MYASTHENIA GRAVIS AND RELATED DISORDERS: EXPERIMENTAL AND CLINICAL ASPECTS

ANNALS OF THE NEW YORK ACADEMY OF SCIENCES

Volume 681

MYASTHENIA GRAVIS AND RELATED DISORDERS: EXPERIMENTAL AND CLINICAL ASPECTS

Edited by Audrey S. Penn, David P. Richman, Robert L. Ruff, and Vanda A. Lennon

The New York Academy of Sciences
New York, New York
1993

The paper used in this publication meets the minimum requirements of American National Standard for Information Sciences—Permanence of Paper for Printed Library Materials, ANSI Z39.48-1984.

Library of Congress Cataloging-in-Publication Data

Myasthenia gravis and related disorders : experimental and clinical aspects / edited by Audrey S. Penn ... [et al.].
p. cm. -- (Annals of the New York Academy of Sciences, ISSN 0077-8923 ; v. 681)
Includes bibliographical references and index.
ISBN 0-89766-755-7 (cloth : alk. paper). -- ISBN 0-89766-756-5 (pbk. : alk. paper)
1. Myasthenia gravis--Congresses. 2. Myasthenia gravis--Molecular aspects--Congresses. 3. Acetylcholine--Receptors--Congresses. 4. Calcium channels--Congresses. I. Penn, Audrey S. II. Series.
[DNLM: 1. Myasthenia Gravis--complications--congresses. W1 AN626YL v.681 1993 / WE 555 M9943 1993]
Q11.N5 vol. 681
[RC935.M8]
500 s--dc20
[616.7'442]
DNLM/DLC
for Library of Congress 93-8239
CIP

BC/PCP

Printed in the United States of America

ISBN 0-89766-755-7 (cloth)

ISBN 0-89766-756-5 (paper)

ISSN 0077-8923

ANNALS OF THE NEW YORK ACADEMY OF SCIENCES

Volume 681
June 21, 1993

MYASTHENIA GRAVIS AND RELATED DISORDERS: EXPERIMENTAL AND CLINICAL ASPECTS[a]

Editors and Conference Organizers
AUDREY S. PENN, DAVID P. RICHMAN, ROBERT L. RUFF, AND VANDA A. LENNON

CONTENTS

[a] This volume is the result of a conference entitled **Myasthenia Gravis and Related Disorders: Experimental and Clinical Aspects,** held in Washington, DC on April 12–15, 1992, by the New York Academy of Sciences and the Myasthenia Gravis Foundation.

Part III. The Nicotinic Acetylcholine Receptor: Structural and Functional Aspects

Part IV. Autoimmunity and the Acetylcholine Receptor

Part V. Autoimmunity and Calcium Channels

Part VI. Symposium on Congenital Myasthenic Syndromes

Part VII. Diagnosis and Treatment in Myasthenia and the Lambert-Eaton Myasthenic Syndrome

Financial assistance was received from:

Supporters

- HOFFMANN-LA ROCHE INC.
- MYASTHENIA GRAVIS FOUNDATION
  - California Chapter
  - Myasthenia Gravis Association, Inc., Detroit
  - Greater Chicago Area Chapter
  - Greater New York Chapter
  - Los Angeles Chapter
  - Myasthenia Gravis Association of Western Pennsylvania
- NATIONAL INSTITUTE OF NEUROLOGICAL DISEASES AND STROKE, NIH

Contributors

- BURROUGHS WELLCOME COMPANY
- MERCK SHARP & DOHME RESEARCH LABORATORIES
- MYASTHENIA GRAVIS FOUNDATION
  - Arizona Chapter
  - Arkansas Chapter
  - Carolinas Chapter
  - Connecticut Chapter
  - Eastern Pennsylvania Chapter
  - Georgia Chapter
  - Greater Dallas/Ft. Worth Chapter
  - Greater Kansas City Chapter
  - New Mexico Chapter
  - Northwest Texas Chapter
  - San Antonio Chapter
  - Virginia Chapter
  - Western New York Chapter
  - Wisconsin Chapter
- NATIONAL INSTITUTE OF ALLERGY AND INFECTIOUS DISEASES, NIH
- RANSOM CONSULTING

The Myasthenia Gravis Foundation sponsors research fellowships and medical symposia; produces literature for patients, doctors, and nurses; and has thirty-eight local chapters that provide patient services, including counseling, newsletters, clinics, and support groups. For more information, please contact the Myasthenia Gravis Foundation, 53 West Jackson, Suite 660, Chicago, Illinois 60604.

Preface

AUDREY S. PENN

Department of Neurology
Columbia University College of Physicians and Surgeons
New York, New York 10032

This volume contains contributions to the eighth international conference on myasthenia gravis entitled Myasthenia Gravis and Related Disorders: Experimental and Clinical Aspects. This conference was the latest in a series cosponsored by the New York Academy of Sciences and the Myasthenia Gravis Foundation, an organization of patients, their families, and other interested lay persons. The goals for both organizations have been to organize meetings and to publish the proceedings, giving the latest information about myasthenia and, more recently, about related disorders of neuromuscular transmission. The conferences have grown from the first informal gathering at the Massachusetts General Hospital in 1947 and the first symposium at the University of Pennsylvania in 1954 to participation by more than 400 scientists and clinicians from over fourteen countries. Subsequent conferences sponsored by the two organizations were held in 1965, 1970, 1975, 1980, and 1986.[a]

Myasthenia gravis is the prototypic human autoimmune disease, inasmuch as both the antigen and mechanisms of immunopathogenesis are known. Studies of this disease and the models produced by deliberate immunization with purified antigen (acetylcholine receptor) have provided insights to analysis of normal and abnormal neuromuscular transmission, to analysis of the antigen, and to analysis of mechanisms of autoimmunity. Much of this information has been presented and discussed at previous conferences. Since the last conference, however, there has been an explosion of new information on the developmental biology of the thymus gland and of T cells, on the molecular biology of T-cell receptors and B-cell rearrangements that result in antibody specificity, and on the biochemistry of glycoprotein receptors. All of this information has an impact on analysis of mechanisms underlying myasthenia. Myasthenia is known to be characterized by a pathological thymus. Thymectomy is an accepted and very effective therapy. Therefore, review of new information on T-cell maturation and selection of T-cell receptors within the thymus that mediate tolerance is directly relevant to myasthenia. This part of the proceedings should afford new insights to molecular and cellular immunologists and pathologists as well as to those studying myasthenia.

Knowledge of the detailed sequence of the acetylcholine receptor is being exploited to study the ion channel and specific T-cell epitopes. Pertinent information can be derived from discussion of other membrane glycoprotein receptors. Of particular interest is the voltage-gated calcium channel of presynaptic terminals at the neuromuscular junction, which is almost certainly the target antigen in the Lambert-Eaton syndrome. The discussion of this protein and its physiology as well as its localization to small-cell carcinomas of the lung should provide insights into the production of autoimmunity. Review of these topics contributes to discus-

[a] The following *Annals* resulted, respectively, from the conferences: Volume 135, Article 1, 1966; Volume 183, 1971; Volume 274, 1976; Volume 377, 1981; and Volume 505, 1987.

sion of current diagnosis and therapy of both forms of the disease with emphasis on progress in attempts to design more specific therapies.

These proceedings allow scientists from a variety of disciplines, including biochemistry, physiology, pharmacology, immunology, and molecular biology to interact with physicians and physician scientists working specifically on these diseases. Indeed, new information on the Lambert-Eaton myasthenic syndrome has confirmed that the antibodies present in some patients react with voltage-gated calcium channels. Accordingly, the isolation, characterization, and toxicology of these proteins, as discussed during the conference, shed new light on the disease and vice versa. Because others with this syndrome have a paraneoplastic disorder, mechanisms responsible for this form are of equal interest. Although extremely rare, the congenital myasthenic syndromes provide unusual insights into abnormalities that can occur in the acetylcholine receptor itself and other proteins, both pre- and postsynaptic, whose absence or malfunction causes neuromuscular dysfunction. The latest information on these disorders is presented.

The disorders of neuromuscular transmission challenge scientists of varied disciplines to see normal structure and function in new ways. The resulting dialogue and cross-fertilization have, in the past, contributed greatly to new information on mechanisms that in turn has challenged both clinicians and scientists to translate this information into newer diagnostic and therapeutic measures. As has become traditional, current diagnosis and both current and potential new therapies are discussed in this volume.

The organizers of this conference—David P. Richman, Robert L. Ruff, Vanda A. Lennon, and I—are grateful to the New York Academy of Sciences, to all of the participants, and to all organizations that gave financial support. In particular, we wish to acknowledge the support of so many of the individual chapters of the Myasthenia Gravis Foundation as well as of the National Lay Board during the fortieth anniversary of the Foundation.

An Immunogenic Self-Peptide for T Cells in Mice with Experimental Myasthenia[a]

QUN XU, MARK AGIUS, ESWARI GUDIPATI,
ROBERT H. FAIRCLOUGH, AND DAVID P. RICHMAN

Department of Neurology and Center for Neurobiology
University of California, Davis
Davis, California 95616

Myasthenia gravis (MG) is a T cell–dependent, B cell–mediated autoimmune disease in which autoantibodies against muscle acetylcholine receptor (AChR) attack the AChRs at the neuromuscular junction. A pivotal role for anti-AChR T cells in the autoimmune process in MG was demonstrated as early as 15 years ago, including some work done in our laboratory.[1]

As an animal model for human autoimmune disease, experimental autoimmune myasthenia gravis (EAMG) is useful for investigating the autoimmune mechanism involved in human disease. This animal model replicates MG in almost every characteristic that has been analyzed.[2] EAMG provides a means of analyzing the role of T cells in regulating autoimmune response in human disease.

MATERIAL AND METHODS

Production of T-Cell Hybridomas

T cells were isolated from the draining popliteal and inguinal lymph nodes of 8–12-week-old C57BL/B6 mice 10 days after subcutaneous injection of 300 μg purified *Torpedo* AChR emulsified in complete Freund's adjuvant. Lymph node cells were initially cultured in serum-free HL-1 Ventrex medium containing 25 μg/mL of the mouse AChR peptide alpha 185-197 (FIG. 1). Five days later responding T-cell blasts were cultured in complete culture medium containing 25 μg/mL of the mouse peptide. Syngeneic irradiated spleen cells were used as a source of antigen-presenting cells every 7–10 days. Hybridomas were generated by fusing the T cells with the mutant thymoma line BW5147. The hybridomas were screened for their response to the mouse peptide and to the corresponding *Torpedo* peptide (FIG. 1) by an IL-2 release assay.

RESULTS AND DISCUSSION

Thirty hybridomas were screened for their response to mouse AChR peptide alpha 185-197 and to *Torpedo* AChR peptide alpha 180-197 by IL-2 release assay.

[a] This work was supported in part by Grants from the National Institutes of Health (NS15462, NS19779, NS24304, NS01226).

mouse AChR peptide alpha185-197	K H W V F Y S C C P T T P
Torpedo AChR peptide alpha 180-197	D Y R G N K H W V Y Y T C C P D T P

FIGURE 1. The amino acid sequences of mouse AChR alpha-subunit peptide 185-197 and *Torpedo* AChR alpha-subunit peptide 180-197. These two peptides differ by three amino acids, two of which are conservative.

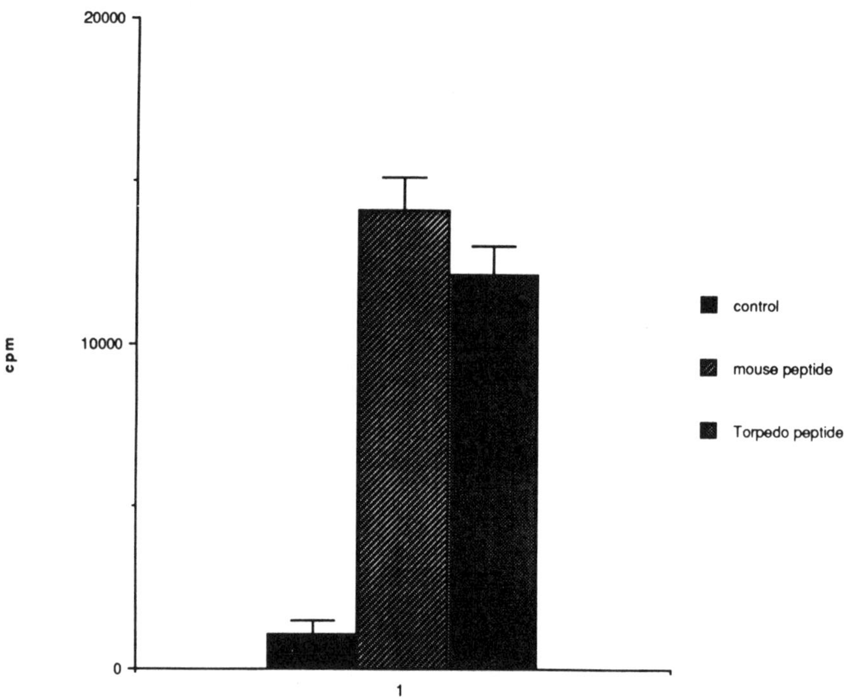

FIGURE 2. Response of hybridoma P14 to mouse and *Torpedo* peptides. IL-2 release was measured according to Kappler *et al.* (1981). Briefly, 10^5 hybridoma cells were added to 10^6 C57B1/6 antigen-presenting cells, with or without antigen, in a total volume of 200 μL. Twenty-four hours later, 100 μL of supernatant was transferred to 5×10^3 IL-2–dependent HT-2 cells in 100 μL cyclic Dulbecco's modified Eagle medium (cDMEM). After an additional 24 hours, the cells were pulsed for 8 hours with 1 μCi [^{3}H]thymidine per well. The cpm thymidine incorporation was calculated as the mean ($\pm$SEM) for triplicate culture. The FIGURE presents the results of the IL-2 release assay for hybridoma P14. The background is 1,000 cpm.

Two of the these hybridomas, P14 and P16, responded both to mouse and *Torpedo* peptides. Results of the IL-2 release assays are presented in FIGURES 2 and 3.

In susceptible mouse strains, for example, C57BL/B6 and SJL, there appears to be a limited number of immunogenic sites on the alpha-subunit, the only subunit studied to date. One of the peptides identified in C57BL/B6 is alpha 183-200.[3] Interestingly, this peptide is immunogenic in Lewis rats as well. All of these results, however, are significantly limited by the fact that the antigen analyzed has been AChR purified from the electric organ of *Torpedo californica*. *Torpedo* AChR is nearly always the immunogen in EAMG but is, obviously, not the autoan-

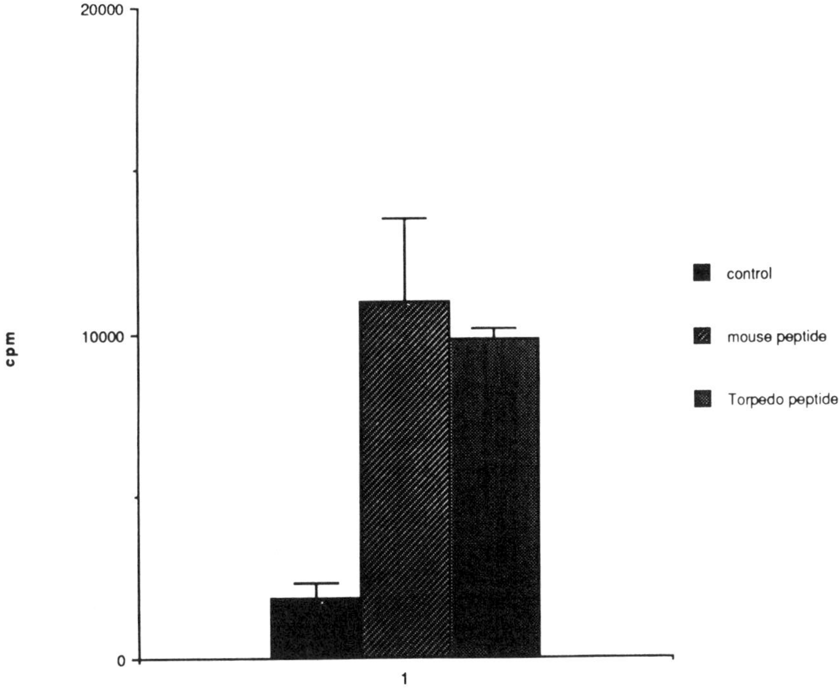

FIGURE 3. Response of hybridoma P16 to mouse and *Torpedo* peptides. The results of IL-2 release assay as in FIG. 2 for hybridoma P16. The background is 1,800 cpm.

tigen, which is isogeneic muscle AChR located at the neuromuscular junction. In terms of the regulation and manipulation of the autoimmune response in EAMG, it is mouse AChR and its peptides that are important. Because of these limitations of the previous studies and because of the importance of an accurate and precise definition of immunogenic peptides for understanding the autoimmune mechanism and developing a specific immunotherapy, we isolated T cells from C57BL/6 mice immunized with *Torpedo* AChR but tested the response of these T cells to mouse AChR peptide alpha 185-197 and to the corresponding *Torpedo* AChR peptide, alpha 180-197. Our ability to readily recall the anti-AChR response with the mouse peptide and, in fact, to isolate T-cell lines by stimulation with this peptide demonstrate that this self-peptide contains a T-cell autoepitope involved in EAMG in this susceptible strain.

Autoimmune diseases have been induced by self-peptide responding T cells in some animal models, such as T cell–mediated experimental allergic encephalomyelitis (EAE). T-cell clones recognizing self-myelin basic protein (MBP) peptide induce the disease in mice, but T-cell clones responding to rat or guinea pig MBP peptides do not, indicating that T-cell recognition of self-MBP was necessary for induction of EAE in mice.[4] Inasmuch as EAMG is a B cell–mediated, T cell–dependent disease, the autoimmune process in EAMG is different from that in EAE. However, it is a reasonable hypothesis that in a chronic disease, such as human MG, the T cells that respond to self-peptide may play a key role in self-perpetuating the autoimmune processing.

T cells responding to self-tissue are normally eliminated by clonal deletion in early life or are suppressed at later times. Why were T cells recognizing self-peptide found in our animal model? As we have demonstrated, the T-cell clones responding to mouse peptide also respond to the corresponding *Torpedo* peptide. The similarity between the two peptides is likely important in the cross-reactivity that appears to provoke the autoimmune response. On the other hand, the small differences between the two peptides (they differ by three amino acids) may be important in the breakdown in tolerance required for the initiation of the immune response to the *Torpedo* AChR in the first place.

REFERENCES

1. RICHMAN, D. P., J. PATRICK & B. G. W. ARNASON. 1976. Cellular immunity in myasthenia gravis: Response to purified acetylcholine receptor and autologous thymocytes. N. Engl. J. Med. **294:** 694–698.
2. LENNON, W. A., J. M. LINDSTROM & M. E. SEYBOLD. 1975. Experimental autoimmune myasthenia: A model of myasthenia gravis in rat and guinea pigs. J. Exp. Med. **141:** 1365–1375.
3. ATASSI, M. Z., B. MULAC-JERICEVIC, T. YOKOI & T. MANSHOURI. 1987. Localization of the functional sites on the alpha chain of acetylcholine receptor. Fed. Proc. Fed. Am. Soc. Exp. Biol. **46:** 2538–2547.
4. ZAMVIL, S. S. & L. STEINMAN. 1990. The T lymphocyte in experimental allergic encephalomyelitis. Annu. Rev. Immunol. **8:** 579–621.

The Thymus and T-Cell Tolerance[a]

JONATHAN SPRENT

Department of Immunology, IMM4A
The Scripps Research Institute
10666 North Torrey Pines Road
La Jolla, California 92037

INTRODUCTION

It is now well accepted that the capacity of the immune system to distinguish self from nonself is largely under the control of T cells. B cells play a subsidiary role because, for the most part, the function of B cells is strictly T-dependent. The purpose of this article is to give a brief overview of how the T-cell repertoire is formed and the various mechanisms that have been invoked to account for self/nonself discrimination. To begin, it is useful to consider the essential features of T-cell specificity.

T-cell Specificity

The vast majority of T cells in the secondary lymphoid tissues carry α-β T-cell receptor (TCR) molecules and coexpress either CD4 or CD8 molecules.[1-3] The specificity of both types of T cells is directed to peptides bound to a groove on major histocompatibility complex (MHC) molecules.[4] For primary immune responses, the complex of peptides and MHC molecules have to be presented by specialized antigen-presenting cells (APC) such as dendritic cells and macrophages.[1,5] MHC molecules come in two forms, termed class I and class II.[6] Class I MHC molecules are found on nearly all cells and display antigen (peptide) to $CD8^+$ ($CD4^-8^+$) cells. Class II molecules show a more restricted tissue distribution and present antigen to $CD4^+$ ($CD4^+8^-$) cells. The α-β TCR molecules on $CD4^+$ and $CD8^+$ cells are not obviously different, and the MHC class specificity of these cells appears to reflect that CD4 and CD8 molecules have binding specificity for MHC molecules: CD4 binds to class II, whereas CD8 binds to class I.

It is now clear that a proportion of T cells express γ-δ TCR.[7] Although these cells are attracting considerable interest, the specificity of γ-δ T cells and the role of these cells in tolerance is still unclear. The remainder of this article deals exclusively with typical α-β^+ T cells.

Superantigens

In addition to recognizing peptide/MHC complexes, α-β^+ T cells have specificity for a unique class of antigens, termed *superantigens*.[8-11] These antigens seem

[a] This work was supported by Grants CA25803, CA38355, and AI21487 from the United States Public Health Service. This is publication no. 7283-IMM from the Research Institute of Scripps Clinic.

to be recognized in intact form rather than as peptides. The immunogenicity of superantigens appears to reflect that these molecules cause T cells and typical APC to stick together by binding jointly to the sides of MHC and TCR molecules. Superantigens bind to the lateral aspect of the β chain of the α-β TCR, and there is little or no contact with the peptide-binding site (which is a product of both the α and β chain). A notable feature of superantigens is that binding to the TCR is Vβ-specific. There are 20–30 members of the Vβ family, and individual superantigens have binding specificity for 1–5 Vβs. This means that each superantigen can be recognized by a very high proportion of T cells, up to 20 percent. Because superantigens bind selectively to MHC class II molecules, T-cell recognition of superantigens is largely restricted to $CD4^+$ T cells.

Superantigens are divided into two categories. Most superantigens are soluble products of bacteria and viruses; bacterial enterotoxins are a classic example, and these molecules cause food poisoning, presumably by stimulating T cells to release large quantities of toxic cytokines. In addition, it has recently become clear that superantigens can exist as cell-surface molecules. These molecules are products of endogenous retroviruses, especially the mouse mammary tumor virus (MMTV), and are expressed predominantly on B cells. Mls and related antigens in mice fall under this category. As discussed later, endogenous superantigens have proved to be an invaluable tool for assessing the mechanism of self-tolerance.

T-cell Differentiation in the Thymus

Although a small proportion of T cells seem to develop by an extrathymic pathway of differentiation, the vast majority of T cells arise in the thymus.[12–14] In discussing thymocyte differentiation it is important to emphasize that only a very small fraction of thymocytes are released to the peripheral lymphoid tissues. Most (>95%) thymocytes die *in situ*. At face value this might seem a very wasteful process. In practice, however, it appears that the widescale destruction of thymocytes is a reflection of a highly efficient mechanism for selecting a T-cell repertoire that is tailor-made for the individual. Ultimately, the death of most thymocytes can be viewed as a reflection of MHC polymorphism. Each individual expresses a unique set of MHC molecules, and the T cells in each individual have to be selected so as to be able to recognize antigen (peptide) presented by these particular self–MHC molecules; for the most part, T cells have specificity for peptides bound to autologous MHC molecules, recognition of peptides presented by foreign (allo) MHC molecules being much less efficient. To cope with the problem of MHC polymorphism, the thymus takes the brute force approach of generating large numbers of immature T cells with specificity for all of the MHC molecules expressed in the species as a whole. From this large immature repertoire, the thymus selects a small proportion of T cells with specificity for the particular self MHC molecules displayed in the thymus. This process is termed positive selection.

Positive selection occurs in the cortical region of the thymus. The cortex consists largely of immature T cells expressing a low level of α-β TCR molecules; these cells are double-positive for CD4 and CD8 molecules and arise from a small subset of cycling $CD4^-8^-$ cells situated under the capsule. Positive selection reflects contact of $CD4^+8^+$ cells with the dense network of cortical epithelial cells. Although the mechanism of positive selection is still unclear, the simplest idea is that $CD4^+8^+$ thymocytes are screened for binding specificity for MHC class I and II molecules expressed on epithelial cells; both the TCR and either CD4 or CD8 contribute to binding. If binding occurs, the T cells receive a protective signal that allows the cells to survive and eventually be exported from the thymus. $CD4^+8^+$ cells failing to bind to epithelial cells do not receive a protective signal, and these cells—the bulk of thymocytes—die from "neglect" within a few days. After positive selection, the selected $CD4^+8^+$ cells rapidly upregulate TCR expression and downregulate CD4 or CD8 to become typical "single-positive" T cells. If selection is directed to class I molecules, the T cell downregulates CD4 and maintains CD8; the reverse applies when selection is directed to class II molecules.

Although positive selection of T cells is now well-documented, the signals and receptor/ligand interactions involved in this process are still poorly understood. For example, the notion that selection involves the delivery of a protective signal is more hypothesis than fact. It is also unclear whether selection is directed to polymorphic MHC epitopes, a range of self-peptides bound in the groove of MHC molecules, or both. This issue is discussed elsewhere in this volume by Janko Nikolić-Žugić.

Some of the T cells undergoing positive selection are presumed to have the requisite "physiological" level of specificity for self-MHC molecules required by postthymic T cells. In the periphery, these cells respond vigorously to MHC molecules containing foreign peptides but ignore MHC molecules lacking these peptides (or containing self-peptides); the T cells are thus self-tolerant. Other T cells undergoing positive selection have higher (above physiological) reactivity to self MHC molecules. These cells are potentially autoreactive and have to be destroyed. As discussed below, the clonal elimination of potentially autoaggressive T cells takes place largely in the thymus.

Negative Selection in the Thymus

Although it has long been argued that self-tolerance induction reflects clonal deletion, unequivocal proof for this notion was obtained only quite recently. The first direct evidence for clonal deletion came from the finding that strains of mice expressing particular MMTV-derived endogenous superantigens show $V\beta$-specific deletion of T cells.[15-17] The key finding was that this deletion is apparent at the level of thymocytes. Similar evidence for clonal deletion of T cells in the thymus has come from studies with TCR-transgenic mice, that is, mice containing a near-monoclonal population of T cells specific for a defined antigen.[14]

The site of negative selection in the thymus is controversial. For Mls and related endogenous superantigens, $V\beta$ deletion appears to occur quite late in ontogeny, at the level of single-positive T cells. With TCR transgenic mice, however, deletion can occur relatively early in differentiation, at or before the $CD4^+8^+$ stage. These two sets of observations are difficult to compare because most of the data on TCR-transgenic mice deal with conventional antigens rather than superantigens. Nevertheless it is of interest that studies with a TCR-transgenic line with dual specificity for a conventional antigen (lymphocytic choriomeningitis virus) and an endogenous superantigen (Mls^a) have shown that deletion occurs early in differentiation for the conventional antigen and late for the superantigen.[18] The simplest explanation for this finding is that endogenous superantigens have a restricted distribution in the thymus, with strong expression in the medulla (the site of mature T cells) but only weak expression in the cortex. Viruses, by contrast, can enter both the cortex and medulla.

With regard to the cell types controlling negative selection, it is well-accepted that typical APC play a decisive role.[13] These cells are bone marrow (BM)–derived and are concentrated in the medulla and the corticomedullary junction. Whether epithelial cells contribute to negative selection is controversial. Although it is generally agreed that thymic epithelial cells are much less tolerogenic than typical APC, there is good evidence that epithelial cells are at least partly tolerogenic.[19,20] The most likely possibility is that epithelial cells are only tolerogenic for high-affinity T cells, tolerance of low-affinity cells being controlled by APC.

Tolerance to Non-MHC Self-antigens

Most of the evidence on self-tolerance induction has come from studies with MHC antigens and endogenous superantigens. Much less is known about tolerance to other self-antigens. In considering this question, it is important to emphasize that T cells are clearly not tolerant to all self-antigens. Thus, it is well-established that certain self-antigens present in the brain, thyroid, and other tissues are strongly immunogenic when injected in association with adjuvant.[21,23] Such immunization induces organ-specific autoimmune disease. Because spontaneous organ-specific autoimmune disease is relatively rare, one has to account for why the self-antigens concerned are generally ignored by the immune system. The simplest possibility is that such antigens, for example, myelin basic protein in the brain (the target for experimental autoimmune encephalomyelitis), are normally sequestered from the immune system: the antigens remain in the tissue concerned and reach the lymphoid tissues in only negligible quantities. Conversely, typical resting T cells remain confined to the lymphoid tissues (including blood and lymph) and rarely enter the extravascular tissue of the brain or thyroid.

In addition to tissue-specific self-antigens, the body contains a multitude of self-antigens that come into continuous contact with T cells during their normal patterns of migration. How are T cells rendered tolerant to these antigens? There are two broad possibilities:

(1) *Self-Tolerance Involves Extrathymic Tolerance Induction*

Although intrathymic tolerance is well-documented, many groups argue that clonal deletion of T cells in the thymus cannot be totally efficient and that small numbers of autoreactive T cells will inevitably reach the periphery. It is also argued that it is unlikely that the full range of self-antigens expressed in the peripheral lymphoid tissues could reach the thymus in sufficient quantities to tolerize all potentially reactive thymocytes. For these reasons, many workers feel that backup mechanisms for tolerance must exist in the post-thymic environment.

At present, induction of anergy is the most popular back-up mechanism for tolerance. To understand anergy, it is important to point out that stimulation of mature resting T cells is considered to require at least two signals: strong cross-linking of the TCR through contact with antigen on APCs and the delivery of "costimulatory" signals from the APC.[1] Professional APC such as macrophages and dendritic cells express high levels of costimulatory signals and are strongly immunogenic for resting T cells. By contrast, cell types that lack these "second signals" are relatively nonimmunogenic. Furthermore, at the level of resting T cells, it is argued that contact with such "nonprofessional" APC induces tolerance. In support of this idea, there is convincing evidence that, under defined conditions *in vitro,* nonprofessional APC induce T cells to enter an anergic state: in the absence of costimulation, the T cells receive an inhibitory signal that makes the cells refractory to subsequent stimulation by professional APC.[24]

Although anergy induction initiated by cells lacking second signals is well-documented *in vitro,* whether this form of tolerance can occur *in vivo* is less clear. The most popular approach for addressing this question is to prepare transgenic mice with MHC genes placed under tissue-specific promoters, for example, the insulin promoter. These mice express MHC antigens in tissue-specific sites, for example, the β cells of pancreatic islets. The question asked here is whether such expression of MHC antigens on nonprofessional APC induces T-cell tolerance. The results of this approach are mixed. Some groups find that the transgenic mice show widespread anergy,[25,26] even in the thymus.[25] Other groups, however, observe little or no tolerance.[27–29] Reaching generalized conclusions from these experiments is thus difficult.

We have searched for anergy induction by nonprofessional APC by transferring T cells to long-term parent → F_1 BM chimeras.[30] When prepared with supralethal irradiation (1300 rad), MHC-different parent → F_1 chimeras are rapidly depleted of virtually all host BM-derived cells, and host MHC expression is limited to low-level expression on endothelial cells and certain other non–BM-derived cells. When normal parental-strain T cells are transferred to these chimeras, the expression of host MHC antigens on host stromal cells does not induce tolerance. On the contrary, some of the donor T cells proliferate vigorously and develop into effector cells with strong antihost reactivity; these data apply to both $CD4^+$ and $CD8^+$ T cells. Our interpretation of this finding is that presentation of antigen by nonprofessional APC under *in vivo* conditions is not intrinsically tolerogenic. The consequences of antigen recognition under these conditions depends on the relative affinity of the T cells. T cells with high affinity for antigen do not require strong costimulation, and these T cells can be stimulated by a wide variety of different cell types including endothelial cells. Cells with lower affinity are heavily dependent on costimulation, and these cells require professional APC for activation. When low-affinity T cells are exposed to antigen on nonprofessional APC *in vivo,* we envisage that the main consequence of this encounter is not tolerance (anergy) induction but simply lack of stimulation.

One of the reasons we doubt that anergy induction by nonprofessional APC *in vivo* is of generalized importance is that *in vitro* studies have shown that anergy induction by nonprofessional APC can be overcome by addition of "bystander" professional APC.[24] Because the latter are numerous *in vivo,* it is difficult to envisage how T cells exposed to antigen on nonprofessional APC, for example, endothelial cells and resting B cells, would fail to receive bystander second signals from adjacent APC. Despite this theoretical objection, there is at least one clear example of tolerance induced *in vivo* by antigen expressed on small B cells (which are generally acknowledged to be nonprofessional APC).[31] Although this is an intriguing observation, it should be mentioned that the mechanism of tolerance induction in this model is still unclear. One possibility is that the T cells are clonally eliminated. In this respect, it is noteworthy that in certain situations exposure of mature T cells to antigen *in vivo* leads to initial expansion of the responding cells, but this phase of proliferation is then followed by widescale elimination of the T cells.[32,33] The physiological significance of this finding has yet to be resolved. One possibility is that the T cells are eliminated as a reflection of overstimulation by professional APC. In support of this idea, it has been found that at least a proportion of the stimulated T cells die in the lymphoid tissues as the result of apoptosis.[33,34] In addition, however, there is evidence that some of the activated T cells simply leave the lymphoid tissues and home to the gut.[34] Such homing might have the dual function of diverting effector T cells to regions where the cells are needed and, at the same time, allow the immune system to purge itself of too many cells of a given specificity. For these reasons it is questionable whether the "clonal elimination" of T cells participating in apparently normal immune responses should be regarded as evidence of a tolerogenic mechanism (even though the end result can be functional tolerance).

A point to emphasize in the above experiments is that, prior to T-cell elimination, many of the activated T cells display anergy. Hence, at least in certain situations, anergy may be a prelude to cell death. Although it is possible that anergy can at times be reversible, the status of anergy as a major mechanism of tolerance *in vivo* is still uncertain. In particular, the view that anergy induction in the post-thymic environment is essential for maintaining self-tolerance is far from convincing.

(2) *Clonal Deletion in the Thymus Is Sufficient to Account for Self-Tolerance*

One of the conceptual problems with the notion that peripheral mechanisms are needed to maintain self-tolerance is that the distinction between self and nonself is blurred. Thus, if inefficient clonal deletion of autoreactive T cells in the thymus were countered by anergy induction in the post-thymic environment, such blunting of the immune response would be expected to apply equally to immune responses directed to nonself antigens. Compromising the immune response to foreign antigens in order to maintain self-tolerance would be potentially dangerous, especially if the response were directed to a virulent pathogen. In view of this problem, it is useful to speculate on whether self/nonself discrimination can be explained entirely in terms of clonal deletion in the thymus. Teleologically, the advantage of this idea is that self and nonself are then clearly defined, and the immune system is free to mount unhindered responses to foreign antigens.

As discussed earlier, self-tolerance does not apply to organ-specific antigens: self-tolerance is restricted to self-antigens that come into frequent contact with T cells. Such antigens also come into contact with other blood-borne cells, notably APC. The central question is whether the movement of APC from the peripheral lymphoid tissues into the thymus is sufficiently extensive to induce efficient tolerance (clonal deletion) of all immature thymocytes reactive to the various self-antigens borne by immigrant APC. Although little is known about the range of self-antigens carried by APC, it is clear that thymic APC do have a rapid turnover and are continuously entering the thymus from the peripheral lymphoid tissues.[35] It is also well-documented that the medullary region of the thymus, the site of APC, is permeable to blood-borne proteins.[36]

Although one can make the case that entry of self-antigens into the thymus is extensive and highly efficient, there is still the objection that tolerance induction in the thymus cannot be foolproof and that autoreactive T cells will inevitably leak through to the periphery. This does not necessarily follow. Indeed, one can argue that evolutionary pressures have designed the thymus to be as efficient as possible in inducing tolerance. APC, especially dendritic cells, form a dense barrier at the corticomedullary junction, and it is difficult to envisage how T cells could avoid confronting this barrier before exiting to the periphery.

Because many self-antigens, for example, certain hormones, are present in only very low levels, one could argue that the concentration of these antigens reaching the thymus is too low to cause tolerance. The point to emphasize here is that immature T cells are exquisitely sensitive to tolerance induction. Thus it is likely that even very low levels of antigens are sufficient to delete high-affinity T cells in the thymus. Cells of lower affinity may escape deletion, but these T cells would be operationally tolerant because the concentration of antigen encountered in the post-thymic environment would be too low to trigger these cells.

Another objection that can be raised is that clonal deletion in the thymus would not apply to cell-bound antigens expressed, for example, in the liver. Although T cells presumably make frequent contact with these antigens in the post-thymic environment, one can argue that the expression of these antigens on nonprofessional APC (liver cells) is simply nonimmunogenic. Imposing active tolerance to these antigens is thus unnecessary.

As argued above, one can make a reasonable case that steady-state self/nonself discrimination at the T-cell level can be explained entirely in terms of clonal deletion in the thymus (combined with sequestration of tissue-specific antigens). Despite the popularity of back-up mechanisms for tolerance, the argument that it is essential to postulate the existence of such mechanisms to account for normal self-tolerance is unconvincing.

Breaking of Self-Tolerance

As discussed earlier, it is clear that T cells are not tolerant to tissue-specific antigens. These antigens are strongly immunogenic when injected in adjuvant, and the lack of autoreactivity to these antigens in the normal animal presumably reflects that such antigens are sequestered from the immune system. It would seem likely, therefore, that development of T-mediated autoimmune disease to these antigens is attributable to a breakdown of antigen sequestration, presumably because antigen somehow gains access to the lymphoid tissues and thus comes into close contact with (nontolerant) T cells.

This notion is probably an oversimplification because one can envisage many situations in which substantial pulses of tissue-specific antigens reach the lymphoid tissues, for example, as the result of trauma, tissue infarction, or infection. Yet organ-specific autoimmune disease at the T-cell level is relatively rare. For this reason, it would seem that contact with self-antigens alone is not sufficient to stimulate autoreactive T cells. Something else is needed. What could this be?

In considering this question it should be borne in mind that most antigens are relatively nonimmunogenic unless injected with adjuvant. Although the mode of action of adjuvants is still unclear, a likely possibility is that these agents amplify the immunogenicity of APC by upregulating the costimulatory activity of these cells. Does it follow, therefore, that all T-cell–mediated responses require the presence of adjuvant? This possibility might seem unlikely because, in contrast to soluble antigens, the immune system generally gives strong primary responses to pathogenic microorganisms without the need for exogenous adjuvant. But is it possible that pathogens contain their own built-in adjuvants? This is conceivable because it has recently become apparent that many microorganisms produce superantigens, for example, streptococcal enterotoxins.[10] As discussed earlier, superantigens are strongly immunogenic for unprimed T cells, but, in addition, superantigens are potent activators of APC and thus display adjuvant activity.[37] It should be mentioned that microorganisms also release other substances with powerful adjuvant activity, for example, lipopolysaccharide (LPS).[38]

Although it has long been argued that autoimmune disease is triggered by infection,[22,23] precisely how infection breaks tolerance at the T-cell level remains controversial. A popular view is that certain microorganisms display antigens that cross-react with self-antigens. Evidence that such molecular mimicry applies at the T-cell level, however, is rather sparse. In light of the finding that many microorganisms, including certain viruses, express superantigens, the question arises whether superantigens are an important contributing factor for induction of autoimmune disease. Although there is little direct evidence available on this question, one can envisage a scenario where superantigens synthesized by pathogens act as adjuvants for tissue-specific self-antigens. As the result of local inflammation, both the self-antigens and the superantigens are released into the bloodstream (or lymph) and enter the lymphoid tissues where they are taken up by APC. After upregulation of costimulatory activity by the superantigens, the APC present the self-antigens to T cells in immunogenic form. After activation, the T cells then home to the site of infection and reencounter the antigens that led to initial T-cell induction. The end result is autoimmunity.

Regulation of Autoimmune Responses

One of the problems with the above model is that it fails to explain why progressive T-mediated autoimmune disease is uncommon. One way around this problem is to invoke immunoregulation. As discussed earlier, one can make the case that normal self/nonself discrimination can be explained entirely in terms of clonal deletion in the thymus, but this only applies when T cells and tissue-specific self-antigens remain compartmentalized. When sequestration of self-antigens fails and overt autoreactivity results, it would obviously be advantageous to have back-up mechanisms for inhibiting the T cells concerned.

Although discussing T-cell suppression is beyond the scope of this article, suffice it to say that there is convincing evidence that, once activated, autoreactive T cells are highly susceptible to various forms of immunoregulation, including the production of inhibitory cytokines, idiotype-specific antibodies, and antigen-specific suppressor T cells.[21] Rapid onset of immunoregulation may explain the rarity of overt autoimmune disease. Thus T-cell response to self-antigens might be quite common, for example, during infection, but remain subclinical because of the subsequent onset of suppression.

In considering the role of immunoregulation in preventing autoimmune disease, it should be mentioned that there are a number of situations in which mice spontaneously develop generalized autoimmune disease (GAD). These situations include thymectomy on day 3 after birth,[39] treatment with cyclosporine after irradiation,[40,41] and thymus-grafting of nude mice;[42] in each of these situations, supplementing the mice with bulk populations of normal T cells prevents the onset of GAD. At face value these findings would seem to imply that the immune system is poised to attack self-antigens but is normally kept in check by "regulator" cells. There is another possibility, however. In each of the models described above, it is notable that the host mice have a marked quantitative deficiency of T cells and are thus prone to infection. According to the scenario considered earlier, one can argue that infection makes the mice susceptible to autoimmunity through release of self-antigens from the sites of infection. These antigens (plus superantigens and LPS) then stimulate autoreactive T cells. Because of the overall paucity of T cells, however, the immune system is unable to mount effective response to pathogens, and infection persists. Inflammation and release of self-antigens continues, therefore, and the autoimmune response to these antigens progresses to the stage of overt autoimmunity (multiorgan autoimmune disease being a reflection of multiorgan infection). This chain of events fails to occur if the T-cell pool is of normal size. Here, the immune system is able to rapidly clear pathogens and thereby prevent the onset of progressive autoimmunity.

The key feature of this model is that regulatory T cells do not act by inhibiting the function of autoreactive T cells: regulation is simply a reflection of an optimal immune response to pathogens resulting in interruption of the release of self-antigens into the lymphoid tissues.

CONCLUDING COMMENTS

The maintenance of self-tolerance is often viewed as being highly complex and dominated by antiidiotypic networks, anergy, and T-suppression. In this article I have tried to make the point that self-tolerance at the resting T-cell level can be explained simply in terms of a combination of clonal deletion in the thymus (for circulating antigens) and antigen sequestration (for tissue-specific antigens). The development of autoimmune disease presumably reflects abnormal release of tissue-specific antigens into lymphoid tissues. T-cell contact with these antigens alone, however, is not sufficient to lead to progressive autoimmune disease. In addition, it would appear that some form of infection is necessary to amplify the immunogenicity of APC presentng the tissue-specific antigens. Superantigens such as bacterial toxins are good candidates for providing the signals needed to upregulate APC function.

This scenario is clearly an oversimplification because it fails to account for the association between autoimmune disease and certain MHC alleles.[22,23] It also ignores the evidence that autoreactive T cells are skewed in terms of Vβ expression[43] and that some forms of autoimmune disease seem to involve the γ-δ subset of T cells responding to heat-shock proteins.[43] At present it is impossible to accommodate all these data into a single unified theory for autoimmune disease.

REFERENCES

1. SPRENT, J. & S. R. WEBB. 1987. Function and specificity of T cell subsets in the mouse. Adv. Immunol. **41:** 39–133.
2. HEDRICK, S. M. 1989. T lymphocyte receptors. *In* Fundamental Immunology, second edition. W. E. Paul, Ed.: 291–314. Raven Press. New York.
3. SHEVACH, E. M. 1989. Accessory molecules. *In* Fundamental Immunology, second edition. W. E. Paul, Ed.: 413–444. Raven Press. New York.
4. CARBONE F. R. & M. J. BEVAN. 1989. Major histocompatibility complex control of T cell recognition. *In* Fundamental Immunology, second edition. W. E. Paul, Ed.: 541–570. Raven Press. New York.
5. STEINMAN, R. M. 1991. The dendritic cell system and its role in immunogenicity. Annu. Rev. Immunol. **9:** 271–296.
6. ROBINSON, M. A. & T. J. KINDT. 1989. *In* Fundamental Immunology, second edition. W. E. Paul, Ed.: 489–539. Raven Press. New York.
7. RAULET, D. H. 1989. The structure, function, and molecular genetics of γ/δ T cell receptor. Annu. Rev. Immunol. **7:** 175–207.
8. JANEWAY, C. A. JR. 1991. Selective elements for the Vβ region of the T cell receptor: Mls and the bacterial toxic mitogens. Adv. Immunol. **50:** 1–53.
9. ACHA-ORBEA, H. & E. PALMER. 1991. Mls—a retrovirus exploits the immune system. Immunology Today **12:** 356–361.
10. MARRACK, P. & J. KAPPLER. 1991. The staphylococcal enterotoxins and their relatives. Science **248:** 705–710.
11. ABE, R. & R. J. HODES. 1989. T cell recognition of minor lymphocyte stimulating (MLS) gene products. Annu. Rev. Immunol. **7:** 683–708.

12. Fowlkes, B. J. & D. M. Pardoll. 1984. Molecular and cellular events of T cell development. Adv. Immunol. **44:** 207–264.
13. Sprent, J. 1989. T lymphocytes and the thymus. *In* Fundamental Immunology, second edition. W. E. Paul, Ed.: 69–94. Reven Press. New York.
14. von Boehmer, H. 1990. Developmental biology of T cells in T cell receptor transgenic mice. Annu. Rev. Immunol. **8:** 531–556.
15. Kappler, J. W., N. Roehm & P. Marrack. 1987. T cell tolerance by clonal elimination in the thymus. Cell **49:** 273–280.
16. MacDonald, H. R., R. Schneider, R. K. Less, R. C. Howe, H. Acha-Orbea, H. Festenstein, R. M. Zinkernagel & H. Hengartner. 1988. T-cell receptor Vβ use predicts reactivity and tolerance to Mls^a-encoded antigens. Nature (London) **332:** 40–45.
17. Kappler, J. W., U. Staerz, J. White & P. C. Marrack. 1988. Self tolerance eliminates T cells specific for Mls-modified products of the major histocompatibility complex. Nature (London) **332:** 35–40.
18. Pircher, H., K. Burki, R. Lang, H. Hengartner & R. M. Zinkernagel. 1989. Tolerance induced in double specific T cell receptor transgenic mice varies with antigen. Nature (London) **342:** 559–561.
19. Ohki, H., C. Martin, C. Corbel, M. Coley & N. M. Le Douarin. 1987. Tolerance induced by thymic epithelial cells in birds. Science **237:** 1032–1035.
20. Gao, E.-K., D. Lo & J. Sprent. 1990. Strong T cell tolerance in parent → F_1 chimeras prepared with supralethal irradiation. J. Exp. Med. **171:** 1101–1121.
21. Cohen, I. R. 1986. Regulation of autoimmune disease: Physiological and therapeutic. Immunol. Rev. **94:** 5–21.
22. Schwartz, R. S. & S. K. Datta. 1989. Autoimmunity and autoimmune disease. *In* Fundamental Immunology, second edition. W. E. Paul, Ed.: 819–866. Raven Press. New York.
23. Sinha, A. A., M. T. Lopez & H. O. McDevitt. 1990. Autoimmune diseases: The failure of self tolerance. Science **248:** 1380–1388.
24. Mueller, D. L., M. K. Jenkins & R. H. Schwartz. 1989. Clonal expansion versus functional clonal inactivation: a costimulatory signalling pathway determines the outcome of T cell antigen receptor occupancy. Annu. Rev. Immunol. **7:** 445–480.
25. Burkly, L. C., D. Lo & R. A. Flavell. 1990. Tolerance in transgenic mice expressing MHC molecules extrathymically on pancreatic cells. Science **248:** 1364–1368.
26. Miller, J. F. A. P., G. Morahan, J. Allison & M. Hoffmann. 1991. A transgenic approach to the study of peripheral T cell tolerance. Immunol. Rev. **122:** 103–116.
27. Bohme, J., K. haskins, P. Stecha, W. van Ewijk, M. LeMeur, P. P. Gerlinger, C. Benoist & D. Mathis. 1989. Transgenic mice with I-A on islet cells are normoglycemic but immunologically intolerant. Science **244:** 1179–1182.
28. Murphy, K. M., C. T. Weaver, M. Elish, P. M. Allen & D. Y. Loh. 1989. Peripheral tolerance to allogeneic class II histocompatibility antigens expressed in transgenic mice: evidence against a clonal deletion mechanism. Proc. Natl. Acad. Sci. USA **86:** 10034–10038.
29. Miller, J., L. Daitch, S. Rath & E. Selsing. 1990. Tissue-specific expression of allogeneic MHC class II molecules induces neither tissue rejection nor clonal inactivation of alloreactive T cells. J. Immunol. **144:** 334–341.
30. Gao, E.-K., H. Kosaka, C. D. Surh & J. Sprent. 1991. T cell contact with Ia antigens on nonhemopoietic cells *in vivo* can lead to immunity rather than tolerance. J. Exp. Med. **174:** 435–446.
31. Eynon, E. E. & D. C. Parker. 1992. Small B cells as antigen-presenting cells in the induction of tolerance to soluble protein antigens. J. Exp. Med. **175:** 131–138.
32. Webb, S. R., C. Morris & J. Sprent. 1990. Extrathymic tolerance of mature T cells: clonal elimination as a consequence of immunity. Cell **63:** 1249–1256.
33. Kawabe, Y. & A. Ochi. 1991. Programmed cell death and extrathymic reduction in Vβ8$^+$ CD4$^+$ T cells in mice tolerant to *Staphylococcus aureas* enterotoxin B. Nature (London) **349:** 245–248.
34. Sprent, J. 1976. Fate of H-2 activated T lymphocytes in syngeneic hosts. I. Fate

in lymphoid tissues and intestines traced with ^{3}H-thymidine, ^{125}I-deoxyuridine and ^{51}Chromium. Cell Immunol. **21:** 278–302.

35. Kampinga, J., P. Nieuwenhuis, B. Roser & R. Aspinal. 1990. Differences in turnover between thymic medullary dendritic cells and a subset of cortical macrophages. J. Immunol. **145:** 1659–1663.
36. Raviola, E. & M. J. Karnovsky. 1972. Evidence for a blood-thymus barrier using electron-opaque tracers. J. Exp. Med. **136:** 466–498.
37. Jupin, C., S. Anderson, C. Damais, J. E. Alouf & M. Parant. 1988. Toxic shock syndrome Toxin 1 as an inducer of human necrosis factors and γ interferon. J. Exp. Med. **167:** 752–761.
38. C. A. Janeway. 1992. The immune system evolved to discriminate infectious nonself from noninfectious self. Immunology Today **13:** 11–16.
39. Sakaguchi, S., T. Takahashi & Y. Nishizuka. 1982. Study on cellular events on postthymectomy autoimmune oophoritis in mice. II. Requirement of Lyt-1 cells in normal female mice for the prevention of oophoritis. J. Exp. Med. **156:** 1577–1586.
40. Glazier, A., P. J. Tutschka, E. R. Farmer & G. W. Santos. 1983. Graft-versus-host disease in cyclosporine A-treated rats after syngeneic and autologous bone marrow reconstitution. J. Exp. Med. **158:** 1–8.
41. Cheney, R. T. & J. Sprent. 1985. Capacity of cyclosporine to induce graft-versus-host disease and impair intrathymic T cell differentiation. Transplant. Proc. **17:** 528–530.
42. Sakaguchi, S. & N. Sakaguchi. 1990. Thymus and autoimmunity: Capacity of the normal thymus to produce pathogenic self-reactive T cells and conditions required for the induction of autoimmune disease. J. Exp. Med. **172:** 537–545.
43. Cohen, I. R. 1991. Autoimmunity to chaperonins in the pathogenesis of arthritis and diabetes. Annu. Rev. Immunol. **9:** 567–590.

Positive Selection of the T-Cell Repertoire Is Affected by Mutations in the Peptide-binding Site of MHC Class I Molecules

JANKO NIKOLIĆ-ŽUGIĆ AND RUBENDRA DYALL

Laboratory of T Cell Development
Immunology Program
Memorial Sloan-Kettering Cancer Center
New York, New York 10021

INTRODUCTION

T cells recognize fragments of foreign antigen bound to self–major histocompatibility complex (MHC)–encoded molecules. In the thymus, these molecules shape the T-cell repertoire by positive and negative selection.[1–4] The $\alpha\beta$ T-cell receptor (TcR) of immature thymocytes must interact with MHC molecules expressed by thymic cortical epithelium for the thymocyte to be positively selected.[1–4] Thymocytes that fail to contact self-MHC will die from neglect.

The class I MHC molecule peptide-binding groove is composed of a "floor" of antiparallel β strands and "walls" of two long α-helices[5] (I. A. Wilson, personal communication). During MHC-restricted recognition of foreign peptides the T-cell receptor most likely contacts the outer face of the α helices and the peptide lying between them.[6–8] Positive selection of the T-cell repertoire occurs in the thymus in the absence of foreign peptides. It is not known what the relative role is of the outer face of the α helices and the peptide binding groove in this selection process. To address this, we compared the ability of a number of H-2K^b variant MHC molecules (TABLE 1) to select a K^b restricted repertoire. Most interest focused on two K^b mutants, H-2K^{bm8} (changes at positions 22, 23, 24, 30) and H-2K^{bm5} (position 116), with substitutions only in the β strands forming the floor of the groove.[9,10] These changes are probably not detectable on the exterior of the molecule, and in fact these variants are serologically indistinguishable from wild-type K^b with a battery of antibodies.[9–13] Structurally, these changes are inaccessible to TcR, and their effects on T-cell recognition are likely to be due to changes in peptide binding.[10] Comparison of the structure of HLA-A2 and HLA-Aw68, with 10 residue differences in amino acids lining the floor and walls of the groove, shows that each substitution causes only very local structural changes.[13] The backbone structure of the two molecules is the same and unaltered residues remain essentially unperturbed. The same principle holds for evolutionary distant molecules, like HLA-A2 and a murine class I K^b molecule (I. A. Wilson, personal communication). For these reasons, the α-helical regions of K^{bm5} and K^{bm8} are expected to be identical to those of K^b. Two other K^b mutants, H-2K^{bm1} (positions 152, 155, 156) and H-2K^{bm3} (positions 77 and 89), have substitutions on the α helices that locally affect serological sites[9] and TcR binding, as well as point into the groove and affect peptide presentation.[10]

C57BL/6 (B6, H-2^b) mice are able to mount an ovalbumin (OVA)-specific CTL response when immunized with syngeneic cells transfected with an OVA vector or with spleen cells carrying OVA.[15,16] This CTL activity is restricted by the H-2K^b class I molecule and is specific for the OVA peptide contained within the fragment 253-276. The response is very heterogeneous inasmuch as analysis of clones and long-term lines has shown that T-cell receptors Vβ5.1, Vβ5.2, Vβ8, Vβ9, and Vβ15 (F. R. Carbone and J. Nikolić-Žugić, unpublished results) as well as other Vβs are represented in the CTL response. In an H-2^b animal, K^b can therefore positively select OVA-specific T cells.

In this report we use the K^b-restricted OVA-specific response to examine the positive selection by various K^{bm} molecules. We provide evidence that self-peptides, presented in the groove of the class I molecule on thymic epithelial cells, are critically involved in positive selection of the OVA-specific T-cell repertoire.

TABLE 1. Characteristics of K^b Mutations

Mutant	Position of Altered Amino Acid	Classification of Residue[10]
K^{bm1}	152 Glu → Ala	ligand/TcR
	155 Arg → Tyr	TcR/ligand
	156 Leu → Tyr	ligand
K^{bm3}	77 Asp → Ser	ligand
	89 Lys → Ala	silent
K^{bm5}	116 Tyr → Phe	ligand
K^{bm8}	22 Tyr → Phe	ligand
	23 Met → Ile	silent
	24 Glu → Ser	ligand
	30 Tyr → Asn	silent

RESULTS AND DISCUSSION

To study the influence of substitutions in the class I molecule on the selection of the K^b-restricted repertoire, $F_1 \rightarrow P$ bone marrow chimeras were constructed by injecting bone marrow cells from (B6.PL-Thyl^a K^b × K^{bm})F_1 mice into lethally irradiated B6 or K^b-mutant mice. When bone marrow–derived T cells had matured in the irradiated hosts (8–12 weeks later), the mice were primed with OVA on H-2^b × H-2^{bm} cells. Spleen cells from these mice were boosted *in vitro* with E.G7-OVA (an EL4 transfectant-synthesizing OVA), F_1 cells coated with OVA peptide 253-276, or with allogeneic H-2^d spleen cells; CTL activity was measured 5 days later. The H-2^d–specific effector cells were typed for their expression of Thy1.1 (indicating bone marrow origin) or Thy1.2 only (indicating host origin). Cells that matured in an irradiated K^{bm1} or K^{bm8} host did not respond to K^b plus OVA (TABLE 2). On the other hand, T cells that matured in a wild-type K^b, K^{bm3}, or K^{bm5} host responded to OVA measured by the specific lysis of E.G7-OVA targets (TABLE 2). Spleen cells from all groups mounted a significant and comparable response to allogenic cells in primary mixed lymphocyte culture, documenting general immunocompetence of chimeric animals (12, and not shown). Seventy to ninety-two percent of the alloreactive CTL activity was sensitive to anti-Thy1.1

TABLE 2. The Ability of K^b-mutant MHC Molecules To Select the CTL Response to OVA Assayed in [$F_1 \rightarrow$ Parent] Radiation Chimeras[a]

	OVA-specific Response		Alloreactive Response		
	Percent Specific Lysis of Targets		Percent Specific Lysis of Targets		
Radiation Chimeras	EL4	E.G7	EL4	P815	Percent Donor* T Cells
Experiment 1					
[(B6.PL × bm1)$F_1 \rightarrow$ bm1]	3/0	0/1	4/2	48/30	92
[(B6.PL × bm3)$F_1 \rightarrow$ bm3]	10/6	41/27	2/0	41/18	91
[(B6.PL × bm5)$F_1 \rightarrow$ bm5]	8/4	37/27	7/4	47/24	83
[(B6.PL × bm8)$F_1 \rightarrow$ bm8]	6/3	10/10	6/3	43/24	74
Experiment 2					
[(B6.PL × bm5)$F_1 \rightarrow$ B6]	8/3	56/47	3/4	70/49	83
[(B6.PL × bm5)$F_1 \rightarrow$ bm5]	7/4	56/42	6/3	77/58	85
[(B6.PL × bm8)$F_1 \rightarrow$ B6]	2/3	58/54	4/2	65/47	78
[(B6.PL × bm8)$F_1 \rightarrow$ bm8]	10/6	7/8	6/4	69/54	70

[a] Radiation chimeras were made by injecting 10^7 F_1 bone marrow cells depleted of T cells into 950-rad lethally irradiated recipients of the indicated type (F_1 bone marrow $\rightarrow$ irradiated host).[1,2] The chimeras were maintained on antibiotic water and were primed after 8–12 weeks by iv injection of 2.5×10^7 irradiated spleen cells containing OVA as described previously.[12] In experiment 1 spleen cells syngeneic with the marrow donor were used to prime, and in experiment 2 B6.PL spleen cells were used. Seven days after priming, spleen cells from the chimeras were stimulated in culture with irradiated (120 Gy) E.G7-OVA cells[13] to boost the OVA-specific response. After 5 days in culture, CTL activity was assayed in a 3-h ^{51}Cr-release assay. Percent specific lysis at two effector : target ratios are shown: 6 : 1 and 2 : 1 in experiment 1, and 100 : 1 and 33 : 1 in experiment 2.

antibody and complement and therefore due to cytotoxic T cell (CTL) of donor (bone marrow) origin. The failure of the [(B6.PL × bm8) $\rightarrow$ bm8] chimeras to respond to OVA shows that changes in the MHC peptide-binding site can profoundly influence positive selection and implies strongly that self-peptides participate in this process.

The nonresponsiveness of the ($F_1 \rightarrow$ bm8) and ($F_1 \rightarrow$ bm1) chimeras to K^b-plus-OVA could be due to tolerance to the mutant K^b molecule. This is unlikely because, as shown in TABLE 3, (B6.PL × bm8)F_1 and (B6.PL) × bm1)F_1 mice do respond to OVA plus K^b. The homozygous mutant mice B6.C-H-2^{bm1} and B6.C-H-2^{bm8}, however, are themselves nonresponders to OVA. The other mutant mice, B6.C-H-2^{bm3} and B6.C-H-2^{bm5}, mount a CTL response to OVA that cross-reacts on OVA plus K^b (TABLE 3). A correlation, therefore, exists between the ability of four K^b mutants to make a CTL response to OVA and their ability to select the OVA-plus-K^b repertoire in radiation chimeras. To study this more closely, we analyzed the ability of the K^b variants to present the K^b-restricted OVA peptide *in vitro*. Several primary (heterogeneous) CTL lines form B6 mice specific for K^b-plus-$OVA_{253\text{-}276}$ were tested for lysis of target cells expressing different class I molecules, and target cells expressing K^{bm1} or K^{bm8} were in no case recognized in the presence of peptide.[12,13] In addition, none of the 11 clones from B6 mice specific for K^b-plus-$OVA_{253\text{-}276}$ could react with OVA-plus-K^{bm1} or K^{bm8} [13] (J. Nikolić-Žugić, unpublished results).

The inability of cells derived from K^{bm1} or K^{bm8} mice to present the OVA peptide suggests a trivial explanation for the failure of such mice to select the response in the ($F_1 \rightarrow$ Parent) chimeras. Although we typed the alloreactive CTL as >70% of bone marrow donor origin, it could be that the cells in the chimeras that present OVA to resting CTL are still exclusively of host origin. If this were the case, then (B6.PL × bm1 → bm1) and (B6.PL × bm8 → bm8) chimeras would have inappropriate presenting cells for the K^b-plus-OVA response. This explanation is unlikely to be true because previous work suggests that antigen-presenting cells turn over rapidly following irradiation,[17] and we always injected the mice with OVA-bearing immunogenic cells that expressed H-2K^b. We did test this possibility, however, by assaying the ability of spleen cells from these chimeras to stimulate anti–H-2K^b CTL. Splenocytes from (B6.PL × bm1 → bm1) and (B6.PL × bm8 → bm8) radiation chimeras were shown to be as effective in stimulating a K^b-specific CTL response as cells from normal F_1 animals (not shown). Thus the nonresponding chimeras do have appropriate H-2K^b–expressing presenting cells.

The H-2K^{bm8} molecule differs from wild-type H-2K^b in four substitutions in a β strand in the floor of the peptide-binding site. Two of these, at positions 22 and 24, point up into the site and contact the bound peptide, whereas the two others are considered silent with 23 pointing down and 30 located away from the peptide or TcR contact area[10] (TABLE 1). Functional studies with the point mutants of the H-2K^{bm8} indicate that the change at position 24 completely disrupts presentation of OVA, and the one at 22 does so to a lesser extent.[18] Mutations at two other positions had no effect on antigen presentation. The inability of K^{bm8} to select the heterogeneous response to K^b-plus-$OVA_{253\text{-}276}$ suggests very strongly that self-peptides associated with the class I molecules of thymic epithelial cells play a critical role in selecting the T-cell repertoire. The idea that tissue-specific self-

TABLE 3. Failure To Respond to OVA in the [$F_1 \rightarrow$ Parent] Radiation Chimeras Is Not Due to Tolerance[a]

Responder	Percent Specific Lysis on Targets: E.G7-OVA	EL4
Experiment 1		
B6	60/41	17/9
(B6 × bm1)F_1	47/30	18/7
(B6 × bm3)F_1	66/53	25/8
(B6 × bm5)F_1	58/46	14/6
(B6 × bm8)F_1	53/36	9/8
Experiment 2		
B6	60/53	0/0
bm1	4/2	2/2
bm3	41/25	3/0
bm5	54/39	3/1
bm8	12/6	23/15

[a] The indicated responder mice were immunized by intravenous injection of 2.5×10^7 1000-rad irradiated syngeneic spleen cells that had been cytoplasmically "loaded" with OVA.[15] Seven days later spleen cells from the primed animals were restimulated in culture with syngeneic spleen cells that had been incubated with a CNBr digest of OVA, 100 μg/mL, and washed. The CNBr fragments of OVA include the K^b-restricted targeting peptide. CTL activity was measured 5 days later and is expressed as percent specific lysis at effector : target ratios 50 : 1 and 17 : 1. The ability of the CTL to lyse E.G7-OVA targets correlated with their ability to lyse syngeneic K^b-mutant targets in the presence of the OVA peptide.

peptides[19,20] or erroneous self-peptides[21] are involved in positive selection has been suggested previously, and in some cases background genes (*i.e.* not the restriction element) have been shown to influence selection.[22–24]

Another explanation for our results is that one or more of the K^{bm8} mutations act by allosterically altering the architecture of the α helices, thereby influencing the MHC class I contact with TcR. Although x-ray crystallography is the only technique that can directly answer this issue, available serologic, structural, and functional (see below) results argue strongly against this explanation. The fact that for four H-$2K^{bm}$ moelcules the ability to positively select the OVA-plus-K^b-specific T-cell repertoire in the thymus correlates with their ability to present the OVA peptide indicates that self-peptides use some form of molecular mimicking to positively select the T-cell repertoire for foreign peptides. Similar effects have been observed in three independent TcR transgenic models.[25–27] The mechanism for this mimicry is suggested by the studies of Garrett *et al.*[14] describing the three-dimensional structure of a second human class I protein, HLA-Aw68. Comparison of the HLA-A2 and HLA-Aw68 structures revealed that MHC polymorphism apparently alters subsites or pockets within the binding cleft, leaving the overall architecture of the molecule unchanged. Such changes clearly alter peptide binding. A unique set of peptides was bound to the HLA-Aw68 molecule inasmuch as the non-HLA–derived electron density in the cleft differed from that found in HLA-A2.[14] It is the specific requirements of the individual subsites within the cleft that give rise to the observed differences in peptide binding, and, probably, peptide conformation.

Modeling of H-$2K^b$ and H-$2K^{bm8}$, according to the HLA-A structure, predicts that the mutations in H-$2K^{bm8}$ are likely to alter an important antigen-contacting pocket, termed pocket B.[28] Recent crystallographic elucidation of the structure of K^b supports the validity of such modeling, as K^b and HLA-A2 appear virtually identical in α_1 and α_2 domain, and only slight changes that could be accounted for by crystal variation are observed in the α_3 domain and β_2-microglobulin molecule.[28a,b] An implication from this modeling is that other parts of the H-$2K^{bm8}$ peptide-binding groove should be identical to that of H-$2K^b$. Using the H-$2K^b$ restricted peptide derived from the vesicular stomatitis virus nucleocapsid protein (VSV $N_{52\text{-}59}$),[29] we could demonstrate that the rest of the H-$2K^{bm8}$ peptide-binding groove is functionally equivalent to H-$2K^b$, because H-$2K^{bm8}$ can efficiently present VSV N to H-$2K^b$– and H-$2K^{bm8}$–restricted T cells (J. Nikolić-Žugić and R. Dyall, unpublished results and TABLE 4). Based on these results, we speculated that the defect in OVA presentation and repertoire selection by H-$2K^{bm8}$ is due to the mutation that renders the critical peptide-binding pocket nonpermissive for binding of OVA-mimicking self-peptide(s) and OVA peptide; the mutation may also change the conformation of the bound peptides to the point at which they are no longer recognized by OVA-specific T cells. The binding of self-peptides is difficult to study, but we studied the interaction of H-$2K^{bm8}$ and OVA peptide. We raised CTL against VSV N and examined the ability of the OVA peptide to competitively inhibit presentation of VSV N by H-$2K^b$ or H-$2K^{bm8}$. An influenza nucleoprotein peptide, $NP_{365\text{-}380}$, which binds to D^b, was used as a control, and this peptide had no effect on either OVA or VSV presentation. $OVA_{253\text{-}276}$ could readily inhibit the interaction of VSV N and H-$2K^b$ but had no effect on VSV N : H-$2K^{bm8}$ interaction (TABLE 4, Exp. 1). When the same experiment, however, was performed with a minimal determinant $OVA_{257\text{-}264}$ (TABLE 4, Exp. 2) or a slightly larger peptide $OVA_{257\text{-}265}$ (not shown), the results were quite different. Both K^b and K^{bm8} targets were prevented from binding VSV peptide, as evidenced by lack of killing by specific CTL. H-$2K^{bm8}$ is therefore capable of binding the OVA peptide. It is still possible that the affinity of the octameric peptide for K^b is higher than for K^{bm8}, but we could not observe this difference in our competition experiments. By that argument, a likely explanation for the results obtained with a longer peptide is that an octameric contaminant, able to efficiently bind to H-2K^b, but not H-2K^{bm8},

TABLE 4. Inhibition of K^b and K^{bm8}-restricted VSV N Presentation by Different OVA Peptides[a]

Presenting Molecule	Competitor	Molar Excess	Percent Relative ^{51}Cr Release
Experiment 1			
H-2K^b	$NP_{365\text{-}380}$	360	95.2
		120	99.8
		40	102.1
		20	97.2
	$OVA_{253\text{-}276}$	360	22.7
		120	56.4
		40	87.9
		20	98.3
H-2K^{bm8}	$NP_{375\text{-}380}$	360	94.9
		120	101.9
		40	100.1
		20	99.9
	$OVA_{253\text{-}276}$	360	99.2
		120	98.7
		40	103.4
		20	99.4
Experiment 2			
H-2K^b	$OVA_{257\text{-}264}$	1250	7.3
		125	7.3
		12.5	12.7
H-2K^{bm8}	$OVA_{257\text{-}264}$	1250	6.0
		125	7.0
		12.5	8.4

[a] An effector CTL line was derived from H-2K^{bm8} mice after the ip immunization with 10^6 plaque-forming units (PFU) of VSV. Target cells expressing H-2K^b or H-2K^{bm8} [13] were preincubated with indicated concentrations of inhibiting peptides for 15 min before the addition of 1 μM (exp. 1) or 0.5 μM (exp. 2) of VSV $N_{53\text{-}63}$ peptide. After 1 h of incubation, ^{51}Cr-labeled targets were washed and incubated for 3 h with anti-VSV CTL line at 20 : 1 effector to target ratio. The results are expressed as:

$$\text{Percent relative } {}^{51}\text{Cr-release} = \frac{\text{Percent specific } {}^{51}\text{Cr-release in the presence of competitor}}{\text{Percent specific } {}^{51}\text{Cr-release without competitor}}.$$

Actual ^{51}Cr-release without inhibitor was between 33% and 83%.

was present in the preparation of $OVA^{253\text{-}276}$. The concentration of this contaminant was probably low enough to reveal the difference in binding. Experiments are in progress to address this issue.

The prediction of the theory of molecular mimicry is that the H-2K^{bm8} molecule must not only present VSV N but also be able to positively select the T-cell repertoire specific for it. TABLE 5 shows that this is indeed the case, as ($F_1 \rightarrow$ bm8) chimeric animals can mount a strong anti–VSV N CTL response that is restricted to H-2K^b, evidenced by the ability of these cells to lyse a VSV N transfectant of EL4.[30]

Therefore, the H-2K^b–restricted antigens, OVA and VSV N, may have different pocket-binding requirements in the peptide-binding site. Both peptides occupy most of the site, because they can compete with each other for binding to H-2K^b, and octa- and nonapeptides bind to K^b in an extended conformation across the entire binding cleft (I. A. Wilson, personal communication). Critical contact for these peptides, however, is probably provided by different binding pockets or contact residues. Peptide motifs necessary for binding to MHC class I molecules were defined by Rammensee's group[31] and were proven correct by crystallography[32] and by predicting peptide epitopes.[33] These motifs involve usually two out of eight or nine residues in a peptide. Positions 3 and 5 and a hydophobic residue (L, I, etc.) at position 8 probably make up the MHC contact site of the K^b-bound peptides, including those of OVA$_{257\text{-}264}$ and VSV N$_{52\text{-}59}$. Good recognition of VSV N and a complete lack of recognition of OVA$_{253\text{-}276}$ and OVA$_{257\text{-}264}$ [12,13] (not shown and TABLE 4) presented by K^{bm8} implicate that certain pockets in the peptide-binding cleft can have profound effects on peptide conformation, as seen by TcR. These issues will be resolved by an x-ray crystallographic analysis of the K^b-OVA complex and subsequent functional experiments.

TABLE 5. Response of $F_1 \rightarrow$ P Chimeras to VSV N Protein[a]

Radiation Chimeras	Percent Specific Lysis EL4	N1[b]
[(B6.PL × bm5)$F_1 \rightarrow$ bm5]	2/1	70/64
[(B6.PL × bm5)$F_1 \rightarrow$ B6]	1/1	69/59
[(B6.PL × bm8)$F_1 \rightarrow$ bm8]	2/1	72/62
[(B6.PL × bm8)$F_1 \rightarrow$ B6]	0/2	67/64

[a] Experiments were performed exactly as described in the legend to TABLE 2, except for immunization of chimeras (10^6 PFU VSV ip) and *in vitro* stimulation (25×10^6 syngeneic F_1 spleen cells coated with VSV N peptide).

[b] N1 cells (N protein transfected EL4 line)[30] were used as targets in ^{51}Cr-release assay.

Either way, the behavior of OVA- and VSV-mimicking self-peptides during the positive selection of the T-cell repertoire specific for these antigens is quite similar to that of OVA and VSV peptides. The alteration of the pocket B in H-2K^{bm8} renders this molecule incapable of OVA presentation and anti-OVA T-cell repertoire selection, but that alteration does not affect presentation of, and repertoire selection for, VSV N$_1$ by H-2K^{bm8}. This additionally demonstrates very local effects of H-2K^{bm8} mutation, rendering the change of helical structure(s) by an allosteric effect even less likely.

Which peptides govern positive selection and how the interplay between the positive and negative selection allow the emergence of broad and nonautoaggressive T-cell repertoire are discussed elsewhere.[34,35] The correlation between antigen presentation and repertoire selection by a given MHC molecule may not be strict for all antigens.[36] A widely accepted fact is that most MHC class I molecules cannot fold correctly and are biochemically unstable at the cell surface without a bound peptide. As these molecules by default carry self-peptides, these peptides will be present at the time of positive selection. All the TcRs that have a "snug fit" for the MHC : self-peptide complex will be critically dependent on the peptide. Whether there are other TcRs that can be selected by a loose interaction, and that are therefore less dependent on peptides, remains to be seen. But other MHC molecules may be stable without the peptide (*e.g.*, L^d [37]), and it is critical to elucidate the requirements for positive selection in this case. In addition, a case that a nonpresenting molecule can mediate positive selection of a class II–restricted transgenic TcR has been recently reported by Kaye *et al.*,[38] but the transgenic TcR heterodimer was engineered from α and β chains that do not naturally form a TcR, and the MHC restriction element for this receptor differed from the positively selecting MHC molecule. It is possible that positive selection on class II molecules happens with a great deal of promiscuity, given the variable length of peptides associated with these molecules.[39] At present, it is not clear what the relative physiological significance of the above[12,25–27,37] observations is. As other defined MHC class I- and class II–restricted antigenic peptides become available, we should be able to determine how general the principles are that apply for the antigens studied here.

SUMMARY

$H\text{-}2K^b$ mutant molecules ($H\text{-}2K^{bm}$) and the $H\text{-}2K^b$–restricted response to OVA and VSV N peptides were used to investigate the influence of polymorphism of structurally defined regions of the MHC class I molecules on intrathymic positive selection of the T-cell repertoire. We show that the positive selection of the T-cell repertoire in the thymus requires the self-peptide to be present in the MHC antigen-binding site. A correlation between the ability of four MHC molecules to present antigenic peptide and to positively select T cells specific for it was noted. The self-peptides involved in positive selection may therefore mimic the foreign peptide during intrathymic selection. A structural correlate of this mimicry may be a similar or identical binding requirement for the antigen-binding pocket(s)/residues of the MHC peptide-binding site.

REFERENCES

1. BEVAN, M. J.1977. Nature (London) **269:** 417–418.
2. ZINKERNAGEL, R. M. et al. 1978. J. Exp. Med. **147:** 882–896.
3. KISIELOW, P. *et al.* 1988. Nature (London) **335:** 730–733.
4. SHA, W. C. *et al.* 1988. Nature (London) **336:** 73–76.
5. BJORKMAN, P. J. *et al.* 1987. Nature (London) **329:** 506–512.
6. DAVIS, M. M. & P. J. BJORKMAN. 1988. Nature (London) **334:** 395–402.
7. CLAVERIE, J. M., A. PROCHNICKA-CHALUFOUR & L. BOUGUELERET. 1989. Immunology Today **107:** 10–14.
8. CHOTHIA, C., D. R. BOSWELL & A. M. LESK. 1988. EMBO J. **7:** 3745–3755.
9. NATHENSON, S. G. *et al.* 1986. Annu. Rev. Immunol. **4:** 471–502.
10. BJORKMAN, P. J. *et al.* 1987. Nature (London) **329:** 512–518.
11. BLUESTONE, J. A. *et al.* 1986. J. Immunol. **137:** 1244–1250.
12. NIKOLIĆ-ŽUGIĆ, J. & M. J. BEVAN. 1990. Nature (London) **344:** 65–67.
13. NIKOLIĆ-ŽUGIĆ, J. & F. R. CARBONE. 1990. Eur. J. Immunol. **20:** 2431.
14. GARRETT, T. P. J. *et al.* 1989. Nature (London) **342:** 692–696.
15. MOORE, M. W., F. R. CARBONE & M. J. BEVAN. 1988. Cell **54:** 777–785.

16. CARBONE, F. R. & M. J. BEVAN. 1990. J. Exp. Med. **171:** 377–387.
17. RON, Y., D. LO & J. SPRENT. 1986. J. Immunol. **137:** 1764–1771.
18. PULLEN, J. A., D. H. HUNT & L. R. PEASE. 1991. J. Immunol. **146:** 2145–2151.
19. MARRACK, P. & J. KAPPLER. 1988. Immunology Today **9:** 308–315.
20. MARRACK, P., J. MCCORMACK & K. KAPPLER. 1989. Nature (London) **338:** 503–505.
21. KOURILSKY, P. & J. M. CLAVERIE. 1989. Cell **56:** 327–329.
22. OGATA, M. *et al.* 1986. J. Immunol. **136:** 1178–1185.
23. SINGER, A. *et al.* 1986. Prog. Immunology **VI:** 60–66.
24. FRY, A. M., M. M. COTTERMAN & L. A. MATIS. 1989. J. Immunol. **143:** 2723–2729.
25. BERG, L. J., G. D. FRANK & M. M. DAVIS. 1990. Cell **60:** 1043–1053.
26. SHA, W. C. *et al.* 1990. Proc. Natl. Acad. Sci. USA **87:** 6168.
27. JACOBS, H. *et al.* 1990. Eur. J. Immunol. **20:** 2333–2337.
28. SAPER, M. A., P. BJORKMAN & D. C. WILEY. 1991. J. Mol. Biol. **219:** 277–319.
28a. FREMONT, D. H. *et al.* 1992. Science **257:** 919–927.
28b. MATSUMURA, M. *et al.* 1992. Science **257:** 927–934.
29. VAN BLEEK, G. M. & S. G. NATHENSON. 1990. Nature (London) **348:** 213.
30. PUDDINGTON, L., M. J. BEVAN, J. K. ROSE & L. LEFRANCOIS. 1986. J. Virol. **60:** 708–715.
31. FALK, K. *et al.* 1991. Nature (London) **351:** 290.
32. JARDETZKY, A. *et al.* 1991. Nature (London) **353:** 326.
33. PAMER, E., J. HARTY & M. J. BEVAN. 1991. Nature (London) **353:** 852–855.
34. NIKOLIĆ-ŽUGIĆ, J. & F. R. CARBONE. 1991. Year Immunol.
35. NIKOLIĆ-ŽUGIĆ, J. 1991. Immunology Today **12:** 65–70.
36. KAST, W. M., L. P. DEWAAL & C. J. M. MELEIF. 1984. J. Exp. Med. **160:** 1752–1766.
37. LIE, W.-R. *et al.* 1990. Nature (London) **344:** 439–441.
38. KAYE, J., N. VASQUEZ & S. M. HEDRICK. 1991. J. Immunol. **148:** 3342–3353.
39. RUDENSKY, A. Y. *et al.* 1991. Nature (London) **353:** 622–627.

Mechanisms of Helper T-Cell Regulation of B-Cell Activity[a]

ROBERT L. COFFMAN

Department of Immunology
DNAX Research Institute
901 California Avenue
Palo Alto, California 94304

THE HETEROGENEITY OF CD4+ T CELLS

It is now recognized that $CD4^+$ or helper T lymphocytes (Th cells) comprise a number of functionally distinct subsets. These subsets, although not readily identified or separated by cell-surface phenotype, are characterized by quite distinct patterns of cytokine production. At least three major subsets are recognized in both mouse and humans: Th1 cells that produce IL-2 and IFN-γ, but not IL-4, IL-5, or IL-10; Th2 cells that produce IL-4, IL-5, and IL-10, but not IL-2 or IFN-γ; and Th0 cells that produce all of these cytokines.[1,2] In addition, a number of other cytokines, such as IL-3, GM-CSF, and TNF-α are produced by all subsets, although not necessarily in similar amounts. Whether all populations of activated Th cells are mixtures of only these subsets, or contain cells with stable phenotypes distinctly different from these three, is still unclear.[2] In this review, we will discuss primarily the mechanisms by which the Th1 and Th2 subsets provide helper functions to B cells, inasmuch as little analysis has been done yet with Th0 cells.

Because nearly all of the regulatory and effector functions of $CD4^+$ T cells are mediated by secreted cytokines, these distinctly different cytokine patterns are causally linked to the functions of each subset rather than being empirically correlated with them. The major exception is the Th contact-mediated activation of resting B cells, which, however, is a property of all three subsets.[3,4] An important feature of the sets of cytokines produced by Th1 and Th2 cells is that they include factors that specifically inhibit effector functions stimulated by the other subset and factors that inhibit the differentiation, proliferation, or function of the opposite subset. This appears to be the basis for the reciprocal and mutually inhibitory relationship between cell-mediated and humoral immunity,[5,6] processes that depend largely on Th1 and Th2 cells, respectively.

Several important tools have become available in the past few years that have permitted a detailed dissection of the mechanisms by which $CD4^+$ Th cells regulate the activity of B cells. One of these tools includes clonal long-term Th lines with defined antigen specificities and patterns of cytokine production. Others include highly purified recombinant cytokines and neutralizing monoclonal antibodies specific for those cytokines. More recently, the definition and cloning of several of the cell-surface molecules that are important for Th-B interactions has had an important impact. Largely for technical reasons, a clearer synthesis of the mechanisms of helper function has emerged from studies in the mouse rather than in humans, and it is primarily this work that will be summarized here.

[a] The DNAX Research Institute is supported by the Schering-Plough Corporation.

B-CELL ACTIVATION AND PROLIFERATION

Nearly all of the functions of $CD4^+$ T cells are mediated by secreted products; the notable exception is the activation of resting B cells.[3,7–9] Although there are many reports in the older literature of Th supernatants and partially purified factors that can cause resting B cells to enter the cell cycle, none have held up under rigorous scrutiny. Common sources of artifact include impure preparations of resting B cells and mycoplasma contamination of Th lines. Certain cytokines, especially IL-4, can induce some very early changes in resting B cells,[10] but full activation and entry into the cell cycle require direct contact with activated T cells. This contact-mediated signal can be delivered by recently activated cells, not resting Th cells; but it does not require direct recognition of the antigen by either the T or B cells, and is it not MHC-restricted.[3,8,9,11,12] Under normal conditions, of course, a Th cell would be activated by recognition of antigen processed and presented by the same B cell that then responds to contact with the activated Th cell.[13] The entire process would appear antigen-dependent and MHC-restricted, although that is true only for the activation of the Th cell.

This contact-mediated signal can be delivered quite efficiently with isolated plasma membranes from activated Th cells.[14] The key cell-surface molecules mediating this interaction have recently been identified as CD40 on B cells and the CD40 ligand on T cells.[15,16] Nearly all splenic B cells will respond to this signal and will proliferate extensively without further additions.[14,17,18] Proliferation can be enhanced, however, with some Th cytokines, especially IL-4 and IL-5. Differentiation to high-rate immunoglobulin-producing plasma cells, however, does not occur in the absence of added cytokines.[3] The principal differentiation factor from Th1 cells is IL-2, whereas the most potent factors from Th2 cells are IL-4 and IL-5.

THE REGULATION OF IMMUNOGLOBULIN-ISOTYPE SWITCHING

In addition to stimulating the overall activation, clonal expansion, and terminal differentiation of B cells, Th cells also regulate isotype switching. The presence of IL-4 during the first 2–3 days after B-cell activation will substantially enhance switching to IgG1 and is absolutely required for switching to IgE.[17,19] Similarly, IFN-γ stimulates switching to IgG2a, and to some extent to IgG3.[20] These factors are mutually antagonistic, such that IFN-γ can completely inhibit IL-4–mediated IgG1 and IgE switching, and IL-4 will inhibit IgG2a and IgG3 switching.[21,22] Some of these activities are apparent from studies with Th clones. Thus, only Th2 cells are able to provide helper function for primary IgE responses, whereas Th1 clones can act as preferential helpers for IgG2a responses and inhibitors of IgE responses.[3] The only other factor shown to induce specific isotype switching in the mouse is TGF-β, which specifically induces IgA switching.[23] TGF-β appears to be produced by Th cells but is made by many other cell types as well.

Both Th subsets stimulate the production of IgG1, the major non-IgM isotype produced in *in vitro* responses, even though IL-4 appears to be the only cytokine known to induce this switch. The explanation comes from the recent demonstration that the contact-mediated activation signal, as delivered by plasma membranes or intact cells, induces germ-line Cγ1 gene transcription, the essential prerequisite step for IgG1 switching.[24] The importance of these two cytokines *in vivo* has been demonstrated in a number of models with recombinant cytokines,[25,26] anticytokine antibodies,[27] and, for IL-4, targeted disruption of the IL-4 gene.[28] The observation that mice with nonfunctional IL-4 genes make significant, although subnormal, IgG1 responses, but no IgE responses, is consistent with the existence of another signal that can induce IgG1, but not IgE, switching. The details of isotype switching in humans have not been elaborated as clearly as in the mouse. The induction of IgE by IL-4[29] and of IgA by TGF-β[30] have been reproduced in human B-cell cultures, however, and the induction of IgE responses only by IL-4–producing human Th clones has been shown.[31,32]

REFERENCES

1. MOSMANN, T. R. & R. L. COFFMAN. 1989. Heterogeneity of cytokine secretion patterns and functions of helper T cells. Adv. Immunol. **46:** 111–147.
2. MOSMANN, T. R., J. H. SCHUMACHER, N. F. STREET, R. BUDD, A. O'GARRA, T. A. FONG, M. W. BOND, K. W. MOORE, A. SHER & D. F. FIORENTINO. 1991. Diversity of cytokine synthesis and function of mouse CD4⁻ T cells. Immunol. Rev. **123:** 209–229.
3. COFFMAN, R. L., B. W. SEYMOUR, D. A. LEBMAN, D. D. HIRAKI, J. A. CHRISTIANSEN, B. SHRADER, H. M. CHERWINSKI, H. F. SAVELKOUL, F. D. FINKELMAN, M. W. BOND & T. R. MOSMANN. 1988. The role of helper T cell products in mouse B cell differentiation and isotype regulation. Immunol. Rev. **102:** 5–28.
4. HODGKIN, P. D., L. C. YAMASHITA, B. SEYMOUR, R. L. COFFMAN & M. R. KEHRY. 1991. Membranes from both Th1 and Th2 cell clones stimulate B cell proliferation and prepare B cells for lymphokine-induced differentiation to secrete Ig. J. Immunol. **147:** 3696–3702.
5. PARISH, C. R. 1972. The relationship between humoral and cell-mediated immunity. Transplant. Rev. **13:** 35–66.
6. LIEW, F. Y. 1982. Regulation of delayed-type hypersensitivity to pathogens and alloantigens. Immunol. Today **3:** 18–23.
7. SNOW, E. C. & R. J. NOELLE. 1987. Thymus-dependent antigenic stimulation of hapten-specific B lymphocytes. Immunol. Rev. **99:** 173–192.
8. WHALEN, B. J., H. P. TONY & D. C. PARKER. 1988. Characterization of the effector mechanism of help for antigen-presenting and bystander resting B cell growth mediated by Ia-restricted Th2. J. Immunol. **141:** 2230–2239.
9. OWENS, T. 1988. A noncognate interaction with anti-receptor antibody-activated helper T cells induces small resting murine B cells to proliferate and to secrete antibody. Eur. J. Immunol. **18:** 395–401.
10. RABIN, E. M., J. OHARA & W. E. PAUL. 1985. B-cell stimulatory factor 1 activates resting B cells. Proc. Natl. Acad. Sci. USA **82:** 2935–2939.
11. TONY, H. P. & D. C. PARKER. 1985. Major histocompatibility complex-restricted, polyclonal B cell responses resulting from helper T cell recognition of antiimmunoglobulin presented by small B lymphocytes. J. Exp. Med. **161:** 223–241.
12. TONY, H. P., N. E. PHILLIPS & D. C. PARKER. 1985. Role of membrane immunoglobulin (Ig) crosslinking in membrane Ig-mediated, major histocompatibility-restricted T cell-B cell cooperation. J. Exp. Med. **162:** 1695–1708.
13. VITETTA, E. S., R. FERNANDEZ-BOTRAN, C. D. MYERS & V. M. SANDERS. 1989. Cellular interactions in the humoral immune response. Adv. Immunol. **45:** 1–106.
14. HODGKIN, P. D., L. C. YAMASHITA, R. L. COFFMAN & M. R. KEHRY. 1990. Separation of events mediating B cell proliferation and Ig production by using T cell membranes and lymphokines. J. Immunol. **145:** 2025–2034.
15. ARMITAGE, R., W. FANSLOW, L. STROCKBINE, T. SATO, K. CLIFFORD, B. MACDUFF, D. ANDERSON, S. GIMPEL, T. DAVISSMITH, C. MALISZEWSKI, E. CLARK, C. SMITH, K. GRABSTEIN, D. COSMAN & M. SPRIGGS. 1992. Molecular and biological characterization of a murine ligand for cd40. Nature **357:** 80–82.
16. NOELLE, R. J., M. ROY, D. M. SHEPHERD, I. STAMENKOVIC, J. A. LEDBETTER & A.

ARUFFO. 1992. A 39-kDa protein on activated helper T cells binds CD40 and transduces the signal for cognate activation of B cells. Proc. Natl. Acad. Sci. USA **89:** 6550–6554.
17. LEBMAN, D. A. & R. L. COFFMAN. 1988. Interleukin 4 causes isotype switching to IgE in T cell-stimulated clonal B cell cultures. J. Exp. Med. **168:** 853–862.
18. RIEDEL, C., T. OWENS & G. J. NOSSAL. 1988. A significant proportion of normal resting B cells are induced to secrete immunoglobulin through contact with anti-receptor antibody-activated helper T cells in clonal cultures. Eur. J. Immunol. **18:** 403–408.
19. LAYTON, J. E., E. S. VITETTA, J. W. UHR & P. H. KRAMMER. 1984. Clonal analysis of B cells induced to secrete IgG by T cell-derived lymphokine(s). J. Exp. Med. **160:** 1850–1863.
20. LEE, F., T. YOKOTA, T. OTSUKA, P. MEYERSON, D. VILLARET, R. COFFMAN, T. R. MOSMANN, D. RENNICK, N. ROEHM, C. SMITH, A. ZLOTNIK & K. ARAI. 1986. Isolation and characterization of a mouse interleukin cDNA clone that expresses B-cell stimulatory factor 1 activities and T-cell- and mast-cell-stimulating activities. Proc. Natl. Acad. Sci. USA **83:** 2061–2065.
21. VITETTA, E. S., A. BOSSIE, R. FERNANDEZ-BOTRAN, C. D. MYERS & K. G. OLIVER. 1987. Interaction and activation of antigen-specific T and B cells. Immunol. Rev. **99:** 193–239.
22. BERGSTEDT-LINDQVIST, S., P. SIDERAS, H. R. MACDONALD & E. SEVERINSON. 1984. Regulation of Ig class secretion by soluble products of certain T-cell lines. Immunol. Rev. **78:** 25–50.
23. COFFMAN, R. L., D. A. LEBMAN & B. SHRADER. 1989. Transforming growth factor β specifically enhances IgA production by lipopolysaccharide-stimulated murine B lymphocytes. J. Exp. Med. **170:** 1039–1044.
24. SCHULTZ, C. L., P. ROTHMAN, R. KÜHN, M. KEHRY, W. MÜLLER, K. RAJEWSKY, F. ALT & R. L. COFFMAN. 1992. T helper cell membranes promote IL-4–independent expression of germ-line Cgamma1 transcripts in B cells. J. Immunol. **149:** 60–64.
25. FINKELMAN, F. D., I. M. KATONA, T. R. MOSMANN & R. L. COFFMAN. 1988. IFN-gamma regulates the isotypes of Ig secreted during *in vivo* humoral immune responses. J. Immunol. **140:** 1022–1027.
26. SAVELKOUL, H. F., B. W. SEYMOUR, L. SULLIVAN & R. L. COFFMAN. 1991. IL-4 can correct defective IgE production in SJA/9 mice. J. Immunol. **146:** 1801–1805.
27. FINKELMAN, F. D., J. HOLMES, I. M. KATONA, J. F. URBAN, JR., M. P. BECKMANN, L. S. PARK, K. A. SCHOOLEY, R. L. COFFMAN, T. R. MOSMANN & W. E. PAUL. 1990. Lymphokine control of *in vivo* immunoglobulin isotype selection. Annu. Rev. Immunol. **8:** 303–333.
28. KUHN, R., K. RAJEWSKY & W. MULLER. 1991. Generation and analysis of interleukin-4 deficient mice. Science **254:** 707–710.
29. PENE, J., F. ROUSSET, F. BRIERE, I. CHRETIEN, J. Y. BONNEFOY, H. SPITS, T. YOKOTA, N. ARAI, K. ARAI, J. BANCHEREAU & J. E. DE VRIES. 1988. IgE production by normal human lymphocytes is induced by interleukin 4 and suppressed by interferons gamma and alpha and prostaglandin E2. Proc. Natl. Acad. Sci. USA **85:** 6880–6884.
30. VAN VLASSELAER, P., J. PUNNONEN & J. DE VRIES. 1992. Transforming growth factor-β directs IgA switching in human B cells. J. Immunol. **148:** 2062–2067.
31. MAGGI, E., G. DEL PRETE, D. MACCHIA, P. PARRONCHI, A. TIRI, I. CHRETIEN, M. RICCI & S. ROMAGNANI. 1988. Profiles of lymphokine activities and helper function for IgE in human T cell clones. Eur. J. Immunol. **18:** 1045–1050.
32. PENE, J., F. ROUSSET, F. BRIERE, I. CHRETIEN, X. PALIARD, J. BANCHEREAU, H. SPITS & J. E. DE VRIES. 1988. IgE production by normal human B cells induced by alloreactive T cell clones is mediated by IL-4 and suppressed by IFN-gamma. J. Immunol. **141:** 1218–1224.

Steps in the Generation of Autoantibodies[a]

URSULA STORB

Department of Molecular Genetics and Cell Biology
University of Chicago
Chicago, Illinois 60637

Myasthenia gravis is an autoimmune disease in which autoantibodies against acetylcholine receptors clearly seem to play a major pathogenetic role (reviewed in ref. 1). The purpose of this paper is to briefly summarize what we know about the events that lead to antibody formation and to point out steps in this process where autoantibodies may arise or where, in the normal case, their appearance may be controlled (TABLE 1).

Antibodies are a product of B lymphocytes.[2] The early phase of the development of B lymphocytes occurs in the bone marrow where an apparently multipotent hemopoietic stem cell differentiates into pro-B cells.[3] These cells become capable of rearranging immunoglobulin (Ig) genes presumably by the activation of the recombinase activating genes, RAG-1 and RAG-2.[4] The process of Ig gene rearrangement requires the V(D)J recombinase, whose components are at present only incompletely understood. Rearrangement also requires that the Ig gene loci are "accessible" to the recombinase.[2,4] Accessibility seems to parallel transcriptional competence of an Ig locus, but the presence of a transcription complex on the target genes for the recombinase does not seem to be necessary.[5] Also, the target DNA must be at least partially undermethylated.[6,7] Although the details of accessibility are not understood, it is generally accepted that, in mouse and humans, heavy (H) and light (L) chain genes are sequentially activated and H chain genes are therefore rearranged before L chain genes.

Antibody specificities that are created by rearrangement of variable (V), diversity (D), and joining (J) elements of heavy chain, or V and J of light chain genes, seem to be entirely random.[2] Thus, there is a priori no inhibition of the appearance of autoantibodies. Antibody specificity created by rearrangement is a result of the particular germline V, (D), and J genes assembled, as well as the alterations occurring at the joining site. Deletions of a few nucleotides at one or both coding ends of a particular joint and the addition of untemplated nucleotides (N regions), in the case of H genes, result in a vastly increased diversity. Inasmuch as these alterations at the DJ, VD, and VJ joints affect the third hypervariable region of the heavy and light chains, they have a major impact on antibody specificity.

Because of the randomness of the joining events, conceivably autoantibodies are constantly produced against most kinds of self-antigens. Once a pre-B cell has acquired both H and L chains, a poorly understood feedback mechanism leads to the cessation of the V(D)J recombinase.[2] Thus, rearrangement stops and no new antibody specificities will arise by this process. Control mechanisms must also exist that cause the deletion or inactivation of B cells that produce autoantibodies. Examples of transgenic mice with autoantibody transgenes demonstrate this con-

[a] This work was supported by NIH Grants AI24780, HD23089, and GM38649.

trol. When mice, which carry transgenic H and L genes with specificity for histocompatibility antigens H-2K^k, do not carry this MHC specificity, most B cells have the transgene encoded Igs on the cell membrane, and antibodies to H-2K^k are present in the serum.[8,9] When the transgene, however, is bred into mice of the H-2K^k type, the mice lack transgene-positive B cells and serum antibodies encoded by the transgenes.[8,9] It appears that deletion of transgene-expressing cells occurs in the bone marrow in the transition from pre-B to B cells.[9]

In contrast to the clonal deletion in the above example, transgenic mice that produce antibodies against lysozyme show clonal B-cell anergy when a lysozyme-encoding transgene is also present.[10] The B cells are not deleted, but they down-regulate the membrane-bound IgM antibodies to lysozyme.[10] In these latter mice, as well as in mice that carry antiviral antibody transgenes, B-cell tolerance against lysozyme and the virus, respectively, can be broken by certain conditions.[11,12] Apparently, T-cell tolerance is crucial in this kind of B-cell anergy.[11-13] A variety of explanations have been considered for the two types of tolerance induction and maintenance, but the issue requires further investigation.[14]

TABLE 1. How Can Autoantibodies Arise or Be Prevented?

+[a]	Rearrangement of Ig genes is random—no prevention of the formation of autoantibodies
	Feedback inhibition of V(D)J recombinase; thus no new autoimmune activities will arise by rearrangement
	Clonal deletion of autoimmune-specific B cells in bone marrow (T)[b]
	Clonal anergy of autoimmune cells (T)
	Virgin B cells in periphery: repertoire selection (T)
+	Somatic hypermutation: random creation of new specificities
	Selection of high-affinity mutants; selection against autoantibodies? (T)
	Expansion of mutants by autoantigen (T)

[a] +: Stages at which new antibody specificities arise.

[b] (T): Stages at which T cells can be involved in an autoantigen-specific way. (T cells are presumably also involved during the rearrangement and somatic hypermutation of Ig genes, but in these events they do not direct or prevent the creation of antibody specificities).

Presumably, in some forms of autoimmunity the process of tolerance induction is compromised so that it fails to control autoantibodies that arise by the random rearrangement process. How these events relate to the normal B-cell repertoire selection is not clear. In mice at least, the pool of peripheral long-lived B lymphocytes is skewed in the expression of certain Vh genes compared with a much more complex repertoire seen in bone marrow pre-B cells.[15]

A second stage in B-cell development when new antibody specificities arise is during the creation of memory B cells. The V regions of both H and L genes are hypermutated in this process, which is ongoing in the germinal centers of peripheral lymphoid tissues.[16] The mutations appear to be entirely random and comprise the V(D)J region as well as some flanking regions, but generally not the C region.[17,18] Apparently the mutations arise in centroblasts in the "dark zone" of the germinal centers.[19] These cells give rise to the so-called centrocytes that accumulate in the "light zone" in association with a dense network of follicular dendritic cells (FDC).

It is assumed that in the light zone the B cells interact with antigens bound to the FDC and that only B cells with high affinities for the antigens present on the FDC survive (presumably by receiving a positive signal). Cells that have mutated their Ig genes to lower affinity of where the genes were made nonfunctional by introduction of a stop codon apparently undergo apoptosis.

Clearly, there must also be a control mechanism that operates after the creation of somatically mutated specificities that eliminates autoreactive B cells. In autoimmune mice with the autosomal recessive mutation, *lpr*, antiself antibodies that arise by somatic hypermutation exist at high levels.[20,21] It was found that such autoantibodies are oligoclonal or monoclonal.[21] B cells producing mutated autoantibodies in *lpr* mice are apparently greatly expanded in numbers by proliferation after the mutation process has stopped.[21] Thus, not only are these cells not eliminated in autoimmunity, they appear to be selected to proliferate. As Weigert *et al.* suggest, the selection occurs by way of antigen stimulation.[21] It is not clear where or how B cells, which produce somatically mutated antibodies with self-specificity, are normally selected against. The *lpr* mouse has a defect in the Fas antigen gene.[22] The Fas protein, a cell-surface molecule expressed on certain T and B cells and other cells, is apparently a receptor for an apoptotic cell signal (the ligand is gld).[22] Interestingly, when a mixture of +/+ and *lpr* Ig allotype-marked bone marrow cells are transferred into *lpr* mice, the autoimmune antibodies are entirely of the *lpr* allotype.[20] This indicates that even in the presence of wild-type T cells, the *lpr* autoimmune B cells survive. It suggests that the lack of selection against these B cells is an intrinsic B-cell problem, presumably due to the absence of the Fas receptor molecule. Thus, presumably in the normal individual self-reactive B cells, which arose by somatic mutation, are eliminated by receiving an apoptotic stimulus from an unknown type of cell (dendritic cell or T cell).

In conclusion, autoimmune antibodies can arise at several stages of B-cell development. Their persistence may be due to B- or T-cell defects or other poorly understood abnormalities of B-cell regulation.

REFERENCES

1. Vincent, A. 1991. Autoimmunity to acetylcholine receptors in myasthenia gravis. Biochem. Soc. Trans. **19:** 180–183.
2. Engler, P. & U. Storb. 1988. Immunoglobulin gene rearrangement. *In* Genetic Recombination. R. Kucherlapati & G. R. Smith, Eds.: 667–700. Am. Soc. Microbiol.
3. Ikuta, K., N. Uchida, J. Friedman & I. L. Weissman. 1992. Lymphocyte development from stem cells. Annu. Rev. Immunol. **10:** 759–783.
4. Schatz, D. G., M. A. Oettinger & M. S. Schlissel. 1992. V(D)J recombination: molecular biology and regulation. Annu. Rev. Immunol. **10:** 359–383.
5. Engler, P., P. Roth, J. Y. Kim & U. Storb. 1991. Factors affecting the rearrangement efficiency of an Ig test gene. J. Immunol. **146:** 2826–2835.
6. Engler, P., D. Haasch, C. A. Pinkert, L. Doglio, M. Glymour, R. Brinster & U. Storb. 1991. A strain-specific modifier on mouse chromosome 4 controls the methylation of independent transgene loci. Cell **65:** 939–947.
7. Hsieh, C.-L. & M. R. Lieber. 1992. CpG methylated minichromosomes become inaccessible for V(D)J recombination after undergoing replication. EMBO J. **11:** 315–325.
8. Nemazee, D. A. & K. Bürki. 1989. Clonal deletion of B lymphocytes in a transgenic mouse bearing anti-MHC class I antibody genes. Nature (London) **337:** 562–566.
9. Nemazee, D. & K. Buerki. 1989. Clonal deletion of autoreactive B lymphocytes in bone marrow chimeras. Proc. Natl. Acad. Sci. USA **86:** 8039–8043.
10. Goodnow, C. C., J. Crosbie, S. Adelstein, T. B. Lavoie, S. J. Smith-Gill, R. A.

BRINK, H. PRITCHARD-BRISCOE, J. S. WOTHERSPOON, R. H. LOBLAY, K. RAPHAEL, R. J. TRENT & A. BASTEN. 1988. Altered immunoglobulin expression and functional silencing of self-reactive B lymphocytes in transgenic mice. Nature (London) **334:** 676–682.

11. GOODNOW, C. C., R. BRINK & E. ADAMS. 1991. Breakdown of self-tolerance in anergic B lymphocytes. Nature (London) **352:** 532–536.
12. ZINKERNAGEL, R. M., S. COOPER, J. CHAMBERS, R. A. LAZZARINI, H. HENGARTNER & H. ARNHEITER. 1990. Virus-induced autoantibody response to a transgenic viral antigen. Nature (London) **345:** 68–71.
13. ADELSTEIN, S., H. PRITCHARD-BRISCOE, T. A. ANDERSON, J. CROSBIE, G. GAMMON, R. H. LOBLAY, A. BASTEN & C. C. GOODNOW. 1991. Induction of self-tolerance in T cells but not B cells of transgenic mice expressing little self antigen. Science **251:** 1223–1225.
14. GOODNOW, C. C. 1992. Transgenic mice and analysis of B-cell tolerance. Annu. Rev. Immunol. **10:** 489–518.
15. GU, H., D. TARLINTON, W. MÜLLER, K. RAJEWSKY & I. FÖRSTER. 1991. Most peripheral B cells in mice are ligand selected. J. Exp. Med. **173:** 1357–1371.
16. STEELE, E. J. 1991. Somatic hypermutation in V-regions. CRC Press. Boca Raton, FL.
17. HACKETT JR., J., B. J. ROGERSON, R. L. O'BRIEN & U. STORB. 1990. Analysis of somatic mutations in κ transgenes. J. Exp. Med. **172:** 131–137.
18. ROGERSON, B., J. HACKETT JR., A. PETERS, D. HAASCH & U. STORB. 1991. Mutation pattern of immunoglobulin transgenes is compatible with a model of somatic hypermutation in which targeting of the mutator is linked to the direction of DNA replication. EMBO J. **10:** 4331–4341.
19. LIU, Y.-J., G. D. JOHNSON, J. GORDON & I. C. MACLENNAN. 1992. Germinal centres in T-cell–dependent antibody responses. Immunology Today **13:**17–21.
20. COHEN, P. L. & R. A. EISENBERG. 1991. *Lpr* and *gld*: single gene models of systemic autoimmunity and lymphoproliferative disease. Annu. Rev. Immunol. **9:** 243–269.
21. SHLOMCHIK, M. J., A. MARSHAK-ROTHSTEIN, C. B. WOLFOWICZ, T. L. ROTHSTEIN & M. G. WEIGERT. 1987. The role of clonal selection and somatic mutation in autoimmunity. Nature (London) **238:** 805–811.
22. WATANABE-FUKUNAGA, R., C. I. BRANNAN, N. G. COPELAND, N. A. JENKINS & S. NAGATA. 1992. Lymphoproliferation in mice explained by defects in Fas antigen that mediates apoptosis. Nature (London) **356:** 314–317.

T-Cell Receptor Genes in Autoimmunity[a]

ARGYRIOS N. THEOFILOPOULOS,
ROBERT S. BALDERAS, ROBERTO BACCALA,
AND DWIGHT H. KONO

Department of Immunology
The Scripps Research Institute
10666 North Torrey Pines Road/IMM3
La Jolla, California 92037

INTRODUCTION

T cells appear to be an absolute requirement for disease expression in almost all autoimmune diseases, as exemplified in appropriate animal models. T cells may participate in autoimmune disease pathogenesis either directly, in the form of cytotoxic cells, as in several organ-specific autoimmune diseases, for example, insulin-dependent diabetes and multiple sclerosis; or indirectly, in the form of helper cells required for autoantibody production and immune complex formation either in the circulation, as in lupus, or on the target organ, as in antireceptor autoimmune diseases, such as myasthenia gravis. Recent efforts to further define the role of T cells in autoimmune disease pathogenesis have focused on two basic questions: What are the characteristics of the inciting autoantigenic epitopes and the responding T-cell clonotypes? and Do specific T-cell receptor (TCR) repertoires predispose to disease, and, if so, do such repertoires reflect unique germline V gene composition or structure, errors in T-cell tolerance mechanisms, and/or other properties of the thymic selection process, such as repertoire shaping by MHC and self-ligands? This article summarizes the various theories of autoimmunity as they relate to the circumvention of T-cell tolerance and addresses issues related to the correlation of autoimmune diseases with genomic composition, polymorphism, and expression of TCR genes.

TCR REPERTOIRE AND SELF-TOLERANCE

Although it had long been known that the antigen receptor of B lymphocytes was membrane-bound immunoglobulin, later shown to be encoded by multiple gene segments for heavy (V, variable; D, diversity; J, joining) and light (V,J) chains, the protein and genes of the T-cell antigen receptor (TCR) remained elusive until the mid-1980s. Through the studies of several investigators (reviewed in refs. 1–3), it was documented that, like the Ig gene loci, the TCR genes are grouped into arrays of coding segments scattered over large tracts of chromosomal DNA.

[a] This is publication number 7310IMM from the Department of Immunology, the Scripps Research Institute, 10666 North Torrey Pines Road/IMM3, La Jolla, California 92037. The author's work cited herein was supported by NIH Grants AR31203, AR39555, and AG09430.

Four genes (α, β, γ, δ) were identified that encode two types of dimeric receptor structures, one termed $\alpha\beta$ and the other $\gamma\delta$. The vast majority of helper and cytotoxic T cells express the $\alpha\beta$ receptor, whereas the $\gamma\delta$ receptor is expressed on a minor subset of peripheral T cells that is, however, enriched in epithelial sites; the exact function and specificity of the limited diversity-displaying $\gamma\delta$ cells remains unknown. The β locus has two C regions of nearly identical structure arrayed in tandem with a cluster of 7 J elements and one D element 5′ to each. The number of Vβ elements varies from ≈24 in mice and rats[1,4] to ≈70 in humans.[1,2,5] The TCR α locus, lacking D segments and having many more J segments (≈50) than the β locus, is unique in that all of the TCR δ coding segments are located within the intervening DNA between the Vα and Jα segments and are deleted in most $\alpha\beta$-bearing T cells. The Vβ and Vα genes are arranged into multiple subfamilies, some being single member and others encompassing several members, that, by convention exhibit >75% homology in their nucleotide sequences. The V(D)J gene segments are rearranged stochastically during T-cell ontogeny to form a functional V region gene, which is then joined to C-region sequences by RNA splicing following transcription. This mechanism, together with random α- and β-chain pairing, imprecise joining, and nucleotide additions at the joining points allows generation of the high degree of TCR diversity required to recognize and respond to a virtually limitless universe of antigens.

The TCR functions as an activating mechanism that triggers the T-cell response to conventional antigens only in the context of cell-surface molecules encoded by the genes of the major histocompatibility complex (MHC).[6] Several studies have shown that T cells do not recognize unprocessed macromolecules but rather recognize antigenic peptides bound to MHC molecules, a characteristic markedly different from that of the B-cell receptor that recognizes discontinuous sequences on soluble globular molecules.[7,8] It has also been found that MHC class I molecules, expressed on most cell types, normally present peptides derived from proteins produced endogenously in the cell, whereas MHC class II molecules, expressed on only a few cell types (B cells, macrophages, and dendritic cells) can effectively present peptides derived from molecules taken from the outside. The elaborate processes by which peptides are created, bind to MHC, and are transported to the cell surface have recently been outlined (reviewed in refs. 9–11). Moreover, techniques have recently been developed by which self- and foreign peptides bound to MHC molecules can be isolated and characterized functionally and biochemically.[12–14] In addition to conventional antigens, a second class of T cell–activating molecules has recently been identified, collectively termed "super-antigens (reviewed in ref. 15). These molecules, products of mouse endogenous retroviral genes or exogenous microbial agents, unlike conventional antigens, bind outside the antigen-binding groove of MHC class II molecules; need, in most instances, not to be processed; and engage TCR clones almost solely on the basis of the Vβ they display.

Although the enormous diversity in T- (and B-) cell receptor structures is of obvious benefit in responding to nonself-molecules, it carries the inherent risk that some of these receptors may exhibit antiself specificity because the receptor repertoire is generated without regard for which antigens are part of self. To avoid this potentially destructive effect, non-germline-encoded learning processes had to be developed whereby self-reactive cells are eliminated or otherwise silenced. Recent studies, primarily in transgenic mice, have shown that a binary process takes place intrathymically during T-cell development (reviewed in refs. 16, 17). On one hand, recognition of self-MHC molecules (with or without self-peptides, depending on TCR affinity) on thymic epithelial cells by T cells expressing appropriate receptors leads to the phenotypic and functional progression of thymocytes from the immature double-positive ($CD4^+8^+$) stage of differentiation to the mature single-positive ($CD4^+8^-$ or $CD8^+4^-$) stage of differentiation; this process has been termed "positive selection." On the other hand, those positively selected cells with high antiself affinity are deleted, presumably in a subsequent step, from the repertoire by apoptosis, yielding tolerance; this process has been termed "negative selection." Additional investigations revealed that for some self-antigens, intrathymic T-cell tolerance is conferred by clonal inactivation rather than deletion.[18] Obviously, not all self-antigens will be present in the thymus and, therefore, mechanisms must be in place to ensure tolerance development in the periphery. Indeed, as experimentally assessed in mice expressing transgenic self-antigens under the influence of certain tissue-specific promoters or following injection of a superantigen, extrathymic T-cell tolerance was documented and found to also be mediated by clonal deletion or anergy, with the nondeletional mechanisms considered to be more likely for peripheral forms of tolerance under physiologic conditions (reviewed in refs. 19, 20).

With regard to activation, deletion, or inactivation following T-cell contact with antigen, several studies have shown that such outcomes depend on various factors, including differentiation stage, physicochemical properties of the antigen, epitope density, and polymeric or monomeric state, amount and mode of presentation, receptor affinity, as well as production of co-stimulatory signals, such as those delivered by cytokines and other activation/adhesion molecules.[21,22] Moreover, the distinction between the two major tolerance mechanisms, that is, deletion versus anergy, has been the subject of considerable debate. Some investigators believe that anergy is simply a transitory state whose final outcome is cell death, whereas others believe that anergic cells are long-lived, can recover from this state, and become fully functional upon appropriate signal circumstances. A third possibility is that anergy or deletion constitutes a hierarchical state with the tolerant cell destined to survive or die, depending on the strength of the inductive stimulus.

CIRCUMVENTION OF T-CELL TOLERANCE AND AUTOIMMUNITY

Review of the literature offers several theories for circumvention of T-cell tolerance and induction of pathologic autoimmune responses. Autoimmune damage resulting from absence of T- (and B-) cell self-tolerance is well-established, as exemplified by autoimmune responses against antigens normally sequestered behind anatomical barriers in the so-called immunologically "privileged" sites. Induction of unresponsiveness requires contact between the antigen and the maturing immune system early in ontogeny; tolerance cannot easily be established if the antigen is sequestered and unavailable to the immune system. The opportunity for an autoimmune response can, therefore, be provided by release of such antigens in the general circulation induced by trauma or inflammation. Examples of autoimmune responses following availability of anatomically sequestered antigens include sympathetic opthalmia following eye injury, orchitis after vasectomy, and experimental allergic encephalomyelitis (EAE) and perhaps multiple sclerosis caused by an immune response to brain myelin basic protein (reviewed in ref. 23).

A corollary hypothesis that addresses molecular rather than anatomic sequestration is that of cryptic self-determinants.[24] It has been suggested that each self- (and foreign) molecule has both dominant and subdominant (or cryptic) T-cell determinants, and the dominant or cryptic behavior of such determinants depends on various factors, such as antigen processing, appropriate sequence motif for efficient peptide-MHC interaction, frequency of corresponding T-cell precursors, and amount and availability of competing molecules. Efficiently processed and presented dominant determinant regions of a self-molecule will act as good tolerance inducers, whereas inefficiently processed and presented subdominant or cryptic determinant regions will be poor tolerogens. Confinement of tolerance to certain dominant determinants of a molecule has clearly been documented in HEL transgenic mice.[25] It has thus been hypothesized that increased availability of a subdominant self-determinant beyond a threshold level induces expansion of the nontolerized T cells and ensuing disease. Although intellectually appealing, the mode by which the postulated increased availability of subdominant self-determinants occurs has not been clearly defined, but microbial agents mimicking such self-determinants may be invoked.

Another concept supported by recent transgenic mouse studies[26,27] holds that autoimmunity may result from activation of self-antigen–binding T cells that, due to low antigen expression, inadequate presentation, or lack of costimulatory signals, were maintained at a quiescent state but were not deleted or anergized (self-antigen–ignoring cells). In one set of experiments, double transgenic mice were constructed with a TCR reactive against the lymphocytic choriomeningitis virus (LCMV) p32-42 glycoprotein (GP) epitope and the LCMV-GP gene expressed in β pancreatic cells under the influence of the insulin promoter. The TCR transgenic thymocytes of such mice were positively, but not negatively, selected, and, as evidenced by the LCMV-GP–specific vigorous proliferative and cytotoxic *in vitro* responses, the spleen T cells were not anergic, although the animals were free of insulitis or diabetes. Insulitis and diabetes, however, resulted when such animals were infected with the appropriate LCMV strain bearing the GP p32-42 determinant but not with a LCMV strain without this determinant. Similar results were reported from studies using mice transgenic for the LCMV-GP only. Thus, self-reactive cytotoxic T cells appear not to be inactivated to all proteins expressed on extrathymic tissues, presumably because the self-antigen–expressing cells (here β cells) do not display enough antigen, MHC class II molecules, or appropriate costimulatory signals. When infection occurs with a virus bearing the self- (or related) molecules, however, effective presentation and costimulation may occur by way of infected B cells or macrophages (class II MHC-expressing cells), leading to activation of the previously quiescent T cells and their subsequent attraction to, and damage of, the organ displaying the structurally similar self-antigen. These results are compatible with the antigenic mimicry hypothesis of autoimmunity (reviewed in ref. 28), particularly as it pertains to infectious agents. Because of the rules imposed upon construction of all coding sequences, identical oligopeptides are often found in totally unrelated proteins, and many peptide fragments of infectious agents are homologous with host proteins, including MHC molecules.[29] High amounts of such foreign peptides may engage and expand the quiescent, nontolerant, self–cross-reacting T cells and thus initiate the autoimmune process.

The concept that autoimmune responses may not be directed against pre-existing self-determinants, but rather against newly created self-determinants for which T- (and B-) cell tolerance has never been established, has also been considered. Such neo-self determinants may, for example, be generated following binding of drugs and other haptenic groups to self-antigens,[30] by antigen-antibody complexing, as in anti-IgG (rheumatoid factor) induction;[31] by gene recombination, conversion, or mutation; or following viral infection. In addition to molecular mimicry or tissue damage and release of sequestered antigens, viruses may promote autoimmunity through induction of new self-peptide determinants brought about by the viral effects on the overall cellular machinery, including transcriptional, translational, and posttranslational events as well as peptide generation and transport. Certain drugs known to have the side effect of inducing autoimmune responses, such as procainamide, hydralazine, and mercury, may also act through similar mechanisms.[30]

Autoimmunity may also result from errors in intrathymic negative selection, a possible example of which may be the development of autoimmunity in irradiated mice reconstituted with bone marrow and then treated with, and withdrawn from, cyclosporin,[32] presumably due to the interference of this drug with negative selection.[33] An additional example may be the development of multiorgan autoimmune manifestations in mice thymectomized early in life, wherein escape to the periphery of self-superantigen–binding Vβ clones was reported.[34]

Polyclonal T-cell activation may also be a contributing or initiating mechanism, particularly in systemic autoimmunity. Such a mechanism is presumably involved in the graft-versus-host model of autoimmunity, wherein stimulated alloreactive donor T cells initiate, by way of soluble factors, the differentiation of recipient B cells into antibody-secreting cells, regardless of whether the receptors of B cells are antiself or anti–non-self.[30] Another possible example is polyclonal stimulation of a large set of T cells reactive with B cell–MHC class II molecule-bound microbial superantigens;[35] such stimulated T cells in close contact with the superantigen-displaying B cells may then induce the latter to produce antibodies, including autoantibodies.

Finally, some investigators hold that disturbances in immunoregulatory T-cell circuits may constitute the primary cause of autoimmune syndromes. Uncontested evidence for abnormalities in idiotypic networks in such diseases are scant or absent. Moreover, there is considerable debate about the existence, in general, and the role, in particular, of suppressor T cells in such diseases. Examples, however, of T-cell subsets or clones inducing or inhibiting disease development upon passive transfer have been reported[36,37] and point to the need for further study in this area.

TCR GENOMIC AND EXPRESSION PROFILES IN AUTOIMMUNE DISEASES

Inasmuch as T cells recognize MHC-bound peptides by way of their TCR, considerable efforts have been made to determine the diversity of T-cell clones engaged in autoimmune responses as well as whether genomic composition, polymorphic variants, and/or expression of certain TCR genes are predisposing factors in autoimmunity.

Genomic Studies

Animal Models of Autoimmunity

With regard to mouse and rat germline TCR repertoires, studies by several investigators indicate a gene pool size of 24 Vβ genes organized into 20 subfamilies, all of which contain a single member, except Vβ3, 5, and 8, which contain 2–3 members.[1,4] Three mouse Vβ haplotypes have been defined based on gene pool size: $V\beta^b$, encompassing most laboratory strains of mice without Vβ genomic deletions, but lacking functional transcripts for Vβ17 and Vβ19;

Vβ^a haplotype (SJL, SWR, C57L, C57BR) with deleted Vβ5, -8, -9, -11, -12, and -13; and Vβ^c (RIIIS) haplotype, which exhibits additional genomic deletions in Vβ6, -15, and -17. Coding region polymorphisms for Vβ1, 3.1, 6, 10, and 15 between Vβ^a and Vβ^b haplotype mice have also been identified, with differences ranging from 1 to 6 amino acids.[38] We[39] and Kotzin *et al.*[40] also identified in NZW mice a deletion beginning at the first exon of Cβ1 and ending at the first exon of Cβ2. Studies by both groups have shown that this genomic deletion plays no apparent role in the (NZBxNZW)F_1 lupus-like disease (reviewed in ref. 41). With regard to mouse Vα, 36 sequences, grouped into 13 subfamilies, have thus far been identified. We[42] have further identified four main Vα genomic haplotypes, termed a, b, c, and d. Lupus and diabetes-predisposed mice were distributed throughout these Vα haplotypes together with several normal strains of mice. Based on these findings, it seems unlikely that autoimmune disease expression in mice is caused by abnormalities encoded in particular Vβ or Vα haplotypes *per se,* or by novel combinations of such haplotypes with MHC alleles. These findings do not preclude, however, the possibility that an otherwise normal haplotype or allele may nonetheless contribute to the unique constellation of factors, allowing the autoimmune response to occur.

Human Autoimmune Diseases

Statistical estimates suggest that the human genome contains about 60 to 70 Vβ genes; thus far, however, only 49 unique sequences have been published.[1,2,5] These genes have been grouped into 23 subfamilies; 9 contain multiple members, and 14 are single member genes. Southern blot analysis of 14 Vβ subfamily gene segments in ≈100 individuals identified only one with a single Vβ gene difference in the germline pool size (reviewed in ref. 1), suggesting the rarity of variation in the number of genes among individuals. With regard to allelic polymorphisms, restriction fragment length polymorphism (RFLP) analysis, sequence data, and family studies have, thus far, tentatively identified polymorphic variants for 10 human Vβs (Vβ1.1, 2.1, 5.1, 6.7, 7.2, 8.2, 12.2, 13.4, 15.1, and 16), of which several resulted in amino acid change, and all but one are coding region-related.[5,43,44] Although some studies have found correlations of Cβ-Vβ or Cα-Vα, genomic polymorphisms with predisposition to certain autoimmune diseases such as multiple sclerosis, myasthenia gravis, insulin-dependent diabetes mellitus (IDDM), Graves' disease, Hashimoto's thyroiditis, lupus, and rheumatoid arthritis, other studies have failed to confirm such associations (reviewed in ref. 45). The frequent occurrence of recombination hot spots in the vicinity of these genes indicates that the genetic distance between the different V gene segments is large. Given the limited polymorphisms in this region, it is therefore unlikely that a particular restriction site marker will be close enough to a biologically relevant allelic variation to reveal an association with autoimmune disease predisposition. Actual cloning and sequencing of the genes will be required to determine whether such associations, if any, exist.

Expression Studies

Animal Models of Autoimmunity

Systemic Lupus Erythematosus (*SLE*). In 1987, Kappler *et al.*[46] reported the first clear-cut evidence that self-superantigen–recognizing T cells are deleted intrathymically from the repertoire prior to exportation of T cells to the periphery, presumably in order to confer tolerance. Subsequently, using anti-Vβ antibodies or RNA probes, a large number of somatically imposed, superantigen-dependent Vβ clonal deletions were identified in the mouse by these and other investigators (reviewed in ref. 15). Furthermore, the endogenous superantigens mediating this effect were shown to be related to the undefined minor lymphocyte stimulating (*Mls*) antigen, and to be actually encoded by mouse mammary tumor virus (MMTV) proviral genes integrated on various chromosomes (reviewed in refs. 15, 47). Kotzin *et al.*,[48] using immunofluorescence studies with certain monoclonal Vβ-specific antibodies, first addressed the issue of the integrity of the self-superantigen–directed clonal deletion process in lupus (NZBxNZW)F_1 and (NZBxSWR)F_1 mice and found that these mice at both young and advanced age exhibited Vβ deletions similar to those of MHC and *Mls*/MMTV-matched normal controls. Using a multiprobe RNase protection assay encompassing all mouse Vβ genes, we[49] also found that lupus-prone strains (NZB, NZW, BXSB, MRL) obeyed the rules of self-superantigen–imposed Vβ clonal deletions. It was, therefore, concluded that systemic autoimmunity in these mice is unlikely to be caused by abnormalities in intrathymic superantigen-related negative selection. These findings, however, do not exclude defects in central tolerance for conventional self-antigens, nor do they address the issue of possible defects in peripheral tolerance as contributors to the disease of these mice.

Helper T-cell clones specific for anti-DNA responses have also been derived from (NZBxSWR)F_1 lupus mice and found to exhibit diverse Vβs.[50] Although the junctional sequences of these clones were also diverse, importantly, those with helper activity for pathogenic cationic IgG anti-DNA autoantibodies displayed a high frequency of negatively charged (anionic) residues (aspartic or glutamic acid) in their CDR3, thereby suggesting that antigenic epitopes with basic charge may be involved in this autoimmune response.

Insulin-dependent Diabetes Mellitus. Class II MHC transgenic nonobese diabetic (NOD) mice (in themselves IE^- and expressing a unique IA), a model of human IDDM, are protected from expression of insulitis and diabetes (reviewed in ref. 51). It was, therefore, suggested that this protective effect was due to cross-tolerance by deletion of certain clones cross-reactive with endogenous superantigens as well as islet cell antigens. In fact, an initial report concluded that Vβ5-bearing clones are responsible for the disease of this mouse.[52] Later studies showed, however, that the class II MHC effect could not be explained on the basis of deletion of diabetogenic clones[53] and that the reported Vβ5 restriction was actually artifactual. Subsequently, studies in both NOD mice[54] and diabetes-prone BB rats[55] have indicated that insulitis is most likely mediated by a diverse Vβ T-cell population.

Experimental Autoimmune Encephalomyelitis (EAE). EAE is an experimentally induced autoimmune demyelinating disease in which $CD4^+$ T lymphocytes responsive to myelin basic protein (MBP) are activated and mediate demyelination and paralysis resembling human multiple sclerosis. Several studies have examined the TCR repertoires used by EAE-susceptible strains of mice (PL/J, B10.PL) and reported utilization of a restricted population of T-cell receptors (Vβ8.2/Vβ13 and Vα2/Vα4) with conserved junctional sequences (reviewed in ref. 56). Furthermore, the response of EAE-susceptible Lewis rats was also found to be restricted and to use Vβ/Vα genes homologous to those in mice, even though the Lewis rat responds to a different portion of the MBP molecule than that of the above mouse strains. These findings were of considerable importance, because it was subsequently shown in mice and rats that antibodies to the predominant TCRs used in EAE could prevent and even cure this disease.[57,58] Moreover, "vaccination" with inactivated encephalitogenic T-cell clones and, more recently, with synthetic peptides containing sequences from encephalitogenic TCRs were reported to prevent or cure EAE.[59,60] A more recent study,[61] however, questions the universality of this latter therapeutic approach because, in this study, Vβ-peptide immunized rats sometimes exhibited more severe, chronic relapsing disease than untreated counterparts.

Collagen-induced Arthritis. It was initially reported that susceptibility to this disease in mice was dependent on certain Vβs, particularly the Vβ6 polymorphic variant of the $V\beta^b$ haplotype mice.[62] Subsequent studies by several laboratories have dispelled this notion.[38,63,64] Moreover, although an enrichment of Vβ6 and Vβ8.2 was noted in arthritic joints of such mice[65] and a therapeutic effect for an anti-Vβ8 antibody was reported,[66] others have found no therapeutic effects for anti-Vβ8 or anti-Vβ6 antibodies in mice[67] or for anti-TCR αβ antibody in rats[68] with established collagen-induced arthritis.

Experimental Myasthenia Gravis (EMG). T cell–immunodominant epitopes within the α subunit of AChR have been mapped for human, mouse, and rat T cells by proliferative responses to synthetic peptides. Ten Lewis rat T-cell clones with reactivity to native AChR and peptides thereof[69] were screened by polymerase chain reaction (PCR) for expression of Vβ gene segments (L. Smith, J Lindstrom, and A. N. Theofilopoulos, in preparation). Four clones that reacted to the peptide α100-116, previously reported to be a T cell–immunodominant epitope in Lewis rats, were found to express four Vβs (4, 8.2, 10 and 17) and three Jβs (1.4, 2.3, 2.6); thus, despite their recognition of a common limited epitope, all use different Vβ gene combinations. Moreover, the Vβs and Jβs in the remaining six clones, exhibiting no peptide reactivity, but a variety of AchR subunit–reactivity patterns, were also diverse with regard to their Vβs. Actual cloning and sequencing of the entire Vβ gene from these clones also failed to show any obvious selection for particular amino acids in the third hypervariable region.

Human Autoimmune Diseases

Multiple Sclerosis. Paralleling the EAE model, studies have been conducted to address TCR clonal restriction in MBP-responding T-cell clones and lines from multiple sclerosis (MS) patients. In examining T cells specific for the MBP immunodominant determinant (residues 82-102),[70] evidence was obtained for preferential usage of the Vβ17 gene segment, but additional Vβs were also found, including Vβ12, 14, 7, and 8. Similar results were obtained examining T-cell lines generated from control subjects. Contrastingly, studies with MBP-specific T-cell clones and lines from several subjects showed dominance of Vβ5.2 and, to a lesser extent, Vβ6.1, and, surprisingly, specificity for different MBP determinants appeared to be independent of this biased TCR Vβ gene use.[71] Although preferential Vβ usage was also noted in MBP-specific T-cell clones derived from normal individuals, the dominant Vβs (Vβ14, 7) differed from those in MS patients. Yet, in another study,[72] MBP-specific T-cell clones from peripheral blood of several MS patients showed a tendency for Vβ restriction, but the dominant Vβs (Vβ15, 14, 12) differed from one individual to another without any apparent relation to the epitope recognized or to the MHC allele. Although no Vα predominance was noted in the above studies with peripheral blood-derived MBP-specific clones, a limited Vα heterogeneity in brain lesions was reported,[73] but again, the dominant Vαs differed from patient to patient. Furthermore, a more recent study[74] identified diverse Vβ and Vα genes in active brain lesions of MS patients, and clear differences in such repertoires between plaques obtained from the same MS case were observed.

Rheumatoid Arthritis. T cells appear to play an important role in the pathology of rheumatoid arthritis (RA), and, therefore, the repertoire of synovium-infiltrating T cells has also been investigated, but the results, as in MS, are diverse. Initial rearrangement studies[75,76] with IL-2–expanded synovial T cells suggested clonal dominance, but the rearrangements were distinct from one individual to another, suggesting utilization of different Vβs. It was noteworthy in these studies that rearrangements observed in T cells from different joints of one individual and in several fragments of one synovium exhibited some similarity in TCR gene rearrangement patterns, suggesting that the same dominant clone was present in several afflicted joints and different areas of an inflamed synovium. Other studies with this approach,[77,78] however, failed to obtain evidence for oligoclonality. Subsequently, specific Vβs were analyzed in T cells from synovial tissue or fluid. One study[79] used PCR to analyze bulk synovial fluid T cells from a few patients and identified expansion of several Vβs, including Vβ2, 4, 6, 7, 8, 13, and 16. A subsequent study[80] using IL-2–propagated synovial fluid T-cell lines, found that Vβ14 predominated in most samples. Cloning and sequencing indicated that the Vβ14$^+$ T-cell population was dominated by a few clones, as indicated by junctional sequence identity. Surprisingly, in these patients, levels of Vβ14 expression in peripheral blood were very low. Based on these findings, the authors hypothesized that a superantigen might initially engage a large proportion of Vβ14$^+$ T cells in peripheral blood, and that a portion of these cells exhibiting cross-reactivity with a self-antigen in the synovium would then be sequestered at that site. Another study[82] identified Vβ14, and also Vβ17 and Vβ3, as the dominant clonotypes in IL-2–propagated RA synovial tissue-derived T-cell populations. Finally, in one study, the TCR Vβ and Vα repertoires in synovial fluid of one RA patient were found to be highly heterogeneous and polyclonal, but some repetitive sequences of Vβ2.1 and Vβ3.1 were noted;[82] in another study,[83] Vβ was found to be heterogeneous, but a patient-specific Vα restriction was noted.

Thyroiditis. Total intrathyroidal T cells from patients with Graves' disease or Hashimoto's thyroiditis have also been analyzed and found to exhibit a somewhat dominant Vα-gene expression, but once again, the predominant Vα gene varied from patient to patient.[84] Studies, however, related to the Vβ repertoires of such cells have shown diverse clonotypic origins.[85,86]

SUMMARY

T cells are primary participants in the pathogenesis of the MHC-dependent autoimmune diseases, and therefore, evidence for association of TCR V-gene repertoires with such disorders has been actively sought. With very few exceptions, no clear-cut evidence for correlation of particular RFLP-defined V-C–region genomic polymorphisms with autoimmune disease predisposition has thus far been demonstrated. With regard to TCR V-gene repertoires engaged in responses to autoantigens, restricted use of certain Vβ and Vα genes in response to myelin basic protein has been documented in animal models. In many spontaneous and experimentally induced animal and human autoimmune diseases, however, the picture is far from clear. Although dominance of certain TCR V genes has been noted, the clonal restrictions are not absolute; they differ from one study to another and from one patient to another. Such variations may be caused by MHC allele-dependent determinant selection mechanisms, secondary T-cell infiltrates in inflammatory sites, different patient populations and stages of disease, or the involvement of different pathogens that, nevertheless, lead to the same clinical entity. Overall, the results indicate that efforts to intervene therapeutically in autoimmune diseases by vaccination with modified T-cell clones, V region–synthetic peptides, or TCR blocking analogues may not be easily applicable. Further studies on the characterization of the specific antigens involved in autoimmune disease pathogenesis is required in order to accurately address the issue of TCR utilization in autoimmune diseases.

REFERENCES

1. WILSON, R. K., E. LAI, P. CONCANNON, R. K. BARTH & L. E. HOOD. 1998. Structure, organization and polymorphism of murine and human T-cell receptor α and β chain gene families. Immunol. Rev. **101:** 149–172.
2. TOYONAGA, B. & T. W. MAK. 1987. Genes of the T-cell receptor in normal and malignant T cells. Annu. Rev. Immunol. **5:** 585–620.
3. BRENNER, M. B., J. L. STROMINGER & M. S. KRANGEL. 1988. The $\gamma\delta$ T cell receptor. Adv. Immunol. **43:** 133–192.
4. SMITH, L. R., D. H. KONO & A. N. THEOFILOPOULOS. 1991. Complexity and sequence identification of 24 rat Vβ genes. J. Immunol. **147:** 375–379.
5. PLAZA, A., D. H. KONO & A. N. THEOFILOPOULOS. 1991. New human Vβ genes and polymorphic variants. J. Immunol. **147:** 4360–4365.
6. ZINKERNAGEL, R. M. & P. C. DOHERTY. 1979. MHC restricted cytotoxic cells: Studies on the biological role of polymorphic major transplantation antigens determining T-cell restriction specificity, function and responsiveness. Adv. Immunol. **27:** 51–177.
7. BABBIT, B., P. M. ALLEN, G. MATSUEDA, E. HABER & E. R. UNANUE. 1985. Binding of immunogenic peptides to Ia histocompatibility molecules. Nature (London) **317:** 359–361.
8. BUUS, S., A. SETTE, S. M. COLON, C. MILES & H. M. GREY. 1987. The relationship between major histocompatibility complex (MHC) restriction and the capacity of Ia to bind immunogenic peptides. Science **235:** 1353–1358.
9. ELLIOT, T. 1991. How do peptides associate with MHC class I molecules? Immunology Today **12:** 386–388.
10. BRACIALE, T. J. & V. L. BRACIALE. 1991. Antigen presentation: Structural themes and functional variations. Immunology Today **12:** 124–129.
11. SADEPH-NASSERI, S. & R. N. GERMAIN. 1992. How MHC class II molecules work: Peptide-dependent competition of protein folding. Immunology Today **13:** 43–46.
12. FALK, K., O. ROTZSCHKE, S. STEVANOVIC, G. JUNG & H.-G. RAMMENSEE. 1991. Allele-specific motifs revealed by sequencing of self-peptides eluted from MHC molecules. Nature (London) **351:** 290–296.
13. RUDENSKY, A., P. PRESTON-HURLBURT, S.-C. HONG, A. BARLOW & C. A. JANEWAY JR. 1991. Sequence analysis of peptides bound to MHC class II molecules. Nature (London) **343:** 622–627.
14. HUNT, D. F., R. A. HENDERSON, J. SHABANOWITZ, K. SAKAGUCHI, H. MICHEL, N. SEVILIR, A. L. COX, E. APPELLA & V. H. ENGELHARD. 1992. Characterization of peptides bound to the class I MHC molecule HLA-A2.1 by mass spectrometry. Science **255:** 1261–1263.
15. HERMAN, A., J. W. KAPPLER, P. MARRACK & A. M. PULLEN. 1991. Superantigens: Mechanism of T-cell stimulation and role in immune responses. Annu. Rev. Immunol. **9:** 745–772.

16. VON BOEHMER, H. & P. KISIELOW. 1990. Self-nonself discrimination by T cells. Science **248:** 1369–1373.
17. BLACKMAN, M., J. KAPPLER & P. MARRACK. 1990. The role of T cell receptor in positive and negative selection of developing T cells. Science **248:** 1335–1341.
18. RAMSDELL, F. & B. J. FOWLKES. 1990. Clonal deletion versus clonal anergy: The role of the thymus in inducing self tolerance. Science **248:** 1342–1348.
19. LO, D., J. FREEDMAN, S. HESSE, R. L. BRINSTER & L. SHERMAN. 1991. Peripheral tolerance in transgenic mice: Tolerance to class II MHC and non-MHC transgene antigens. Immunol. Rev. **122:** 87–102.
20. SCHONRICH, G., U. KALINKE, F. MOMBURG, M. MALLISON, A.-M. SCHMITT-VERHULST, B. MALISSEN, G.-J. HAMMERLING & B. ARNOLD. 1991. Down-regulation of T cell receptors on self-reactive T cells as a novel mechanism for extrathymic tolerance induction. Cell **65:** 293–304.
21. SPRENT, J., E.-K. GAO & S. R. WEBB. 1990. T cell reactivity to MHC molecules: Immunity versus tolerance. Science **248:** 1357–1363.
22. SCHWARTZ, R. H. 1990. A cell culture model for T lymphocyte clonal anergy. Science **248:** 1349–1356.
23. THEOFILOPOULOS, A. N. 1987. Autoimmunity. *In* Basic and Clinical Immunology. D. P. Stites, J. D. Stobo & J. V. Wells, Eds. **6:** 128–166. Lange Medical Publications. New York.
24. GAMMON, G., E. SERCARZ & G. BENICHOU. 1991. The dominant self and the cryptic self: Shaping the autoreactive T-cell repertoire. Immunology Today **12:** 193–195.
25. CIBOTTI, R., J. M. KANELLOPOULOS, J.-P. CABANIOLS, O. HALLE-PANENKO, K. KOSMATOPOULOS, E. SERCARZ & P. KOURILSKY. 1991. Tolerance to a self-protein involves its immunodominant but does not involve its subdominant determinant. Proc. Natl. Acad. Sci. USA **89:** 416–420.
26. OHASHI, P. S., S. OEHEN, K. BUERKI, H. PIRCHER, C. T. OHASHI, B. ODERMATT, B. MALISSEN, R. M. ZINKERNAGEL & H. HENGARTNER. 1991. Ablation of "tolerance" and induction of diabetes by virus infection in viral antigen transgenic mice. Cell **65:** 305–317.
27. OLDSTONE, M. B. A., M. NERENBERG, P. SOUTHERN, J. PRICE & H. LEWICKI. 1991. Virus infection triggers insulin-dependent diabetes mellitus in a transgenic model: Role of anti-self (virus) immune response. Cell **65:** 319–332.
28. OLDSTONE, M. B. A. 1987. Molecular mimicry and autoimmune disease. Cell **50:** 819–820.
29. OHNO, S. 1991. Many peptide fragments of alien antigens are homologous with host proteins, thus canalizing T-cell responses. Proc. Natl. Acad. Sci. USA **88:** 3065–3068.
30. GOLDMAN, M., P. DRUET & E. GLEICHMANN. 1991. T_H2 cells in systemic autoimmunity: Insights from allogeneic diseases and chemically-induced autoimmunity. Immunology Today **12:** 223–227.
31. NEMAZEE, D. A. 1985. Immune complexes can trigger specific, T cell–dependent, autoanti-IgG antibody production in mice. J. Exp. Med. **161:** 242–256.
32. SOROKIN, R., H. KIMURA, K. SCHODER, D. H. WILSON & D. B. WILSON. 1986. Cyclosporin-induced autoimmunity. Conditions for expressing disease, requirement for intact thymus, and potency estimates of autoimmune lymphocytes in drug-treated mice. J. Exp. Med. **164:** 1615–1625.
33. GAO, E.-K., R. CHENEY, O. KANAGAWA & J. SPRENT. 1988. Abnormal differentiation of thymocytes in mice treated with cyclosporin A. Nature (London) **336:** 176–179.
34. SMITH, H., I. M. CHEN, R. KUBO & K. S. K. TUNG. 1989. Neonatal thymectomy results in a repertoire enriched in T cells deleted in adult thymus. Science **245:** 749–752.
35. FRIEDMAN, S. M., D. N. POSNETT, J. R. TUMANG, B. C. COLE & K. CROW. 1991. A potential role for microbial superantigens in the pathogenesis of systemic autoimmune disease. Arthritis Rheum. **34:** 468–480.
36. COHEN, I. R. 1986. Regulation of autoimmune disease: Physiological and therapeutic. Immunol. Rev. **94:** 5–21.
37. PAWRIE, F. & D. W. MASON. 1990. OX-22^{high} $CD4^+$ T cells induce wasting disease with

multiple organ pathology, prevention by the OX-22^{low} subset. J. Exp. Med. **172:** 1701–1708.

38. SMITH, L. R., A. PLAZA, P. A. SINGER & A. N. THEOFILOPOULOS. 1990. Coding sequence polymorphisms among Vβ T cell receptor genes. J. Immunol. **144:** 3234–3237.
39. NOONAN, D. J., R. KOFLER, P. A. SINGER, G. CARDENAS, F. J. DIXON & A. N. THEOFILOPOULOS. 1986. Delineation of a defect in T cell receptor β genes of NZW mice predisposed to autoimmunity. J. Exp. Med. **163:** 644–653.
40. KOTZIN, B. L., V. L. BARR & E. PALMER. 1985. A large deletion within the T-cell receptor beta-chain gene complex in New Zealand White mice. Science **229:** 167–171.
41. THEOFILOPOULOS, A. N., R. KOFLER, P. A. SINGER & F. J. DIXON. 1989. Molecular genetic of murine lupus. Adv. Immunol. **46:** 61–108.
42. SINGER, P. A., R. J. MCEVILLY, R. S. BALDERAS, F. J. DIXON & A. N. THEOFILOPOULOS. 1988. T cell receptor Vα haplotype characterization in normal and autoimmune laboratory mouse strains. Proc. Natl. Acad. Sci. USA **85:** 7729–7733.
43. BACCALA, R., D. H. KONO, S. M. WALKER, R. S. BALDERAS & A. N. THEOFILOPOULOS. 1991. Genomically imposed and somatically modified human thymocyte Vβ gene repertoires. Proc. Natl. Acad. Sci. USA **88:** 2908–2912.
44. POSNETT, D. J. 1990. Allelic variations of human TCR V gene products. Immunology Today **11:** 368–373.
45. HILLERT, J. & O. OLERUP. 1992. Germ-line polymorphism of TCR genes and disease susceptibility—fact or hypothesis? Immunology Today **13:** 47–49.
46. KAPPLER, J., N. ROEHM & P. MARRACK. 1987. T cell tolerance by clonal elimination in the thymus. Cell **49:** 273–280.
47. ACHA-ORBEA, H. & E. PALMER. 1991. *Mls*—a retrovirus exploits the immune system. Immunology Today **12:** 356–361.
48. KOTZIN, B. L., J. W. KAPPLER, P. C. MARRACK & L. R. HERRON. 1989. T cell tolerance to self antigens in New Zealand hybrid mice with lupus-like disease. J. Immunol. **143:** 89–94.
49. SINGER, P. A. & A. N. THEOFILOPOULOS. 1990. T cell receptor Vβ repertoire expression in murine models of SLE. Immunol. Rev. **118:** 103–127.
50. ADAMS, S., P. LEBLANC & S. K. DATTA. 1991. Junctional region sequences of T-cell receptor β-chain genes expressed by pathogenic anti-DNA autoantibody-inducing helper T cells from lupus mice. Possible selection by cationic autoantigens. Proc. Natl. Acad. Sci. USA **88:** 11271–11275.
51. SMILEK, D. E., C. B. LOCK & H. O. MCDEVITT. 1990. Antigen recognition and peptide-mediated immunotherapy in autoimmune disease. Immunol. Rev. **118:** 37–50.
52. REICH, E.-P., R. S. SHERWIN, O. KANAGAWA & C. A. JANEWAY JR. 1989. An explanation for the protective effect of the MHC class II I-E molecule in murine diabetes. Nature (London) **341:** 326–328.
53. BOHME, J., B. SCHUHBAUR, O. KANAGAWA, D. BANOIST & D. MATHIS. 1990. MHC-linked protection from diabetes dissociated from clonal deletion of T cells. Science **249:** 293–295.
54. NAKANO, N., H. KIKUTANI, H. NISHIMOTO & T. KISHIMOTO. 1991. T cell receptor V gene usage of islet beta cell-reactive T cells is not restricted in non-obese diabetic mice. J. Exp. Med. **173:** 1091–1097.
55. PRUD'HOMME, G. J., T.-L. LONG, D. C. BOCARRO, R. S. BALDERAS & A. N. THEOFILOPOULOS. 1991. Analysis of pancreas-infiltrating T cells in diabetic NOD mice: Fusion with BW5147 yields a high frequency of islet-reactive hybridomas. Autoimmunity **10:** 285–289.
56. HEBER-KATZ, E. & H. ACHA-ORBEA. 1989. The V-region disease hypothesis: Evidence from autoimmune encephalomyelitis. Immunology Today **10:** 164–169.
57. STEINMAN, L. 1991. The development of rational strategies for selective immunotherapy against autoimmune demyelinating disease. Adv. Immunol. **49:** 357–379.
58. ZALLER, D. M., G. OSMAN, O. KANAGAWA & L. HOOD. 1990. Prevention and treatment of murine experimental allergic encephalomyelitis with T cell receptor Vβ-specific antibodies. J. Exp. Med. **171:** 1943–1955.

59. OFFNER, H., G. A. HASHIM & A. A. VANDENBARK. 1991. T cell receptor peptide therapy triggers autoregulation of experimental encephalomyelitis. Science **251:** 430–432.
60. HOWELL, M. K., S. T. WINTERS, T. OLEE, H. C. POWELL, D. J. CARLO & S. W. BROSTOFF. 1990. Vaccination against experimental allergic encephalomyelitis with T cell receptor peptides. Science **246:** 668–670.
61. DESQUENNE-CLARK, L., T. R. ESCH, L. OTVOS JR. & E. HEBER-KATZ. 1991. T-cell receptor peptide immunization leads to enhanced and chronic experimental allergic encephalomyelitis. Proc. Natl. Acad. Sci. USA **88:** 7219–7223.
62. HAQQI, R. M., S. BANERJEE, M. A. BEHLKE, G. DUNGUN, D. Y. LOH, J. STUART, H. S. LUTHRA & C. S. DAVID. 1988. Vβ6 gene of T cell receptor may be involved in type II collagen-induced arthritis in mice (abstract) FASEB J. **2:** A661.
63. ANDERSSON, M., T. J. GOLDSCHMIDT, E. MICHAELSSON, A. LARSSON & R. HOLMDAHL. 1991. T-cell receptor Vβ haplotype and complement component C5 play no significant role for the resistance to collagen-induced arthritis in the SWR mouse. Immunology **73:** 191–196.
64. SPINELLA, D. G., J. R. JEFFERS, R. A. REIFE & J. M. STUART. 1991. The role of C5 and T-cell receptor Vβ genes in susceptibility to collagne-induced arthritis. Immunogenetics **34:** 23–27.
65. HAQQI, T. M., G. D. ANDERSON, S. BANERJEE & C. S. DAVID. 1992. Restricted heterogeneity in T-cell antigen receptor Vβ gene usage in the lymph nodes and arthritic joints of mice. Proc. Natl. Acad. Sci. USA **89:** 1253–1255.
66. CHIOCCHIA, G., M.-C. BOISSIER & C. FOURNIER. 1991. Therapy against murine collagen-induced arthritis with T cell receptor Vβ-specific antibodies. Eur. J. Immunol. **21:** 2899–2905.
67. GOLDSCHMIDT, T. J., L. JANSSON & R. HOLMDAHL. 1990. *In vivo* elimination of T cells expressing specific T-cell receptor Vβ chains in mice susceptible to collagen-induced arthritis. Immunology **69:** 508–514.
68. YOSHINO, S. & L. G. CLELAND. 1992. Depletion of α/β T cells by a monoclonal antibody against the α/β T cell receptor suppresses established adjuvant arthritis, but not established collagen-induced arthritis, in rats. J. Exp. Med. **175:** 907–915.
69. FUJII, Y. & J.LINDSTROM. 1988. Specificity of the T cell immune response to acetylcholine receptor in experimental autoimmune myasthenia gravis. Response to subunits and synthetic peptides. J. Immunol. **140:** 1830–1837.
70. WUCHERPFENNIG, K. W., K. OTA, N. ENDO, J. G. SEIDMAN, A. ROSENZWEIG, H. L. WEINER & D. A. HAFLER. 1990. Shared human T cell receptor Vβ usage to immunodominant regions of myelin basic protein. Science **248:** 1016–1019.
71. KOTZIN, B. L., S. KARUTURI, Y. K. CHOU, J. LAFFERTY, J. M. FORRESTER, M. BETTER, G. E. NEDWIN, H. OFFNER & A. A. VANDENBARK. 1991. Preferential T-cell receptor β-chain variable gene use in myelin basic protein-reactive T-cell clones from patients with multiple sclerosis. Proc. Natl. Acad. Sci. USA **88:** 9161–9165.
72. BEN-NUN, A., R. S. LIBLAU, L. COHEN, D. LEHMANN, E. TOURNIER-LASSERVE, A. ROSENZWEIG, Z. JINGWU, J. C. M. RAUS & M.-A. BACH. 1991. Restricted T-cell receptor Vβ gene usage by myelin basic protein-specific T-cell clones in multiple sclerosis: Predominant genes vary in individuals. Proc. Natl. Acad. Sci. USA **88:** 2466–2470.
73. OKSENBERG, J. R., S. STUART, A. B. BEGOVICH, R. B. BELL, H. A. ERLICH, L. STEINMAN & C. C. A. BERNARD. 1990. Limited heterogeneity of rearranged T-cell receptor Vα transcripts in brains of multiple sclerosis patients. Nature (London) **345:** 344–346.
74. WUCHERPFENNIG, K. W., J. NEWCOMBE, H. LI, C. KEDDY, M. L. CUZNER & D. A. HAFLER. 1992. T cell receptor Vα-Vβ repertoire and cytokine gene expression in active multiple sclerosis lesions. J. Exp. Med. **175:** 993–1002.
75. STAMENKOVIC, I., M. STEGAGNO, K. A. WRIGHT, S. M. KRANE, E. P. AMENTO, R. B. COLVIN, R. J. DUQUESNOY & J. T. KURNICK. 1988. Clonal dominance among T-lymphocyte infiltrates in arthritis. Proc. Natl. Acad. Sci. USA **85:** 1179–1183.
76. VAN LAAR, J. M., A. M. M. MILTENBURG, M. J. A. VERDONK, M. R. DAHA, R. R. P.

De Vries, P. J. Van Den Elsen & F. C. Breedveld. 1992. T-cell receptor β-chain gene rearrangements of T-cell populations expanded from multiple sites of synovial tissue obtained from a patient with rheumatoid arthritis. Scand. J. Immunol. **35:** 187–194.

77. Cooper, S. M., D. L. Dier, K. D. Roessner, R. C. Budd & J. A. Niclas. 1991. Diversity of rheumatoid synovial tissue T cells by T cell receptor analysis. Arthritis Rheum. **34:** 537–546.
78. Duby, A. D., A. K. Sinclair, S. L. Osborne-Lawrence, W. Zeldes, L. Kan & D. A. Fox. 1989. Clonal heterogeneity of synovial fluid T lymphocytes from patients with rheumatoid arthritis. Proc. Natl. Acad. Sci. USA **86:** 6206–6210.
79. Sottini, A., L. Imberti, R. Gorla, R. Cattaneo & D. Primi. 1991. Restricted expression of T cell receptor Vβ, but not Vα, genes in rheumatoid arthritis. Eur. J. Immunol. **21:** 461–466.
80. Paliard, S., S. G. West, J. A. Lafferty, J. R. Clements, J. W. Kappler, P. Marrack & B. L. Kotzin. 1991. Evidence for the effects of a superantigen in rheumatoid arthritis. Science **253:** 325–329.
81. Howell, M. D., J. P. Diveley, K. A. Lundeen, E. Esty, S. T. Winters, D. J. Carlo & S. W. Brostoff. 1991. Limited T-cell receptor β-chain heterogeneity among interleukin 2 receptor-positive synovial T cells suggests a role for superantigens in rheumatoid arthritis. Proc. Natl. Acad. Sci. USA **88:** 10921–10925.
82. Uematsu, Y., H. Wege, A. Straus, M. Ott, W. Bannwarth, J. Lanchburg, G. Panayi & M. Steinmetz. 1991. The T-cell–receptor repertoire in the synovial fluid of a patient with rheumatoid arthritis is polyclonal. Proc. Natl. Acad. Sci. USA **88:** 8534–8538.
83. Bucht, A., J. R. Oksenberg, S. Lindblads, A. Gronberg, L. Steinman & L. Klareskog. 1992. Characterization of T-cell receptor $\alpha\beta$ repertoire in synovial tissue from different temporal phases of rheumatoid arthritis. Scand. J. Rheumatol. **35:** 159–165.
84. Davies, T. F., A. Martin, E. S. Concepcion, P. Graves, L. Cohen & A. Bennun. 1991. Evidence of limited variability of antigen receptors in intrathyroidal T cells in autoimmune thyroid disease. N. Engl. J. Med. **325:** 238–244.
85. Davies, T. F., A. Martin, E. S. Concepcion, P. Graves, N. Lahat, W. L. Cohen & A. Ben-Nun. 1992. Evidence for selective accumulation of intrathyroidal T lymphocytes in human autoimmune thyroid disease based on T cell receptor V gene usage. J. Clin. Invest. **89:** 157–162.
86. Lipoldova, M., M. Londei, B. Brubeck-Lobenstein, M. Feldmann & M. J. Owen. 1989. Analysis of T-cell receptor usage in activated T-cell clones from Hashimoto's thyroiditis and Graves' disease. J. Autoimmunity **2:**1–13.

The Thymus in Myasthenia Gravis[a]

HARTMUT WEKERLE

Max-Planck-Institut für Psychiatrie
D-8033 Martinsried, Germany

HISTORY

Myasthenia gravis (MG) is more than just a neuromuscular disease. Obviously, disabling disturbance of neuromuscular transmission is the clinical hallmark of the disease. In addition, however, MG is characterized by the occurrence of humoral autoantibodies binding to the postsynaptic acetylcholine receptor (AChR), by the linkage of disease susceptibility to certain HLA haplotypes, and by characteristic changes of the thymus.[1]

There was never a shortage of hypotheses trying to explain the pathogenesis of MG. In the 1960s, three concepts were in vogue. First, Simpson proposed that the thymus might be involved in MG.[2] In the author's words, this concept was "based on clinical data indicating a multisystem disease with resemblances to systemic lupus erythematosus, transmission to the newborn, response to corticosteroids, and pathological changes in muscle and thymus suggestive of immunological disease."[3] Then Van de Velde and Friedman,[4] and Van der Geld and Strauss[5] independently drew attention to the humoral cross-reactivity of MG serum immunoglobulins with skeletal muscle and thymic myoid cell determinants. They were the first to propose primary intrathymic autosensitization against a muscle-specific autoantigen shared by skeletal and myoid cells. At that time, the dichotomy of T and B cells was yet unknown, and AChR as target structures of MG autoantibodies remained to be established. Finally, the third early thymus/MG hypothesis was Goldstein's thymitis concept.[6] It built on experimental work suggesting an autoimmune inflammatory reaction within the thymus leading to the release of a hormone-like, nonimmunoglobulin substance that interfered with neuromuscular transmission.

The world of MG research was changed by two reports that revealed the pathogenic role of autoantibodies against the AChR of the neuromuscular endplate. Patrick and Lindstrom described rabbits that had been immunized with AChR isolated from *Torpedo californica* electric organs and that then developed fatigability and muscle weakness. The disease could be reversed by treatment with Prostigmin.[7] The second seminal experiment was carried out by Toyka and Drachman, who transferred experimental autoimmune myasthenia gravis (EAMG) to immunosuppressed mice by transferring large quantities of MG patient–derived IgG preparations. By then it was clear that anti-AChR antibodies could cause MG-like disease, EAMG, and that very similar (auto-)antibodies were circulating in the blood of MG patients. But now, what of the thymus? How could production of a humoral autoantibody be linked to the central organ of T lymphocytes?

[a] Support for our research came from the Max-Planck-Society and the Deutsch Forschungsgemeinschaft.

HYPOTHESIS

Our own interest in MG was kindled by an unexpected *in vitro* observation. In an attempt to decipher the complexities of lymphocyte-stroma interactions during T-cell differentiation, thymic stroma cells were dissociated and grown as monolayers. These cultures were used to test the effect of thymic stroma on differentiation and function of (autologous) T lymphocytes.[8] Surprisingly, not only "conventional" stromal components (epithelium, fibroblasts, and phagocytes) grew in these cultures. In addition, colonies of small cells formed that proliferated rapidly, and, having reached a certain density, fused to form huge, elongated giant cells. In long-standing cultures, these elongated cells contracted spontaneously, which disclosed beyond any doubt, their identity as myogenic cells.[9]

The number of myogenic colonies varied drastically between different strains of mouse, with C3H/f mice displaying a maximum, and C57BL/6 having a minimum, of *in vitro* thymic myogenesis. Screening a panel of major histocompatibility complex (MHC) congenic strains suggested that the MHC has a role in determining frequency of myogenic precursors in the thymus. These myogenic cells express ample amounts of AChR, which was either distributed evenly or concentrated in "hot spots" on the myotube membrane.[10,11] Myogenesis in cultures of human thymus exhibited the same features.[12]

Our observation of myogenesis in thymus stroma cultures had succeeded previous work that had shown that thymus T cells could be specifically autoimmunized against autochthonous stroma cells.[13] Could it be that autoreactive thymic T cells could be activated also against myogenic cells differentiating within the thymic stroma? If so, one could postulate that the thymus is involved in two steps of the initial phase of the myasthenogenic autoimmune pathogenesis. We proposed that the thymus contains stem cells inducible to myogenesis and that actual induction depends on genetic control (including the MHC) and suitable developmental stimuli. Differentiation of thymic myoid cells is a necessary, but not sufficient, condition to trigger MG. To initiate MG, conditions must be created that allow local T cells to recognize topically produced AChR. Local antigen-presenting cells should take up AChR-containing material from myoid cells and present it in the molecular context of MHC class II determinants to specific T cells. Once activated, the AChR-reactive T cells could disseminate into the periphery and recruit AChR-responsive B cells to produce the pathogenic autoantibodies.[14]

AChR-SPECIFIC T LYMPHOCYTES IN THE MYASTHENIC THYMUS

Obviously, the intrathymic pathogenesis concept of MG predicts that MG thymuses contain an enhanced proportion of AChR-specific, autoreactive T lymphocytes. AChR-specific T cells in an MG patient should be more frequent than in the donor patient's peripheral blood, and AChR frequencies of MG thymus and blood should be higher than the frequencies in healthy donors' organs. Recent progress in T lymphocyte–culture technology allowed us to directly test whether indeed the thymic T cells contain an enhanced proportion of AChR-reactive T cells. It has been known that in some, but not all MG patients, peripheral blood lymphocytes (PBL) may respond to AChR isolated from electric fish organs. The "primary" *in vitro* reactions, measured either by cytokine release[15] or by proliferation assays,[16,17] were quite low and definitely variable. It was a substantial advance when Hohlfeld *et al.* succeeded for the first time to isolate AChR-specific

T cells from the blood of MG patients and to establish monospecific T-cell lines.[18] The AChR-specific lymphocytes were characterized as $CD4^+$ T cells, which recognize their AChR-autoantigen epitope in the molecular context of MHC class II products, mainly DR molecules.[19] Importantly, this work further showed that the AChR-specific T cells were helper lymphocytes capable of enhancing the production of anti-AChR autoantibodies by MHC-compatible B lymphocytes.[20]

Using a similar technology, Melms *et al.* investigated the AChR reactivity of T cells in thymuses removed from MG patients for therapeutical indications. Confirming the early studies mentioned above, the primary proliferation responses of MG PBL were low against *Torpedo* AChR but fairly high to a control foreign T-cell antigen, tetanus toxoid (TT). In stark contrast, T cells from all MG thymuses examined responded equally well to AChR as to TT. Furthermore, the methods used allowed the isolation of AChR-specific T-cell lines from all thymus samples, but only from part of the same donors' PBL.[21] Very similar results were obtained by Sommer *et al.*, who screened both MG thymuses with hyperplasia and thymomas.[22] The thymus-derived T-cell lines differed in no way from Hohlfeld's PBL lines. They all expressed the $CD4^+$ $CD8^-$ T-cell subset phenotype and were restricted by MHC class II determinants.

All data indicate that, in MG, the thymus shelters a relatively high number of AChR-specific, autoimmune T lymphocytes that seem capable of helping AChR-specific B cells produce pathogenic autoantibodies. At this point, however, the question is left open of whether the autoreactive T cells were produced within the thymus or whether they immigrated from the periphery. It should be noted in this context that very recent experiments established that AChR-specific T cells are by no means exclusively found in MG patients but are also present in healthy donors.[23–25]

THE THYMIC MICROENVIRONMENT IN MG

Hyperplastic MG thymuses differ from normal, age-matched organs in several respects. The medulla of hyperplastic thymuses has features that very closely resemble the histology of peripheral immune organs. Perhaps the most prominent change is the formation of germinal centers.[26,27] These seem to be localized often within perivascular spaces, perhaps contributing to a second important change: the disruption of the intercellular matrix structures positioned around medullary blood vessels.[28–31]

Germinal centers are the sites of active humoral immune responses. In germinal centers B cells interact with local follicular dendritic cells (FDC) and helper T lymphocytes leading to maturation of high-affinity immunoglobulin. In germinal centers, immunoglobulin H chain usage is switched to T cell–dependent isotypes. Furthermore, somatic mutations are selected that result in an improved antigen binding, and, finally, differentiation to secretory plasma cells is initiated within germinal centers.[32] All of these events supposedly take place also in germinal centers of MG hyperplastic thymuses, and they may involve B cells specific for AChR determinants. In fact, the production of anti-AChR autoantibodies by thymic B lymphocytes has been demonstrated by Vincent *et al.*[33]

B-cell responses are related more closely to the effector phase of MG than to the initiation phase. More recent immunocytochemical investigations revealed, however, intercellular formations that could well reflect very early cellular responses of the MG pathogenesis. In a very detailed set of immunocytochemical investigations, Kirchner *et al.* identified the three cellular components required in an initial myasthenogenic response. They demonstrated myoid cells expressing AChR in the normal and MG thymus alike. In the normal human thymus, myoid cells are often located within the medullary areas containing mature T cells, but rarely, if ever, within germinal centers.[34] The numbers of myoid cells seem to be high in hyperplastic thymuses with or without germinal centers, and low in atrophic, possibly "burnt-out" thymuses.[30] Conspicuous are the changes of the cellular microenvironment enclosing the myoid cells. Many of these myoid cells are in very close contact with strongly MHC class II–expressing interdigitating cells, the most efficient antigen-presenting cells of the thymus.[21,34] Myoid cells do not seem to express MHC class II determinants on their own membranes,[34,35] but Schluep *et al.* found thymus medullary cells containing AChR-like material and expressing membrane markers typical for APC.[35]

The thymus of myasthenic patients thus holds all elements theoretically required for activation of AChR-specific autoimmune T cells, which could trigger the pathogenesis of the disease. There are autoantigen-producing cells (myoid cells) and autoantigen-presenting cells (interdigitating cells, IDC) in close contact, and these complexes are located within immunocompetent, mainly $CD4^+$ $CD8^-$, T lymphocytes. Yet, it remains open thus far whether these structures are *primary* or, rather, *secondary* events in the pathogenesis.

THYMUS IN EXPERIMENTAL AUTOIMMUNE MYASTHENIA GRAVIS

EAMG is often termed the experimental model of human MG. This is not the entire truth. Like MG, EAMG is caused by autoantibodies binding to AChR of the neuromuscular junction. Unlike MG, however, the pathogenic autoantibodies in EAMG are either induced by immunization with AChR in adjuvant[7] or they are passively transferred.[36,37] Hence, EAMG represents the effector phase, but not the initiation phase of MG.

Examination of thymus changes in EAMG would possibly shed light on the causal involvement of the thymus in the immune pathogenesis of MG. In case of secondary involvement, the thymus of EAMG animals could have lesions similar to those in human MG. Meinl *et al.* investigated the thymuses of rats with acute or with chronic EAMG. Acute EAMG was produced by transferring MAb 35, a monoclonal antibody binding to the main immunogenic region of the AChR α-chain.[38] Chronic EAMG was induced by immunizing the rats with AChR purified from the electric organs of *Torpedo californica* and emulsified in complete Freund's adjuvant.

Immunocytochemistry was used to identify myoid cells, IDC, and lymphocyte subsets in the EAMG thymuses. The structure of healthy rat and human thymuses was very similar. As expected, cortex and medulla could be readily distinguished. Myoid cells were seen often at the corticomedullary junction and also in the thymic medulla proper. In both species. IDC were distributed throughout the medulla, without any particular, nonrandom association with another cell type, such as myoid cells. Finally, silver staining of the thymic reticulin fiber network confirmed the concentric arrangement of connective tissue around medullary blood vessels enframing the thymic perivascular space.

None of the changes described as typical for human MG thymus hyperplasia were ever seen in EAMG rat thymus specimens. Germinal centers were searched for using monoclonal antibodies specific for B cells as well as for FDC. B cells

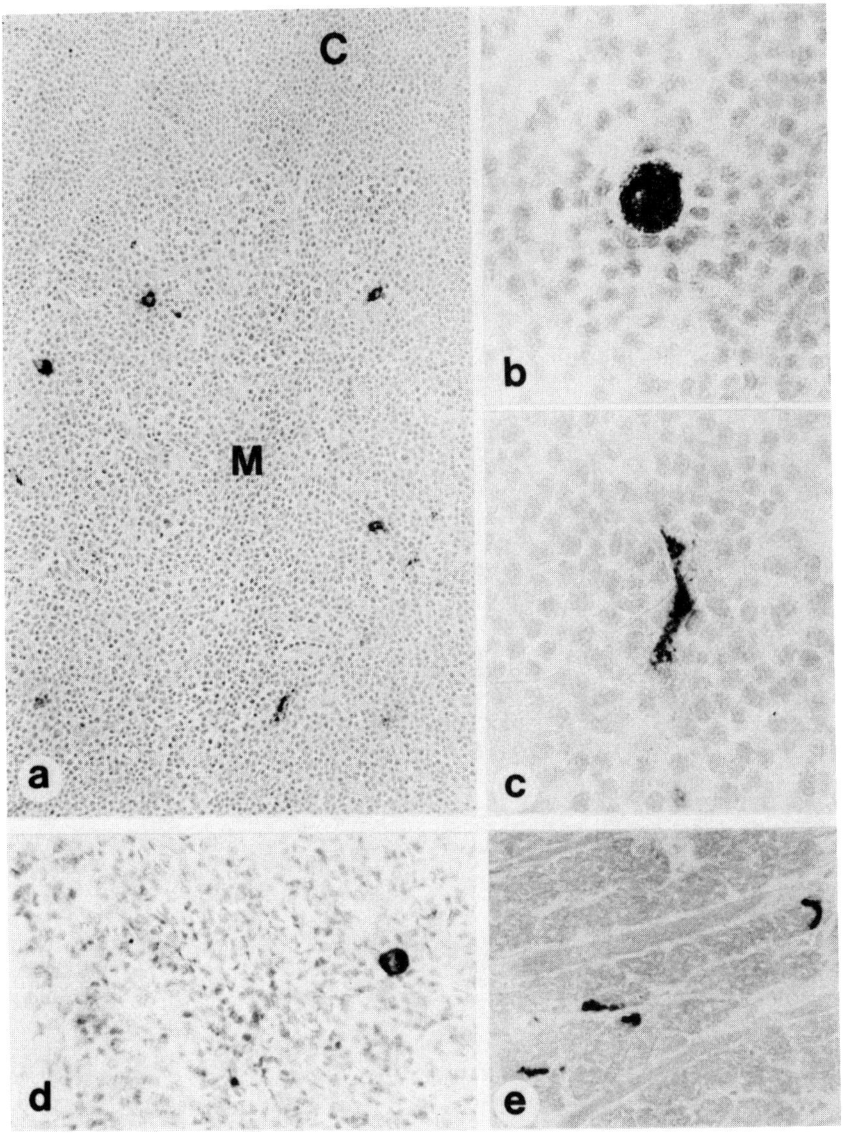

FIGURE 1. EAMG thymuses do not form germinal centers. B-cell arrangements in a lymph node follicle (d) and in the thymic medulla of a normal (b), and EAMG rat (actively induced, chronic EAMG, a). Immunostaining with MAb KiB1R. Follicular dendritic cells (FDC) staining germinal centers in lymph node (e) are not demonstrable in chronic EAMG thymus medulla (c). MAb KiM4R stains, however, some perivascular cells. (Meinl *et al.*[48] With permission from the *American Journal of Pathology*.)

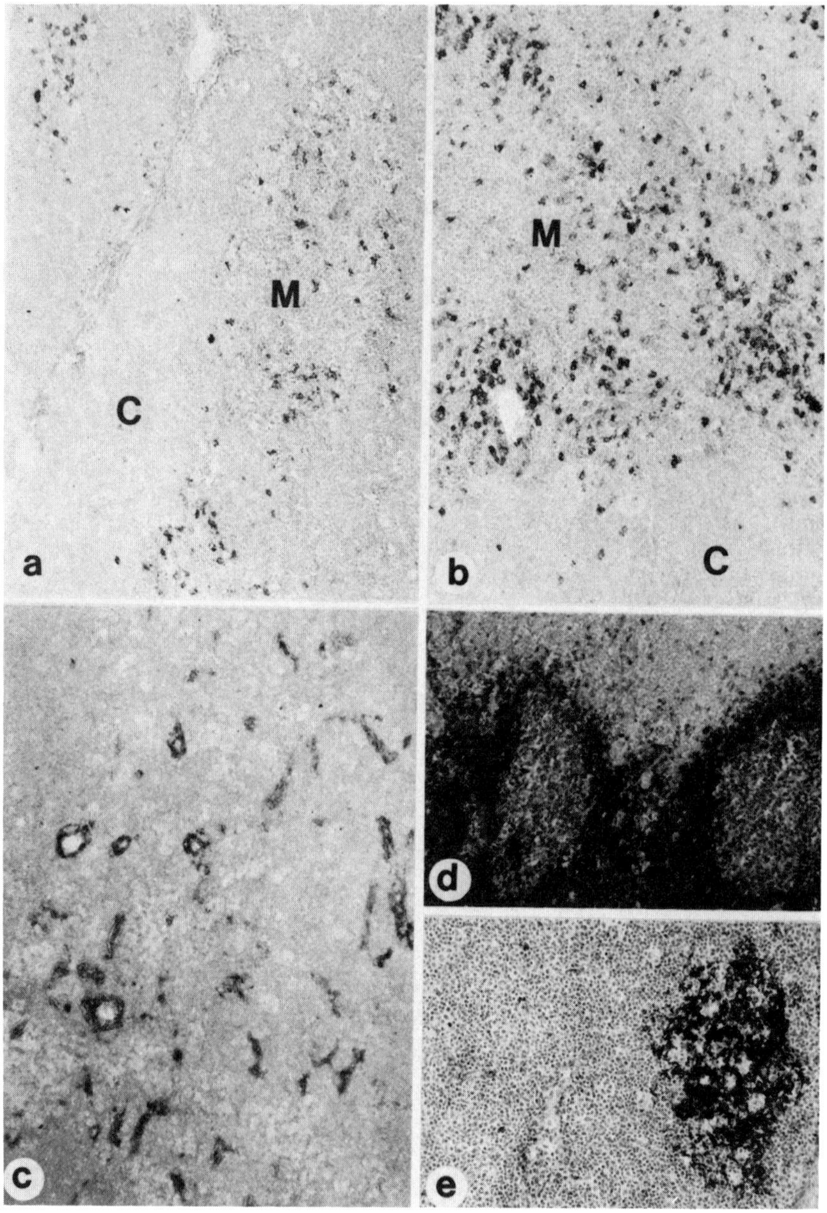

FIGURE 2. Myoid cells in rat thymus. Immunostaining with antitroponin MAb in thymic medulla of normal (a) and chronic EAMG rat (b,c). AChR expression (MAb 35) on myoid cell of a rat with chronic EAMG (d) and on end-plate of control skeletal muscle (e). (Meinl *et al.*[48] With permission from the *American Journal of Pathology*.)

were observed in all thymus specimens regardless of normal or EAMG rat origin. Most of these were scattered throughout the medulla, and another few were located in the cortex as well. In no case did we note B-cell accumulations suggestive of primary or secondary follicle formation. In support of these findings, none of our normal or EAMG thymuses contained any demonstrable FDC, which can be considered highly sensitive indicators of germinal centers (FIG. 1).[39] Equally undisturbed by EAMG were the perivascular reticulin structures of the medullary perivascular space. Even in thymuses involuted in the course of either peracute transfer EAMG, or long-standing chronic EAMG, no discontinuity of the reticulin meshwork was noted.

Finally, the microenvironment of the myoid cells was examined and found to be unaffected by EAMG. Distribution and number of myoid cells in EAMG thymuses were indistinguishable from those in normal control tissue. Furthermore, no association with IDC or other thymic antigen-presenting cells was noted (FIG. 2). The absence of cellular reactions around thymic myoid cells was especially noteworthy in rats with acute, passively transferred disease, which commonly had marked inflammatory infiltrations around peripheral neuromuscular end-plates ("lymphorrhagias").

All changes typical for human MG thymuses are missing in the thymuses of rats with actively induced or passively transferred EAMG. This could mean that rodent thymuses are principally resistant to MG-like changes or that the autoimmune process was too fresh to lead to secondary thymic lesions. Both arguments are refuted by the striking thymus changes seen in spontaneous mouse autoimmune syndromes like autoimmune diabetes mellitus of NOD mice[40] and murine lupus erythematosus.[41] Thus thymus changes related to autoimmunity can happen in rodents, but they did not in EAMG. One may cautiously conclude that the thymic changes of the MG thymus are related to spontaneous development of the disease; for example, they are involved in the initiation of MG.

CONCLUSION AND SUMMARY

The hypothesis that MG starts in the thymus is hard to formally prove. Several predictions made based on this concept can, however, be examined. Among these, we tested whether the thymus contains the cells required for initiation of an AChR-specific T-cell reaction. Indeed, an increased proportion of AChR-specific T cells are demonstrable, along with AChR-producing (myoid cells) and -presenting cells (IDC). That the changes typical of the MG thymus are primary rather than secondary to the immunopathogenesis of MG is suggested by histological analysis of the thymus in EAMG. In EAMG, where a myasthenogenic immune response is initiated outside the thymus, the characteristic thymus changes are not seen. Furthermore, very recent work involving transplantation of MG thymus fragments to immunodeficient SCID mice[42,43] indicated that an intact autoimmune microenvironment is required to maintain production of pathogenic autoantibodies. In conclusion, several experimental observations made during the past few years support the thymic pathogenesis of MG. They have not formally proven this concept, but they further stress the importance of studying the thymus as a model microenvironment for initial phases of human autoimmune disease. The generally accepted success of early thymectomy in therapy of MG associated with thymus hypertrophy[44–47] further strengthens this contention.

ACKNOWLEDGEMENTS

I am obliged to my colleagues in Freiburg, Würzburg, and Munich for fruitful collaboration.

REFERENCES

1. Schönbeck, S., S. Chrestel & R. Hohlfeld. 1990. Int. Rev. Neurobiol. **32:** 175–200.
2. Simpson, J. A. 1960. Scott. Med. J. **5:** 410.
3. Simpson, J. A., P. O. Behan & H. M. Dick. 1976. Ann. N.Y. Acad. Sci. **274:** 382–389.
4. Van de Velde, R. L. & N. B. Friedman. 1966. J. Am. Med. Assoc. **198:** 287–288.
5. Van der Geld, H. W. R. & A. J. L. Strauss. 1966. Lancet **i:** 57–60.
6. Goldstein, G. 1968. Lancet **ii:** 119–122.
7. Patrick, J. & J. Lindstrom. 1973. Science **180:** 871–872.
8. Wekerle, H., I. R. Cohen & M. Feldman. 1973. Eur. J. Immunol. **3:** 745–748.
9. Wekerle, H., B. Paterson, U.-P. Ketelsen & M. Feldman. 1975. Nature **256:** 493–494.
10. Wekerle, H. & U.-P. Ketelsen. 1977. Lancet **i:** 678–680.
11. Wekerle, H., U.-P. Ketelsen, A. D. Zurn & B. W. Fulpius. 1978. Eur. J. Immunol. **8:** 579–582.
12. Kao, I. & D. B. Drachman. 1977. Science **195:** 74–75.
13. Cohen, I. R. & H. Wekerle. 1972. Science **176:** 1324–1325.
14. Wekerle, H., R. Hohlfeld, U.-P. Ketelsen, J. R. Kalden & I. Kalies. 1981. Ann. N.Y. Acad. Sci. **377:** 455–475.
15. Abramsky, O., A. Aharonov, C. Webb & S. Fuchs. 1975. Clin. Exp. Immunol. **19:** 11–16.
16. Richman, D. P., J. Patrick & B. G. W. Arnason. 1976. N. Engl. J. Med. **294:** 694–698.
17. Conti-Tronconi, B. M., M., Morgutti, A. Sghirlanzoni & F. Clementi. 1979. Neurology **29:** 496–501.
18. Hohlfeld, R., K. V. Toyka, K. Heininger, H. Grosse-Wilde & I. Kalies. 1984. Nature (London) **310:** 244–246.
19. Hohlfeld, R., B. Conti-Tronconi, I. Kalies, J. Bertrams & K. V. Toyka. 1985. J. Immunol. **135:** 2393–2399.
20. Hohlfeld, R., I. Kalies, B. Kohleisen, K. Heininger, B. M. Conti-Tronconi & K. V. Toyka. 1986. Neurology **36:** 618–621.
21. Melms, A., B. C. G. Schalke, T. Kirchner, H. K. Müller-Hermelink, E. Albert & H. Wekerle. 1988. J. Clin. Invest. **81:** 902–908.
22. Sommer, N., N. Willcox, G. C. Harcourt & J. Newsom-Davis. 1990. Ann. Neurol. **28:** 312–319.
23. Salvetti, M., S. Jung, S.-F. Chang, H. Will, B. C. G. Schalke & H. Wekerle. 1991. Ann. Neurol. **29:** 508–516.
24. Sommer, N., G. C. Harcourt, N. Willcox, D. Beeson & J. Newsom-Davis. 1991. Neurology **41:** 1270–1276.
25. Melms, A., G. Malcherek, U. Hern *et al.* 1992. Ann. Neurol. **31:** 311–318.
26. Castleman, B. & E. H. Norris. 1949. Medicine (Baltimore) **28:** 27–58.
27. Alpert, L. I., A. Papatestas, A. Kark, R. S. Osserman & K. Osserman. 1971. Arch. Pathol. **91:** 55–61.
28. Berrih, S., W. Savino & S. Cohen. 1985. J. Histochem. Cytochem. **33:** 655–664.
29. Bofill, M., G. Janossy, N. Willcox, M. Chilosi, J. K. Trejdosiewicz & J. Newsom-Davis. 1985. Am. J. Pathol. **119:** 462–473.
30. Kirchner, T., B. Schalke, A. Melms, T. v. Kügelgen & H. K. Müller-Hermelink. 1986. Virchows Arch. B **52:** 237–257.
31. Karttunen, T. 1987. Virchows Arch. A **411:** 245–252.
32. Liu, Y.-J., G. D. Johnson, J. Gordon & I. C. M. MacLennan. 1991. Immunology Today **13:** 17–21.

33. Vincent, A., G. K. Scadding, H. C. Thomas & J. Newsom-Davis. 1978. Lancet **i:** 305–307.
34. Kirchner, T., F. Hoppe, B. Schalke & H. K. Müller-Hermelink. 1988. Virchows Arch. B **54:** 295–302.
35. Schluep, M., N. Willcox, A. Vincent. G. K. Dhoot & J. Newsom-Davis. 1987. Ann. Neurol. **22:** 212–222.
36. Toyka, K. V., D. B. Drachman, A. Pestronk & I. Kao. 1975. Science **190:** 397–399.
37. Lennon, V. A., M. E. Seybold, J. M. Lindstrom, C. Cochrane & R. Ulevitch. 1978. J. Exp. Med. **146:** 973–983.
38. Tzartos, S. J., S. Hochschwender, P. Vasquez & J. Lindstrom. 1987. J. Neuroimmunol. **15:** 185–194.
39. Wacker, H. H., H. J. Radzun, V. Mielke & M. R. Parwaresch. 1987. J. Leukocyte Biol. **41:** 70–77.
40. Savino, W., C. Boitard, J.-F. Bach & M. Dardenne. 1991. Lab. Invest. **64:** 405–417.
41. De Vries, M. J. & W. Hijmans. 1967. Immunology **12:** 179–192.
42. Schönbeck, S., F. Padberg, A. Marx, R. Hohlfeld & H. Wekerle. 1993. Transplantation of myasthenia gravis thymus to SCID mice. Ann. N.Y. Acad. Sci. This volume.
43. Schönbeck, S., F. Padberg, R. Hohlfeld & H. Wekerle. 1992. J. Clin. Invest. **90:** 245–250.
44. Lanska, D. J. 1990. Neurology **40:** 1828–1829.
45. Hatton, P. D., J. T. Diehl, B. D. T. Daly *et al.* 1989. Ann. Thorac. Surg. **47:** 838–840.
46. Mulder, D. G., M. Graves & C. Herrmann. 1989. Ann. Thorac. Surg. **48:** 551–555.
47. Jaretzki, A., A. S. Penn, D. S. Younger *et al.* 1988. J. Thorac. Cardiovasc. Surg. **95:** 747–757.
48. Meinl, E., W. E. F. Klinkert & H. Wekerle. 1991. Am. J. Pathol. **139:** 995–1008.

The Pathological Basis of Thymoma-associated Myasthenia Gravis

H. K. MÜLLER-HERMELINK, A. MARX, K. GEUDER, AND TH. KIRCHNER

Institute of Pathology
University of Würzburg
Josef-Schneider-Strasse 2
8700 Würzburg, Germany

Almost exactly one hundred years ago, Hermann Hoppe (1892), a young American neuropathologist from Cincinnati, mentioned for the first time the coincidence of myasthenia gravis (MG) and a mediastinal tumor mass at the meeting of the Society for Psychiatry and Neurological Diseases in Berlin.[1] Soon after this observation it became evident that the coincidence of MG and mediastinal tumor was not by mere chance. Much of the literature and our own material have shown that thymoma is found in about 10% of the MG patients and that MG is detected in 15 to 80% of thymoma patients. The percentage depends, to some extent, on whether the statistical evaluation is derived from hospitals with a major interest in neuromuscular disease or from thoracic surgery units.[2–5]

In the past, different hypotheses have been formulated to explain the extraordinary association between the two diseases (TABLE 1 and refs. 5–7), but based on recent findings, most of them seem to be quite unlikely.[8] There is, for instance, no indication of a common genetic error or risk of predisposition for the two diseases.

Furthermore, thymic lymphofollicular hyperplasia probably does not progress to thymoma, inasmuch as both MG-associated thymic alterations clearly differ with regard to MHC determinants, age, and sex distribution.[9,10] There are also no observations that a T cell–mediated antitumor response, as can be assumed for the paraneoplastic Lambert-Eaton syndrome in small lung cancer,[11] might provoke MG.[7] Thus, it is extremely difficult to find any anti–AChR reactivity in mature T cells or activated T cells derived from thymoma *in vitro*.[12] We have found strong evidence, however, that the thymoma-associated MG could be caused by aberrant selection processes occurring in the thymoma as a neoplastic mimicry of the normal thymic function.

Pathohistological features of thymic epithelial tumors are very heterogeneous (FIG. 1), and in recent years we have proposed a new classification of thymic tumors that relates tumor features to the organization of microenvironments of the normal thymus.[5,13] Using this classification we show that medullary thymoma, mixed thymoma, cortical thymoma, and well-differentiated thymic carcinoma comprise organotypic thymic features that are lost in other thymic carcinomas of high-grade malignancy. Most interestingly, these heterogeneous types of epithelial tumors not only exhibit differences with respect to invasiveness and prognosis but also significant differences with regard to a development of paraneoplastic MG.[5] Thus MG association is low for medullary, mixed, and predominantly cortical thymoma; it is high for cortical thymoma and well-differentiated thymic carcinoma; and MG is generally not detected in patients with other types of thymic carcinoma (TABLE 2). Most of our studies on the pathogenesis of paraneoplastic MG are based on the tumors with high MG association, namely cortical thymomas and well-

TABLE 1. Hypotheses for Thymoma-association with MG

Susceptibility to MG and to thymoma by the same genetic error
Progression of thymic lymphofollicular hyperplasia to thymoma
Induction of an antitumor and paraneoplastic autoimmune reaction by tumor-associated antigens
Lack of negative selection in thymoma
False-positive selection or autosensitization in thymoma

differentiated thymic carcinomas. We do not know at present whether similar pathogenic mechanisms are also effective in the group of tumors with low MG association. There are, however, some indications of why MG is not found in other types of thymic carcinomas and in tumors of extrathymic origin.

A hallmark of all thymic tumors associated with MG is a presence of CD1-positive immature thymocytes showing an intense interaction with neoplastic epithelial cells of the cortical type.[5,14–17] These epithelial cells correspond to the thymic nurse cells that are highly involved in processes of positive T-cell selection in the thymus.[18,19] Intratumorous lymphoepithelial interactions of CD1-positive thymocytes are a constant feature of cortical thymomas and well-differentiated thymic carcinomas associated with MG. They are not found in other types of thymic carcinomas or in tumors of extrathymic sites.

IS AChR EXPRESSED BY THYMOMA EPITHELIAL CELLS?

If positive selection, normally occurring in the thymic cortex, also takes place in the interaction of CD1-positive thymocytes and thymoma epithelial cells, we may speculate on an abnormal presence of immunogenic peptides related to the main autoantigen in MG, namely the acetylcholine receptor (AChR) on the neoplastic epithelial cells. Therefore we started to study the genomic organization and transcription of the AChR alpha-, beta-, gamma-, and delta-subunit in thymic epithelial tumors. Restriction analysis of genomic DNA showed that the organization of the AChR subunit genes in these tumors is identical to the findings in normal tissues.[20,21]

Looking at the level of transcription dot blot hybridization revealed a presence of mRNA of the alpha-, beta-, gamma-, and delta-subunit as well as a very rare

TABLE 2. MG Association of Thymic Epithelial Tumors

Tumor Type	Total Cases	MG Association (Percent)
Medullary thymoma	9	33
Mixed thymoma	31	39
Predominantly cortical thymoma	12	33
Cortical thymoma	65	66
Well-differentiated thymic carcinoma	26	77
Other types of thymic carcinoma	12	0

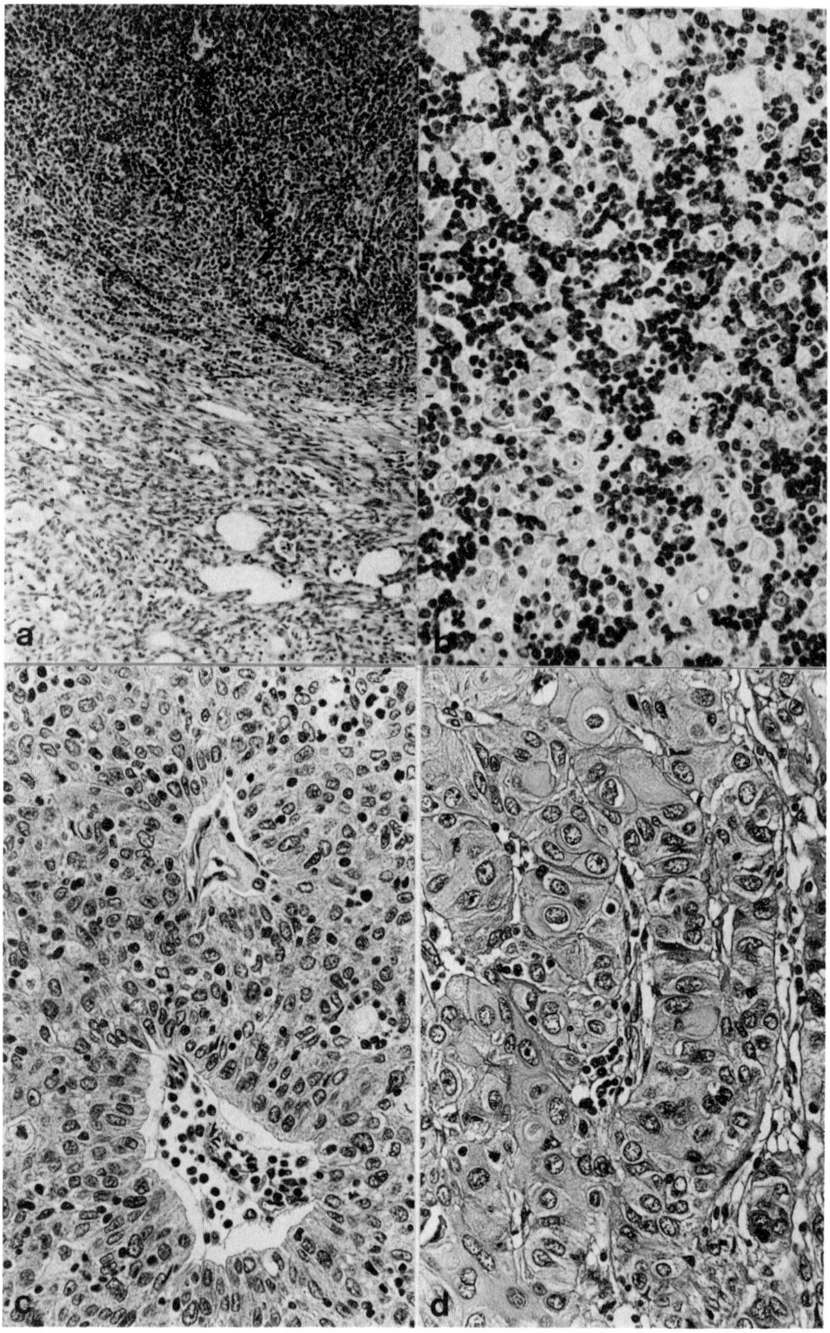

occurrence of the epsilon-subunit mRNA in normal thymus and MG-associated thymic lymphofollicular hyperplasia–containing myoid cells,[22] which were previously identified to bear AChR by immunohistochemistry.[23,24] Dot and Northern blot analysis, however, did not show transcription of any AChR-subunit genes in thymic epithelial tumors, in which myoid cells are generally absent.[21] A recent investigation by Hara *et al.*[25] confirms these results of Northern and dot blot hybridization analysis. Using a very sensitive PCR technique, however, they were able to detect transcripts of the alpha-AChR subunit in thymoma epithelial cells. This most interesting and provocative fact has also been found in our recent studies. We established a thymoma cDNA library cloned into the lambda gt 11 bacteriophage. Plaque screening was done by a synthetic oligonucleotide corresponding to a cDNA sequence of the cytoplasmic part of the AChR alpha chain. Among a total number of 2×10^6 clones, we found two individual clones expressing the alpha-AChR subunit (FIG. 2). Therefore thymoma epithelial cells apparently transcribe the AChR-alpha subunit to a very low extent and perhaps only in individual cells and/or temporarily. The question remains, however, whether a complete and functional receptor is produced by these epithelial cells. Again the answer to this question is *no*: no functional acetylcholine-gated channels or glycine-gated and GABA-gated channels, which show sequence homology with the AChR, could be detected on thymoma epithelial cells by the most sensitive patch-clamp technique, whereas acetylcholine-triggered currents were regularly present in the TE 671 cell line (TABLE 3 and ref. 26).

ARE IMMUNOGENIC AChR EPITOPES PRESENT IN THYMIC EPITHELIAL TUMORS?

For the abnormal intratumorous selection of thymocytes with specificity for the AChR, it could be sufficient that thymoma epithelial cells express only relevant immunogenic AChR epitopes rather than the complete receptors. Using monoclonal antibodies to denatured AChRs, a very immunogenic cytoplasmic epitope of the alpha-subunit (VICE-alpha), which has been defined by Tzartos and Remoundos,[27] was consistently found in a few epithelial cells of the normal thymic medulla and in many neoplastic epithelial cells of cortical thymoma and well-differentiated thymic carcinoma.[24,28] VICE-alpha is best detected by the monoclonal antibody (MAb) 155 of Tzartos *et al.*[29,30] It was originally thought to comprise residues 371–378 and is more recently localized to residues 373–380 of the cytoplasmic part of the AChR-alpha subunit.[27] The expression of VICE-alpha on neoplastic epithelial cells of thymomas and well-differentiated thymic carcinomas

---

FIGURE 1. Heterogeneity of thymic epithelial tumors. **a:** MG-associated, mixed thymoma composed of a medullary-type differentiation with spindle cells (lower half of the picture) and a cortical-type differentiation with many thymocytes (upper half of the picture). **b:** MG-associated cortical thymoma, typically rich in immature thymocytes intermingled with large, pale, and polygonal cortical-type epithelial cells. **c:** MG-associated well-differentiated thymic carcinoma exhibiting the prevalence of epithelial cells with little atypia, epidermoid differentiation palisading around perivascular spaces, and the preserved occurrence of few immature thymocytes. **d:** Moderately differentiated thymic carcinoma (not associated with MG) with moderate atypia of epithelial cells and almost complete lack of immature thymocytes.

FIGURE 2. Occurrence of AChR alpha-subunit transcripts in thymoma. Detection of the alpha-subunit cDNA in two clones from a thymoma cDNA library by Southern blot hybridization, using the AChR alpha-subunit cDNA as a probe.[20–22] Lane 1: lambda gt11 bacteriophage only; lanes 2 and 3: positive clones; lanes 4 and 5: irrelevant clones (negative controls); lane 6: AChR alpha-subunit CDNA insert of plasmid pTE1.1 (provided by R. Schoepfer).

can be correlated to the clinical presence or absence of paraneoplastic MG (TABLE 4). Binding of the MAb 155 to VICE-alpha is seen in most of the cortical thymomas and well-differentiated thymic carcinomas that exhibit high MG associations. Predominantly cortical thymomas take an intermediate position, however, and most of the medullary and mixed thymomas with low MG association are negative. Moreover, 29 of 35 thymic epithelial tumors with MG association expressed VICE-alpha, whereas 16 of 17 tumors without MG association were negative.

TABLE 3. Screening for Ion Channels in Thymic and Thymomatous Epithelial Cells from Short-term Cultures by Patch-Clamp Analyses (Whole-cell recording) According to Siara *et al.*[26]

	Percentage of Positive Cells		
Cell Origin	ACh-gated Channel	Glycine-gated Channel	GABA-gated Channel
Thymus 11398	0%	0%	0%
Thymus 12773	0%	0%	0%
Thymus 13778	0%	0%	0%
Thymoma 1494	0%	0%	0%
Thymoma 10433	0%	0%	0%
Thymoma 18633	0%	0%	0%
Thymoma 15977	0%	0%	0%
TE 691 cell line	100%	0%	0%

Our first immunohistochemical studies already indicated that the binding of the MAb 155 to VICE-alpha is apparently not related to the presence of AChRs in thymoma epithelial cells.[24] Therefore, we wanted to get more insight into proteins or peptides relevant for this cross-reactivity. For these studies we constructed an oligonucleotide probe of the AChR-alpha chain amino acids 371–378, which are relevant for the binding of MAb 155.[29,30] Using this oligonucleoticle, we probed EcoRI-digested genomic DNA of thymomas and muscle. Four restriction fragments, one of more than 14,000, and the others of about 7000, 3500, and 1500 bp, were detected.[21] As the 14,000 bp fragment is due to the EcoRI restriction fragment of the AChR-alpha subunit gene, one can conclude that there could be three additional genomic loci that might code for proteins with VICE-alpha but that are different from the AChR. Finally, Western blots from tissue extracts of thymomas revealed a non-AChR protein with a molecular mass of 153 kDa that is labeled by MAb 155.[31,32] Based on these findings it seems likely that one of the hallmarks of MG-associated thymoma is the expression of AChR epitopes, especially those that are related to VICE-alpha. This is not a *de novo* expression, because VICE-

TABLE 4. Immunohistochemical VICE-alpha Detection by the MAb 155 in Different Types of Thymic Epithelial Tumors and in Tumors with and without MG

	MAb 155 Binding	
Tumor Type	Positive	Negative
Medullary thymoma	0	1
Mixed thymoma	1	13
Predominantly cortical thymoma	7	6
Cortical thymoma	14	1
Well-differentiated thymic carcinoma	8	0
Moderately differentiated thymic carcinoma	0	1
Tumors with MG	29	6
Tumors without MG	1	16

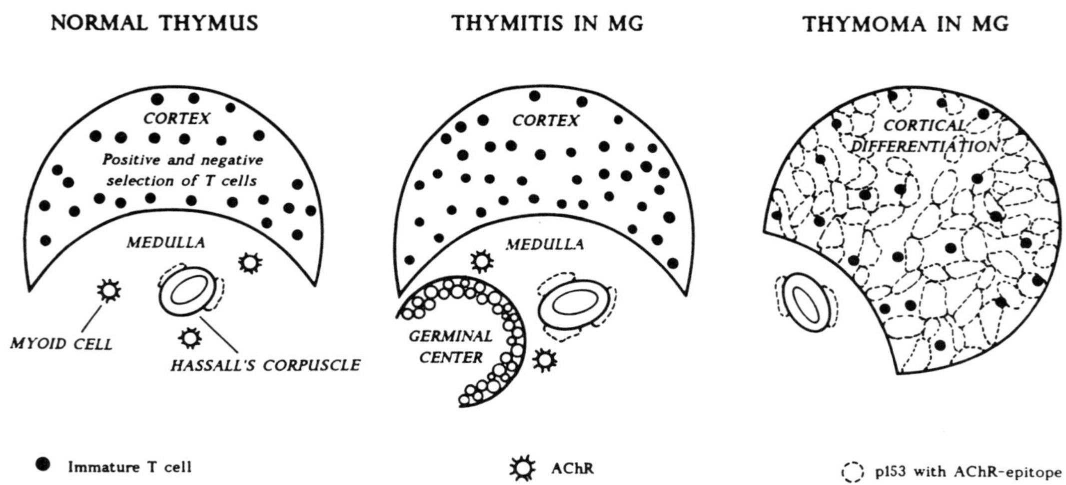

FIGURE 3. Scheme showing the different expressions of the AChR epitope–bearing protein (p153) and of AChRs in normal thymus, thymitis in MG, and thymoma in MG. Normal thymus and thymitis in MG contain AChR-expressing myoid cells and few epithelial cells with p153 exclusively in the medulla. Thymoma in MG exhibits no AChR-expressing myoid cells and shows an aberrant hyperexpression of p153 on cortical-type epithelial cells next to immature thymocytes.

alpha occurs also on a few medullary epithelial cells of the nontumorous thymus. Thymomas exhibit, however, a hyperexpression as well as an aberrant expression of VICE-alpha in the microenvironment of cortical epithelial differentiation (FIG. 3). Of course, it is now tempting to speculate that such aberrant expressions of AChR epitopes cause activation rather than production of tolerance of receptor-specific T cells, or that false-positive selection of AChR-sensitive T cells might finally cause MG in thymoma patients.[33]

SIGNIFICANCE OF THYMOMA-STRIATED MUSCLE AND OTHER CROSS-REACTING EPITOPES

The finding of cross-reacting epitopes between striated muscle and thymus epithelium is not new. It is well known that, in addition to antibodies to the AChR, thymoma-associated autoimmunity is also characterized by autoantibodies to skeletal muscle antigens. Some of these antibodies have also been shown to stain thymoma epithelial cells: Gilhus *et al.*[34] and Dardenne *et al.*[35] have raised sera or a monoclonal antibody defining epitopes shared by muscle and thymic epithelial cells. Thus we were challenged to test the hypothesis of whether VICE-alpha, shared by the AChR and thymoma epithelial cells, is also present in skeletal muscle proteins. In a cooperative study with Mary Osborn, we finally succeeded in showing that VICE-alpha is also expressed by a major protein of skeletal muscle,[36] namely fast troponin I.[37] So far autoantibodies to fast troponin I have not been tested in MG patients in comparison to AChR autoimmunity. It might turn out, however, that the autoimmunity to the AChR is only the symptomatic finding of an autoimmune attack induced by a single immunogenic peptide for T cells. Up to now a definite proof for the immunogenic relevance of VICE-alpha in paraneoplastic MG is still lacking, inasmuch as the 153 kDa thymoma protein with the VICE-alpha is not yet sequenced and available for testing. Recently, however, we found another example of an abnormal protein expression in the cortical microenvironment of thymomas, which might soon enable a further testing of our hypothesis. Twenty-two different thymomas were investigated with antibodies against neurofilament epitopes. Epitopes of the 68 kDa and 160 kDa neurofilaments were only present in the neoplastic epithelial cells of cortical thymoma and thymic carcinoma, whereas medullary and mixed thymoma were negative.[38] In addition, the frequency of autoantibodies against axons of human fetal spinal cord and dorsal ganglia was tested in patients with thymic epithelial tumors, and most interestingly the antiaxonal antibodies were detected much more frequently in patients with thymic epithelial tumors than in myasthenic patients without thymoma.

In conclusion, these findings favor the hypothesis that aberrant expressions of protein epitopes in thymic epithelial cells, especially those with cortical thymic differentiation, might cause false-positive selection or autosensitization of immature T cells, which could stimulate the production of autoantibodies after an export to the periphery.

REFERENCES

1. HOPPE, H. H. 1892. Ein Beitrag zur Kenntnis der Bulbärparalyse. Berl. Klin. Wschr. **29:** 332–336.
2. SOUADJIAN, J. V., P. ENRIQUEZ, M. N. SILVERSTEIN & J.-M. PEPIN. 1974. The spectrum of diseases associated with thymoma. Arch. Intern. Med. **134:** 370–374.

3. ROSAI, J. & G. D. LEVINE. 1976. Tumors of the thymus. *In* Atlas of Tumor Pathology, Sec Ser, Fasc 13. Armed Forces Institute of Pathology. Washington DC.
4. OTTO, H. F. 1984. Pathologie des Thymus. Springer. Berlin.
5. KIRCHNER, TH. & H. K. MÜLLER-HERMELINK. 1989. New approaches to the diagnosis of thymic epithelial tumors. In Progress in Surgical Pathology, 10. C. M. Fenoglio-Preiser, M. Wolff & F. Rilke, Eds.: 167–186. Field and Wood Inc. Philadelphia.
6. WILLCOX, N., M. SCHLUEP, M. A. RITTER, N. J. SCHUURMAN, J. NEWSOM-DAVIS & B. CHRISTENSSON. 1987. Myasthenic and nonmyasthenic thymoma. An expansion of a minor cortical epithelial cell subset? Am. J. Pathol **127:** 447–460.
7. HOHLFELD, R. 1990. Myasthenia gravis and thymoma: paraneoplastic failure of neuromuscular transmission. Lab. Invest. **62:** 241–243.
8. WILLCOX, N. 1992. The Third Euromyasthenia Conference: Meeting report and update on myasthenia research. Autoimmunity **11:** 209–212.
9. COMPSTON, A. X., A. VINCENT, J. NEWSOM-DAVIS & J. R. BATCHELOR. 1980. Clinical, pathological, HLA-antigen and immunological evidence for disease heterogeneity in myasthenia gravis. Brain **103:** 597–601.
10. SCHALKE, B. C. G., H. G. MERTENS, H. LINDNER, TH. KIRCHNER, A. MARX, H. K. MÜLLER-HERMELINK, A. ANDREAS-ZIETZ & E. ALBERT. 1992. Myasthenia Gravis: Immungenetik Thymusmorphologie und klinischer Verlauf. *In* Neuromuskuläre Erkrankungn. K. Kunze, A. Arlt & G. Thaysen, Ed.: 157–164. G. Fischer Verlag. Stuttgart, Jena, New York.
11. MORRIS, C. S., M. M. ESIRI, A. MARX & J. NEWSOM-DAVIS. 1992. Immunocytochemical characteristics of small cell lung carcinoma associated with the Lambert-Eaton myasthenic syndrome. Am. J. Pathol. **140:** 839–845.
12. SOMMER, N., N. WILLCOX, G. C. HARCOURT & J. NEWSOM-DAVIS. 1990. Myasthenic thymus and thymoma are selectively enriched in acetylcholine receptor-reactive T cells. Ann. Neurol. **28:** 312–319.
13. MARINO, M. & H. K. MÜLLER-HERMELINK. 1985. Thymoma and thymic carcinoma. Relation of thymoma epithelial cells to the cortical and medullary differentiation of the thymus. Virchows Arch. Pathol. Anat. **407:** 119–149.
14. CHAN, W. C., C. S. ZATAARI, S. TABEI, M. BIBB & R. K. BRYNES. 1984. Thymoma: An immunohistochemical study. Am. J. Clin. Pathol. **82:** 160–166.
15. MOKHTAR, N., S. M. HSU, R. P. LAD, B. F. HAYNES & E. S. JAFFE. 1984. Thymoma: Lymphoid and epithelial components mirror the phenotype of normal thymus. Hum. Pathol. **15:** 378–384.
16. VAN DER KWAST, T. H., E. VAN VLIET, E. M. CRISTEN, W. VAN EWIJK & R. O. VON DER HEUL. 1985. An immunohistochemical study of the epithelial and lymphoid components of six thymomas. Hum. Pathol. **16:** 1001–1008.
17. MÜLLER-HERMELINK, H. K., M. MARINO & G. PALESTRO. 1986. Pathology of thymic epithelial tumors. *In* Current Topics in Pathology Vol. 75. The Human Thymus: Histophysiology and Pathology. H. K. Müller-Hermelink, Ed.: 207–268. Springer. Berlin.
18. WEKERLE, H. & U.-P. KETELSEN. 1980. Thymic nurse cells—Ia bearing epithelium involved in T-lymphocyte differentiation? Nature **283:** 402–404.
19. MARRACK, P., J. MCCORMACK & J. KAPPLER. 1989. Presentation of antigen, foreign major histocompatibility complex proteins and self by thymus cortical epithelium. Nature **338:** 503–505.
20. GEUDER, K. I., R. SCHOEPFER, TH. KIRCHNER, A. MARX & H. K. MÜLLER-HERMELINK. 1989. The gene of the alpha-subunit of the acetylcholine receptor: Molecular organisation and transcription in myasthenia-associated thymomas. Thymus **14:** 179–186.
21. GEUDER, K. I., A. MARX, V. WITZEMANN, B. SCHALKE, TH. KIRCHNER & H. K. MÜLLER-HERMELINK. 1992. Genomic organization and lack of transcription of the nicotinic acetylcholine receptor subunit genes in myasthenia gravis–associated thymoma. Lab. Invest. **66:** 452–458.
22. GEUDER, K. I., A. MARX, V. WITZEMANN, B. SCHALKE, K. V. TOKA, T. KIRCHNER & H. K. MÜLLER-HERMELINK. 1992. Pathogenetic significance of fetal-type acetyl-

choline receptors on thymic myoid cells in myasthenia gravis. Dev. Immunol. **2:** 69–75.

23. SCHLUEP, M., N. WILLCOX, A. VINCENT, G. K. DHOOT & J. NEWSOM-DAVIS. 1987. Acetylcholine receptors in human thymic myoid cells *in situ:* An immunohistological study. Ann. Neurol. **22:** 212–222.
24. KIRCHNER, T., S. TZARTOS, F. HOPPE, B. SCHLAKE, H. WEKERLE & H. K. MÜLLER-HERMELINK. 1988. Pathogenesis of myasthenia gravis. Acetylcholine receptor–related antigenic determinants in tumor-free thymuses and thymic epithelial tumors. Am. J. Pathol. **130:** 268–280.
25. HARA, Y., S. UENO, T. UEMICHI, N. TAKAHASHI, S. YORIFUJI, Y. FUJII & S. TARUI. 1991. Neoplastic epithelial cells express alpha-subunit of muscle nicotinic acetylcholine receptor in thymomas from patients with myasthenia gravis. FEBS Lett. **279:** 137–140.
26. SIARA, J., R. RÜDEL & A. MARX. 1991. Absence of acetylcholine-induced current in epithelial cells from thymus glands and thymomas of myasthenia gravis patients. Neurology **41:** 128–131.
27. TZARTOS, S. & S. REMOUNDOS. 1992. Precise epitope mapping of monoclonal antibodies to the cytoplasmic side of the acetylcholine receptor alpha-subunit. Dissecting a potentially myasthenogenic epitope. Eur. J. Biochem. **207:** 915–922.
28. KIRCHNER, T., F. HOPPE, H. K. MÜLLER-HERMELINK, B. SCHALKE & S. TZARTOS. 1987. Acetylcholine receptor epitopes on epithelial cells of thymoma in myasthenia gravis. Lancet **1:** 218.
29. TZARTOS, S., L. LANGEBERG, S. HOCHSCHWENDER, L. W. SWANSON & J. LINDSTROM. 1986. Characteristics of monoclonal antibodies to denatured *Torpedo* and to calf acetylcholine receptors: species, subunit and region specificity. J. Neuroimmunol. **10:** 235–253.
30. TZARTOS, S., A. KOKLA, S. L. WALGRAVE & B. M. CONTI-TRONCONI. 1988. Localization of the main immunogenic region of human muscle acetylcholine receptor to residues 67–76 of the alpha-subunit. Proc. Natl. Acad. Sci. USA **85:** 2899–2903.
31. MARX, A., T. KIRCHNER, F. HOPPE, R. O'CONNOR, B. SCHALKE, S. TZARTOS & H. K. MÜLLER-HERMELINK. 1989. Proteins with epitopes of the acetylcholine receptor in epithelial cell cultures of thymomas in myasthenia gravis. Am. J. Pathol. **134:** 865–877.
32. MARX, A., G. O'CONNOR, K. I. GEUDER, F. HOPPE, B. SCHALKE, S. TZARTOS, I. KALIES, T. KIRCHNER & H. K. MÜLLER-HERMELINK. 1990. Characterization of a protein with an acytylcholine receptor epitope from myasthenia gravis–associated thymomas. Lab. Invest. **62:** 279–285.
33. MARX, A., K. I. GEUDER, R. SCHOEPFER, S. TZARTOS, U. KRISTOFFERSSON, B. SCHALKE, T. KIRCHNER & H. K. MÜLLER-HERMELINK. 1991. Analysis of the acetylcholine receptor epitope-bearing protein p153 in thymomas favors "false-positive T-cell selection" as a Mechanism of Paraneoplastic Myasthenia gravis. *In* Lymphatic Tissues and *In Vivo* Immune Responses. B. A. Imhof, S. Berrih-Aknin & S. Ezine, Eds.: 577–583. Marcel Dekker, Inc. New York, Basel, Hong Kong.
34. GILHUS, N. E., J. A. AARLI, B. CHRISTENSSON & R. MATRE. 1984. Rabbit antiserum to citric acid extract of human skeletal muscle staining thymomas from myasthenia gravis patients. J. Neuroimmunol. **7:** 55–64.
35. DARDENNE, M., W. SAVINO & J. F. BACH. 1987. Thymomatous epithelial cells and skeletal muscle share a common epitope by a monoclonal antibody. Am. J. Pathol. **126:** 194–198.
36. MARX, A., M. OSBORNE, S. TZARTOS, K. I. GEUDER, B. SCHALKE, W. NIX, T. KIRCHNER & H. K. MÜLLER-HERMELINK. 1992. A striational muscle antigen and myasthenia gravis–associated thymomas share an acetylcholine-receptor epitope. Dev. Immunol. **2:** 77–83.
37. OSBORN, M., A. MARX, T. KIRCHNER, S. TZARTOS, U. PLESSMAN & K. WEBER. 1992. A shared epitope in the acetylcholine receptor alpha subunit and fast troponin I of skeletal muscle: Is it important for myasthenia gravis? Am. J. Pathol. **140:** 1215–1223.
38. MARX, A., T. KIRCHNER, A. GREINER, H. K. MÜLLER-HERMELINK, B. SCHALKE & M. OSBORN. 1992. Neurofilament epitopes in thymoma and antiaxonal autoantibodies in myasthenia gravis. Lancet **339:** 707–708.

Transplantation of Myasthenia Gravis Thymus to SCID Mice[a]

SIMONE SCHÖNBECK,[b,c] FRANK PADBERG,[b,e]
ALEXANDER MARX,[d] REINHARD HOHLFELD,[b,c] AND
HARTMUT WEKERLE[b]

[b]*Department of Neuroimmunology*
Max-Planck Institute
8033 Planegg-Martinsried, Germany

[c]*Department of Neurology*
Klinikum Grosshadern
University of Munich
8000 Munich 71, Germany

[d]*Department of Pathology*
University of Würzburg
D-8700 Würzburg, Germany

INTRODUCTION

The classical passive transfer model of myasthenia gravis (MG) by way of antibodies to the acetylcholine receptor (AChR) represents the effector phase of the disease solely.[1] This model does not allow studies on the role of the thymus. Indeed, even in actively induced experimental MG,[2,3] secondary changes in the thymus of recipient mice could not be observed.[4] This provides further evidence that the afferent phase of the autoimmune response against the AChR is initiated in the thymus (see Wekerle, this volume). To examine the role of the thymus in the immunopathogenesis of MG, thymus tissue was transplanted into severe combined immunodeficient (SCID) mice. This model allowed the study of the role of the thymus in the initial phase of MG.

This paper will first give an introduction to the use of SCID mice as a model for human diseases and then discuss different possibilities for reconstituting SCID mice with human grafts. Our model will then be demonstrated and discussed in detail.

THE "HUMANIZED" SCID MOUSE MODEL

The C.B.-17 *scid*/*scid* mouse (SCID mouse) carries a spontaneously arising autosomal recessive mutation leading to severe combined immunodeficiency.[5] Although the bone marrow of SCID mice appears normal, these mice are lympho-

[a] This work was supported by the Max-Planck Society and the Deutsche Forschungsgemeinschaft.

[e] This work is part of F. Padberg's medical thesis performed at the Ludwig Maximilians Universität, Munich, Germany.

penic, and their thymus size is only between 1 and 10% of normal. The SCID thymocytes are CD3-, CD4-, and CD8-negative.[6] The serum contains less than 1 μg/mL immunoglobulin (Ig). SCID mice are unable to mount an immune response against T-cell independent or T-cell dependent antigens, and they are incapable of rejecting allogeneic cells. Pre–B cells carrying the B-cell marker B220 and cytoplasmic IgM are also not detectable.[5]

In contrast to the lymphoid lineage, the differentiation and function of cells of the myeloid lineage do not seem to be affected by the *scid* mutation. It could be demonstrated that the *scid* mutation results in a defective VDJ recombinase system, which is used by the antigen-specific receptors of B cells and T cells.[7]

The arrest of lymphocyte differentiation, however, is not absolute in SCID mice. Mainly depending on the conditions of animal maintenance, between 2 and 23% of adult SCID mice have detectable serum Ig levels, reaching 30–47% in mice that are at least one year older.[8] The latter mice have not only functional Ig-secreting B cells but also functional T cells. Most old SCID mice possess functional T or B lymphocytes, regardless of whether they contain serum immunoglobulin.

Reconstitution of SCID mice with elements of the human immune system has been accomplished by two routes. To produce human peripheral blood leukocyte (hu-PBL) SCID mice, normal hu-PBL were injected into the peritoneal cavity of SCID mice.[9] By this route PBL from patients with systemic lupus erythematosus[10] and autoimmune thyroiditis[11] were transferred to SCID mice where they continued to produce autoantibodies.

There seem to be some limitations in the use of hu-PBL SCID mice for the study of human autoimmune diseases. First, when Epstein-Barr virus (EBV)-positive donors are used, EBV-transformed lymphocytes, and sometimes even aggressive B-cell lymphomas arise in SCID mice *in vivo*.[12] Second, some of the hu-PBL SCID develop signs of graft-versus-host (GvH) disease.[13] Third, the normal microenvironment required for an immune response is obviously not maintained in mice injected with single cell suspensions. Preformed plasma cells that continue to produce Ig have a life span between 3 days and 3 months.[14] Secreted IgG has a half-life of 10 days in SCID mice.[15]

In another approach, SCID mice are engrafted with human hematolymphoid organs, for example, fetal liver or bone marrow, thymus, and peripheral lymphoid organs.[16] The organs are transplanted under the renal capsule. The idea is to provide an intact human microenvironment with normal human progenitor cells. In these models EBV transformants are not observed. GvH reactions usually do not occur. Whereas originally only fetal lymphoid organs were felt to be suitable for transplantation, Barry *et al.* showed that human postnatal thymus could also be engrafted successfully. In this approach, however, only 56% of mice showed viable grafts after 1–3 months.[17]

TRANSPLANTATION OF THYMUS FROM MYASTHENIA GRAVIS PATIENTS INTO SCID MICE

In order to investigate the role of the thymus *in vivo,* we transplanted small pieces of MG thymus under the kidney capsule of SCID mice. A second group of mice received thymocytes as single cell suspensions (1×10^8 cells per mouse injected intraperitoneally (ip)). The experimental procedures are schematically outlined in FIGURE 1. Between 5–10 mice were transplanted or injected from one thymus. Every two weeks, blood was taken from the lateral tail vein, and antibodies

(Abs) against the human AChR were measured by an α-bungarotoxin (BGT)–binding RIA. The time kinetics of the anti-AChR Ab titers can be seen in FIGURE 2. Ab titers in mouse sera were measured for 22 weeks posttransplantation. The thymus was obtained from a patient with a high preoperative anti-AChR Ab titer (32 nmol/L). The left side (FIG. 2A) represents the titers after injection of 1×10^8 thymocytes per mouse ip; the right side (FIG. 2B) represents titers after transplantation of an intact piece of thymus (approximately $1–2 \times 10^6$ cells per graft). Peaking 3 weeks after injection of thymocytes, anti-AChR Ab concentrations of up to 12 nmol/L could be detected (FIG. 2A). When an intact thymic microenvironment

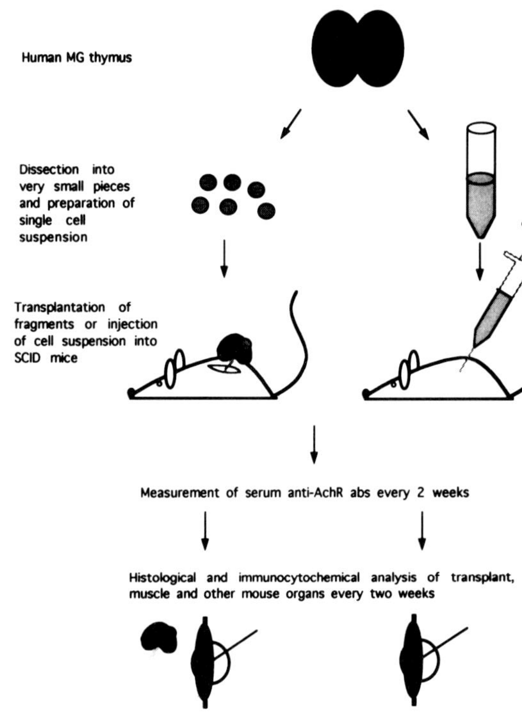

FIGURE 1. Experimental design.

was transplanted, the kinetics of anti-AChR Ab titers were different. As can be seen in FIGURE 2B, Abs were barely detectable in the first 2–3 weeks but rose markedly thereafter. After the peak level (also approximately 12 nmol/L) was reached, at about 11 weeks posttransplantation, the titers began to fall. In principle, the same observation can be made when a thymus from a patient with a low preoperative anti-AChR Ab titer is transplanted. in that case, Abs are hardly detectable after injection of thymocytes but rise markedly 3–4 weeks after transplantation of intact tissue.[18]

The difference in the time kinetics between the two groups is remarkable for

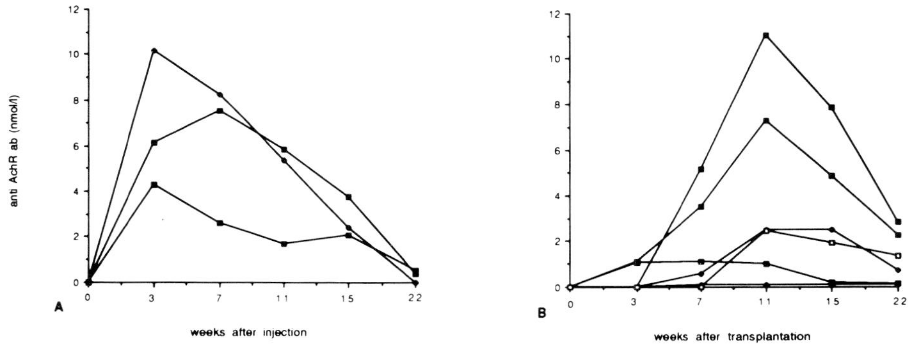

FIGURE 2. Kinetics of anti-AChR Ab titers after ip injection of 1 × 10^8 thymocytes from MG thymus into SCID mice **(A)** and transplantation of a 1 mm^3 piece of MG thymus tissue **(B)**. Each graph represents a group of mice that had received pieces from the same thymus. Pieces of MG thymus were injected **(A)** or transplanted **(B)** from a patient with a high preoperative anti-AChR Ab titer. Anti-AChR Abs were measured by the α-BGT binding assay and given in nmol/L. On the horizontal axis, the weeks after injection (transplantation) are indicated.

two reasons. First, the peak level of anti-AChR Abs was approximately the same (11 nmol/L) in the mice transplanted with intact thymus and in the group injected with 100 times more suspended cells. Second, the different time of peak suggests that Ab secretion by the intact thymus reflects a complex interactive immune response, whereas in the other group, only preformed plasma cells secreted immunoglobin. This latter effect is similar to *in vitro* experiments, showing that dissociated thymocytes can produce anti-AChR Abs for a short while. The transplanted intact tissue, however, seems to be a potent autoimmune microenvironment.

ANTI-AChR Abs BIND TO THE SCID MOUSE MUSCLE END-PLATE

To investigate whether human anti-AChR Abs are present not only in the serum of the transplanted mice but also at the muscle end-plate, cryostat sections of muscle were double stained with fluorescein (FITC)-labeled α-BGT and rhodamine-labeled antibodies to human immunoglobulin. FIGURES 3A and 3B demonstrate the presence of human Ig (red) at several muscle end-plates (green). Because in MG and EAMG one of the primary effector mechanisms of AChR breakdown is the local activation of complement, we attempted to immunolocalize complement at SCID end-plates. A polyclonal anti-mouse C3 antibody (green) was applied together with ant-human Ig (red). As can be seen in FIGURES 3C and 3D, mouse complement is indeed detectable at SCID muscle end-plates. Control mice transplanted with normal human thymus showed no binding human anti-AChR Abs or mouse complement (not shown).

ANTI-AChR Abs REPRESENT ONLY A MINOR PART OF HUMAN Abs PRODUCED IN TRANSPLANTED SCID MICE

In addition to AChR-specific human autoantibodies, total human and murine IgG and IgM concentrations were measured at regular intervals. The results are given in TABLE 1. When the titers of anti-AChR Abs (for better comparison given here in ng/mL) are compared with total human IgG and IgM in the sera, only 0.1% to 0.01% of the Abs are AChR-specific. This is in accordance with *in vitro* experiments and supports the observation that the majority of B cells and plasma cells present in the hyperplastic MG thymus are not specific for the AChR.

All mice were also tested for murine Ig before transplantation and at regular intervals thereafter. Only mice without detectable murine Ig were transplanted. After transplantation of MG thymus, most of the SCID mice (70%) became leaky. Because this high percentage of leakiness was not observed in mice engrafted with normal human thymus (20–30%), or in untransplanted (10%) mice, we assume that the leakiness is the result of continuous stimulation by the hyperplastic thymus, that is, by lymphokine secretion.

ALL ELEMENTS OF AN MG THYMUS ARE PRESENT IN THE TRANSPLANTS UNTIL AT LEAST 16 WEEKS POSTTRANSPLANTATION

At regular intervals, usually every 2 to 3 weeks, one mouse was sacrificed, and the transplant was examined histologically and immunohistochemically. FIGURES

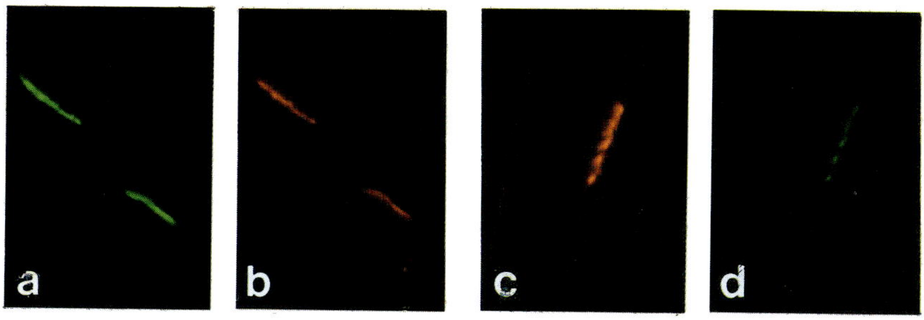

FIGURE 3. Staining of muscle end-plates of SCID mice with FITC-α-BGT **(a)**, rhodamine-labeled anti-human IgG (**b** and **c**), and mouse complement 3′ **(d)** 7 weeks after transplantation of human MG thymus (280 ×).

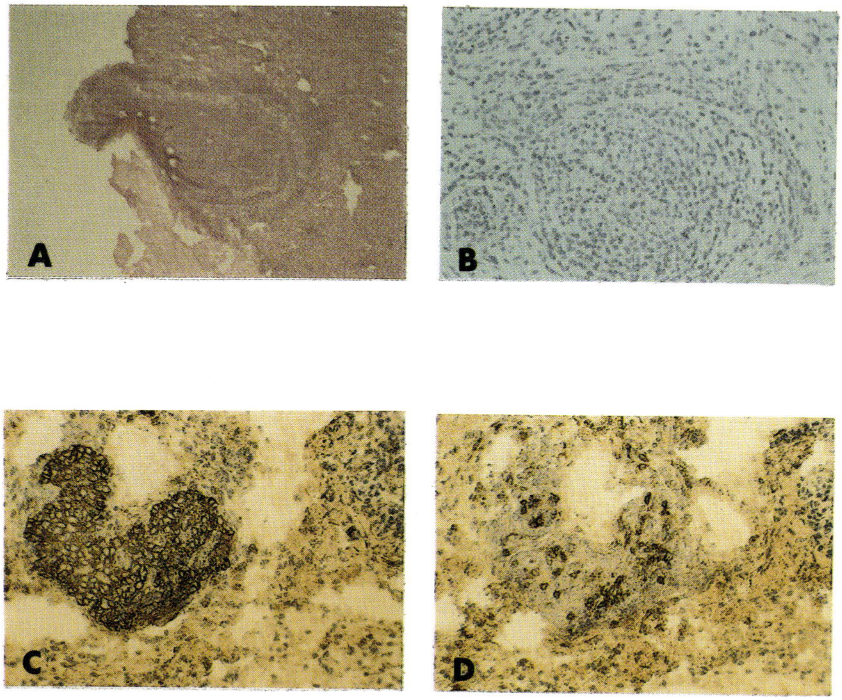

FIGURE 4. Histological and immunocytochemical analysis of an MG thymus graft placed under the kidney capsule of SCID mice. **A:** Overview of a kidney with an MG thymus graft (20 ×). **B:** Overview of a germinal center as typically seen in lymphofollicular hyperplasia of the MG thymus (75 ×). Staining on **C** and **D** was performed on sequential sections on a thymus graft 16 weeks after transplantation. **C:** Anticytokeratin 8/18; indirect peroxidase staining (160 ×). **D:** Anti-CD4; indirect peroxidase staining (160 ×).

4A and 4B show a 4-week-old graft in hematoxylin and eosin staining. FIGURE 4A is an overview of the transplant surrounded by mouse kidney and a fibrous capsule at the interface between thymus and kidney. The size of the fibrous capsule varies considerably between different kidneys. In transplants older than 12 weeks, sometimes only small areas of lymphoid tissue are present inside the capsule. FIGURE 4B demonstrates a germinal center as it is typically found in the MG thymus. It contains HLA DR-positive cells inside and is negative for cytokeratin.[18] Staining for cytokeratin 8/18 and CD4 is shown in serial sections in a transplant 16 weeks after engraftment (FIGURES 4C and D). As expected, staining for CD1, a marker for T-cell precursors in the thymic cortex, was essentially negative (not shown). CD4- and CD8-positive T cells (and B cells) and cytokeratin, however, were still present.

TABLE 1. Human and Murine Immunoglobulin Concentrations after Transplantation of MG Thymus to SCID Mice

	Sample	AChR Ab (ng/mL)[a]	Human IgG⁺ IgM (ng/mL)	Mouse IgG + IgM (ng/mL)
Two weeks after transplantation	1	<0.3	NT[b]	NT
	4	<0.3	0	0
	5	<0.3	2070	713
	6	<0.3	2320	0
Four weeks after transplantation	3	15.12	10580	1218
	4	<0.3	8040	0
	5	<0.3	6800	797
	6	4.29	9980	60

[a] A better comparison is the concentration given in ng/mL instead of nmol/L.
[b] Not tested.

PERSPECTIVES

This new animal model supports the central role of the thymus in the pathogenesis of MG. It should be useful to investigate further the thymic elements that are critical in the pathogenesis of MG. Our goals for future research are to investigate not only the hyperplastic MG thymus but also thymomas, and to manipulate the thymic implants in order to dissect the elements that are necessary and sufficient for an immune response against the AChR.

ACKNOWLEDGEMENTS

We thank Professor Spelsberg, Dr. Dienemann, and Professor Sebening for kindly providing thymus tissue; Professor Thierfelder for the SCID mice; Professor Fateh-Moghadam, Dr. Wick, and Ms. Herbeger for measuring anti-AChR antibody titers; and Professor Byron Waksman for helpful discussions.

REFERENCES

1. Toyka, K. V., D. B. Drachman, A. Pestronk & I. Kao. 1975. Myasthenia gravis. Passive transfer from man to mouse. Science **190:** 397–399.
2. Patrick, J. & J. Lindstrom. 1973. Autoimmune response to AchR. Science **180:** 871–872.
3. Lennon, V. A., J. M. Lindstrom & M. E. Seybold. 1975. Experimental autoimmune myasthenia gravis: a model of myasthenia gravis in rats and guinea pigs. J. Exp. Med. **141:** 1365–1375.
4. Meinl, E., W. E. F. Klinkert & H. Wekerle. 1991. The thymus in myasthenia gravis: changes typical for the human disease are absent in experimental autoimmune myasthenia gravis of the Lewis rat. Am. J. Pathol. **139:** 995–1008.
5. Bosma, G. C., R. D. Custer & M. J. Bosma. 1983. A severe combined immunodeficiency mutation in the mouse. Nature **301:** 527–530.
6. Custer, R. P., G. C. Bosma & M. J. Bosma. 1985. Severe combined immunodeficiency (SCID) in the mouse. Am. J. Pathol. **120:** 464–477.
7. Lieber, M. R., J. E. Hesse, S. Lewis, G. C. Bosma, N. Rosenberg, M. Mizuuchi, M. J. Bosma & M. Gellert. 1988. The defect in murine severe combined immune deficiency: Joining of signal consequences but not coding sequences in V(D)J recombination. Cell **55:**7–16.
8. Bosma, G. C., M. Fried, R. P. Custer, A. Carroll, D. M. Gibson & M. J. Bosma. 1988. Evidence of functional lymphocytes in some (leaky) SCID mice. J. Exp. Med. **167:** 1016–1033.
9. Mosier, D. E., R. J. Gulizia, S. M. Baird & D. B. Wilson. 1988. Transfer of a functional human immune system to mice with severe combined immunodeficiency. Nature **335:** 256–259.
10. Duchosal, M. A., P. J. McConahey, C. A. Robinson & F. J. Dixon. 1990. Transfer of human systemic lupus erythematosus in severe combined immunodeficient (SCID) mice. J. Exp. Med. **172:** 985–988.
11. Macht, L., N. Fukuma, K. Leader, D. Sarsero, C. A. S. Pegg, D. I. W. Phillips, P. Yates, S. M. McLachlan, C. Elson & B. Rees Smith. 1991. Severe combined immunodeficient (SCID) mice: A model for investigating human thyroid autoantibody synthesis. Clin. Exp. Immunol. **84:** 34–42.
12. Okano, M., Y. Taguchi, H. Nakamine, S. J. Pirruccello, J. R. Davis, K. W. Beisel, K. L. Kleveland, W. G. Sanger, R. R. Fordyce & D. T. Purtilo. 1990. Characterization of Epstein-Barr Virus–induced lymphoproliferation derived from human peripheral blood mononuclear cells transferred to severe combined immunodeficient mice. Am. J. Pathol. **137:** 517–522.
13. Krams, S. M., K. Dorshkind & M. A. Gershwin. 1990. Generation of biliary lesions after transfer of human lymphocytes into severe combined immunodeficient (scid) mice. J. Exp. Med. **170:** 1919–1930.
14. MacLennan, I. C. M. & D. Gray. 1986. Antigen driven selection of virgin and memory B cells. Immunol. Rev. **91:** 61–86.
15. Hassan, M. S., M. Abedi-Valugerdi, G. Lefranc, L. Hammarström & C. I. E. Smith. 1991. Biological halflife of normal and truncated human IgG3 in SCID mice. Eur. J. Immunol. **21:** 1319–1322.
16. McCune, J. M., R. Namikawa, H. Kaneshima, L. D. Shultz, M. Lieberman & I. L. Weissman. 1988. The SCID-hu mouse: murine model for the analysis of human hematolymphoid differentiation and function. Science **241:** 1632–1639.
17. Barry, T. S., D. M. Jones, C. B. Richter & B. F. Haynes. 1991. Successful engraftment of human postnatal thymus in severe combined immune deficient (SCID) mice: differential engraftment of thymic components with irradiation versus anti-asialo GM-1 immunosuppressive regimens. J. Exp. Med. **173:** 167–180.
18. Schönbeck, S., F. Padberg, R. Hohlfeld & H. Wekerle. 1992. Transplantation of thymic autoimmune microenvironment to SCID mice: A new model of myasthenia gravis. J. Clin. Invest. **90:** 245–250.

Molecular Analysis of Intrathymic Nicotinic Acetylcholine Receptor

L. M. WHEATLEY, D. URSO, Y. ZHENG, E. LOH, AND A. I. LEVINSON

Department of Medicine
The University of Pennsylvania School of Medicine
Philadelphia, Pennsylvania 19104

INTRODUCTION

Antibodies directed against the nicotinic acetylcholine receptor alpha-chain (nAChRα) at the myoneural junction clearly cause the clinical signs and symptoms of myasthenia gravis (MG). The thymus appears to play a special role in the pathogenesis of MG. There is thymic pathology in the majority of patients with MG, 70% having thymic hyperplasia with formation of germinal centers, and 10% having epithelial cell thymomas.[1] B and T cells reactive to nAChRα can be recovered from these thymic specimens.[2] Thymectomy frequently leads to clinical improvement.[1] Over the years, intrathymic localization of nAChRα has fueled speculation on the role of the thymus in the pathogenesis of the disease. There is still some question about which thymic cells express nAChRα and whether the structure is exactly the same as that found at the myoneural junction. The polymerase chain reaction (PCR) with its exquisite sensitivity allows these questions to be addressed. Accordingly, the studies reported here were aimed at evaluating the primary sequence of thymic nAChRα. Preliminary attempts at localization were also made.

METHODS

Amplification Scheme

The nAChRα is a pentameric structure composed of five subunits in the stoichiometry of α_2, β, δ, γ, or ε.[3] The α-subunit contains the acetylcholine binding site, is the most conserved across species,[2] and is the target for the majority of anti-AChR antibodies as well as the source of T-cell epitopes.[4–8] For these reasons, it was decided to concentrate on this subunit. An amplification strategy was devised that would produce cDNA encompassing the coding sequence for the entire extracellular portion of the α-chain including the ligand-binding site and main immunogenic regions, as well as portions of the transmembrane and cytoplasmic domains. Total RNA was reverse transcribed (RT) using a 3′ primer that corresponds to nucleotides 891-917 of the mouse muscle sequence. One-tenth of the reverse transcription reaction was then used in a PCR reaction employing a 3′ primer nested immediately inside the one used in the RT (871-890) altered to encode a Sal I restriction site, and a 5′ primer composed of the first 27 bp of exon 2, again altered to contain an Eco R I restriction recognition sequence (FIG. 1). Two microliters of the first PCR reaction was used in subsequent reamplification cycles using the same primers. RNA treated in an identical manner, except for the omission of reverse transcriptase, was used as a negative control in all experiments.

RNA: Sources and Preparation

The thymus and kidneys of 6–8-week-old BALB/k and SJL mice were harvested using separate sets of instruments to gain access to the thorax and abdomen and to remove each organ. This was done to minimize the possibility of contaminating specimens with muscle cells. Some of these thymuses were bisected and half of the thymus passed through a stainless steel sieve to separate the stroma (retained by the mesh) from the thymocytes (fall through). Muscle was harvested from the leg.

SV-40–transformed mouse thymic cell lines were obtained (from Drs. S. Faas and B. Knowles, Wistar Institute, Philadelphia; manuscript submitted for publication) that were categorized by ultrastructural characteristics, monoclonal antibody staining, and cytokine production as being of the following lineages: cortical reticular epithelial cells, interdigitating dendritic cells, medullary epithelial cells, and subcapsular cortical epithelial cells. The human rhabdomyosarcoma cell line TE671 (ATCC #HT 139) was used as a positive control.

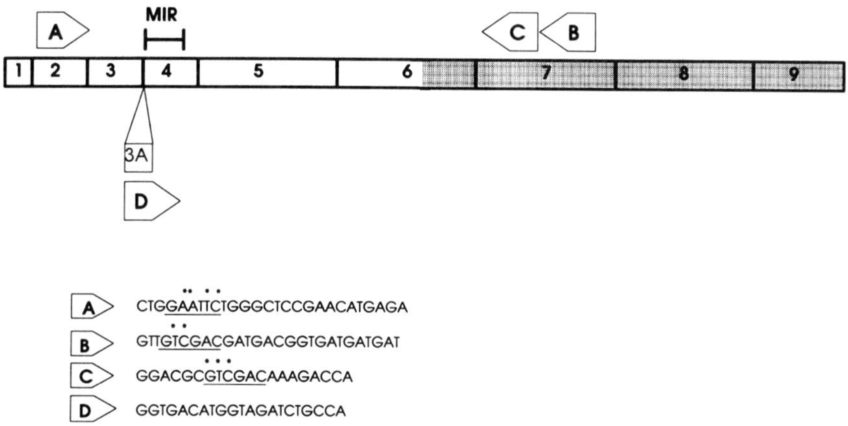

FIGURE 1. A schematic representation of the alpha-subunit of nicotinic acetylcholine receptor amplification procedure. A newly described exon (3A) is depicted along with the nine exons that constitute this molecule. The open rectangles represent the extracellular portion of the receptor subunit; the stippled area represents the transmembrane and cytoplasmic regions. The segment encoding the main immunogenic region (MIR) is indicated. The lettered arrows indicate where the oligonucleotide primer sequences anneal. In the primer sequences, the letters marked with asterisks are alterations in the native sequence to accommodate restriction enzyme sequences. The restriction enzyme sites are underlined. (Wheatley *et al.*[22] With permission from *The Journal of Immunology*.)

Human specimens were obtained after surgical resection for symptomatic MG and from one patient with thymoma but without myasthenia. The thymuses from the myasthenic patients all showed germinal-center hyperplasia. As with the mouse thymuses, some of the human thymuses were also separated into thymocytes and stroma.

Total RNA was isolated from the above samples using acid guanidium thiocyanate.[9] One to five micrograms was used in each RT.

Analysis

PCR products were assessed for specificity by Southern hybridization with a full-length (1.7 kb) nAChRα cDNA (provided by Dr. J. Merlie, Washington University, St. Louis). The probe was labeled with [^{32}P]dCTP and allowed to hybridize overnight at 42°C in 50% formamide, 2X Denhardts, 5X SSC, 0.2% SDS, and 150 μg/mL salmon sperm DNA. The blots were washed three times in 2X SSC at room temperature for 5 minutes followed by two washes in 0.2X SSC and 0.1% SDS at 65°C for 30 minutes each.

Reamplified PCR products were digested at the engineered restriction sites and ligated into Bluescript SK$^+$ (Strategene, La Jolla), which were then used to transform JM109. Clones carrying the cDNA were identified by restriction enzyme digestion and were sequenced using the Sequenase 2.0 kit (United States Biochemical, Cleveland).

RESULTS

Southern Hybridization

A band of appropriate size (880 bp) was found to hybridize specifically with our nAChRα probe when the source of the RNA was from whole-mouse thymus, whereas the kidney was negative under the same exposure conditions. When the

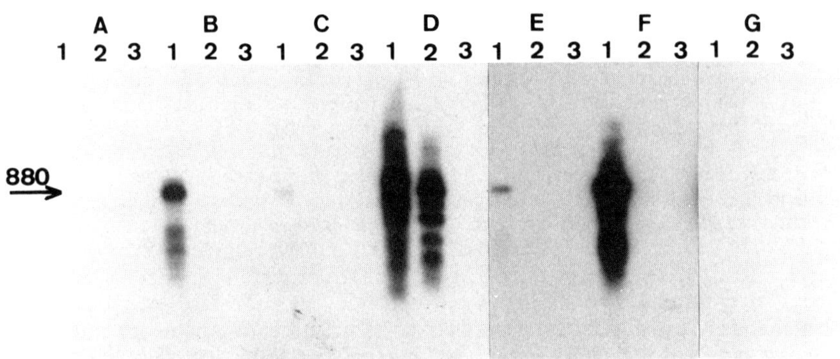

FIGURE 2. PCR of murine cells. Southern hybridization of tissue and cell line PCR products. Each letter represents a different tissue source for PCR. A, murine thymocyte suspension; B, whole murine thymus; C, murine thymic medullary epithelial cells; D, TE671; E, murine thymic cortical reticular epithelial cells; F, murine muscle; G, murine kidney. In the lanes marked 1, the PCR was performed with our standard primers. In lanes marked 2, the 5′ oligonucleotide was derived from exon 3A. In lanes marked 3, the RNA was not reverse transcribed prior to the PCR. (Wheatley *et al.*[22] With permission from *The Journal of Immunology*.)

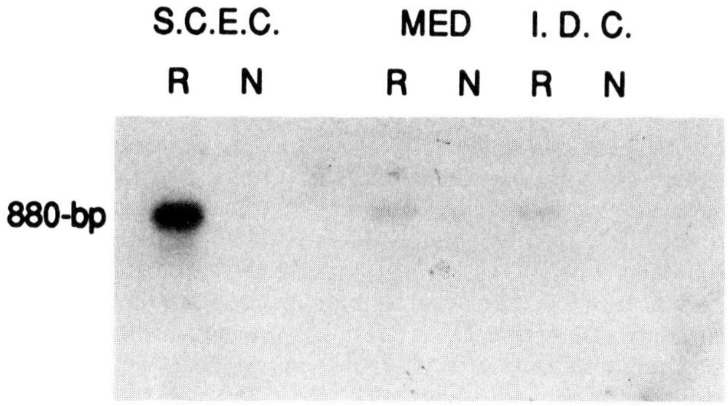

FIGURE 3. PCR SV40 cell lines. Southern hybridization of the SV40-transformed mouse thymic subcapsular cortical cell line (S.C.E.C.), the medullary epithelial cell line (MED), and the interdigitating dendritic cell (I.D.C.) line. The lanes labeled R have been reverse transcribed and amplified per our standard protocol. The lanes labeled N are PCRs of total RNA that have not been reverse transcribed.

starting material was RNA from mouse thymocytes, no specific bands could be detected even at a time when the unfractionated half of the same thymuses produced specific bands (FIG. 2). A new exon named 3A has recently been described in human muscle. The length of the cDNAs generated from the mouse specimens was too short to contain this exon. Nevertheless in an effort to increase the probability of locating this exon, if it existed in mouse tissue, a new primer derived from the 5′ end of exon 3A was employed in the PCR. No cDNA was generated from mouse thymus or muscle but was detected in our positive control, the human rhabdomyosarcoma cell line, TE671 (FIG. 2).

To further approach receptor localization, RNA was generated from SV-40–transformed murine thymic cell lines. All cell lines tested, which include interdigitating dendritic cell, subcapsular cortical epithelial cell, and medullary epithelial and cortical reticular epithelial cell lines, were positive on Southern hybridization for the specific 880 bp band. Again, no cDNA was generated when the 3A 5′ primer was substituted for the standard 5′ primer located in exon 2 (FIGURES 2 and 3).

In the cDNA generated from the myasthenic thymuses and the thymoma, the 880 bp was detected. In addition, an indistinct but specifically hybridizing band was seen in 955 bp (FIG. 4). This band would correspond in size to a cDNA transcript containing the 75 bp exon 3A. PCR using the 5′ primer developed from exon 3A also generated specifically hybridizing bands. Clones generated for sequencing from unfractionated human thymus specimens were screened with a ^{32}P-labeled 3A oligonucleotide primer. 3A was found to be present in 16% of clones screened.

Sequencing

cDNA clones corresponding to the major bands from whole mouse thymus and the medullary and cortical reticular epithelial cell lines were sequenced and found to have an identical sequence with mouse muscle over the region examined. In addition, in cDNA prepared from thymus, mouse muscle, and the TE671 cell line, multiple bands shorter in length than the predicted 880 bp were also found to specifically hybridize with our nAChRα probe. Several of the smaller bands from TE671 were isolated and successfully cloned. The sequences of some of these bands reveal segmental losses consistent with alternate splicing. In these clones, there is a cytidine/thymidine-rich region 5′ to the putative splice site and an adenosine/guanosine pair immediately adjacent to the splice site. The reading frame is maintained through this splice junction. Not all of the clones, however, have this type of consensus recognition sequence. Some have 5–6 bp identical sequences on each side of the join. These clones may represent artifacts of the amplification. Incomplete cDNAs might anneal at these sequences and thereby serve as primers for strand completion with loss of intervening sequences (FIG. 5).

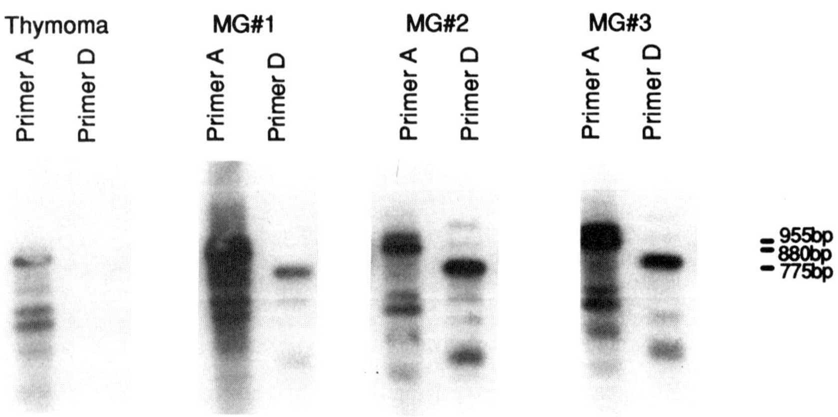

FIGURE 4. PCR of human thymic samples. Southern hybridization of human thymus specimens. The source of the specimens are from three patients with MG and thymic hyperplasia, and from one patient with thymoma, but without MG. The lanes labeled A were PCRs performed using our standard 5′ oligonucleotide primer. The lanes labeled D are PCR products generated using the 5′ oligonucleotide derived from exon 3A as the 5′ primer.

Clones from the cDNA of each of the unfractionated human thymuses were also sequenced and found to share nucleotide identity with human muscle. Two of these sequences contained exon 3A.

DISCUSSION

The issue of intrathymic expression of nAChR has been a controversial one (TABLE 1). Since the thymus was discovered to contain binding sites for α-bungarotoxin,[10] there has been speculation about whether the thymus expresses AChR or an AChR-like protein, and if so, where such a protein might be located and whether

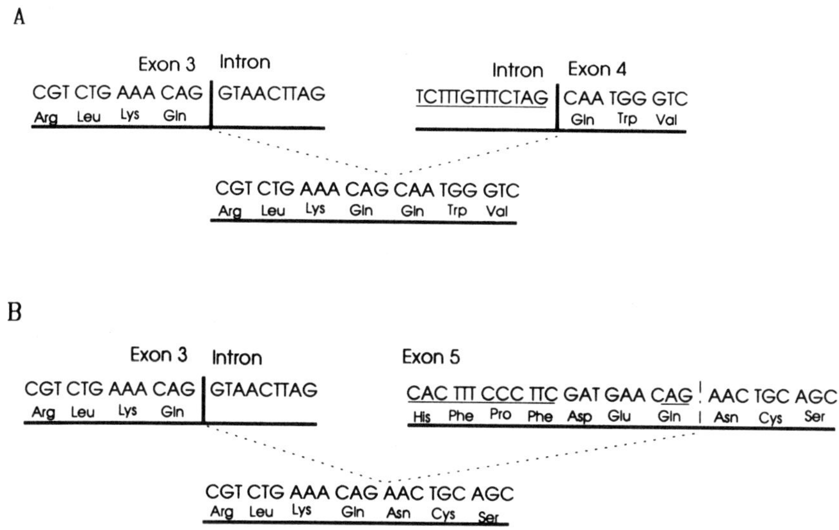

FIGURE 5. An example of a shorter transcript of TE671 consistent with alternate splicing. Section A illustrates the usual splicing of exon 3 to exon 4 with the deletion of an intron. Section B shows the joining seen in a 650 bp band sequenced from TE671. Note that exon 3 is spliced into the middle of exon 5 in an area preceded by a run of pyrimidines, and an AG pair is found immediately adjacent to the putative splice site. The joining causes the loss of exon 4 and the first 135 bp of exon 5 but does not shift the reading frame. (Wheatley *et al.*[22] With permission from *The Journal of Immunology.*)

it would be important in the pathogenesis of MG. Myoid cells, epithelial cells, and thymocytes have all been reported to express AChR-like molecules.[11-16] Most investigators have approached this question by using MAb and immunohistochemical staining. The results of these studies have not unequivocally shown, however, whether the target of MAb is AChR or a cross-reactive protein. For example, Marx *et al.* reported binding of a MAb with specificity for the cytoplasmic portion

TABLE 1. A History of Intrathymic AChR

1977	Engel *et al.*: α-BGT binding in the epithelial cell compartment
1978	Kao and Drachman: AChR binding to cultured thymus
1980	Fuchs *et al.*: anti-AChR binding to thymocytes
1987	Schluep *et al.*: anti-AChR binding to myoid cells *in situ*
1989	Geuder *et al.*: no α-chain transcript in MG thymoma by Northern analysis
1990	Marx *et al.*: α-chain cytoplasmic epitope binding to cross-reactive protein in MG thymoma and epithelial cells cultured from same
1990	Nelson and Conti-Tronconi: α-, β-, γ-, and δ-binding proteins from bovine thymus
1991	Mihovilovic and Roses: α_3 neuronal transcript in thymus; positive for nAChRα on very long exposure
1991	Hara *et al.*: no nAChR α-chain transcript in MG thymoma and hyperplastic thymus by Northern analysis; α-chain positive on PCR of MG thymoma, hyperplastic thymus, and neoplastic epithelial cell lines.

of AChR in myasthenic thymomas and in cortical epithelial cell lines derived from these thymomas.[15] On further studies, the target of this binding was determined to be a non-AChR 153 kDa protein that expressed a cross-reactive epitope.[17]

Several groups have reported that nAChR is expressed on myoid cells in myasthenic and normal thymuses.[12,13] One of the most systematic studies indicated that MAbs to the extracellular domain of the nAChRα bind almost exclusively to myoid cells.[14] In a recent study, binding of MAbs to the α-, β-, δ-, and γ-subunits of the receptor was detected in bovine thymus, suggesting that a complete extrajunctional nAChR is expressed.[18] The site of tissue expression was not determined, however. Taken together, these studies indicate that a complete extrajunctional form of receptor is likely expressed in normal thymus, predominantly on myoid cells.

To further approach the question of intrathymic nAChR expression, investigators have resorted to molecular approaches. These techniques avoid the pitfalls of identifying cross-reactive non-AChR proteins and provide the wherewithal to determine primary structure when only small amounts of protein are available for analysis. Geuder *et al.* did not detect nAChR mRNA on Northern analysis.[19] Northern hybridization of the RNA samples did reveal α_3 neuronal nAChR in normal human thymus in a study by Mihovilovic and Roses. Using an nAChRα probe, these investigators also detected a 2.3 kb band in thymus RNA on extremely long exposures.[20]

We used PCR in our analysis of intrathymic nAChR. We focused on nAChRα because this chain contains B-cell and T-cell epitopes that appear to be important in the pathogenesis of MG. PCR provided the added sensitivity to identify expression of this scant autoantigen at the molecular level. Moreover, it made it possible to clone and sequence nAChRα expressed in thymus and thereby determine the nucleotide sequence and, by inference, the protein sequence.

Results of studies carried out with mouse thymic specimens are highlighted by the finding of nAChRα mRNA expression in whole thymus preparations but not in single cell suspensions. These results tentatively localize mRNA expression to stromal populations of cells. Although the stromal populations were not directly identified, our findings of nAChRα mRNA expression in subcapsular cortical, cortical reticular and medullary epithelial cell lines, and dendritic cell lines raise the possibility that these thymic cellular constituents contribute to stromal-cell expression.

Southern hybridization of amplified nAChRα products from whole thymus, the thymic epithelial cell lines, and the positive control TE671 also revealed smaller bands than expected. Sequencing of the smaller bands suggests that some of them may represent an alternatively spliced product not previously identified on screening cDNA libraries, whereas others may be the result of a PCR artifact. In favor of alternate splicing is the appearance of a cytidine/thymidine-rich region 5′ to the putative splice junction and an AG pair immediately adjacent to the splice site in several of our clones. This type of sequence is frequently observed at splice junctions.

We determined that the sequence of the major 880 bp band detected on Southern analysis of the nAChRα product amplified from whole thymus tissue, thymic cortical reticular epithelial, and medullary epithelial cell lines was identical to that of the receptor expressed at the myoneural junction. No evidence for expression of an isoform containing the recently described exon 3A was found in any of the mouse RNA specimens, in line with the original report that suggested that this isoform was expressed only in human tissues.[21]

Our studies of thymic specimens from MG patients with thymic hyperplasia and one patient who had thymoma, but not MG, provide new insights into nAChR mRNA expression in human thymic tissue. Similar to our studies in the mouse, a number of specifically hybridizing bands smaller than the 880 bp band were seen on Southern analysis of the amplified nAChRα product. Unlike the mouse studies, however, a larger 955 bp band, approximating the size of transcripts containing exon 3A, was also detected. Further evidence for the expression of this receptor isoform was provided by the studies in which an exon 3A primer was substituted for our standard 3′ primer in the PCR. Sequencing of the 880 and 975 bands revealed sequences identical to the respective isoforms reported in muscle.[21] The smaller bands have not been sequenced as yet to determine if they might represent additional isoforms of this receptor expressed in human tissue.

In summary, we have detected nAChRα mRNA expression in normal mouse thymus and in thymic tissue from patients with MG, and thymoma without MG. Furthermore, we have detected nAChRα mRNA in a number of thymic stromal cell lines, including subcapsular cortical epithelial, cortical reticular epithelial and medullary epithelial, and dendritic cell lines. It is significant that the nAChRα products from all of these sources are identical in sequence to their muscle counterparts. Inasmuch as the products encompass the reported nAChRα T-cell epitopes as well as B-cell epitopes, our data support the hypothesis that intrathymic nAChRα expression is involved in the pathogenesis of MG. If our finding of nAChRα expression in these stromal cell lines can be extrapolated to expression on their normal thymic tissue counterparts, our results raise the possibility that aberrant selection of the T-cell repertoire by thymic cells coexpressing MHC class II proteins and nAChRα might contribute to the development of autoreactive T cells in patients with myasthenia gravis.

REFERENCES

1. WEKERLE, H. & H. K. MULLER-HERMELINK. 1986. The thymus in myasthenia gravis. *In* The human thymus. Histopathology and pathology. H. K. Muller-Hermelink, Ed.: 179–206. Springer. Berlin, Heidelberg, New York, Tokyo.
2. LEVINSON, A. I., B. ZWEIMAN & R. P. LISAK. 1987. Immunopathogenesis and treatment of myasthenia gravis. J. Clin. Immunol. **7:** 187–197.
3. KARLIN, A. 1991. Explorations of the nicotinic acetylcholine receptor. Harvey Lect. **85:** 71–107.
4. RALSTON, S., V. SARIN, H. L. THANH, J. RIVIER, J. L. FOX & J. M. LINDSTROM. 1987. Synthetic peptides used to locate the α-bungarotoxin binding site and immunogenic regions on α-subunits of the nicotinic acetylcholine receptor. Biochemistry **26:** 3261.
5. TZARTOS, S. J., A. KOKLA, S. L. WALGRAVES & B. M. CONTI-TRONCONI. 1988. The main immunogenic region is localized within residues 63-80 of the α-subunit. Proc. Natl. Acad. Sci. USA **85:** 2899–2903.
6. ZHANG, Y., M. SCHLUEP, S. FRUTIGER, G. J. HUGHES, M. JEANNET, A. STECK & T. BARKAS. 1990. Immunologic heterogeneity of autoreactive T lymphocytes against the nicotinic acetylcholine receptor in myasthenic patients. Eur. J. Immunol. **20:** 2577–2583.
7. OSHIMA, M., T. ASHIZAWA, M. S. POLLACK & M. Z. ATASSI. 1990. Autoimmune T cell recognition of human acetylcholine receptor: the sites of T cell recognition in myasthenia gravis on the extracellular part of the α subunit. Eur. J. Immunol. **20:** 2563–2569.
8. FUJI, Y. & J. LINDSTROM. 1988. Specificity of the T cell immune response to acetylcholine receptor in experimental autoimmune myasthenia gravis. J. Immunol. **140:** 1830–1837.
9. CHOMCZYNSKI, P. & N. SACCHI. 1987. Single-step method of RNA isolation by acid guanidinium thiocyanate-phenol-chloroform extraction. Anal. Biochem. **162:** 156–159.

10. Engel, W. K., J. L. Trotter, D. E. McFarlin & C. L. McIntosh. 1977. Thymic epithelial cell contains acetylcholine receptor. Lancet **1:** 1310–1311.
11. Hara, Y., S. Ueno, T. Uemichi, N. Takahashi, S. Yorifuji, Y. Fujii & S. Tarui. 1991. Neoplastic epithelial cells express α-subunit of muscle nicotinic acetylcholine receptor in thymomas from patients with myasthenia gravis. FEBS Lett. **279:** 137–140.
12. Kao, I. & D. B. Drachman. 1978. Thymic muscle cells bear acetylcholine receptors: possible relation to myasthenia gravis. Science **195:** 74–75.
13. Werkerle, H., U.-P. Ketelson, A. D. Zurn & B. W. Fulpius. 1978. Intrathymic pathogenesis of myasthenia gravis: transient expression of acetylcholine receptors on thymus-derived myogenic cells. Eur. J. Immunol. **8:** 579–582.
14. Schluep, M., N. Willcox, A. Vincent, G. K. Dhoot & J. Newsom-Davis. 1987. Acetylcholine in human thymic myoid cells *in situ:* an immunologic study. Ann. Neurol. **22:** 212–222.
15. Marx, A., T. Kirchner, F. Hoppe, R. O'Conner, B. Schalke, S. Tzartos & H. K. Muller-Hermelink. 1989. Proteins with epitopes of the acetylcholine receptor in epithelial cell cultures of thymomas in myasthenia gravis. Am. J. Pathol **134:** 865–877.
16. Fuchs, S., I. Schmidt-Hopfeld, G. Tridente & R. Tarrab-Hazdai. 1980. Thymic lymphocytes bear a surface antigen which cross-reacts with acetylcholine receptor. Nature **287:** 162–164.
17. Marx, A., R. O'Connor, K. I. Geuder, F. Hoppe, B. Schalke, S. Tzartos, I. Kalies, T. Kirchner & H. K. Muller-Hermelink. 1990. Characterization of a protein with an acetylcholine receptor epitope from myasthenia gravis–associated thymomas. Lab. Invest. **62:** 279–286.
18. Nelson, S. & B. Conti-Tronconi. 1990. Adult thymus expresses an embryonic nicotinic acetylcholine receptor-like protein. J. Neuroimmunol. **29:** 81–92.
19. Geuder, K. I., R. Schoepfer, T. Kirchner, A. Marx & H. K. Muller-Hermelink. 1989. The gene of the α-subunit of the acetylcholine receptor: Molecular organization and transcription in myasthenia-associated thymomas. Thymus **14:** 179–186.
20. Mihovilovic, M. & A. Roses. 1991. Expression of mRNAs in human thymus coding for the $\alpha 3$ subunit of a neuronal acetylcholine receptor. Exp. Neurol. **111:** 175–180.
21. Beeson, D., A. Morris, A. Vincent & J. Newsom-Davis. 1990. The human muscle nicotinic acetylcholine receptor α-subunit exists as two isoforms: a novel exon. EMBO J. **9:** 2101–2106.
22. Wheatley, L. M., D. Urso, K. Tumas, J. Maltzman, E. Loh & A. I. Levinson. 1992. Molecular evidence for the expression of nicotinic acetylcholine receptor alpha-chain in mouse thymus. J. Immunol. **148:** 3105–3109.

Nicotinic Neuronal Acetylcholine Receptor α-3 Subunit Transcription in Normal and Myasthenic Thymus[a]

MIRTA MIHOVILOVIC,[b] CHRISTINE HULETTE,[c]
JACQUELYN MITTELSTAEDT,[b] CAROL AUSTIN,[b] AND
ALLEN D. ROSES[b,d]

[b]*Department of Medicine*
Division of Neurology
[c]*Department of Pathology*
[d]*Department of Neurobiology*
Duke University Medical Center
Durham, North Carolina 27710

The thymus expresses antigens known to bind α-elapid toxins[1-3] and antibodies[4,5] that recognize neuromuscular-type acetylcholine receptors (AChRs). Immunohistological studies indicate that these thymic AChR-related antigens are expressed in both thymic epithelial[1,5] and thymic myoid cells.[2,4,5] It is possible to differentiate at least two populations of these thymic "cholinergic" antigens based on their immunological reactivity towards different antineuromuscular-type AChR monoclonal antibodies. The best characterized cholinergic antigen is expressed in the myoid cells and appears to conform antigenically to a neuromuscular-type thymic AChR.[4,5] In addition, an antigen expressed only in thymomas from myasthenia gravis (MG) patients and that cross-reacts with an antineuromuscular-type AChR antibody has been detected.[6] The existence of these cholinergic antigens, together with the thymic anomalies that are associated with MG,[7-9] have given support to the notion that intrathymic immunization with a cholinergic antigen could be involved in the triggering and/or maintenance of an autoimmune attack in myasthenia gravis.

In an effort to further characterize the reported cholinergic antigens, we screened human thymic cDNA libraries with cDNA probes, encoding the subunits of the mouse neuromuscular AChR. This approach led to the isolation of human thymic cDNA clones encoding the α-3 subunit of a neuronal-type nicotinic AChR.[10] This transcript is known to be expressed in both ganglia[11] and the central nervous system.[12]

Although it is widely accepted that the main target of the autoimmune attack in MG is the neuromuscular AChR, the disease manifests itself as a heterogeneous disorder in which anti-AChR titers do not correlate with disease activity.[13-15] About 10% of patients do not have detectable circulating anti-AChR antibodies, and antibodies to other muscle antigens are frequently found in sera from MG patients.[16-19] The MG syndrome has occasionally been associated with periph-

[a] This work was supported by the NIH Program Project P01-NS-26630 (MAPV), a Clinical Research Grant (M. Mihovilovic) from the Muscular Dystrophy Association (MDA1), and M01-RR-30, National Center for Research Resources, General Clinical Research Centers Program from the National Institutes of Health.

eral[20-22] and central nervous system involvement.[23,24] Antibodies able to bind to antigens of neural origen[25,26] and, in some cases, to a neural α-bungarotoxin binding component[27] are present in MG sera. This latter observation suggests antibody cross-reactivity with neuronal AChR or AChR-related antigens.

The role of a neuronal-type AChR in thymus has yet to be determined; it is possible, however, that it participates as a transducer or modulator of signals delivered to the thymus by the autonomic nervous system. The thymus receives both sympathetic and parasympathetic innervation;[28-31] this innervation is crucial for normal thymic development[32,33] and may influence thymic function.[34] The identification of a number of neurotransmitters in thymus, including adrenalin, vasoactive intestinal peptide, met-enkephalin, and acetylcholine[28-31] indicates that a tight neuroregulatory control exists in the thymus. Adrenergic, cholinergic, and peptidergic nerve networks run through the subcapsular cortex and the corticomedullary junction.[35] Prothymocytes that undergo T cell–receptor rearrangement within the cortex localize to these same areas.[36,37] Thus immature thymocytes are located within a highly innervated thymic microenvironment regulated by a network of neuroactive ligands and their receptors. This suggests that innervation may play an important role in thymocyte maturation and that regulation of this network may have important implications for thymic function.

We have analyzed the thymic transcription of the alpha-3 subunit of a neuronal AChR in normal and MG thymi as part of our effort to define the role of putative thymic neuronal AChR in myasthenia gravis.

METHODS

Tissues and Cells

All tissues were quick frozen in liquid nitrogen and stored at −80°C. Preparations enriched in thymocytes were obtained by teasing thymic tissue under sterile conditions.[38] The released cells were collected in phosphate buffered saline (PBS) and filtered through gauze, and the thymocyte-enriched fraction was recovered from the interface of lymphocyte separation media (LSM, Organon Teknika Corp., Durham, NC). This fraction was washed with 5 volumes of PBS, and the cells were collected by centrifugation at 2000 × *g* for 10 minutes. Pellets were quick frozen and stored at −80°C. Microscopic examination of the thymocyte-enriched fraction showed that more than 95% of the cells were morphologically consistent with lymphoid cells.

Northern Blot Analysis

Analysis was performed as previously reported[10] employing poly(A)$^+$ RNA preparations. The α-3 cDNA clone, 72NTBSZ, encoding the mature human subunit starting at amino acid #10 and the β-amyloid cDNA (2-alpha), encoding approximately 800 bp of the carboxy terminus of the β-amyloid peptide, were used for the analysis. Amyloid clone 2-alpha was isolated from a λgt11 human frontal cortex cDNA library. It hybridizes to a 3.5 kb transcript encoding the β-amyloid precursor protein.[39]

To compare the abundance of α-3 and β-amyloid precursor transcripts present in different thymic poly(A)$^+$ preparations, the autoradiographic bands corresponding to the 3.0 kb α-3 and 3.5 kb β-amyloid transcripts (see FIG. 1), obtained from duplicate blots, were scanned at 600 nm in a Gilson Response spectrophotometer (Oberlin, OH) adapted with a scanning platform. The integrated area below each absorbancy peak was taken as a measure of the abundance of the corresponding transcripts. Normal thymic RNA preparations in each blot were run as controls.

Ribonuclease Protection Assays

The assays were performed as indicated in reference 40, using radiolabeled cRNA probes produced through runoff synthesis, in the presence of [P^{32}] UTP from linearized Bluescript SK^+ clones. cRNA probes were hybridized with the poly$(A)^+$ RNA samples. cRNA-protected fragments were analyzed electrophoretically in 5% acrylamide denaturing gels. To determine the relative abundance of the different protected fragments, autoradiographic scanning of individual samples was performed at 600 nm as described above.

Polymerase Chain Reaction Analysis

Plasmids were purified employing QIAGEN columns (QIAGEN, Chatsworth, CA). Genomic DNAs were purified from human lymphocytes through phenol/chloroform extractions.[41]

Polymerase chain reactions (PCRs) were run as previously described,[40] employing either 1 ng of cloned cDNA or 125 ng of genomic DNA in a final volume of 0.025 milliliter. Cycling conditions were 9 min of denaturation at 94°C, followed by 15 cycles of 3 min at 94°C, 3 min at 55°C, and 3 min at 72°C and continuing with 15 cycles of 3 min at 94°C, 3 min at 37°C, and 3 min at 72°C.

Histochemistry

Sections of thymomas and hypertrophic MG thymi were processed through xylene, embedded in paraffin, and 5 μm sections were cut and mounted on Probe-On, glass slides (Fisher, Pittsburg, PA). Representative sections were stained with hematoxylin and eosin. Immunostaining for the amyloid precursor protein (APP) was performed using a mouse monoclonal antibody to the APP amino terminus (Boheringer-Mannhein, Indianapolis, IN) diluted 1:2 after pretreatment of the sections with pronase (Biomeda Corp., Foster City, CA) and 99% formic acid for 20 minutes.[42] The Streptavidin-Biotin Universal Detection System (Biomeda Corp., Foster City, CA) was used, and peroxidase activity was visualized using diaminobenzidine (Sigma, St. Louis, MO).

RESULTS AND DISCUSSION

Transcription of the α-3 AChR Subunit Gene in Normal and MG Thymus

Transcripts encoding for the α-3 subunit of a neuronal AChR are found in both normal and MG hypertrophic thymus.[10] To expand our original observations and to investigate the expression of the thymic α-3 AChR subunit transcripts in MG

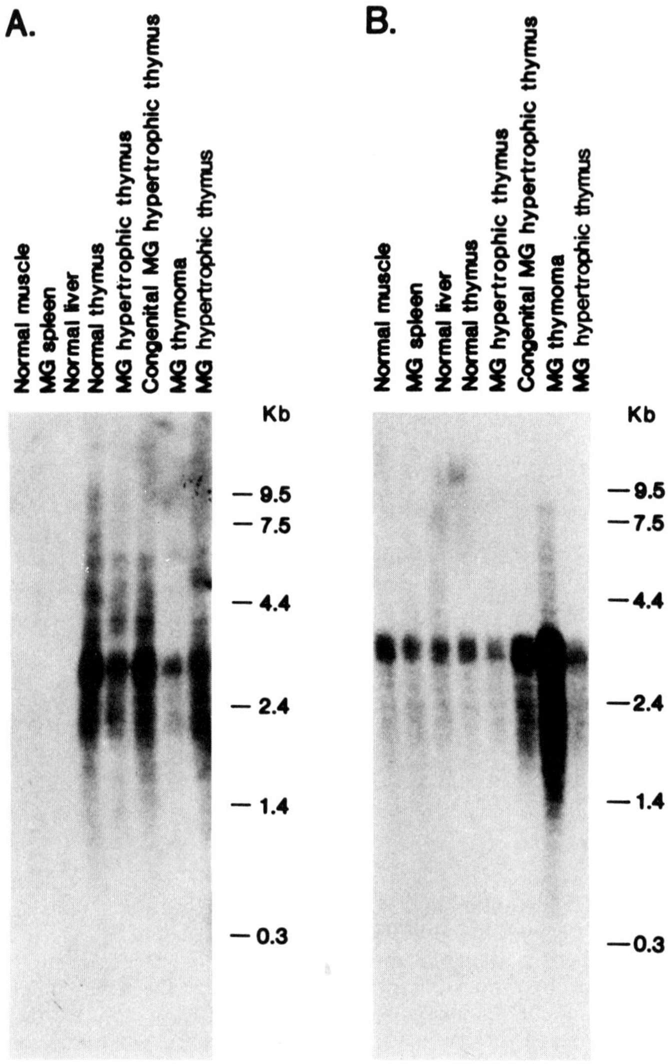

FIGURE 1. Hybridization of nicotinic neuronal AChR α-3 subunit and β-amyloid cDNA probes to different lymphoid and nonlymphoid tissues. Blotted RNA species were hybridized with radiolabeled **(A)** α-3 cDNA probe and **(B)** β-amyloid probe as indicated in METHODS. Blots in **A** and **B** are identical. 2.0 μg of poly(A)$^+$ RNA was applied per lane.

thymic tissues, identical blots of poly(A)$^+$ thymic RNA preparations from MG and normal thymic tissues were prepared and hybridized with either a β-amyloid or an α-3 cDNA probe. To determine the tissue specificity of the α-3 transcript expression, poly(A)$^+$ RNA purified from other lymphoid tissues (spleen, peripheral lymphocytes) and nonlymphoid tissues (muscle, liver) were also included in the analysis. Only thymic tissues expressed the α-3 transcripts (FIG. 1); α-3 transcriptional levels were similar for the hyperplastic and the normal thymi. MG thymomas, however, presented an overall decrease in these levels, except for one case (TABLE 1, #5) in which α-3 transcription was normal. In this case, however, β-amyloid transcription was the highest observed for all of the preparations analyzed. Thus, the ratio of α-3 to β-amyloid hybridization for all thymomas investigated was lower than that observed in the normal thymus and MG-hypertrophic thymus.

β-amyloid precursor transcription is ubiquitous, occurring in almost all tissues tested.[39] Our studies have confirmed this; in addition we have shown that thymomas express either normal or higher than normal levels of this transcript. The pathological significance of this relative overexpression and the corresponding underexpression of α-3 transcripts are not known but may be the result of the expansion of the thymic epithelium that is the neoplastic component in thymomas (TABLE 1 and ref. 43). Expression of the β-amyloid precursor protein by the thymic epithelia has been confirmed immunohistochemically in the two thymomas (#1 and #5) and the two MG hypertrophic thymi studied.

To investigate whether α-3 and β-amyloid precursor expression were confined to a specific thymic cell population, poly(A)$^+$ RNA was prepared from thymic cellular fractions enriched in thymocytes or thymic stroma and epithelia. There was a tendency for the amyloid transcripts to be preferentially associated with the thymic stromal fraction that is rich in epithelial cells, whereas the α-3 transcripts distributed equally well between the two fractions (data not shown). The thymocyte-enriched fraction represents a mixed cell population that contains cells that are easily released from the thymic lobes by rupture of the thymic epithelia and stroma. Wright's staining of a typical preparation shows that at least 95% of the cells are thymocytes, but spindle-like cells are also present. In this context, we are not able to determine which type of cell most likely expresses the α-3 transcripts.

Based on the extent of hybridization of the α-3 cRNA probes to thymic poly(A)$^+$ RNA preparations in the ribonuclease protection assays (see below), we have estimated that α-3 transcripts correspond to about 0.02% of the thymic poly(A)$^+$ RNA species present in the normal thymus; however, we have not been able to clearly demonstrate the presence of α-3 transcripts in peripheral lymphocytes of

TABLE 1. Comparison of α-3 AChR Subunit and β-Amyloid Precursor Protein Transcription in Thymomas: Correlation with Histological Findings

Thymomas/Diagnosis	Level of α-3 Transcription[a]	Level of β-Amyloid Transcription[a]	Main Histological Findings
1/MG	low	high	Spindle-cell thymoma
2/MG	low	high	Mixed thymoma with abundant epithelial cells
3/non-MG	low	normal	Mixed thymoma
4/MG	traces	normal	Involuted thymoma with fibrosis
5/MG	normal	high	Mixed well-differentiated thymoma

[a] Level of transcription is estimated from the autoradiographic scanning of the 3.0 kb α-3 and the 3.5 kb β-amyloid cDNA bands resolved in Northern blots of normal thymus and thymoma poly(A)$^+$ RNA preparations. The signal intensity obtained with normal thymus poly(A)$^+$ RNA preparation is taken as normal. Signals whose intensities are decreased by 50% or more are considered below normal levels. Signals whose intensities are increased by 50% or more are considered above normal levels.

MG or non-MG patients, employing Northern blot analysis[10] or the more sensitive ribonuclease protection assay (not shown). These results then suggest that α-3 thymic transcripts are most likely expressed in a nonthymocyte cell. On the other hand, if thymocytes do express α-3 transcripts, our findings could be explained by intrathymic apoptosis of these thymocytes or by down-regulation of α-3 transcription through thymocyte development. The possibility that the methods employed are not sensitive enough to detect α-3 transcripts derived from a minor fraction of peripheral lymphocytes cannot be excluded.

To define the thymic cells with active transcription of the α-3 gene, an *in situ* analysis is necessary.

Analysis of Thymic cDNA Clones Encoding for the α-3 Subunit of a Nicotinic Neuronal AChR

PCR and sequencing analysis of 15 cDNA clones that encode for the α-3 subunit of a neuronal AChR have shown the existence of three 3′end noncoding variant cDNA forms. These are schematized in FIG. 2. The consensus polyadenylation signal sequence (AATAAA) is present in the normal cDNA clone and in the variant form #1 (clone 42NTBSZ). This consensus sequence is located 18 and 15 bp upstream from the poly(A) tail of the normal and variant cDNA, respectively. By contrast, variant from #2 (clone 72NTBSZ) does not contain this canonical sequence or its variants.[44]

Current sequence information indicates variability arising from alternative splicing of the α-3 gene with variant sequences starting at position 1445 (1600 bp in FIG. 2) and 1490 (1645 bp in FIG. 2) of the normal α-3 transcript. Generation of 3′end variants from a single primary transcript has been proposed to account for cDNA clones encoding the α-4 neuronal AChR,[45] with alternative splicing occurring within the coding region. Our findings indicate that a similar mechanism operates for the generation of the α-3, 3′end noncoding variants.

The stability of mRNA depends on sequences found in its 3′end noncoding region.[44] Thus the characterization of multiple 3′ α-3 noncoding variants suggests that, in thymus, α-3 transcripts differing in their stability are being produced. mRNAs responsible for the transient expression of peptides (growth factors, oncogene products, ref 44) have very short half-lives and carry in their 3′ noncoding regions the decay consensus AUUUA sqeuence. This sequence mediates the targeting of mRNA for degradation.[46] Variant form #1 carries this AUUUA sequence 119 bp and 130 bp from position 1490, possibly contributing to the low levels of expression for this transcript (see below). Two out of 15 isolated clones correspond to this variant. The 3′end of variant #1 has been found to be associated with both the variant clone 42NTBSZ, as shown in FIG. 2, and with cDNA sequences corresponding to the normally spliced α-3 transcript. Preliminary evidence, based on ribonuclease protection assays employing 3′ noncoding cRNA probes suggests that all 3′ variant transcripts are expressed in both the normal and the MG thymi (not shown). The existence of these variants suggests that there is tight regulation at the transcriptional level for the α-3 AChR subunit gene.

Variant form #1 (FIG. 2) in addition has a 122 bp sequence spliced into its coding region. Sequence comparison of this variant with the well-conserved sequences of the rat and chicken α-3 gene[57] indicates that the 122 bp fragment is located between exons 5 and 6 of the normal transcript and arises through alternative splicing. This finding was further confirmed through genomic and cDNA PCR analysis (see below). Alternative splicing has been reported for the rat β-1 subunit,[47] the rat

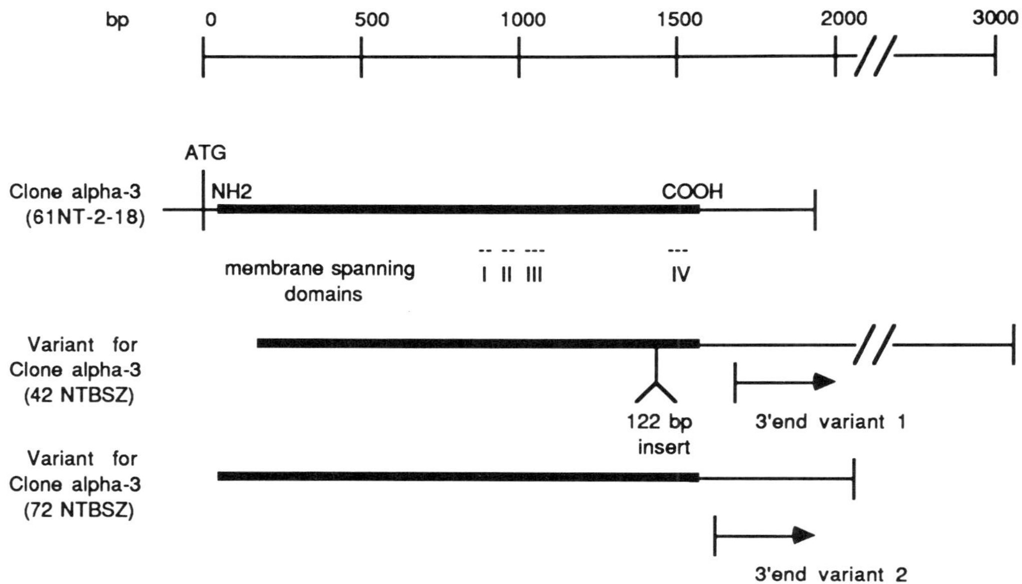

FIGURE 2. Schematic representation of thymic cDNA clones encoding for the nicotinic neuronal AChR α-3 subunit. The length of each depicted clone is represented by a line. The bold line represents the coding sequence for the subunit. Arrows indicate the site in which 3′end variants diverge from the normal cDNA sequence (see text).

α-4 subunit,[45] and for the human α-1 subunit.[48] The 122 bp α-3 variant sequence corresponds to an antisense Alu sequence that not only shifts the reading frame of the exon 6 but also introduces a stop codon located at positions 37 through 39 of the insert. If translated, the transcript will encode for a truncated peptide whose biological and immunogenic activity will differ from that of the normal α-3 subunit (see below). This finding confirms the observation that multiple transcripts encoding for variant AChR peptides are generated through alternative splicing and suggests that the regulation at the transcriptional level influences the surface expression of the AChRs. Within the coding sequence, variant #1 clone 42NTBSZ (except for its 122 bp insert) has 100% sequence identity with the normal α-3 cDNA clone 61NT-2-18,[10] indicating that the normal and variant cDNAs are derived from the same gene.

PCR Analysis of the Exon 5/Exon 6 Junction of the Gene Encoding the Nicotinic Neuronal AChR α-3 Subunit

To define more clearly the relationship between exon 5, exon 6, and the 122 bp Alu insert, a PCR analysis of the exon 5-exon 6 junction was performed using genomic DNA and normal and variant cDNAs as controls. The results are presented in FIGURE 3. The expected PCR products are schematized in the FIGURE. The results indicate that intervening sequences do exist between exon 5 and the 122 bp Alu sequence, and between the 122 bp Alu sequence and exon 6. Thus the α-3 122 bp variant appears to arise through alternative splicing of the α-3 gene and not from a pseudogene.

Transcriptional Analysis for the Normal and 122 bp Variant Form of the Thymic Nicotinic Neuronal AChR α-3 Subunit

To investigate the expression of the α-3 variant #1 transcripts in thymus, we carried out ribonuclease protection assays employing cDNA-derived (cRNA) probes that extend from exon 5 through exon 6 of both the α-3 normal and the 122 bp variant cDNA clones. The cRNA probes employed in the assay are schematized in FIGURE 4. The data obtained with the normal α-3 cRNA probe demonstrate protection of the expected 379 bp cRNA species that extends from exon 5 through exon 6 of the normal transcript. In addition, the exon 5–derived 259 bp cRNA fragment is protected, indicating that partially spliced RNA transcripts are present in the RNA thymic preparations. These data were reproducible under a variety of hybridization temperatures that ranged from 42°C to 70°C in independent experiments (not shown).

The results obtained with both normal and MG hypertrophic thymus poly(A)$^+$ RNA preparations confirmed the Northern blot analysis showing that the normal and the MG hypertrophic thymus express equivalent amounts of the α-3 transcripts. Likewise, in the two thymomas investigated (#1 and #2 in TABLE 1), it was found that α-3 transcript thymic expression was greatly reduced (not shown). Except for cRNA species that were protected by either the normal or the MG hypertrophic thymic RNA, no other protected cRNA fragments were evident for these two thymomas.

In contrast to the normal cRNA probe, protection of the 496 bp cRNA variant sequence and a corresponding 381 bp sequence, representing a partially spliced exon 5/122 bp Alu species, was only observed under high stringency conditions (hybridization temperatures equal or higher than 55°). Hybridization temperatures of 45°C resulted in preferential protection of the 122 bp Alu sequence and limited protection of exon 5 (259 bp) and exon 6 (115 bp) cRNA sequences (not shown). This observation is likely the result of competition between the abundant homolo-

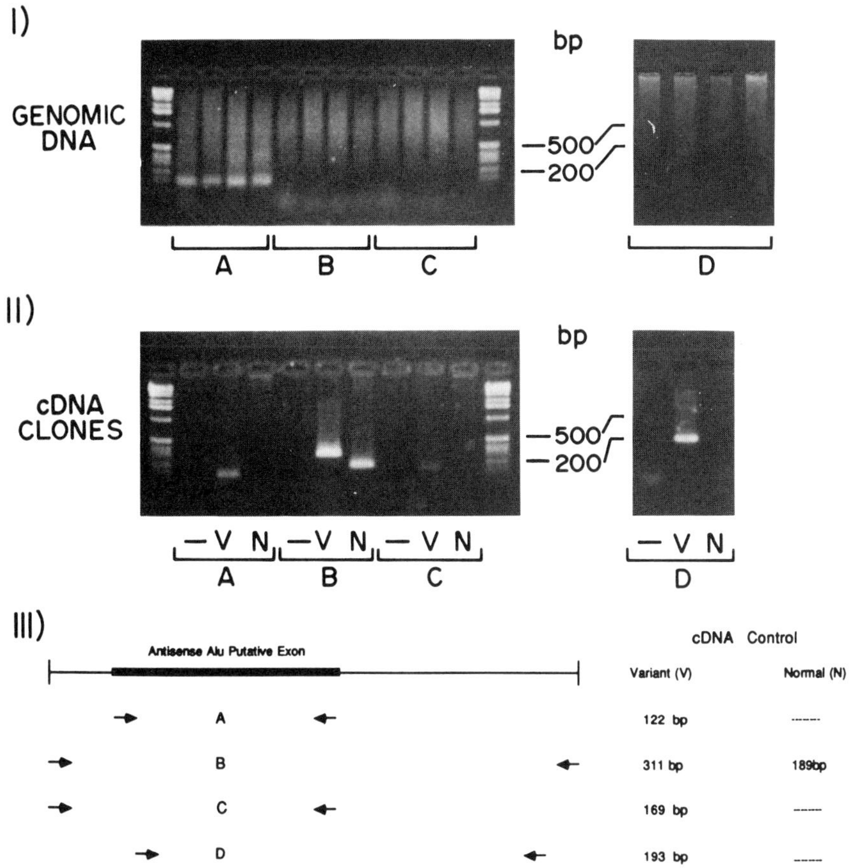

FIGURE 3. PCR analysis of the exon 5/exon 6 junction of the gene encoding for the nicotinic neuronal AChR α-3 subunit. **I:** PCR analysis was performed as indicated in METHODS employing 4 genomic DNA samples. **II:** cDNA clones 61NT-2-18 encoding for the normal sequence (N) and 42NTBSZ encoding for the 122 bp variant #1 sequence (V) were employed as controls. **III:** Primers used in the analysis are represented by arrows. The primer pair combinations used in the assay are schematized (A,B,C,D), and the length of the expected PCR products are entered in the columns to the right.

gous Alu sequences known to be present within partially spliced poly(A)$^+$ RNAs and the α-3 transcripts for binding to the variant cRNA probe. The high stringency conditions necessary to hybridize the variant α-3 496 bp cRNA to thymic RNA indicates that this protection is specific and that the variant transcript represents

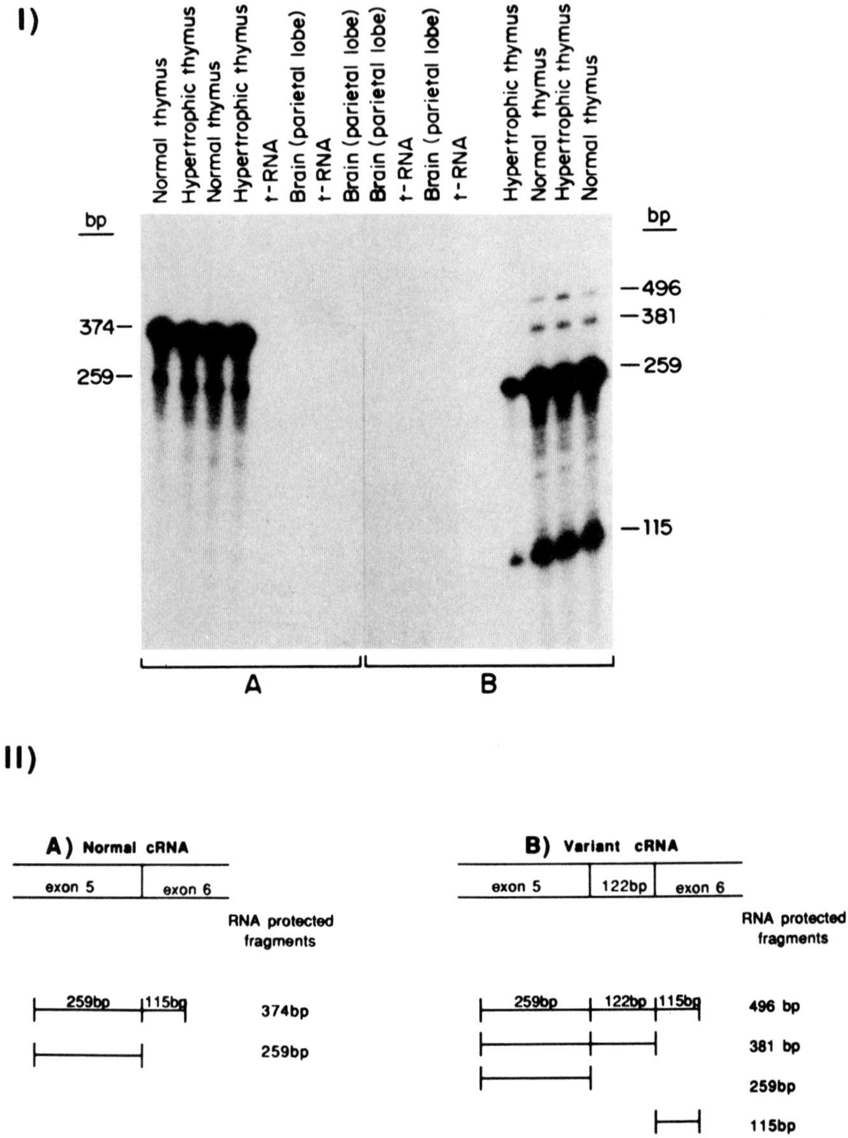

FIGURE 4. Ribonuclease protection assay of cRNA probes encoding for normal and variant forms of the nicotinic neuronal AChR α-3 subunit in normal and MG hypertrophic thymus. **I:** Ribonuclease protection assays were performed as indicated in METHODS. The eight left lanes represent data obtained with the normal cRNA probe. The right lanes represent data obtained with the variant cRNA probe. Lane 5 from right is blank. Except for lane 4 from the right, which represents data employing 2.0 μg of poly(A)$^+$ RNA, all others represent protection by 8.0 μg of poly(A)$^+$ RNA. **II:** Characteristics of the cRNA probes employed and of the cRNA fragments protected in the assay. Schematic representation of the cRNA-protected species, which are electrophoretically resolved in the assay.

a minor α-3 transcript normally expressed in thymus. Under high stringency conditions we have estimated, as deduced from the intensity of the autoradiographic bands, that approximately 2 to 6% of the α-3 transcripts expressed in thymus correspond to this variant form. This figure is lower than the frequency with which the variant cDNAs are represented in the corresponding thymic cDNA library (two out of 15 clones or 13% of the isolated clones correspond to this α-3 variant.) It is likely that the competition between Alu sequences and the variant RNA species for hybridization with the 498 bp cRNA probe results in less full-length probe protection and in underestimation of this α-3 variant transcription in thymus.

The α-3 122 bp variant, if translated, will generate a truncated peptide lacking the complete fourth transmembrane domain (M4) for this subunit and 4 amino acids upstream of this domain. In addition, the truncated peptide will contain 12 amino acids at its carboxy terminus, representing a peptide (MESCSVTQNGVQ) with a mean hydrophobicity index of −0.4.[49] Due to the hydrophilic characteristics of the dodecapeptide, it is unlikely that it possesses the minimal structural requirements to contribute to an M4 functional domain.[50]

Using cDNA manipulations and Xenopus oocytes to express the neuromuscular type *Torpedo californica* AcChR, Tobimatsu *et al.*[50] have reported that a deletion of 9 out of 19 amino acids in the hydrophobic fragment M4 of the α-1 subunit eliminates the biological function of the peptide. By analogy, the homologue α-3 variant may lack biological activity.

The significance of the Alu sequence splicing within the α-3 coding region of the normal α-3 transcript is not known. Our observation follows other examples in which an Alu sequence is found spliced into mature mRNA.[51-53] The two best documented examples of Alu splicing involve an antisense Alu sequence.[51,52] In these cases, Alu splicing arises from either a genomic point mutation[51] or represents alternative splicing arising from one gene.[52] For the autosomal recessive ornithine δ-aminotransferase deficiency, the splicing has been shown to be associated with a point mutation. This mutation generated an additional donor splice site and activated a cryptic acceptor splice site at the Alu 3′end. The translated product corresponded to an inactive truncated peptide.[51] On the other hand, alternative splicing of the decay-accelerating factor, known to inhibit the amplification of the complement cascade, has been shown to generate two variant forms of nontruncated peptides that differ in their amphipathic characteristics and correspond to the soluble and membrane-bound forms of this factor.[52]

The level of transcription for the α-3 122 bp variant is similar in two normal and two MG hypertrophic thymi tested, indicating that the splicing is not mediated by a genomic mutation but arises from the normal gene. This observation follows the prediction of Mitchell *et al.*[51] In their report they point out that the consensus anti-Alu sequence[54] carries both cryptic acceptor and potential donor sites that could mediate antisense Alu splicing.

From a functional point of view, alternative splicing of exons carrying an in-frame stop codon has been associated with developmental regulation. This regulatory role has been well defined for the SxI drosophila sex determination gene that produces an active peptide only in females[55] and for the rat glutamic acid decarboxylase gene that appears to encode an inactive product during early fetal development.[56] Whether or not the expression of the variant #1 transcript is subjected to metabolic or developmental control is a matter of speculation. if translated within a specific thymic compartment, however, the corresponding variant peptide may have important immunological and metabolic consequences in terms of intrathymic antigen presentation and AChR function.

SUMMARY

Thymic transcription of the α-3 subunit of the AChR was studied through sequencing and PCR analysis of thymic cDNA clones, Northern blotting, and ribonuclease protection assays. This analysis revealed at least three, 3'end sequence variants for the α-3 subunit as well as a variant that results from the alternative splicing of an antisense 122 bp Alu sequence between exons 5 and 6 of the normal transcript. The spliced Alu sequence not only shifts the exon 6 reading frame but also carries an in-frame stop codon. If translated, this variant transcript would produce a truncated peptide lacking the fourth transmembrane domain of the subunit and carrying a carboxy terminus dodecapeptide not found in any other known AChR subunit sequence. The putative variant subunit may lack biological activity and should differ antigenically from its normal counterpart. In comparing the normal, the MG hypertrophic, and the MG thymoma for transcription of the α-3 subunit and its 122 bp variant, it was found that there were no qualitative or quantitative changes in α-3 transcript expression in the MG hypertrophic thymi. Thymomas, however, showed an overall decrease in α-3 transcription and a comparative increase in β-amyloid precursor transcription. The decrease in the levels of α-3 transcription in thymomas may be related to the proliferation of thymic epithelial cells.

REFERENCES

1. ENGEL, W. K., J. L. TROTTER, D. E. MCFARLIN & M. L. MCINTOSH. 1977. Lancet **i:** 1310–1311.
2. KAO, I. & D. B. DRACHMAN. 1977. Science **195:** 74–75.
3. KAWANAMI, S., H. KAMEI, J. OITA, M. KUROKAWA, Y. UCHIDA & K. HAYASHI. 1987. Neurology **234:** 207–210.
4. SCHLUEP, M., N. WILLCOX, A. VINCENT, G. K. DHOOT & J. NEWSOM-DAVIS. 1987. Ann. Neurol. **22:** 212–222.
5. KIRCHNER, T., S. TZARTOS, F. HOPPE, B. SCHALKE, H. WEKERLE & H. K. MULLER-HERMELINK. 1988. Am. J. Pathol. **130:** 268–280.
6. O'CONNORS, M. R., K. I. GEUDER, F. HOPP, B. SCHALKE, S. TZARTOS, I. KALIES, T. KIRCHNER & H. K. MULLER-HERMELINK. 1990. Lab. Invest. **62:** 279–285.
7. CHILOSI, M., A. IANNUCCI, L. FIORE-DONATI, G. TRIDENTE, M. PAMPANIM, G. PIZZOLO, M. RITTER, M. BOFILL & G. JANOSSY. 1986. J. Neuroimmunol. **11:** 191–204.
8. BOFILL, M., G. JANOSSY, N. WILLCOX, M. CHILOSI, L. K. TREJDOSIEWICZ & J. NEWSOM-DAVIS. 1985. Am. J. Pathol. **119:** 462–473.
9. SAVINO, W., H. EMONARD & J-A. GRIMAUD. 1988. J. Neuroimmunol. **18:** 223–230.
10. MIHOVILOVIC, M. & A. D. ROSES. 1991. Exp. Neurol. **111:** 175–180.
11. BOYD, R. T., M. H. JACOB, S. COUTURIER, M. BALLIVET & D. K. BERG. 1988. Neuron **1:** 495–502.
12. DENERIS, E. S., J. BOULTER, J. CONNOLLY, E. WADA, K. WADA, D. GOLDMAN, L. W. SWANSON, J. PATRICK & S. HEINEMANN. 1989. Clin. Chem. **35:** 731–737.
13. LINDSTROM, J. M., M.L E. SEYBOLD, V. A. LENNON, S. WHITTINGHAM & D. D. DUANE. 1976. Neurology **26:** 1054–1059.
14. ROSES, A. D., C. W. OLANOW, M. W. MCADAMS & R. J. M. LANE. 1981. Neurology **31:** 220–224.
15. DRACHMAN, D. B., R. H. ADAMS, M. SC. LORRAINE, F. JOSIFEK AND S. G. SELF. 1982. N. Engl. J. Med. **307:** 769–775.
16. VAN DER GELD, H. W. R. & A. J. L. STRAUSS. 1966. Lancet **i:** 57–60.
17. GILHUS, N. E., J. A. AARLI & R. MATRE. 1983. J. Neuroimmunol. **5:** 239–249.
18. GILHUS, N. E., J. A. AARLI & R. MATRE. 1984. Neurology **34:** 246–249.
19. MEHL, V. S. & R. W. LANG. 1984. J. Neuroimmunol. **6:** 347–360.

20. Regev, I., N. Bornstein, R. Carasso & Y. Vardi. 1982. Acta Neurol. Scand. **65:** 681–682.
21. Bogousslavsky, J., F. Regli, A. M. Doret, B. W. Fulpius, B. Ostinelli, T. Rabinowicz & J. Ruzicka. 1983. Eur. Neurol. **22:** 301–306.
22. Kimura, S. & A. Nezu. 1989. Brain & Dev. **11:** 429–432.
23. Papazian, O. 1976. Neurology **26:** 311–316.
24. Tucker, D. M., D. P. Roeltgen, D. Wann & R. I. Wertheimer. 1988. Neurology **38:** 1173–1177.
25. Muller, K. M. I. 1989. J. Neurol. Sci. **93:** 263–275.
26. Xiao, B.-G., C.-Z. Lu, B. Hojeberg & H. Link. 1991. J. Neurol. Sci. **105:** 118–123.
27. Gotti, C., A. Ogando & F. Clementi. 1989. Neurosci. **32:** 759–767.
28. Williams, J. M. & D. L. Felten. 1981. Anat. Rec. **199:** 531–539.
29. Felten, D. L., S. Y. Felten, S. L. Carlson, J. A. Olschowska & S. Livnat. 1985. J. Immunol. **135:** 755s–765s.
30. Bulloch, K. 1987. *In* The Neuro-Immune-Endocrine Connection. C. S. Cotman, R. E. Brinton, A. Galaburda, B. McEwen & D. M. Schneider, Eds.: 34–43. Raven Press. New York.
31. Magni, F., F. Bruschi & K. Marianne. 1987. Brain Res. **424:** 379–385.
32. Bulloch, K. 1985. *In* Neural Modulation of Immunity. R. Guillemin, M. Cohen & T. Melnechuk, Eds.: 110–141. Raven Press. New York.
33. Bulloch, K. & W. Pomerantz. 1984. J. Comp. Neurol. **228:** 57–68.
34. Rozzman, T. L. & W. H. Brooks. 1985. J. Neuroimmunol. **10:** 59–72.
35. Kendall, M. D. & A. A. Al-Shawaf. 1991. Brain, Behav. Immun. **5:** 9–28.
36. Penit, C. & F. Vasseur. 1988. J. Immunol. **140:** 3315–3323.
37. Boyd, R. L. & P. Hugo. 1991. Immunol. Today **12:** 71–79.
38. Denning, S. M., D. T. Tuck, K. H. Singer & B. F. Haynes. 1987. J. Immunol. **138:** 680–686.
39. Tanzi, T. E., J. F. Gusella, P. C. Watkins, G. A. P. Bruns, P. St. George-Hyslop, M. L. Van Keuren, D. Patterson, S. Pagan, D. M. Kurnit & R. L. Neve. 1987. Science **235:** 880–884.
40. Mihovilovic, M., C. Hulette & J. Gilbert. 1993. Methods Neurosci. **12.** In press.
41. Kunkel, L. M., U. Tantravahi, M. Elsenhard & S. A. Latt. 1982. Nucleic Acids Res. **10:** 1557–1578.
42. Vinters, H. V., W. M. Pardridge, D. L. Secor & N. Ishii. 1988. Am. J. Pathol. **133:** 150–162.
43. Lampert, I. A. & R. A. Ritter. 1989. *In* Thymus Update 1. The Microenvironment of the Human Thymus. M. D. Kendall & M. A. Ritter, Eds.: 5–25. Harwood Academic Publishers. London, Paris, New York, Melbourne.
44. Proudfoot, N. J. & E. Whitelaw. 1988. *In* Transcription and Splicing. B. D. Haines & D. M. Glover, Eds.: 97–129. IRL Press. Oxford, Washington, D.C.
45. Goldman, D., E. Deneris, W. Luyten, A. Kochhar, J. Patrick & S. Heinemann. 1987. Cell **48:** 965–973.
46. Malter, J. S. 1989. Science **246:** 664–666.
47. Goldman, D. & K. Tamai. 1989. Nucleic Acids Res. **17:** 3049–3056.
48. Beeson, D., A. Morris, A. Vincent & J. Newsom-Davis. 1990. EMBO J. **9:** 2101–2106.
49. Kyte, J. & R. F. Doolittle. 1982. J. Mol. Biol. **157:** 105–132.
50. Tobimatsu, T., Y. Fujita, K. Fukuda, K-i. Tanaka, Y. Mori, T. Konno, M. Mishina & S. Numa. 1987. FEBS Lett. **222:** 56–62.
51. Mitchell, G. A., D. Labuda, G. Fontaine, J. M. Saudubray, J. P. Bonnefont, S. Lyonnet, L. C. Brody, G. Steel, C. Obie & D. Valle. 1991. Proc. Natl. Acad. Sci. USA **88:** 815–819.
52. Caras, I. W., M. A. Davitz, L. Rhee, G. Weddell, D. W. Martin Jr. & V. Nussenzweig. 1987. Nature **325:** 545–548.
53. Lundwall, A. B., R. A. Wetsel, T. Kristensen, A. S. Whitehead, D. E. Woods, R. C. Ogden, H. R. Colten & B. F. Tack. 1985. J. Biol. Chem. **260:** 2108–2112.
54. Labuda, D. & S. Striker. 1989. Nucleic Acids Res. **17:** 2477–2491.

55. BELL, L. R., E. M. MAINE, P. SCHEDL & T. W. CLINE. 1988. Cell **55:** 1037–1046.
56. BOND, R. W., R. J. WYBORSKI & D. I. GOTTLIEB. 1990. Proc. Natl. Acad. Sci. USA **87:** 8771–8775.
57. NEF, P., C. ONEYSER, C. ALLIOD, S. COUTURIER & M. BALLIVET. 1988. EMBO J. **7:** 595–601.

High IL-6 Gene Expression and Production by Cultured Human Thymic Epithelial Cells from Patients with Myasthenia Gravis

S. COHEN-KAMINSKY,[a] R. M. DELATTRE,[b]
O. DEVERGNE,[b] I. KLINGEL-SCHMITT,[a] D. EMILIE,[b]
P. GALANAUD,[b] AND S. BERRIH-AKNIN[a]

[a]*CNRS URA-1159*
Hôpital Marie Lannelongue
92350 Le Plessis-Robinson, France

[b]*INSERM U131*
92140 Clamart, France

Interleukin-6 (IL-6) is a multifunctional cytokine that regulates immune responses, acute-phase reactions, and hematopoiesis. In addition, a number of clinical and experimental observations strongly suggested that IL-6 could play a role in the pathogenesis of systemic autoimmune diseases in which an overproduction of IL-6 could induce polyclonal B-cell activation, abnormal differentiation of B lymphocytes, and autoantibody production.[1] We have recently shown a synergistic induction of IL-6 production and gene expression in normal human thymic epithelial cells (TEC) by lipopolysaccharide (LPS) and cytokines such as IL-1β and TNF-α.[2] Inasmuch as IL-6 is involved in T-cell activation, growth, and differentiation, as well as in maturation of activated B cells to Ig-producing cells, we thought that IL-6 production by TEC may be related to the thymic abnormalities found with a high frequency in MG patients, namely thymic hyperplasia and the presence of lymphoid follicles containing B cells.[3]

Using a secondary TEC culture system providing >90% TEC purity and similar techniques as in our previous work,[3] we investigated IL-6 production by MG-cultured TEC (from three patients with hyperplastic thymus) compared to control TEC (from four infant thymuses). IL-6 activity was measured by the hybridoma growth factor biological activity. As shown in FIGURE 1, LPS and cytokines (IL1-β, TNF-α), either alone or in combination, enhance or superinduce IL-6 production by MG TEC to levels dramatically higher than that seen in controls. We also studied the lymphoepithelial cell contact as a stimulating agent for IL-6 production. For this purpose autologous thymocytes were frozen at the time of primary TEC culture and then added in the secondary cultures at various amounts. Autologous thymocytes stimulated control and MG TEC in a time- and dose-dependent way. In addition, MG TEC presented a higher sensitivity to the stimulation of IL-6 production by autologous thymocytes compared to controls. *In situ* hybridization analysis of cytocentrifuged cells with an mRNA antisense probe specific for human IL-6 and labeled with ^{35}S (for technical details see ref. 2) confirmed these results and assessed that TEC (and not thymocytes) were actually the IL-6–producing cells. Computer-aided quantification of IL-6 mRNA levels (FIG. 2) indicated that both the numbers of cells expressing IL-6 mRNA and the amounts of cytoplasmic IL-6 mRNA per cell were increased upon stimulation. The cell-surface molecules (in particular adhesion molecules) that could be involved in signaling of IL-6 production through the lymphoepithelial cell contact are under investigation.

These results suggest that MG TEC present an IL-6 hypersecretion phenotype compared to controls. The induction of an IL-6 overexpression in TEC could be of pathophysiological relevance in MG, leading to activation and expansion of autoreactive T and B cells known to be present in MG thymus.[4,5] Consistent with this hypothesis, by *in situ* hybridization on thymic pathological sections, we recently demonstrated a higher number of IL-6 mRNA expressing cells as compared to controls, with an interesting localization close to the lymphoid follicles.[6] Furthermore in transgenic mice carrying the human IL-6 genomic gene fused with the human Ig heavy chain enhancer where abnormal IL-6 gene expression occurs, a massive polyclonal plasmocytosis in thymus, lymph node, and spleen, with hypergammaglobulinemia accompanying auto-antibody production, has been generated.[7] Further studies on the molecular mechanisms at the gene level that induce abnormal IL-6 gene expression and the mechanisms of IL-6 action through its receptor would provide critical understanding of the diseases where disregulated production of IL-6 is associated with polyclonal B-cell abnormalities or autoimmune reactions (reviewed in ref. 1), including myasthenia gravis.

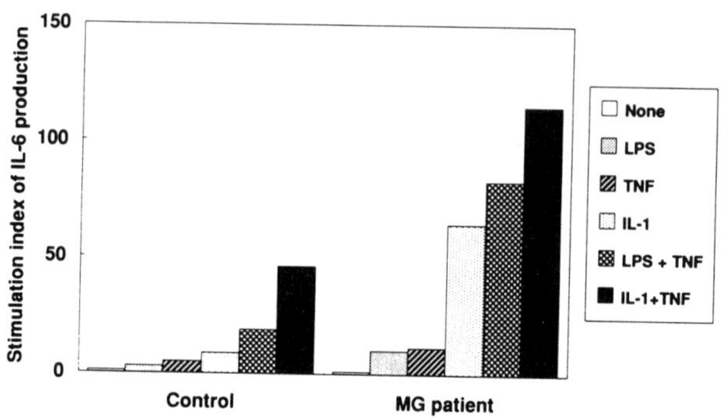

FIGURE 1. Effect of LPS and cytokines on MG TEC IL-6 production compared to control TEC. TEC in secondary cultures were stimulated with the indicated agents. LPS, TNFα, and IL-1β were used at concentrations of 10 μg/mL, 100 ng/mL, 1 ng/mL, respectively. When agents were used in combination, TNFα was added 2 hours before LPS, as described.[2] The amounts of IL-6 were measured in 48 h supernatants. Stimulation indexes are ratios between IL-6 amounts produced under stimulation and those produced spontaneously. Note the synergistic action of TNFα combined either to LPS or IL-1β and the high stimulation indexes of IL-6 production observed with the patient. Results are from one experiment and are representative of three independent experiments.

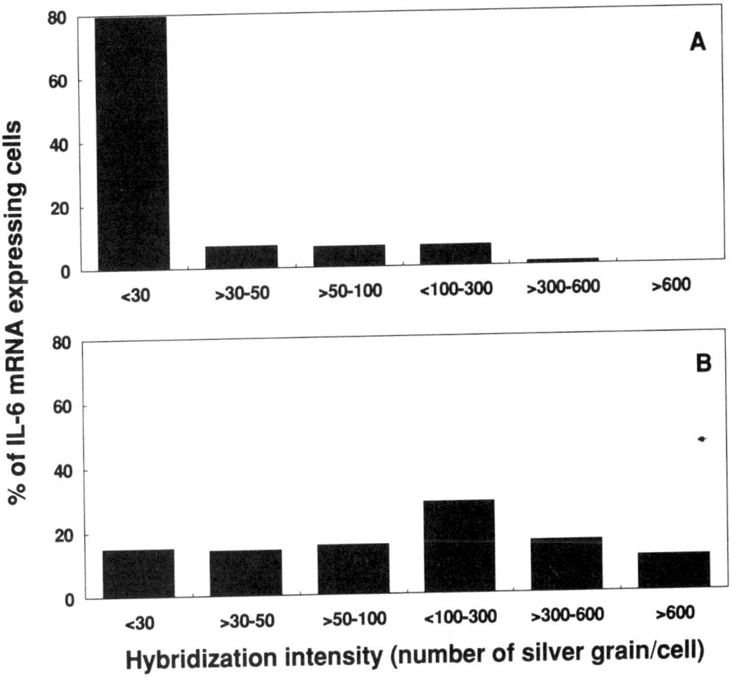

FIGURE 2. Quantitative analysis of IL-6 mRNA expression in stimulated MG TEC. TEC from secondary culture were incubated without stimulating factor **(A)** or with TNFα + LPS **(B)**. TNFα (100 ng/mL) was added 2 h before LPS (10 μg/mL). After 24 h, TEC were trypsinized cytocentrifuged, and then hybridized with [^{35}S]labeled antisense IL-6 RNA probe. The sense probe was used as a negative control and showed a maximum of 8 silver grains on the surface corresponding to one TEC. Computer-aided quantification of silver grains was carried out by an image analysis technique using the "Starwise Grains" program (IMSTAR, Paris, France), as previously described.[2] Note that both the cell number of IL-6 expressing cells and the amounts of cytoplasmic mRNA per cell are increased after stimulation as assessed by the progressive shift of IL-6 mRNA expressing cells towards high-grade hybridization intensity.

REFERENCES

1. Hirano, T. & T. Kishimoto. 1991. *In* Molecular Autoimmunity. N. Talal, Ed.: 177–194. Raven Press. Harcourt Brace Jovanovitch Publishers. London.
2. Cohen-Kaminsky, S., R. M. Delattre, O. Devergne, P. Rouet, D. Gimond, S. Berrih-Aknin & P. Galanaud. 1991. Cell Immunol. **138:** 79–93.
3. Berrih-Aknin, S., E. Morel, F. Raimond, D. Safar, C. Gaud, J. P. Binet, P. Levasseur & J. F. Bach. 1987. Ann. N.Y. Acad. Sci. **505:** 50–70.
4. Berrih-Aknin, S., S. Cohen-Kaminsky, V. Lepage, D. Neumann, J. F. Bach & S. Fuchs. 1991. J. Autoimmunity **4:** 135–153.
5. Leprince, C., S. Cohen-Kaminsky, S. Berrih-Aknin, B. Vernet der Garabedian, D. Treton, P. Galanaud & Y. Richard. 1990. J. Immunol. **145:** 2115–2122.
6. Emilie, D., M. C. Crevon, S. Cohen-Kaminsky, M. Peuchmaur, O. Devergne & S. Berrih-Aknin. 1991. Hum. Pathol. **22:** 461–468.
7. Suematsu, S., T. Matsuda, K. Aozasa, S. Akira, N. Nakano, S. Ohno, J. I. Miyazaki, K. I. Yamamura, T. Hirano & T. Kishimoto. 1989. Proc. Natl. Acad. Sci. USA **86:** 7547–7551.

In Situ Production of Interleukins in Hyperplastic Thymus from Myasthenia Gravis Patients

D. EMILIE,[a] O. DEVERGNE,[a] S. COHEN-KAMINSKY,[b]
S. BERRIH-AKNIN,[b] AND P. GALANAUD[a]

[a]*INSERM U131*
32 rue des Carnets
92140, Clamart, France

[b]*CNRS URA 1159*
Le Plessis-Robinson, France

Cells within the thymus of patients suffering from myasthenia gravis (MG) produce antiacetylcholine receptor autoantibodies, and this production is T cell–dependent. Moreover, numerous studies have outlined the role of accessory cells in the differentiation of B lymphocytes. Both T-cell and macrophage help to the antibody response can be provided at least partly by the production of cytokines. In the present work we asked whether a production of lymphokines and monokines occurs *in situ* in the thymus of patients presenting an MG associated with a hyperplastic thymus.

We analyzed by *in situ* hybridization the expression of four interleukin genes (IL-1β, IL-6, IL-2, and interferon-γ) in seven thymuses displaying a follicular hyperplasia obtained from patients with myasthenia gravis. As controls, six normal thymuses obtained during thoracic surgery were hybridized in parallel. No positive cell was detected with any of the four probes in these latter tissues. By contrast, IL-1β– and IL-6–producing cells were detected in the seven hyperplastic thymuses from MG patients. Positive cells were present in similar amounts and with similar distributions, suggesting that the same cells produced both monokines. Positive cells were mainly located in perifollicular areas of hyperplastic thymuses and in the connective structures emerging from the septa, at the site of cortex disruption (FIG. 1). The comparison of *in situ* hybridization and immunohistochemical results using an anti-CD68 monoclonal antibody (mAb) and an anticytokeratin mAb suggested that thymic epithelial cells and/or perifollicular macrophages were responsible for this production, as they were similarly distributed as monokine-producing cells.[1] By contrast, the large macrophages that were present inside germinal centers and that are involved in the phagocytosis of B lymphocytes dead at this site by apoptosis did not produce the monokines.

IL-2–producing cells were detected in perifollicular areas and to a lesser extent inside follicles. They were clearly outnumbered by CD25-positive cells that were similarly distributed.[1] Such a distribution of IL-2–producing cells and of CD25-positive cells with respect to follicle architecture was not specific for hyperplastic thymuses, as it was also found in lymph nodes from infected patients, displaying a benign follicular hyperplasia.[2] Despite the expression of these molecular and immunohistochemical markers of T-cell activation, interferon-γ–producing cells were extremely rare in myasthenic thymuses. This virtual absence of interferon-γ–producing cells in MG contrasts with a number of other immunological reac-

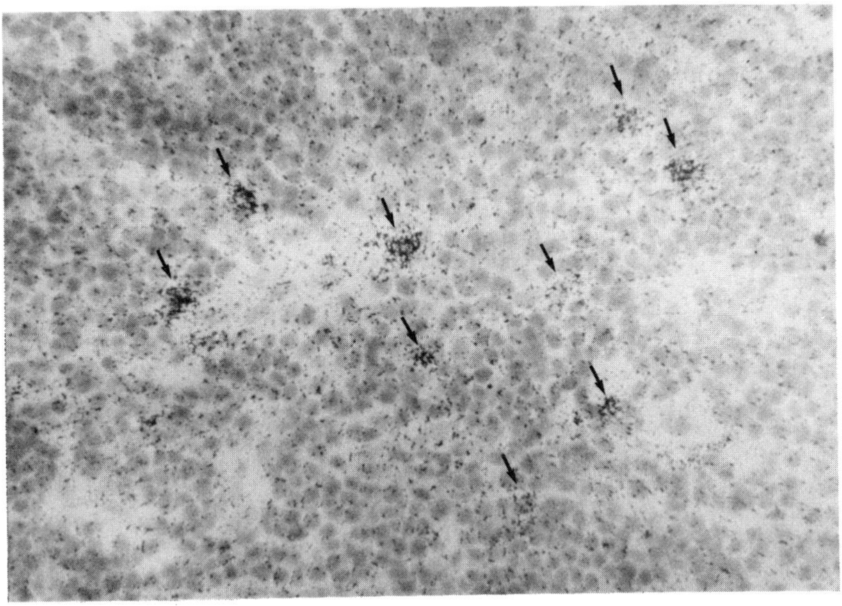

FIGURE 1. IL-1β–producing cells in perifollicular areas of hyperplastic thymuses from MG patients.

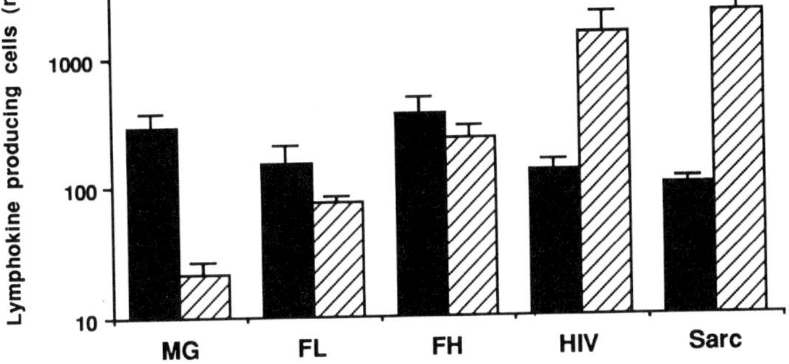

FIGURE 2. Enumeration of lymphokine genes in various immunopathological situations. MG: myasthenia gravis (n = 7) (ref. 1); FL: follicular lymphomas (n = 21) (ref. 3); FH: follicular hyperplastic lymph nodes (n = 8) (ref. 2); HIV: HIV-replicating lymph nodes (n = 8) (ref. 2); Sarc: sarcoidosis lymph nodes (n = 10) (ref. 4).

tions, in which interferon-γ–producing cells are at least as numerous as IL-2–producing cells (FIG. 2).

These findings indicate that the *in situ* autoantibody production developing in thymuses from MG patients may be supported by a local production of both monokines and IL-2, which themselves reflect the activation of T lymphocytes, macrophages, and epithelial cells. Although the cytokine production that occurs in these thymuses shares a number of similarities with other nonautoimmune responses, MG is characterized by a unique underexpression of the interferon-γ gene, which suggests either the involvement of a peculiar T-cell subpopulation or the existence of regulatory pathways responsible for the down-regulation of the interferon-γ gene expression in this situation.

REFERENCES

1. EMILIE, D. *et al.* 1991. *In situ* production of interleukins in hyperplastic thymuses from myasthenia gravis patients. Hum. Pathol. **22:** 461–468.
2. EMILIE, D. *et al.* 1990. Production of interleukins in HIV-1 replicating lymph nodes. J. Clin. Invest. **86:** 148–159.
3. PEUCHMAUR, M. *et al.* 1991. Interleukin-2 and interferon-γ production in follicular lymphomas. Am. J. Clin. Pathol. **95:** 55–62.
4. DEVERGNE, O. *et al.* 1992. Production of cytokines in sarcoid lymph nodes: Preferential expression of IL-1β and interferon-γ genes. Hum. Pathol. **23:** 317–323.

MYF-4 Does Not Mediate AChR Receptor Subunit mRNA Expression in Thymic Tissues[a]

HENRY J. KAMINSKI,[b,f] ROBERT A. FENSTERMAKER,[d] FADI W. ABDUL-KARIM,[c] JULIE CLAYMAN,[e] AND ROBERT L. RUFF[b]

Department of [b]Neurology, [c]Pathology, [d]Neurosurgery, and [e]Surgery
Cleveland Veterans Affairs Medical Center
University Hospitals of Cleveland
Case Western Reserve University School of Medicine
Cleveland, Ohio 44106

Evidence suggests the thymus is involved in the autoimmune process of myasthenia gravis.[1] Myoid cells have been found to be present in normal and neoplastic thymus, and muscle-specific proteins, including skeletal muscle nicotinic acetylcholine receptors (AChR), or antigenically similar proteins, have been identified.[1] AChR expression in muscle appears to be regulated to a great extent by members of the MyoD gene family, a group of DNA binding proteins.[2,3] Myf-4 is a member of this gene family, which when transfected into cultured fibroblasts stimulates expression of muscle-specific proteins and conversion of cells to a muscle-like phenotype.[4] We found expression of the α- and ε-subunit genes of the AChR in thymic tissue.[5] Here we investigate whether Myf-4 may play a role in the expression of AChR subunit mRNA and in the differentiation of myoid cells in the thymus.

METHODS AND MATERIAL

RNA was extracted from paraffin-embedded surgical specimens of thymus (age of specimens ranged from 2 to 11 years, see TABLE 1), reverse-transcribed (RT), and DNA-amplified by polymerase chain reaction (PCR) in the presence of [α^{32}P]dCTP using the method of Stanta and Schneider.[6] For Myf-4 RT-PCR the forward primer was 5′-CTTGAGGACAAGGGGGCTGGG-3′ (bases 201-220), and the reverse primer was 5′-ACCTTGGGCAGCCGCTGGTT-3′ (complementary to bases 409-390).[4] PCR reaction mixtures were separated by polyacrylamide gel electrophoresis and autoradiography was performed. RNA samples that were not subjected to RT were included in PCR reactions to serve as negative controls.

All myasthenic persons were grade IV Osserman clinical grades at the time of thymectomy. Four had elevated titers of AChR antibodies. The time from onset

[a] Dr. Kaminski is supported by a Career Development Award from the Department of Veterans Affairs and National Eye Institute Grant EY 09186.

[f] Address for correspondence: Dr. Henry J. Kaminski, Department of Neurology 127W, Cleveland Veterans Affairs Medical Center, 10701 East Blvd., Cleveland, Ohio 44106.

of myasthenia to thymectomy ranged from two months to ten years. All nonmyasthenics presented with a mediastinal mass. Only rare myoid cells were appreciated in the pathological specimens. For control tissues a two-year old paraffin-embedded muscle specimen that had been obtained from a patient who had undergone leg amputation for ischemic vascular disease was used. The muscle was histologically normal.

RESULTS

In the specimens from the 10 patients studied, β-actin had been identified in 8 patients, indicating the presence of amplifiable RNA. In all 8 of these patients, ε- and α-subunit mRNA was found, but no γ-subunit transcripts were identified. The expected fragment from Myf-4 amplification was identified from RNA extracted from paraffin-embedded human muscle but was not detected from thymic tissue. The results of our previous report[5] and the present study are summarized in TABLE 1.

TABLE 1. Patient Characteristics

Patient	Age	Sex	Amplified mRNA	Pathology
Nonmyasthenic				
1	79	F	α-, ε-subunits β-actin	thymoma, epithelial predominant
2	76	M	none	thymoma, epithelial predominant
3	67	F	α-, ε-subunits β-actin	thymoma, epithelial predominant
4	45	F	α-, ε-subunits β-actin	thymoma, lymphocytic predominant
5	55	F	α-, ε-subunits β-actin	thymic hyperplasia
Myasthenic				
6	53	F	α-, ε-subunits β-actin	thymic hyperplasia
7	45	F	α-, ε-subunits β-actin	thymic hyperplasia
8	29	M	α-, ε-subunits β-actin	thymic hyperplasia
9	57	F	α-, ε-subunits β-actin	thymic hyperplasia
10	48	M	none	atrophic thymus, predominantly fat

DISCUSSION

The AChR is a pentameric protein of four subunits: α-, β-, δ-, and either a γ- or a ε-subunit. In the adult mammal the γ-subunit is replaced by the ε-subunit.

Kobayashi *et al.*[7] found expression of all AChR subunit mRNAs in cultured thymic myoid cells, and Hara *et al.*[8] found α-subunit mRNA in thymomas of myasthenics. We identified α- and ε-subunit transcripts in thymic tissue of various histopathologies of myasthenics and nonmyasthenics but did not find γ-subunit mRNA.[5] These investigations contrast with those of others who did not find α-subunit mRNA in thymic tissue.[9,10] The greater sensitivity of PCR may explain these inconsistencies.

The transcriptional control of muscle-specific genes in thymus is not understood. We searched for Myf-4 mRNA in thymic tissue to determine if it might be involved in expression of the AChR subunit genes in thymus or be involved in myoid cell development, but we were unable to detect Myf-4 transcripts. Myogenin, the rat homologue of Myf-4, may particularly influence γ-subunit expression.[2] Therefore, the absence of Myf-4 message is consistent with the lack of γ-subunit expression. Other MyoD family members[3] or other mechanisms[11] may mediate α- and ε-subunit gene expression in pathological thymus. In normal adult murine thymus, Grounds *et al.*,[12] using *in situ* hybridization, found scattered cells, primarily in thymic medulla, which expressed MyoD and myogenin but did not stain with antimuscle antibodies. These cells were presumed to be myoid cells. Normal, hyperplastic, and neoplastic thymuses may differ in their expression of MyoD mRNA and the mechanisms by which muscle-specific transcripts are regulated.

SUMMARY

We investigated whether a MyoD gene family member plays a role in the expression of AChR or AChR-like proteins in human thymus. We amplified from thymic tissue, mRNA of subunits of the fetal- and adult-type AChR, and Myf-4, a gene product that appears to regulate AChR expression. RNA extracted from paraffin-embedded thymic tissue of five myasthenics and five nonmyasthenics was subjected to reverse transcription followed by polymerase chain reaction using primers specific for these mRNAs. Thymic RNA from eight of ten patients contained transcripts of the α- and ε-subunits (specific for the adult-type AChR). Presence of these transcripts did not correlate with thymic pathology or clinical presentation. No γ-subunit (specific for the fetal-type AChR) or Myf-4 transcripts were found. Our results indicate that mRNA for the adult-type AChR is expressed in thymic tissue and is expressed by mechanisms not involving Myf-4. Thymic AChR subunits may be the primary epitope that initiates the immune response in MG against the adult-type AChR.

REFERENCES

1. SCHONBECK, S., S. CHRESTEL & R. HOHLFELD. 1990. Myasthenia gravis: prototype of the antireceptor autoimmune diseases. Int. Rev. Neurobiol. **32:** 175–200.
2. NUMBERGER, M., I. DHRR, W. KUES, M. KOENEN & V. WITZEMANN. 1991. Different mechanisms regulate muscle-specific AChR γ- and ε-subunit gene expression. EMBO J. **10:** 2957–2964.
3. TAPSCOTT, S. & H. WEINTRAUB. 1991. MyoD and the regulation of myogenesis by helix-loop-helix proteins. J. Clin. Invest. **87:** 1133–1138.
4. BRAUN, T., E. BOBER, G. BUSCHHAUSEN-DENKER, S. KOTZ, K. H. GRZESCHIK & H. H. ARNOLD. 1989. Differential expression of myogenic determination genes in muscle cells: possible autoactivation by the Myf gene products. EMBO J. **8:** 3617–3625.

5. KAMINSKI, H. J., F. W. ABDUL-KARIM, R. A. FENSTERMAKER, J. CLAYMAN & R. L. RUFF. 1992. Acetylcholine receptor mRNA expression in neoplastic thymic tissue. Neurology **10:** 308.
6. STANTA, G. & C. SCHNEIDER. 1991. RNA extracted from paraffin-embedded human tissues is amenable to analysis by PCR amplification. Biotechniques **11:** 304–308.
7. KOBAYASHI, T., M. MISHINA, M. MICHIKAWA, Y. KAMEDA, Y. MATSUMOTO, T. TSUKAGOSHI & I. KAMO. 1990. α-, β-, δ-, γ-, and ε-subunit–specific mRNAs of acetylcholine receptors (AChRs) express in cultured rat thymus myoid cells. J. Neurol. Sci. **98** (Suppl): 299.
8. HARA, Y., S. UENO, T. UEMICHI, N. TAKAHASHI, S. YORIFUJI & Y. FUJII. 1990. Neoplastic epithelial cells express α-subunit of the muscle nicotinic acetylcholine receptor in thymomas from patients with myasthenia gravis. FEBS Lett. **279:** 137–140.
9. MIHOVILOVIC, M. & A. D. ROSES. 1991. Expression of mRNAs in human thymus coding for the $\alpha3$ subunit of a neuronal acetylcholine receptor. Exp. Neurol. **111:** 175–180.
10. GEUDER, K., R. SCHOEPFER, T. KIRCHNER, A. MARX & H. K. MULLER-HERMELINK. 1989. The gene of the α-subunit of the acetylcholine receptor: molecular organization and transcription in myasthenia-associated thymomas. Thymus **14:** 179–186.
11. SHIMIZU, N., E. DIZON & R. ZAK. 1992. Both muscle-specific and ubiquitous nuclear factors are required for muscle-specific expression of the myosin heavy-chain β gene in cultured cells. Mol. Cell. Biol. **12:** 619–630.
12. GROUNDS, M. D., K. L. GARRET & M. W. BEILHARZ. 1992. The transcription of MyoD1 and myogenin genes in thymic cells. Exp. Cell Res. **198:** 357–361.

Myasthenia Gravis–Associated Thymic Epithelial Tumors Express Neurofilaments and Are Associated with Antiaxonal Autoimmunity

ALEXANDER MARX, THOMAS KIRCHNER,
AXEL GREINER, BERTHOLD SCHALKE,[a] AND
HANS KONRAD MÜLLER-HERMELINK

Pathologisches Institut
and
[a]Neurologische Klinik der Universität Würzburg
Josef-Schneider-Strasse 2
8700 Würzburg, Germany

Thymic epithelial tumors (TET) are a heterogeneous group of neoplasms that have partly retained the unique thymic functions for homing and maturation of T cells.[1] This might cause an abnormal intratumorous T-cell selection and the frequent paraneoplastic autoimmunity associated with TET.[2,3] In fact, the occurrence of autoantibodies against the acetylcholine receptor (AChR) could be correlated with an expression of AChR epitopes[2,3] by neoplastic thymic epithelial cells. Evidence, however, that an aberrant protein expression in the T cell selecting microenvironment of TET does, in fact, induce autoimmunity is still lacking. As a model to test this hypothesis, we report here on an expression of neurofilament (NF) epitopes in thymic epithelial tumors and on the occurrence of antiaxonal autoantibodies in the majority of patients with thymoma-associated MG.

Twenty-two TET from patients with and without MG, five cases of thymitis with lymphoid follicular hyperplasia from MG patients, and eight non-MG control thymuses were investigated by immunohistochemistry on frozen sections using monoclonal antibodies (mAbs) against the 68 kDa (mAb NR4), 160 kDa (mAbs BF10 and NN18) and 200 kDa neurofilament (mAb NE52; all from Boehringer, Mannheim, Germany). Human antiaxonal autoantibodies in sera diluted 1 : 10 in PBS/0.5% BBSA were detected by immunofluorescence on frozen sections of fetal human spinal cord (FIG. 1b).[4]

Whereas in thymitis and normal thymus the mAbs decorated very few scattered cells only in the thymic medulla, a strong expression of NF epitopes on large numbers of neoplastic epithelial cells was encountered in all thymomas of cortical type (FIG. 1a) and in thymic carcinomas (n = 12) but not in medullary and mixed thymomas (n = 10). By double immunofluorescence, NF immunoreactivity was found to be confined to keratin-positive epithelial cells.

As summarized in TABLE 1, the frequency of antiaxonal autoantibodies was significantly higher in patients with TET than in age-matched controls. There was no significant difference between controls and MG patients without TET. Although we do not yet know the proteins to which these autoantibodies bind, the expression of NF epitopes in neoplastic thymic epithelial cells may possibly explain this novel type of paraneoplastic autoimmunity.

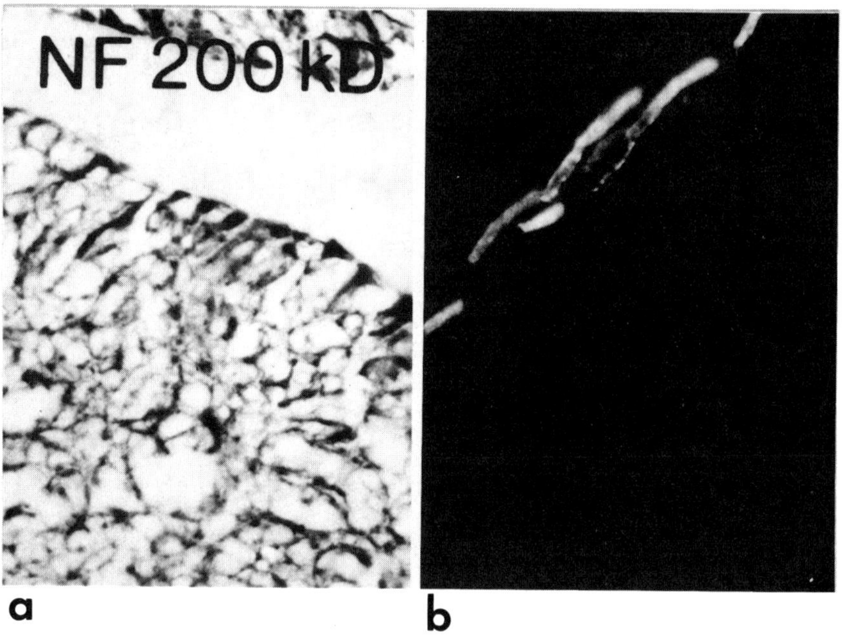

FIGURE 1. a: Binding of anti-200 kDa neurofilament mAb NE52 to a cortical thymoma (immunoperoxidase, × 250). **b:** Staining of human fetal spinal cord axons by serum autoantibodies from a thymoma patient with MG (immunofluorescence, × 250).

CONCLUSIONS

Because neurofilament expression was confined to thymomas of cortical type and thymic carcinomas, but absent in medullary and mixed thymomas, the results support our concept for a classification of thymomas distinguishing a medullary versus cortical line of differentiation.[1] Because the expression of the 68 kDa NF epitope was highly associated with invasive growth (8/9 cases), the 68 kDa NF may serve as a prognostic parameter for the evaluation of thymic epithelial tumors. Based on the demonstration of antiaxonal autoimmunity in patients with thymic epithelial tumors (TABLE 1), it will now be reasonable to test whether

TABLE 1. Immunoreactivity of Sera (1 : 10) from MG Patients (n = 20) and Controls (n = 50) with Human Spinal Cord Axons

	Immunofluorescence	
	Positive	Negative
MG patients with thymoma[a]	7	3
MG patients without thymoma	2	8
Controls without MG[a]	3	47

[a] The difference is significant with $X^2 = 14.01$, $p < 0.01$.

autoaggressive T or B cells are directed against intratumorous neurofilaments. If such a correlation exists, it would strongly suggest that an aberrant protein expression in neoplastic thymic epithelial cells can induce autoimmunity. This in turn would support our concept that the aberrant expression of nonreceptor proteins with an AChR-epitope in thymic epithelial tumors could be the trigger of anti-AChR autoimmunity in paraneoplastic myasthenia gravis.[2,3]

REFERENCES

1. Kirchner, T. & H. K. Müller-Hermelink. 1989. New approaches to the diagnosis of thymic epithelial tumors. *In* Progress in Surgical Pathology. C. M. Fenoglio-Preisser, M. Wolff & F. Rilke, Eds.: **10:** 167–186. Field and Wood Inc. Philadelphia.
2. Kirchner, T., S. Tzartos, F. Hoppe, B. Schalke, H. Wekerle & H. K. Müller-Hermelink. 1988. Pathogenesis of myasthenia gravis: acetylcholine receptor–related antigenic determinants in tumor-free thymuses and thymic epithelial tumors. Am. J. Pathol. **130:** 268–280.
3. Marx, A., R. O'Connor, K. I. Geuder, F. Hoppe, B. Schalke, S. Tzartos, I. Kalies, T. Kirchner & H. K. Müller-Hermelink. 1990. Characterization of a protein with an acetylcholine receptor epitope from myasthenia gravis–associated thymomas. Lab. Invest. **62:** 279–286.
4. Marx, A., T. Kirchner, A. Greiner, H. K. Müller-Hermelink, B. Schalke & M. Osborn. 1992. Neurofilament epitopes in thymoma and antiaxonal autoimmunity in myasthenia gravis. Lancet **339:** 707–708.

Thymus Histology and Acetylcholine Receptor Antibodies in Generalized Myasthenia Gravis

JOEL J.-F. OGER

University of British Columbia
2211 Wesbrook Mall
Vancouver, British Columbia, Canada

Although the etiology of generalized myasthenia gravis (GMG) is unknown, the pathogenesis includes reduced nicotinic acetylcholine receptor (AChR) as a result of an autoimmune attack.[1,2] When human AChR is used as antigen, most patients with GMG have serum antibodies.[3,4] Failure to detect antibodies in clinically unquestionable GMG has been a nagging problem encountered in 10–15% of patients with GMG.[5–8] As the assay described by Lindstrom[9] was improved, circulating antibodies are now found in up to 95% of a selected group of patients with generalized MG.[10] Characterization of the relationship between AChR antibody status and thymus histology might improve our understanding of the pathogenesis of MG. Thus we have reviewed the 54 patients who had thymectomy for GMG at University of British Columbia Hospitals in 1983–1991.

METHODS

GMG patients were included on two criteria: (1) weakness of ocular muscles, together with bulbar or limb muscles, and decrement or increased jitter, and a positive response to a Tensilon® test; and (2) AChR antibodies (AChRAb) measured in our laboratory before thymectomy. We used the staphylococcus A precipitation of human antigen.[13] AChRAb were considered to be present if cpm precipitated by serum and Staphylococcus A exceeded the mean plus 3 SD of cpm precipitated by 5 sera that had been randomly selected among 40 healthy adult controls. Thymus slides were reviewed by a single pathologist, unaware of the clinical or antibody status of the patients according to standard criteria.[11] Interestingly, review of these slides resulted in reclassification of seven cases; six were downgraded from follicular hyperplasia to normal thymus.

RESULTS

Among 54 patients with generalized MG, lymphoid follicular hyperplasia (LFH) was found in 27 (50%), a normal thymus in 19 (35%), and thymoma in 8 (15%). AChRAb were detected in 44 (81.5%). Ten patients (18.5%) had no antibodies. In 7 of the 10 seronegative patients, the thymus was normal (70%). By contrast, the thymus was normal in only 12 of 44 seropositive patients (27.2%); this difference was significant ($p < 0.01$). All 8 patients with thymoma had AChR antibodies.

DISCUSSION

Our experience with 54 consecutive thymectomies for generalized MG confirmed the association between thymic pathology and AChRAb; however, in contrast with expectations, we also found that only 60% of patients had both abnormal histology and circulating antibodies. In this series the frequency of thymic abnormalities and antibody positivity was comparable to that found by others who also used human AchR as antigen[7,9,10,11] (see TABLE 1). We found no blocking antibodies in the 10 seronegative patients.

We found 10 patients with no circulating AChRAb. Seven had a normal thymus and only 3 had LFH. There are arguments, however, for the presence of antibodies in "seronegative" MG: IgG and complement deposit at the neuromuscular junction of seronegative patients.[12,13] Mice injected with IgG from seronegative MG have block conduction.[14,15] Neonatal MG has been transmitted by one seronegative mother.[16] Thus a proportion of seronegative MGs are probably antibody-mediated. Are they all? When neither circulating antibodies nor an abnormal thymus are found, there is little evidence to support the autoimmune nature of the disease. We found 7 such patients in our cohort of 54 (13%). The existence of these 7 normothymic seronegative generalized MG patients suggests that there may be still unidentified "neuromuscular blocking molecules."[6]

TABLE 1. Thymus Histology and Percentage of Seronegativity for Each Histological Type

Author	No. of Patients	Type	Thymoma	Hyperplasia	Involution
Verma and Oger[8]	41	Generalized	14% (0%)[a]	50% (8%)	36% (33%)
Evoli *et al.*[17]	263	Generalized	33% (2%)	44% (7%)	22% (40%)
Vincent *et al.*[6]	42	Mixed[b]	23% (NA)[c]	60% (NA)	17% (NA)
Hankins *et al.*[18]	47	Generalized	20% (NA)	58% (NA)	21% (NA)
Berrih *et al.*[7]	70	Mixed	16% (0%)	51% (50%)	31% (13%)

[a] In parentheses: Proportion of seronegative patients in histological subgroup.
[b] Mixed groups included ocular and generalized myasthenia patients.
[c] Not available in published material.

Patients with seronegative generalized myasthenia are a small but important group that is heterogeneous. We believe the normothymic seronegative patients should be separated from seronegative patients with thymic hyperplasia. In seronegative normothymic MG, nonantibody factors or antibodies that do not bind to AChR could be responsible for impaired neuromuscular transmission.

REFERENCES

1. DRACHMAN, D. B., R. N. ADAMS, J. L. F. JOSIFEK & S. G. SELF. 1982. N. Engl. J. Med. **307:** 769–775.
2. VINCENT, A. 1980. Physiol. Rev. **60:** 756–824.
3. LEFVERT, A. K., K. LBERGSTROM, G. MATELL *et al.* 1978. J. Neurol. Neurosurg. Psychiatry **41:** 394–403.

4. LIMBURG, P. D., T. H. THE, E. HUMMEL-TAPPEL *et al.* 1983. J. Neurol. Sci. **58:** 357–370.
5. SOLIVEN, B. C., D. J. LANGE, A. S. PENN *et al.* 1988. Neurology **38:** 514–517.
6. VINCENT, A., J. NEWSOM-DAVIS, P. NEWTON & N. BEACK. 1983. Neurology **33:** 1276–1282.
7. BERRIH, S., M. EVELYNE, C. GAUD *et al.* 1984. Neurology **34:** 66–71.
8. P. VERMA & J. OGER. 1992. Neurology **42:** 586–589.
9. LINDSTROM, J. 1977. Clin. Immunol. Immunopathol. **7:** 36–43.
10. OGER, J., R. KAUFMAN & K. BERRY. 1987. Can. J. Neurol. Sci. **14:** 297–302.
11. ROSAI, J. 1989. Mediastinum. H. Ackermann's Surgical Pathology. C. V. Mosby. Toronto. 348–370.
12. TSUJIHATA, M., T. YOSHIMURA, A. SATOH *et al.* 1989. Neurology **39:** 1359–1363.
13. GARABEDIAN, B. V., E. MOREL & J. F. BACH. 1986. J. Neuroimmunol. **12:** 65–74.
14. DRACHMAN, D. B., S. DE SILVA, D. RAMSAY & A. PESTRONK. 1987. Ann. N.Y. Acad. Sci. **505:** 90–105.
15. MOSSMAN, S., A. VINCENT & J. NEWSOM-DAVIS. 1986. Lancet **i:** 116–119.
16. MELBER, D. 1988. N. Engl. J. Med. **319:** 996.
17. EVOLI, A., E. BARTOCCIONI, A. P. BATOCCHI *et al.* 1989. Clin. Invest. Med. **12:** 104–109.
18. HANKINS, J. R., R. F. MAYER, SATTERFIELD *et al.* 1985. Ann. Surg. **201:** 618–625.

Imaging Ligand Binding Sites on the *Torpedo* Acetylcholine Receptor[a]

ROBERT H. FAIRCLOUGH[b], ROBERT JOSEPHS, AND
DAVID P. RICHMAN

Department of Neurology
University of Chicago
5841 South Maryland Avenue
Chicago, Illinois 60637

Department of Molecular Genetics and Cell Biology
University of Chicago
920 East 58th Street
Chicago, Illinois 60637

Department of Neurology
University of California, Davis
2315 Stockton Boulevard, Room 5308
Sacramento, California 95817

INTRODUCTION

Electron microscopy and Fourier image analysis of tubular arrays of *Torpedo* acetylcholine receptor (AChR) have generated two models[1,2] of mass distribution for the complex of five protein subunits of this ligand-gated ion channel. Studies employing electron cryomicroscopy determined the structure to 18 Å resolution. This was sufficient to trace individual subunits through most of the height of the receptor.[1] However, no secondary structures or individual amino acid side chains were detected.

In the case of bacteriorhodopsin, Henderson *et al.*[3] used electron cryomicroscopy of two-dimensional crystals to generate a three-dimensional density map of this integral membrane protein at 3.5 Å resolution in the plane of the membrane and 10 Å perpendicular to the plane. This map allowed identification of particular amino acid side chains as well as the main secondary structural elements of the protein. Certainly electron cryomicroscopy may be expected to generate similar views of secondary structural motifs in the AChR and possibly a view of individual amino acid side chains in selected regions of the AChR.

Whereas the current AChR data provide less than atomic resolution models, they do provide data useful for determining structural and functional features of interest. For instance, we are mapping specific AChR ligand-binding sites and particular amino acid side chains onto the current three-dimensional models. We are also mapping the distribution of the five polypeptide subunits in these models to resolve the dilemma of AChR subunit organization.

[a] This work was supported by National Institutes of Health Grant NS 24304 and RR01209.

[b] Current address: Department of Neurology, University of California, Davis, 274 Briggs Hall, Davis, CA 95616.

RESULTS AND DISCUSSION

Models of Ion Channel Gating

Certain residues in the second stretch of hydrophobic amino acids (M-2), in each subunit, have been implicated as lining the walls of the AChR ion channel. Chimeric calf/*Torpedo* AChRs constructed to have *Torpedo* residues only in the M-2 region of the δ subunit have ion conductances similar to the native *Torpedo* AChR.[4] Chimeras constructed with calf residues only in the M-2 region of δ have ion conductances similar to the calf AChR[4]. Photo-affinity labeling with noncompetitive open channel blockers, chlorpromazine and triphenylmethyl phosphonium, results in modification of α-ser 248[5,6] and the homologous residues in β, γ, and δ,[5,7–10] and residues homologous to α-leu 251 in β and γ.[7,8] Site-specific mutation of residues homologous to α-ser 248 and 252 and α-leu 251 affect both single channel conductance and equilibrium binding of QX-222, an open channel blocker.[10–13] Finally, site-specific mutations of the glu and gln residues, homologous to α-glu 241 that lies to the amino-terminal side of the hydrophobic residues in M-2, give rise to AChRs with significantly altered mean channel conductances.[14] The current rectification patterns and effects of Mg^{2+} on the conductance of these mutated receptors indicates a cytoplasmic localization for the gln and glu's.[14]

In experiments designed to locate Ca^{2+} type–binding sites, we have probed AChR-enriched membranes with Tb(III) ions and used anomalous X-ray scattering from these ions to locate them in a small angle X-ray diffraction experiment.[15] Small angle X-ray diffraction experiments with centrifugally oriented membranes produce electron density profiles of the membranes perpendicular to the bilayer planes (see FIG. 1, solid curve). A major feature of the profiles obtained from AChR-enriched membranes includes two peaks separated by 41 Å, which represent the electron-rich phosphate head groups on opposite sides of the bilayer. Between the two peaks is a trough that corresponds to the hydrocarbon interior of the bilayer. The asymmetric protrusion of electron density on one side of the bilayer corresponds to the synaptic head of the AChR. Anomalous scattering experiments employ at least two incident X-ray energies: one just below and one at the L_3 absorption edge of the Tb(III) ion. This effectively tunes the scattering power of Tb(III) into and out of the diffraction pattern and allows one to determine by Fourier difference mapping the distribution of Tb(III) density relative to the electron density across the membrane. The results of the refinement of the difference Fourier mapping is presented in FIGURE 1 as the dashed curve. Two Tb(III) sites, separated by 21–22 Å, lie between the phosphate head regions of the membrane electron-density profile (see arrows in FIG. 1). We hypothesize that Tb(III) in these sites requires a ring of carboxylate residues from the ion channel wall lining. In particular, the glu and gln residues flanking the amino terminal end of the five subunits' M-2 segments, which have been implicated in controlling the AChR ion channel conductance, are attractive candidates for building the Tb(III) binding site closest to the cytoplasmic side phosphates.

Unwin has proposed that the conserved motif of aligned small polar and large hydrophobic residues along the putative channel lining segments of different channel superfamilies strongly implicates these residues in similar structural paradigms working by way of similar mechanisms to open and close the channels.[16] The structural motif for the N-terminal residues of the M-2 segments lining the channel is the α-helix[16] (FIG. 2A). If the five subunits' M-2 helices pack to form a pore, it may be analogous to riboflavin synthase, which contains such a bundle involving helices from five subunits packed with a right-handed twist.[17] The packing is stabilized by having a small residue from one helix contacting a large residue from a neighboring helix.[18] Unwin proposes that the open channel pore (FIG. 2B) is lined with backbone carbonyl and amide groups along with the small polar residues from one of the ridges along the purported M-2 helix (S in FIG. 2A). Unwin then proposes a mechanism for closing the gate that involves concerted displacements of the helices, moving the large hydrophobic residues on an adjacent helical ridge (L in FIG. 2A) into the pore in the closed channel state[14] (FIG. 2C). The steric block imposed by these residues is invoked to explain the tight seal formed by the AChR in this state,[16] preventing not only the passage of ions but also of small,

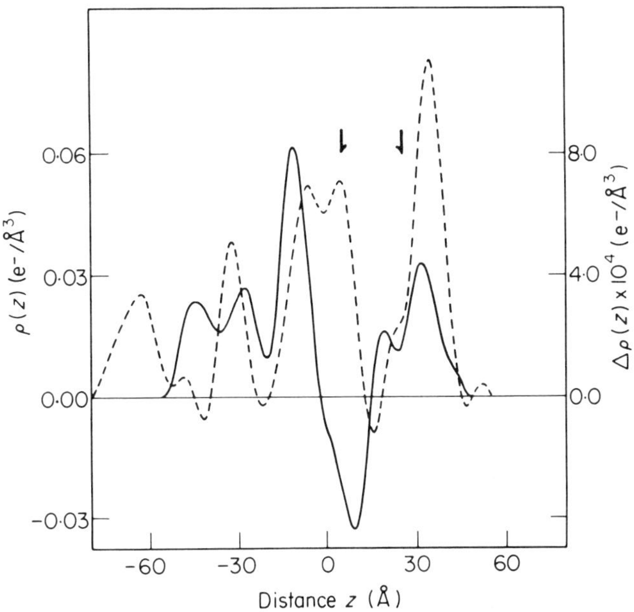

FIGURE 1. Electron density profile perpendicular to AChR-enriched membranes (solid line) and the distribution of Tb(III) ions (dashed line) determined by anomalous scattering.[15] The two arrows call attention to two Tb(III) sites in the hydrocarbon region of the bilayer spaced 21–22 Å apart. The extracellular side of the membrane is to the left of the FIGURE, and the cytoplasmic side is to the right.

uncharged molecules.[19] Many features of this mechanism are attractive, but the stability of the M-2 bundle with the hydrophobic ridges in the pore is possibly so great as to prohibit the facile transition to the open state (FIG. 2B), with only the free energy of agonist binding to drive the transition.

We would like to propose a second model of channel gating that involves a third ridge (A in FIG. 2A) that may exist on the other side of the polar ridge. The A ridge is amphiphilic with large hydrophobic groups on the interior and charged/polar glu or gln residues at the N-terminal end of the helical ridge. In a helical bundle with this third ridge facing the pore, a ring of carboxylates would stabilize

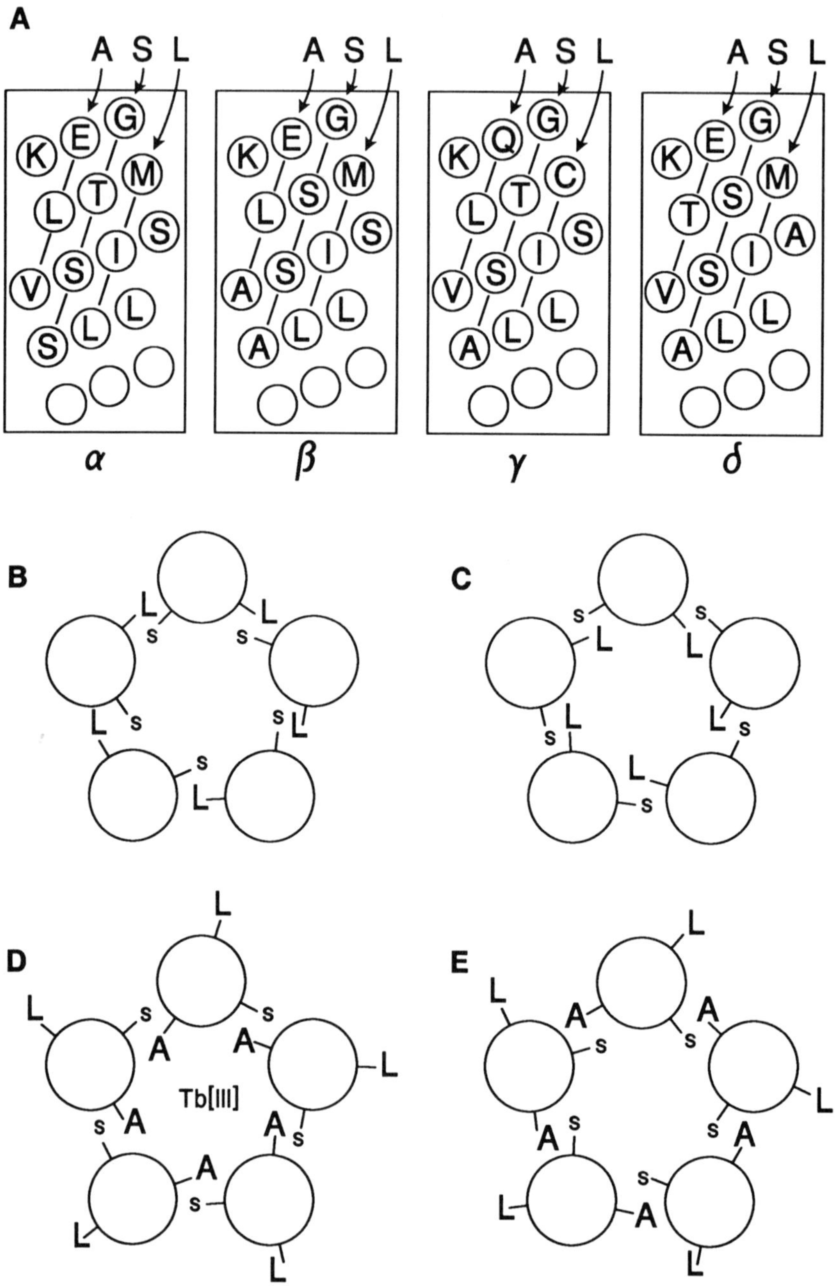
A
A S L
K E G L T M V S I S S L L
α
A S L
K E G L S M A S I S A L L
β
A S L
K Q G L T C V S I S A L L
γ
A S L
K E G T S M V S I A A L L
δ
B
C
D
Tb[III]
E

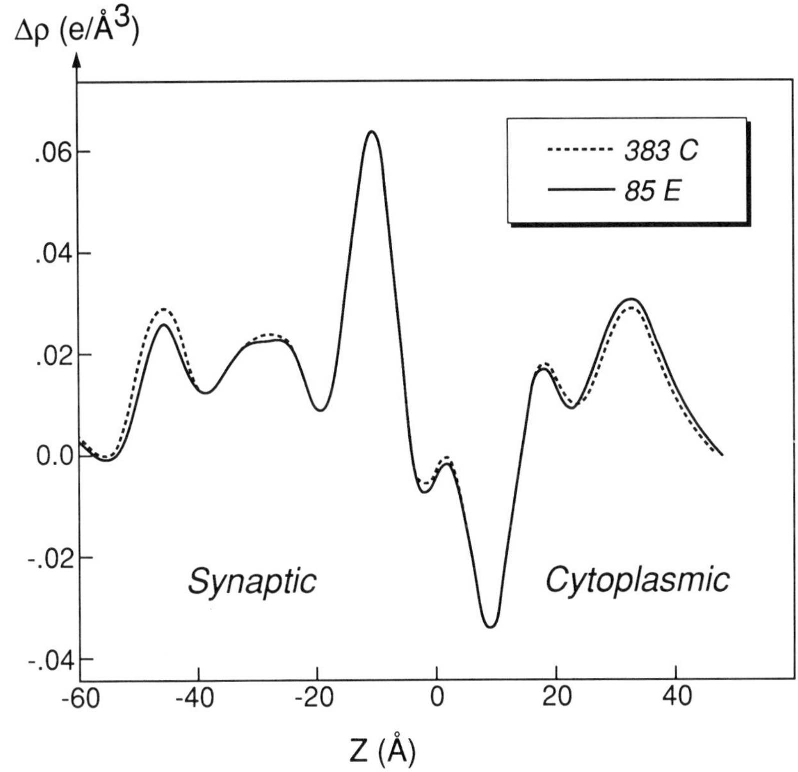

FIGURE 3. Electron density profile perpendicular to AChR-enriched membranes decorated with two different monoclonal antibodies. 383C produces additional electron density at the top of the extracellular region of the AChR, and 85E produces additional electron density on the cytoplasmic side of the membrane.

FIGURE 2. Helical models of the N-terminal residues of M-2 and flanking regions of the AChR subunits. In **A** is presented a helical net diagram of these residues for each subunit. The most N-terminal residue under consideration is at the top of each FIGURE and is on the cytoplasmic side of the membrane. The amino acid side chains form ridges along the indicated lines that are designated as A for amphiphilic, S for small polar, and L for large hydrophobic. **B** is a model of the five M-2 helices forming a water-filled open pore, after Unwin.[16] The view is from the cytoplasmic side of the AChR. Of note are the packing of the ridges of the helices, putting the small polar side chain ridge closest to the pore. **C** is a model for a closed channel pore presented by Unwin.[16] **D** is an alternative model for a closed channel pore stabilized by (1) a multivalent cation chelated by the ring of carboxylates of the N-terminal flanking residues of M-2, and (2) the hydrophobic residues to the interior along the A ridge of the helices. **E** represents an alternative open pore model to the one proposed by Unwin.[16]

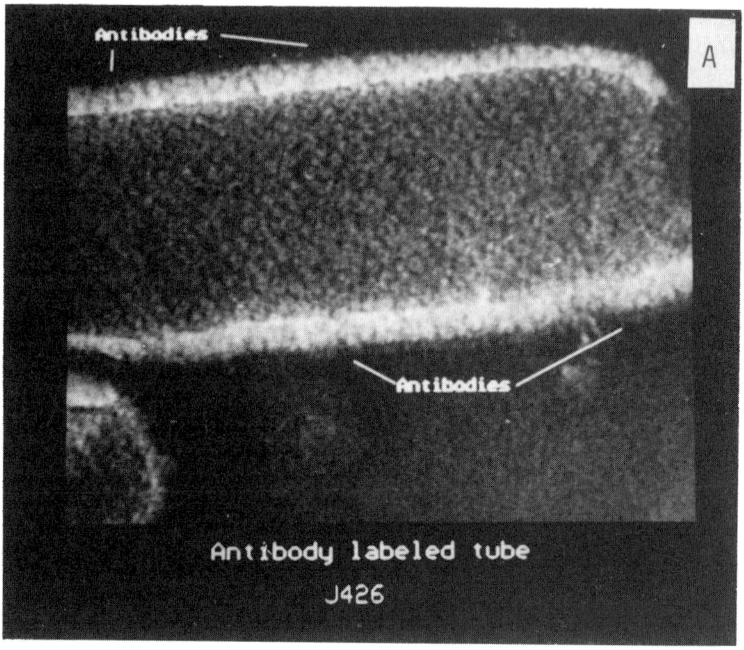
A
Antibodies
Antibodies
Antibody labeled tube
J426

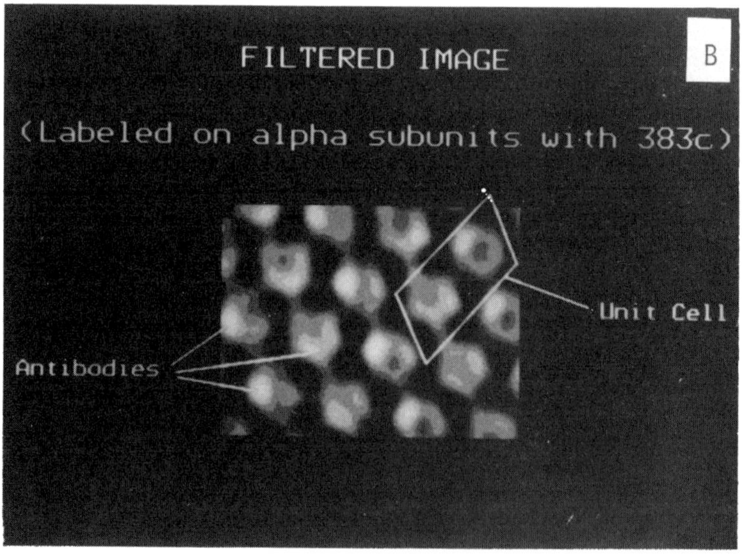
B
FILTERED IMAGE
(Labeled on alpha subunits with 383c)
Unit Cell
Antibodies

the closed state as long as a multivalent cation were situated in the center (FIG. 2D). Such an amphipathic pore seal would better explain the barrier to small uncharged molecules, as well as to ions, than the pure hydrophobic seal. Furthermore the free energy of agonist binding is sufficient to release bound Ca(II)[20] or Tb(III),[21] which could destabilize this closed arrangement of the five-helical bundle. Concerted displacements of the helices driven by mutual repulsion of the carboxylates could move the interior hydrophobic residues out of the pore and return them to helical interfaces in contact with the ridge of small polar residues (FIG. 2E), similar to the open state described before (FIG. 2B). We would propose as a candidate for the desensitized state the state initially described by Unwin as the closed state (FIG. 2C).

Localization of α-Cys (192-193)

To open the channel gate, agonists bind to regions of the AChR near α-cys (192-193), the residues affinity alkylated by 4-(*N*-maleimido)benzyltrimethylammonium iodide (MBTA).[22] Where is this region on the AChR model? Small angle X-ray diffraction and electron microscopy place α-bungarotoxin (α-BTX), which blocks agonist binding to this site, at the top of the synaptic protrusion.[23] Criado *et al.*[24] and subsequently Pedersen *et al.*,[25] however, have localized by way of monoclonal antibodies (MAbs), a cytoplasmic exposed segment of the α-subunit between residues 156 and 179 that forces cys (192-193) to be very close to the level of the synaptic-side phosphate head groups. For example, if the polypeptide chain extending from 179 to 192 forms a β-strand, one would expect the distance subtended to be 42 Å, just long enough to traverse the lipid bilayer, thereby placing cys (192-193) very close to the level of the phosphate head groups on the synaptic side of the membrane.

To resolve where cys (192-193) resides, we have employed MAb 383C, which binds to the AChR with a stoichiometry of one antibody per two toxin sites (Fairclough *et al.*, in preparation). This antibody completely inhibits α-BTX binding, blocks carbamylcholine binding to one site, and has little effect on *d*-tubocurarine (TC) binding.[26] 383C binds to the α-subunit in Western blots and peptide scans to peptide α-(187-199) (Fairclough *et al.*, in preparation). 383C marks the location on the AChR of α-(192-193). The fact that this antibody binds to AChR in membranes is not consistent in itself with α-(192-193) being located near the synaptic-side phosphate head groups. Small angle X-ray diffraction analysis, in fact, finds 383C well up on the synaptic protrusion as seen in FIGURE 3 (Fairclough *et al.*, in preparation). Finally, FIGURE 4A shows a two-dimensional array of AChR rosettes decorated with 383C (Fairclough *et al.*, in preparation). Fourier filtration of this electron micrograph produces the result displayed in FIGURE 4B (Fairclough *et al.* in preparation) in which the epitope is visualized as the light stain excluding spots on the rosettes. Clearly α-(192-193) is located up on the synaptic head accessible to this antibody for at least one of the two α-subunits.

FIGURE 4. Electron micrograph **(A)** and the Fourier filtered image **(B)** of one side of an AChR tubular array decorated with 383C. The electron micrograph was taken of the array negatively stained with 2% phosphotungstic acid at an electron optical magnification of 39,000. The filtered image has light regions on the rosettes that correspond to areas of increased stain exclusion from the presence of the antibody-combining site.

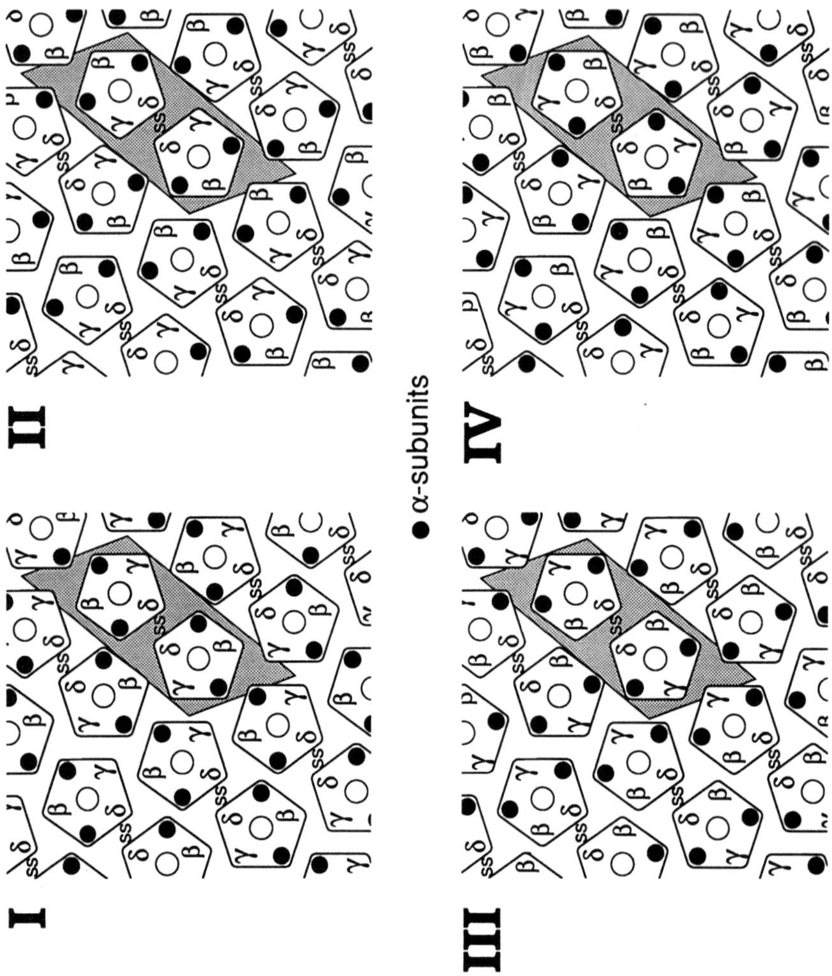
I
II
III
IV
● α-subunits

The filtration of the electron micrograph indicates two sites on the AChR to which this antibody binds, one of which has a higher occupancy than the other. The toxin-blocking characteristic of this MAb suggests both arms of the MAb are attached, whereas the carbamylcholine binding data suggest the second arm may not be attached. Studies with 383C Fab fragments are now in progress.

AChR Subunit Organization

Electron cryomicroscopy of AChR arrays has enabled visualization of five subunits arranged with pseudo-pentagonal symmetry surrounding the central pore of the AChR rosette.[27] Several groups have addressed the question of the subunit organization, and all seem to agree that the two α-subunits are not nearest neighbors.[28–33] That leaves one of β, γ, or δ to be the unique subunit between the α's on one side with the other two subunits on the other side. Two electron microscopic[28,31] studies using different AChR preparations, but both marking δ and α, find that δ is not the unique subunit between α's. The four remaining possible arrangements are illustrated in FIGURE 5. Model I was proposed on the basis of direct chemical cross-linking of subunits[29] and on the basis of toxin cross-linking to AChR subunits and the affect of receptor-specific ligands upon the pattern of toxin cross-linking.[30] An immunoelectron microscopic study localized three extracellular protein ligands on two-dimensional crystalline arrays of the AChR: α-BTX, Fab 35, and wheat germ agglutinin (WGA).[31] α-BTX and Fab 35 label the toxin/agonist site and the main immunogenic region (MIR) of the α-subunit, respectively. WGA binds uniquely to a polysaccharide on δ. The imaging study finds that WGA and Fab 35 bind very closely to one another, suggesting that the MIR rather than the toxin site is near the α/δ interface. Also this study identifies δ by the site of closest approach of two rosettes across a two-fold axis in the array. The conclusion is that δ is not the unique subunit between α's and that model I is favored over model II because of the relative positions of α and δ in the rosette. This study also located a cytoplasmic determinant on β: Fab 111. The extra stain exclusion was found between the two α-subunits. The consequence of this localization is additional data that favors model I and argues against model II.

Another electron microscopic study focused on detergent-solubilized AChR dimers cross-linked by way of the δ-subunits or the β-subunits. The α-subunits were labeled in the rosettes with biotinylated cobra toxin and visualized in the electron micrographs with avidin.[28] Analysis of scanning transmission electron micrographs of δ-δ and β-β cross-linked rosettes decorated with toxin/biotin/avidin led to the conclusion that neither δ nor β was the unique subunit between α's, and hence γ must be the one.[28]

A study of pairwise protein complexes formed by cells coexpressing mouse α/γ-, α/δ-, or α/β-subunits[32] revealed that each of γ and δ forms a binary complex with α, whereas β does not. The α/γ- and α/δ-complexes can bind carbamylcholine, whereas α alone cannot. Furthermore the α/γ-complex binds TC with a higher

FIGURE 5. Four models of subunit organization that are being tested by probes for different subunits and subunit interfaces. Models I and II have β as the unique subunit between α's. They differ by an interchange of γ and δ. Models III and IV have γ as the unique subunit between α's. They differ by an interchange of β and δ.

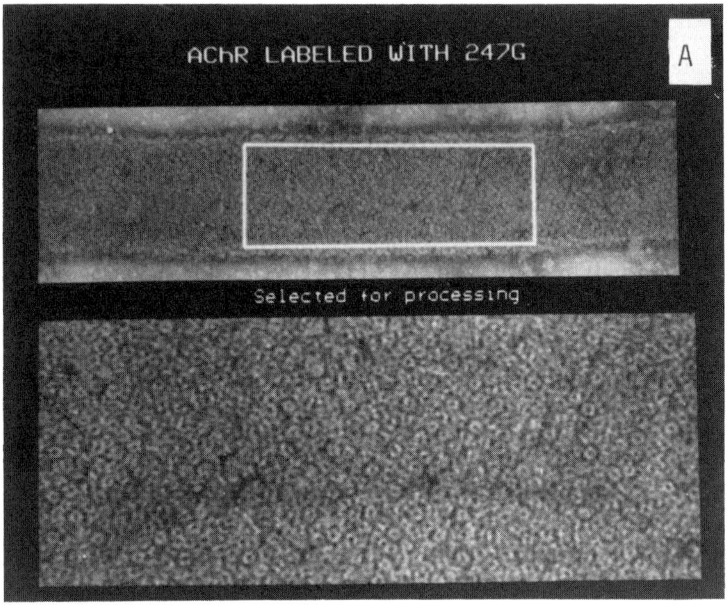
AChR LABELED WITH 247G
A
Selected for processing

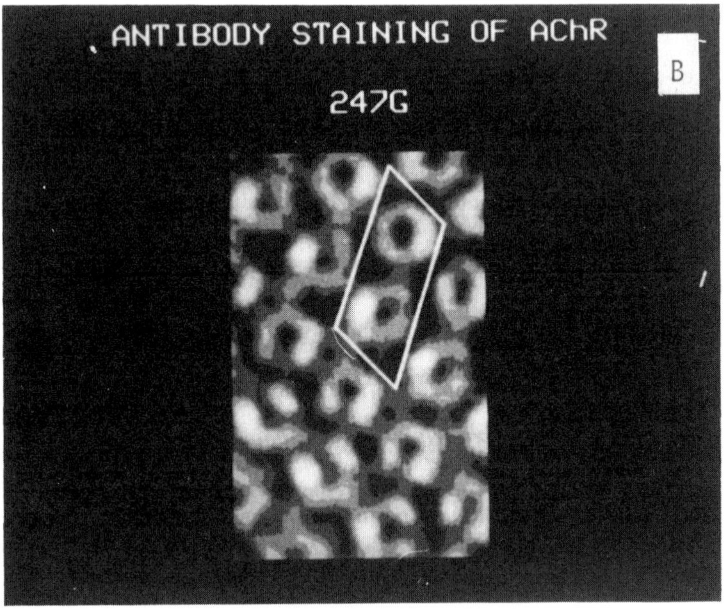
ANTIBODY STAINING OF AChR
B
247G

affinity than the α/δ-complex, and α alone does not bind TC at all. In the rosette, one α-subunit must bear rotational symmetry to the other. It is likely that the same face of the α-subunit complexes with γ or δ to form the two agonist sites and the high and low affinity TC binding sites, respectively. This requires either γ or δ to be the unique subunit between the α's, and because δ has been ruled out by two previous studies,[28,31] γ is the favored subunit here.[32]

Cohen and Pedersen also find a high-affinity TC site at the α/γ interface by specific photoaffinity labeling of these subunits with [^{3}H]TC, and a low-affinity TC site at the α/δ interface.[33] A higher concentration of TC is required for incorporation of TC into this site. They propose that γ is the unique subunit between α's, but using the α/δ orientation about the rosette as seen by Kubalek *et al.*,[31] they favor model IV.[33]

Given the divergence of thinking on the subunit organization, it is clear that mapping the distribution of each subunit within the three-dimensional model would be a start to sorting out this dilemma. The approach we are taking is to use multiple extracellular probes, MAbs or their Fab fragments, directed to each subunit in the immunoelectron microscopic paradigm using AChR tubular arrays. Additional mapping will use cytoplasmic side-directed probes to determine the subunit organization on that side of the membrane and how it compares to the extracellular organization. Initially we have used negatively stained specimens of MAb-decorated AChR arrays for the three-dimensional reconstructions of these complexes. We also have initiated electron cryomicroscopy studies as well. This technique should allow one to access the structural changes that occur upon binding the probes.

The preliminary results of this approach include the mapping in projection of the epitopes for MAb 383C, discussed previously, and for MAb 247G (Fairclough *et al.*, in preparation). The map of the 383C epitope is presented in FIGURE 4B, which is a Fourier filtration of the electron micrograph in FIGURE 4A. The localization of the light regions of stain exclusion on the rosettes corresponds to the region marked by α-BTX in the Kubalek *et al.* study.[31] This localization is inconsistent with the distribution of α-subunits predicted by model III, which has the two α-subunits in a rosette along a line roughly parallel to the short axis of the unit cell. It is consistent with models I and IV, which have the line between the α-subunits roughly perpendicular to the short axis of the unit cell.

The second MAb imaged is 247G. This MAb inhibits the binding of 50% of α-BTX and inhibits the binding of TC to the high-affinity site. Carbamylcholine binds to 247G/AChR complexes as well as to AChR alone, but 247G inhibits the agonist-induced opening of the ion channel. Because 247G does not bind to any AChR subunit on a Western blot, but does interfere with the high-affinity TC binding site, it is possibly a marker for the α/γ interface that some feel constitutes the high-affinity TC binding site.[32,33]

The result of the Fourier filtration of the better preserved side of the tube in FIGURE 6A is presented in FIGURE 6B (Fairclough *et al.*, in preparation). The

FIGURE 6. Electron micrograph **(A)** and the Fourier filtered image of one side **(B)** of an AChR tubular array decorated with antibody 247G. The filtered image **(B)** presents a unit cell outlined with the following dimensions: a = 77 Å, the short side; b = 162 Å, the long side; and γ = 114°, the angle between a and b. Note the light regions on the rosettes corresponding to the antibody-combining site.

filtered image of the decorated array also contains light regions on the rosettes that indicate increased stain exclusion reflecting the regular presence of the MAb-combining site. Most of the antibody is not apparent in the filtered image because flexibility between antibody domains places much of the MAb mass in different locations on each receptor. Fourier filtration eliminates signal from nonperiodic elements of the array.

From the filtration in FIGURE 6B, 247G binds to a unique site on the AChR not statistically distributed between the two agonist binding sites. Furthermore, taking 247G as a marker for the α/γ-interface, the position of additional stain exclusion overlaps a region of the unit cell not identified by any previous marker[31] and assigned by default to be the γ-subunit in model I. This localization favors model I over model IV, which has β in the location labeled by 247G.

Many questions remain in attempts to define the subunit organization of the AChR. The mapping of additional MAb epitopes, however, with different specificities on each subunit holds out the possibility of a more complete description and understanding of the subunit distribution than localization of any single probe alone. Apart from a total three-dimensional atomic resolution model, much can be learned from electron microscopic studies of complexes of AChR arrays with MAb or other probes specific to different segments of each subunit.

ACKNOWLEDGMENT

Synchrotron X-rays were provided by the Stanford Synchrotron Radiation Lab. We would like to thank Jennifer Holly, Suzanne Black, and Leon Gross for their input on the electron microscopic studies, and George Twaddle, Dr. Eswari Gudipati, Michele Mandala, and Carol Fair for their input on MAb characterization and AChR membrane preparation. We also thank Dr. Mark McNamee for a critical reading of the manuscript.

REFERENCES

1. UNWIN, N., C. TOYOSHIMA & E. KUBALEK. 1988. J. Cell Biol. **107:** 1123–1138.
2. MITRA, A. K., M. P. MCCARTHY & R. M. STROUD. 1989. J. Cell Biol. **109:** 755–774.
3. HENDERSON, R., J. M. BALDWIN, T. A. CESKA, F. ZEMLIN, E. BECKMANN & K. H. DOWNING. 1990. J. Mol. Biol. **213:** 899–929.
4. IMOTO, K., C. METHFESSEL, B. SAKMANN, M. MISHINA, Y. MORI, T. KONNO, K. FUKUDA, M. KURASAKI, H. BUJO, Y. FUJITA & S. NUMA. 1986. Nature **324:** 670–674.
5. HUCHO, F., W. OBERTHUR & F. LOTTSPEICH. 1986. FEBS Lett. **205:** 137–142.
6. GIRAUDAT, J., J.-L. GALZI, F. REVAH, J.-P. CHANGEUX, P.-Y. HAUMONT & F. LEDERER. 1989. FEBS Lett. **253:** 190–198.
7. GIRAUDAT, J., M. DENNIS, T. HEIDMANN, P.-Y. HAUMONT, F. LEDERER & J.-P. CHANGEUX. 1987. Biochemistry **26:** 2410–2418.
8. REVAH, F., J.-L. GALZI, J. GIRAUDAT, P.-Y. HAUMONT, F. LEDERER & J.-P. CHANGEUX. 1990. Proc. Natl. Acad. Sci. USA **87:** 4675–4679.
9. OBERTHÜR, W., P. MUHN, H. BAUMANN, F. LOTTSPEICH, B. WITTMANN-LIEBOLD & F. HUCHO. 1986. EMBO J. **5:** 1815–1819.
10. GIRAUDAT, J., M. DENNIS, T. HEIDMANN, J.-Y. CHANG & J.-P. CHANGEUX. 1986. Proc. Natl. Acad. Sci. USA **83:** 2719–2723.
11. LEONARD, R. J., C. G. LABARCA, P. CHARNET, N. DAVIDSON & H. A. LESTER. 1988. Science **242:** 1578–1581.

12. REVAH, F., D. BERTRAND, J.-L. GALZI, A. DEVILLERS-THIERY, C. MULLE, N. HUSSY, S. BERTRAND, M. BALLIVET & J.-P. CHANGEUX. 1991. Nature **353:** 846–849.
13. BERTRAND, D., A. DEVILLERS-THIERY, F. REVAH, J.-L. GALZI, N. HUSSY, C. MULLE, S. BERTRAND, M. BALLIVET & J.-P. CHANGEUX. 1992. Proc. Natl. Acad. Sci. USA **89:** 1261–1265.
14. IMOTO, K., C. BUSCH, B. SAKMANN, M. MISHINA, T. KONNO, J. NAKAI, H. BUJO, Y. MORI, K. FUKUDA & S. NUMA. 1988. Nature **335:** 645–648.
15. FAIRCLOUGH, R. H., R. C. MIAKE-LYE, R. M. STROUD, K. O. HODGSON & S. DONIACH. 1986. J. Mol. Biol. **189:** 673–680.
16. UNWIN, N. 1989. Neuron **3:** 665–676.
17. LADENSTEIN, R., M. SCHNEIDER, R. HUBER, H.-D. BARTUNIK, K. WILSON, K. SCHOTT & A. BACHER. 1988. J. Mol. Biol. **203:** 1045–1070.
18. CHOTHIA, C., M. LEVITT & D. RICHARDSON. 1981. J. Mol. Biol. **145:** 215–250.
19. HUANG, L. M., W. A. CATTERALL & G. EHRENSTEIN. 1978. J. Gen. Physiol. **71:** 397–410.
20. CHANG, H. W. & E. NEUMANN. 1976. Proc. Natl. Acad. Sci. USA **73:** 3364–3368.
21. RUBSAMEN, H., A. T. ELDEFRAWI, M. E. ELDEFRAWI & G. P. HESS. 1978. Biochemistry **17:** 3818–3825.
22. KAO, P. N., A. J. DWORK, R. J. KALDANY, M. I. SILVER, J. WIDEMAN, S. STEIN & A. KARLIN. 1984. J. Biol. Chem. **259:** 11662–11665.
23. FAIRCLOUGH, R. H., J. FINER-MOORE, R. A. LOVE, D. KRISTOFFERSON, P. J. DESMEULES & R. M. STROUD. 1983. Cold Spring Harbor Symp. Quant. Biol. **48:** 9–20.
24. CRIADO, M., S. HOCHSCHWENDER, V. SARIN, J. L. FOX & J. LINDSTROM. 1985. Proc. Natl. Acad. Sci. USA **82:** 2004–2008.
25. PEDERSEN, S. E., P. C. BRIDGMAN, S. D. SHARP & J. B. COHEN. 1990. J. Biol. Chem. **265:** 569–581.
26. MIHOVILOVIC, M. & D. P. RICHMAN. 1987. J. Biol. Chem. **262:** 4978–4986.
27. BRISSON, A. & P. N. T. UNWIN. 1985. Nature **315:** 474–477.
28. KARLIN, A., E. HOLTZMAN, N. YODH, P. LOBEL, J. WALL & J. HAINFELD. 1983. J. Biol. Chem. **258:** 6678–6681.
29. KISTLER, J., R. M. STROUD, M. W. KLYMKOWSKY, R. A. LALANCETTE & R. H. FAIRCLOUGH. 1982. Biophys. J. **37:** 371–383.
30. HAMILTON, S. L., D. R. PRATT & D. C. EATON. 1985. Biochemistry **24:** 2210–2219.
31. KUBALEK, E., S. RALSTON, J. LINDSTROM & N. UNWIN. 1987. J. Cell Biol. **105:** 9–18.
32. BLOUNT, P. & J. P. MERLIE. 1989. Neuron **3:** 349–357.
33. PEDERSEN, S. E. & J. B. COHEN. 1990. Proc. Natl. Acad. Sci. USA **87:** 2785–2789.

Molecular Evolution of the Binding Site of the Acetylcholine Receptor[a]

SARA FUCHS,[b] DORA BARCHAN,[b]
SYLVIA KACHALSKY,[b] DRORIT NEUMANN,[b]
MIRIT ALADJEM,[b] ZVI VOGEL,[c] MICHAEL OVADIA,[d]
AND ELAZAR KOCHVA[d]

Departments of [b]Chemical Immunology and [c]Neurobiology
The Weizmann Institute of Science
Rehovot 76100, Israel

[d]Department of Zoology
Tel-Aviv University
Tel-Aviv, Israel

The nicotinic acetylcholine receptor (AChR), which is known as the major autoantigen in myasthenia gravis and its animal model, experimental autoimmune myasthenia gravis, is a well characterized ligand-gated ion channel. It is an integral membrane glycoprotein that forms an oligomeric complex composed of four homologous subunits present in a molar stoichiometry of $\alpha_2\beta\gamma\delta$ with an overall molecular mass of 250,000 daltons.[1-3] The complete amino acid sequence of the receptor subunits has been derived from DNA clones of *Torpedo,* chick, calf, mouse, and human receptors.[4-8]

Among the subunits of AChR, the α-subunit has been shown to contain the cholinergic binding site.[9,10] Affinity-labeling experiments indicated that within the α-subunit the binding site is in close proximity to a sulfhydryl group,[1] thus focusing attention to the four cysteines (residues 128, 142, 192, and 193) present in the extracellular portion of the α-subunit. The first two cysteines (128 and 142) are common to all AChR subunits, whereas the other two (192 and 193) are present exclusively in the α-subunit. α-Bungarotoxin (α-BTX), which binds to nicotinic AChR with very high affinity at or near the cholinergic binding site, has been most instrumental in the analysis of the ligand binding site of AChR. Nevertheless, it should be pointed out that there are some cases of nicotinic AChRs in which the structural elements required for acetylcholine binding are not identical to those required for α-BTX binding. For instance, neuronal nicotinic AChRs[11] as well as snake muscle AChRs[12] are not affected by α-BTX, although they retain characteristic cholinergic pharmacology.

Several experimental approaches have been employed in an attempt to identify the ligand binding site in AChR. Earlier work from our laboratory, using proteolytic fragmentation of the α-subunit along with probing the fragments with antibodies to synthetic peptides and with α-BTX, led to the localization of the ligand binding site within a fragment of the α-subunit, which contains the two tandem cysteines

[a] The research described in this paper was supported by Grants from the Association Francaise Contre les Myopathies, the Los Angeles Chapter of the Myasthenia Gravis Foundation, the Israeli Academy of Sciences, and the Leo and Julia Forscheimer Center for Molecular Genetics at the Weizmann Institute.

at positions 192 and 193, and not the first two cysteines.[13,14] Later, by employing synthetic peptides, we showed that residues 185–196 of the α-subunit contain the essential components of the ligand binding site of AChR.[14–16] This segment contains the two tandem cysteine residues, at positions 192 and 193, which are unique to the α-subunit and are present at this position in all α-subunits of muscle and neuronal AChRs thus far sequenced. We have shown that these two cysteines as well as tryptophan, at position 187, are required for retaining toxin binding.[16]

Studies by other groups using cyanogen bromide-derived fragments of affinity-labeled α-subunit,[17] proteolytic fragmentation,[18] synthetic peptides,[19–21] and genetic constructs[22,23] have also localized the cholinergic binding site in close proximity or contiguous to the two tandem cysteine residues 192 and 193. Recent experiments with other acetylcholine-derived affinity labels[24,25] confirmed the putative localization of the binding site and demonstrated that other regions in the extracellular domains of the α-subunit and possibly in other subunits as well, may be in close contact with the ligand. The role of specific amino acid residues in ligand binding was studied by several groups, by replacing various amino acid residues, in either synthetic peptides or in genetically engineered peptides.[26,27]

TABLE 1. Snake and Mongoose Are Resistant to α-BTX Injection

Species	Dose,[a] μg of α-BTX/g of Body Weight	Death, Min after Injection
Snake	10	survived
Lizard	10	30
Mongoose	0.3	survived
Rabbit	0.3	90
Mouse	0.1	10
	0.1[b]	10

[a] α-BTX (in 0.5 mL of phosphate-buffered saline) was injected into snake and lizard (ip), into mongoose and rabbit (im), and into mice (iv).

[b] α-BTX was incubated with snake or mongoose serum (1 mL) for 30 min before injection into mice.

In an attempt to analyze in detail the structure of the cholinergic binding site of AChR and to elucidate the structural requirements for agonist versus α-BTX binding, we have been studying the ligand binding domains of nonclassical muscle AChRs of animals that are resistant to α-bungarotoxin. We have also used immunological reagents for further analysis and characterization of the ligand binding site of AChR. In the following, we will describe the structural and immunological analysis of the ligand binding region of muscle AChR from three species of snakes and from a mongoose (mammalia), all of which are resistant to α-neurotoxins and whose receptors do not bind α-BTX.

THE BINDING SITE OF SNAKE AChR

Some elapid snakes were shown to be resistant to the antagonistic effects of α-BTX, whereas their AChR is sensitive to agonist activation.[12] As a first step towards understanding the molecular basis of this phenomenon, we examined

```
        CCACACCTTAATTTGGAGGCATAGTCCCATAATGAGAGCACAGAGGGAAACATTGAGACTACACAAACAGCTGTAAACCCAGGTTCTGCG   90

        TGAGTGCGACCTCTAAAATAAAGGCTTCTTTGGGGGAAACTATCTCCTGCTTTGATCGATCTGGAAATACCTGCAGCAGGTGGGTTCTGC  180

        CGCATTACTACCGGTTTGGCGCCCGAAAAACTTTGTGCACCTGATGAGTCCGTGCTTGTGCGCATTTGCACTTAAAACAGTCTGCGCATG  270

        CACACAACATTAAAAAGGAATAAAAATACATTTTTTTCAATGAAGATTGTTCTGAGCATGTGCAGAAAAGAAAATAAGGATGGCGCTGCG  360

        AGAATAAAACCCGGTATGGCCACGTGTCCGCCAGAGTTGCTACCTATTTGGCTGTATCAGACCAAATAGGTAGGAACCCACCCATCTGCC  450

        CTGCAGATGTTACAATTAGCTTCAACCAAAAAAAGAAAGAGAGAGACGGCTGGCCCCCAGTGGCAAGAATATTGTGGAGAAACAGCTACG  540

        TTGGCATCTTTGGGCACGGGGAGCCCTACTCCTTGTCCAATGTTTTCAGAGTAGTAGGGCTATTGGAACCAGGCTGCAAAAACCCGAATG  630

        ACCACTAGATGGGAACAAACGTGTTGGAGATCCAGTATCTCCTTGCTGGCAAGTCTAGCTATTGTCCCATCTTGGCAGCTCTGATGTAAT  720

                                                                                              A  D
        AACTCATACCCAGCAGGAGCCAATGGGTTAAAGTGTATTAAAATTCGATGCTATTCTTATGACTCTTTTCCCCCCCCTTTTTAGTGCAGA  810
                                                                                             |||||
                                                                                             GCAGA
                                                                                              A  D

         G  I  F  A  I  D  Q  F  T  K  V  L  L  N  Y  T  G  H  I  T  W  N  P  P  A  I  F  K  S  Y
Natrix  TGGAATTTTTGCTATCGACCAGTTCACCAAAGTTCTTCTCAACTATACAGGACATATCACCTGGAATCCACCAGCCATCTTCAAGAGCTA  900
         ||    ||||| || | | | |||||||| || || ||  |||| || || || ||||||||||  || |||||||||| || |||||
Mouse   CGGCGACTTTGCCATTGTCAAATTCACCAAGGTGCTCCTGGACTACACCGGCCACATCACCTGGACACCGCCAGCCATCTTTAAAAGCTA
         G  D  F  A  I  V  K  F  T  K  V  L  L  D  Y  T  G  H  I  T  W  T  P  P  A  I  F  K  S  Y

         C  E  I  I  V  T  H  F  P  F  D  E  Q  N  C  S  M  K  L  G  T  W  T  Y  D  G  N  V  V  D
Natrix  CTGTGAAATTATAGTCACCCACTTCCCCTTTGACGAGCAAAACTGTAGCATGAAGCTGGGAACTTGGACTTACGATGGCAATGTGGTTGA  990
        |||||| || || ||||| ||||| |||||  || ||||| ||||| ||||||||||||||| || ||||| || || |||  |||||| | 
Mouse   CTGTGAGATCATTGTCACTCACTTTCCCTTCGATGAGCAGAACTGCAGCATGAAGCTGGGCACCTGGACCTATGACGGCTCTGTGGTGGC
         C  E  I  I  V  T  H  F  P  F  D  E  Q  N  C  S  M  K  L  G  T  W  T  Y  D  G  S  V  V  A

         I  S  P
Natrix  CATCTCCCCGGTAGGCTTTAGTTAGGTAGCTGTTTTT  1027
        |||   ||||
Mouse   CATTAACCCG
         I  N  P
```

FIGURE 1. Nucleotide sequence of *Natrix* HincII genomic fragment. Homology to mouse AChR α-subunit cDNA and deduced amino acid sequences are depicted.[28]

whether the nonpoisonous water snake *Natrix tessellata* was also resistant to α-BTX. Intraperitoneal administration of α-BTX into water snakes in concentrations several orders of magnitude higher than the LD_{50} (LD = lethal dose) for other species did not result in snake death (TABLE 1).[28] Similar weight-equivalent doses were lethal in lizards, which are phylogenetically related to the snake and were reported to be sensitive to α-BTX.[12] We ruled out the possibility that resistance of the snake to α-BTX was due to neutralizing factors in the serum, inasmuch as preincubation of snake serum with α-BTX (10 μg) did not abolish its lethal effect upon subsequent injection into mice.[28]

A genomic clone from *Natrix* genomic DNA, which hybridized to a mouse AChR α-subunit cDNA probe, was first isolated.[28] Sequence analysis of this 1 kb clone revealed significant homology (75%) to the mouse α-subunit cDNA, on a stretch of 200 nucleotides (FIG. 1). According to the homology, the sequence AGTG at nucleotides 803–806 and CGGT at nucleotides 999–1002 are possible candidates for acceptor and donor splice sites, respectively. The deduced amino acid sequence between these two putative splice sites revealed considerable homology with residues 96–160 of the mouse AChR α-subunit. This includes the two cysteine residues at positions 128 and 142 of the α-subunit, which are present in all the four subunits. The homology was much higher with the α-subunit than with the other subunits.

By employing Northern blot analysis, a 3.7 kb AChR α-subunit transcript was identified in both the water snake (*Natrix tessellata*) and cobra (*Naja naja atra*), using either cloned genomic snake sequences or the mouse α-subunit cDNA as probes.[28] This analysis confirmed that the genomic clone isolated codes for portions of the snake AChR α-subunit. The size of the snake α-subunit mRNA (3.7 kb) is higher than that reported for the *Torpedo,*[4] mouse,[8] and human[29] α-subunit mRNA. It is of interest to point out that an mRNA of a size similar to that of snake has been detected for the calf α-subunit and was found to contain a long 3′ untranslated region.[5]

In order to identify the region in the snake that in other species, such as *Torpedo* and mouse, was shown to contain essential components of the α-BTX binding site (*i.e.*, the segment containing the tandem cysteines 192 and 193), we have employed the polymerase chain reaction (PCR) for specific amplification of the cDNA encompassing this region. We have cloned and sequenced the domain of the α-subunit from *Natrix* and cobra AChR (a 300 bp fragment, corresponding to amino acid residues 119–222), which contains the four extracellular cysteines (128, 142, 192, and 193) of the α-subunit.[28] Sequence analysis of the amplified 300 bp fragments of *Natrix* and cobra revealed 72% homology to the mouse α-subunit cDNA, on the level of nucleotides and 78% homology at the deduced amino acids.[28] The presence of two cysteines at positions 192 and 193 verifies that this fragment corresponds to the snake AChR α-subunit. Most differences between the snake and mouse sequences concentrate on the vicinity of these tandem cysteines and might contribute to the unique pharmacological properties of the snake receptor.

We have extended our study to a third snake, the sand boa (*Eryx jaculus*), and cloned and sequenced a 250 bp PCR fragment corresponding to amino acid residues 122–205 (FIG. 2). This nonvenomous snake was chosen because we thought that it might bind α-BTX as has been reported for another boa, *Boa constrictor,*[12] and therefore may differ in the binding site from that of the snakes that do not bind α-BTX. Nevertheless, binding experiments with muscle Triton extracts from this boa did not reveal any significant binding to α-BTX, as is the case with the water

```
        122                 *                                       140     *
        A   I   F   K   S   Y   C   E   I   I   V   T   Y   F   P   F   D   E   Q   N   C
Boa     GCC ATC TTC AAA AGC TAT TGT GAA ATT ATA GTC ACC TAT TTC CCC TTT GAT GAG CAA ACC TGC
Cobra   GCC ATC TTC AAA AGC TAT TGT GAA ATT ATA GTC ACC TAT TTC CCC TTT GAT GAG CAA ACC TGC
        A   I   F   K   S   Y   C   E   I   I   V   T   Y   F   P   F   D   E   Q   N   C

                                150                                     160
        S   M   K   L   G   T   W   T   Y   D   G   T   V   V   A   I   Y   P   E   G   P
Boa     AGT ATG AAG CTG GGA ACT TGG ACT TAC GAT GGC ACC GTG GTC GCC ATC TAC CCG GAA GGT CCC
Cobra   AGT ATG AAG CTG GGA ACT TGG ACT TAC GAT GGC ACC GTG GTT GCC ATC TAC CCG GAA GGT CCC
        S   M   K   L   G   T   W   T   Y   D   G   T   V   V   A   I   Y   P   E   G   P

                            170                                     180
        R   P   D   L   S   K   H   M   Q   S   G   E   W   T   L   K   D   Y   R   G   F
Boa     CGT CCA GAT TTG AGT AAG CAC ATG CAG AGT GGA GAA TGG ACA TTG AAA GAT TAC CGA GGT TTT
Cobra   CGT CCA GAT TTG AGT AAG TAC ATG CAG AGT GGA GAA TGG ACA TTG AAA GAT TAC CGA GGT TTT
        R   P   D   L   S   K   Y   M   Q   S   G   E   W   T   L   K   D   Y   R   G   F

                        190         *   *                           200
        W   H   S   V   N   Y   S   C   C   L   D   T   P   Y   L   D   I   T   Y   H   F
Boa     TGG CAC TCG GTG AAC TAT TCT TGT TGC CTT GAT ACG CCG TAT CTT GAT ATC ACC TAC CAC TTT
Cobra   TGG CAC TCG GTG AAC TAT TCT TGT TGC CTT GAT ACG CCG TAT CTT GAT ATC ACC TAC CAC TTT
        W   H   S   V   N   Y   S   C   C   L   D   T   P   Y   L   D   I   T   Y   H   F
```

FIGURE 2. Alignment of nucleotide and deduced amino acid sequences for the sand boa PCR fragment of the AChR α-subunit, as compared with the corresponding cobra fragment. Amino acid residues are numbered from 122 to 205, corresponding to their position in the mouse AChR α-subunit. Cysteine residues are marked with an asterisk.

snake and boa (FIG. 3). Sequence analysis of the boa fragment revealed a very high similarity to the cobra (FIG. 2) and water snake respective fragments.[28] Inasmuch as the sand boa does not display a different pharmacological specificity from that of the other two snakes, it is not surprising that the sequence of its presumed binding site, close to cysteines 192 and 193, is identical to that of the other toxin-resistant snakes.

Alignment of the snake's PCR-derived sequences and those of the respective segments of muscle AChR α-subunits of other species indicates that the snake sequence is highly homologous to other muscle and not to neuronal AChR, which also does not bind α-BTX (FIG. 4).[28] Among the various muscle AChRs, the snake receptor exhibits some significant differences, most of which occur in the vicinity of cysteines 192 and 193. Several amino acids that are identical in distant species, such as *Torpedo,* chick, mouse, calf, and human are different in snake. Of these,

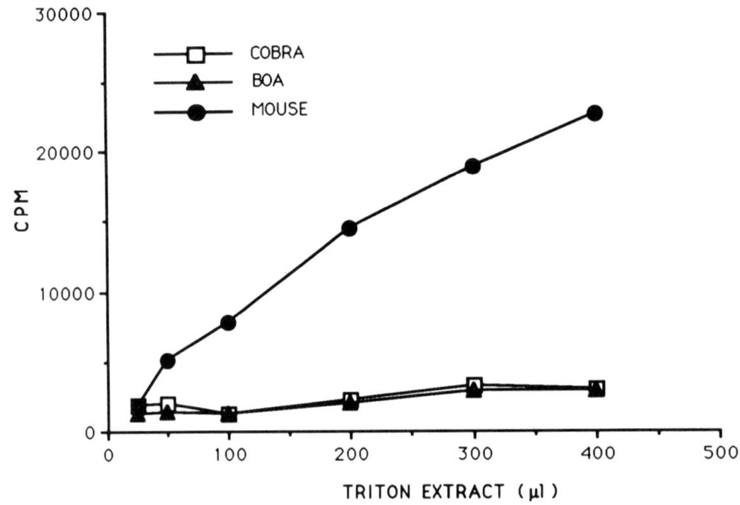

FIGURE 3. Binding of [^{125}I]α-BTX to cobra, boa, and mouse muscle Triton extracts.

substitution of tryptophan 184 for phenylalanine, lysine 185 for tryptophan, and proline 184 for leucine are in close proximity to the two tandem cysteines and might be relevant for binding properties. In addition, of particular interest is Trp 187, which is conserved in *Torpedo,* chick, mouse, and calf but substituted for Ser in the snake as well as in human. We have previously shown that chemical modification of this tryptophan in the synthetic peptide 185–196 of the *Torpedo* α-subunit abolished its binding to α-BTX.[16] Moreover, the synthetic dodecapeptide, corresponding to residues 185–196 of the human α-subunit, which does not contain tryptophan 187, did not bind α-BTX.[16] Finally, the PCR-derived fragments of all three snakes contain a putative *N*-glycosylation site in the binding-site domain at position 189 (Asn 189). This glycosylation site is not present in other known α-subunits. Hence, it is possible that the snake AChR α-subunit might be

```
                            140                 160                   180                200
                  *             *                                                  **
                             .                  .                     .                   .
Cobra       AIFKSYCEIIVTYFPFDEQNCSMKLGTWTYDGTVVAIYPEGPRPDLSNYMQSGEWTLKDYRGFWHSVNYSCCLDTPYLDITYHF
Natrix      ...........................R...........................A............................
Boa         ................................................H...................................
Torpedo     ............H....Q...T....I......K.S.S..SD.....TF.E....VM.....WK.W.Y.T..P...........
Chick       .................Q...............M.V.N..SD......F.E....VM.....WK.W.Y.A..P...........
Mouse       ............H...................S....N..SDQ....NF.E....VI.EA..WK.W.F....PT..........
Calf        ............H...................S..V.N..SDQ.....F.E....VI.ES..WK.W.F.A..PS..........
Human       ............H...................S....N..SDQ.....F.E....VI.ES..WK...T....P...........
```

FIGURE 4. Comparison of amino acid residues 122–205 of AChR α-subunit from three snakes and other species.

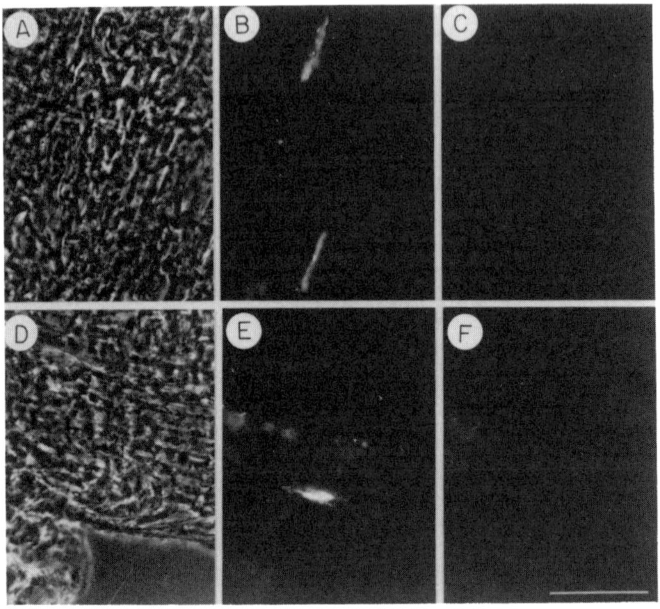
A
B
C
D
E
F

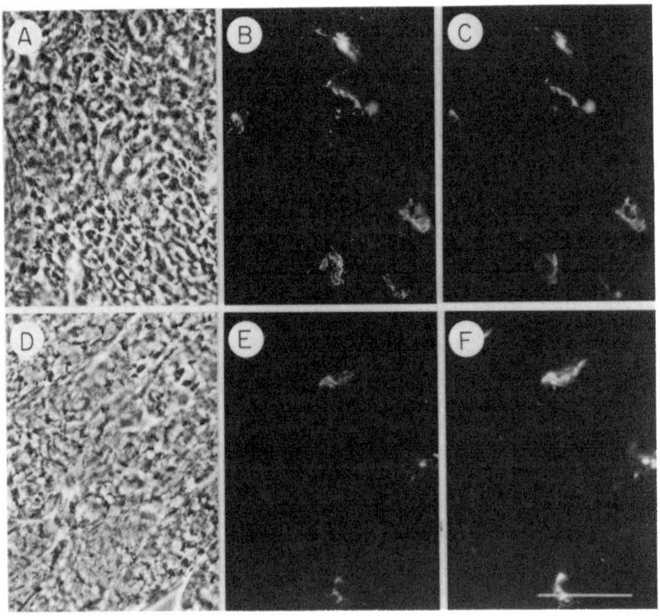
A
B
C
D
E
F

more glycosylated than that of the other AChRs. These changes or part of them may explain the lack of α-BTX binding to snake AChR.

THE BINDING SITE OF THE MONGOOSE AChR

We have noted that some of the sequence differences between the snake and other species' AChR could be specific to the group. Therefore, we have extended our study to a mammal (mongoose) that is resistant to neurotoxins[30] and includes snakes in its diet. We first examined the toxic effect of α-BTX in the mongoose as compared with rabbit and mouse (TABLE 1).[31] Intramuscular administration of α-BTX into mongoose in amounts of 0.3–2 μg/g of body weight did not kill the mongoose, whereas 0.1 and 0.3 μg/g of body weight were found lethal in mice and rabbits, respectively. As has been observed for snakes, the resistance of mongoose to α-BTX was not due to neutralizing factors in the blood serum, because preincubation of mongoose serum with α-BTX did not abolish its toxic effect upon subsequent injection into mice.

We then showed that similarly to the snake, mongoose AChR also does not bind α-BTX. Thus, Triton extracts of mongoose muscle did not bind [^{125}I]α-BTX, whereas extracts of mouse muscle exhibited a specific binding that could be displaced by unlabeled α-BTX or by *d*-tubocurarine.[31]

Histochemical experiments using diaphragm sections demonstrated that mongoose AChR does not bind α-BTX. Acetylcholinesterase and AChR were visualized at the endplate regions, whereas α-BTX did not bind to these regions.[31] AChR in the mongoose end-plate was visualized by two different AChR-specific antibodies (FIG. 5). Both antibodies stained the mongoose end-plate, whereas no staining was observed with α-BTX. By contrast, both the anti-AChR antibodies and α-BTX stained the end-plate regions in sections of the rat diaphragm (FIG. 5).[31]

The results described above suggest that mongoose AChR differs from other, toxin-binding, AChRs in its binding site, as was demonstrated for the snake AChR. We therefore cloned and sequenced a mongoose cDNA fragment that includes the binding-site domain (*i.e.*, the segment from the α-subunit containing the tandem cysteines 192 and 193). The polymerase chain reaction was used to amplify the cDNA fragment, which contains the binding-site region from mongoose single-stranded cDNA. The resulting 250 bp fragment hybridized to the mouse AChR α-subunit cDNA. Sequence analysis of this amplified mongoose fragment (FIG. 6) revealed high homology with the respective mouse fragment, corresponding to amino acid residues 122–205 of the α-subunit (homology of 89% in nucleotides

---

FIGURE 5. Staining of AChR in mongoose (top) and rat (bottom) diaphragm sections. Diaphragm sections were incubated with two types of purified AChR-specific antibodies (2 μg/mL). **A–C:** rabbit antidenatured *Torpedo* AChR antibody;[42] **D–F:** rabbit antipeptide 351–368 of the human AChR α-subunit.[43] Sections were then incubated with a mixture of tetramethylrhodamine (TMR)–α-BTX and FITC-conjugated goat anti-rabbit IgG. **A** and **D** show phase contrast images. **B** and **E** show FITC fluorescence, demonstrating AChR staining with both types of antibodies. **C** and **F** show TMR fluorescence of the same fields, demonstrating that TMR–α-BTX did not stain the mongoose end-plate (top) and that α-BTX staining colocalized with that of AChR in the rat end-plate (bottom). Bar represents 50 μm.

```
          122                     *       140                                     140     *
          A   I   F   K   S   Y   C   E   I   I   V   T   H   F   P   F   D   E   Q   N   C
Mongoose  GCC ATC TTC AAA AGC TAC TGT GAG ATC ATC GTC ACC CAC TTT CCC TTT GAT GAA CAG AAC TGC
Mouse     GCC ATC TTT AAA AGC TAC TGT GAG ATC ATT GTC ACT CAC TTT CCC TTC GAT GAG CAG AAC TGC
          A   I   F   K   S   Y   C   E   I   I   V   T   H   F   P   F   D   E   Q   N   C

                                      150                                     160
          S   M   K   L   G   T   W   T   Y   D   S   S   V   V   V   I   N   P   E   S   D
Mongoose  AGC ATG AAG CTG GGT ACC TGG ACC TAT GAC AGC TCT GTG GTT GTC ATC AAC CCG GAA AGC GAC
Mouse     AGC ATG AAG CTG GGC ACC TGG ACC TAT GAC GGC TCT GTG GTG GCC ATT AAC CCG GAA AGT GAC
          S   M   K   L   G   T   W   T   Y   D   G   S   V   V   A   I   N   P   E   S   D

                                  170                                     180
          Q   P   D   L   S   N   F   M   E   S   G   E   W   V   I   K   E   A   R   G   W
Mongoose  CAA CCT GAC CTA AGC AAC TTC ATG GAA AGC GGA GAG TGG GTG ATC AAG GAG GCC CGG GGC TGG
Mouse     CAG CCC GAC CTG AGT AAC TTC ATG GAG AGC GGG GAG TGG GTG ATC AAG GAA GCT CGG GGC TGG
          Q   P   D   L   S   N   F   M   E   S   G   E   W   V   I   K   E   A   R   G   W

                              190     *   *                           200
          K   H   N   V   T   Y   A   C   C   L   T   T   H   Y   L   D   I   T   Y   H   F
Mongoose  AAG CAC AAT GTG ACC TAC GCC TGC TGC CTC ACC ACC CAC TAC CTG GAC ATC ACC TAC CAC TTC
Mouse     AAG CAC TGG GTG TTC TAC TCC TGC TGC CCC ACC ACT CCC TAC CTG GAC ATC ACC TAC CAC TTC
          K   H   W   V   F   Y   S   C   C   P   T   T   P   Y   L   D   I   T   Y   H   F
```

FIGURE 6. Alignment of nucleotide and deduced amino acid sequences for the mongoose PCR fragment of the AChR α-subunit, as compared with the corresponding mouse fragment. Amino acid residues are numbered from 122 to 205, corresponding to their position in the mouse AChR α-subunit. Cysteine residues are marked with an asterisk.

and 92% in amino acids). The mongoose segment contains the four cysteines at positions 128, 142, 192, and 193, thus verifying that it corresponds to the AChR α-subunit. Interestingly, five of the seven amino acid differences between the mouse and the mongoose fragments concentrate in the vicinity of the tandem cysteines in a stretch of eleven amino acid residues (residues 187–197) and most probably contribute to the inability of the mongoose receptor to bind α-BTX. Three of these five differences are at positions 187, 189, and 194, where major substitutions take place also in the snake AChR; one of them (187) creates a potential *N*-glycosylation site in the mongoose AChR.

PHARMACOLOGICAL AND IMMUNOLOGICAL SPECIFICITY OF THE SNAKE- AND MONGOOSE-EXPRESSED AChR FRAGMENTS

The cloned fragments corresponding to amino acid residues 122–205 of the mongoose, snake, and mouse AChR were expressed in bacteria employing a pET 8C–derived expression vector.[31] The expressed protein fragments were localized in the insoluble pellet, in inclusion bodies. The pellet could bc solubilized in 8 M urea or 6 M guanidine hydrochloride and could be maintained in solution in 4 M urea. Chromatography in SDS-polyacrylamide gel revealed that the expressed protein fragments have the expected molecular weight of 8 kDa and constitute the major protein in the pellet (FIG. 7). Antibodies against a synthetic peptide corresponding to residues 143–158 of the *Torpedo* AChR α-subunit stained all three fragments (FIG. 7 right), indicating that the expressed fragments are indeed from the AChR α-subunit. Overlay of the blotted proteins with $[^{125}I]\alpha$-BTX showed that the toxin binds only to the mouse fragment and not to snake or mongoose fragments (FIG. 7).[31]

Despite the unique pharmacological properties of the snake and mongoose (*i.e.*, toxin binding), the vairous expressed binding-site domains were immunologically cross-reactive, when analyzed with several polyclonal and monoclonal AChR-specific antibodies.[31,32] The snake-, mongoose-, and mouse-expressed domains (residues 122–205), as well as the *Torpedo* and mouse α-subunit, reacted with antibodies raised against gel-purified expressed domain from the mongoose or mouse α-subunit. Similar cross-reactivity was observed also in immunoprecipitation experiments with soluble fragments obtained following *in vitro* transcription and translation, using rabbit reticulocyte lysates.[32] Of special interest is the cross-reactivity with monoclonal antibody 5.5 (mcAb 5.5). This particular monoclonal antibody was developed in our laboratory and was shown to be a highly specific reagent directed against the AChR binding site, in a conformation-dependent manner.[33] In previous reports, we have demonstrated that mcAb 5.5 competes with cholinergic agonists and antagonists, including α-neurotoxins, in receptor binding, and acts directly as a cholinergic antagonist in blocking acetylcholine-induced sodium transport and modifying ion channel properties.[33–35] Our study demonstrated that the α-subunit domains from snake, mongoose, and mouse, expressed in a cell-free system or in bacteria, all recognize mcAb 5.5,[32] suggesting that depsite the physical vicinity or overlapping of the binding sites for acetylcholine and α-BTX, there may be distinct structural requirements for each of them.

DISCUSSION AND CONCLUDING REMARKS

Elucidation of the ligand binding site of AChR is of an utmost importance for understanding of the different physiological functions of this receptor and, in general, of ligand-receptor interactions. We have analyzed the binding site of AChR from animals that are resistant to α-neurotoxins and demonstrated that AChR in these animals have a modified binding site. The mongoose is particularly

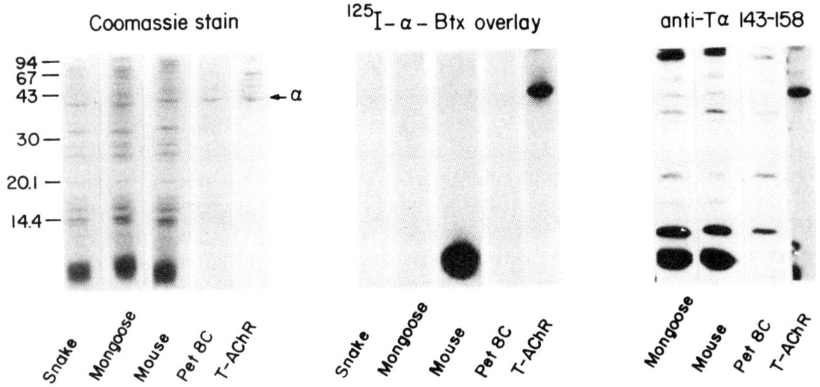

FIGURE 7. Analysis of expressed protein fragments. The expressed protein fragments or purified *Torpedo* AChR were resolved by polyacrylamide gel electrophoresis. The gels were stained for proteins (left panel) or blotted and overlayed with [^{125}I]α-BTX, or with anti-peptide 143–156, followed by [^{125}I]protein A.

interesting because it is very homologous to other mammals and differs from them only in its resistance to α-BTX. Analysis of the binding-site domain of the mongoose AChR indicated that five out of the seven amino acid differences between the mouse and mongoose α-subunit fragment (residues 122–205) cluster in a short stretch of eleven amino acids (residues 187–197) encompassing the two tandem cysteines 192 and 193 at the binding site (FIG. 6). The substitutions are at positions 187 (Trp to Asn), 189 (Phe to Thr), 191 (Ser to Ala), 194 (Pro to Leu), and 197 (Pro to His). Alignment of the mongoose-derived sequence at the ligand binding region with those of other species, including the snake and one neuronal receptor, is depicted in FIGURE 8. Three of the mongoose substitutions are at the same positions (187, 189, and 194), at which major substitutions take place also in the snake. Position 194, adjacent to the tandem cysteines is Pro in the toxin-binders, whereas in the mongoose and snakes it is Leu. This substitution may result in a major conformational change, as proline can form a β-bend or turn. Such a conformational change presumably will not affect the interaction between the receptor and its natural ligand, acetylcholine, but can interfere with the binding of the much larger polypeptide antagonist, α-BTX. It should be noted that in the neuronal AChR, which does not bind α-BTX,[36] as well as in the α7 α-BTX–binding subunit from chick brain,[37,38] there is no proline residue adjacent to the tandem cysteines (corresponding to Pro 194 in the muscle α-subunit). Nevertheless, the

```
           180                  200
                     **
Mongoose   EARGWKHNVTYACCLTTHYLD
Human      .S.....S...S..PD.P...
Calf       .S.....W.F....PS.P...
Mouse      .......W.F.S..P..P...
Chick      DY.....W.Y....PD.P...
Torpedo    DY.....W.Y.T..PD.P...
Snake      DY..FW.S.N.S...D.P...

Rat(PC12)  DAV.TYNTRK.E..AEI-.P.
```

FIGURE 8. Interspecies comparison of amino acid residues 180–200 of AChR α-subunit. Sequence sources are as referred to in Neumann *et al.*[28] All sequences were compared to mongoose.[31]

brain α-BTX–binding protein has a proline residue three residues following the tandem cysteines, which may play a similar role to that played by Pro 194 in the muscle toxin-binding AChRs.

The other two major substitutions are from aromatic residues, at positions 187 and 189 in toxin-binders, to nonaromatic residues in the mongoose and snake. Position 189, which is Tyr or Phe in the toxin-binders, is changed to Thr in the mongoose, Asn in the snakes, and Lys in all neuronal AChRs. At position 187, Trp in the toxin-binding AChRs has been substituted by Asn in the mongoose and Ser in the snake. That Trp 187 is important for toxin-binding is supported by the observation that its chemical modification in the synthetic *Torpedo* peptide 185–196 eliminated the binding activity to α-BTX, and that the homologous human synthetic dodecapeptide with Ser at 187 and Thr at 189 did not bind α-BTX either.[16] A longer synthetic peptide of the human sequence 32-mer (residues 173–204) was reported to bind α-BTX 150 times weaker than the homologous *Torpedo* peptide.[39] Nevertheless, intact human AChR binds α-BTX, and it is possible that conformational factors not present in the synthetic peptides or that other domains in the receptor or the toxin molecule contribute to the toxin-binding activity of the human receptor.

Asn 187 in the mongoose and Asn 189 in the snake AChR α-subunit are both putative *N*-glycosylation sites. It is not known yet whether these asparagines are glycosylated in the intact mongoose and snake receptors, and if they are, whether such a glycosylation contributes to toxin resistance. It should be noted that the nonglycosylated mongoose and snake protein fragments expressed in *E. coli* (FIG. 7), as well as the respective nonglycosylated synthetic peptides (residues 185–196), do not bind α-BTX. It is possible, however, that glycosylation that adds a bulky group in the binding-site domain may provide additional protection toward the toxin without affecting acetylcholine binding. Thus, it seems that Pro at position 194, an aromatic residue (Tyr or Phe) at 189, and Trp at 187 are strong requirements for toxin-binding, as mutations at these three positions in the mongoose and snake AChR abolish their binding to α-BTX. It is still not clear whether all three changes are required and what the contribution of each is in conferring toxin resistance.

The different interaction of α-BTX and mcAb 5.5 with the mouse and mongoose domains suggests that these two reagents recognize different moieties within the cholinergic binding site. The cholinergic binding site is designed to recognize acetylcholine, a small molecule, and it is possible that selective pressure against toxin recognition could design a sequence that recognized acetylcholine but not α-BTX. Our results demonstrate that the amino acid substitutions involved in conferring toxin resistance did not abolish the interaction with mcAb 5.5. Thus, mcAb 5.5 is directed against an overlapping, but not identical, structure within the cholinergic binding site. These results suggest that the specific requirements for the various reagents reacting with this site, such as α-BTX, mcAb 5.5, and acetylcholine, could be distinct from each other. Determination of the specific residues required for mcAb 5.5 recognition, and comparison with the known specificity of acetylcholine binding site, could provide further insight into this question.

Finally, although mongoose and snake AChR do not bind α-BTX, they still retain their cholinergic properties and, therefore, amino acid residues, which are essential for binding of acetylcholine, should be conserved also in these receptors. Indeed, both the mongoose- and snake-binding site domains contain the amino acids that were shown to be labeled by dimethyl-aminobenzene diazonium fluoroborate (DDF),[24,40] maleimido benzyltrimethylammonium (MBTA),[17] and lophotoxin[41] at the same positions as other muscle and neuronal receptor α-subunits. These include Tyr 190, Cys 192 and 193, and Tyr 198. The aromatic residues Trp 149 and Tyr 151, which were reported to be labeled by DDF,[24,40] are also conserved in the mongoose and snake. In conclusion, our study indicates that the requirements for acetylcholine binding are not sufficient for α-BTX binding. Though both bind primarily to the same site in AChR, additional structural elements are necessary for α-BTX binding, and those can be manipulated by genetic pressure without the loss of the major physiological function, which is acetylcholine binding.

REFERENCES

1. KARLIN, A. 1980. *In* The Cell Surface and Neuronal Functions. C. W. Cotman, G. Poste & G. L. Nicolson, Eds.) 191–260. Elsevier/North Holland, Amsterdam.
2. CONTI-TRONCONI, B. M. & M. A. RAFTERY. 1982. Annu. Rev. Biochem. **51:** 491–530.
3. CHANGEUX, J. P., A. DEVILLERS-THIERY & P. CHEMOUILLI. 1984. Science **225:** 1335–1345.
4. NODA, M., H. TAKAHASHI, T. TANABE, M. TOYOSATO, Y. FURUTANI, T. HIROSE, M. ASAI, S. INAYAMA, T. MIYATA & S. NUMA. 1982. Nature **299:** 793–797.
5. NODA, M., Y. FURUTANI, H. TAKAHASHI, M. TOYOSATO, T. TANABE, S. SHINIZU, S. KIKYOTANI, T. KAYANO, T. HIROSE, S. INAYAMA & S. NUMA. 1983. Nature **305:** 818–823.

6. DEVILLERS-THIERY, A., J. GIRAUDAT, M. BENTABOULET & J. P. CHANGEUX. 1983. Proc. Natl. Acad. Sci. USA **80:** 2067–2071.
7. CLAUDIO, T., M. BALLIVET, J. PATRICK & S. HEINEMANN. 1983. Proc. Natl. Acad. Sci. USA **80:** 1111–1115.
8. BOULTER, J., W. LUYTEN, E. EVANS, P. MASON, M. BALLIVET, D. GOLDMAN, S. STENGELIN, G. MARTIN, S. HEINEMANN & J. PATRICK. 1985. J. Neurosci. **5:** 2545–2552.
9. HAGGERTY, J. G. & S. C. FROHNER. 1981. J. Biol. Chem. **256:** 8294–8297.
10. GERSHONI, J. M., E. HAWROT & T. L. LENTZ. 1983. Proc. Natl. Acad. Sci. USA **80:** 4973–4977.
11. PATRICK, J. & W. STALLCUP. 1977. J. Biol. Chem. **252:** 8629–8633.
12. BURDEN, S. Y., M. C. HARTZELL & D. YOSHIKAMI. 1975. Proc. Natl. Acad. Sci. USA **72:** 3245–3249.
13. NEUMANN, D., J. M. GERSHONI, M. FRIDKIN & S. FUCHS. 1985. Proc. Natl. Acad. Sci. USA **82:** 3490–3493.
14. FUCHS, S., D. NEUMANN, A. SAFRAN, M. SOUROUJON, D. BARCHAN, M. FRIDKIN, J. M. GERSHONI, R. MANTEGAZZA & S. PIZZIGHELLA. 1986. Ann. N.Y. Acad. Sci. **505:** 256–271.
15. NEUMANN, D., D. BARCHAN, A. SAFRAN, J. M. GERSHONI & S. FUCHS. 1986. Proc. Natl. Acad. Sci. USA **83:** 3008–3011.
16. NEUMANN, D., D. BARCHAN, M. FRIDKIN & S. FUCHS. 1986. Proc. Natl. Acad. Sci. USA **83:** 9250–9253.
17. KAO, P., A. DWORK, R. KALDANY, M. SILVER, J. WEIDEMAN, S. STEIN & A. KARLIN. 1984. J. Biol. Chem. **259:** 11662–11665.
18. PEDERSEN, S. E., E. B. DREYER & J. B. COHEN. 1986. J. Biol. Chem. **261:** 13735–13743.
19. WILSON, P. T., T. L. LENZ & E. HAWROT. 1985. Proc. Natl. Acad. Sci. USA **82:** 8790–8794.
20. MULAC-JERICEVIC, B. & M. Z. ATASSI. 1986. FEBS Lett. **199:** 68–74.
21. GOTTI. C., G. MAZZOLA, R. LONGHI, D. FORNASARI & F. CLEMENTI. 1987. Neurosci. Lett. **82:** 113–119.
22. BARKAS, T., A. MAURON, B. ROTH, C. ALLIOD, S. J. TZARTOS & M. BALLIVET. 1987. Science **235:** 77–80.
23. GERSHONI, J. M. 1987. Proc. Natl. Acad. Sci. USA **84:** 4318–4321.
24. DENNIS, M., J. GIRAUDAT, F. KOTZYBA-HIBERT, M. GOELDNER, C. HIRTH, J. Y. CHANG, C. LAZURE, M. CHRETIEN & J. P. CHANGEUX. 1988. Biochemistry **27:** 2346–2357.
25. CHANGEUX, J. P. 1990. Trends Pharmacol. Sci. **11:** 485–492.
26. TZARTOS, S. J. & M. S. REMOUNDOS. 1990. J. Biol. Chem. **256:** 21462–21.
27. OHANA, B. & J. M. GERSHONI. 1990. Biochemistry **29:** 6409–6415.
28. NEUMANN, D., D. BARCHAN, M. HOROWITZ, E. KOCHVA & S. FUCHS. 1989. Proc. Natl. Acad. Sci. USA **86:** 7255–7259.
29. MISHINA, M., T. TOBIMATSU, K. IMOTO, K. TANAKA, Y. FUJITA, K. FUKUDA, M. KURASAKI, H. TAKAHASHI, Y. MORIMOTO, T. HIROSE, S. INAYAMA, T. TAKAHASHI, M. KUNO & S. NUMA. 1985. Nature **313:** 364–369.
30. OVADIA, M. & E. KOCHVA. 1977. Toxicon **15:** 541–548.
31. BARCHAN, D., S. KACHALSKY, D. NEUMANN, Z. VOGEL, M. OVADIA, E. KOCHVA & S. FUCHS. 1992. Proc. Natl. Acad. Sci. USA **89:** 7717–7721.
32. KACHALSKY, S., M. ALADJEM, D. BARCHAN & S. FUCHS. 1993. FEBS Lett. In press.
33. MOCHLY-ROSEN, D. & S. FUCHS. 1981. Biochemistry **20:** 5920–5924.
34. SOUROUJON, M. C., D. MOCHLY-ROSEN, A. S. GORDON & S. FUCHS. 1983. Muscle & Nerve **6:** 303–311.
35. GOLDBERG, G., D. MOCHLY-ROSEN, S. FUCHS & Y. LASS. 1983. J. Membr. Biol. **76:** 123–128.
36. BOULTER, J., K. EVANS, D. GOLDMAN, G. MARTIN, D. TRECO, S. HEINEMANN & J. PATRICK. 1986. Nature (Lond.) **319:** 368–374.
37. SCHOEPFER, R., W. G. CONROY, P. WHITING, M. GORE & J. LINDSTROM. 1990. Neuron **5:** 35–48.

38. Couturier, S., D. Bertrand, J. M. Matter, M. C. Hernandez, S. Bertrand, N. Millar, S. Valera, T. Barkas & M. Ballivet. 1990. Neuron **5:** 847–856.
39. Wilson, P. T. & T. L. Lentz. 1988. Biochemistry **27:** 6667–6674.
40. Galzi, J. L., F. Revah, A. Bessis & J. P. Changeux. 1991. Ann. Rev. Pharmacol. **31:** 37–72.
41. Abramson, S. N., Y. Li, P. Culver & P. Taylor. 1989. J. Biol. Chem. **264:** 12666–12672.
42. Bartfeld, D. & S. Fuchs. 1977. FEBS Lett. **77:** 214–218.
43. Fuchs, S., D. Neumann, A. Safran, S. Pizzighella, R. Mantegazza, M. P. Daniels & Z. Vogel. 1987. Biochemistry **26:** 4611–4616.

Physostigmine and Neuromuscular Transmission[a]

ALFRED MAELICKE, T. COBAN, A. SCHRATTENHOLZ, B. SCHRÖDER, S. REINHARDT-MAELICKE, A. STORCH, J. GODOVAC-ZIMMERMANN, CHRISTOPH METHFESSEL,[b] E. F. R. PEREIRA,[c] AND EDSON X. ALBUQUERQUE[c]

Laboratory of Molecular Neurobiology
Institute of Physiological Chemistry and Pathobiochemistry
Johannes-Gutenberg-University Medical School
Duesbergweg 6, D-6500
Mainz, Germany

[b]Bayer AG
Leverkusen, Germany

[c]Department of Pharmacology and Experimental Therapeutics
University of Maryland Medical School
Baltimore, Maryland

INTRODUCTION

Myasthenia gravis (MG) is an autoimmune disease that affects postsynaptic nicotinic acetylcholine receptors (nAChRs) of the neuromuscular junction.[1] It has been shown that the density distribution of end-plate nAChRs of myasthenic patients is much lower than that of healthy suspects, resulting in a significant decrease in the junctional sensitivity to acetylcholine (ACh).[2,3] Electrophysiological and ultrastructural evidence suggest that this is due to the presence of anti-nAChR immunoglobulins in myasthenic sera.[4–6] The diagnosis and therapy of MG still rely on the acute and chronic use of carbamates,[7,8] which are believed to improve cholinergic transmission by inhibition of hydrolysis of acetylcholine by acetylcholinesterase. Typically, carbamates with a quaternary amino group, such as neostigmine, were employed to affect neuromuscular transmission,[9] whereas those with a ternary amino group, such as physostigmine (eserine, Phy), act predominantly at central cholinergic synapses.[10]

Although it is widely accepted that the anticholinesterase properties of carbamates are responsible for their therapeutic efficacy in the treatment of MG, recent studies by our laboratories suggest that the direct action of carbamates on nAChRs may be of similar if not even higher importance. Direct effects of carbamates on nAChR channel activity have been reported repeatedly in the literature. Katz and Miledi[11] observed as early as 1977 that Phy displays agonistic action at muscle end-plates. Bloch and Stallcup[12] reported in 1979 an agonistic action of neostigmine

[a] This work was supported by Grants from the Deutsche Forschungsgemeinschaft (to A. Maelicke; Ma 599/12), the Fonds der Chemischen Industrie (to A. Maelicke), the U.S. Army Medical Research and Development Command Contract DAMD-17-88-C-8119 (to E. X. Albuquerque), and the National Institutes of Health (NS29296; to E. X. Albuquerque).

at cultured mammalian muscles. We demonstrated in 1985 that some carbamates, at concentrations as low as 0.1 μM, are capable of inducing single channel currents with amplitudes and open times typical for the nAChR channel at end-plates of single muscle fibers isolated from the interosseal muscle of the frog *Rana pipiens*.[13] At higher concentrations, the same carbamates act as open channel blockers.[14,15]

Recently we have observed that the direct action of carbamates on nAChRs is accompanied by an unexpected pharmacology. By studying ligand-induced time-resolved cation flux into sealed vesicles of *Torpedo marmorata* electrocytes, we showed that the Phy-induced channel activity was not blocked by established nicotinic competitive antagonists.[16] Moreover, Phy was capable of activating the nAChR channel even when the receptor was preincubated with desensitizing concentrations of ACh.[17] These data suggested that Phy may act by way of binding site(s) distinct from those for acetylcholine and established cholinergic agonists and competitive antagonists. Therefore, in addition to the general significance of these findings for our understanding of cholinergic transmission, they may offer novel therapeutic approaches to diseases manifested in a malfunction of peripheral or central nAChR ion channels.

In this report, we have brought together some of the key findings of our previous studies. In addition, we show that neither the channel-activating action of Phy, nor its peculiar pharmacology, is confined to a single (exotic) tissue but rather is a general property of nAChRs from many species and tissues. We also discuss the structural location of the Phy binding site in regard to the ACh site on the α-polypeptide of the nAChR, and we propose consideration of the action of some carbamates and other nAChR ligands in research and medical applications.

METHODS

Biochemical Studies

[Benzene ring-(n)-^{3}H](-)physostigmine hemisulphate (1.69 Ci/mmol = 62.5 GBq/mmol) was purchased from Amersham Buchler; benzoquinonium was a gift from Sterling-Whinthrop. All other chemicals were from established sources (see previous publications from our laboratories).

nAChR-rich membrane vesicles were prepared from *Torpedo marmorata* electric tissue according to Duguid and Raftery,[18] with the modifications described by Reinhardt *et al.*[19] If required, further purification was achieved by pH 11 extraction as described by Neubig *et al.*[20] The receptor concentration of the purified membrane suspension was approximately 15 μM in terms of ACh binding sites, at a protein concentration of five milligrams per millilitre.

Photolabeling of nAChR by Phy was achieved as described elsewhere.[21] Briefly, nAChR-enriched *Torpedo* membrane fragments (TMF) (approx. 2 μM in terms of ACh binding sites) were incubated with 50 μM [^{3}H]Phy, 10 mM Tetram, 10 mM sodium fluoride, 1 mM 8-azido ATP, and 10 mM ouabain in 100 mM Tris/HCl, pH 8.5, in a total volume of 0.5–1.0 mL for 1 h at 4°C in the dark. The mixture was then flashed twice (flash duration, 4 ms) at maximal intensity (approx. 2000 J) employing a Bläsing IM 7000 photolysis instrument (Raytest, Essen) equipped with a XFS6 285/1000Q quartz lamp. To reduce irradiation damage of the protein, the light was passed through a Schott WG 295 cut-off filter before reaching the quartz cuvette (1 cm light path) containing the reaction mixture.

The subunit specificity of Phy labeling was determined by SDS-PAGE followed by semidry blotting onto Trans-Blot nitrocellulose membranes (BIO-RAD) and autoradiography. The [^{3}H]Phy-labeled nAChR α-subunit was purified by preparative PAGE, concentrated by ultrafiltration, and submitted to cleavage by trypsin (Boehringer). The reaction was stopped by the addition of phenylmethylsulfonyl fluoride (PMSF). The cleavage mixture was submitted to tricine/SDS-PAGE as described by Schägger *et al.*,[22] using a 16.5% (6% cross-linker) separating gel containing 6 M urea. After electrophoresis, the gels were electroblotted onto polyvinylidenefluoride (PVDF) membranes (Immobilon-P, Millipore), and the radioactively labeled peptides were identified by autoradiography. The labeled band(s) were cut out, and the N-terminal amino acid sequence was determined directly from the PVDF membrane. Automated Edman degradation was carried out on an Applied Biosystems pulsed liquid-phase sequencer (ABI 477A).

Ion flux studies were carried out as described previously.[16] Briefly, purified membrane vesicles from *Torpedo marmorata* were loaded with the fluorophore PTS prior to rapid mixing in a stopped-flow fluorometer with a solution containing Cs^+ and a nAChR channel–activating ligand. Activation of the channel causes influx of Cs^+ into the vesicles and concomitant heavy metal quenching of the 1,3,6,8-pyrene tetrasulfonate (PTS) fluorescence. Time-resolved studies of PTS fluorescence quenching, therefore, represent the kinetics of cation influx through *Torpedo* nAChR. Blocking of channel activity was achieved by incubation of the PTS-loaded vesicles with appropriate concentrations of antagonist prior to rapid mixing with a channel-activating drug.

Electrophysiological Studies

Single channel studies of *Torpedo* nAChR were performed by C. Methfessel using giant membrane vesicles prepared by a procedure based on that of Riquelme *et al.*[23] Artificial liposomes were obtained by sonication of a mixture of 50 mg Azolectin (Serva) and 5 mg cholesterol suspended in 1 mL H_2O. The lipid suspension was mixed with *Torpedo* membrane fragments at a ratio (v/v) of 9 : 1 employing a glass potter followed by sonication. Tetram was added to a final concentration of 50 micromoles. After gel permeation centrifugation through Sephadex G25M, a cycle of dehydration and rehydration[23] was performed. Giant liposomes were collected with a Pasteur pipet and applied to the patch-clamp chamber.

Cell Culture of Hippocampal Neurons

The technique for obtaining neuronal hippocampal cultures has been described elsewhere.[24] Briefly, pregnant rats (17–19 days) were anesthetized with dry ice and sacrificed by cervical dislocation. The cerebral hemispheres were removed from 6–8 fetuses and placed in cold physiological solution. Their hippocampi were dissected, minced, and incubated in trypsin (0.25%) for 30 min at 35.5°C. Mechanical dissociation was achieved with fluxes of culturing solution consisting of modified Eagle medium (MEM, GIBCO), 10% fetal calf serum, 10% inactivated horse serum, glutamine (1 mM), and DNAse (40 μg/mL). The dissociated cells were plated on petri dishes coated with collagen. The dishes were incubated in the water-saturated atmosphere of 10% CO_2/90% air at 35°C, and 24 hours later the culturing solution was replaced with another lacking DNAse and fetal calf serum. After 7 days, 5-fluoro-2′-deoxyuridine/uridine (FDUR), an inhibitor of base cell growth, was added to the culture medium for 24 hours. We used 6 to 60-day-old cultures for patch-clamp experiments.

The procedure to obtain myoball cultures was adapted from that described by Giller *et al.*[25] for mouse tissues. Briefly, hind limb muscles were removed from 1–2-day-old neonatal rats and minced and incubated in trypsin (0.25%) for 30 min at 35.5°C. Then, a procedure similar to that described above was followed to dissociate mechanically the muscle fibers. The dissociated cells were plated on petri dishes coated with collagen. The dishes were incubated in the water-saturated atmosphere of 10% CO_2/90% air at 35°C, and 24 hours later the culturing medium was replaced by another lacking DNAse and fetal calf serum. After seven days, FDUR was added to the culture medium for 24 hours to arrest the proliferation of fibroblasts. Patch-clamp experiments were performed on 1–2-week-old cultured cells.

Patch-clamp Recordings and Data Analysis

Single-channel currents were recorded from outside-out patches excised either from cultured hippocampal neurons or from cultured myoballs using the high-resolution patch-clamp techniques.[26] The external bath solution consisted of (in mM) NaCl-165, KCl-5, $CaCl_2$-2 glucose-10, and HEPES − 5 (pH = 7.2; 330 mosM); and the internal pipet solution consisted of (in mM) CsCl − 80, CsF − 80, EGTA − 10, and HEPES − 10 (pH = 7.3; 340 mosM). A perfusion system consisting of multibarrel glass capillaries connected to several syringes was used to apply the pharmacological agents to the patches. This system allowed a fast exchange of solutions and, thereby, a better quantification of fast effects of the drugs to be tested. When the outside-out patches were obtained from the cultured neurons, the micropipette with the patch was placed inside one of the capillaries through which external solution containing atropine (1 μM), 2-amino-5-phosphono-valeric acid (50 μM), and tetrodotoxin (TTX) (0.3 μM) was delivered. A similar procedure was followed after obtaining the outside-out patches from the cultured myoballs, except that the external solution contained only TTX (0.3 μM). After ensuring that there were no "spontaneous" channel openings, the micropipette was moved to the other capillaries from which the pharmacological agents to be tested were delivered. All recordings were made at room temperature (20–25°C) using an LM-EPC-7 patch-clamp system (List Electronics). Following a modulation, the data were stored on video-cassette tape for later computer analysis. The data were filtered at 3 kHz (Bessel), digitized at 12.5 kHz, and analyzed on an IBM AT microcomputer. The IPROC-2 program[27] was used to analyze the single channel currents in order to obtain the amplitude, frequency of openings, and mean open, closed, and burst times. The open events were considered valid when the amplitude decreased to beyond 50% of the amplitude estimated. The burst times represent all of the openings separated from the adjacent opening by an interval greater than 6 ms (in recordings obtained from outside-out patches excised from cultured hippocampal neurons) or 1 ms (in recordings obtained from cultured myoballs).

RESULTS AND DISCUSSION

In this work, a novel aspect regarding the activation and function of the nAChR has been unveiled. Our present studies demonstrate that Phy is the prototype of

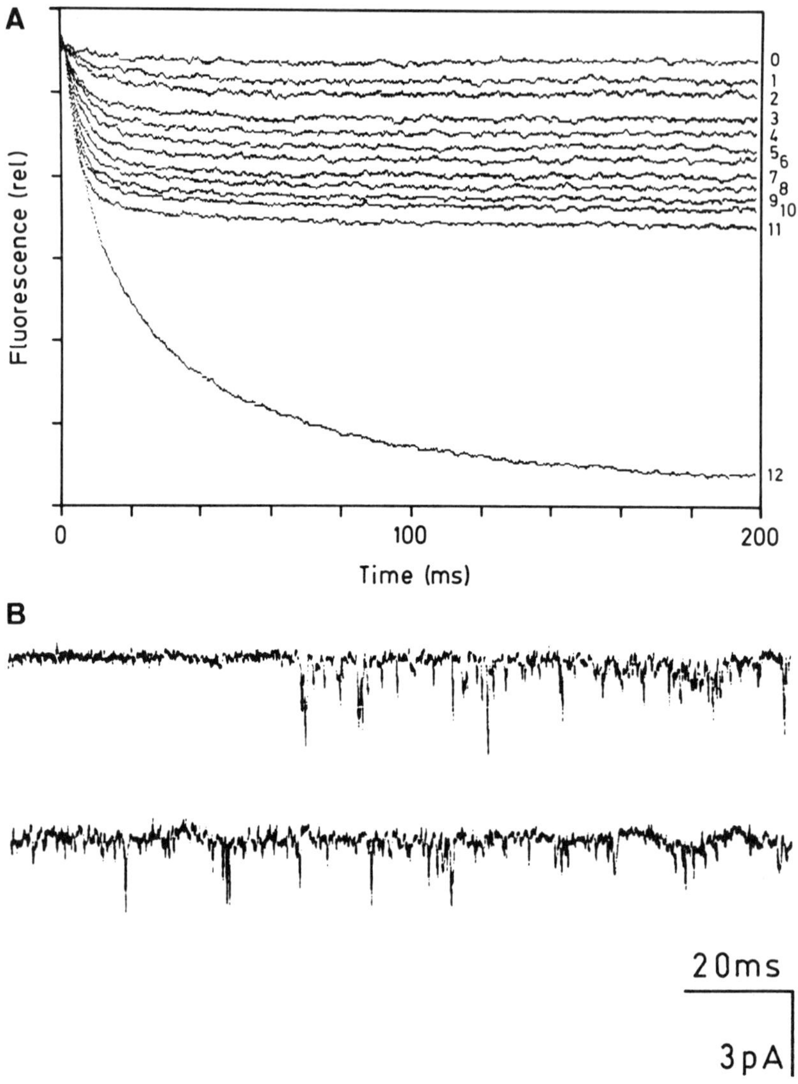

FIGURE 1. Activation by physostigmine of nAChR channels from *Torpedo marmorata* electric tissue. **A:** *Torpedo* membrane vesicles loaded with 1,3,6,8-pyrene tetrasulfonate (PTS) and suspended in Na^+/HEPES buffer were rapidly mixed with Cs^+/HEPES buffer in the absence (trace 0) and presence of (−)-physostigmine (traces 1–11). The change in fluorescence was recorded as a function of time. Final concentrations of Phy in the reaction mixture were 0, 62.5 μM, 125 μM, 250 μM, 500 μM, 750 μM, 1 mM, 1.5 mM, 2 mM, 3 mM, 4 mM, 5 mM, and 6 mM, respectively. Half-maximal response was observed at 400–800 μM of Phy. As a control, the dye-loaded vesicles were disrupted by the addition of 2% dimethylsulfoxid so that the total amount of vesicle-entrapped dye was exposed to the heavy metal quencher (trace 12). **B:** Samples of single-channel current recordings from reconstituted giant vesicles containing *Torpedo* nAChR. The concentration of Phy in the bath solution was 100 μM at a membrane potential of −80 mV.

a new class of channel-activating ligand at the nAChR. Using several preparations, we have demonstrated biochemically and electrophysiologically that Phy can activate the nAChR of several species and tissues by binding to sites distinct from those for acetylcholine.

Torpedo electrocyte membranes, representing a tissue particularly rich in nAChR, are advantageous for biophysical and biochemical studies of nicotinic cholinergic transmission.[28] Employing time-resolved ion flux studies (FIG. 1A) and single-channel current recordings (FIG. 1B), it was demonstrated that *Torpedo* nAChR is susceptible to the channel-activating action of Phy. This action of the drug is independent of its carbamate function, as eseroline induces channel activation with the same characteristics and potency.

The unusual pharmacology of channel activation by Phy is demonstrated in FIGURE 2. Uninhibited Cs^+ flux is observed even after preincubation of the membrane vesicles with saturating concentrations of α-bungarotoxin (α-BGT) (FIG. 2A), that is, under conditions at which ACh did not activate the nAChR channel. Similarly, channel activation by Phy is not inhibited by the ACh-antagonistic monoclonal antibody WF6[29] and low molecular weight ACh antagonists such as *d*-tubocurarine, gallamine, and hexamethonium.[16] By contrast, the monoclonal antibody FK1, which was raised against the *Torpedo* nAChR, blocks the channel-activating action of Phy (FIG. 2B), with little, if any, effect on either ACh binding to the nAChR or ACh-induced channel activation.

Preincubation of *Torpedo* membrane vesicles with a large concentration of ACh leads to desensitization of the nAChR, as is shown in FIGURE 2C by the inability of a pulse of ACh to induce channel activity. Under identical experimental conditions, Phy was still able to induce influx of Cs^+, the total amplitude of induced fluorescence quenching and the kinetics of ion flux remaining practically unchanged as compared to control experiments in the absence of ACh (FIG. 2C).[17]

Photoaffinity labeling with radioactively labeled Phy demonstrated that only the α-subunit of *Torpedo* nAChR is involved in Phy binding (FIG. 3A). By means of epitope mapping studies[30] employing a set of synthetic peptides, each of approximately 20 amino acids in length, covering overlappingly the whole primary structure of the *Torpedo* nAChR α-subunit, the region α118-137 was identified as containing major element(s) of the Phy binding site (FIG. 3B).[31] Purification on a preparative scale of radioactively labeled α polypeptide, proteolytic digestion and gel electrophoretic separation of the labeled peptide, and Edman degradation identified lys-125 as an element within or in close proximity to the Phy binding site. The binding site for ACh has been mapped to a region around the adjacent cysteines α192 and α193,[32-35] with other attachment points further upstream in the α subunit N-terminal region. None of the studies reported so far has indicated the region around lys-125 as containing essential elements of the ACh binding site. Thus, the binding sites for ACh and Phy reside in the N-terminal extracellular region of the same subunit of the nAChR, albeit in separate locations. This agrees with the distinct pharmacology of the two drugs reported in FIGURE 2.

Residue 125 of the α-subunit is located just a few positions in front of the disulfide-bridged loop Cys-128-Cys-142, containing a prominent consensus sequence in all nAChR polypeptides sequenced and the fairly common *N*-glycosylation site at Gln-141. Lys-125 (and its nearest neighbors) is conserved in most nAChR α polypeptides, but is conspicuously absent in non-α polypeptides.[36] These homology considerations prompted us to investigate other nAChR species as to whether they also bind physostigmine.

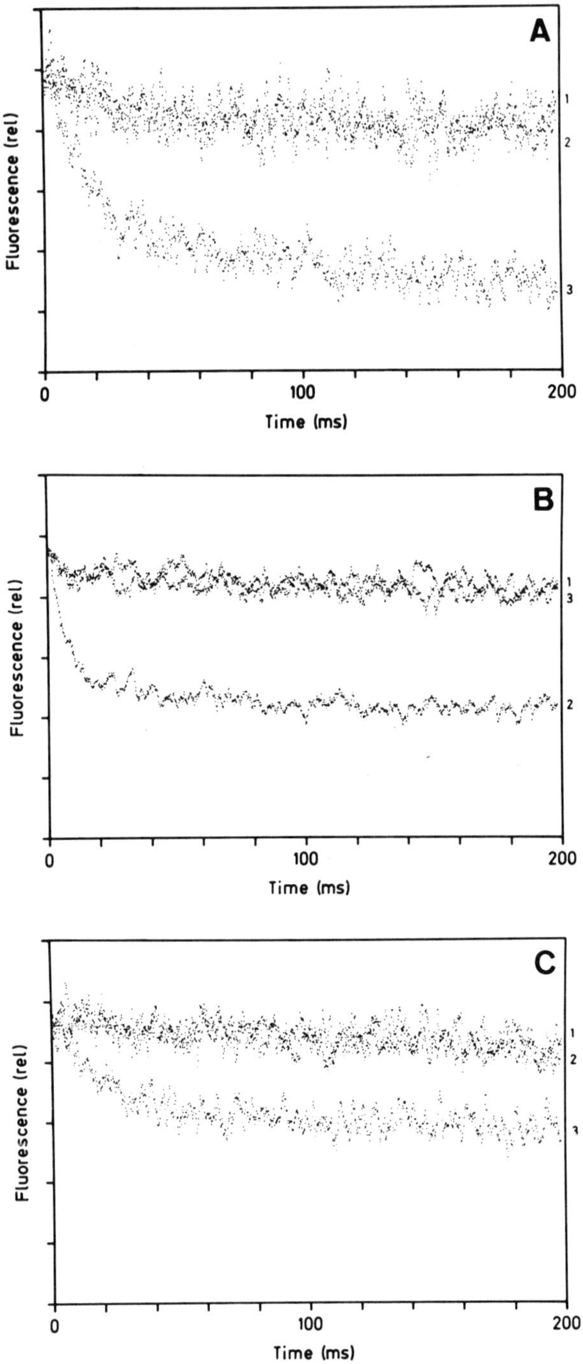
A
Fluorescence (rel)
1
2
3
0
100
200
Time (ms)
B
Fluorescence (rel)
1
3
2
0
100
200
Time (ms)
C
Fluorescence (rel)
1
2
3
0
100
200
Time (ms)

Interestingly, we have found that Phy activates the mammalian neuronal and muscle nAChR channel. The general significance of the observation originally obtained with frog muscle[13] and the *Torpedo* system[16] was tested by electrophysiological techniques applied to a variety of other cholinergic tissues. We have recently described the presence of nAChRs sensitive to α-BGT on cultured fetal rat hippocampal neurons.[24,37,38] Also, we have proven that methyllycaconitine (MLA), a potent neurotoxin extracted from the seeds of *Delphinium brownii,* is an important probe to study this receptor. MLA was shown to be a highly selective, potent, and reversible nicotinic antagonist at neuronal nAChRs sensitive to α-BGT.[39] Here, in outside-out patches excised from cultured hippocampal neurons, we have demonstrated that Phy is capable of activating the ion channel of neuronal nAChR (Fig. 4). The current-voltage relationship revealed that the single channel currents activated by Phy had at least two conductance values, 23 and 46 pS, the 46 pS single channel currents being the most predominant. Further analysis of the kinetic properties of the single channel currents evoked by Phy displayed that the mean open time of the currents was shortened, whereas the mean burst time was elongated when the concentrations of Phy were increased from 1 to 5 μM (TABLE 1). These findings are in agreement with Phy being an open channel blocker. We also recorded single channel currents activated by Phy in outside-out patches obtained from cultured rat myoballs (data not shown). The Phy agonist action on the nAChRs of either rat myoballs, rat hippocampal neurons, or frog muscles preserved a similar pharmacology as the Phy-induced response on the *Torpedo* preparation. In all these nicotinic systems, the channel activation by Phy could not be blocked by competitive nicotinic antagonists (FIG. 5). Following a pattern similar to that seen with vesicles of *Torpedo,* benzoquinonium, a competitive and noncompetitive blocker of the nAChR, was able to block the agonist actions of Phy on outside-out patches obtained from the hippocampal neurons. Benzoquinonium decreased the frequency of Phy-activated single-channel currents and shortened their lifetimes. These effects were voltage- and concentration-dependent (FIG. 6 and TABLE 2). These data altogether suggest that we have discovered a general property of nAChRs.

It is, therefore, tempting to speculate on the physiological significance of the newly discovered binding site at the nAChR as a site involved in some kind

FIGURE 2. Pharmacology of *Torpedo* nAChR channel activation by physostigmine. **A:** Preincubation with saturating concentration of α-BGT does not inhibit nAChR channel activation by Phy. PTS-loaded *Torpedo* membrane vesicles were incubated with 200 μM α-BGT for 30 min prior to rapid mixing with Cs^+ Hepes buffer containing 200 μM Phy. Trace 1: "leakage flux" in the absence of an activating ligand; trace 2: 20 μM ACh; trace 3: 200 μM Phy. α-BGT, at the experimental conditions used, is capable of blocking channel activation by ACh but not by Phy. **B:** Inhibition of Phy-induced Cs^+ influx by monoclonal antibody FK1. Under the same experimental conditions as described above for α-BGT, saturating concentrations of the antibody FK1 were applied, and the Phy-induced Cs^+ influx was measured. Trace 1: leakage flux; trace 2: Phy-induced flux in the absence of FK1; trace 3: blockade of Phy-induced flux by FK1. **C:** Physostigmine activates desensitized *Torpedo* nAChR. Dye-loaded membrane vesicles were incubated for 15 min with 200 μM acetylcholine (ACh) followed by rapid mixing with an equal volume of Cs^+ buffer with (trace 2) or without (trace 1) 100 μM ACh. Only leakage flux was observed, suggesting that the majority of channels remained closed due to desensitization. When ACh-pretreated membrane vesicles were rapidly mixed with Phy (200 μM), however, uninhibited Cs^+ influx was observed (trace 3).

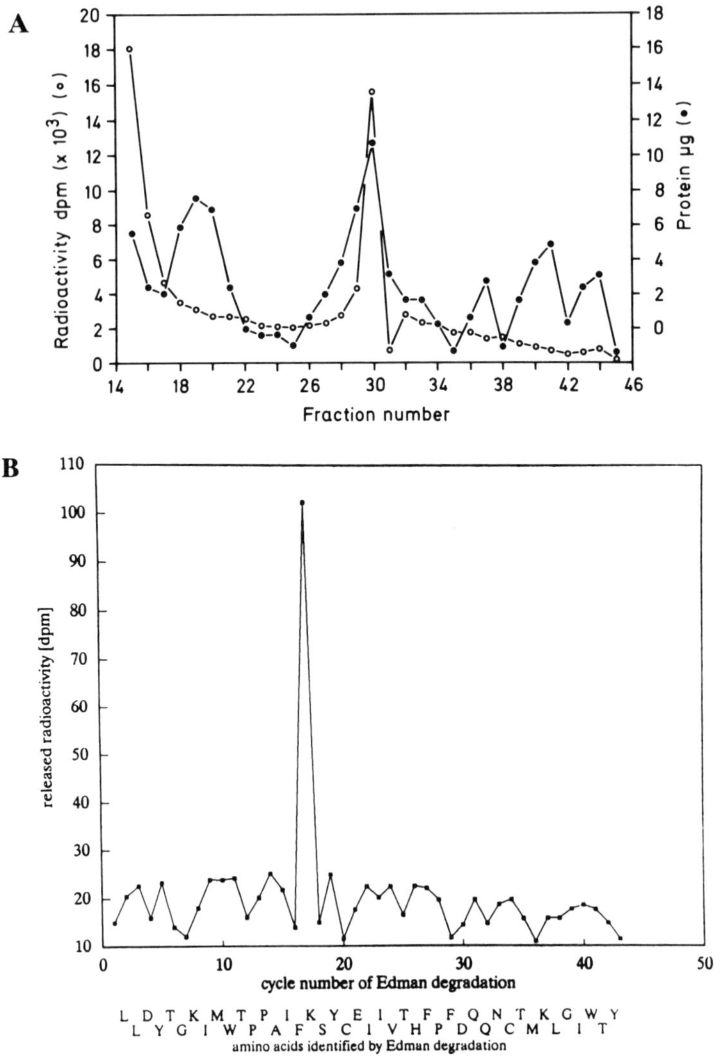

FIGURE 3. Localization of the binding site of physostigmine within the primary structure of *Torpedo* nAChR. **A:** Separation of [^{3}H]Phy-labeled *Torpedo* nAChR α-polypeptide by preparative SDS-PAGE. A preparative batch of *Torpedo* membrane fragments was photolabeled with [^{3}H]Phy, and the reaction mixture was submitted to preparative SDS-PAGE. Of the radioactivity associated with protein bands, the majority was found in the first peak of eluted protein, corresponding to the nAChR α-polypeptide. **B:** Automated Edman degradation of the [^{3}H]Phy-labeled peptide fragment. The radioactive peptide identified by autoradiography of the high resolution SDS gel (see **A**) was blotted onto a PVDF membrane, and the N-terminal amino acid sequence was determined directly from the membrane employing a pulsed liquid-phase sequencer. The following sequence was determined:

LLDYTGKIMWTPPAIFKSYCEIIVTHFPFDQQNCTMKLGIWTY

It corresponds to residues 109-151 of the primary structure of *Torpedo* nAChR α-subunit. As shown by the elution profile, only cycle 17 of the Edman degradation, corresponding to Lys-125, was associated with a significant release of radioactivity.

A

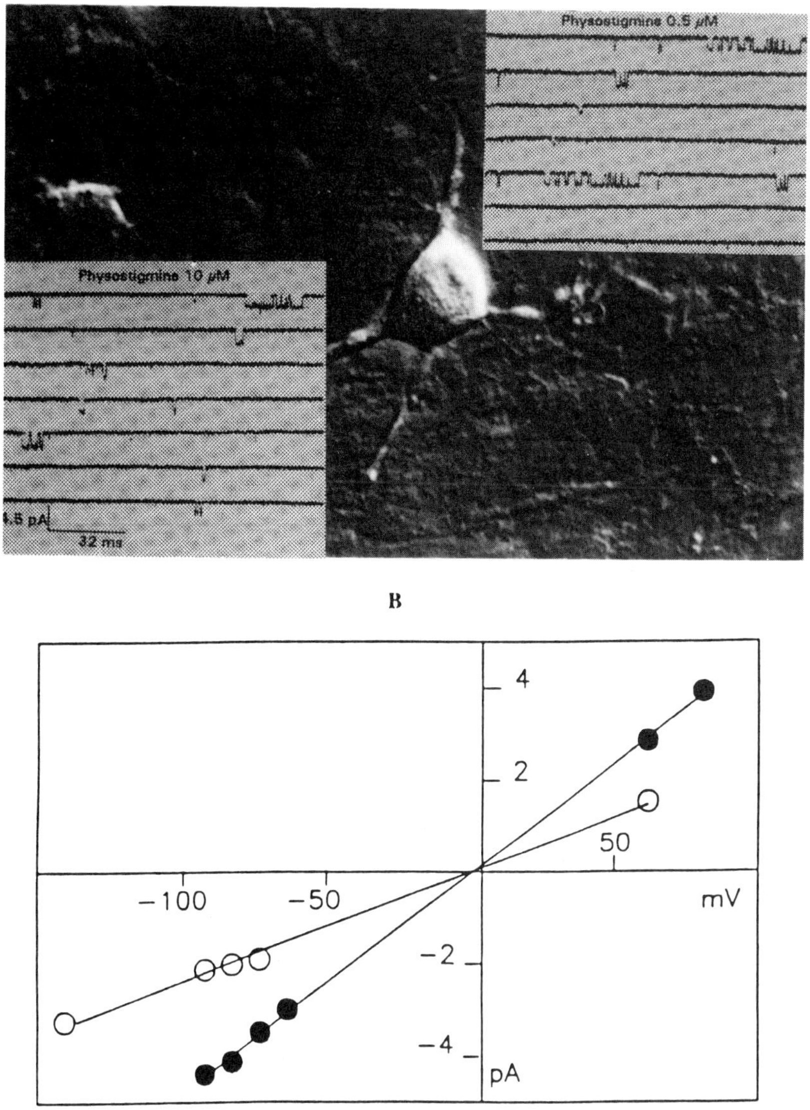

FIGURE 4. **A:** Plate showing a fetal rat hippocampal neuron grown in culture for 15 days from which an outside-out patch was excised. Insets show single channel currents activated by 0.5 and 10 μM Phy at −80 mV. **B:** Current voltage relationship of the Phy-evoked single channel currents in the outside-out patches obtained from fetal rat hippocampal neurons grown in culture for 15–25 days.

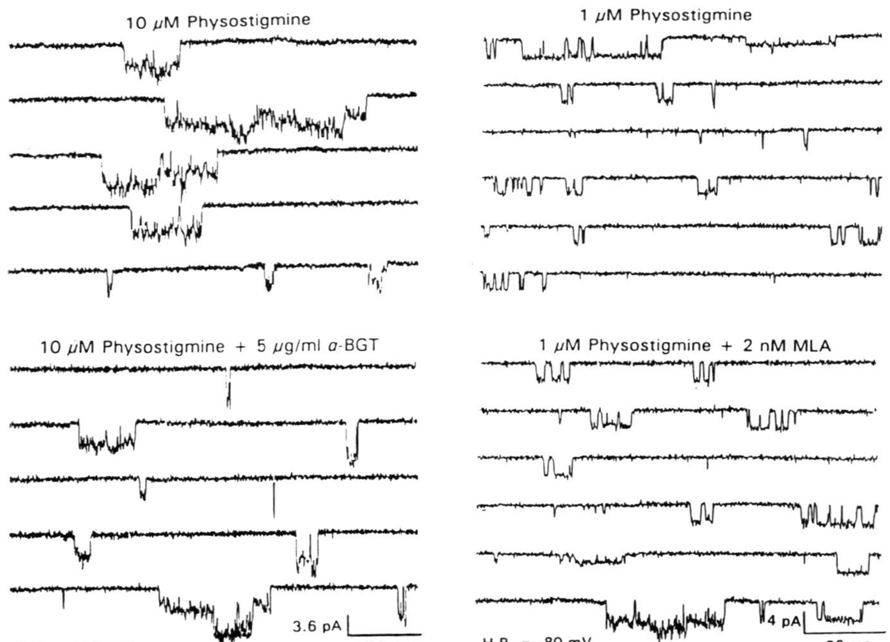

FIGURE 5. Left column: Single-channel currents activated by 10 μM Phy in an isolated frog muscle fiber under cell-attached conditions were not blocked after 1.5 h of incubation of the preparation with 5 μg/mL α-BGT. Right column: Single-channel currents activated by 1 μM Phy in an outside-out patch excised from a cultured hippocampal neuron were not blocked by 2 nM MLA, a nicotinic antagonist that blocks completely, reversibly, and instantaneously the (+) anatoxin-α–evoked single channel currents at 1 pM.[38] H.P. = holding potential.

TABLE 1. Kinetic Parameters of the 46 pS Single Channel Currents Activated by Phy (1.0 and 5.0 μM) in Outside-out Patches Excised from Cultured Fetal Rat Hippocampal Neurons at -80 mV[a]

[Physostigmine]	τ_{open} (ms)	τ_{closed} (ms)	τ_{burst} (ms)
1 μM	0.20 ± 0.05	0.17 ± 0.05	1.19 ± 0.15
	1.93 ± 0.10	1.46 ± 0.25	5.32 ± 0.42
5 μM	0.09 ± 0.02	0.10 ± 0.04	1.05 ± 0.30
	1.11 ± 0.21	1.55 ± 0.24	10.2 ± 1.80

[a] Results are mean ± SEM of three experiments.

of regulatory process for the activation of the nAChR. In the context of the topic of this symposium, it is particularly significant that the Phy-induced channel activity is independent of the state of the ACh binding site, whether blocked by antagonist or desensitized by ACh itself. Compounds with the properties of physostigmine, therefore, may be useful in overcoming the adverse effects of a direct blockade, for example, by antibodies, of nAChR channel activation by way of the classical ACh sites. As the involvement of direct blocking antibodies in the development of myasthenic symptoms has neither been proven nor disproven, our observations provide new incentives to thoroughly analyze this possibility. Finally, it is worthwhile to mention that we have already identified several other drugs, the majority of them without the capability of active serine alkylation, that activate the nAChR channel by way of the novel regulatory site described in this communication.

SUMMARY

Single channel studies carried out in cultured rat myoballs and cultured hippocampal neurons, and ion flux studies performed on *Torpedo* electrocyte membrane vesicles, showed that physostigmine (Phy), a well-established acetylcholinesterase inhibitor, interacts directly with nicotinic acetylcholine receptors (nAChR). Low

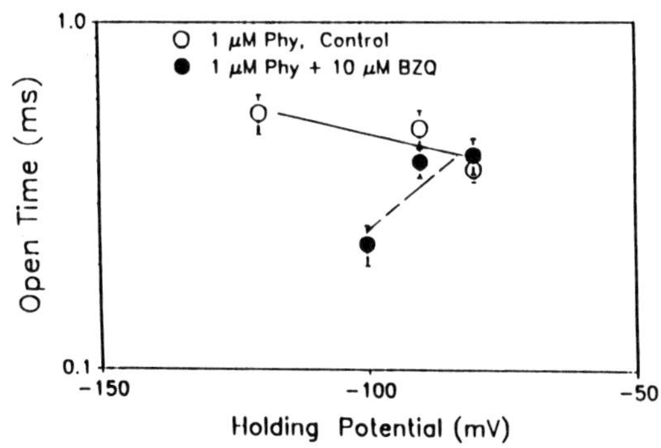

FIGURE 6. The plot shows the voltage-dependent effect of 10 μM benzoquinonium on the 23-pS single-channel currents activated by 1 μM Phy. The results represent mean ± SEM of three experiments.

concentrations (0.1 μM) of Phy activate the receptor integral channel, whereas higher concentrations blocked the channel in its opened state. In contrast to channel activation by acetylcholine (ACh) and classical cholinergic agonists, however, Phy was capable of activating the nAChR channel even when the ACh binding sites were blocked by competitive antagonists, such as α-neurotoxins and *d*-tubocurarine, or when the nAChR was desensitized by preincubation with high concentrations of ACh. The binding site at which Phy binds and activates the nAChR was mapped. It was located within the N-terminal extracellular region of the α-polypeptide, in close proximity to the binding site of the natural transmitter. These data identify a novel binding site at nAChRs from many species and tissues that may be involved in receptor regulatory processes.

TABLE 2. Blockade of Phy-evoked Single Channel Currents in Outside-out Patches Excised from Cultured Hippocampal Neurons by Benzoquinonium[a]

1 μM Phy + X μM BZQ	τ_{open} (ms)	Frequency of channel openings (% of control)
X = 0.0	0.61 ± 0.06	100
X = 0.1	0.63 ± 0.04	57 ± 3
X = 1.0	0.46 ± 0.02	63 ± 5
X = 10	0.39 ± 0.03	15 ± 4

[a] The TABLE shows the concentration-dependent effects of benzoquinonium on the frequency and mean open time of 28 pS single channel openings induced by 1 μM Phy at a fixed holding potential (−100 mV).

ACKNOWLEDGMENTS

The synthetic peptides employed in the epitope mapping studies of antibody FK1 were generously provided by Dr. B. M. Conti-Tronconi (St. Paul, MN). We thank Gerhard Goellner for help in the preparation and analysis of some *Torpedo* membrane preparations. The technical assistance by Birgit Graeff, Barbara Marrow, and Mabel A. Zelle is gratefully acknowledged. We thank Drs. Ch. Franke (Munich) and B. W. Krüger (Bayer AG, Leverkusen) for helpful discussions.

REFERENCES

1. PATRICK, J. & J. LINDSTROM. 1973. Autoimmune response to acetylcholine receptor. Science **280:** 871–872.
2. FAMBROUGH, D. M., D. B. DRACHMAN & S. SATYAMURTI. 1973. Neuromuscular junction in myasthenia gravis: decreased acetylcholine receptors. Science **182:** 293–295.
3. ALBUQUERQUE, E. X., J. E. RASH, R. F. MAYER & J. R. SATTERFIELD. 1976. An electrophysiological and morphological study of the neuromuscular junction in patients with myasthenia gravis. Exp. Neurol. **51:** 536–563.
4. RASH, J. E., E. X. ALBUQUERQUE, C. S. HUDSON, R. F. MAYER & J. R. SATTERFIELD. 1976. Studies on human myasthenia gravis: Electrophysiological and ultrastructural evidence compatible with antibody attachment to the acetylcholine receptor. Proc. Natl. Acad. Sci. USA **73:** 4584–4588.

5. SAHASHI, K., A. G. ENGEL, J. M. LINDSTROM, E. H. LAMBERT & V. A. LENNON. 1978. Ultrastructural localization of immune complexes (IgG and C_3) at the end-plate in experimental autoimmune myasthenia gravis. J. Neuropathol. Exp. Neurol. **37:** 212–213.
6. RASH, J. E., C. S. HUDSON, R. F. MAYER, W. F. GRAHAM, T. J. A. JOHNSON & F. D. GIDDINGS. 1981. Freeze-fracture, labelled-replica, and electrophysiological studies of junctional fold destruction in myasthenia gravis and experimental autoimmune myasthenia gravis. Ann. N.Y. Acad. Sci. **377:** 38–59.
7. WALKER, M. B. 1934. Treatment of myasthenia gravis with physostigmine. Lancet **1:** 1200–1201.
8. ENGEL, A. G., E. H. LAMBERT & T. SANTA. 1973. Study on long-term anticholinesterase therapy: Effects on neuromuscular transmission and on motor end-plate fine structure. Neurology **23:** 1273–1281.
9. HERMANN JR., C. 1988. Myasthenia gravis. *In* Conn's Current Therapy, R. E. Rakel, Ed.: 794–800. Saunders. Philadelphia.
10. NILSSON, L., A. ADEM, J. HARDY, B. WINBLAD & A. NORDBERG. 1987. Do tetrahydroaminoacridine (THA) and physostigmine restore acetylcholine release in Alzheimer brains via nicotinic receptors? J. Neural Transm. **70:** 357–368.
11. KATZ, B. & R. MILEDI. 1977. A reexamination of curare action at the motor endplate. Proc. R. Soc. Lond. B. **196:** 59–72.
12. BLOCH, R. J. & W. B. STALLCUP. 1979. Agonist action of neostigmine on acetylcholine receptors of cultured mammalian muscle. Brain Res. **172:** 378–381.
13. SHAW, K.-P., Y. ARACAVA, A. AKAIKE, D. L. RICKETT & E. X. ALBUQUERQUE. 1985. The reversible cholinesterase inhibitor physostigmine has channel-blocking and agonist affects on the acetylcholine receptor-ion channel complex. Mol. Pharmacol. **28:** 527–538.
14. AKAIKE, A., S. R. IKEDA, N. BROOKES, G. J. PASCUZZO, D. K. RICKETT & E. X. ALBUQUERQUE. 1984. The nature of the interactions of pyridostigmine with the nicotinic acetylcholine receptor-ionic complex channel II. Patch clamp studies. Mol. Pharmacol. **25:** 102–112.
15. ALBUQUERQUE, E. X., A. AKAIKE, K.-P. SHAW & D. L. RICKETT. 1984. The interaction of anticholinesterase agents with the acetylcholine receptor-ionic channel complex. Fund. Appl. Toxicol. **4:** S27–S33.
16. OKONJO, K. O., J. KUHLMANN & A. MAELICKE. 1991. A second pathway of activation of the nicotinic acetylcholine receptor ion channel. Eur. J. Biochem. **200:** 671–677.
17. KUHLMANN, J., K. O. OKONJO & A. MAELICKE. 1991. Desensitization is a property of the cholinergic binding region of the nicotinic acetylcholine receptor, not of the receptor-integral ion channel. FEBS Lett. **279:** 216–218.
18. DUGUID, J. R. & M. A. RAFTERY. 1973. Fractionation and partial characterization of membrane particles form *Torpedo californica* electroplax. Biochemistry **12:** 3593–3597.
19. REINHARDT, S., H. SCHMIADY, B. TESCHE & F. HUCHO. 1984. Functional and structural analysis of acetylcholine receptor-rich membranes after negative staining. FEBS Lett. **173:** 217–221.
20. NEUBIG, R. R., N. D. BOYD & J. B. COHEN. 1982. Conformations of *Torpedo* acetylcholine receptor associated with ion transport and desensitization. Biochemistry **21:** 3460–3467.
21. SCHRATTENHOLZ, A., J. GODOVAC-ZIMMERMANN, H.-J. SCHÄFER, E. X. ALBUQUERQUE & A. MAELICKE. 1992. Photoaffinity labeling of *Torpedo* acetylcholine receptor by the reversible cholinesterase inhibitor physostigmine. Proc. Natl. Acad. Sci. USA. Submitted.
22. SCHÄGGER, H. & G. VON JAGOW. 1987. Tricine-sodium dodecyl sulfate polyacrylamide gel electrophoresis for the separation of proteins in the range from 1 to 100 kDa. Anal. Biochem. **166:** 368–379.
23. RIQUELME, G., E. LOPEZ, L. M. GARCIA-SEGURA, L. A. FERRAGUT & L. M. GONZALES-ROS. 1990. Giant liposomes: A model system in which to obtain patch-clamp recordings of ion channels. Biochemistry **29:** 11215–11222.

24. Aracava, Y., S. S. Deshpande, K. L. Swanson, H. Rapoport, S. Wonnacott, G. Lunt & E. X. Albuquerque. 1987. Nicotinic acetylcholine receptor in cultured neurons from the hippocampus and brain stem of the rat characterized by single channel recording. FEBS Lett. **222:** 63–70.
25. Giller, Jr., E. L., J. H. Neale, P. N. Bullock, B. K. Schrizer & J. P. G. Nelson. 1977. Choline acetyltransferase activity of spinal cord cell cultures increased by co-culture with muscle and by muscle conditioned medium. J. Cell Biol. **74:** 16–19.
26. Hamill, O. P., A. Marty, E. Neher, B. Sakmann & F. J. Sigworth. 1981. Improved patch-clamp techniques for high resolution current recording from cells and cell-free membrane patches. Pfluegers Arch. **391:** 85–100.
27. Sachs, F., J. Neil & N. Barkakati. 1982. The automated analysis of data from single ionic channels. Pfluegers Arch. **395:** 331–349.
28. Raftery, M. A., S. Blanchard, J. Elliot, P. Hautig, H. P. Moore, U. Quast, M. Schimerlic, V. Witzemann & W. Wu. 1979. Properties of *Torpedo californica* acetylcholine receptor. Adv. Cytopharmacol. **3:** 159–182.
29. Fels, G., R. Plümer-Wilk, M. Schreiber & A. Maelicke. 1986. A monoclonal antibody interfering with binding and response of the acetylcholine receptor. J. Biol. Chem. **261:** 15746–15754.
30. Conti-Tronconi, B. M., F. Tang, B. M. Diethelm, S. R. Spencer, S. Reinhardt-Maelicke & A. Maelicke. 1990. Mapping of a cholinergic binding site by means of synthetic peptides, monoclonal antibodies and α-bungarotoxin. Biochemistry **29:** 6221–6230.
31. Schröder, B., S. Reinhardt-Maelicke, S. Diethelm, B. M. Conti-Tronconi & A. Maelicke. 1992. Mapping of the binding site for physostigmine within the primary structure of the *Torpedo* nAChR α-subunits. FEBS Lett. In press.
32. Kao, P., A. Dwork, R. Kaldany, M. Silver, J. Wideman, S. Stein & A. Karlin. 1984. Identification of the α-subunit half-cystein specifically labeled by an affinity reagent for the acetylcholine receptor-binding site. J. Biol. Chem. **259:** 11662–11665.
33. Dennis, M., J. Giraudat, F. Kotzyba-Hibert, M. Goeldner, C. Hirth, J.-Y. Chang, C. Lazure, M. Chretien & J.-P. Changeux. 1988. Amino-acids of the *Torpedo marmorata* acetylcholine-receptor α-subunit labeled by a photoaffinity ligand for the acetylcholine binding-site. Biochemistry **27:** 2346–2357.
34. Galzi, J.-L., F. Revah, D. Black, M. Goeldner, C. Hirth & J.-P. Changeux. 1990. Identification of a novel amino acid α-tyrosine 93 within the cholinergic ligand-binding sites of the acetylcholine receptor by photoaffinity labeling. Additional evidence for a three-loop model of the cholinergic ligand-binding sites. J. Biol. Chem. **265:** 10430–10437.
35. Conti-Tronconi, B. M., B. M. Diethelm, X. Wu, F. Tang, T. Bertazzon, B. Schröder, S. Reinhardt-Maelicke & A. Maelicke. 1991. α-Bungarotoxin and the competing antibody WF6 interact with different amino acids within the same cholinergic subsite. Biochemistry **30:** 2575–2584.
36. Maelicke, A. 1988. Structure and Function of the Nicotinic Acetylcholine Receptor. *In* Handbook of Experimental Pharmacology. V. P. Whittaker, Ed. **86:** 267–313. Springer-Verlag. Berlin and Heidelberg.
37. Albuquerque, E. X., A. C. S. Costa, M. Alkondon, K.-P. Shaw, A. S. Ramoa & Y. Aracava. 1990. Functional properties of nicotinic and glutamatergic receptors. J. Recep. Res. **11:** 603–625.
38. Alkondon, M. & E. X. Albuquerque. 1991. Initial characterization of the nicotinic acetylcholine receptors in rat hippocampal neurons. J. Recep. Res. **11:** 1001–1021.
39. Alkondon, M., E. F. R. Pereira, S. Wonnacott & E. X. Albuquerque. 1992. Blockade of nicotinic currents in hippocampal neurons defines methyllycaconitine as a potent and specific receptor antagonist. Mol. Pharmacol. **41:** 802–808.

Degradation of Acetylcholine Receptors at Vertebrate Neuromuscular Junctions[a]

MIRIAM M. SALPETER,[b] JACOPO ANDREOSE,[c]
JAMES P. O'MALLEY,[b] RUFENG XU,[b]
GUIDO FUMAGALLI,[c] AND TERJE LOMO[d]

[b]*Department of Neurobiology and Behavior*
Cornell University
Ithaca, New York 14853

[c]*Department of Pharmacology*
Center for the Study of Peripheral Neuropathy
University of Milan
Milan, Italy

[d]*Institute of Neurophysiology*
Oslo, Norway

BACKGROUND ON R_s AND R_r AChRs

Recent studies from our laboratories on AChR degradation suggest the presence of two populations of junctional AChR in vertebrate muscle. These two populations differ not only in their degradation rate, but also in the response of that degradation rate to denervation, reinnervation, and muscle electrical activity. AChRs of the first population (which we will refer to as slowly degrading AChRs, or R_s) are predominantly synthesized in innervated muscle. After insertion into the postsynaptic membrane, their degradation half-life ($t_{1/2}$) can be modulated reversibly, being 3–5 days in denervated and 8–11 days in innervated or reinnervated muscle.[1–4] AChRs of the second population (which we will refer to as rapidly degrading AChRs, or R_r) are inserted into the postsynaptic membrane of denervated muscle and have a $t_{1/2}$ of ~1 day.[5] Unlike the R_s, reinnervation of denervated muscle does not slow the degradation rate of the R_r once they are in the muscle membrane.[6]

In an extensive collaboration involving investigators from Norway, Italy, and the United States, we have undertaken a series of studies designed to understand the differences in the regulation of the degradation rates of these two AChR populations. We are separately examining embryonic muscle in tissue culture as well as innervated and denervated adult muscle *in vivo* or in organ culture. The variables examined include reinnervation of denervated muscle, exogenous factors such as dibutyryl cAMP (dB-cAMP), and muscle activity, both spontaneous and induced by electrical stimulation. The studies to be described have all been done using rodent muscles, including rat soleus, mouse sternomastoid, and mouse

[a] This work was supported by Grants from the NIH, NS09315 and GM10422 (to M. M. Salpeter); A.S.I. Grant 91-RS-35 and Theleton-Italy project: Nerve-muscle interactions (to G. Fumagalli); the Norwegian Research Council for Science and Humanities (to T. Lomo); short-term fellowship from the European Neurobiology Network (to J. Andreose); and an MDA postdoctoral fellowship (to J. P. O'Malley).

diaphragm muscles for the *in vivo* or organ culture studies, and rat leg muscles for the tissue culture studies. Although some of the detailed degradation half lives and time dependence of postdenervation changes vary slightly for the different muscles, the emerging principles are amazingly similar.

For adequate background we will first describe the events occurring in an adult muscle in response to denervation (FIG. 1). The normal adult neuromuscular junction (nmj) contains a population of R_s that has a degradation $t_{1/2}$ of ~8 to 11 days. If the muscle is denervated and the R_s are labeled at the same time (so that only receptors present in the membrane before denervation can be seen by radioactive assays), the degradation $t_{1/2}$ for the first 3 to 10 days after denervation (the exact time depending on the muscle used) is ~10 days and thus similar to that of innervated nmjs. After this initial period the R_s accelerate to a $t_{1/2}$ of about 3 to 5 days, which is maintained until they are fully degraded. If the nerve, however, is allowed to regenerate before the R_s AChRs are fully degraded, these accelerated receptors become stabilized back to a $t_{1/2}$ of ~10 days.[3] The stabilization of the accelerated R_s can be reproduced in organ culture by adding membrane-permeable cAMP analogues to the organ culture bath.[7]

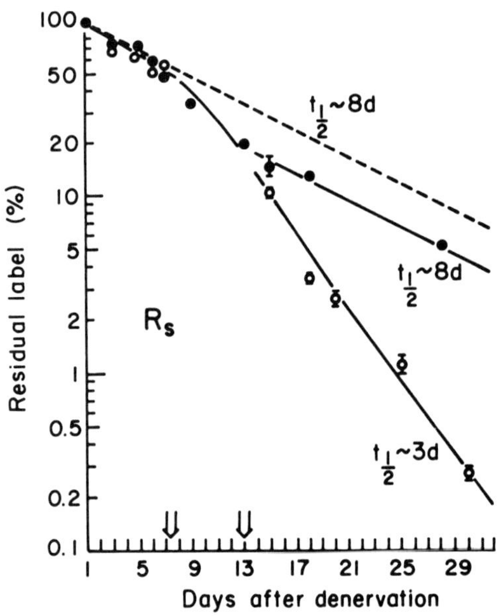

FIGURE 1. Degradation rate of R_s AChRs depends on innervation. R_s AChRs of mouse sternomastoid muscle were saturated with [^{125}I]α-BGT prior to denervation. Dashed line gives the degradation rate of R_s AChRs in control innervated muscle ($t_{1/2}$ ~ 8 d). If muscle is denervated and not allowed to reinnervate, the degradation of the R_s accelerates to a $t_{1/2}$ of about 3 d (○). If, however, the nerve is crushed and allowed to regenerate (which occurred between days 7 to 14 in this study, as marked by arrows), the prelabeled accelerated R_s AChRs are restabilized (●). Adapted from Salpeter *et al.*[3]

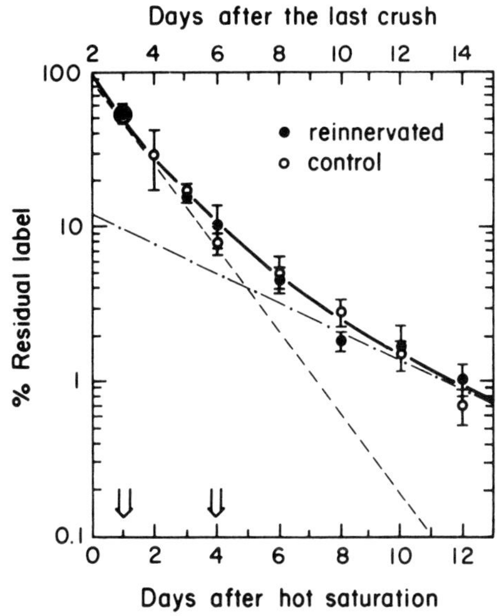

FIGURE 2. Degradation rate of R_r AChRs is not affected by reinnervation. Mouse sternomastoid muscles were denervated by being crushed every second day for 14 days; the nerve was then either allowed to regenerate (●) or prevented from regenerating (○). All AChRs were saturated with nonradioactive α-BGT four days after the first crush and with $[^{125}I]\alpha$-BGT two days after the last crush, before reinnervation began. Reinnervation of the muscle occurs between days 3 and 6 after the last crush in this multiple crush experiment (marked by arrows). No significant difference was seen between the degradation rates of the reinnervated versus the nonreinnervated control. Note that neither degradation curve can be fit to a single exponential but consists of two populations: about 10% of the AChRs (–·–·) have a $t_{1/2}$ of ~3 d (*i.e.* like the accelerated form of the R_s), and about 90% of the AChRs (–––) have a $t_{1/2}$ of ~1 d and are R_r. The solid line is the sum of these two components and gives a good fit to the experimental data. (The curve was not extended long enough to establish whether any stabilization occurred in the small accelerated R_s component after reinnervation.) (S. L. Shyng & M. M. Salpeter.[6] With permission from the *Journal of Neuroscience*.)

Once the R_s are degraded in a denervated nmj, they are predominantly replaced by the rapidly degrading AChRs (R_r; $t_{1/2}$ ~ 1–2 d). In addition to the predominant R_r population, about 10% of the junctional AChRs at the denervated nmj are accelerated R_s with a $t_{1/2}$ of ~3–5 days.[6] Thus both R_r (~90%) and R_s (~10%) are made in denervated muscle, but in the absence of nerve, the R_s are in the accelerated form.

The junctional R_r have entirely different stability properties from that of the R_s (FIG. 2). They degrade with a $t_{1/2}$ of ~1–2 days, as embryonic AChRs do, and are not stabilized by either reinnervation or dB-cAMP.[7] If chronic electrical stimulation is applied to long-term denervated muscles, in which the junctional population is predominantly R_r, a stable population of AChRs appears in the membrane.[8] In light of our previous inability to stabilize R_r, as well as our unpublished experiments to be described below, we interpret these results to mean that during the period of electrical stimulation, the synthesis of R_r is down-regulated and the R_r are replaced by stabilized R_s.

R_r AND R_s IN TISSUE CULTURE CELLS

As indicated above, denervated nmjs contain predominantly R_r but have an ~10% accelerated R_s population (*i.e.* $t_{1/2}$ ~ 3–5 days).[6] Our unpublished results

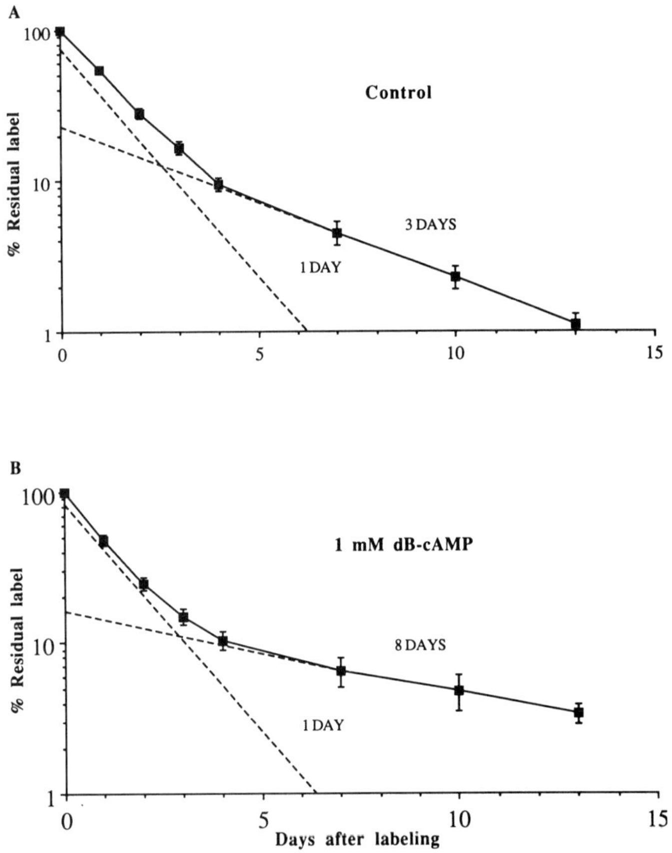

FIGURE 3. Effect of dB-cAMP on AChR degradation in primary rat myotubes in culture. AChRs in control myotubes **(A)** or myotubes treated with dB-cAMP were saturated with nonradioactive α-bungarotoxin (α-BGT) on day 3 of treatment, followed by $[^{125}I]\alpha$-BGT one day later. The degradation of these AChRs inserted between days 3 and 4 was followed for 13 days. A two-component degradation curve was seen with 80% of the AChRs degrading with a $t_{1/2}$ of 1 day (in both **A** and **B**) and 20% degrading with a $t_{1/2}$ of 3 days in control myotubes **(A)** and with a $t_{1/2}$ of 8 days in dB-cAMP–treated muscles **(B)**. These results show that dB-cAMP stabilizes the R_s but not the R_r population of AChRs in cultured myotubes.

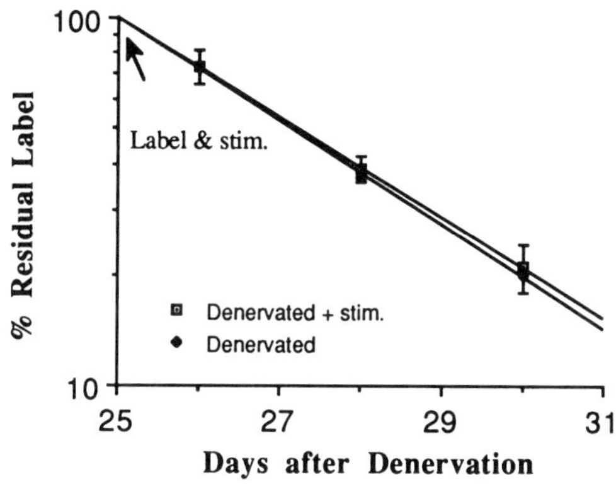

FIGURE 4. Electrical stimulation does not affect R_r degradation rate. Twenty-five days after denervation of rat soleus muscles (when most of the R_s has degraded and was replaced by R_r), the end-plate AChRs were labeled to saturation with [^{125}I]α-BGT. Electrodes were implanted, and the right soleus muscle was electrically stimulated until the muscles were removed. End-plate specific radioactivity was measured by γ-counting. Each time point represents the mean ± SD of 4 animals. The $t_{1/2}$ of the R_r AChRs in the rat soleus muscle was 2.1 days on the denervated muscles and 2.2 days on the denervated and stimulated soleus muscles, indicating that the R_r degradation rate cannot be modulated by muscle activity. The simulation consisted of a train of 60 stimuli at 100 Hz, once every 60 seconds. Similar results were obtained using a different stimulation pattern (a train of 200 stimuli at 20 Hz every 15 seconds).

indicate that a similar phenomenon is seen in rat primary myotubes from newborn pups grown in tissue culture (FIG. 3). The exponential degradation curve of AChRs in cultured muscles has two components. In addition to the R_r ($t_{1/2} \sim 1$ day) usually reported for these cells, which constitute 80 to 90% of the labeled AChRs, there is also a 10 to ~20% slow component ($t_{1/2} \sim 3$ days), which is like the accelerated R_s seen in denervated junctions. As in denervated muscle, the relatively small R_s component of myotube AChRs becomes stabilized to a $t_{1/2}$ of ~10 days in response to cAMP analogues. cAMP analogues, however, do not stabilize the R_r population, which retains its $t_{1/2}$ of 1 day (O'Malley, J. P., L. L. Rubin, and M. M. Salpeter, unpublished).

ROLE OF ELECTRICAL ACTIVITY IN STABILIZING AChRs

Our findings on the role of electrical activity are to some extent consistent with previous results but also contain unexpected aspects. If the R_r in long-term denervated soleus muscles are labeled prior to initiation of *in vivo* electrical stimulation, the $t_{1/2}$ value remains unchanged (FIG. 4). Thus, as expected from the fact that R_r are not stabilized in response to reinnervation or cAMP analogues, the R_r are also not stabilized by electrical stimulation.

The response of the R_s to electrical stimulation is more complicated. These results are summarized in FIGURE 5. If the R_s are labeled just prior to denervation, and electrical stimulation is initiated immediately, then the acceleration of the R_s is prevented, just as reported by Brenner and Rudin.[9] These results are consistent with previous reports that muscle paralysis leads to acceleration of the R_s.[10] We find, however, that after the R_s have already accelerated, electrical stimulation (unlike reinnervation or cAMP analogues) cannot reverse the process.[10a]

A similar phenomenon is seen in the tissue-cultured myotubes. In these myotubes spontaneous contractile activity begins at around day 6 in culture.

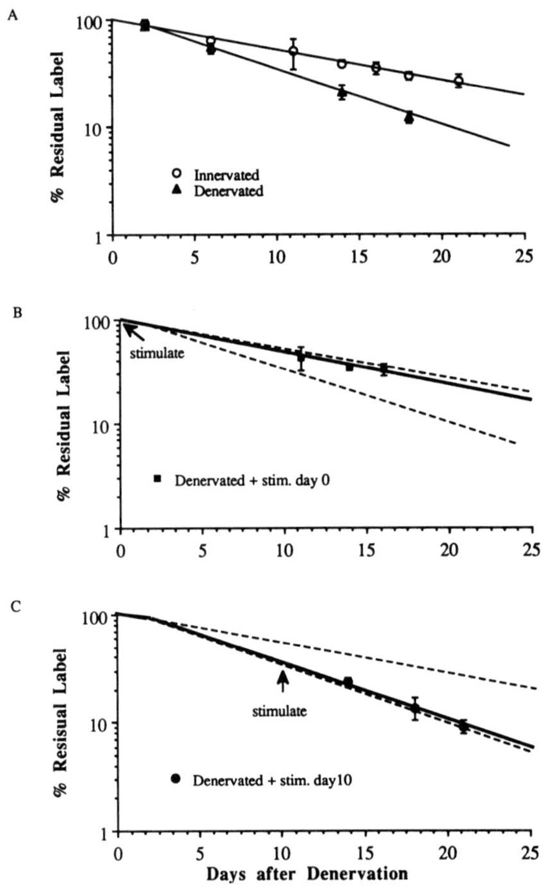

FIGURE 5. Effects of electrical stimulation on R_s degradation rate. R_s AChRs of rat soleus muscles were labeled by injection of [^{125}I]BGT at time of denervation. End-plate–specific radioactivity was measured by γ-counting. Each time point represents the mean ± SD of 3 to 8 muscles. **A:** Original R_s of denervated muscle (▲) initially degrade with a $t_{1/2}$ indistinguishable from AChRs that control innervated end-plates (○) ($t_{1/2}$ ~ 10 days) and that accelerate by 5 days after denervation to a $t_{1/2}$ of ~5 days. **B:** If electrical activity is initiated at time of denervation (arrow), the degradation rate of the R_s (■) is prevented from accelerating. **C:** Once acceleration has taken place, however, electrical stimulation (arrow) of the muscle (one train of 60 stimuli at 100 Hz every 100 seconds) does not restabilize the R_s degradation rate (●).

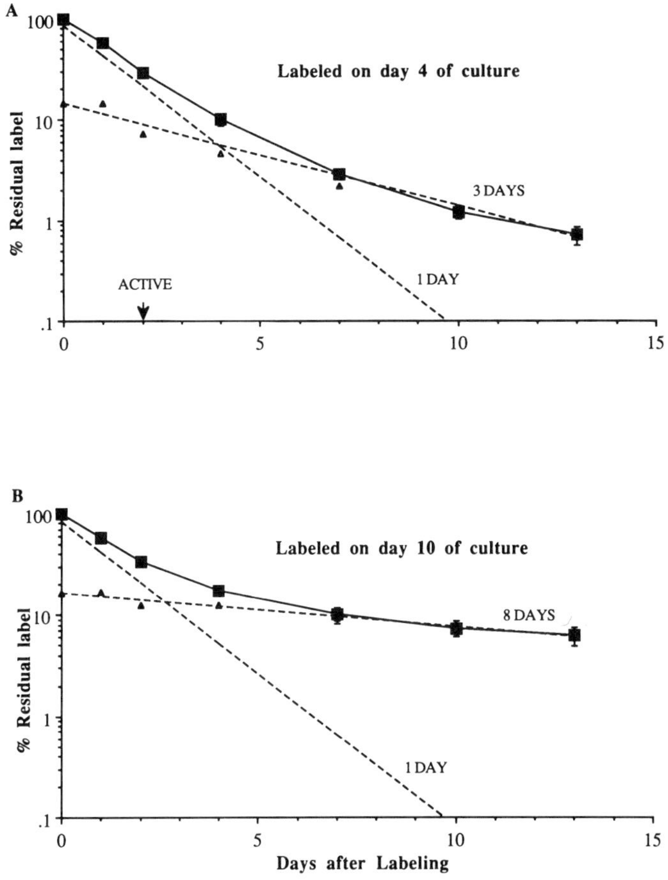

FIGURE 6. Decay of AChRs produced in inactive **(A)** and active **(B)** rat myotubes. The total AChR population was labeled to saturation with $[^{125}I]\alpha$-BGT on day 4 **(A)** and day 10 **(B)** of culture (between which times the myotubes become spontaneously active; see arrow in **A**). The decay of the receptors was measured as in FIG. 3. Values shown are the means ± SEM of 5 experiments.

If all surface AChRs are labeled on day 4 prior to the onset of activity, the degradation curve looks as described above with an 80% R_r component ($t_{1/2}$ ~ 1 d) and a 20% accelerated R_s component ($t_{1/2}$ ~ 3 d). The R_s retain their 3 d $t_{1/2}$ even after muscle activity begins (FIG. 6A). The muscle activity does not stabilize the R_s already in the membrane. If, however, all surface AChRs are labeled on day 10, after onset of activity, the R_s have a stable $t_{1/2}$ of ~10 days, whereas the $t_{1/2}$ of the R_r is unaffected and remains at 1 day (FIG. 6B) (O'Malley *et al.*, unpublished). Thus the R_s AChRs produced in myotubes prior to onset of muscle activity decay at a different rate than those produced after activity begins, even at the same myotube age in culture (days 10–17, FIG. 7). This would indicate that stabilization of the R_s AChR is not a property of the age of the cell. The presence of R_r with unaltered degradation half-lives at either time provides an internal control and confirms our claim that R_r AChRs are not stabilized by muscle activity.

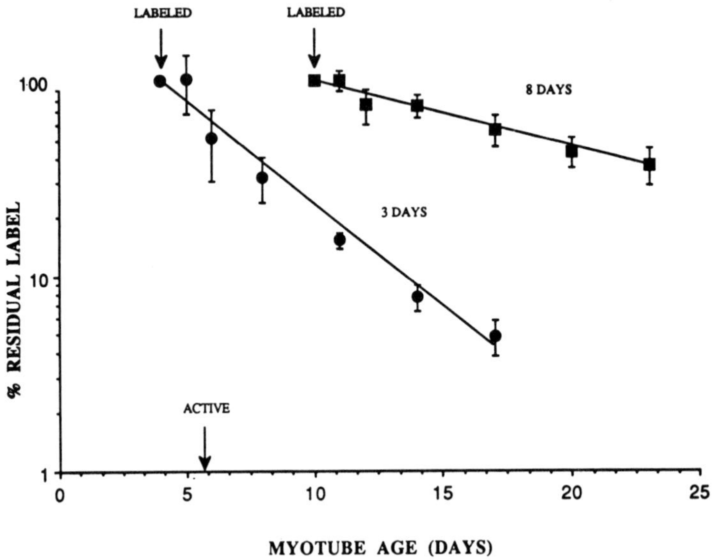

FIGURE 7. Decay rates of R_s AChRs inserted before and after day 6, *i.e.* the initiation of spontaneous myotube activity. The R_r ($t_{1/2}$ ~ 1 day) component was subtracted from the data shown in FIG. 6, and the R_s component was plotted (expressed as a percentage of the original R_s population). Values shown are the means ± SEM of 5 experiments and show that the stabilization of the R_s component depends on when the AChRs are inserted relative to initiation of muscle activity (see FIG. 6) and not on the age of the myotube.

DISCUSSION

It is still premature to propose a unifying theory of how AChR degradation is regulated in muscle. It is clear, however, that R_r and R_s behave differently from each other, and that whereas R_s can be modulated in the muscle membrane, R_r cannot, at least not by any manipulations tried by us. Furthermore, in the modulation of R_s degradation, electrical or spontaneous muscle activity does not fully mimic the nerve or cAMP and must act by a mechanism different from that of cAMP.

The question of what distinguishes R_r from R_s has been raised previously.[7,11] One appealing suggestion centers around the possibility that R_r is the embryonic or γ-containing AChR, whereas R_s is the mature or ε-containing AChR. The time course of appearance, however, of the ε subunit,[12,13] relative to when AChRs are stabilized,[14–16] argues against accepting the suggestion that γ and ε subunits are involved in regulating AChR degradation. The answer to what distinguishes R_r from R_s must wait further clarification.

We propose the following hypotheses to account for the presence of stable AChRs in active or innervated muscle. During muscle activity or after innervation, the synthesis of the R_r is down-regulated and, as the membrane-bound R_r degrade, they are replaced by stable R_s. The behavior, however, of any R_r remaining in the membrane is unaltered. The modulation of R_s degradation appears to involve cAMP-dependent phosphorylation, although it is not yet clear what is being phosphorylated. One possibility is that R_s may be inserted into muscle membrane in its stable form and that in an inactive membrane it accelerates. The acceleration of the R_s could be due to a change in the phosphorylation state either of the AChR, the AChR-related proteins, other kinases, or transcription factors. Treatment of muscle cells by cAMP analogues, but not muscle activity, can then reverse this process, which results in stabilization of the R_s. Studies on the relationship between muscle activity and phosphatase activity (now in progress) should help clarify this model.

REFERENCES

1. BEVAN, S. & J. H. STEINBACH. 1983. Denervation increases the degradation rate of acetylcholine receptors at end-plates *in vivo* and *in vitro*. J. Physiol. **336:** 159–177.
2. LEVITT, T. A. & M. M. SALPETER. 1981. Denervated endplates have a dual population of junctional acetylcholine receptors. Nature **291:** 239–241.
3. SALPETER, M. M., D. L. COOPER & T. LEVITT-GILMOUR. 1986. Degradation rates of acetylcholine receptors can be modified after they are inserted into the postjunctional plasma membrane of the vertebrate neuromuscular junction. J. Cell Biol. **103:** 1399–1403.
4. STANLEY, E. & D. DRACHMAN. 1981. Denervation accelerates the degradation of junctional acetylcholine receptors. Exp. Neurol. **73:** 390–396.
5. SHYNG, S.-L. & M. M. SALPETER. 1989. Degradation rate of acetylcholine receptors inserted into denervated vertebrate neuromuscular junctions. J. Cell Biol. **108:** 647–651.
6. SHYNG, S. L. & M. M. SALPETER. 1990. Effect of reinnervation on the degradation rate of junctional acetylcholine receptors synthesized in denervated skeletal muscles. J. Neurosci. **10**(12): 3905–3915.
7. SHYNG, S. L., R. XU & M. M. SALPETER. 1991. Cyclic AMP stabilizes the degradation of original junctional acetylcholine receptors in denervated muscle. Neuron **6:** 469–475.
8. FUMAGALLI, G., S. BALBI, A. CANCIANO, & T. LØMO. 1990. Regulation of turnover and number of acetylcholine receptors at neuromuscular junctions. Neuron **4:** 563–569.
9. BRENNER, H. R. & W. RUDIN. 1989. On the effect of muscle activity on the end-plate membrane in denervated mouse muscle. J. Physiol. **410:** 501–512.
10. AVILA, O. L., D. B. DRACHMAN & A. PESTRONK. 1989. Neurotransmission regulates stability of acetylcholine receptors at the neuromuscular junction. J. Neurosci. **9** (8): 2902–2906.
10a. ANDREOSE, J., R. XU, T. LØMO, M. M. SALPETER & G. FUMAGALLI. 1993. J. Neurosci. In press.
11. GU, Y., A. FRANCO JR., P. D. GARDNER, J. B. LANSMAN, J. R. FORSAYETH & Z. W. HALL. 1990. Properties of embryonic and adult muscle acetylcholine receptors transiently expressed in COS cells. Neuron **5:** 147–157.
12. BRENNER, H. R., V. WITZEMANN & B. SAKMANN. 1990. Imprinting of acetylcholine receptor messenger RNA accumulation in mammalian neuromuscular synapses. Nature **344:** 544–547.

13. Gu, Y. & Z. W. Hall. 1988. Characterization of acetylcholine receptor subunits in developing and in denervated mammalian muscle. J. Biol. Chem. **263**(26): 12878–12885.
14. Michler, A. & B. Sakmann. 1980. Receptor stability and channel conversion in the subsynaptic membrane of the developing mammalian neuromuscular junction. Dev. Biol. **80:** 1–17.
15. Reiness, C. G. & C. B. Weinberg. 1981. Metabolic stabilization of acetylcholine receptors at newly formed neuromuscular junctions in rat. Dev. Biol. **84**(2): 247–254.
16. Steinbach, J. H., J. Merlie, S. Heinemann & R. Bloch. 1979. Degradation of junctional and extrajunctional acetylcholine receptors by developing rat skeletal muscle. Proc. Natl. Acad. Sci. USA **76**(7): 3547–3551.

cDNA and Genomic Clones Encoding the Human Muscle Acetylcholine Receptor

D. BEESON, A. VINCENT, A. MORRIS, M. BRYDSON,
L. JACOBSON, F. BAGGI, S. JEREMIAH,[a] S. POVEY,[a]
AND J. NEWSOM-DAVIS

Neurosciences Group
Institute of Molecular Medicine
John Radcliffe Hospital
Oxford, OX3 9DU, United Kingdom

[a]*MRC Human Biochemical Genetics Unit*
The Galton Laboratory
University College London
Wolfson House, 4 Stephenson Way
London NW1 2HE, United Kingdom

The autoimmune response in myasthenia gravis (MG) is directed against the human muscle acetylcholine receptor (AChR). Detailed study of the role of the AChR in MG and hereditary myasthenia (HM) is restricted by the low level of AChR (<2 pmol/g) found in human muscle. We have isolated genomic and cDNA clones encoding the human muscle AChR α-, β-, γ-, δ-, and ε-subunits and are using the clones for expression of the AChR subunits. Here, in TABLE 1 we summarize information gained from the primary sequence of the cDNAs and show the chromosomal localization of the genes;[2] in FIGURE 1 we demonstrate their translation in rabbit reticulocyte lysates.

The isolation of genomic clones for the α- and γ-subunits, and cDNA clones for the α- and δ-subunits have been described by other groups.[3–6] Our isolation of α-subunit cDNA clones revealed two forms of α-subunit mRNA, one of which contains the sequence for a novel exon encoding an additional 25 amino acids (aa) between aa58 and aa59 of the mature α-subunit.[7] Translation of both mRNA forms to generate two α-subunit protein isoforms of 437 and 462 aa has been demonstrated in the human muscle cell line TE671.[8]

It has been reported that two different β-subunit mRNAs exist in rat muscle generated by alternative splice junctions at the third coding exon.[9] We have isolated eight cDNAs covering the analogous region of the human β-subunit; none of these contain a similar nine nucleotide insertion. cDNA clones encoding the human γ-subunit were derived from muscle of an amputated leg of a diabetic with peripheral neuropathy. The muscle obtained had the characteristics of denervated muscle, having both elevated levels of α-bungarotoxin–binding and γ-subunit mRNA. Sequencing of several γ-subunit cDNAs showed their sequences to be, as predicted, from genomic clones[4] and to encode a subunit containing 495 amino acids.

By sequence analogy with other AChR subunits, we predict the mature ε-subunit has 473 amino acids. Somewhat surprisingly the ε-subunit gene is not closely linked to the γ and δ on chromosome 2 but is located on chromosome 17.

TABLE 1. Human AChR[a]

	Subunit				
	α	β	γ	δ	ε
Chromosome	2	17	2	2	17
Chromosomal localization	2q24–2q32	17p11–17p12	2q33–2qter	2q33–2qter	—
Amino acids in mature protein	437 462	478	495	495	473
Relative molecular mass	49,694 52,403	54,447	55,740	56,772	52,752
Isoelectric point	6.45 6.28	5.28	6.07	6.68	4.88
Potential extracellular *N*-glycosylation sites [Asn-X-Ser or Asn-X-Thr]	Asn 141[b]	Asn 141	Asn 30 Asn 141	Asn 76 Asn 143 Asn 169	Asn 65 Asn 141
cAMP dependent protein kinase site [Arg-Arg-X-Ser]	—	—	—	354	351
Number of potential protein kinase C sites Ser or Arg-X-Thr or Lys	5	4	3	8	4

[a] The Peptidesort program was used to generate data shown in this TABLE.
[b] Numbering is for the 437 aa isoform of the α-subunit.

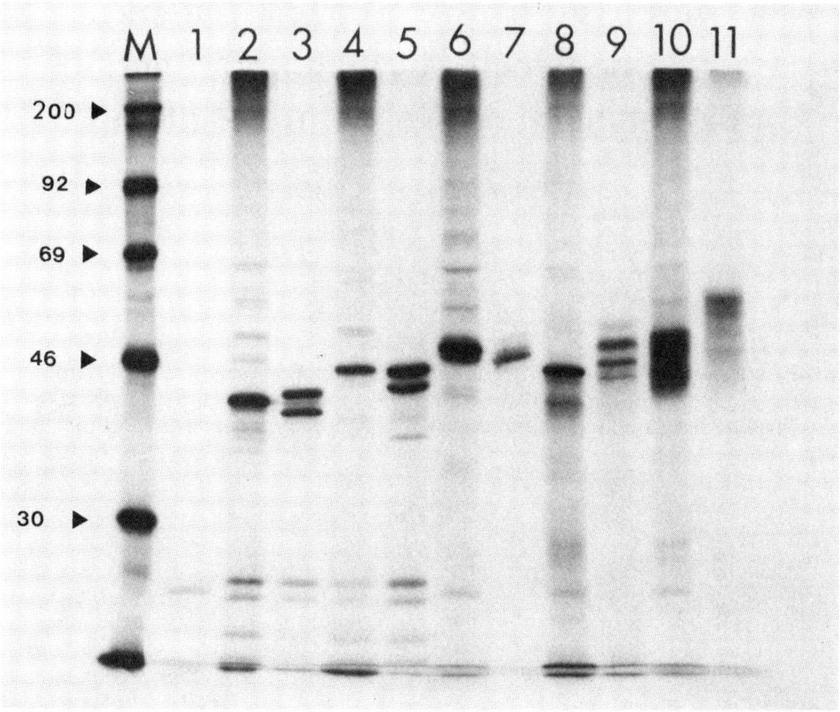

FIGURE 1. Translation in rabbit reticulocyte lysates. 5′-capped RNA transcripts coding for the human AChR α-, β-, γ-, and δ-subunits were generated using the SP6 polymerase system. Individual subunit RNAs were translated using nuclease-treated rabbit reticulocyte lysates either with or without the addition of canine pancreatic microsomal membranes (CPMM). [^{35}S]methionine-labeled products were analyzed by SDS/PAGE followed by autoradiography. Subunit RNA added to lysates are as follows. M: molecular weight markers; lane 1: no RNA; lane 2: α437; lane 3: α437 + CPMM; lane 4: α462; lane 5: α462 + CPMM; lane 6: β; lane 7: β + CPMM; lane 8: γ; lane 9: γ + CPMM; lane 10: δ; lane 11: δ + CPMM. α437 and α462 are the two α-subunit isoforms.

REFERENCES

1. DEVEREUX, J., P. HAEBERLI & O. SMITHIES. 1984. Nucleic Acids Res. **12:** 378–395.
2. BEESON, D., S. JEREMIAH, L. WEST, S. POVEY & J. NEWSOM-DAVIS. 1990. Ann. Hum. Genet. **54:** 199–208.
3. NODA, M., Y. FURUTANI, H. TAKAHASHI *et al.* 1983. Nature **305:** 818–823.
4. SHIBAHARA, S., T. KUBO, H. PERSKI, H. TAKAHASHI, M. NODA & S. NUMA. 1985. Eur. J. Biochem. **146:** 15–22.
5. LUTHER, M., R. SCHOEPFER, P. WHITING *et al.* 1989. J. Neurosci. **9:** 1082–1096.
6. SCHOEPFER, R., M. LUTHER & J. LINDSTROM. 1988. FEBS Lett. **226:** 235–240.
7. BEESON, D., A. MORRIS, A. VINCENT & J. NEWSOM-DAVIS. 1990. EMBO. J. **9:** 2101–2106.
8. MORRIS, A., D. BEESON, L. JACOBSON, F. BAGGI, A. VINCENT & J. NEWSOM-DAVIS. 1991. FEBS Lett. **295:** 116–118.
9. GOLDMAN, D. & K. TAMAI. 1989. Nucleic Acids Res. **17:** 3049–3056.

Stereo View of Conformation and Myasthenogenicity in Synthetic Peptides of Acetylcholine Receptor α-Subunit[a]

MASAHARU TAKAMORI

Department of Neurology
Kanazawa University School of Medicine
13-1, Takara-machi
Kanazawa, Japan

Structural models for the recognition of acetylcholine receptor (AChR) synthetic peptides by B cells expected at β-turn, and T cells expected at α-helix, were designed to seek a conformation that would approach the myasthenogenic architecture.

Synthetic Peptides Used as Immunogens

Artificially formed peptides (model peptides) were synthesized by variously linking putative B-cell epitope sequences (*Torpedo californica* AChR α190–195 as ACh-binding site recognized by blocking antibody,[1] or α67–76 as "main immunogenic region"[2] recognized by binding antibody) with putative T-cell epitope sequence (α107–116, adapted to Lewis strain rats under MHC class II restriction[3]). Intervation of the artificial sequence, Asn-Pro-Gly-Gly (NPGG),[4] in natural sequence segments was also designed to potentially adopt a β turn in antigenic peptides. They are listed in the left column of TABLE 1.

Antibody Assays and Microelectrode Study in Peptide-immunized Rats (TABLE 1)

The model peptide, α107–116 · NPGG · α190–195, was potently immunogenic in Lewis rats to provoke the antipeptide antibody reactive with native AChR in the manner of "blocking," and to induce EAMG as confirmed by reduced miniature end-plate potential (MEPP) amplitudes. This showed more potent myasthenogenicity than the natural sequence peptide, α183–200.[1] The model peptide, α107–116 · α67–76 · α107–116, was potently immunogenic in Lewis rats to provoke the antipeptide antibody reactive with native AChR in the manner of "binding," and to induce EAMG as confirmed by reduced MEPP amplitudes.

Anti-peptide Antibody Assays in Rats with EAMG Induced by Immunization with Torpedo *AChR* (TABLE 2)

In seven EAMG rats, anti-(α107–116 · α190–195 · α107–116) and anti-(α107–116 · NPGG · α67–76) antibodies showed high titers and were correlated

[a] This work was supported by Grants from the Japanese Ministries of Education, Science and Culture (03454239), and Health and Welfare.

TABLE 1. Antipeptide and Antinative AChR Antibodies, and MEPP Amplitudes in Peptide-immunized Rats[a]

Control	Antipeptide Antibody (cpm) (Written Under Each, n = 15)	Anti-rat AChR Antibody: Binding Antibody (pmol/mL) 7 ± 2	Anti-rat AChR Antibody: Blocking Antibody (percent) 0.3 ± 0.05	MEPP Amplitudes (mV) 0.71 ± 0.128 (n = 60 End-plates; 20 End-plates in Each Tested Rat)
Natural sequence peptides				
Torpedo α183–200	20136 ± 5893 (7/7) (control: 242 ± 57)	13 ± 8 (4/7)	0.1 ± 0.09 (0/7)	0.52 ± 0.149 (4/7)
B-cell epitopes				
Torpedo α190–195	247 ± 7 (5/5) (control: 154 ± 11)	4 ± 2 (0/5)	0.1 ± 0.07 (0/5)	0.68 ± 0.185 (0/5)
Torpedo α67–76	858 ± 438 (5/5) (control: 222 ± 20)	3 ± 3 (0/5)	0.3 ± 0.1 (0/5)	0.67 ± 0.201 (0/5)
T-cell epitope				
Torpedo α107–116	189 ± 13 (5/5) (control: 132 ± 12)	2 ± 3 (0/5)	0.1 ± 0.10 (0/5)	0.70 ± 0.147 (0/5)
Model peptides including B- and T-cell epitopes (*Torpedo* sequence)				
α107–116 · α190–195	2690 ± 297 (5/5) (control: 228 ± 11)	5 ± 2 (1/5)	0.1 ± 0.06 (0/5)	0.67 ± 0.118 (1/5)
α107–116 · NPGG · α190–195	1879 ± 228 (5/5) (control: 215 ± 18)	23 ± 12 (4/5)	0.2 ± 0.04 (0/5)	0.45 ± 0.156 (4/5)
α107–116 · α190–195 · α107–116	6755 ± 1380 (5/5) (control: 210 ± 24)	8 ± 3 (2/5)	0.2 ± 0.08 (0/5)	0.63 ± 0.133 (2/5)
α107–116 · α67–76	1259 ± 694 (7/7) (control: 243 ± 13)	3 ± 3 (0/7)	0.2 ± 0.1 (0/7)	0.67 ± 0.221 (0/7)
α107–116 · NPGG · α67–76	3485 ± 1598 (8/8) (control: 241 ± 16)	3 ± 2 (0/8)	0.3 ± 0.10 (2/8)	0.64 ± 0.154 (2/8)
α107–116 · α67–76 · α107–116	1424 ± 890 (10/10) (control: 250 ± 7)	3 ± 2 (0/10)	0.6 ± 0.1 (5/10)	0.60 ± 0.198 (5/10)

[a] Values represent the mean ± SD. No. of "positive"/No. of tested rats is inserted in parentheses ("positive" in the antibody assays indicates the titer above the control mean + 2 SD; "positive" in the MEPP amplitude indicates the reduction of mean amplitude, estimated as significant by $p < 0.05$ vs. the control by one-way analysis of variance and Bonferroni multiple comparisons).

TABLE 2. Antipeptide Antibodies (cpm) in Rats with EAMG Induced by Immunization with *Torpedo californica* AChR[a]

	Anti-(α190–195)	Anti-(α67–76)	Anti-(α107–116)	Anti-(α183–200)	Anti-(α107–116 · α190–195)
Control (n = 15)	154 ± 11	222 ± 20	132 ± 12	242 ± 57	228 ± 11
Tested rats (n = 7)	274 ± 11	616 ± 151	164 ± 10	465 ± 12	383 ± 53
	Anti-(α107–116 · NPGG · α190–195)	**Anti-(α107–116 · α190–195 · α107–116)**	**Anti-(α107–116 · α67–76)**	**Anti-(α107–116 · NPGG · α67–76)**	**Anti-(α107–116 · α67–76 · α107–116)**
Control (n = 15)	215 ± 18	210 ± 24	243 ± 13	241 ± 16	250 ± 7
Tested rats (n = 7)	685 ± 193	773 ± 194	826 ± 99	2134 ± 766	350 ± 52

[a] Values represent the mean ± SD.

with antinative AChR blocking and binding antibody titers (data are not shown), respectively. Therefore, the synthetic peptide may assume an antigenic conformation that differs from the antigenic conformation stimulating the induction of EAMG.

Stereo View of Conformation in Synthetic Peptides

Model peptides, highly immunogenic as demonstrated in the experiments, formed a rigid structure composed of β-turn and α-helix as revealed by the computer graphics of conformation drawn by interproton distance geometry based on two-dimensional nuclear magnetic resonance.[5]

REFERENCES

1. TAKAMORI, M., S. OKUMURA, M. NAGATA & H. YOSHIKAWA. 1988. Myasthenogenic significance of synthetic α-subunit peptide 183–200 of *Torpedo californica* and human acetylcholine receptor. J. Neurol. Sci. **85:** 121–129.
2. TZARTOS, S. J., A. KOKLA, S. L. WALGRAVE & B. M. CONTI-TRONCONI. 1988. Localization of the main immunogenic region of human muscle acetylcholine receptor to residues 67–76 of the α subunit. Proc. Natl. Acad. Sci. USA **85:** 2899–2903.
3. FUJII, Y. & J. LINDSTROM. 1988. Specificity of the T cell immune response to acetylcholine receptor in experimental autoimmune myasthenia gravis. Response to subunits and synthetic peptides. J. Immunol. **140:** 1830–1837.
4. TAKAMORI, M., S. OKUMURA, K. KOMAI & R. SATAKE. 1990. Conformational modification enhances myasthenogenicity in synthetic peptide of acetylcholine receptor α-subunit. J. Neurol. Sci. **99:** 219–227.
5. BRAUN, W. & N. GŌ. 1985. Calculation of protein conformations by proton-proton distance constraints. A new efficient algorithm. J. Mol. Biol. **186:** 611–626.

Interaction of Antiacetylcholine Receptor Monoclonal Antibodies with the Acetylcholine Receptor[a]

QUN XU, ROBERT H. FAIRCLOUGH, AND
DAVID P. RICHMAN

Department of Neurology and the Center for Neurobiology
University of California, Davis
Davis, California 95616

A preponderance of the antibody response in myasthenia gravis (MG) and experimental autoimmune myasthenia gravis (EAMG) is directed against a region of the acetylcholine receptor (AChR) called the main immunogenic region (MIR) localized on the alpha-subunit at a site 30 Å distant from the alpha-bungarotoxin binding site also localized on the alpha-subunit (FIG. 1). The archetype of an MIR-directed monoclonal antibody (mAb) is mAb 35,[1] developed by Tzartos and Lindstrom (TABLE 1). This antibody induces passive transfer EAMG in rats. It does not impair carbamylcholine (CbCho)-induced $^{22}Na^{+}$ conductance through the AChR, but rather its pathogenic effect appears to be mediated through inflammatory mechanisms.

Although antibodies directed to regions near the agonist binding site represent a smaller population of the antibodies in MG and EAMG, they can have profound functional effects on neuromuscular transmission by disrupting binding of acetylcholine to AChR. Such an antibody is mAb 387D,[2] which completely inhibits the binding of CbCho to the receptor (TABLE 1).

Given the likely spatial separation of the two antibody epitopes defined by mAb 35 and 387D, we asked whether the binding of one of these antibodies could influence the binding of the other.

METHODS

An ELISA was employed to study inhibition of binding of each mAb, labeled with biotin, by the other mAb. Membrane-bound *Torpedo* AChR complexes with unlabeled inhibiting mAb, at various stoichiometries, were formed and then titrated with biotin-labeled testing mAb. Binding of labeled testing mAb was determined by addition of alkaline phosphatase avidin and then the substrate, *p*-nitrophenyl phosphate. The extent of the reaction is determined by the OD_{410}.[3]

[a] This work was supported in part by Grants from the National Institutes of Health (NS15462, NS19779, NS24304, NS01226), the Muscular Dystrophy Society, and the Amyotrophic Lateral Sclerosis Society of America.

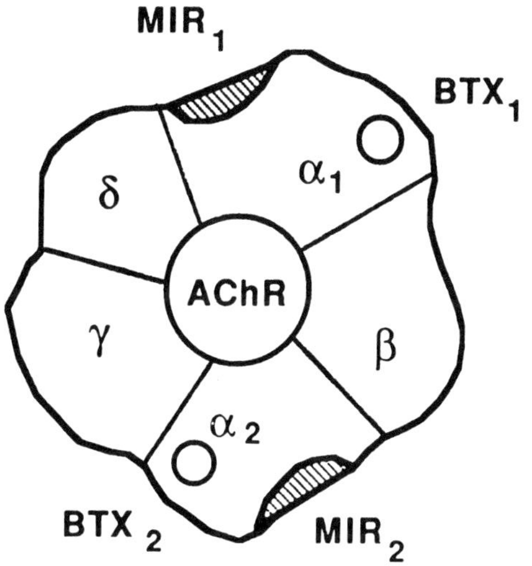

FIGURE 1. Acetylcholine receptor (AChR) molecule viewed from the synaptic side with borders arbitrarily drawn between subunits. AChR is a pentameric glycoprotein composed of two alpha-subunits. The main immunogenic region (MIR) on each alpha-subunit is located at least 30 Å from the alpha-bungarotoxin (BTX) binding site.

RESULTS AND DISCUSSION

For biotin-labeled mAb 35, which binds to the MIR of the AChR, titration of 1 mAb 387D : 1 AChR complexes resulted in a 100-fold decrease in mAb 35 affinity. By preparing, predominantly, complexes of 2 mAb 387D : 1 AChR, mAb 35 affinity was reduced by only 10-fold. Thus, it appears that when each alpha-subunit within a single AChR molecule is bound by a different mAb 387D molecule (2 mAbs : 1 AChR stoichiometry), the effect on the mAb 35 epitopes is less. To test the hypothesis that the greater effect seen with stoichiometry of 1 mAb : 1 AChR, that is, 1 mAb/2 alpha-subunits, is related to intramolecular cross-linking of the 2 alpha-subunits within a single AChR by two arms of one mAb 387D, we examined the effect on mAb 35 binding of the Fab fragment of mAb 387D, which is incapable of cross-linking. At the 2 Fab fragments of 387D : 1 AChR stoichiometry, which is equivalent to a 1 : 1 stoichiometry of intact mAb, the effect was identical to the smaller reduction in affinity observed with the 2 mAb 387D : 1 AChR stoichiometry.

TABLE 1. Properties of mAb 387D and mAb 35

	mAb 387D	mAb 35
Blocked by BTX[a]	+	−
Blocked by CbCho	−	−
Blocks CbCho	+	−
Anti-MIR	−	+

[a] alpha-bungarotoxin.

Therefore, it seems likely that the major effect of mAb 387D on the mAb 35 epitope occurs as a result of intramolecular cross-linking.

REFERENCES

1. TZARTOS, S. J. & J. M. LINDSTROM. 1980. Monoclonal antibodies used to probe acetylcholine receptor structure: Localization of the main immunogenic region and detection of similarities between subunits. Proc. Natl. Acad. Sci. USA **77:** 755–759.
2. MIHOVILOVIC, M. & D. P. RICHMAN. 1987. Anti-acetylcholine receptor monoclonal antibodies as probes of the acetylcholine receptor cholinergic binding site. J. Biol. Chem. **262:** 4978–4986.
3. AGIUS, M. A., C. J. GEANNOPOULOS, R. H. FAIRCLOUGH & D. P. RICHMAN. 1988. Monoclonal anti-idiotypic antibodies directed against myasthenia-inducing anti-acetylcholine receptor monoclonal antibodies: Preponderance of nonparatope-directed antibodies affecting antigen binding. J. Immunol. **140:** 62–68.

Characterization of the Epitope of an Antiacetylcholine Receptor Antibody That Inhibits Fifty Percent of Alpha-Bungarotoxin Binding[a]

QUN XU, GEORGE M. TWADDLE, DAVID P. RICHMAN, AND ROBERT H. FAIRCLOUGH

Department of Neurology
and
the Center for Neuroscience
University of California, Davis
Davis, California 95616

The anti-acetylcholine receptor (AChR) monoclonal antibody (MAb) 247G binds to the AChR with a stoichiometry of 1 MAb for 2 α-bungarotoxin (α-BTX) sites. It inhibits a maximum of 50% of α-BTX binding to AChR.[1] This MAb blocks activation of the cholinergic ion channel without affecting agonist binding or agonist-induced desensitization.[2] These observations indicate that MAb 247G stabilizes a conformational state of the AChR in which cholinergic transitions leading to desensitization have been uncoupled from those leading to activation of the channel. We have further analyzed the characteristics of the 247G epitope as a means of investigating the structure and function of the AChR.

METHODS

Titration of Membrane-bound Torpedo *AChR with Labeled MAb 247G*

Purified MAb 247G was labeled with the hydroxysuccinimide ester of [^{3}H]biotin or 5- (and -6)-carboxytetramethylrhodamine using standard methods[3] and separated from free reagent by gel filtration. Labeling stoichiometry was determined

[a] This work was supported by National Institues of Health Grant NS 24304.

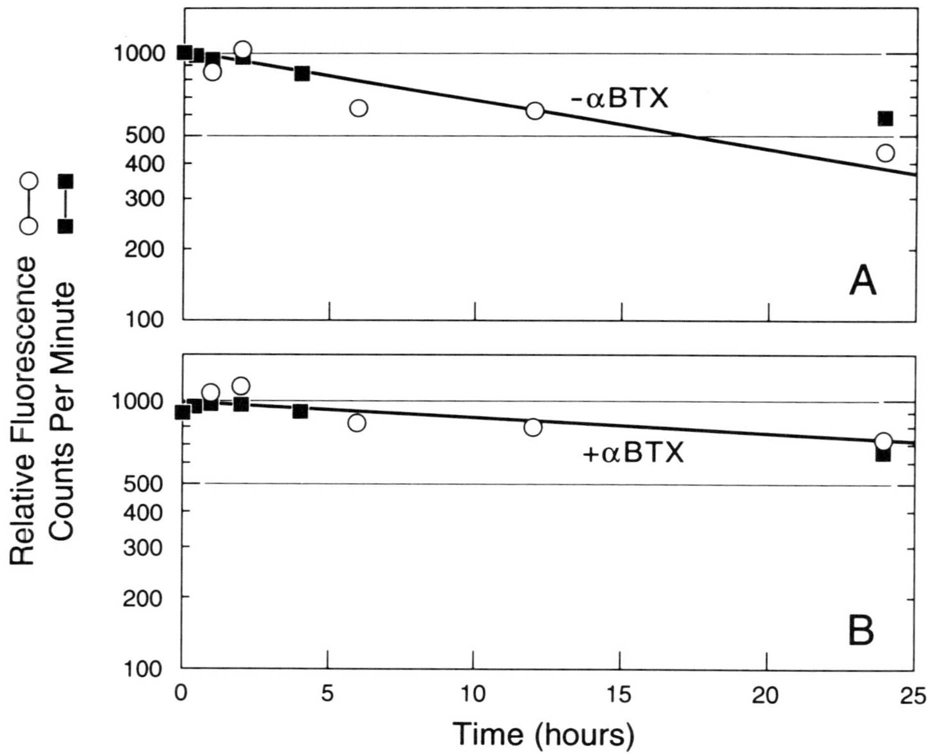

FIGURE 1. Time course of dissociation of MAb 247G : AChR complexes in the absence **(A)** and presence **(B)** of half-saturating amounts of α-BTX. The lifetime of the complexes in the absence of α-BTX is 25 hours, which increases to 80 hours in the presence of α-BTX.

by liquid scintillation counting or optical density measurement and a Lowry assay of the labeled material.

Labeled MAb 247G was incubated with membrane-bound AChR for 3 hours. Complexes were separated from free MAb by sedimentation in an Airfuge at 20 psi for 5 minutes, followed by three washes with phosphate buffer.

Dissociation of MAb 247G : AChR Complexes

Complexes of membrane-bound AChR with [^{3}H]-labeled or rhodamine-labeled MAb 247G were resuspended in a 100-fold excess of unlabeled MAb 247G, and dissociation of the complexes was followed by pelleting aliquots at various time intervals and observing the cpm of ^{3}H or the fluorescence pelleting with the membranes in the absence and presence of α-BTX.

RESULTS AND DISCUSSION

Titration of membrane-bound *Torpedo* AChR with [^{3}H]biotin-labeled MAb 247G revealed a stoichiometry of 1 MAb per AChR with a dissociation constant of 0.5 nM. We studied the rate of dissociation of 247G : AChR complexes, as a means of determining whether the stoichiometry of 1 MAb per AChR can be explained by MAb cross-linking of two epitopes on a single AChR. The lifetime of the complexes, 25 hours (FIG. 1), is the same as that obtained with Fab fragments of a MAb that binds with similar affinity, suggesting an absence of cross-linking. Thus, it appears that MAb 247G binds to a single AChR site and blocks α-BTX binding to that site. In the presence of half-saturating amounts of α-BTX, however, the lifetime of 247G : AChR complexes is increased from 25 to 80 hours (FIG. 1), suggesting that the 247G epitope is altered by binding of α-BTX to the AChR. Our hypothesis is that binding of α-BTX to 247G : AChR complexes is occurring at the toxin site not blocked by 247G and is inducing a structural change in the 247G epitope, which overlaps the blocked toxin site. These observations make MAb 247G a useful probe of structural changes induced by AChR agonists and antagonists.

REFERENCES

1. MIHOVILOVIC, M. & D. P. RICHMAN. 1984. Modification of alpha-bungarotoxin and cholinergic ligand-binding properties of *Torpedo* acetylcholine receptor by a monoclonal antiacetylcholine receptor antibody. J. Biol. Chem. **259:** 15051–15059.

2. MIHOVILOVIC, M. & D. P. RICHMAN. 1987. Anti-acetylcholine receptor monoclonal antibodies as probes of the acetylcholine receptor cholinergic binding site. J. Biol. Chem. **262:** 4978–4986.
3. BAYER, E. A. & M. WILCHEK. 1978. The avidin-biotin complexes as a tool in molecular biology. Trends Biochem. Sci. **3:** N257.

Influence of Immunological Fine-specificity on the Induction of Experimental Myasthenia Gravis[a]

KEITH A. KROLICK, PATRICIA A. THOMPSON, THOMAS E. ZODA, AND TRAI-MING YEH

Department of Microbiology
University of Texas Health Science Center at San Antonio
San Antonio, Texas 78284

INTRODUCTION

Myasthenia gravis (MG) and its animal model, experimental autoimmune myasthenia gravis (EAMG), are characterized by neuromuscular dysfunction that results in an ease of fatigability and relative weakness of voluntary muscles.[1–3] A common feature in each case is a T lymphocyte–dependent serum autoantibody response against the acetylcholine receptor (AChR); the binding of anti-AChR autoantibodies to AChR at the neuromuscular junction causes the observed impairment of neuromuscular transmission. Although it has long been noted that the major feature of MG responsible for the neuromuscular dysfunction observed in patients is the presence of circulating serum antibodies reactive with the AChR, attempts to correlate serum anti-AChR antibody concentrations with disease severity have been disappointing in some patients.[4–7] In other words, some severely ill patients have marginally detectable anti-AChR antibody titers, whereas certain other patients with very high antibody titers demonstrate marginally detectable disease symptoms. Even in light of the various tests of proposals to explain the lack of clear relationships between antibody titer and disease induction, no clear relationships to disease severity have yet been determined.

The main premise on which our studies have been based is that some, but not all, anti-AChR antibodies produced in MG patients are directly responsible for disease induction; a certain proportion (differing from patient-to-patient) of the anti-AChR antibodies produced has no direct relationship to disease (FIG. 1). If in different MG patients different subsets of the total clonotypic spectrum of possible anti-AChR antibodies were produced (some dominated by antibody species with disease-causing potential and some dominated by benign antibody species), one would expect that differences in the severity of neuromuscular impairment would result. The ability to predict disease severity, therefore, would be dependent on the ability to identify and measure levels of the disease-causing antibody subset; measurement of total anti-AChR antibody titers would likely be misleading unless the relevant subset composed the majority of the antibody.

Finally, complex immunoregulatory phenomena, probably T cell–mediated, also influence the production and characteristics of the anti-AChR antibodies. Buried within the complexities of immunopathology and immunoregulation is an

[a] This research is currently supported by NIH Grants NS28172 and NS24954 and a Grant from the Muscular Dystrophy Association.

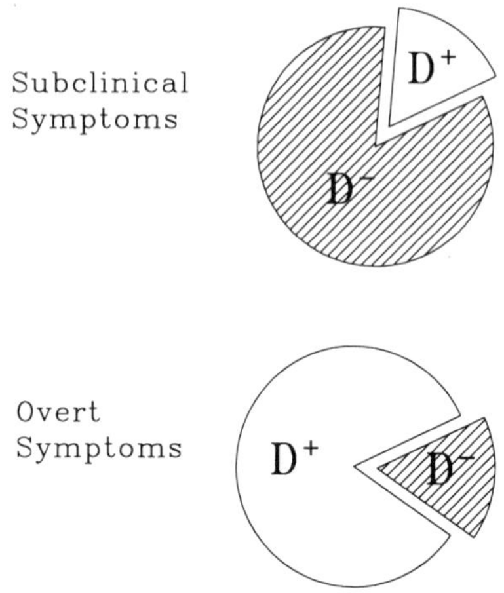

FIGURE 1. Total anti-AChR antibody. Hypothetical differences in the proportions of disease-causing and benign anti-AChR antibodies produced by two MG patients that demonstrate differences in disease symptom severity. The explanation proposed here is that although both patients may have the same serum titer of anti-AChR antibodies, only that patient with a supra-threshold titer of the disease-causing antibody subset would demonstrate severe neuromuscular impairment.

understanding of the quantitative and qualitative relationships between the anti-AChR antibody response and the induction of AChR-dependent neuromuscular disease symptoms. Patient-to-patient variability with regard to disease severity may be explained by immunoregulatory activities that determine which anti-AChR antibodies are expressed/produced from within the potential anti-AChR antibody repertoire of fine specificities. These activities may involve major histocompatibility complex (MHC)–selected immunoregulatory T-cell activities that steer the anti-AChR antibody response. Therefore, in searching for an understanding of disease mechanisms that lead to an ability to predict disease state and progress, one must include not only parameters associated with the fine-specificity repertoire of disease-causing antibodies directly, but also parameters associated indirectly with the regulation of that particular subset of the total anti-AChR response.

STRAIN-DEPENDENT EAMG SUSCEPTIBILITY IN INBRED RATS

Data obtained from this laboratory concerning disease-susceptible Lewis rats suggest that characteristics dictated by the clonal B-cell origin of anti-AChR antibodies dramatically influences resulting myopathological effects. Isoelectric focusing (IEF) analyses can be useful for assessing the clonotypic origins, heterogeneity, isotype, and relative antigen binding avidity of serum anti-AChR antibodies.[8–14] Immunopathological differences associated with different clonotypic species of anti-AChR antibodies can thus be identified, giving a different perspective on causative disease mechanisms than do studies of the bulk serum antibody response. Comparisons have been performed[13] to evaluate the characteristics of the anti-AChR antibodies produced in two genetically different rat strains:[15] AChR-immune Lewis rats that demonstrate easily detectable neuromuscular dysfunction and AChR-immune Wistar Furth (WF) rats that demonstrate significant resistance to the induction of antibody-mediated muscle dysfunction (FIG. 2). (In experiments described, both decreased relative tension of contraction and increased sensitivity to curare [seen as decreased RT_{50} values] reflect diminished AChR-dependent muscle contractile function compared to normal muscles[16,17]).

With regard to the complexity of IEF banding patterns (reflecting the number of AChR-reactive B-cell clones making antibody), both Lewis and WF rats express similar clonotypic antibody heterogeneity and isotype preference (FIG. 3). Because antibody-isotype differences do not appear to provide a clear explanation for disease sensitivity in the two rat strains, perhaps antibody-binding avidity plays a role in determining the myasthenogenicity of the antibodies produced in each case. Consistent with this notion, some antibody clonotypes from the two rat strains are

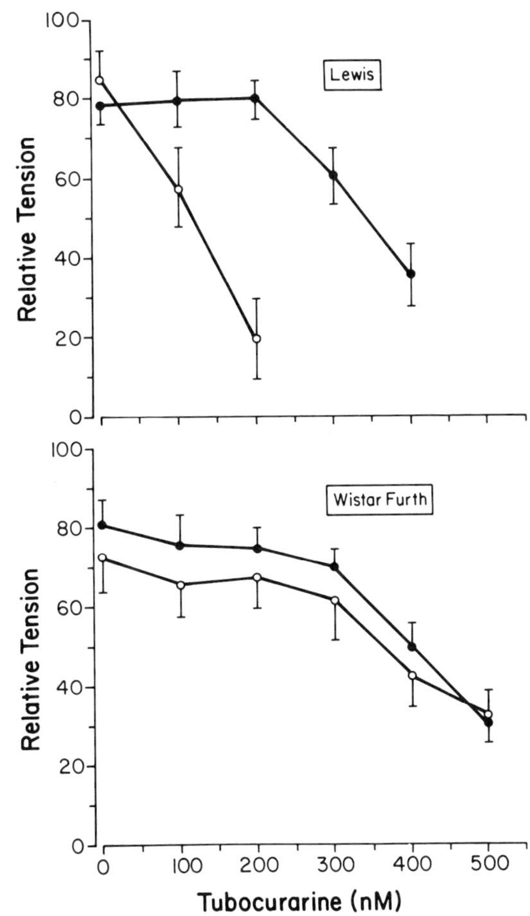

FIGURE 2. Effect of AChR immunization on the AChR-dependent contractile function of EDL muscles from Lewis and WF rats (as indicated). Rats of each strain were either injected (sc) 4 weeks before with 50 μg AChR in complete Freund's adjuvant (CFA) (○) or left unimmunized (●). Values represent the mean (±SEM for 4 rats) relative tension of EDL muscles produced following a 3.35 V stimulation in the presence of increasing doses of tubocurarine. (Zoda *et al.*[13] With permission from the *Journal of Immunology*.)

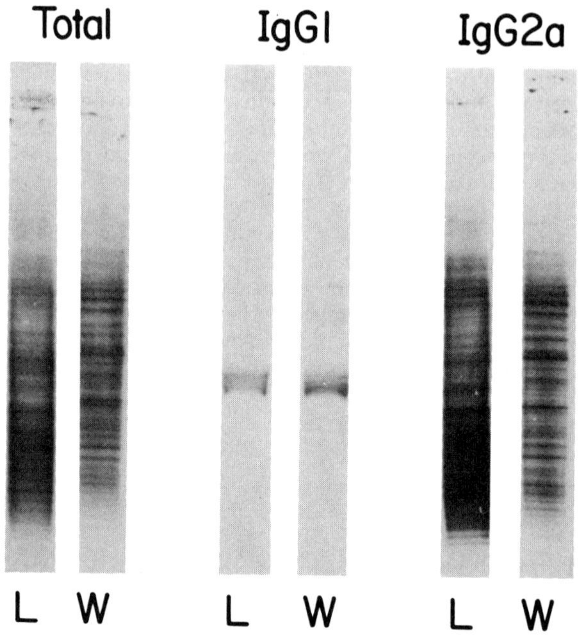

FIGURE 3. Isoelectric focusing analysis of antiserum from one representative Lewis (L) or Wistar Furth (W) rat immunized 4 weeks before with AChR. AChR-specific antibody bands immunoblotted onto AChR-coated nitrocellulose paper were probed for total, IgG1, or IgG2a, as indicated. (Zoda *et al.*[13] With permission from the *Journal of Immunology*.)

not comigrating in pH gradients, indicating a lack of identity. There does not appear, however, to be a clear distinction between Lewis anti-AChR antibodies and WF antibodies with regard to binding avidity either; the relative binding avidities appear to be quite similar between Lewis and WF anti-AChR antibodies.[13] Nevertheless, inasmuch as the responses of both rat strains are highly dominated by the same IgG subclass (IgG2a) with no apparent allotypic differences,[18] it is likely that many of the differences in net charge and pI between Lewis and WF antibodies are the result of variable-region amino acid differences; this could indicate differences in antibody fine specificity that reflect differences in resulting AChR perturbations.

The main conclusion derived from this information is that although the antibodies produced following a single AChR immunization of Lewis and WF rats are of similar clonotypic heterogeneity and avidity, and dominated by the same IgG subclass, only Lewis rats go on to demonstrate impaired AChR-dependent neuromuscular function. There are two explanations for this difference in disease induction: (1) the antibodies produced by the two rat strains may be of overlapping but nonidentical fine specificities and therefore express different degrees of myasthenogenicity, or (2) the antibodies produced by both strains include specificities that are myasthenogenic, but the biochemistry or presentation of junctional AChR in WF rats somehow differs from AChR in Lewis rats, resulting in resistance to antibody-mediated effects. This second possibility is less likely with the demonstration of impaired muscle function in WF rats after the passive transfer of preformed anti-AChR antibodies from Lewis donor rats (TABLE 1).

Following multiple injections of AChR, WF rats do finally demonstrate AChR-dependent neuromuscular dysfunction (FIG. 4). There are at least two explanations that should be considered with regard to this observation. First, the early WF responses to AChR involve antibodies with no disease-causing activity; upon hyperimmunization, however, antibody variable region gene mutations may occur that result in shifts in antibody-binding fine specificity and/or avidity; acquisition of these new antibody specificities to the expressed anti-AChR repertoire could result in increased disease-causing potential. A second, but not mutually exclusive, explanation for the disease breakthrough observed in WF rats is that the disease-causing fine specificities exist in the WF germ line but are initially expressed at very low frequency and/or are somehow immunosuppressed (T cell–mediated?); upon hyperimmunization, eventual activation and expansion of disease-causing B-cell clones results in the production of enough anti-AChR antibody to exceed the threshold needed to result in impaired muscle function as described in FIGURE 1. A simple disease versus antibody titer relationship suggests that this disease-causing threshold may exist (FIG. 5). It is also of note, however, that the expression of disease symptoms in hyperimmunized WF rats is transient (<10 weeks) (FIG. 4), although the total anti-AChR antibody titer remains virtually constant during the same period of time. This may indicate that the production of the small subset of anti-AChR antibodies responsible for disease symptoms in WF rats may continue to be restricted by immunoregulatory factors. Changing IEF banding patterns seen over the multiple immunization timetable are also consistent with this possibility and may provide a more sensitive way of monitoring relationships between anti-AChR antibody production and AChR-dependent muscle dysfunction.[13]

TABLE 1. Passive Transfer of Anti-AChR Antibodies from Lewis Rats Can Cause Impaired Muscle Function in Wistar Furth Rats

Amount of Ab Transferred[a]	RT_{50} (nM)[b] Lewis	WF
0	365 ± 30	437 ± 23
200	350 ± 6	400[c]
400	196 ± 37	140 ± 39
800[d]	<100	<100

[a] Micrograms of IgG fraction of AChR-immune Lewis rat serum transferred by intravenous injection.

[b] Concentrations of tubocurarine required to inhibit relative contractile tension of EDL muscles by 50%. Mean values (±SEM) represent determinations of muscles from rats of the indicated strain (3–6 rats/group). Evaluations of AChR-dependent contractile function of EDL muscles from many healthy and AChR-immunized rats (>200 each) have indicated that healthy EDL muscles demonstrate RT_{50} values of 350–450 nM curare, whereas AChR-immune rats demonstrate RT_{50} values of 100–200 nM curare.[9–13,16,17]

[c] One rat tested at this dose of antibody transferred.

[d] Both rat strains showed visible signs of muscle dysfunction following transfer of 800 μg of Lewis antibody. RT_{50} values for these animals were not calculated because relative tensions were determined to be decreased by about 50–70% prior to the addition of any curare. Exposure to 100 nM curare resulted in no detectable contractile function (RT = 0).[13]

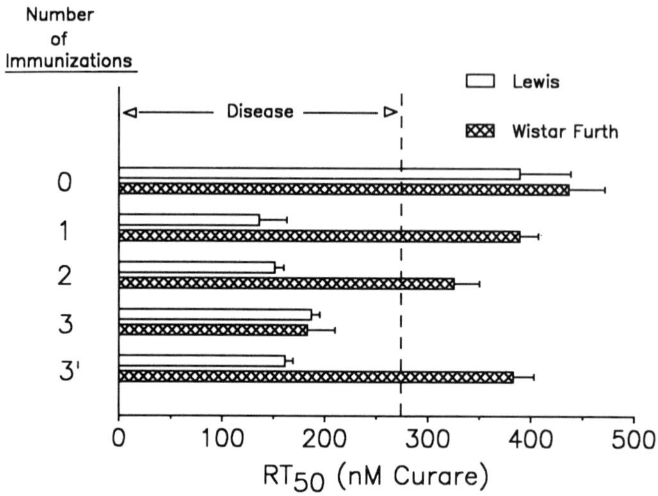

FIGURE 4. Disease induction in Lewis rats (open bars) and Wistar Furth rats (hatched bars) following multiple AChR immunizations. The dose of curare necessary to inhibit contractile force by 50% (RT_{50}) is thought to reflect the level of functional AChR expression and has been shown to be a sensitive reproducible indicator of antibody-mediated AChR perturbations.[16,17] The vertical dashed line has been arbitrarily placed to indicate the lowest RT_{50} value ever determined for EDL muscles from healthy control rats (*i.e.*, 270 nM curare); in reality, after greater than 200 assessments, control values are most frequently found in the 350–500 nM range and diseased values in the 100–250 nM range. Groups of four rats of each strain were injected (sc) with 50 μg nAChR emulsified in CFA, followed by 1 or 2 additional AChR booster injections given in saline. Three weeks following each injection, a group from each rat strain was sacrificed and tested for AChR-dependent muscle contractile function. The number of immunizations indicated is the sum of total injections given; 3′ indicates that rats were given 3 injections and allowed to wait 10 weeks, instead of the usual 3–5 weeks, before analysis of muscle function. (Zoda *et al.*[13] With permission from the *Journal of Immunology*.)

DIFFERING DISEASE-CAUSING POTENTIALS ASSOCIATED WITH ANTI-AChR ANTIBODIES FROM EAMG-SUSCEPTIBLE RATS

Conformation Dependence of Important AChR Epitopes

From the studies described above comparing disease-sensitive Lewis rats to disease-resistant Wistar Furth rats, a particular set of anti-AChR antibodies prominently produced by Lewis rats, but absent in WF rats, appears to be required for induction of neuromuscular disease. Other studies from this and other laboratories have been reported demonstrating that, in fact, Lewis anti-AChR antibodies are not themselves all equally capable of causing impaired AChR-dependent neuromuscular function. For instance, certain key studies by other investigators[19–22] demonstrate the importance of the three-dimensional conformation of AChR with regard to anti-AChR antibody fine specificity for AChR epitopes; that is, in experimental systems of MG, serum antibody responses to denatured AChR versus

AChR of native conformation are found to differ with regard to subsequent induction of disease.

In addition, our own studies demonstrate clonotypic distinctions between disease-causing and non–disease-causing anti-AChR antibodies in Lewis rats immunized with either conformationally intact AChR or reduced and SDS-denatured AChR.[11] Reactivity of antibodies directed against linear (as opposed to conformational) epitopes often are incapable of eliciting disease symptoms; however, although reactivity of serum antibodies with AChR epitopes demonstrating native conformation appears to be a prerequisite for inducing disease, it is not necessarily sufficient. Thus, as expected, challenge with native AChR (nAChR) results in the production of serum antibodies reactive with nAChR; these antibodies are oligoclonal, dominated by the rat IgG2a subclass, and capable of interfering with normal AChR-dependent muscle contraction as reflected by measurements of AChR-dependent muscle contractile tension and sensitivity to the AChR antagonist, curare. Antibodies produced as a result of immunization with denatured AChR (dAChR) but *cross-reactive with nAChR,* are similar, but not identical, to those produced directly against nAChR; dAChR-stimulated antibodies are skewed to the acidic end of the pH gradient (in contrast to the more alkaline nature of antibodies stimulated with native AChR) and are clearly incapable of causing detectable impairment of AChR-dependent muscle contractile function (summarized in TABLE 2). Therefore, the difference in disease-causing potential is demonstrated despite the observation that immunization with either nAChR or dAChR can generate similar levels of circulating antibodies that react with the native form of AChR and that both antibody responses express a similar isotype distribution

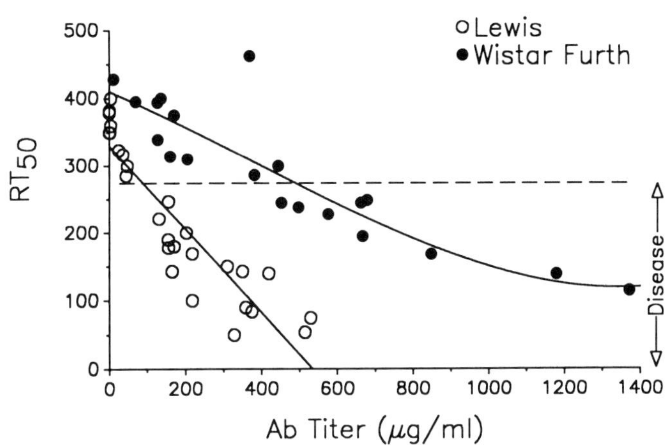

FIGURE 5. Relationship of curare sensitivity (RT_{50}), demonstrated by EDL muscles from WF rats, with serum anti-AChR antibody concentration. All rats were given an initial injection (sc) of 50 μg of AChR in CFA; some rats (groups of 4 rats each) were given 1 or 2 additional injections (sc) of 50 μg AChR in buffered saline at approximately 2 weeks and 4 weeks, respectively, following the first injection. Disregarding the exact immunization protocol, rats were then ranked according to resulting anti-AChR antibody titers and plotted as shown above, against curare sensitivity (RT_{50}). (Zoda *et al.*[13] With permission from the *Journal of Immunology*.)

TABLE 2. Immunization of Lewis Rats with AChR of Native versus Denatured Conformation

	nAChR Immunization	dAChR Immunization
Anti-AChR antibody titer with reactivity versus nAChR	High	High
Clonotypic antibody heterogeneity	High	High
Antibody avidity	Medium to High	Medium to High
Isotype expression	IgG2a dominated	IgG2a dominated
Disease induction	+ + + +	—

and levels of clonotypic heterogeneity. Further evidence of immunopathological differences comes from the intravenous administration of nAChR-induced versus dAChR-induced antibodies into immunologically naive rats; the inability of transferred anti-dAChR antibodies to interfere with AChR function is observed even though antibody binding can be demonstrated to extracellular AChR of native conformation (TABLE 3).[11] Finally, antibodies stimulated by immunization with nAChR and then adsorbed against dAChR-Sepharose to remove antibodies that recognize linear antigenic determinants retain the disease-causing subset of antibody.

Coexpression of Benign and Disease-causing Anti-AChR Antibodies in Lewis Rats

Rat and mouse monoclonal antibody studies from several laboratories have been designed to examine disease-associated antibody characteristics and have, in fact, demonstrated the differential abilities of various AChR-specific monoclonal antibodies to passively transfer EAMG.[23-27] Although the use of monoclonal antibody techniques to study the immune response to particular antigenic determinants has proven extremely useful, there still exists some confusion concerning relation-

TABLE 3. Passive Transfer of EAMG with Anti-AChR Antibodies: Requirement of Binding by Antibody to AChR of Native Conformation

Primary Antibody Tranferred[a]	Secondary Antibody Transferred[b]	Resulting RT_{50}[c]
None	None	372 ± 51
Anti-nAChR	None	110 ± 12
Anti-dAChR	None	365 ± 22
Anti-dAChR (TNP-conjugated)	None	377 ± 14
None	Anti-TNP	389 ± 10
Anti-dAChR (TNP-conjugated)	Anti-TNP	175 ± 22

[a] Groups of 3 rats each were injected with 500 μg of indicated affinity-purified antibody.

[b] Rats were injected with 1 mg of affinity-purified rat antitrinitrophenol antibodies 4 days after primary antibody transfer.

[c] The dose of curare necessary to inhibit contractile force by 50% (RT_{50}) is thought to reflect the level of functional AChR expression and has been shown to be a sensitive reproducible indicator of antibody-mediated AChR perturbations.[16,17] Values are means ± SEM determined 48 hr following antibody transfers.

ships between antibody characteristics and their disease-causing potential. Such confusion may have arisen from the inherent biases produced by the inability to fuse, clone, and expand B cells representing the entire expressed antibody repertoire contributing to the polyclonal anti-AChR antibody response or biases introduced by screening strategies based on favored antibody specificities. An alternative strategy, presented in a previous report,[9] involves the identification of disease-causing clonal B-cell products (AChR-reactive antibodies present in immune rat serum using methods that are independent of the fine specificity, frequency, tissue distribution, and fusability of the responding B cells. Unfortunately, regardless of the method of antibody isolation, direct identification of important AChR determinants will always be difficult, considering the clear requirement of native AChR conformation for the binding of important disease-causing anti-AChR anti-

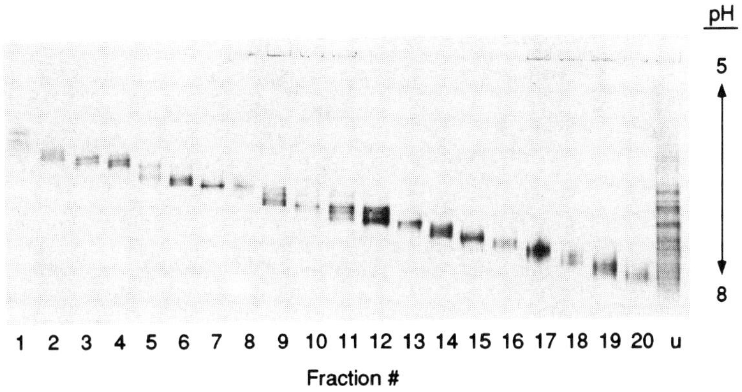

FIGURE 6. The polyclonal rat anti-AChR antibody response can be fractionated by preparative isoelectric focusing (pIEF) allowing the recovery of clonotypically restricted AChR-reactive antibodies. pIEF was performed as described in reference 9. Recovered protein from each fraction was then rerun on analytical IEF gels to assess purity and maintenance of binding activity against AChR. Numbers indicate the relative position in the pH gradient from which each fraction was recovered (increasing from the acidic to the alkaline range). Unfractionated antiserum is indicated by "u". (Thompson & Krolick.[9] With permission from *Clinical Immunology and Immunopathology*.)

bodies.[11] Therefore, the use of conventional approaches to epitope mapping (*e.g.*, synthetic peptide-binding studies and Western blot analysis) may give falsely underestimated information because these methods require antigen denaturation.

Our approach to assessing the heterogeneity and disease-causing potential of the anti-AChR antibodies generated in rats upon immunization with AChR is to purify antibodies directly from serum by preparative isoelectric focusing (pIEF) techniques (FIG. 6).[28] Following electrofocusing, each purified pIEF fraction containing certain clonotypic antibody species can be passively transferred into naive recipient rats and tested for its ability to induce symptoms of impaired AChR-dependent neuromuscular function. The goal of such studies is to identify the disease-causing subset of anti-AChR antibodies from among the entire spectrum of

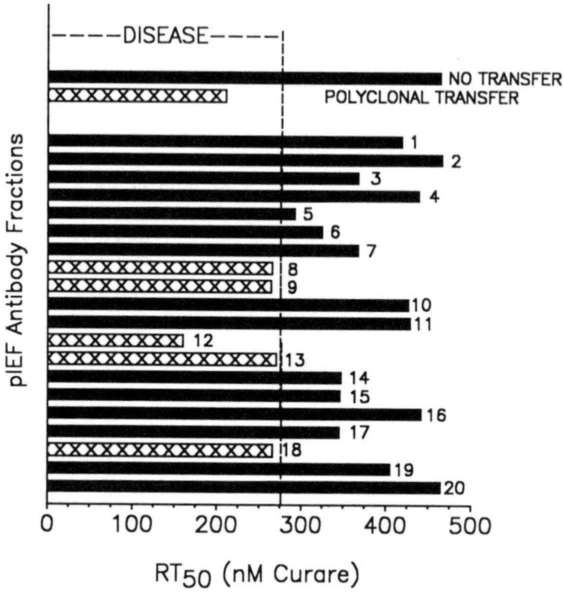

FIGURE 7. Passive transfer of pIEF fractions containing anti-AChR antibodies of equivalent binding activities revealed differences in the ability to induce AChR-dependent neuromuscular impairment. Approximately 500 μg of specific anti-AChR antibody (normalized to give equivalent binding activity) contained in each pIEF fraction described in FIG. 6 was injected (iv) into pairs of receipient Lewis rats. Forty-eight hr later the recipient rats were sacrificed, and EDL muscles were evaluated for impaired contractile function. Data are expressed as mean RT_{50} values; individual RT_{50} values of each pair generally varied by $<10\%$. This dose of curare, necessary to inhibit contractile force by 50%, is thought to reflect the level of functional AChR expression and has been shown to be a sensitive reproducible indicator of antibody-mediated AChR perturbations.[16,17] The vertical line that divides pIEF fractions into disease-causing fractions (hatched bars) and benign fractions (solid bars) has been arbitrarily placed to indicate the lowest RT_{50} value ever determined for EDL muscles from healthy control rats (*i.e.*, 270 nM curare); in reality, after greater than 200 assessments, control values are most frequently found in the 350–500 nM range and diseased values in the 100–250 nM range. (Thompson & Krolick.[9] With permission from *Clinical Immunology and Immunopathology*.)

AChR-reactive antibody clonotypes in Lewis rats with EAMG and to characterize individual antibody species capable of contributing to neuromuscular dysfunction.

Following transfer of antibodies obtained from the pIEF-fractionated sera of AChR-immune Lewis rats to immunologically naive, healthy recipient rats, two main subsets of AChR-specific antibody can be identified (FIG. 7). One subset, representing about one-third of the expressed clonotypic antibody repertoire, is capable of directly perturbing AChR-dependent neuromuscular function following transfer. A second subset demonstrates no detectable ability to induce disease symptoms following transfer. Although the anti-AChR antibodies in these experiments, produced by immunization with *Torpedo* AChR, are not 100% cross-reactive with rat AChR,[9] the inability of some antibody fractions to perturb AChR function cannot be explained by their inability to react with AChR of mammalian origin (TABLE 4). Furthermore, the ability of certain pIEF fractions to transfer disease symptoms does not correspond with a particular antibody isotype (although the response is dominated by IgG2a) and does not depend solely on high relative binding avidity (benign reactivities of high relative binding avidity are also observed). Nonetheless, an anti-AChR antibody subset can be directly identified and purified from immune serum that is likely to contain reactivities that are most directly responsible for neuromuscular disease symptoms demonstrated by rats with EAMG.

As an interesting sidelight of this study, a well-characterized monoclonal antibody (MAb35)[27,29,30] with reactivity against a main immunogenic region (MIR) of AChR and with potent disease-causing potential, appears to have a serum equivalent (FIG. 8). Interestingly, this comigrating clonotypic species of serum antibody is of the IgG1 subclass (in general, a poorly represented subclass in the anti-AChR response). Nonetheless, mAb35 represents an antibody with a potentially important fine specificity for a conformational epitope (mAb35 will not bind to denatured AChR), which can be used to compare and evaluate other AChR reactivities of unknown specificity. Although data from binding inhibition studies using mAb35 suggest that recognition of the MIR may be important with regard to disease induction for some species of antibodies produced in the anti-AChR response, this fine specificity does not correlate in every case (FIG. 9); some of the benign pIEF antibody fractions are very effective inhibitors of the mAb35/AChR interaction but are themselves incapable of directly inducing neuromuscular disease symptoms. Equally important is that certain disease-causing pIEF antibody fractions are inferior inhibitors of the binding interaction between mAb35 and AChR; this is not consistent with the notion that a single immunodominant epitope is uniquely responsible for inducing antibody-dependent neuromuscular dysfunction. Similar conclusions have been reached in studies that examined a broad panel of human anti-AChR antibodies in which it was concluded that an MIR of AChR may not always be the primary stimulus of antibody production in MG.[31]

In conclusion, the expressed/produced Lewis rat anti-AChR antibody repertoire was found to be composed of at least two main subsets of antibody, one subset capable of directly perturbing AChR function and another subset not capable of directly perturbing AChR function. Examinations of IEF banding patterns, demonstrated by other sources of anti-AChR antibodies that have no detectable disease-causing potential (*e.g.*, antibodies produced by immunization of rats with

TABLE 4. Relationships between Disease-Causing Potential and Antibody Reactivity with Mammalian AChR[a]

	Disease-Causing Fractions	Benign Fractions
Rat AChR$^+$	8, 9, 12, 13, 18	5[b], 10, 11, 14, 15, 16, 17, 19, 20
Rat AChR$^-$	None	1, 2, 3, 4, 6, 7

[a] Each pIEF antibody fraction was passively transferred into naive recipient rats and scored for the ability to directly interfere with AChR-dependent contractile function of EDL muscles as described in FIG. 7. Fractions that were determined to contain disease-causing AChR reactivity or benign AChR reactivity were then evaluated for their ability to bind AChR of mammalian origin (*i.e.*, B_3CH_1 cells). Data are presented as fraction numbers assigned to each of these four groups.

[b] RT_{50} (292 nM curare) nearly reduced into disease-causing range (see FIG. 7).

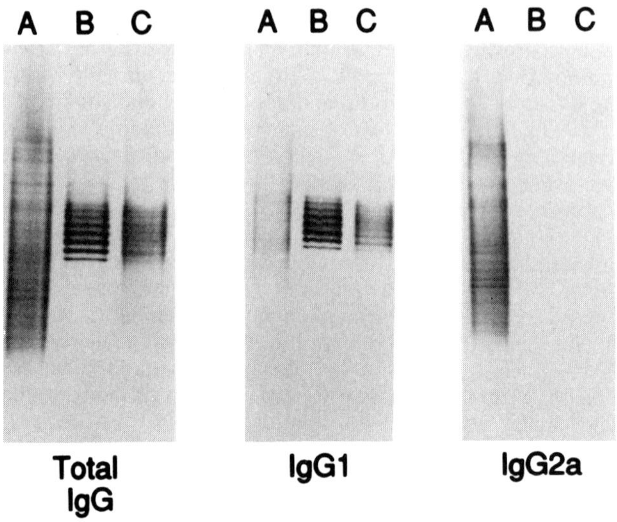

FIGURE 8. The IgG1 fraction of AChR-immune rat serum contains a dominant antibody reactivity that comigrates with MAb35 in IEF analysis. Immune serum was adsorbed to, and IgG1 eluted from, an anti-rat IgG1 affinity column. The purified Ig was then evaluated by IEF and probed for isotype as indicated for each panel. In each panel, lane A is unfractionated AChR-immune serum, lane B is MAb35, and lane C is the IgG1 fraction with immune serum obtained by affinity purification. (Thompson & Krolick.[9] With permission from *Clinical Immunology and Immunopathology*.)

denatured AChR[11] or a rat strain that demonstrates resistance to disease induction[13]), reveal an apparent absence of a strong contribution by disease-causing clonotypes, but maintenance of many of the benign antibody clonotypes. This offers a possible explanation for the lack of disease induction following immunization under these circumstances.

Role of Helper T Lymphocytes in Determining Disease Severity

The anti-AChR antibody response in EAMG is polyclonal at the B-cell level, as described above, and is T cell–dependent. Even in the absence of the outbred genetics of human patient populations, direct demonstrations of the importance of helper T-cell specificity in EAMG is difficult, complicated by restricted antigen recognition governed by the genes of the MHC. Thus, with regard to AChR-stimulated T-cell responsiveness, it has been shown in other laboratories that different strains of mice and rats demonstrate different patterns of MHC-dependent reactivity for AChR and panels of AChR peptides;[32–44] these MHC-determined differences in the expressed/activated T-cell repertoire may be the basis for differences in the severity of disease observed, although this is totally unclear at this time. Studies from this laboratory point to T-cell reactivity directed at a restricted region of the alpha subunit as having the greatest effectiveness for stimulating anti-AChR antibody production[10,12,45] and eventual neuromuscular dysfunction.

T-cell specificity has been evaluated in our laboratory for its influence on the clonotypic heterogeneity and disease-causing potential of antibodies against the AChR.[11] Based on observations presented above, that the anti-AChR antibody response is oligoclonal/polyclonal and that only a subset of these antibodies are able to directly interfere with AChR-dependent function, a question posed has concerned whether restricting the activated AChR-specific helper T-cell repertoire would subsequently result in a similar restriction in the antibody repertoire. Furthermore, would such a T-cell restricted anti-AChR antibody response continue to elicit neuromuscular disease symptoms?

Adoptive transfer studies using T cells from rats immunized with the purified AChR subunits (α, β, γ, and δ) combined with B cells from donor rats immunized

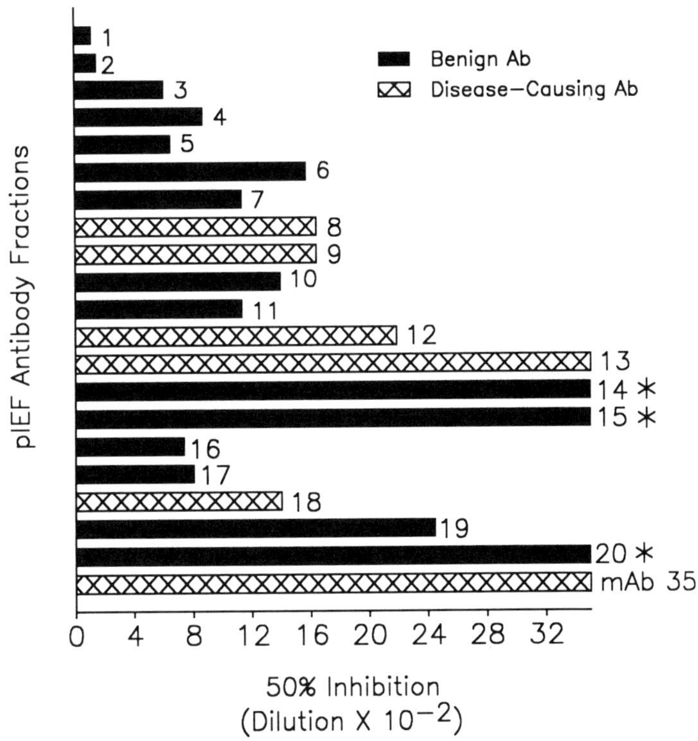

FIGURE 9. pIEF fractions demonstrated different abilities to inhibit MAb35 from binding to AChR. Serial dilutions of each pIEF fraction (each normalized to be of equivalent AChR binding activity) were mixed with a fixed solution of [^{125}I]AChR (2 × 10^7 cpm/μg/tube). Following a one hour incubation, each mixture was added in duplicate to the wells of a microtiter plate coated with MAb35. Data are presented as the dilution of the inhibiting pIEF fraction required to reduce by 50% the [^{125}I]AChR binding to the MAb35-coated wells. Fifty percent inhibition of MAb35 binding to AChR by MAb35 itself is indicated. As scored in FIG. 7, disease-causing pIEF fractions are shown as hatched bars, and benign pIEF fractions are shown as filled bars. (Thompson & Krolick.[9] With permission from *Clinical Immunology and Immunopathology*.)

with the intact AChR suggest that T-cell specificity for antigen may influence the exact subset of antibody clonotypes produced; however, total antibody titers are also affected by the T cells that are used. The level of circulating antibody clearly influences the degree of disease observed. Further ambiguity left by these studies involves the observation that regardless of the AChR subunit specificity of the T cells, all recipient animals demonstrate AChR-dependent muscle dysfunction if adequate anti-AChR antibody titers are reached. Thus, a possible conclusion is that if the importance of T-cell specificity is to be fully appreciated, the activated anti-AChR T-cell repertoire must be further restricted.

In studies reported by others,[37] the rat anti-AChR immune response is shown to be dominated by T-cell reactivity directed toward a small number of biochemical regions. The region including the alpha subunit amino acid residues at positions 100–116 demonstrates prominence in the T-cell reactivity to AChR. Furthermore, significant cross-reactivity is shown to exist for this alpha subunit sequence and determinants associated with the other AChR subunits. For these reasons it appears likely that the (α100–116) T-cell determinant(s) plays an important role in the anti-AChR antibody response and probably in generating disease symptoms. Thus, we exploited these observations to ask questions concerning the immunogenicity of the α100–116 peptide with regard to the regulation of the production of the disease-causing subset of AChR-reactive antibody.[12]

Two main observations made concerning the role of T cells reactive with the α100–116 peptide follow. First, although regulation of the anti-AChR antibody response by peptide-reactive helper T cells results in antibody titers that are lower than responses regulated by T cells reactive with the whole AChR, the clonotypic heterogeneity of the antibodies examined by IEF are similar to those regulated by unrestricted AChR-reactive T cells. This suggests that the spectrum of antibody fine specificities produced is also likely similar and of similar disease-causing potential. Second, disease-causing potential of the anti-AChR antibodies produced with the help of peptide-reactive T cells can be verified by adoptive transfers followed by tests of AChR-dependent muscle function. Thus, T cells reactive with the small (17 amino acids) synthetic peptide are capable of helping virtually all possible anti-AChR antibody clonotypes (FIG. 10) including those involved in inducing neuromuscular disease (FIG. 11).

The virtually complete expression of the entire antibody repertoire reactive with AChR, in spite of the highly restricted T-cell reactivity, was not predicted. These studies are based on the assumption that if the T-cell specificity repertoire is restricted to a very small subset of total possible reactivities that the antibody specificity repertoire would reflect this by demonstrating similar restrictions. Clearly this is not the case. The ability of a single small immunogenic region to stimulate T cells that effectively help all AChR-reactive B cells is difficult to imagine if the various proposed antigen-bridging mechanisms of T-cell to B-cell signaling are correct.[46] It is likely, however, that the region associated with α100–116 or other homologous regions are represented more than once. There are two alpha subunits per AChR monomer, each containing the (α100–116) sequence. Moreover, there appears to be significant T-cell cross-reactivity between the alpha-derived peptide and the other three AChR subunits.[37,45] This may also help explain previous observations, using AChR-reactive T hybridomas, in which over 50% of a panel of 32 clones tested demonstrated stimulated IL-2 production by *in vitro* challenge with all four purified AChR subunits.[47]

Therefore, differences in disease severity caused by manipulations of the T-cell specificities involved may have only a quantitative basis. That is, in studies performed thus far, influences of T-cell specificity on disease severity in rats appear to be directly related to the levels of circulating anti-AChR antibody attained, not influences on the selection of particular B-cell specificities. The effectiveness by which T cells transmit helper signals certainly determines the frequency, but not necessarily the specificity, of B cells activated. It is noteworthy, however, that as in other systems[48] T-cell helper activity may not correlate directly with other T-cell activities such as proliferation. Thus, a single amino acid substitution in the (α100–116) peptide (proline for threonine at position 106) results in a virtually

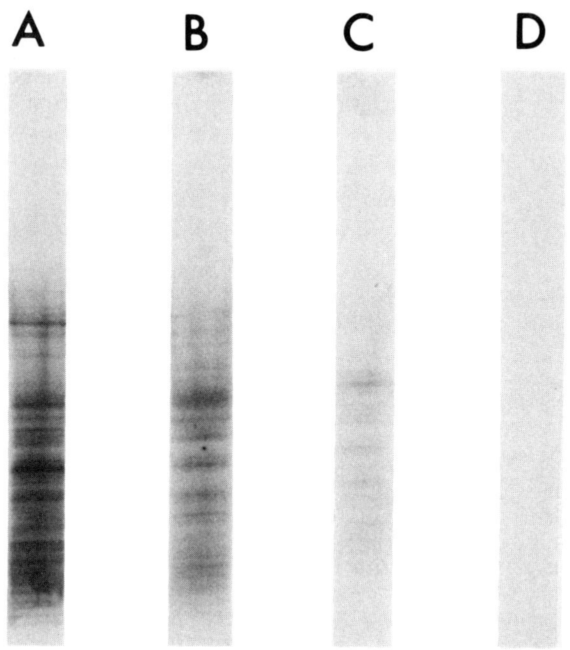

FIGURE 10. Banding pattern in IEF gels of anti-AChR antibodies. Antibodies in pooled antisera were obtained from recipient rats (4–7 rats/group) after they received helper T cells from donor rats immunized with CFA emulsions containing 50 μg of either AChR or AChR peptides as indicated below. All rats were injected with 50 μg AChR in saline 4–6 weeks prior to analysis. Lane A: AChR-immune T cells; lane B: peptide (α100–116)-immune T cells; lane C: T cells from rats injected with CFA only; lane D: serum from original T-cell donor rats immunized with peptide (α100–116) in CFA. Electrofocused serum from normal nonimmune rats demonstrated no AChR reactivity in this assay. (Yeh & Krolick.[12] With permission from the *Journal of Immunology*.)

unaffected ability to stimulate proliferation, whereas nearly all ability to stimulate the help of AChR-reactive B cells with subsequent disease induction is lost. This may imply that at least two T cells with differing fine specificities can react with the (α100–116) peptide and cross-react with the native AChR. One T cell can proliferate when challenged with the original peptide and the cross-reactive proline-substituted peptide but is not capable of providing effective help to B cells. The second T cell demonstrates effective help to B cells and may or may not be easily stimulated to proliferate by the peptides.

Taking all of these observations together, it is unlikely that the specificity of T-cell reactivities will be a clear predictor of subsequent disease severity or the key to future immunotherapies. T-cell reactivity (through proliferation) without careful evaluation of its potential helper activity may be misleading with regard to antibody responses that follow. Even T cells confirmed to have relevant helper activity may demonstrate promiscuous selectivity with regard to the B-cell clones activated. Furthermore, from patient-to-patient, the prominence of particular antigenic determinants in T-cell activation is likely to vary,[49–51] dependent on mechanisms related to antigen-processing events and the role of the MHC in the recognition of antigen by T cells. Thus, the factor more likely to remain constant in its role in disease induction is the fine specificity of anti-AChR antibodies required for interference of AChR function.

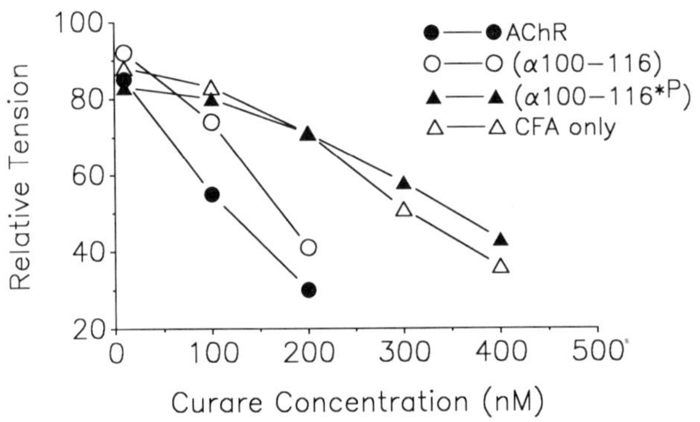

FIGURE 11. Curare sensitivity of EDL muscles from the adoptive recipient rats described in reference 12. Values represent the mean relative tension of muscle contraction obtained from each group of rats (4–7 rats/group) and are plotted versus increasing tubocurarine concentrations. As indicated, T-cell donors were rats injected with CFA emulsions of AChR (●), the (α100–116) peptide (○), the (α100–116*P) peptide (▲), or CFA only (△). (Yeh & Krolick.[12] With permission from the *Journal of Immunology*.)

FINAL CONCLUSIONS

The importance of bridging the observations described above to speculations concerning the lack of correlation between antibody titers and disease severity in MG patients is obvious. The value of being able to directly predict and monitor the disease state by the simple assessment of circulating anti-AChR antibody is only as great as the accuracy of the prediction. Clearly, if the proportion of the total expressed/produced anti-AChR antibody repertoire that has disease-causing potential differs from patient to patient, then assessment of the total antibody titer becomes meaningless unless a particular patient produces disease-causing reactivities that make up a major portion of the total titer. By contrast, a patient could produce a very high anti-AChR antibody titer but remain in relatively good health because the vast majority of the autoantibody is biologically benign. Of course, the key to exploiting the appreciation for antibody subsets with differing disease-causing potential will be to create probes that would allow the easy monitoring of the relevant reactivities. These probes may invovle anti-idiotypic antibodies that specifically recognize idiotopes that are selectively associated with disease-causing anti-AChR antibodies and that are preferably underrepresented on antibodies that lack disease-causing capability. Alternatively, molecular probes that identify expressed complementarity-determining regions associated with activated lymphocytes obtained from MG patients might very precisely assess the contribution made by the disease-causing antibody subset. Identification of all characteristics associated with the antibodies responsible for impaired muscle function would likely lead to a more precise understanding of relationships between antibody titer and disease severity, as well as the opportunity to more directly evaluate disease mechanisms.

REFERENCES

1. Vincent, A. 1980. Immunology of acetylcholine receptors in relation to myasthenia gravis. Physiol. Rev. **60:** 756.
2. Drachman, D. B. 1981. The biology of myasthenia gravis. Annu. Rev. Neurosci. **4:** 195.
3. Lindstrom, J., D. Shelton, Y. Fujii. 1988. Myasthenia gravis. Adv. Immunol. **42:** 233.
4. Lindstrom, J., M. E. Seybold, V. A. Lennon, S. Whittingham & D. D. Duane. 1976. Antibody to acetylcholine receptor in myasthenia gravis: Prevalence, clinical correlates, and diagnostic value. Neurology **26:** 1054.
5. Tindall, R. S. A. 1981. Humoral immunity in myasthenia gravis: Clinical correlations of anti-receptor antibody avidity and titer. Ann. N.Y. Acad. Sci. **377:** 316–331.
6. Drachman, D. B., R. N. Adams, L. F. Josifek & S. G. Self. 1982. Functional activities of autoantibodies to acetylcholine receptors and the clinical severity of myasthenia gravis. N. Engl. J. Med. **307:** 769.
7. Hohlfeld, R., R. Sterz, I. Kalies, H. Wekerle & K. Peper. 1983. Experimental myasthenia: Lack of correlation between the autoantibody titer and the reduction of acetylcholine-controlled ionic channels measured at functioning endplates. Muscle and Nerve **6:** 160.
8. Brown, R. M. & K. A. Krolick. 1988. Clonotypic analysis of the antibody response to the acetylcholine receptor in experimental autoimmune myasthenia gravis. J. Neuroimmunol. **19:** 205.
9. Thompson, P. A. & K. A. Krolick. 1992. Acetylcholine receptor-reactive antibodies in experimental autoimmune myasthenia gravis differing in disease-causing potential: Subsetting by preparative isoelectric focusing. Clin. Immunol. Immunopathol. **62:** 199.
10. Yeh, T. & K. A. Krolick. 1987. Influence of T cell specificity on the heterogeneity and disease-causing capability of antibody against the acetylcholine receptor. J. Neuroimmunol. **17:** 17.
11. Yeh, T. & K. A. Krolick. 1989. Clonotypic analysis of anti-acetylcholine receptor antibodies produced against native and denatured antigen. J. Neuroimmunol. **24:** 133.
12. Yeh, T. M. & K. A. Krolick. 1990. T cells reactive with a small synthetic peptide of the acetylcholine receptor can provide help for a clonotypically heterogeneous antibody response and subsequently impaired muscle function. J. Immunol. **144:** 1654.
13. Zoda, T., T. M. Yeh & K. A. Krolick. 1991. Clonotypic analysis of anti-acetylcholine

receptor antibodies from EAMG-sensitive Lewis rats and EAMG-resistant Wistar Furth rats. J. Immunol. **146:** 663.

14. BROWN, R. M. & K. A. KROLICK. 1988. Selective idiotype suppression of an adoptive secondary anti-acetylcholine receptor antibody response by immunotoxin treatment prior to transfer. J. Immunol. **140:** 893.
15. BIESECKER, G. & D. KOFFLER. 1988. Resistance to experimental autoimmune myasthenia gravis in genetically inbred rats: Association with decreased amounts of *in situ* acetylcholine receptor-antibody complexes. J. Immunol. **140:** 3406.
16. OLSBERG, C. A., L. C. MAXWELL, T. M. MIKITEN & K. A. KROLICK. 1987. Analysis of contractile properties of muscles from rats immunized with purified acetylcholine receptor. J. Neuroimmunol. **14:** 253.
17. THOMPSON, P. A., R. A. BARHON & K. A. KROLICK. 1991. Repetitive nerve stimulation vs. twitch tension in rats with EAMG. Muscle and Nerve **15:** 94.
18. GUTMAN, G. A. 1986. Rat immunoglobulin allotypes. *In* Genetics and Molecular Immunology. D. M. Weir, Ed.: Chapter 98. Blackwell Scientific Publications, Oxford.
19. BARTFELD, D. & S. FUCHS. 1977. Immunological characterization of an irreversibly denatured acetylcholine receptor. FEBS Lett. **77:** 214.
20. BARTFELD, D. & S. FUCHS. 1978. Specific immunosuppression of experimental autoimmune myasthenia gravis by denatured acetylcholine receptor. Proc. Natl. Acad. Sci. USA **75:** 4006.
21. BARTFELD, D. & S. FUCHS. 1979. Fractionation of antibodies to acetylcholine receptor according to antigenic specificity. FEBS Lett. **105:** 303.
22. LINDSTROM, J. M., V. A. LENNON, M. E. SEYBOLD & S. WHITTINGHAM. 1976. Experimental autoimmune myasthenia gravis and myasthenia gravis: Biochemical and immunochemical aspects. Ann. N.Y. Acad. Sci. **274:** 254–274.
23. LENNON, V. & E. LAMBERT. 1980. Myasthenia gravis induced by monoclonal antibodies to acetylcholine receptor. Nature **285:** 238.
24. LENNON, V. A. & E. H. LAMBERT. 1981. Monoclonal antibodies to acetylcholine receptors: Evidence for a dominant idiotype and requirement of complement for pathogenicity. Ann. N.Y. Acad. Sci. **377:** 77–96.
25. GOMEZ, C. & D. RICHMAN. 1983. Anti-acetylcholine receptor antibodies directed against the α-bungarotoxin binding site induce a unique form of experimental myasthenia. Proc. Natl. Acad. Sci. USA **80:** 4089.
26. GOMEZ, C. & D. RICHMAN. 1985. Monoclonal anti-acetylcholine receptor antibodies with differing capacities to induce experimental autoimmune myasthenia gravis. J. Immunol. **135:** 234.
27. TZARTOS, S., S. HOCHSCHWENDER, P. VASQUEZ & J. LINDSTROM. 1987. Passive transfer of experimental autoimmune myasthenia gravis by monoclonal antibodies to the main immunogenic region of the acetylcholine receptor. J. Neuroimmunol. **15:** 185.
28. THOMPSON, P. A. & K. A. KROLICK. 1991. Subsetting of acetylcholine receptor-reactive antibodies by preparative isoelectric focusing. Prep. Biochem. **21:** 229.
29. TZARTOS, S. J., D. E. RAND, B. L. EINARSON & J. M. LINDSTROM. 1981. Mapping of surface structures of electrophorus acetylcholine receptor using monoclonal antibodies. J. Biol. Chem. **256:** 8635.
30. TZARTOS, S. J., A. KOKLA, S. L. WALGRAVE & B. M. CONTI-TRONCONI. 1988. Localization of the main immunogenic region of human muscle acetylcholine receptor to residues 67–76 of the α subunit. Proc. Natl. Acad. Sci. USA **85:** 2899.
31. LENNON, V. A. & G. E. GRIESMANN. 1989. Evidence against acetylcholine receptor having a main immunogenic region as target for autoantibodies in myasthenia gravis. Neurology **39:** 1069.
32. FUCHS, S., D. NEVO, R. TARRAH-HAZDAI & I. YAAR. 1976. Strain differences in the autoimmune response of mice to acetylcholine receptors. Nature **263:** 329.
33. CHRISTADOSS, P., V. A. LENNON & C. DAVID. 1979. Genetic control of experimental autoimmune myasthenia gravis in mice. I. Lymphocyte proliferative response to acetylcholine receptor is under H-2–linked Ir gene control. J. Immunol. **123:** 2540.
34. CHRISTADOSS, P., J. LINDSTROM, R. MELVOLD & N. TALAL. 1985. Mutation at I-A beta chain prevents experimental autoimmune myasthenia gravis. Immunogenetics **21:** 33.

35. BERMAN, P. W. & J. PATRICK. 1980. Linkage between the frequency of muscular weakness and loci that regulate immune responsiveness in murine experimental myasthenia gravis. J. Exp. Med. **152:** 507.
36. BIESECKER, G. & D. KOFFLER. 1988. Resistance to experimental autoimmune myasthenia gravis in genetically inbred rats: Association with decreased amounts of *in situ* acetylcholine receptor-antibody complexes. J. Immunol. **140:** 3406.
37. FUJII, Y. & J. LINDSTROM. 1988. Specificity of the T cell immune response to acetylcholine receptor in experimental myasthenia gravis. Response to subunits and synthetic peptides. J. Immunol. **140:** 1830.
38. ZHANG, Y., T. BARKAS, M. JUILLERAT, B. SCHWENDIMANN & H. WEKERLE. 1988. T cell epitopes in experimental myasthenia gravis of the rat: Strain-specific epitopes and cross-reaction between two distinct segments of the α chain of the acetylcholine receptor (*Torpedo californica*). Eur. J. Immunol. **18:** 551.
39. YOKOI, T., H. MULAC-JERICEVIC & M. Z. ATASSI. 1987. T lymphocyte recognition of acetylcholine receptor. Localization of the full T cell recognition profile on the extracellular part of the α chain of *Tropedo californica* acetylcholine receptor. Eur. J. Immunol. **17:** 1697.
40. ATASSI, M. Z., M. BISERKA, T. YOKOI & Y. MANSHOURI. 1988. Localization of the functional sites on the α chain of acetylcholine receptor. Fed. Proc. Fed. Am. Soc. Exp. Biol. **46:** 2538.
41. LENNON, V. A., D. J. MCCORMICK, E. H. LAMBERT, G. E. GRIESMANN & M. Z. ATASSI. 1985. Region of peptide 125–147 of acetylcholine receptor alpha subunit is exposed at neuromuscular junction and induces experimental autoimmune myasthenia gravis, T cell immunity, and modulating autoantibodies. Proc. Natl. Acad. Sci. USA **82:** 8805.
42. HOHLFELD, R., I. KALIES, F. HEINZ, J. R. KALDEN & H. WERKELE. 1981. Autoimmune rat T lymphocytes monospecific for acetylcholine receptor: Purification and fine specificity. J. Immunol. **126:** 1355.
43. INFANTE, A. J., P. A. THOMPSON, K. A. KROLICK & K. A. WALL. 1991. Determinant selection in murine experimental autoimmune myasthenia gravis: Effect of the bm12 mutation on T cell recognition of acetylcholine receptor epitopes. J. Immunol. **146:** 2977.
44. INFANTE, A. J., H. LEVOCOVITZ, K. A. WALL, P. A. THOMPSON & K. A. KROLICK. 1992. Preferential use of a T cell receptor Vβ gene by acetylcholine receptor-reactive T cells from myasthenia gravis susceptible mice. J. Immunol. **148:** 3385.
45. KROLICK, K. A. & O. E. URSO. 1986. Influence of T cell specificity on the antibody response to the acetylcholine receptor. J. Neuroimmunol. **13:** 75.
46. CELADA, F. & E. E. SERCARZ. 1988. Preferential pairing of T-B specificities in the same antigen: The concept of directional help. Vaccine **6:** 94.
47. TAMI, J. A., O. E. URSO & K. A. KROLICK. 1987. T cell hybridomas reactive with the acetylcholine receptor and its subunits. J. Immunol. **138:** 732.
48. ARTHUR R. & D. MASON. 1986. T cells that help B cell responses to soluble antigen are distinguishable from those producing interleukin 2 on mitogenic or allogeneic stimulation. J. Exp. Med. **163:** 774.
49. HOHLFELD, R., K. V. TOYKA, K. HEININGER, H. GROSSE-WILDE & I. KALIES. 1984. Autoimmune human T lymphocytes specific for acetylcholine receptor. Nature **310:** 244.
50. HOHLFELD, R., K. V. TOYKA, L. L. MINER, S. L. WALGRAVE & B. M. CONTI-TRONCONI. 1988. Amphipathic segment of the nicotinic receptor alpha subunit contains epitopes recognized by T lymphocytes in myasthenia graves. J. Clin. Invest. **81:** 657.
51. MELMS, A., S. CHRESTEL, B. C. SCHALKE, H. WEKERLE, A. MAURON, M. BALLIVET & T. BARKAS. 1989. Autoimmune T lymphocytes in myasthenia gravis. Determination of target epitopes using T lines and recombinant products of the mouse nicotinic acetylcholine receptor gene. J. Clin. Invest. **83:** 785.

T-Helper Epitopes on Human Nicotinic Acetylcholine Receptor in Myasthenia Gravis[a]

LUCIA MOIOLA,[b] MARIA PIA PROTTI,[b,e]
ANGELO A. MANFREDI,[b,e] MI-HA YUEN,[b,c]
JAMES F. HOWARD JR.,[d] AND
BIANCA M. CONTI-TRONCONI[b,c]

[b]*Department of Biochemistry*
University of Minnesota
1479 Gortner Avenue
St. Paul, Minnesota 55108

[c]*Department of Pharmacology*
University of Minnesota
435 Delaware Street
St. Paul, Minnesota 55108

[d]*Department of Neurology*
University of North Carolina at Chapel Hill
Chapel Hill, North Carolina 27599

In both human myasthenia gravis (MG) and in its experimental model, experimental autoimmune myasthenia gravis (EAMG), induced in experimental animals by immunization with purified nicotinic acetylcholine receptor (AChR),[1] the synthesis of anti-AChR antibodies, which are high-affinity IgGs, requires intervention of specific T helper (Th) cells.[2,3] An important role for Th cells in MG and EAMG pathogenesis is supported by the findings that EAMG can be prevented or ameliorated by treatment with anti-Ia or anti-CD4 antibodies,[4,5] and that the only obvious and early effect on the anti-AChR response of thymectomy—a staple in the treatment of MG—is an immediate and substantial decrease of anti-AChR reactivity of circulating T cells.[6] Anti-AChR Th (CD4$^+$) cells are present in the blood and thymus of MG patients, and they can be propagated *in vitro* from those tissues.[7–14]

Because of the pathogenic relevance of anti-AChR Th cells for MG development, we have sought for several years to characterize their functional properties and to determine their epitope repertoire on the AChR molecule—namely, the sequence regions of the AChR and, when possible, the individual residues that form epitopes recognized by the autoimmune CD4$^+$ cells. These studies, and parallel ongoing studies on the molecular interactions occurring between the Th

[a] The studies reported here were partially supported by the NINCDS Grant NS 29919, a Grant from the Muscular Dystrophy Association of America, and the U.S. NIDA program project Grant 5P01-DA05695 (to B. M. Conti-Tronconi). L. Moiola and A. A. Manfredi are recipients of MDA Postdoctoral Fellowships. M. P. Protti was the recipient of the MDA Sidney Blackmer Award.

[e] Permanent address: Laboratory of Immunology, H. S. Raffaele, School of Medicine, Universita' degli Studi di Milano, Milano, Italy.

epitope on the AChR sequences and the restricting MHC class II molecules, should ultimately yield a molecular picture of the MHC/peptide complexes that activate the anti-AChR Th cells.

Towards these aims we have carried out studies that, according to the type of anti-AChR Th cells studied, can be grouped as follows: (1) studies of long-term polyclonal anti-AChR Th cells propagated by stimulation with *Torpedo* AChR; (2) studies of long-term polyclonal anti-AChR Th cell lines propagated by the use of overlapping synthetic peptides, corresponding to the complete sequence of human AChR subunits; (3) studies of unselected $CD4^+$ cells from MG patients' blood; and (4) studies of Th cell lines specific for individual AChR epitopes.

The above approaches have different advantages and shortcomings, as will be discussed in the following sections. The major difference between the usage of AChR-specific Th cell lines and that of unselected blood $CD4^+$ cells is that propagation of AChR-specific Th cell lines on one hand results in a Th cell population highly enriched in AChR-specific Th cells; indeed if their propagation is carried on for enough cycles, all the resulting Th cells may be specific for AChR epitopes.[10]

Such enrichment allows detailed and controlled studies of the recognition and structure of individual AChR epitopes and production of monoclonal AChR-specific Th lines[11-14] necessary for studies of the structure of the corresponding autoimmune T-cell receptors. On the other hand, selective clonal enrichment or loss of Th cells specific for particular epitopes may occur during propagation of the lines and may result in Th cell populations that do not reflect faithfully the original anti-AChR Th repertoire. This risk is obviously avoided by using the total, unselected $CD4^+$ population from the patients' blood, which, however, has the disadvantage that inasmuch as the AChR specific cells are a small minority of the total $CD4^+$ population, their *in vitro* responses to AChR antigens/epitopes may be marginal. Indeed these responses can be detected only in the most severely affected patients.[14]

The combined results of these different approaches allowed definition of several functional properties of the autoimmune anti-AChR Th cells and identification of a large part of their epitope repertoire.

Long-term AChR-Specific Autoimmune T-Cell Lines Propagated by Stimulation with Torpedo *AChR Are CD4⁺ Helper Cells That Recognize DR-Restricted AChR Epitopes, Mostly on the α-Subunit*

Due to its easy purification in large amounts, *Torpedo* electric tissue AChR (TAChR) was first used for propagation *in vitro* of the autoimmune anti-AChR T cells.[2] Its use was justified by its structural similarity with mammalian muscle AChR[15,16] and by earlier studies that demonstrated that AChRs from both *Torpedo* and *Electrophorus* electric tissue could induce an *in vitro* response of T cells from MG patients' blood.[17-19]

Hohlfeld *et al.*[2] first reported propagation of long-term T-cell lines by cycles of stimulation with TAChR and interleukin-2 (IL-2). In collaboration with Klaus Toyka's and Reinhard Hohlfeld's laboratory, we carried out studies using long-term AChR-specific T-cell lines propagated with purified, native TAChR.[3,7,20,21] Because *Torpedo* and human AChRs have substantial sequence dissimilarities,[15,16] the use of TAChR for stimulation and propagation of Th cell lines yields a population enriched only in those Th cells recognizing epitopes conserved in human and *Torpedo* AChRs, and this causes slow growth and a biased epitope repertoire of the T-cell lines obtained. In spite of these limitations, several important conclusions were reached. The T-cell lines obained were AChR-specific: they responded vigorously and in a dose-dependent manner to TAChR,[7] and cross-reacted with human muscle AChR antigens.[2,21] Further, as expected from T cells that recognize denatured segments of the antigen bound to the presenting MHC molecule, they were stimulated equally well by native or fully denatured TAChR.[7] This finding had important practical ramifications, because it justified the use of denatured forms of the AChR—like purified TAChR subunits[7] or synthetic sequences of human muscle AChR (see below)—to study the epitope specificity of the autoimmune T cells.

Recognition of AChR epitopes was consistently restricted by DR molecules.[20] The lines had uniform $CD3^+CD4^+CD8^-$ phenotype[21] and functioned *in vitro* as T helper cells, that is, when co-cultured with B cells from the same patients, they efficiently stimulated synthesis/secretion of anti-AChR antibodies.[3]

Due to the slow growth rate of TAChR-propagated Th cell lines, we could carry out limited investigations of the parts of the AChR molecule forming their epitopes.[7,21] We first challenged the Th lines with purified TAChR subunits.[7] Consistently, the lines recognized most strongly or exclusively the α-subunit. Although preferential recognition of the TAChR α-subunit may occur because this subunit is the most conserved between *Torpedo* and human muscle AChRs,[15,16] and stimulation with TAChR propagates only the Th clones recognizing conserved epitopes, still these results indicated that the AChR α-subunit contains important epitopes for Th sensitization, in addition to the main immunogenic region (MIR): that is, the B epitope, or set of largely overlapping B epitopes, which dominates the anti-AChR antibody response in human MG and rat EAMG,[1] which is also on the α-subunit. Although the MIR is conserved in *Torpedo* and human muscle AChRs, the epitopes recognized by Th cell lines propagated with TAChR did not overlap the MIR region, because complexes of TAChR with high-affinity anti-MIR antibodies stimulated the Th lines as efficiently as TAChR alone.[7]

Two TAChR-propagated Th cell lines were tested for their epitope specificity using a small panel of overlapping synthetic peptides, corresponding to the amino terminal region of the α-subunit of human muscle AChR.[21] Different peptides were recognized by the autoreactive T cells of the two patients. One patient, of HLA-DR 1,4 haplotype, recogized two overlapping synthetic sequences of the human α-subunit, α 7-22 and α 19-34. The other patient, of HLA-DR 3, 8, recognized the synthetic sequence α 1-14. Due to the low number of AChR-specific Th cells obtained by TAChR propagation, testing of the lines with other synthetic AChR sequences for further definition of their epitope repertoire could not be done.

Synthetic Human AChR Sequences Are Efficiently Recognized by the Autoimmune Th Cells

Ideally, human AChR should be used to propagate and study the autoimmune Th cells of MG patients. Unfortunately that is not practical, because purification of suitable amounts of AChR from human muscle is difficult if at all possible, although the recent success of genetically engineered human AChR sequences corresponding to large segments of the α-subunit for studies of Th cells indicates that this is a viable approach for propagation of Th lines.[22] Irrespective of the method used for propagation of AChR-specific T-cell lines, identification of their epitope repertoire necessitates the use of large panels of short AChR sequences, screening the complete sequences of the AChR subunits. Although large panels of short biosynthetic sequences can in principle be genetically engineered, chemical synthesis is presently the easiest and least labor intensive method for their production.

We have exploited the use of short (twenty residue) overlapping peptides corresponding to the complete sequences of the human muscle AChR α-, β-, γ-, and δ-subunits (FIG. 1) for propagation and testing of anti-AChR T cells (refs. 10–13 and Moiola and Conti-Tronconi, in preparation). The demonstration that anti-AChR cells from MG patients recognize denatured AChR as efficiently as native AChR[20] (FIG. 2) strongly suggested that denatured synthetic AChR sequences would be good antigens for T-cell stimulation. This possibility was verified by studying the response of $CD4^+$ cells from the blood of MG patients to a pool of thirty-two overlapping synthetic peptides corresponding to the complete sequence of the human AChR α-subunit (α-pool) (FIG. 1A).[23] $CD4^+$ cells from MG patients responded to the α-pool to an extent that correlated with the severity of their symptoms. During acute phases of MG the $CD4^+$ response was vigorous, and it occurred early (three days) upon exposure to the α-pool (FIG. 3A), indicating the existence of activated, AChR-specific $CD4^+$ cells in the patients' blood. The presence of circulating activated $CD4^+$ cells was directly demonstrated by FACS analysis, which revealed high numbers of activated (DR^+) $CD4^+$ cells (~4% of the total T cells). In patients clinically improved after immunosuppressive treatments, the anti-AChR response of blood $CD4^+$ cells could be detected after more prolonged exposure to the α-pool (FIG. 3B), consistent with stimulation *in vitro* of AChR-specific $CD4^+$ cells that were inactive *in vivo*. Presumably, the activated $CD4^+$ cells detected in acutely ill patients had been eliminated in the recovered patients as a result of the immunosuppressive treatment. FACS analysis of the phenotype of blood lymphocytes from patients after their clinical improvement did not reveal activated $CD4^+$ cells. The delayed but substantial response to the α-pool of these patients, which reached a maximum after seven days and was elicited by relative high doses of α-pool (FIG. 3B), suggests that inactive, AChR-specific Th cells were still present, recruited into active division *in vitro* by prolonged exposure to the α-pool. In MG patients that were in a stable situation of minimal symptoms or remission, no $CD4^+$ response to the α-pool could be detected.

In conclusion, that study[23] clearly indicated that pools of synthetic peptides corresponding to parts, or in principle to all, of the human muscle AChR sequence are useful antigens for studying the AChR-specific response of autoimmune T cells. In a broader perspective, the use of synthetic sequences for the study of the function and epitope repertoire of autoimmune T cells should be feasible for all autoimmune diseases for which the antigen and its primary structure are known.

Propagation of Long-term Th Cell Lines by Stimulation with Pools of Synthetic Sequences of the Human Muscle AChR

Prompted by the results of the study summarized above, we used pools of overlapping synthetic peptides corresponding to the complete sequences of the human AChR α-, β-, γ-, and δ-subunits to propagate *in vitro* AChR-specific Th cell lines from MG patients (TABLE 1) and healthy controls.

All the T lines propagated with peptide pools had uniform $CD3^+$ $CD4^+$ $CD8^-$ phenotype and recognized the AChR epitope in a DR-restricted fashion, because they could be stimulated and propagated by the use of DR-matched antigen-presenting cells (APC). They strongly recognized the peptide pool used for their propagation (FIG. 4) and cross-reacted well with purified mammalian muscle AChR

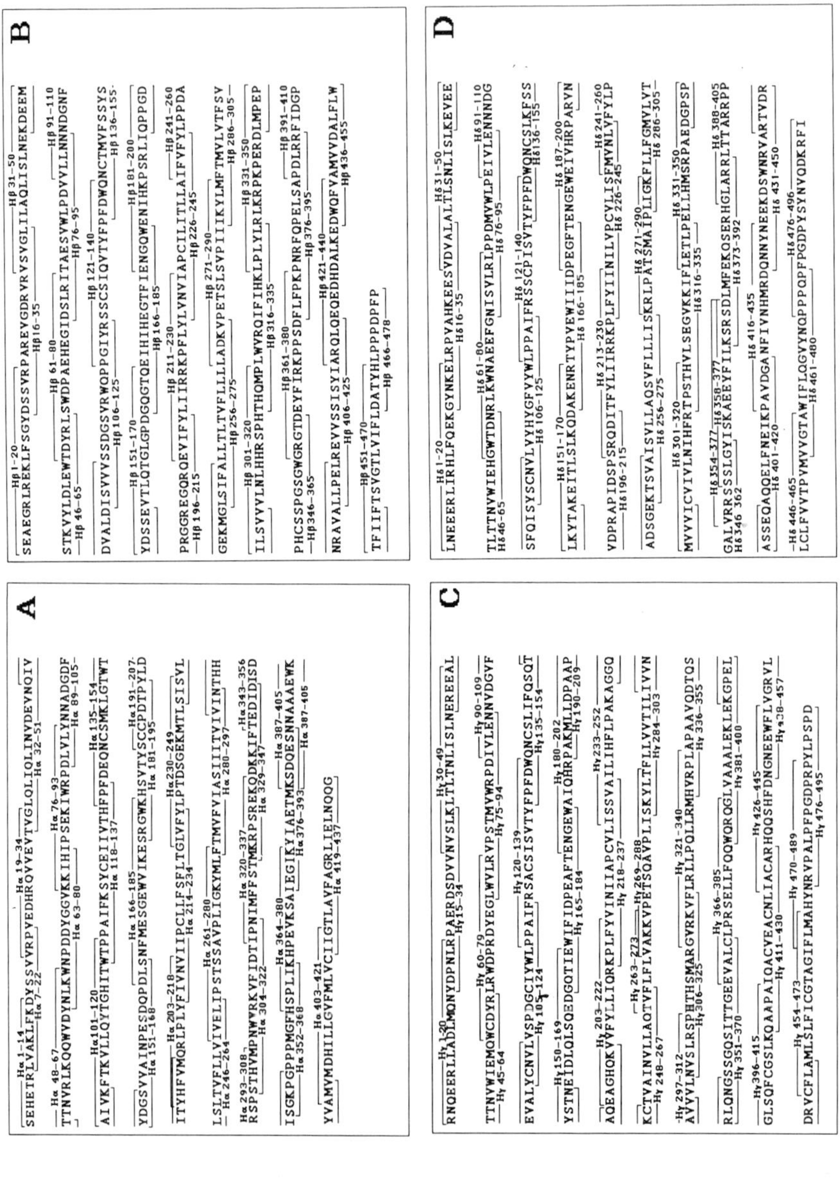

FIGURE 1. Synthetic peptides used; they are aligned on the sequence of the corresponding human AChR subunits (panel A: α-subunit; panel B: β-subunit; panel C: γ-subunit; panel D: δ-subunit).

from bovine muscle (*e.g.*, see FIG. 5). Bovine muscle AChR is a good substitute for the human AChR, which cannot be purified in suitable amounts, because its subunits are highly homologous to those of human AChR.[15,16] Some lines cross-reacted with TAChR: they presumably recognize epitopes within conserved sequence regions in human and *Torpedo* AChRs.[7,11] The T lines cross-reacted minimally or not at all with peptide pools corresponding to subunits other than that used for their propagation (FIG. 5),[12,13] as expected for polyclonal lines containing a mixture of Th cells recognizing epitopes either nonconserved between the different AChR subunits or formed by partially conserved regions of the homologous AChR subunits.[15,16]

Most lines showed, at stages of their propagation that differed from line to line, some degree of reactivity with autologous or DR-matched APC, so that they were stimulated by the sole presence of DR-matched APC.[10] Autologous class II molecule(s) seem the target of this recognition, because it occurred even when APC from a new DR-matched donor were used for the first time. The anti-APC response of some T-cell lines developed over some weeks, reached a peak, and

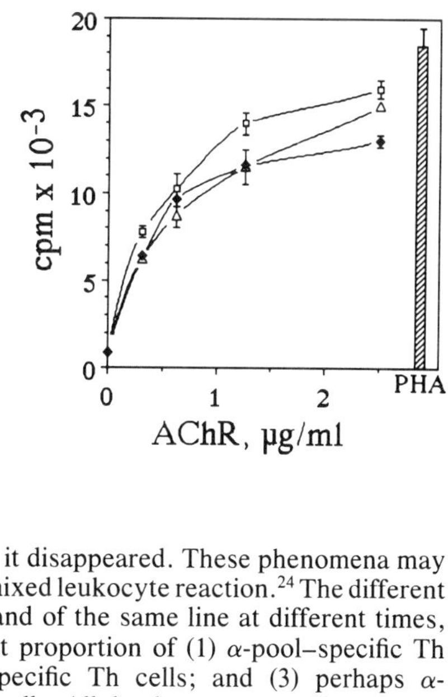

FIGURE 2. Reactivity of AChR-specific Th cells, propagated with cycles of stimulation with native, membrane-bound TAChR, to native membrane-bound TAChR (black triangles), SDS denatured and reduced TAChR (open squares), and SDS denatured TAChR (open triangles). The maximum proliferation induced by PHA (slashed bar PHA) is indicated. T-cell proliferation is expressed as [^{3}H]thymidine incorporation (average cpm ± SD of triplicate cultures). The dose dependence of the T-cell stimulation is the same for the different forms of TAChR used, and in all cases the maximum AChR stimulation is close to that induced by PHA.

then decreased in its intensity. Sometimes it disappeared. These phenomena may be similar to those causing the autologous mixed leukocyte reaction.[24] The different pattern of reaction of the different lines, and of the same line at different times, may be due to the presence of a different proportion of (1) α-pool–specific Th cells; (2) autoreactive, class II CD4$^+$-specific Th cells; and (3) perhaps α-pool–specific but also class II–reactive Th cells. All the three cases outlined above have been demonstrated in experimental animals.[25]

Polyclonal subunit-specific anti-AChR Th cell lines can be easily propagated in large amounts, and they have been successfully used to identify the repertoire of epitopes recognized by the autoreactive Th cells in MG, as will be described below. Structural elucidation of such T-cell epitopes, and possibly identification of sequence segments immunodominant for Th sensitization (see below), could be an important step for devising targeted immunosuppressive therapies.[26] Furthermore, the possibility of propagating from any given patient the autoreactive T clones is a useful step toward the use of such Th cells as "vaccines" to stimulate specific suppressive regulatory responses,[26] as has been done for experimental autoimmune encephalomyelitis (EAE). In EAE, vaccination with Th cells specific for the autoantigen induces sensitization of $CD8^+$ cells, specific for epitopes unique to the autoreactive Th cells[27–33] or shared by all activated Th cells,[34] and resistance to EAE induction or improvement of existing symptoms. Finally, clonal libraries of anti-AChR Th cells obtained from the polyclonal lines by limiting dilution[2] will be an essential tool for investigating the genetics of the anti-AChR autoimmune response and the structure of the T-cell receptor (TcR) of the autoreactive Th cells involved in AChR recognition.

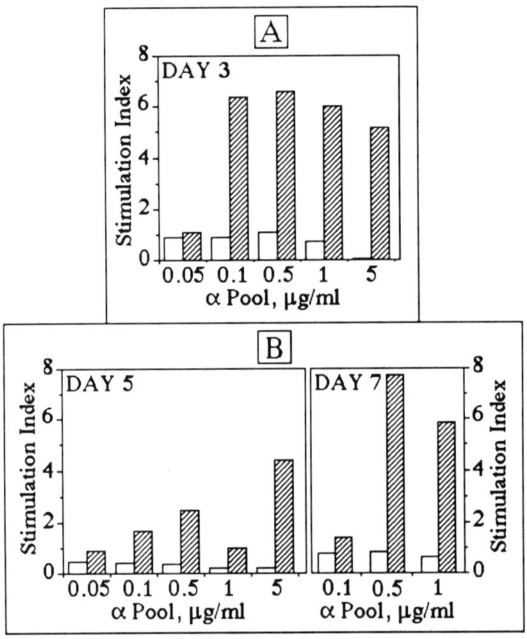

FIGURE 3. A: Reactivity of unselected PBMC (white bar) and of $CD8^+$-depleted, $CD4^+$-enriched cells (hatched bar) from severely affected MG patients after 3 days of culture with α-pool. **B:** Reactivity of unselected PBMC (white bar) and of $CD8^+$-depleted, $CD4^+$-enriched cells (hatched bar) from recovering patients after 5 and 7 days of culture with α-pool. Results are expressed as stimulation index, that is, the ratio between the [^{3}H]thymidine uptake (average cpm of triplicate cultures) in the presence of α pool and the absence of any stimulation. (Modified from Protti *et al.*[23])

Epitope Repertoire of Anti-AChR Th-Cell Lines: Identification of Immunodominant Regions for T-Cell Sensitization on the α-, β-, and γ-Subunits

The AChR sequence regions recognized by four anti-α, five anti-β, four anti-γ, and three anti-δ autoimmune long-term T-cell lines were determined by challeng-

TABLE 1. Characteristics of the Patients

Patient	Age, Sex	Disease Class[a]	Anti-AChR Ab Titer (nM)	Thymectomy	Immunosuppressive Therapy[b]	HLA-DR haplotype	Th lines established			
							anti-α	anti-β	anti-δ	anti-γ
S.D.	36 y, F	2	neg.	+	Pred	DR1,DRw14(6)	+	+	+	+
S.R.	59 y, M	2	4.22	+	Pred	DRw17(3),DR7	+	+	+	−
R.J.	73 y, M	2	3.55	−	Pred	DRw15(2),DR7	+	NA[c]	−	+
T.C.	48 y, F	2	1.78	+	Pred	DRw15(2),DR4	+	+	−	+
D.P.	89 y, M	2	1.25	−	Aza	DR1,DR4	+	−	+	+
M.J.	45 y, F	2	252.00	+	Pred	DRw15(2),DR12(5)	NA[c]	+	NA	NA
M.P.	25 y, F	4	22.50	+	Pred	DRw8,DR9	NA[c]	+	NA	NA

[a] See refs. 11–13 and Moiola and Conti-Tronconi, unpublished.
[b] Pred, prednisone; Aza, azathioprine.
[c] NA, Not attempted.

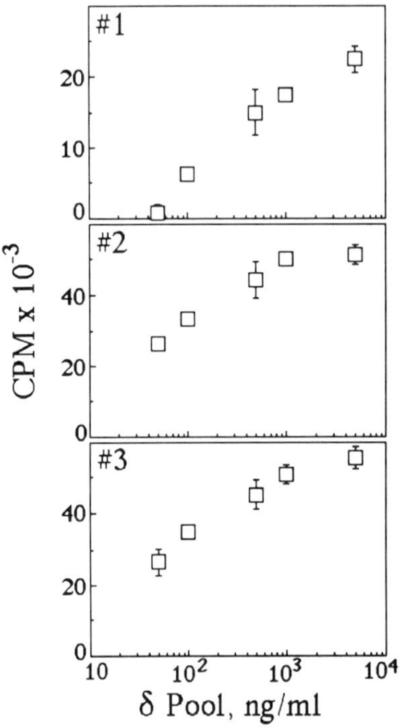

FIGURE 4. Reactivity of δ-pool–specific T-cell lines to increasing concentrations of δ-pool (0.05 to 5 μg/mL). T-cell proliferation is expressed as [^{3}H]thymidine incorporation (average cpm ± SD of triplicate cultures). (Modified from Protti *et al.*[12])

ing the lines with the individual synthetic peptides screening the corresponding AChR subunit. Each line had a individual pattern of peptide recognition that was consistent during the propagation of the line (*e.g.*, see FIG. 6). Some synthetic sequences of the α-, β-, and γ-subunit were recognized by most or all lines, irrespectively of their HLA-DR haplotype. Four immunodominant regions were identified for the α-[11,14] (FIG. 6), three on the β- (Moiola and Conti-Tronconi, unpublished), and two on the γ-subunit (Protti, Manfredi, Moiola, and Conti-Tronconi, unpublished). Therefore, although the Th response against the AChR in MG is polyclonal and polyepitopic, defined small regions of the AChR sequence seem to be the most frequent target of autoimmune Th cell recognition, a situation that could be exploited for development of epitope-specific immunosuppressive procedures, as was done for EAE.[26]

Because the peptides used were twenty residues long, and class II–restricted epitopes may be as short as 9–14 residues,[35–38] each immunodominant synthetic sequence could accommodate several nested epitopes, as described for immunodominant regions of myelin basic protein.[39] The immunodominance of particular sequence regions might be due to their easier cleavage and release during processing of the antigen or their folding into a conformation more agreeable for DR binding. Alternatively, a single truly immunodominant T epitope might exist, recognized in the context of different restricting DR elements, as described for other antigens.[40–42] Both alternatives—immunodominant regions or immunodominant epitopes—might occur for antigen recognition by human Th cells because of the promiscuous nature of the binding of human DR molecules to peptide epitopes: most or all human DR molecules bind a wide spectrum of peptide sequences, and about 25% of the peptide sequences tested bound to any DR molecule.[43]

Epitope Repertoire of Unselected CD4⁺ Cells from the Blood of MG Patients: Identification of a Large Number of T Epitopes on Human AChR and Verification of the Existence of Sequence Regions Immunodominant for T-Cell Sensitization

In addition to the immunodominant regions and other individual epitopes identified by the use of the long-term subunit-specific anti-AChR T-cell lines described above, other parts of the AChR sequence may be involved in anti-AChR response,

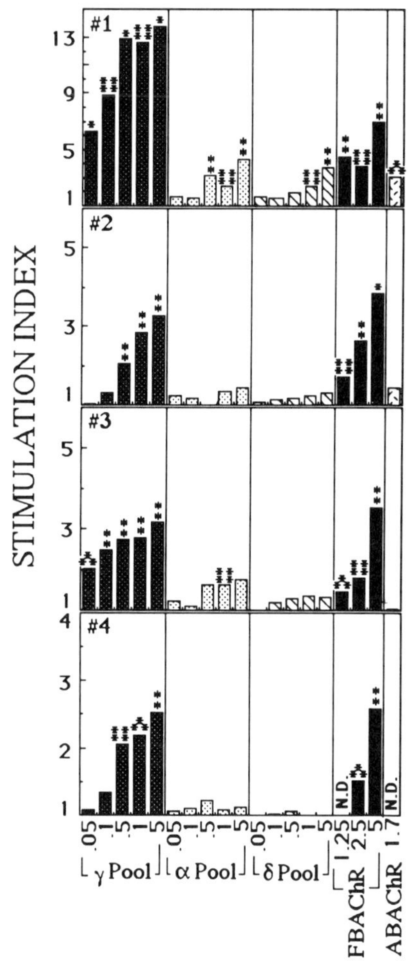

FIGURE 5. Reactivity of γ-pool–propagated T-cell lines to increasing concentrations of γ-pool (0.05 to 5 μg/mL) and to fetal bovine (FBAChR) and adult bovine AChR (ABAChR) purified by α-bungarotoxin affinity chromatography. Results are expressed as stimulation index, that is, the ratio between the [^{3}H]thymidine uptake (mean of the cpm of triplicate cultures) in the presence of the antigen and the absence of any stimulation. Asterisks indicate responses significantly higher than unstimulated proliferative rate (*$p < 0.001$, **$p < 0.005$, ***$p < 0.01$, ****$p < 0.025$). (Protti *et al.*[13] With permission from *Neurology*.)

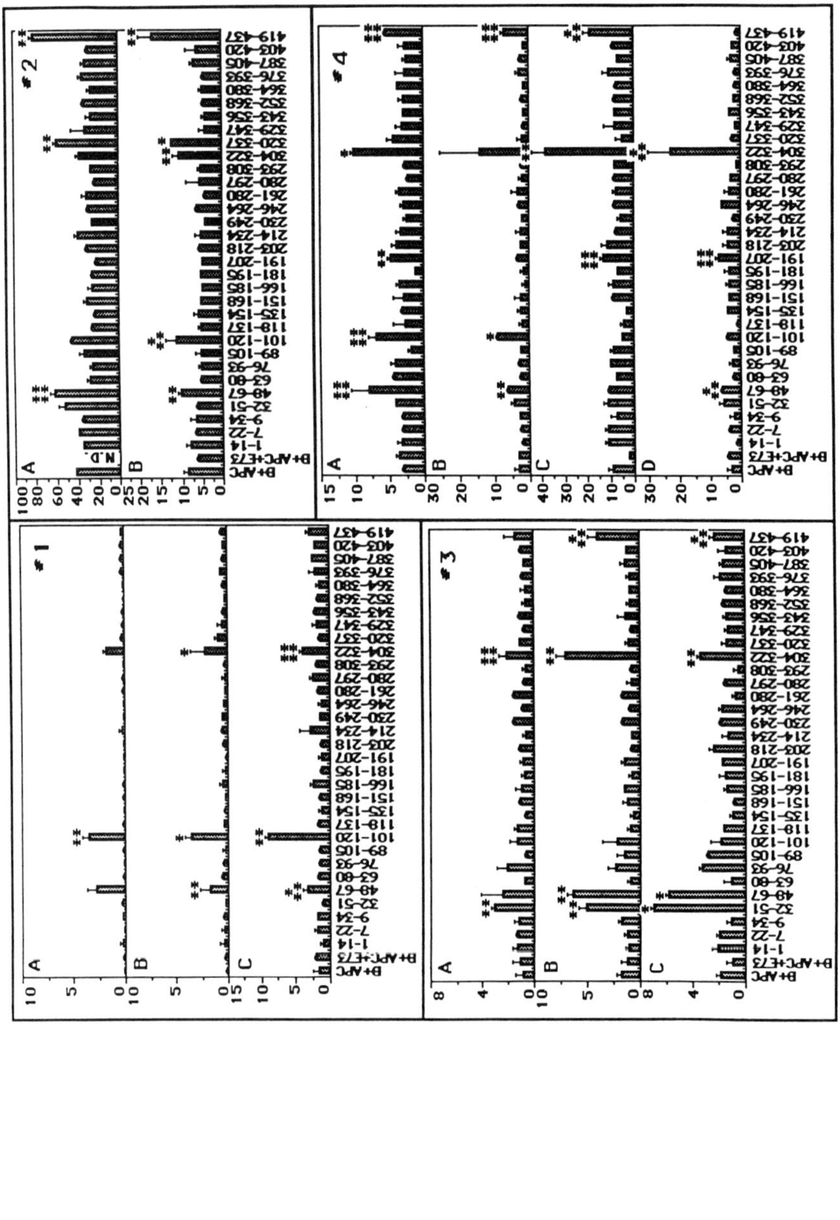

#1
#2
#3
#4
A
B
C
D
N.D.
B+APC
B+APC+E73
1-14
7-22
9-34
32-51
48-67
63-80
76-93
89-105
101-120
118-137
135-154
151-168
166-185
181-195
191-207
203-218
214-234
230-249
246-264
261-280
280-297
293-308
304-322
320-337
329-347
343-356
352-368
364-380
376-393
387-405
403-420
419-437

undetected by long-term anti-AChR $CD4^+$ cell lines because of clonal loss during their propagation or because of the very limited size of the patient population from which the lines were derived (TABLE 1). To properly define the epitope repertoire recognized on the AChR by the autoimmune anti-AChR Th cells, it is necessary to study a relatively large number of MG patients and to use experimental conditions designed to avoid any clonal loss/enrichment.

To obtain a more comprehensive assessment of the epitope repertoire of anti-AChR autoimmune Th cells in MG patients and to verify the existence of the immunodominant sequence region, we tested $CD8^+$-depleted, unselected $CD4^+$ T cells and/or total peripheral blood mononuclear cells (PBMC) from the blood of a relatively large population of MG patients (twenty-two patients), using individual overlapping synthetic peptides, twenty residues long, and screening the complete sequence of the human AChR α-, γ-, and δ-subunits (ref. 14 and Manfredi, Protti, Howard, and Conti-Tronconi, submitted). Several patients were tested twice, once when they suffered from severe symptoms and once during a period of remission. In agreement with our previous studies on the response to the AChR α-subunit of $CD8^+$-depleted, $CD4^+$-enriched cells,[23] only the most severely affected patients recognized AChR epitopes, and they were mainly young women. The extent of the $CD4^+$ cell response was facilitated by removal of the $CD8^+$ cells (see below). Several epitopes were identified on each AChR subunit (ref. 14 and Manfredi, Protti, Howard and Conti-Tronconi, submitted).

A relationship between severity of the disease and reactivity of the circulating $CD4^+$ cells to AChR epitopes was confirmed by experiments in which blood $CD4^+$ cells from four patients were tested twice, once when they were severely affected and once after the clinical situation had improved, or before worsening of the symptoms.[14] In all cases the $CD4^+$ cells recognized several peptides only when the patients were severely affected. After stable improvement of the symptoms, the $CD4^+$ reactivity to the synthetic AChR sequences was markedly reduced or absent.

Most epitopes identified by the use of the long-term T lines (see above) were also detected by the blood $CD4^+$ cells of one or more of the patients used for these studies, which included most of the patient donors of anti-AChR $CD4^+$ lines. The existence of immunodominant regions on the α- and γ-subunits was also confirmed, because those synthetic sequences elicited significant response in a large fraction of HLA-DR disparate patients tested. A few new immunodominant sequence regions were identified, in most cases overlapping the immunodominant sequences discussed above. For example, on the α-subunit sequence, in addition to the previously identified immunodominant sequences α 48-67, α 101-120, α 304-322, and α 419-437, the synthetic sequences α 118-137 (overlapping α 101-120) and 320-337 (overlapping α 304-322) were also recognized with an unusual high frequency (FIG. 7). Two new immunodominant regions, twenty residues long, were identified in the δ-subunit. Several other synthetic sequences were recognized by individual patients.

FIGURE 6. Anti-AChR Th cell lines from four MG patients consistently recognized the same individual synthetic peptides, after several restimulations with the α-pool. The first test was performed after a satisfactory enrichment in anti-AChR–specific cells was reached. B, C, and D refer to successive tests done at one- to two-week intervals. Asterisks indicate responses significantly higher than unstimulated proliferative rate (*$p < 0.001$, **$p < 0.005$, ***$p < 0.01$, ****$p < 0.025$). T-cell proliferation is expressed as $[^3H]$thymidine incorporation (average cpm $\pm$ SD of triplicate cultures). (Originally published in the *Proceedings of the National Academy of Sciences USA*.[11])

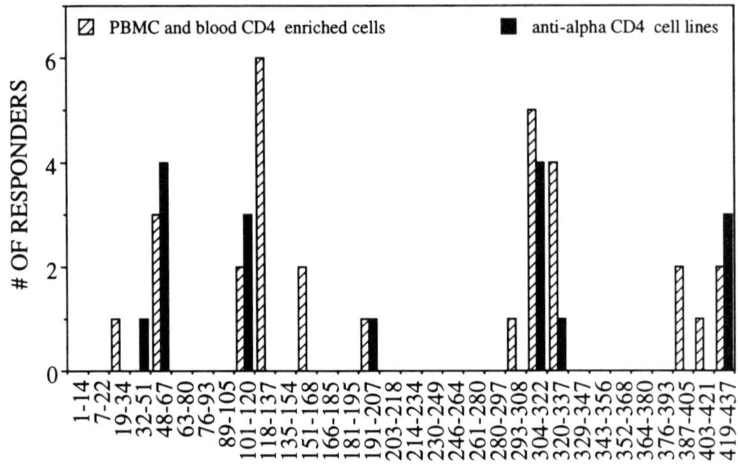

FIGURE 7. Alpha-subunit peptide sequence. Frequency with which the different α-subunit peptides were recognized by unselected CD8$^+$-depleted, CD4$^+$-enriched cells from the blood of eight MG patients (slashed columns)[14] and by long-term anti-AChR Th cell lines propagated from four MG patients by stimulation with α-pool.[11] (Manfredi *et al.*[14] With permission from *Neurology*.)

Important constituent elements of the MIR, the epitope or set of largely overlapping epitopes which dominates the antibody response in MG and rat EAMG[1], are within residues 67-76 of the AChR α-subunit.[44,45] Crucial constituent elements of the cholinergic binding site, which also may induce formation of anti-AChR antibodies in MG patients,[46] are within residues α 187-194 and possibly α 55-74.[47,48] Only one patient recognized a peptide, α 191-207,[14] that marginally overlaps one of these sequence regions, in good agreement with the proposed "complementarity" of T and B epitopes,[49] that is, that T and B cells recognize nonoverlapping sets of epitopes, as demonstrated in different experimental systems.[49] This may occur because (1) the mechanisms of antigen recognition of the B and T cells are very different (surface areas of the native proteins for the B cells, cleaved and denatured short sequence regions, "embedded" onto a suitable MHC molecule for the T cells), and (2) antigen-specific B cells efficiently bind their specific antigen by way of their surface antibodies, process it, and present the Th cell epitopes derived from degradation of the antigen/antibody complexes, where presumably the region of the antigen protected by the bound antibody may be resistant to processing and presentation.[49,50]

In conclusion, these studies on the T repertoire of unselected anti-AChR Th cells clearly indicated that whereas a restricted set of AChR sequence regions is almost invariably involved in the sensitization of autoimmune T cells, a relatively large number of T epitopes are recognized by any individual patient, thus indicating that at least during the later phases of the anti-AChR autoimmune response, when the myasthenic symptoms are obvious, the whole AChR molecule is involved in the autoimmune response (TABLE 2). This does not exclude that the anti-AChR sensitization may be initiated by an altered form of AChR, as it occurs in experimental systems, or by the mechanism of molecular mimicry between one small region of the AChR molecules and an epitope of bacterial or viral origin, inasmuch as it has been proposed that once tolerance to a multideterminant antigen is broken for one epitope, as a result of a molecular mimicry or altered self-protein, responsiveness to multiple epitopes may occur.[51] Comparison between AChR sequences, identified in our studies as forming Th epitopes in MG patients, and the known protein sequences of existing databases did not allow us to reach final conclusions on these matters, because strong sequence similarity with bacterial or microbial protein—including proteins from common pathogens—could be found for some epitope sequences, and also for some control AChR sequences, that were never recognized by our patients. On the other hand, the immunodominant AChR sequence regions did not have any striking similarity with known microbial proteins.

The existence of a relatively large Th repertoire for any individual patient raises doubts as to the feasibility of epitope-specific immunosuppressive procedures. Two considerations may assuage such misgivings. First, it has been demonstrated that both in EAMG and in EAE only few defined T and/or B epitopes cause sensitization of pathogenic T or B cells, whereas antibodies or T-cell responses against other parts of the AChR or of the myelin basic protein do not induce appearance of symptoms.[1,52,53] If this is the case also in human MG, identifications of the AChR T epitopes recognized by the Th cells that cooperated with B cells

TABLE 2. T-cell Epitopes on the α-subunit of Human AChR[a]

Residues	Reference	Residues	Reference
Region α 1-38		Region α 257-269	
α 1-34	21	α 257-269	9,88
α 12-27	87		
α 19-34	14	Region α 293-308	
α 23-38	87	α 293-308	14
Region α 48-71		Region α 304-337	
α 48-67	11,14	α 304-322	11,14
α 56-71	87	α 310-327	88
		α 320-337	11,14
Region α 85-143			
α 85-142	8	Region α 351-368	
α 89-104	87	α 351-368	88
α 101-120	11,14		
α 111-126	87	Region α 387-405	
α 118-137	14	α 387-405	14
α 125-143	9		
		Region α 403-437	
Region α 146-181		α 403-421	14
α 146-162	87	α 419-437	11,14
α 151-168	14		
α 158-174	87		
α 169-181	88		
Region α 182-212			
α 182-198	87		
α 191-207	14		
α 194-210	87		
α 195-212	9,88		

[a] We list the sequence regions of the α-subunit comprising epitopes identified in the studies listed in the Reference column. Some regions include overlapping sequence segments identified as part of a T epitope. It is possible that some overlapping sequences contain the same epitope, within the region of overlap.

for production of highly pathogenic antibodies may reveal that very few of the identified AChR epitopes are involved in MG pathogenesis. Such pathogenic T epitopes may well be the immonodominant regions recognized by most or all of the MG patients. The possibility of "reconstructing" the human anti-AChR response in the SCID mice (see relative chapter in this volume) may be an excellent tool for dissecting the pathogenic potential of different AChR T and B epitopes.

Anti-AChR CD4+ T Cells Exist in Healthy Controls

Th cells against autoantigens have been demonstrated in normal subjects.[54–56] PBMC from normal subjects may respond to synthetic sequences and biosynthetic human segments of the AChR α-subunit,[57] although they did not respond to the complete TAChR molecule.[19,23] On the other hand, specific T lines could be derived from healthy subjects by stimulation with TAChR.[22,57] AChR-specific T-cell clones could be obtained from healthy subjects using a biosynthetic fragment of mouse AChR α-subunit or a full-length biosynthetic human α-subunit, but much less frequently than from MG patients.[58] In another study, specific T lines could be derived from all controls tested by stimulation with biosynthetic fragments of the murine AChR α-subunit.[59] Propagation of specific Th cell lines from healthy controls using pools of synthetic peptides corresponding to the complete sequence of human AChR subunits was seldom successful (refs. 10–13 and Protti, Manfredi, Dalton, Moiola, Howard, and Conti-Tronconi, submitted). A $CD4^+$ δ-subunit–specific line was propagated for a short time from a normal subject[12] and $CD4^+$ cell lines specific for the γ-subunit could be propagated for a short time from three of six healthy controls (ref. 13 and Protti, Manfredi, Dalton, Moiola, Howard, and Conti-Tronconi, submitted). Of the latter, one responded detectably to the γ-pool, but not to any individual peptide, suggesting that the specific $CD4^+$ T cells interacted with γ-subunit epitopes with an affinity too low to allow *in vitro* detection of any peptide epitope. The other two γ-specific lines recognized some individual peptides but could not be propagated for longer than 4–5 weeks, consistent with inefficient recognition of their epitopes (Protti, Manfredi, Dalton, Moiola, Howard, and Conti-Tronconi, submitted). Therefore, although potentially autoreactive $CD4^+$ T cells recognizing AChR sequences may exist in healthy subjects, Th cells recognizing AChR epitopes with high affinity, which can be propagated *in vitro* for long periods of time by stimulating with AChR antigens, are uniquely characteristic of MG patients.

The results summarized above clearly indicate that potential anti-AChR autoimmune Th cells exist in the healthy population and that, as Zhang and coworkers pointed out, "immunoregulatory imbalance rather than the presence of autoreactive T-cell clones is responsible for the pathogenesis of MG."[58] Autoreactive T cells have been demonstrated in many experimental systems where the host does not normally develop autoimmune disease,[60] including EAMG.[61] BALB/c mice develop EAMG upon injection of rat AChR, which is presumably very similar, if not identical, to the autologous murine AChR;[62] in this strain, $CD4^+$ T cells exist, which recognize the murine sequence region α 304-322.[61] The presence of $CD4^+$ cells reactive to self-antigens does not indicate a failure of tolerance, because the many autoreactive cells that are deleted from the T-cell repertoire during maturation within the thymus are probably those with optimal affinity for the self-antigen/MHC complex.[60] T cells with a low-affinity antigen receptor may escape clonal deletion, but because of the similarity between the mechanism of thymic clonal deletion and of peripheral T-cell activation—which both require optimal formation of the ternary complex among the T-cell antigen receptor (TCR), the MHC restricting element, and the T epitope—the autoreactive T cells may never be activated at the periphery, due to the low affinity of their TCRs.[51–60] The potentially autoreactive T cells can be activated under experimental conditions and cause development of autoimmune responses.

Anti-AChR Th Cells Specific for Embryonic Muscle AChR Exist in Many Myasthenic Patients, Suggesting That the Anti-AChR Autoimmune Response May Initiate in a Tissue Other than Innervated Skeletal Muscle

Mammalian muscle AChR exists in two forms, the expression of which is developmentally regulated.[63,64] Embryonic or denervated muscle expresses AChR formed by α-, β-, γ-, and δ-subunits. Upon innervation, the synthesis of the γ-subunit is turned off and a gene encoding a different, homologous ε-subunit is activated in its stead.[63,64] The two forms of muscle AChR have different functional, pharmacologic, and metabolic properties.[63–65]

We could propagate T-cell lines from four of five patients by the use of a pool of synthetic peptides corresponding to the embryonic γ-subunit.[13] The $CD4^+$ T lines obtained responded to the γ-pool vigorously and reproducibly (FIG. 5). A substantial response to the γ-pool frequently occurred at doses as low as 20 nM of each peptide, indicating that the T-cell receptor of these autoimmune Th cells interacts with the MHC molecules/peptide complex with high affinity, that is, in the nanomolar range. These lines strongly recognized AChR from embryonic mammalian muscle and reacted less or not at all with adult mammalian muscle AChR[13] (FIG. 5). The specificity of these lines for embryonic AChR was confirmed by identification of the sequence regions of the γ-subunit forming their epitopes (Protti, Manfredi, Dalton, Moiola, Howard, and Conti-Tronconi, submitted). For all lines, some or all of the synthetic sequences they recognized were unique to the γ-subunit (*i.e.* they are not conserved in the other AChR subunits). Studies on the AChR epitope repertoire of unselected $CD4^+$ cells from MG patients' blood confirmed that most patients recognize γ-subunit epitopes. Because embryonic AChR is not expressed in normal innervated muscle, the existence of embryonic AChR-specific Th cells in a large fraction of MG patients suggests that the primary anti-AChR Th sensitization may occur in a tissue other than skeletal muscle and possibly within the thymus, which expresses an AChR molecule similar or identical to embryonic AChR (see below). Antibodies specific for embryonic AChR may also occur in MG patients.

The Thymus Expresses an AChR-like Protein Similar or Identical to Embryonic Muscle AChR, Which May Be the Primary Antigen in MG

In MG patients, the thymus is frequently hypertrophic and may contain a thymoma.[66,67] The role played by the thymus in the pathogenesis of MG is still uncertain, but Wekerle *et al.*[68] proposed that the anti-AChR sensitization starts within the thymus, which in MG patients frequently contains anti-AChR B cells and T cells.[8,69–72] Immunological cross-reactivity between a thymus component and muscle AChR has been shown.[73–75] The thymus contains binding sites for α-bungarotoxin (α-BTX) that specifically recognizes AChRs from peripheral tissues.[76–78] In addition, cultured cells from adult human thymus can differentiate into striated muscle fibers that bear AChR.[79–81]

We previously demonstrated that the α-BTX–binding component from thymus is related both structurally and immunologically to a true AChR[82] because it contains four peptides with apparent molecular masses of 40 kDa, 51 kDa, 56 kDa, and 66 kDa, which cross-react with antisera raised against TAChR. We called this α-BTX–binding component thymus AChR-like protein (T-AChR-LP).

Inasmuch as antibodies and Th cells specific for embryonic AChR epitopes are present in MG patients, it was important to determine if the T-AChR-LP was similar to embryonic or adult AChR. The former is not expressed in adult (innervated) skeletal muscle and, at least in bovine muscle, the corresponding mRNA is absent at birth.[83] The presence within the thymus of a form of AChR normally absent from other tissues could be related to the breakdown in tolerance that causes MG. Toward this aim, peptides corresponding to unique sequence segments of bovine γ- and ε-subunits were synthesized and used to raise subunit-specific antibodies. These antibodies and other monoclonal and polyclonal antibodies against the α-, β-, and δ-subunits were used to identify the complement of subunits present in preparations of purified T-AChR-LP from bovine thymus.[84] Subunits immunologically corresponding to the embryonic muscle AChR α-, β-, γ-, and δ-subunits were present. Therefore a protein similar or identical to embryonic muscle AChR is expressed in normal thymus, lending credence to the notion that a primary AChR sensitization within the thymus is one of the first steps in the pathogenesis of MG.

The Response of CD4⁺ Cells to AChR Epitopes Is Inhibited by CD8⁺ Cells That Recognize—in a Class I MHC–Restricted Fashion–an Activation Antigen of CD4⁺ Cells

When we tested the response to AChR synthetic sequences of unselected T cells from the blood of MG patients, we found that in most cases the total PBMC population did not respond to AChR antigens, whereas the $CD8^+$-depleted, $CD4^+$-enriched population had an active response (FIG. 3).[10] This suggested the existence of inhibitory $CD8^+$ T cells. This possibility had been proposed previously to explain the finding that $CD4^+$ T-cell depletion enhances anti-AChR antibody secretion by Pokeweed mitogen–stimulated B cells from MG patients.[85] AChR-specific T-suppressor cells have been cloned from animals immunized with TAChR,[86] which inhibited the response of AChR-specific T cells *in vitro* and the immune response to injected TAChR *in vivo*. Circumstantial evidence of an immunoregulatory circuit involving $CD8^+$ cells in MG is the frequent finding of activated $CD8^+DR^+$ cells in the blood of MG patients.[10] The presence of activated $CD8^+$ cells in the blood correlated with the favorable development of the symptoms.[10] In EAE, immunosuppressive circuits involving $CD8^+$ cells specific for idiotypes or activation antigens expressed by the antimyelin $CD4^+$ cells (or in general by activated $CD4^+$ cells) exist and have been exploited for development of specific immunosuppression of the antimyelin $CD4^+$ response by vaccination with activated $CD4^+$ cells or with synthetic sequences of their TcRs, and stimulation of immunosuppressive $CD8^+$ cells.[28–34] The demonstration of the potential for similar $CD8^+$ circuits in MG strongly suggests the feasibility of similar immunosuppressive strategies. We have therefore investigated if the activated $CD8^+$ cells present in the blood of MG patients recognize antigens expressed by the anti-AChR $CD4^+$ cells. Towards this aim, AChR-specific $CD4^+$ blasts from a given myasthenic patient were irradiated and used as an antigen to detect a specific response in microproliferation assays, using as responder $CD4^+$-depleted, $CD8^+$-enriched cells from the blood of the same patients. Consistently, a clear proliferative response of the $CD8^+$ cells was detected. The response was directed against an antigen expressed by activated CD4$^+$ blasts, because irradiated blood CD4$^+$ cells from the same patients did not stimulate proliferation of the CD8$^+$ cells. The response was MHC-restricted, because the response was not induced by irradiated heterologous anti-AChR CD4$^+$ blasts and was inhibited by anti-MHC class I antibodies (Yuen and Conti-Tronconi, in preparation). Similar studies carried out in healthy controls recently vaccinated against tetanus toxin or diphteria toxin indicate that appearance of CD8$^+$ cells recognizing activation antigens expressed by activated antigen-specific CD4$^+$ cells, and able to inhibit CD4$^+$ cell proliferation by a noncytotoxic mechanism, is a general immunoregulatory mechanism that seems to operate during phases of intense CD4$^+$ activation (Yuen and Conti-Tronconi, in preparation).

These results clearly indicate that induction of CD8$^+$ cells against antigen-activated CD4$^+$ cells is part of the normal immunoregulatory circuits that operate in the control of active CD4$^+$ responses. These results also support the possibility that immunization with activated anti-AChR CD4$^+$ cells may be used to control the anti-AChR autoimmune response. At the very least, stimulation of these CD8$^+$ immunoregulatory subsets could be conceivably used to help recovery during the acute myasthenic crisis.

SUMMARY

The synthesis of AChR antibodies requires intervention of AChR-specific Th cells. Because of the paucity of anti-AChR Th cells in the blood of myasthenia gravis (MG) patients, direct studies of these autoimmune cells in the blood are seldom possible. Propagation *in vitro* of anti-AChR T cells from MG patients by cycles of stimulation with AChR antigens selectively enriches and expands the autoimmune T-cell clones, allowing investigation of their function and epitope specificity. *Torpedo* electroplax AChR was initially used for propagation of anti-AChR T-cell lines. Those studies demonstrated the feasibility of *in vitro* propagation of AChR-specific T cells. These are bona fide CD4$^+$ Th cells, which stimulate production *in vitro* of anti-AChR antibodies by B cells of myasthenic patients and recognize equally well denatured and native AChR, suggesting the usefulness of synthetic human AChR sequences as antigens for propagation of the autoimmune Th cells.

We used pools of overlapping synthetic peptides, corresponding to the complete sequences of the human AChR α-, β-, γ-, and δ-subunits, to propagate AChR-specific Th cells from the blood of MG patients. The AChR sequence regions forming epitopes recognized by the autoimmune T cells were determined by challenging the lines with individual synthetic peptides, 20 residues long, screening the AChR subunit sequences. Although each line had an individual pattern of epitope recognition—as expected from their different HLA-DR haplotype—some peptides were recognized by most or all the CD4$^+$ T-cell lines, irrespective of their DR haplotype. The existence of immunodominant regions of the AChR sequence was verified by investigating the response of unselected CD4$^+$ cells from the blood of a relatively large number of MG patients to the individual peptides screening the human α-, γ-, and δ-subunit sequences. Those studies confirmed that each patient has an individual pattern of peptide recognition. The studies also identified a large number of T epitopes of the human AChR and verified the existence of sequence regions immunodominant for T-helper sensitization, because a limited number of sequence regions, including all those immunodominant for the T-helper lines, were recognized by most patients.

Anti-AChR CD4$^+$ T lines could be propagated from some healthy controls only for a brief period of time. They recognized AChR sequences poorly, suggesting a low affinity of their T-cell receptors for the corresponding AChR epitopes.

Most MG patients have CD4$^+$ cells specific for the embryonic form of muscle AChR, which in skeletal muscles disappears after innervation, suggesting that initiation of the anti-AChR immune response may occur in a tissue(s) other than innervated skeletal muscle, and possibly within the thymus, which expresses a protein similar or identical to embryonic muscle AChR.

The response of CD4$^+$ cells to AChR epitopes is inhibited by CD8$^+$ cells, which recognize—in a class I MHC–restricted fashion—a structure (structures) present on antigen-activated CD4$^+$ cells.

REFERENCES

1. Lindstrom, J., D. Shelton & Y. Fujii. 1988. Adv. Immunol. **42:** 233–284.
2. Hohlfeld, R., K. V. Toyka, K. Heininger, H. Grosse-Wilde & I. Kalies. 1984. Nature **310:** 244–24.
3. Hohlfeld, R., I. Kalies, B. Kohleisen, K. Heininger, B. M. Conti-Tronconi & K. V. Toyka. 1986. Neurology **36:** 618–621.
4. Waldor, M. K., S. Sriram, H. O. McDevitt & L. Steinman. 1983. Proc. Natl. Acad. Sci. USA **80:** 2713–2717.
5. Christados, P. & M. J. Dauphinee. 1988. J. Immunol. **137:** 2437–2440.
6. Morgutti, M., B. M. Conti-Tronconi, A. Sghirlanzoni & F. Clementi. 1979. Neurology **29:** 734–738.
7. Hohlfeld, R., K. V. Toyka, S. J. Tzartos, W. Carson & B. M. Conti-Tronconi. 1987. Proc. Natl. Acad. Sci. USA **84:** 5379–5383.
8. Melms, A., B. C. G. Schalke, T. Kirchner, H. K. Muller-Hermelink, E. Albert & H. Wekerle. 1988. J. Clin. Invest. **81:** 902–908.
9. Harcourt, G. C., N. Sommer, J. Rothbard, H. N. A. Wilcox & J. Newson-Davis. 1988. J. Clin. Invest. **82:** 1295–1300.
10. Protti, M. P., A. A. Manfredi, C. Straub, X. D. Wu, J. F. Howard Jr. & B. M. Conti-Tronconi. 1990. J. Immunol. **144:** 1711–1720.
11. Protti, M. P., A. A. Manfredi, C. Straub, X. D. Wu, J. F. Howard Jr. & B. M. Conti-Tronconi. 1990. Proc. Natl. Acad. Sci. USA **87:** 7792–7796.
12. Protti, M. P., A. A. Manfredi, C. Straub, X. D. Wu, L. Moiola, J. F. Howard, Jr. & B. M. Conti-Tronconi. 1991. J. Immunol. **146:** 2253–2261.
13. Protti, M. P., A. A. Manfredi, C. Straub, J. F. Howard Jr & B. M. Conti-Tronconi. 1991. Neurology **41:** 1809–1814.
14. Manfredi, A. A., M. P. Protti, X. D. Wu, J. F. Howard Jr. & B. M. Conti-Tronconi. 1992. Neurology **42:** 1092–1100.
15. Stroud, M. B., M. P. McCarthy & M. Shuster. 1990. Biochemistry **29**(50): 11009–11023.
16. Claudio, T. 1989. Frontiers in Molecular Biology: Molecular Neurobiology Volume. D. M. Glover & B. D. Harnes, Eds.: 63–142. IRL Press. Oxford.
17. Abramsky, O., A. Aharonov, C. Webb & S. Fuchs. 1975. Clin. Exp. Immunol. **19:** 11–16.
18. Richman, D. P., J. P. Antel, J. W. Patrick & B. G. W. Arnason. 1979. Neurology **29:** 291–296.
19. Conti-Tronconi, B. M., M. Morgutti, A Sghirlanzoni & F. Clementi. 1979. Neurology **29:** 496–501.
20. Hohlfeld, R., B. M. Conti-Tronconi, I. Kalies, J. Bertrams & K. V. Toyka. 1985. J. Immunol. **135:** 2393–2399.
21. Hohlfeld, R., K. V. Toyka, L. L. Miner, S. L. Walgrave & B. M. Conti-Tronconi. 1988. J. Clin. Invest. **81:** 657–660.

22. MELMS, A., G. MALCHEREK, U. GERN, H. WIETHOLTER, C. A. MULLER, R. SCHOEPFER & J. LINDSTROM. 1992. Ann. Neurol. **31:** 311–318.
23. PROTTI, M. P., A. A. MANFREDI, C. STRAUB, J. F. HOWARD JR. & B. M. CONTI-TRONCONI. 1990. J. Immunol. **144:** 1276–1281.
24. KOTANI, H., S. TAKADA, Y. VEDA, Y. MURAKAWA, N. SUZUKI & T. SAKANE. 1984. Clin. Exp. Immunol. **56:** 390–395.
25. DOSREIS, G. A. & E. M. SHEVAC. 1985. Eur. J. Immunol. **15:** 446–450.
26. STEINMAN, L. & R. MANTEGAZZA. 1990. FASEB J. **4:** 2726–2731.
27. BEN-NUN, A., H. WEKERLE & I. R. COHEN. 1981. Nature **292:** 60–61.
28. SUN, D., Y. QIN, J. CHLUBA, J. T. EPPLEN & H. WEKERLE. 1988. Nature **332:** 843–845.
29. LIDER, O., T. RESHEF, E. BERAUD, A. BEN-NUN & I. R. COHEN. 1988. Science **239:** 181–183.
30. VANDENBARK, A. A., G. HASHIM & H. OFFNER. Nature 1989. **341:** 541–544.
31. HOWELL, M. D., S. T. WINTERS, T. OLEE, H. C. POWELL, D. J. CARLO & S. W. BROSTOFF. Science 1989. **246:** 668–670.
32. OFFNER H., G. HASHIM & A. A. VANDENBARK. 1991. Science **251:** 430–432.
33. WRAITH, D. C., D. E. SMILEK, D. J. MITCHELL, L. STEINMAN & H. O. MCDEWITT. 1989. Cell **59:** 247–255.
34. LOHSE, A. W., F. MOR, N. KARIN, I. R. COHEN. 1989. Science **244:** 820–822.
35. NICHOLAS, J. A., M. E. LEVELY, M. A. MITCHELL & C. W. SMITH. 1989. J. Immunol. **143:** 2790–2796.
36. LAMBERT, L. E. & E. R. UNANUE. 1989. J. Immunol. **143:** 802–807.
37. BRETT, S. J., K. B. CEASE, C.-A. S. OUYANG & J. A. BERZOFSKY. 1989. J. Immunol. **143:** 771–779.
38. MICHALEK, M. T., B. BENACERRAF & K. L. ROCK. 1989. Proc. Natl. Acad. Sci. USA **86:** 3316–3320.
39. SAKAI, K., A. A. SINHA, D. J. MITCHELL, S. S. ZAMVILL, J. B. ROTHBARD, H. O. MCDEVITT & L. STEINMAN. 1988. Proc. Natl. Acad. Sci. USA **85:** 8608–8612.
40. BRACIALE, T. J., M. T. SWEETSER, L. A. MORRISON, D. J. KITTLESEN & U. L. BRACIALE. 1989. Proc. Natl. Acad. Sci. USA **86:** 277–281.
41. GOOD, M. F., D. PONIBO, K. A. QUAKYI, E. M. RILEY, R. A. HOUGHTEN, A. MENON, D. W. ALLING, J. A. BERZOFSKY & L. H. MILLER. 1988. Proc. Natl. Acad. Sci. USA **85:** 1199–1203.
42. CLERICI, M., N. I. STOCKS, R. A. ZAJAC, R. N. BOSWELL, D. C. BERNSTEIN, D. L. MANN, G. M. SHEARER & J. A. BERZOFSKY. 1989. Nature **339:** 383–385.
43. O'SULLIVAN, P., J. SIDNEY, E. APPELLA, L. UACKER, L. PHILLIPS, S. M. COLON, C. MILES, R. W. CHESNUT & A. SETTE. 1990. J. Immunol. **145:** 1799–1808.
44. BELLONE, M., F. TANG, R. MILIUS & B. M. CONTI-TRONCONI. 1989. J. Immunol. **143:** 3568–3579.
45. SAEDI, M. S., R. ANAND, W. G. CONROY & J. LINDSTROM. 1990. FEBS Lett. **267:** 55–59.
46. BESINGER, U. A., K. V. TOYKA, M. HOMBERG, K. HEININGER, R. HOHLFELD & A. FATEH-MOGHADAM. 1983. Neurology **33:** 1316–1321.
47. CONTI-TRONCONI, B. M., F. TANG, B. M. DIETHELM, S. R. SPENCER, S. REINHARDT-MAELICKE & A. MAELICKE. 1990. Biochemistry **29:** 6221–6230.
48. MCLANE, K. E., X. D. WU, B. DIETHELM & B. M. CONTI-TRONCONI. 1991. Biochemistry **30:** 4925–4934.
49. BERZOFSKY, J., S. J. BRETT, H. Z. STREICHER & H. TAKAHASHI. 1988. Immunol. Rev. **106:** 5–31.
50. MANCA, F., D. FENOGLIO, A. KUNKL, C. CAMBIAGGI, G. LIPIRA & F. CELADA. 1988. Immunology Today **9:** 300–303.
51. GAMMON, G. & E. E. SERCARZ. 1989. Nature **342:** 183–185.
52. SAKAI, K., S. S. ZAMVIL, D. J. MITCHELL, S. HODGKINSON, J. B. ROTHBARD & L. STEINMAN. 1989. Proc. Natl. Acad. Sci. USA **86:** 9470–9474.
53. OFFNER, H., M. VAINIENE, D. P. GOLDON, W. J. MORRISON, R. Y. WANG, G. A. HASHIM & A. A. VANDENBARK. 1991. J. Immunol. **146:** 4165–4172.
54. MARTIN, R., D. JARAQUEMADA, M. FLERLAGE, J. RICHERT, J. WHITAKER, E. O. LONG, D. E. MCFARLIN & H. MCFARLAND. 1990. J. Immunol. **145:** 540–548.

55. Pette, M., K. Fujita, B. Kitze *et al.* 1990. Neurology **40:** 1770–1776.
56. Sommer, N., G. C. Harcourt, N. Willcox, D. Beeson & J. Newsom-Davis. 1991. Neurology **41:** 1270–1273.
57. Salvetti, M., S. Jung, S-F. Chang, H. Will, B. C. G. Schalke & H. Wekerle. 1991. Ann. Neurol. **29:** 508–516.
58. Zhang, Y., M. Schleup, S. Frutiger, G. J. Hughes, M. Jaenner, A. Steek & T. Barkas. 1990. Eur. Immunol. **20:** 2577–2590.
59. Melms, A., S. Chrestel, B. C. G. Schalke, H. Wekerle, A. Mauron, M. Ballivet & T. Barkas. 1989. J. Clin. Invest. **83:** 285–289.
60. Gammon, G. & E. E. Sercarz. 1990. Clin. Immunol. Immunopathol. **56:** 287–297.
61. Bellone, M., N. Ostlie, S. Lei & B. M. Conti-Tronconi. 1991. Eur. J. Immunol. **21:** 2303–2310.
62. Granato, D. A., B. W. Fulpius & J. F. Moody. 1976. Proc. Natl. Acad. Sci. USA **73:** 2872–2876.
63. Hall, Z. W., V. Gorin, L. Silberstein & C. Bennett. 1985. J. Neurosci. **5:** 730–734.
64. Schuetze, S. M. 1986. Trends Neurosci. **9:** 386–388.
65. Gu, Y. & Z. Hall. 1988. Neuron **1:** 117–125.
66. Kumura, J. & M. W. Van Allen. 1967. Neurology **17:** 413–417.
67. Cuenoud, S., T. E. W. Feltkamp, B. W. Fulpius & J. G. H. Oosterhuis. 1980. Neurology **30:** 301–305.
68. Wekerle, H., R. Hohlfed, U.-P. Ketelsen, J. Kalden & I. Kalies. 1981. Ann. N.Y. Acad. Sci. **377:** 455–476.
69. Mittag, T., P. Kornfeld, B. A. Tormay & C. Woo. 1974. N. Engl. J. Med. **294:** 691–694.
70. Kamo, I., S. Furukawa, A. Tada, Y. Mano, Y. Iwasaki, T. Furuse, N. Ho, K. Hayaski & E. Satoyoshi. 1982. Science **215:** 995–997.
71. Fujii, Y., Y. Monden, K. Nakahara, J. Hashimoto & Y. Kawashima. 1984. Neurology **34:** 1182–1186.
72. Lisak, R. P., A. I. Levinson, B. Zweiman & M. J. Kornestein. 1986. J. Immunol. **137:** 1221–1225.
73. Aharonov, A., R. Tarab-Hazdai, O. Abramsky & S. Fuchs. 1975. Proc. Natl. Acad. Sci. USA **72:** 1456–1459.
74. Ueno, S., K. Wada, M. Takahashi & S. Tarui. 1980. Clin. Exp. Immunol. **42:** 463–469.
75. Fuchs, S., H. Schmidt-Hopfeld, G. Tridente & R. Tarrab-Hadzai. 1980. Nature **287:** 162–164.
76. Engel, W. K., J. L. Trotter, D. E. McFarlin & C. L. McIntosh. 1977. Lancet **i:** 1310–1311.
77. Kao, I. & D. B. Drachman. 1977. Science **195:** 74–75.
78. Kawanami, S., H. Kamei, J. Oita, M. Kurokawa, Y. Uchida & K. Hayashi. 1987. J. Neurol. **234:** 207–210.
79. Feltkamp-Vroom, T. 1966. Lancet **i:** 1320–1321.
80. Van de Velde, R. K. L. & N. B. Friedman. 1966. J. Am. Med. Assoc. **198:** 287–288.
81. Wekerle, H., B. Paterson, U. P. Ketelsenand & M. Feldman. 1975. Nature **256:** 493–494.
82. Kawanami, S., B. Conti-Tronconi, J. Racs & M. A. Raftery. 1988. J. Neurol. Sci. **87:** 195–209.
83. Mishina, M., T. Takai, K. Imoto, Noda, H. Takahashi & S. Numa. 1986. Nature **321:** 406–411.
84. Nelson, S. & B. M. Conti-Tronconi. 1990. J. Neuroimmunol. **29:** 81–92.
85. Lisak, R. P., C. Laramore, A. I. Levinson, B. Zweiman, A. R. Moskovitz & A. Whitte. 1984. Neurology **34:** 802–805.
86. Sinigaglia, F., C. Gotti, P. R. Castagnoli & F. Clementi. 1984. Proc. Natl. Acad. Sci. USA **81:** 7569–7573.
87. Oshima, M., T. Ashizawa, M. S. Pollack & M. Z. Atassi. 1990. Eur. J. Immunol. **20:** 2563–2568.
88. Brocke, S., S. Brautbar, T. Steinman, O. Abramsky, J. Rothbard, D. Neumann, S. Fuchs & E. Moses. 1989. J. Clin. Invest. **82:** 1894–1898.

Approaches for Studying the Pathogenic T Cells in Autoimmune Patients[a]

N. WILLCOX, F. BAGGI, A-P. BATOCCHI,[b] D. BEESON, G. HARCOURT, S. HAWKE, L. JACOBSON, H. MATSUO,[c] A-M. MOODY, N. NAGVEKAR, M. NICOLLE, B. ONG,[d] N. PANTIC, J. NEWSOM-DAVIS, AND A. VINCENT

Neuroscience Group
Institute of Molecular Medicine
John Radcliffe Hospital
Oxford OX3 9DU, United Kingdom

INTRODUCTION

Most workers agree that the production of the pathogenic IgG autoantibodies against the acetylcholine receptor (AChR) is likely to be helper T cell–dependent and that the induction of these T cells is probably a crucial step in autosensitization.[1] The arguments are most compelling when applied to the thymoma found in about 10% of myasthenia gravis (MG) patients. It is typically a tumor of cortical epithelial cells that retains the characteristic thymic behavior of generating enormous numbers of (polyclonal) maturing T cells but usually contains extremely few B cells.[2] The epithelial cells express some striated muscle antigens[3] and perhaps one (cytoplasmic) epitope of the AChR[4], to both of which antigens the patients produce autoantibodies. So perhaps these epitopes autosensitize developing T cells *in situ* that are then exported to the periphery. If they eventually encounter a recognizable epitope there, they may then initiate an autoantibody response;[4] such a series of coincidences might explain the old clinical observation that MG symptoms sometimes only appear months or years after removal of the thymoma.[5]

In the 50–60% of early-onset MG patients who have no thymoma, the initiating site is not known, but some suspect that rare myoid cells in the thymic medulla might play a similar role,[6,7] especially as they express the fetal type of AChR that these patients' autoantibodies so often prefer.[8] If they do, that might explain the "hyperplastic" infiltration into the medulla by lymph node–type T-cell areas and germinal centers that is so prevalent in these patients.[8]

To study these inductive cellular events in MG patients, culture responses

[a] This work was supported by the Wellcome Trust, the Sir Jules Thorn and the A. J. Flory Charitable Trusts, the Muscular Dystrophy Group, the Medical Research Councils of Great Britain and Canada, the Neurological Institute "C. Besta" (Milano), and the Italian Telethon.

[b] Present address: Institute of Neurology, Catholic University, 00168 Rome, Italy.

[c] Present address: First Department of Internal Medicine, Nagasaki University School of Medicine, 852 Japan.

[d] Present address: Department of Medicine, National University Hospital, Singapore 0511.

driven by specific antigens are essential (using lymphoid cells from patients and controls). Presumably, *in vivo* in the patient, the pathogenic T cells must recognize antigen in minute doses. How they do so is a fascinating challenge in itself, especially as T cells are traditionally tested against relatively high antigen concentrations *in vitro*. Human AChR was thought to be prohibitively scarce, so the source of antigen for such work is a major concern in our laboratory and others and is, indeed, the central theme of this review. We must point out now that the α-subunit is thought to be the main target of helper T cells[9] but that even this most conserved subunit is only 80% identical in humans and *Torpedo*.[10]

Theoretically possible approaches are summarized in TABLE 1; each has its advantages and drawbacks, some of which are featured below. Others are self-evident. For example, the choice of approach may influence the antigen processing involved; different cell types may prove to process differently, and cell lines, in particular, may behave distinctively. Because of the nature of the response *in vivo* (and our ignorance of its initiation), it may be very important to compare B-cell, muscle-type, and "professional" processing (*e.g.* by macrophages). When synthetic peptides are used, most of these steps will probably be by-passed, though some peptides may still be trimmed.

We have mainly used recombinant (r) human AChR α-subunit, as close as possible to the full-length sequence,[11] but initially starting with r α37-429$^+$ and only more recently progressing to r α1-437$^+$. This material is inevitably heavily denatured during its expression, extraction, and purification from the *E. coli* host, so that it is not recognizable by antibodies to native AChR, including those in MG sera. Fortunately, T cells mainly respond to sequential epitopes (in the form of peptide fragments), and such preparations are quite suitable for them. Shorter recombinant polypeptides (*e.g.* r α37-181) have proved invaluable in locating these epitopes ("mapping") to particular regions,[11] as others have also found with recombinant human antigen.[12]

In the course of cloning and expressing the human α-subunit, we have observed a previously unrecognized optional insert (P3A) of 25 amino acids (aa) between positions 58 and 59 of the originally published sequence.[13] We have more recently shown that this is expressed both in human rhabdomyosarcoma TE671 cells[14] and in human muscle AChR, but not in the cow or lower vertebrates.

RESULTS AND DISCUSSION

Thymoma Cell Suspensions

Thymoma cell suspensions were used in our initial attempts to generate CD4$^+$ T-cell lines because of the strong arguments outlined in the INTRODUCTION. We initiated lines with r α37-429$^+$ in a total of 10 MG patients using autologous irradiated peripheral blood lymphocytes (PBL) as antigen-presenting cells (APC). We were able to establish predominantly CD4$^+$ lines from four of these patients. Only with one could the epitope be mapped (using shorter recombinant constructs) to the 80-113 region (FIG. 1a); that was confirmed by specific stimulation with synthetic peptides p75-114, p72-90, and p62-90, the longest peptide being consistently the most potent. This line grows slowly and was only "cloned" with great difficulty (10 cells/well). In contrast to most other lines (see below), we have so far failed to identify any single HLA-restricting haplotype or even isotype. Monoclonal antibodies to the three class II isotopes—HLA-DR, -DQ, and -DP—all block the responses almost completely (FIG. 1b). Hybrid DQ or DQ/DP

TABLE 1. Advantages and Drawbacks of the Potential Sources of Antigen

Source	Nature of antigen processing	Potential difficulties with: Amounts obtainable	Purity	Contaminating cell antigens	Other drawbacks
Natural AChR					
electric fish	depends on the APC chosen	–[a]	–	–	Sequence differences
mammalian muscle	depends on the APC chosen	+ +	+ +	+ +	Internal epitopes exposed
cell lines (*e.g.* TE671) ± transfection with human class II	muscle/myoid-type?	+ +	+	+ (pre-imm)[b]	Diversity of class II alleles and adhesion molecules
Recombinant AChR or subunits					
transfected into class II$^+$ cell lines (*e.g.* autologous EBV LCL)[c]	B cell type	+ +	+	+ (pre-imm) (*e.g.* EBV)	Diversity of class II alleles; bystander stimulation
expressed in: yeasts	depends on the APC chosen	±	+	+ (pre-imm)	Abnormal posttranslation modifications—$(CHO)_n$ and S-S bonds
E. coli	depends on the APC chosen	±	+ + +	+ + + (pre-imm)	Internal epitopes exposed
Synthetic peptides					
selected; pooled overlapping (*e.g.* 1-20, 16-36, 30-50); pooled hyperlapping (*e.g.* 1-20, 2-22, 4-24)	limited: arbitrarily chosen start and stop points ≡ unnatural processing sites	–	±	–	**Theoretical danger of:** biases towards some epitopes or against others; selection of low-avidity T cells at high [peptide]; competition between epitopes/T cells

[a] Minus means no difficulty; + + +, great difficulty.
[b] pre-imm: cell donors expected to be immune to other cellular antigens.
[c] Epstein-Barr virus–induced lymphoid cell lines.

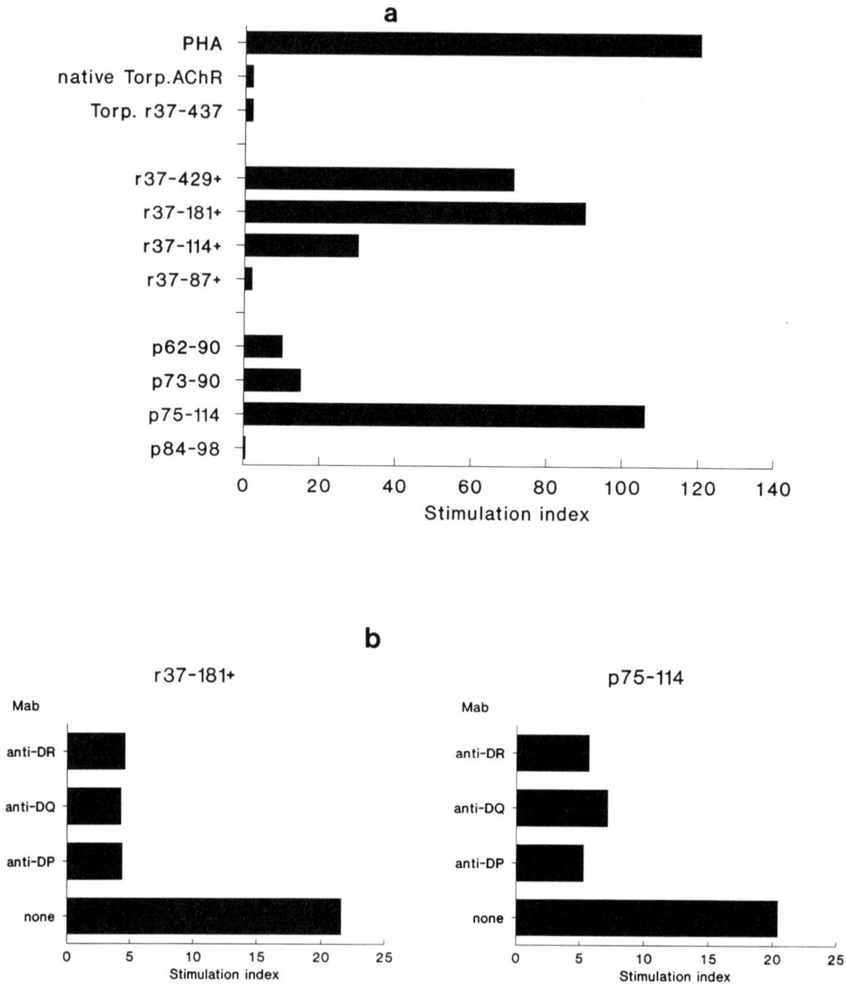

FIGURE 1. Characterization of a T-cell line raised from MG thymoma lymphocytes. **a:** Epitope mapping used recombinant (r) polypeptides and synthetic peptides (p). **b:** Blocking of responses by monoclonal antibodies (Mabs) to HLA-DR (L-243), -DQ (Genox 3.53), and -DP (B7-21), all at a dilution of 1 in 80. Such blocking by all three antibodies is highly unusual (see FIG. 5). Results are stimulation indices from standard proliferation assays pulsed with [^{3}H]thymidine.

molecules seem realistic candidates because the thymoma epithelial cells are strongly HLA-class II–positive,[2] and such aberrations might be expected in them. They might in turn explain some of the difficulties in selecting specific lines from thymoma cells using autologous PBL as APC, though these could also reflect distinctive cytokine requirements of maturing thymic T cells. We were able to identify the T-cell receptor genes used by this T-cell line from its cDNA using anchor PCR,[15] and the results support its clonality (not shown).

Donors without Thymoma

T-Cell Lines

Starting from PBL, we initiated T-cell lines (as above) from six early-onset patients, MG1-6.[11] One of them was established about six months after onset of MG symptoms from patient MG3, and its epitope proved to include part of the P3A insert plus some of the adjacent residues[16,17] (59–65) (FIG. 2, top). This line and its subclones were restricted by HLA-DQw5.[16] A second line, established six months later, showed very similar behavior, though greater heterogeneity. After a further six months of increasing symptoms (requiring corticosteroid treatment), the patient was thymectomized and, interestingly, at this stage there was no detectable response to this epitope in either PBL or thymic T cells. The new lines established from both were restricted by HLA-DR7, but their epitopes could not be mapped with synthetic peptides, as with the lines from patients MG1-2 and MG4-6; nearly all of these "unmapped" lines responded only to preparations of full-length, and not shorter, polypeptides (see below).

Establishing lines from healthy controls has proved no more difficult, and two out of eight have been successfully mapped. One responded to p62-90, and another from control BA responded to p51-65 (FIG. 2, bottom), both in the context of HLA-DRw53. This latter epitope spans positions 58-59 and so is totally disrupted by the presence of the P3A insert (FIG. 2, bottom). Interestingly, there is no response to *Torpedo* AChR or its recombinant α-subunit, despite great sequence homology; perhaps it lacks a nearby processing site that enables these epitopes to be generated from human antigen. Possibly also they might be presented by B cells specific for main immunogenic region (MIR) sites just downstream (aa67-76); the lines from MG3 and BA both seem to be capable of responding to AChR from human muscle (*e.g.* FIG. 2, bottom).[17]

Disease-implicated T Cells

To select for the most disease-implicated T cells, we have also raised T-cell lines from hyperplastic thymus cells of several young MG patients, who show a strong HLA-B8-DR3 association. The only example where mapping was successful came from a 15-year-old DR3/4 patient (P1) thymectomized about six months after onset of symptoms; to enrich activated cells we used the low-density fraction of her cryopreserved thymus cell suspension.[18] The epitope mapped to p144-156 (FIG. 3a), and stimulation was consistently even stronger with the longer spanning peptide 138-167. These "PM-A" cells also respond to p147-163, so the core of their epitope is apparently 147-156, which is almost identical to that of the 147-162 response selected by Atassi and colleagues with pooled synthetic peptides.[19] Our PM-A line was unresponsive to *Torpedo* AChR or its recombinant α-subunit, which differs chiefly by a Val to Lys substitution at 155; this one change was enough to ablate recognition by these T cells (FIG. 3a). Their requirements of the presenting class II molecules were highly exacting too. Only DR4 subtypes with glycine (rather than valine) at β-chain position 86 could present (*i.e.* Dw4 and Dw14.2): natural variants with changes at positions 74, or 70 + 71 also gave negative results.[18]

Hints that these particular T cells were pathogenic in this patient include (1) their isolation in at least two parallel cultures of low-density thymus cells[18] taken when symptoms were flaring; (2) their responsiveness to extremely low con-

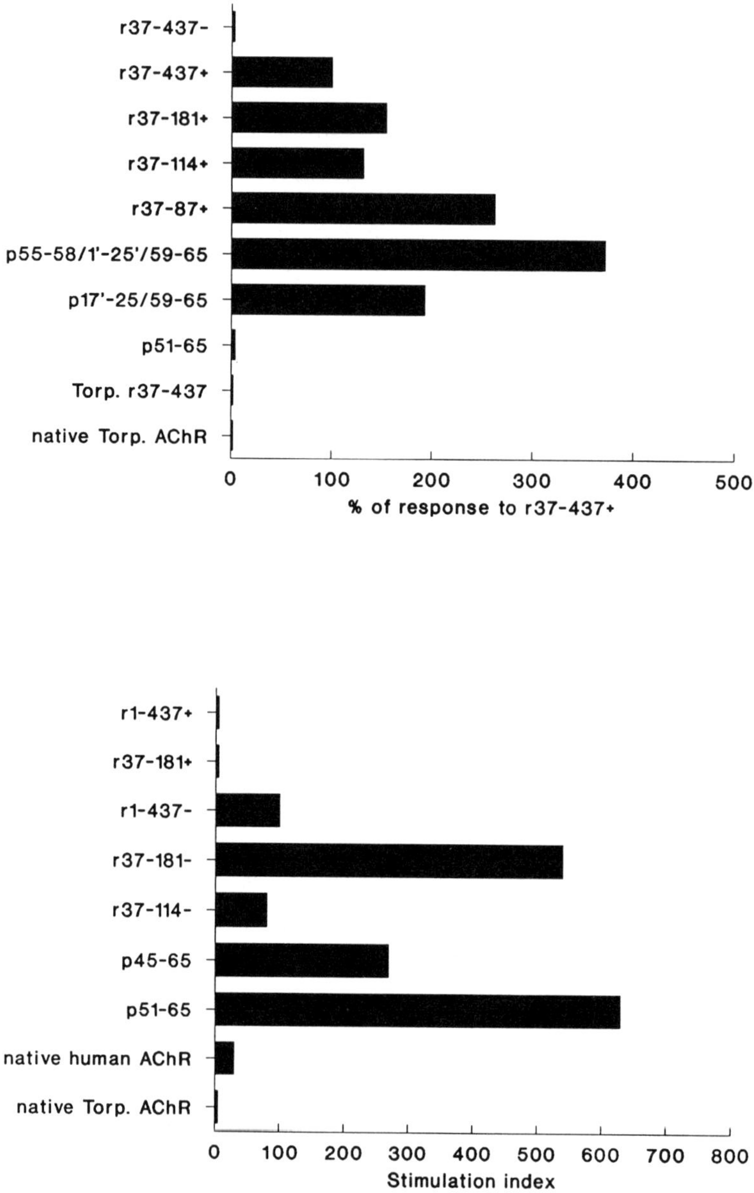

FIGURE 2. Mapping of T-cell epitopes close to the P3A insertion site (α58-59). Top: A clone from patient MG3 only responds to recombinant polypeptides that include the P3A insert 1′–25′ (r α37-87$^+$ etc.). Bottom: A line from control BA only responds to antigens lacking the P3A insert (r α1-437$^-$ etc.). Responses to AChR from human muscle are variable; shown here is the mean of 5 experiments (3 of which were positive).

centrations of soluble or solid-phase native human AChR from muscle (FIG. 3b)[18,20]—despite its glycosylation at Asn[141] and its 128-142 disulphide bond; (3) their recognition of virtually all recombinant or synthetic constructs that include the 138-167 sequence; and (4) efficient presentation by cell types as various as macrophages, EBV cell lines, and mouse L cells[18] or CHO cells or even TE671 cells transfected to express DR4 molecules of the correct subtype. Interestingly, while these TE671 cells can specifically stimulate if peptide is added, they can also present in the absence of any added antigen (FIG. 3c), presumably because they process their endogenous α-subunit appropriately. Having done so, they are usually killed (perhaps involving interferon-γ), as indicated by their reduced [^{3}H]thymidine uptake on co-culture with these T cells[21] (FIG. 3c). Evidently this epitope is generated after processing by myoblast (myoid?)-like cells, B cells, and

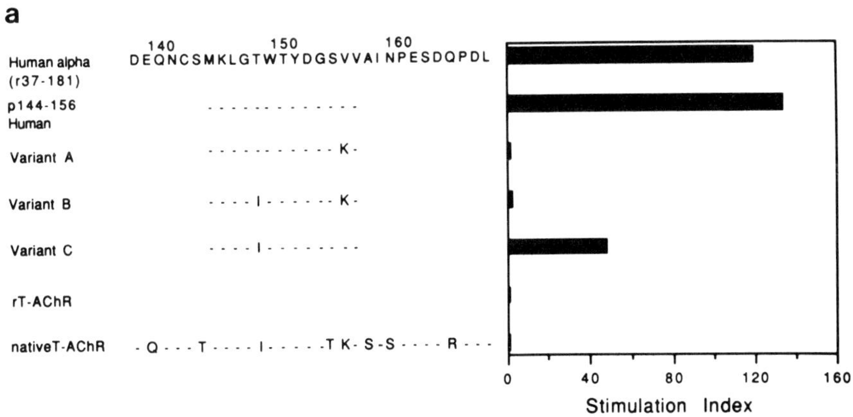

FIGURE 3. Responsiveness of the PM-A T-cell line (derived from a hyperplastic MG thymus) **(a)** to the human and *Torpedo* α144-156 sequences; **(b)** to human muscle AChR either in solution or captured onto Dynabeads by specific mouse Mabs; **(c)** to p144-156 presented by Mitomycin C–treated TE-671 cells either transfected to express HLA-DR4;Dw4 or not (wt). Untreated TE 671 cells take up much more [^{3}H]thymidine, unless PMA T cells are added, in which case uptake is greatly reduced, even in the absence of added peptide. In both cases, specific recognition required DR4; **(d)** to either p144-156 or r α37-181 is abolished by 5 days of preculture with soluble HLA-DR4 complexed with p138-167. (Part **a** from Ong *et al.*[18])

other more professional APC. To confirm their potential pathogenicity, we are currently assaying the ability of these T cells to "help" compatible responder B cells to produce anti-AChR antibodies in response to specific antigen, as well as studying their cytokine profiles.

This line and four of its subclones show the same pattern of specificity and the same VJα and VDJβ T-cell receptor usage, so that, in this case, it was possible rigorously to define each face of the ternary complex of T-cell recognition.

We have begun to exploit these findings by using the *in vitro* responses of these T cells as a model system for testing one strategy for inhibiting them that might prove applicable in patients *in vivo*.[22,23] The peptide 138-167 binds with relatively high affinity to solubilized DR4 : Dw4 molecules. Preincubation of the T cells with the resulting complexes in the absence of APC pre-empts subsequent

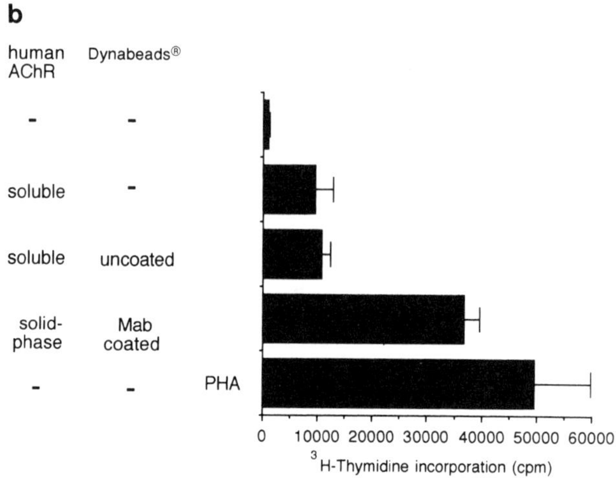

FIGURE 3b. See legend on page 225.

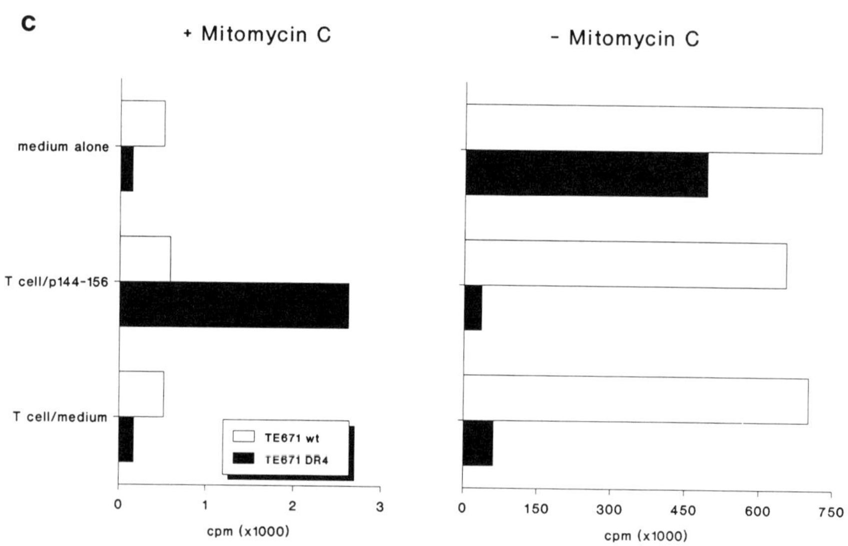

FIGURE 3c. See legend on page 225.

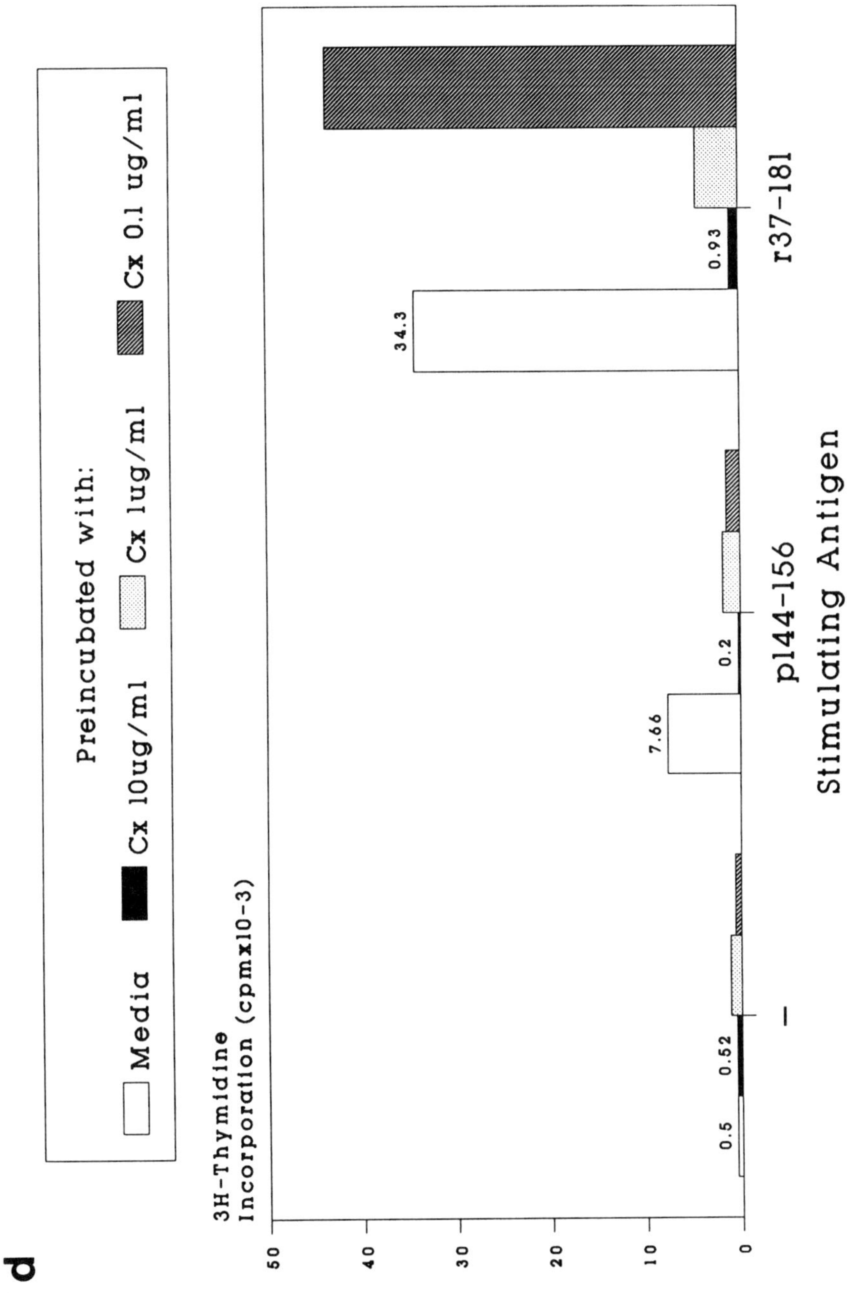

FIGURE 3d. See legend on page 225.

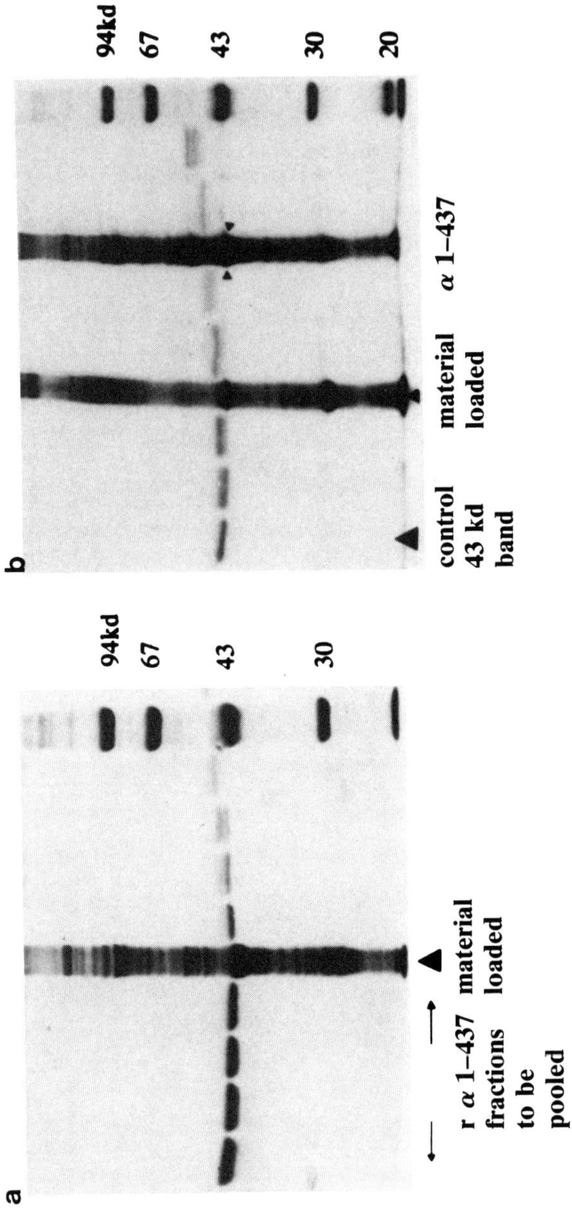

FIGURE 4a and **b.** See legend on page 229.

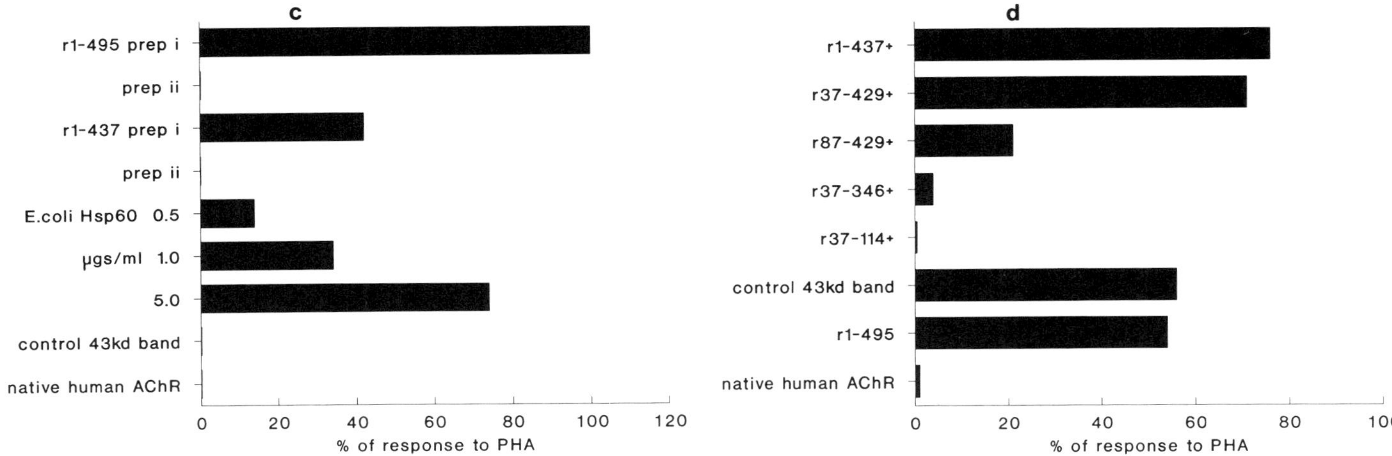

FIGURE 4. Contamination of recombinant AChR α1-437 with *E. coli* components. **a:** Isolation of r α1-437 (solubilized from inclusion bodies) by preparative SDS gel electrophoresis. **b:** Parallel isolation of control 43 kDa band from the same region of a much more heavily loaded separation of r α37-114 (too small to be visible on this photograph). **c:** Response of a T-cell line raised against r γ1-495 (from a healthy control) to *E. coli* heat shock protein (Hsp) 60 present in some r α1-437 and r γ1-495 preparations. **d:** An HLA-DR3–restricted T-cell clone raised from patient MG1 against r α37-429$^+$ responds instead to the control 43 kDa band isolated in **b**. **e:** A contaminant recognized by the clone C5 adsorbs nonspecifically to Dynabeads® but can be removed by 0.25% SDS washes. By contrast, AChR r α1-437, bound specifically by polyclonal antibodies, resists such washing and continues to stimulate PM-A T cells.

responses to a challenge with an optimal dose of antigen plus APC (FIG. 3d), presumably because the complexes engage only the T cell receptors with no accessory signaling. Similar inhibition is seen at 1000 times lower molarity of complex than of free peptide.[23] It is highly epitope-specific; complexes with an irrelevant peptide or none at all are totally inactive, and T cells that recognize other antigens are unaffected. A similar approach is very effective *in vivo* in preventing experimental allergic encephalomyelitis in mice.[22]

Unmapped T-Cell Lines

The first unmapped T cells we tested were unresponsive to r *Torpedo* α-37-k37 and thus appeared to be human AChR-specific. They gave minimal or undetectable responses to native AChR, however, and at least some eventually proved to recognize comigrating contaminants derived from the *E. coli* in which the recombinant polypeptides were produced (FIG. 4). The use of monoclonal and

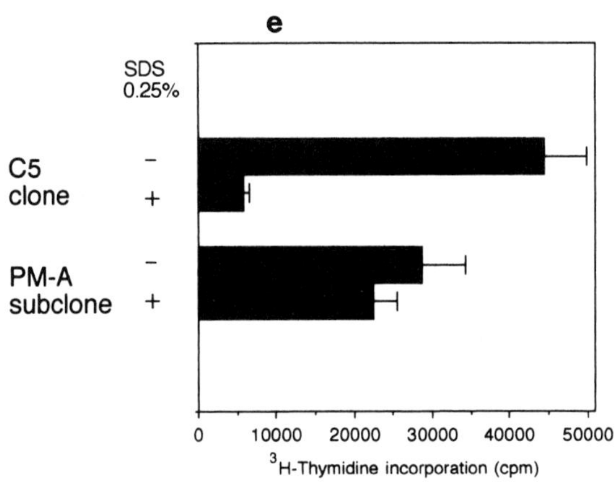

FIGURE 4e. See legend on page 229.

affinity-purified antibodies (*e.g.* to synthetic peptides) to capture antigens onto immunomagnetic beads made quality control testing much more sensitive and rigorous (FIG. 4e). Responses to such impurities should have been no surprise, even after meticulous separation, as most donors have presumably been reacting to *E. coli* throughout life and as unsuspected low-level contaminants are often revealed with great sensitivity after immunizing with supposedly pure antigens. Apparently, there are several different components that can each elicit highly specific class II–restricted T cells; one of them seems to be a heat-shock protein (or a contaminant thereof, FIG. 4c). Thus the AChR-specificity of T cells selected with recombinant material can only be definitively established with an independent source of antigen (preferably synthetic peptide), as underlined by our failure in some cases to detect responses to r *Torpedo* α-subunit prepared similarly.

Nevertheless, we still favor initiating T-cell lines with recombinant antigen, either (1) initially plating the cells at limiting numbers and splitting each well[24] at an early stage to restimulate one third of the cells with the original antigen, one third with an independent form of antigen (*e.g.* intact AChR), and one-third with dummy *E. coli* antigen preparations (all plus APC), and only expanding those that appear to be specific; or (2) initiating lines in greater bulk with intact AChR, and later restimulating with full-length recombinant subunits in the hope of selecting only those T cells that recognize both.

Use of Pooled Synthetic Peptides

Use of pooled synthetic peptides, covering the entire α-subunit sequence, is an alternative way of avoiding some of these problems that has been advocated by Atassi and Conti-Tronconi and their colleagues;[19,25] both laboratories report the added advantage that few (if any) lines are obtained from healthy controls. Because of the potential problems mentioned in TABLE 1, however, we decided to evaluate this strategy thus:

(1) We tested the four lines illustrated in FIGURES 1–3 for responsiveness to pools of synthetic α-subunit peptides (kindly donated by Dr. B. Conti-Tronconi). The thymoma-derived line responded moderately, probably to the constituent peptide 76-93; the line recognizing p51-65 (FIG. 2b) responded very well as its equivalent peptide in the pool is p48-67. As the P3A insert is not represented in the pool, it was no surprise that the lines and clones from MG3 (FIG. 2b) were not stimulated. Similarly, the line recognizing p144-156 also failed to respond, probably because the closest peptide (135-154) stops prematurely, and aa 155 is evidently critical for recognition by this line (FIG. 3a). So the success rate here was two out of four of these $CD4^+$ T-cell lines.

(2) We also raised lines against a pool of our own longest synthetic peptides; it covers the entire α-subunit sequence in only 19 peptides. Most of them are long enough to permit some natural processing or trimming (average length 33 aa, overlapping by a mean of about 8 at each end), but also short enough to bind directly to empty class II molecules. We successfully raised T-cell lines (mainly $CD4^+$) from three of eight MG patients and four of five controls. In one of each, the T cells recognized several different peptides, some of them transiently (FIG. 5a), so much so that patient LF's line ceased growing altogether after about nine weeks. With the remaining lines, by contrast, three other peptides predominated—p75-115, p138-167, and p309-344 (TABLE 2). Responses to p75-115 tended to overtake those to p138-167 later in their evolution, whereas usually those to p309-344 were persistently high.[26] Each was somewhat "promiscuous" but had a preferred restriction element; for p309-344, this was DR3 when it was available (FIG. 5b); otherwise it was once DRw53 and once DR15. For p75-115 it was DR4 (but once DRw53); for p138-167, again it was usually DR4 (but also DR10 once). Some of the DR4-restricted lines showed intriguing preferences for subtypes of DR4.[27] Most interestingly, both control BA (FIG. 5a) and patient PM failed to yield T cells with similar specificities to those of their recombinant antigen-selected lines described above, even though the pool included peptides suitable for each. Also patient PM's line recognized p75-115 in conjunction with DRw53 rather than DR4 (FIG. 5b), even though her previous line had been so clearly DR4-restricted.[18]

Even more strikingly, none of these lines has ever responded to any dose of full-length recombinant or native human AChR—a success rate of 0/9—even though we tried to maximize the sensitivity of these systems (*e.g.* with solid-phase–concentrated antigens). Most of the lines have also failed to respond to the shorter recombinant constructs that include their epitope; the exception that proves this rule occurred when we cleaved an α265-366 fusion protein with trypsin.

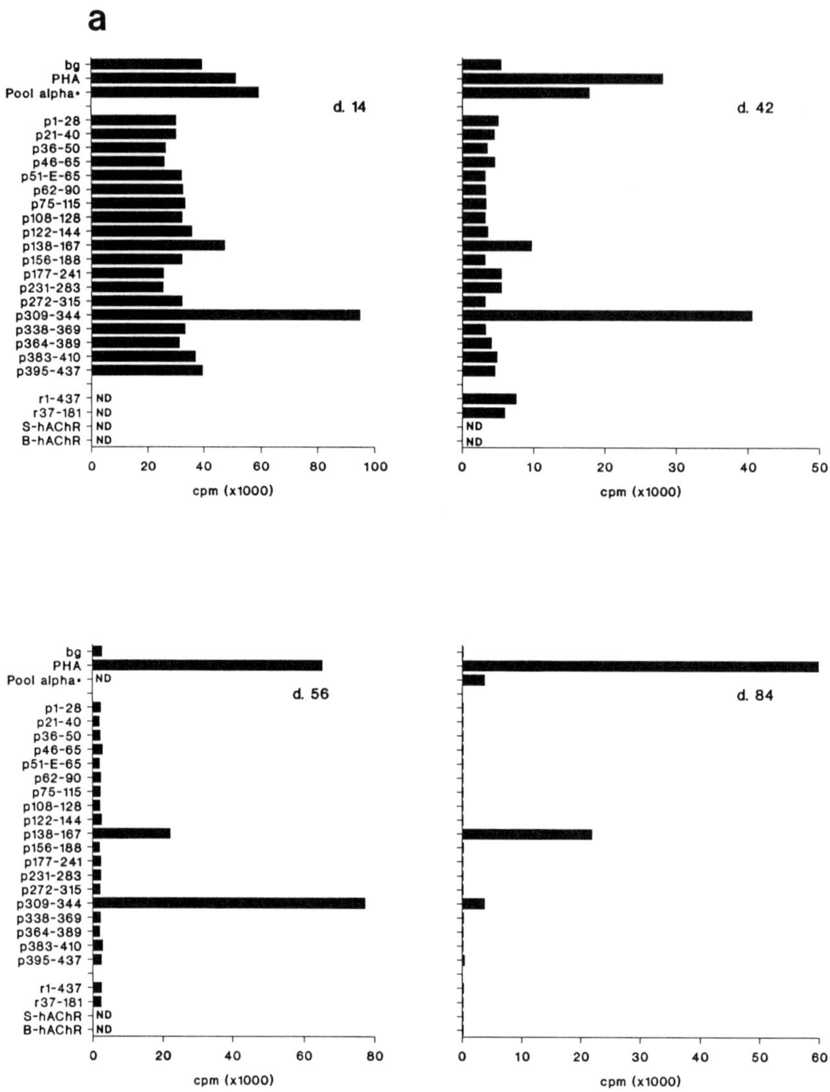

FIGURE 5a. See legend on page 233.

The resulting digest stimulated one of two p309-344–specific lines strongly,[26] showing that sufficient quantities of antigen were available, but that its processing was the limiting factor: when that was overcome artificially (with a pancreatic enzyme), the epitope was unmasked. In other cases, unresponsiveness to recombinant constructs might result from disruption of the epitope at internal sites as they are processed, disruption that is only by-passed when preformed peptides are offered

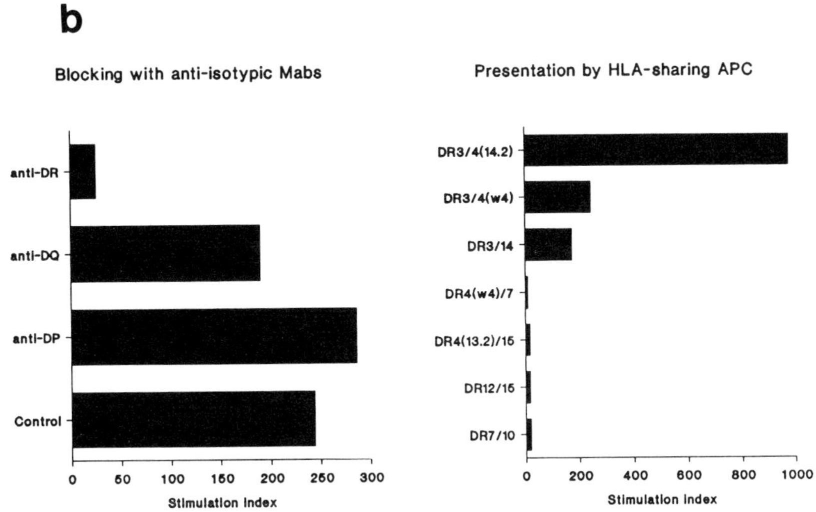

FIGURE 5. Characterization of T-cell lines raised against pooled peptides. **a:** Serial responses of a T-cell line from a healthy control raised against pooled peptides. Evolution of responses from control BA assayed at sequential restimulations. Initially, few of the constituent peptides stimulated above high backgrounds, but, later, responses to p138-167 and p309-344 predominated; the late waning of the latter is not typical. **b:** The response of the line from MG patient PM to p309-344 is blocked by anti-DR Mab and is mainly restricted by HLA-DR3 rather than -DR4.

to empty class II molecules on the APC surface, from selection of low-avidity T cells at high peptide molarities,[28] or from competition by other nonstimulatory peptides, possibilities that we are now beginning to test.

SUMMARY

Our provisional conclusions from this work are as follows. (1) For screening responses of established lines, native human AChR is not prohibitively scarce, especially if it is concentrated onto beads,[20] and class II–transfected TE671 cells may be useful too;[21] both may give vital evidence of AChR-specificity, but it is still crucial to confirm that with synthetic peptides. (2) For mapping epitopes, panels of full-length and shorter recombinant human polypeptides, and of synthetic peptides, are invaluable complementary material: longer peptides tend to stimulate particularly strongly. (3) Initial selection with pooled synthetic peptides can easily generate interesting lines from both patients and controls, but they may depend on the artificial processing sites that are an inevitable consequence of arbitrarily chosen start and stop points. Of course, these might conceivably be employed in unusual antigen-presenting cells (such as thymic myoid cells), so we cannot totally dismiss such "cryptic" epitopes. This system can sometimes select T cells responding to "natural" epitopes too, as now reported for tetanus toxin.[29] Neverthe-

TABLE 2. Summary of T-Cell Lines with Successfully Mapped Epitopes

Selecting Antigen	Source	Proportion of Lines Mapped	Epitope	HLA Class II Restriction	Respond to: Recombinant (α1-437 or Shorter)	Respond to: Human Muscle AChR
Recombinant human AChR α-subunit	MG PBL	1/6	P3A insert +p59-65	DQw5	+ + +	+ variable
	MG thymoma	1/4	p75-115	?	+ + +	?
	MG hyperplastic thymus	1/12	p138-167	DR4; Dw14.2	+ + +	+ +
	Control PBL	2/8	p51-65	DRw53	+ + +	+ variable
			p62-90	DRw53	+ + +	ND[a]
pooled AChR α-subunit peptides	MG PBL	3/8	p75-115	usually DR4	—	—
	Control PBL	4/5	p138-167	usually DR4		
			p309-344	usually DR3		

[a] Not done.

less, for these and other reasons, at present, we strongly favor using the longest human recombinant material possible, because it is apparently processed more naturally. This must be combined with rigorous screening for reactivity to *E. coli*–derived contaminants plus concomitant mapping of epitopes as above. Use of intact AChR for initiating lines may yet become feasible. (4) The T cells thus isolated and characterized so far are proving to be heterogeneous in the epitopes and presenting class II molecules they recognize, and in their T-cell receptor gene usage. It is premature to claim key myasthenogenic epitopes or clonotypes, but HLA-DR3 and the linked -DQw2 do not appear to monopolize presentation. (5) Assessing the disease-relevance of these T cells is a separate problem, highlighted by their apparent similarity in healthy controls. In the meantime, to test their potential pathogenicity, we are assaying their cytokine profiles and ability to help specific antibody production *in vitro*. In the hope that they do prove to be relevant, we are also using some of them to test possible therapeutic strategies that might prove applicable in the patients.

ACKNOWLEDGMENTS

We are extremely grateful to all the patients and controls for their blood samples, to Ms. E. Goodger for taking many of them, and also to Dr. Douglas Young for *E. coli* Hsp 60.

REFERENCES

1. Hohlfeld, R. & K. V. Toyka. 1985. Strategies for the modulation of neuroimmunological disease at the level of autoreactive T-lymphocytes. J. Neuroimmunol. **9:** 193–204.
2. Willcox, N., M. Schluep, M. A. Ritter, H. J. Schuurman, J. Newsom-Davis & B. Christensson. 1987. Myasthenic and nonmyasthenic thymoma. An expansion of a minor cortical epithelial cell subset? Am. J. Pathol. **127:** 447–460.
3. Gilhus, N.-E., J. A. Aarli, B. Christensson & R. Matre. 1984. Rabbit antiserum to a citric acid extract of human skeletal muscle staining thymomas from myasthenia gravis patients. J. Neuroimmunol. **7:** 55–64.
4. Marx, A., R. O'Connor, K. I. Geuder, F. Hoppe, B. Schalke, S. Tzartos, S. I. Kalies, T. Kirchner & H. K. Müller-Hermelink. 1990. Characterization of a protein with an acetylcholine receptor epitope from myasthenia gravis-associated thymomas. Lab. Invest. **62:** 279–286.
5. Namba, T., N. G. Brunner & D. Grob. 1978. Myasthenia gravis in patients with thymoma, with particular reference to onset after thymectomy. Medicine (Baltimore) **57:** 411–433.
6. Wekerle, H. & U-P. Ketelsen. 1977. Intrathymic pathogenesis and dual genetic control of myasthenia gravis. Lancet **i:** 678–680.
7. Kao, I. & D. B. Drachman. 1977. Thymic muscle cells bear acetylcholine receptors: possible relation to myasthenia gravis. Science **195:** 74–75.
8. Willcox, N. & A. Vincent. 1988. Myasthenia gravis as an example of organ-specific autoimmune disease. *In* B Lymphocytes in Human Disease. G. Bird & J. E. Calvert, Eds.: 469–506. Oxford University Press.
9. Hohlfeld, R., K. V. Toyka, S. J. Tzartos, W. Carson & B. M. Conti-Tronconi. 1987. Human T-helper lymphocytes in myasthenia gravis recognize the nicotinic receptor alpha subunit. Proc. Natl. Acad. Sci. USA **84:** 5379–5383.
10. Kubo, T., M. Noda, T. Takai, T. Tanabe, T. Kayano, S. Shimizu, K. Tanaka, H. Takahashi, T. Hirose, S. Inayama, R. Kikuno, T. Miyata & S. Numa.

1985. Primary structure of δ subunit precursor of calf muscle acetylcholine receptor deduced from cDNA sequence. Eur. J. Biochem. **149:** 5–13.

11. NEWSOM-DAVIS, J., G. HARCOURT, N. SOMMER, D. BEESON, N. WILLCOX, & J. B. ROTHBARD. 1989. T-cell reactivity in myasthenia gravis. J. Autoimmunity **2**(Suppl.): 101–108.
12. MELMS, A., G. MALCHEREK, U. GERN, H. WIETHÖLTER, C. A. MÜLLER, R. SCHOEPFER & J. LINDSTROM. 1992. T cells from normal and myasthenic individuals recognize the human acetylcholine receptor: heterogeneity of antigenic sites on the α-subunit. Ann. Neurol. **31:** 311–318.
13. BEESON, D., A. MORRIS, A. VINCENT & J. NEWSOM-DAVIS. 1990. The human muscle nicotinic acetylcholine receptor alpha-subunit exists as two isoforms: a novel exon. EMBO J. **9:** 2101–2106.
14. MORRIS, A., D. BEESON, L. JACOBSON, F. BAGGI, A. VINCENT, & J. NEWSOM-DAVIS. 1991. Two isoforms of the muscle acetylcholine receptor alpha-subunit are translated in the human cell line TE671. FEBS Lett. **295:** 116–118.
15. MOSS, P. A. H., R. J. MOOTS, W. M. C. ROSENBERG, S. J. ROWLAND-JONES, H. C. BODMER, A. J. MCMICHAEL & J. I. BELL. 1991. Extensive conservation of α and β chains of the human T-cell antigen receptor recognizing HLA-A2 and influenza A matrix peptide. Proc. Natl. Acad. Sci. USA **88:** 8987–8990.
16. HARCOURT, G. C., A.-P. BATOCCHI, D. M. W. BEESON, N. PANTIC, A-M. MOODY, H. N. A. WILLCOX, A. VINCENT & J. NEWSOM-DAVIS. Definition of the ternary recognition complex of a $CD4^+$ T cell clone restricted by HLA-DQw5. In preparation.
17. BATOCCHI, A.-P., G. HARCOURT, D. BEESON, N. PANTIC, S. HAWKE, N. WILLCOX & J. NEWSOM-DAVIS. 1993. Differential recognition by T cells of the $P3A^+$ and $P3A^-$ isoforms of the human acetylcholine receptor alpha-subunit. Ann. NY Acad. Sci. This volume.
18. ONG, B., N. WILLCOX, P. WORDSWORTH, D. BEESON, A. VINCENT, D. ALTMANN, J. S. S. LANCHBURY, G. C. HARCOURT, J. I. BELL & J. NEWSOM-DAVIS. 1991. Critical role for the Val/Gly^{86} HLA-DRβ dimorphism in autoantigen presentation to human T cells. Proc. Natl. Acad. Sci. USA **88:** 7343–7347.
19. OSHIMA, M., T. ASHIZAWA, M. S. POLLACK & M. Z. ATASSI. 1990. Autoimmune T cell recognition of human acetylcholine receptor: the sites of T cell recognition in myasthenia gravis on the extracellular part of the α subunit. Eur. J. Immunol. **20:** 2563–2569.
20. HAWKE, S., N. WILLCOX, G. HARCOURT, A. VINCENT & J. NEWSOM-DAVIS. 1992. Stimulation of human T cells by sparse antigens captured on immunomagnetic particles. J. Immunol. Methods. **155:** 41–48.
21. BAGGI, F., M. NICOLLE, A. VINCENT, H. MATSUO, N. WILLCOX & J. NEWSOM-DAVIS. Presentation of endogenous acetylcholine receptor epitope by an MHC class II-transfected human muscle cell line to a specific $CD4^+$ T cell clone from a myasthenia gravis patient. Submitted to J. Neuroimmunol.
22. SHARMA, S. D., B. NAG, X.-M. SU, D. GREEN, E. SPACK, B. R. CLARK & S. SRIRAM. 1991. Antigen-specific therapy of experimental allergic encephalomyelitis by soluble class II major histocompatibility complex–peptide complexes. Proc. Natl. Acad. Sci. USA **88:** 11465–11469.
23. NICOLLE, M. W., A. VINCENT, S. SHARMA, B. NAG, N. WILLCOX & J. NEWSOM-DAVIS. 1993. An *in vitro* model for disease-specific immunotherapy in myasthenia gravis (MG) using soluble MHC class II bound to AChR-derived peptide. Ann. NY Acad. Sci. This volume.
24. PETTE, M., K. FUJITA, B. KITZE, J. N. WHITAKER, E. ALBERT, L. KAPPOS & H. WEKERLE. 1990. Myelin basic protein-specific T lymphocyte lines from MS patients and healthy individuals. Neurology **40:** 1770–1776.
25. PROTTI, M. P., A. A. MANFREDI, C. STRAUB, J. F. HOWARD JR. & B. M. CONTI-TRONCONI. 1990. Immunodominant regions for T helper-cell sensitization on the human nicotinic receptor α subunit in myasthenia gravis. Proc. Natl. Acad. Sci. USA **87:** 7792–7796.
26. MATSUO, H., A.-P. BATOCCHI, S. HAWKE, G. C. HARCOURT, L. JACOBSON, A. VIN-

CENT, N. WILLCOX & J. NEWSOM-DAVIS. 1992. Recognition of cryptic epitopes by T cell lines selected with pooled synthetic acetylcholine receptor α subunit peptides. In preparation.

27. MATSUO, H., K. PILE, N. WILLCOX, B. P. WORDSWORTH, A. VINCENT & J. NEWSOM-DAVIS. 1992. Preferential co-recognition of DR4 subtypes by T cells selected with acetylcholine receptor peptides. In preparation.
28. SCHILD, H., M. NORDA, K. DERES, K. FALK, O. RÖTZSCHKE, K.-H. WIESMÜLLER, G. JUNG & H.-G. RAMMENSEE. 1991. Fine specificity of cytotoxic T lymphocytes primed *in vivo* either with virus or synthetic lipopeptide vaccine or primed *in vitro* with peptide. J. Exp. Med. **174:** 1665–1668.
29. MOIOLA, L., M. P. PROTTI, A. A. MANFREDI, M-H. YUEN, J. F. HOWARD JR. & B. M. CONTI-TRONCONI. 1993. T-helper epitopes on human nicotinic acetylcholine receptor in myasthenia gravis. Ann. N.Y. Acad. Sci. This volume.

Differences in Acetylcholine Receptor-Antibody Interactions between Extraocular and Extremity Muscle Fibers[a]

KENICHIRO ODA

National Institute of Neuroscience
National Center of Neurology and Psychiatry
Tokyo 187, Japan

The extraocular muscle (EOM) is highly vulnerable in people with myasthenia gravis.[1-3] EOMs are involved at some time during the course of the disease in almost every patient, usually during the first year; and in 50–65% of all patients with myasthenia gravis (MG), the initial symptoms were found in the EOMs. Among 10–14% of all myasthenia gravis patients, the disease remained localized to the EOMs for more than two years. This type of manifestation is referred to as ocular myasthenia gravis (OMG).[4] Furthermore, patients afflicted with the generalized disease (generalized MG (GMG)) often continue to exhibit ocular symptoms, even after the general symptoms have been in remission. So far, many studies have revealed the essential role of the antiacetylcholine receptor (AChR) antibody in the pathogenesis of MG.[5] It has been shown, however, that a high proportion (20–50%) of people with OMG are without detectable antibodies.[3,6,7] At present it is not clear why EOMs are frequently involved in MG patients and why a high proportion of OMG patients have no detectable antibodies to AChR. The following hypothesis has been proposed by several authors.[8,9] The AChR molecules of EOM are antigenically distinct from that of limb muscle, and the antibody of OMG might be targeted to certain epitopes unique to ocular muscle AChRs. As such, the antibodies directed against those particular epitopes cannot be detected by conventional radioimmunoassay (RIA) as long as the limb receptor is used as the antigen. To test this hypothesis, the following three investigations were conducted. The anatomical characteristics of human EOM end-plates were examined. Then the specific antibodies that bind to unique epitopes on them were located. Finally, by means of passive transfer, whether or not the antibodies directed against the EOM end-plates have a functional activity that causes a reduction in the number of AChR at the end-plates was determined.

MORPHOLOGY OF HUMAN EOM END-PLATE

EOMs are highly organized striated muscles that control the very precise gradations in the mobility of the eye. They differ from limb muscles in that they

[a] This work was supported in part by a Research Grant (NCNP 1-2-06) from the National Center of Neurology and Psychiatry (NCNP) of the Ministry of Health and Welfare, Japan.

have a high innervation ratio, more variation in fiber types, and unique physiological features.[10] Human EOM fibers have two different types of motor end-plates: en plaque and en grappe types, identified mainly by acetylcholinesterase (AChE) staining.[11–13] Little information, however, is available regarding AChR distribution and the precise innervation pattern of human EOM fibers. To clarify the characteristics of human EOM end-plate morphology, the following study was conducted.[14]

Fresh materials from surgical speciments were used. Bundles of inferior oblique (IO) muscles were obtained from patients undergoing myectomy of the IO muscles for correction of overaction. In the teased muscle fiber preparations, two morphological properties were focused upon: the motor innervation pattern, which was visualized using combined AChE and silver staining, and the AChR distribution on muscle fibers, which was visualized by rhodamine-conjugated α-bungarotoxin (α-BGT).[15] We found that there was one single innervated fiber type and two multiple innervated types (FIGURES 1–4).

Single Terminal Type (Plaque-like End-plate) (FIG. 1)

A single large compact or lobulated end-plate was located on a large-diameter fiber. This end-plate had the same appearance as that of limb muscles. Each muscle fiber of this type was singly innervated by a thick axon. This type constituted around 70% of all muscle fibers.

Multiterminal Type I (Bead-like End-plate) (FIG. 2)

Numerous small end-plates were located on small diameter fibers and were irregularly spaced along the fibers. Considerable variability was encountered in the distance between these small end-plates, with extremes in the range of 2–50 μm. Those small clusters actually contained the AChRs, which were visualized using rhodamine-conjugated α-BGT (FIG. 2b). A group of small end-plates was connected together, like a chain of beads by a thin axon, running along the muscle fibers. This type constituted about 25% of all fibers.

Multiterminal Type II (Grape-like End-plate) (FIG. 3)

Several endings were regularly spaced on a single muscle fiber. Each end-plate was composed of small AChE-positive particles with a "grape-like" appearance (FIG. 3). The muscle fiber of this type was innervated by several different axons emanating from different separate locations on the nerve trunk. Those axons terminated on the muscle fiber at regular intervals, and there were none that ran along the fiber to connect each end-plate region as noted in the multiterminal type I fibers (see FIG. 4). The proportion of this type in the muscle fibers was approximately 5 percent.

Previous studies of human EOM fibers demonstrated single and multiple innervated end-plates. We demonstrated here that the multiterminal type could be divided into two subgroups (FIG. 4; I and II) according to the innervation profile. This classification was achieved by means of the teasing fiber technique, which enabled us to observe the entire surface of a given muscle fiber over a much longer distance compared with the standard histology using frozen or formalin-fixed sections.

Physiologically, singly innervated fibers of EOM correspond to fast and twitch extremity fibers. These fibers are characterized by the ability of the muscle membrane to propagate action potentials and by the fact that the fibers can contract synchronously in response to each motor nerve impulse. On the other hand, multiterminal fibers can be divided into slow-tonic fibers and intermediate fibers

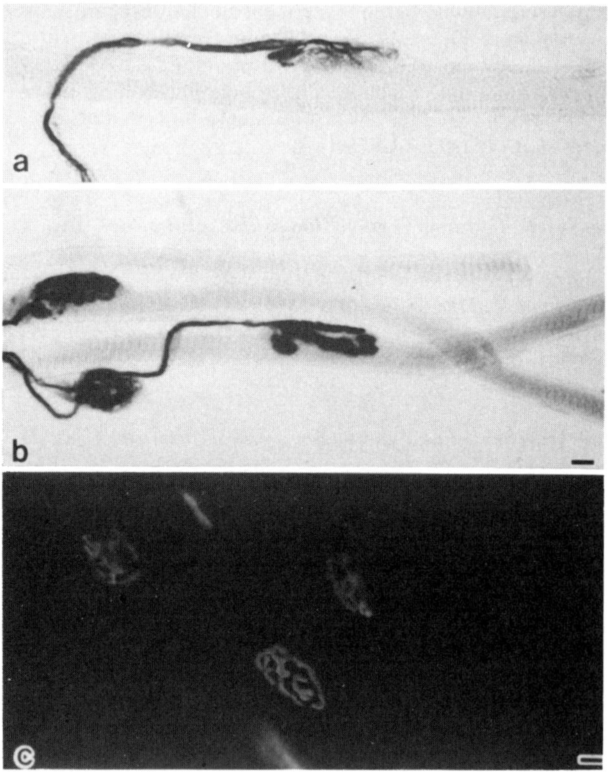

FIGURE 1. Motor innervation pattern and AChR distribution of human extraocular muscle fibers with single terminal end-plates. Combined AChE and silver staining **(a, b)**. Rhodamine-conjugated α-bungarotoxin staining **(c)**. Large lobulated end-plates are localized to a portion of individual muscle fibers, and each end-plate is innervated by a thick myelinated axon (bar = 15 μm).

(named because they share properties of both twitch and tonic fibers). Slow fibers are distinguished by several characteristics;[16] their inability to propagate action potentials and twitch in response to a single volley, their slow isotonic shortening speed, and their ability to develop a maintained contracture in depolarizing solutions. In addition, slow and tonic fibers have extremely slow synaptic current kinetics when compared with twitch fibers. This property, coupled with the pres-

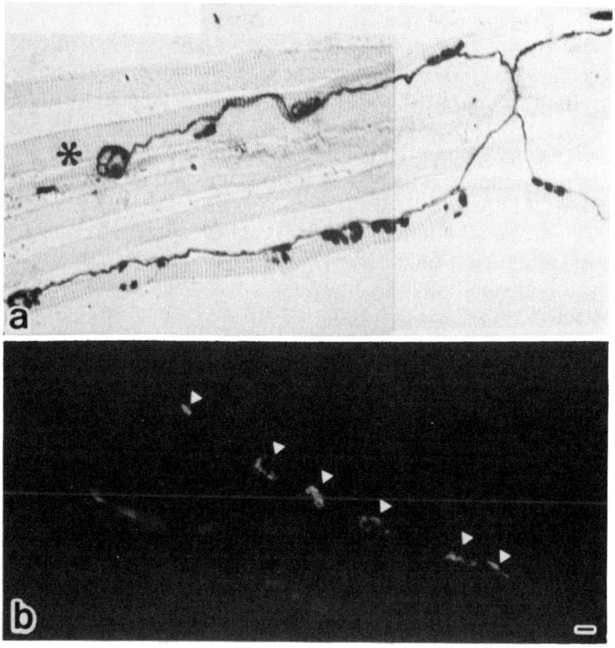

FIGURE 2. Multiterminal muscle fiber type I. **a:** Combined AChE and silver staining. The numerous small end-plates are scattered as a chain of beads by a thin axon, running along the muscle fibers, branching and terminating in small endings. **b:** Rhodamine-conjugated α-bungarotoxin staining. Those small endings contain AChRs (arrowheads) (bar = 15 μm).

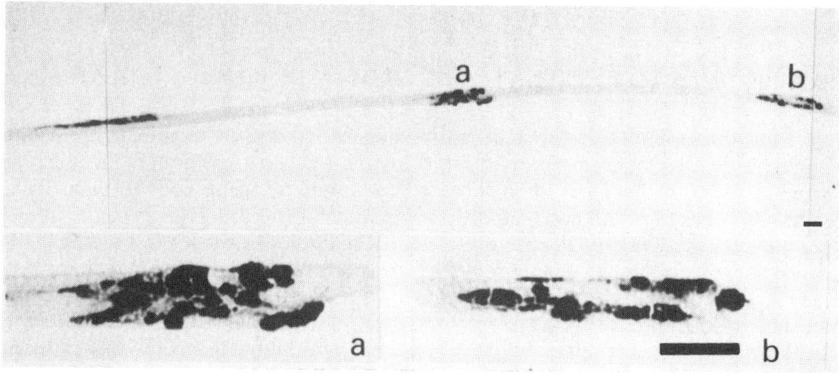

FIGURE 3. Multiterminal muscle fiber type II. AChE staining. End-plates are regularly spaced on a single muscle fiber. At higher magnification of the lower panel, each end-plate **(a, b)** is composed of several small AChE-positive particles (bar = 30 μm).

ence of end-plates distribution along the entire length of the fiber, causes a relatively uniform depolarization of tonic fibers in response to nerve stimulation.[9] EOMs provide a clear example of the divergent physiological roles of fast and slow fibers, where fast fibers generate the saccadic movements needed to position the eyes within the visual field and slow fibers subserve fine positional control and visual tracking.[17]

There are two AChR subtypes in mammalian skeletal muscle.[18] Their subunit composition, antigenicity, ion channel properties, and localization along the muscle fiber are closely linked and are regulated mainly by neural factors.[19] Recently, unique channel properties of multiterminal type receptors have been characterized using single-channel recordings.[17,20,21] Taken together, these anatomical, biochemical, and physiological findings strongly suggest that EOM fibers, especially multiterminal muscle fibers, have a unique AChR molecular subtype, and, if this so, unique epitopes expressed upon it could provide particular antigen sources in persons with myasthenia gravis.

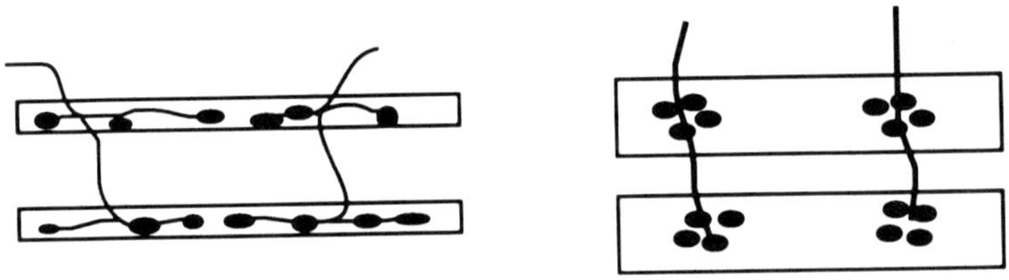

FIGURE 4. Schematic illustration of innervation patterns and the corresponding AChR distribution of human EOM muscle fibers with multiterminal end-plates. Left, polyneuronal innervation with ultraterminal branching (type I); right, polyneuronal innervation without ultraterminal branching (type II).

ANTIBODIES TO END-PLATES OF HUMAN EXTRAOCULAR MUSCLE

In order to detect the specific antibodies directed against human EOM end-plates, we collected sera from 32 OMG patients whose symptoms had been restricted to EOM for more than three years. First, by using conventional RIA,[6] antibody titers to AChR were directly compared using extracts from either human leg muscles or EOMs as the source of AChR antigen. As shown in FIGURE 5, the titer values were higher when assayed with AChR isolated from EOM than those measured with limb muscle AChR. Among 17 sera that contained no detectable antibody to leg muscle AChR, 7 had measurable titer values when measured with EOM AChR. These results confirmed previous reports that the specificity of AChR antibodies to EOM AChR differed from that in limb muscles[7,22] and indicated the presence of antibody subpopulations that bind to unique epitopes of AChR molecules in EOM.

To obtain direct evidence of the presence of EOM end-plate–specific antibod-

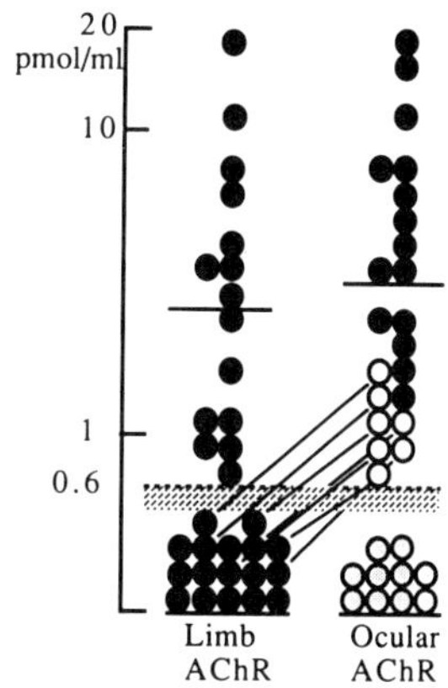

FIGURE 5. Titer values of AChR antibody obtained from the sera of ocular MG patients. Human limb and EOM were used as sources of AChR (normal: below 0.6 pmol/mL).

ies, 17 OMG sera were selected upon the basis of no detectable reactivity with limb muscle AChRs as shown in FIGURE 5, inasmuch as they were expected for candidates who may have had specific antibodies to EOM end-plates. The assay method was as follows. Fresh human EOMs were fixed in methanol and successively incubated with rhodamine-conjugated α-BGT, optimally diluted test sera, and then fluorescein-labeled goat anti-human IgG. The stained muscle fibers were teased and photographed with filters appropriate to show either rhodamine or fluorescein. Twelve of 17 sera tested contained antibodies to EOM end-plates (TABLE 1). These sera were divided into three groups according to the staining pattern. The first four sera contained antibodies to both single and multiterminal

TABLE 1. Antibodies to EOM End-plate

Patient	Anti-single terminal	Anti-multiterminal	Titer of RIA[a] pmol/mL
OMG-1	1:320	1:320	3.4
OMG-2	1:320	1:320	4.2
OMG-3	1:80	1:320	5.3
OMG-4	1:80	1:80	4.8
OMG-5	—	1:320	1.2
OMG-6	—	1:320	2.3
OMG-7	—	1:40	0
OMG-8	—	1:80	0
OMG-9	—	1:80	0
OMG-10	—	1:320	0
OMG-11	—	1:80	0
OMG-12	—	1:160	0
OMG-13	—	—	2.2
OMG-14	—	—	0
OMG-15	—	—	0
OMG-16	—	—	0
OMG-17	—	—	0

[a] RIA with human EOM as antigen.

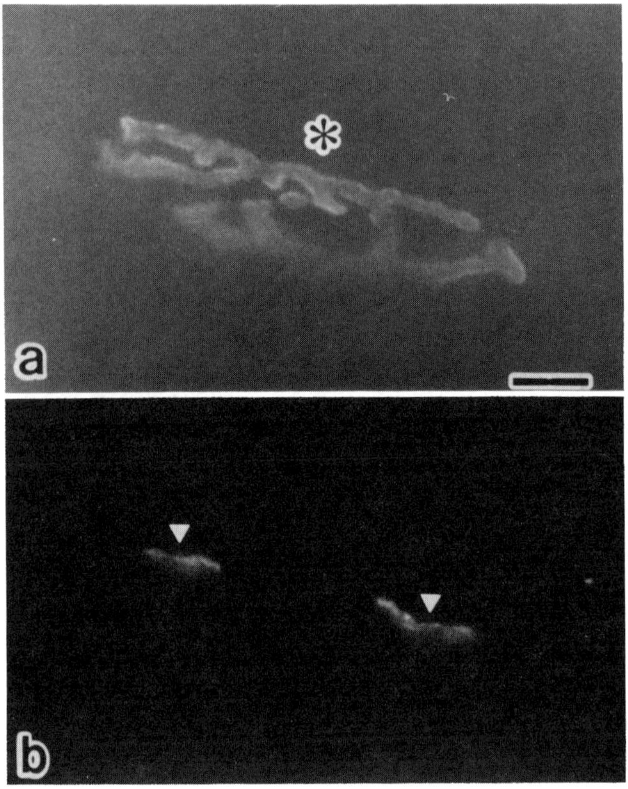

FIGURE 6. EOM fibers successively incubated with rhodamine-conjugated α-BGT, OMG-1 serum (1 : 80), and fluoresceinated anti-human IgG. Teased muscle fibers were photographed with filters appropriate to show fluorescein. The serum of OMG-1 has reactivity with the end-plates of both singe (**a:** *) and multiterminal (**b:** arrowheads) types.

end-plates (FIG. 6). Antibodies in this group were also detectable with the standard RIA using the EOM extract. The next eight sera contained antibodies exclusive to multiterminal end-plates (FIG. 7). These antibodies could not be detected in 6 out of 8 of these sera using RIA. In the third group, except for OMG-13, we could not detect any antibodies using either immunofluorescent method (IF) or RIA. As shown in the first group of the sera, even within a given serum, the titer values between single and multiterminal types were not identical. This difference is thought to provide more evidence for the presence of a different antigenicity between the single-type end-plate and the multiterminal-type end-plate.

We ensured that the sera reactive with EOM end-plates contained the antibodies directed against the AChR (reported elsewhere[8]). When EOM fibers were exposed to the sera reactive to specific end-plate types, about 30–40% of membranous AChRs at the respective end-plates became complexed with IgG and were extractable from the muscle fibers.

There are several reasons why the EOM-specific antibody could be detected using IF. First, as fresh species-matched materials were obtained, the fine antigenic determinants on the receptor were preserved that might have been lost during the detergent solubilization required in RIA. Second, as AChR molecules at the end-plate formed a cluster and were packed at very high densities compared with that presented in the extracts for the standard RIA in solution,[23] even low-affinity antibody subpopulations were detectable using the teased muscle fibers as the substrate.

PASSIVE TRANSFER

To address the issue of whether the antibodies to EOM end-plates in people with ocular myasthenia gravis could cause an AChR loss at end-plates *in vivo,* passive transfer studies were conducted following the protocol established by several workers.[24,25]

Preparation of IgG and Injection of Mice

Sera that contained antibodies that cross-reacted with EOM end-plates of mice were obtained from 7 OMG patients (OMG-1, 3, 5, 6, 7, 13, and 14 in TABLE 1). Control sera were obtained from two normal individuals. One stored serum, used as a positive control, was obtained from a patient with severe generalized MG (GMG). It contained a high titer antibody to mouse junctional AChR and caused a substantial reduction in AChR density at the end-plates of the recipient mice.[25] The immunoglobulin G (IgG) fraction was isolated from each sera by ammonium

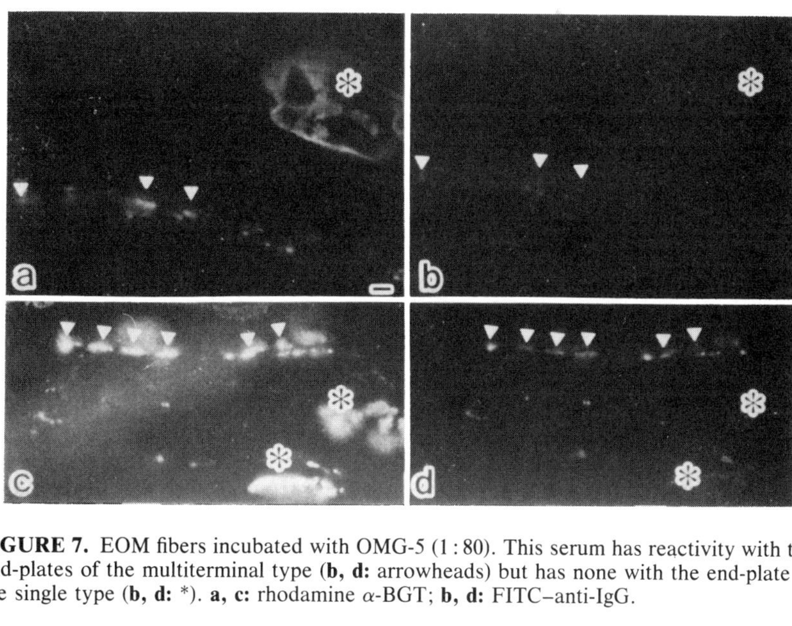

FIGURE 7. EOM fibers incubated with OMG-5 (1 : 80). This serum has reactivity with the end-plates of the multiterminal type (**b, d:** arrowheads) but has none with the end-plate of the single type (**b, d:** *). **a, c:** rhodamine α-BGT; **b, d:** FITC–anti-IgG.

sulphate precipitation and ion-exchange chromatography. These fractions were stored at 4°C at a concentration of 15 mg/mL. Each IgG was injected intraperitoneally into sets of 3 BALB/c mice at 15 mg/day for 7 days. Animals were given a single dose of cyclophosphamide on day 1 to prevent an immune response against the human IgG. Twenty-four hours after the final injections the animals were anesthetized and sacrificed. The extensor digitorum muscles (EDLs) and EOMs were removed.

AChR loss at the end-plate of different-type muscle fibers in the recipient mice was assessed by several methods as follows.

Autoradiography

The use of [^{125}I]α-BGT autoradiography allowed examination, at the light microscopic level, of the distribution of high-density patches of AChR.[26] FIGURE 8a shows the distribution of AChR at the end-plates of EDL muscle fibers obtained from a mouse injected with normal IgG. A discrete cluster of silver grains in the emulsion overlying the end-plate region arose from [^{125}I]α-BGT bound to the underlying AChRs. By contrast, the end-plate region was difficult to visualize in a mouse with the IgG from a GMG (FIG. 8b). Paucity of silver grains reflected the deficiency of specific binding sites for [^{125}I]α-BGT. Furthermore, a radiometric assay revealed that the average number of binding sites for end-plates of EDL from the mice treated with normal IgG and mice with IgG from a GMG were 2.8×10^7 and 0.56×10^7, respectively. As the validity of this protocol of the passive transfer experiment was ensured, the EOM fibers of the recipient mice were examined. When injected IgGs contained both anti-single terminal (anti-S) and anti-multiterminal (anti-M) antibody, the grain density at the end-plates of both singly and multi-innervated muscle fibers of the recipient mouse was significantly decreased (FIG. 9). When IgGs containing anti-M antibody only were injected, selective reduction of the AChR density was observed at the multiterminal end-plates, leaving the AChR density of the single-type end-plate normal (FIG. 10). These results indicated that AChR loss at end-plates of different types was induced by the respective antibody subpopulations that might be mediated through mechanisms similar to those that have been proposed by studies using seropositive GMG IgG, such as accelerated degradation of AChR.[5] While extending this passive transfer study, we encountered some anomalies, in which correlation between the relative AChR loss at end-plates of different types and the composition of antibody subpopulations was low. Thus, we assessed the AChR density at end-plates of different types quantitatively.

Fluorometric Microscopy

Inasmuch as single and multiple innervated muscle fibers are intermingled in the EOM, the standard radiometric method was unable to function, in which the specified end-plate regions containing AChRs labeled with [^{125}I]α-BGT are excised and counted in a gamma counter. To overcome this, we developed a fluorometric method using EOM fibers from recipient mice incubated with rhodamine-conjugated α-BGT. The intensity of the fluorescence at the end-plate was measured using a fluorometric microscope (Olympus, OSP1). In this system, the digitized signal was processed by the image analyzer. To ensure the validity of this method, we first measured the fluorescent intensity at end-plates of EDL muscle prepara-

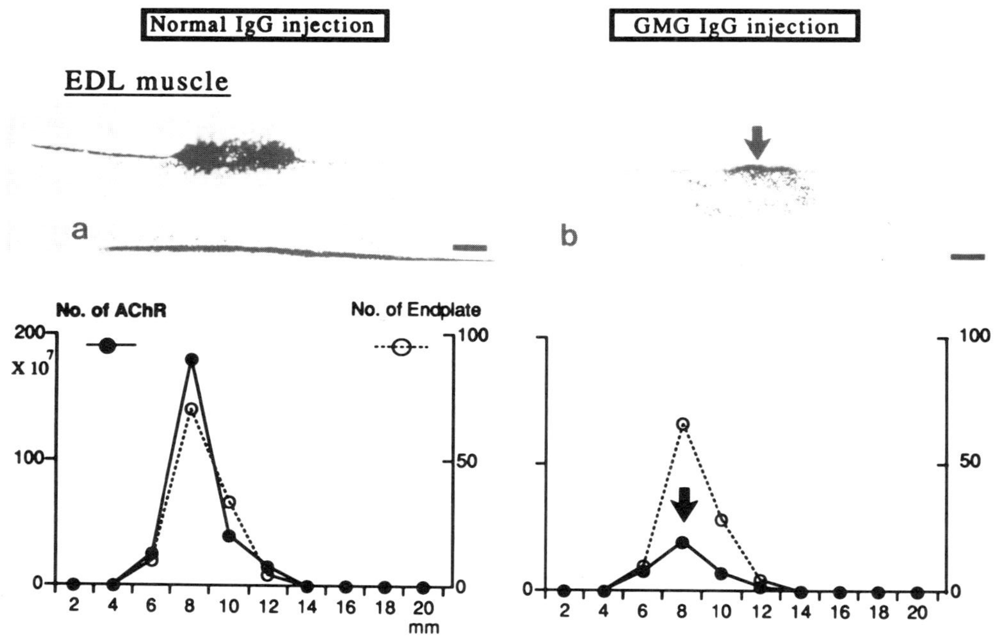

FIGURE 8. Autoradiographs showing sites of [^{125}I]α-BGT of EDL muscle fibers from mice injected with normal IgG **(a)** and GMG IgG **(b)**. **a:** A discrete high-density cluster of silver grains is located in the end-plate region (exposure: 4 days). **b:** A faint cluster of silver grains can be seen in the end-plate region. Distribution of α-BGT binding sites (closed circles) and number of AChE-positive end-plates (open circles) along the length of EDL muscle fibers (in lower panels). There is a significant loss of α-BGT binding sites per end-plate (arrow) of the mouse injected with GMG IgG when compared with those of the mouse with normal IgG.

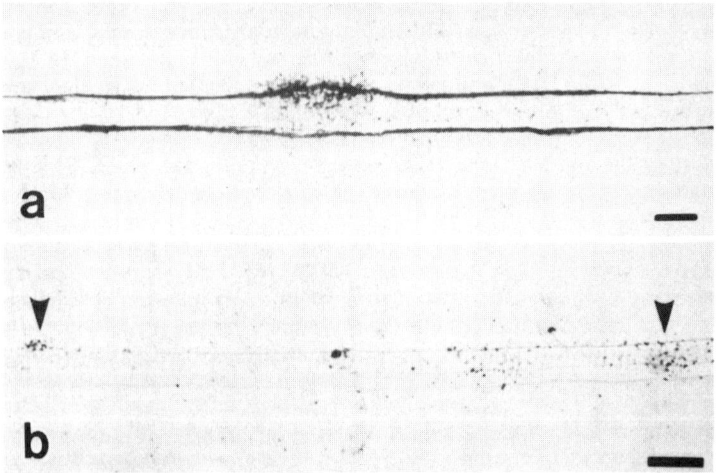

FIGURE 9. Autoradiographs of EOM fibers from a mouse injected with OMG-1 IgG. **a:** EOM fiber with a single terminal (bar = 20 μm). **b:** EOM fiber with multiterminals. The grain density at end-plates of both types are decreased (bar = 10 μm).

tions, which were preincubated with varying concentrations of unconjugated α-BGT, and found that there was an inverse correlation between the signal intensity and the dose of the free toxin added (Oda, unpublished data).

As shown in FIGURE 11, the signal intensity was measured by positioning the fixed pinhole onto the individual fluorescent-labeled end-plate of the multiterminal type in a mouse injected with normal IgG (FIG. 11a) and with OMG-5 IgG, which contained anti-M antibody (FIG. 11b). The left panels show the respective histograms. In the mouse treated with normal IgG, the percent values of intensity were concentrated at around 100 (mean ± SE; 104 ± 35, 100 end-plates). By contrast, in the mouse with OMG IgG, the signal intensity was significantly lower (34 ± 15, N = 100) compared with normal controls. FIGURE 12, in which the results of the passive transfer are assessed with the fluorometric method, indicates the following. In most instances, there was a global correlation between the profile of AChR loss in different types of end-plates and the relative composition of the antibody subgroup that received the injection. Groups of mice injected with anti-M antibody all showed a significant reduction in AChR at the end-plates of M-type fibers. The same was true for the mice with anti-S antibody, in which the single-type AChR decreased. This correlation indicates that antibodies specific to EOM end-plates are functionally active and are responsible for the loss of AChRs at EOM end-plates. In two instances, OMG-6 and OMG-14, there was poor correlation between the AChR loss and the profiles of the antibody subpopulations as suggested by autoradiography. This discrepancy might be explained by the different sensitivity of the immunofluorescent method compared with passive transfer. The efficiency of passive transfer for investigating the presence of circulating antibodies at low concentrations has been reported.[27] The third point is that as observed in OMG-3, 6, and 7, the AChR density of single-terminal fibers of EOM is less than that of EDL muscle. This result suggests that even within a single-terminal end-plate, there are different levels of susceptibility to the AChR molecules confronting the antibody attack. This result also raises the possibility that although single-terminal AChRs of both EDL and EOM may bind antibody equivalently, the metabolism of receptors, such as aggregation, endocytosis, or exocytosis, may proceed at various rates in different muscles because of the influences of intrinsic factors in the muscle fibers. Some of our recent data support this possibility.

Dystrophin-related Protein

Recently, dystrophin-related protein (DRP) was discovered by Love *et al.*[28] This protein has high sequence homology with dystrophin but is encoded by an autosomal gene. The immunohistochemical study revealed that the DRP is highly concentrated at end-plates, with only very weak staining of extrajunctional membranes.[29–32] Therefore, this protein may anchor the cytoplasmic domain of AChRs and stabilize them in the postsynaptic membrane.[33] We obtained an anti-DRP polyclonal antibody (a gift from Dr. M. Takemitsu), which was raised against a

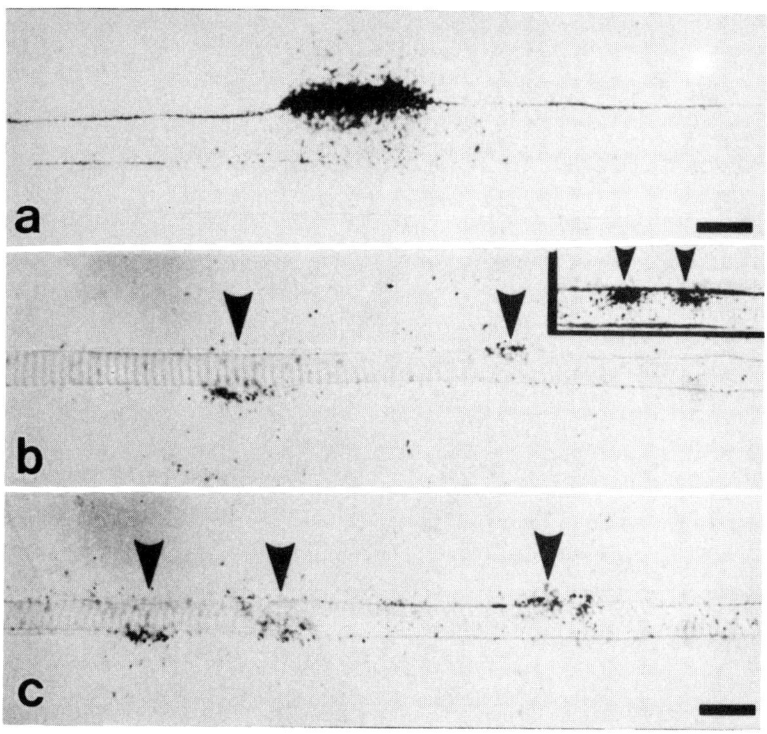

FIGURE 10. Autoradiographs of muscle fibers from a mouse injected with OMG-5. **a:** EOM fiber with a single terminal; **b,c:** EOM fiber with multiterminals. There is a selective reduction in the silver grain density at the multiterminal end-plates (arrowheads). Inset in **b** shows an autoradiograph of multiterminal end-plates from the mouse injected with normal IgG.

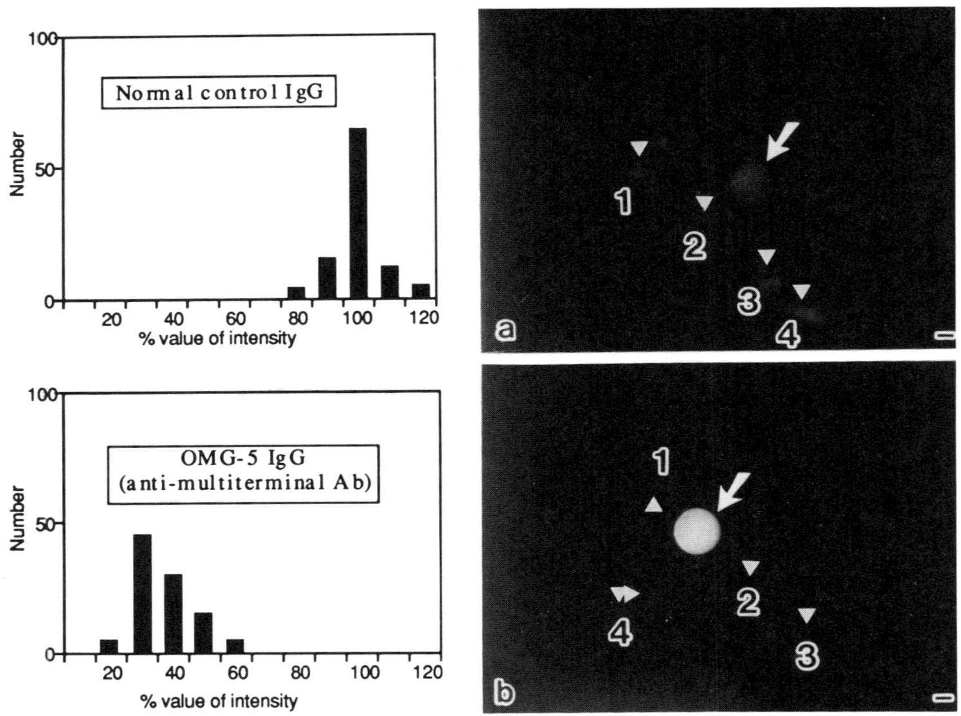

FIGURE 11. Microscopic fluorometry. The intensity of the fluorescence at an individual end-plate (arrowheads and numbering) can be measured by positioning the pinhole (arrows). **a:** EOM fibers with multiterminal end-plates from a mouse injected with normal IgG and from a mouse with OMG-5 IgG (**b**). The histograms show that the percent value of the intensity of signal in **b** is much lower compared with that of **a** (bar = 20 μm).

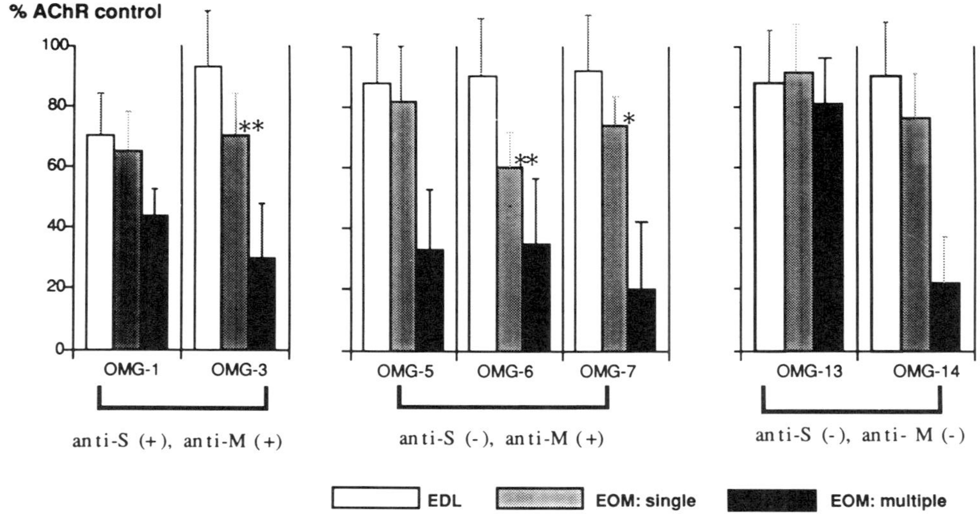

FIGURE 12. Experimental results from injected mice using the fluorometric method. Each column is composed of three different muscles, EDL (open bars ± SD), EOM with single terminal (stippled bars ± SD), and EOM with multiterminal terminals (solid bars ± SD). They were obtained from a group of three mice given the same IgG. There is a significant difference in mean value between EDL and EOM, with a single terminal in the groups of the mice injected with OMG-3,6, and 7. * $p < 0.05$, ** $p < 0.01$.

synthetic peptide corresponding to DRP.[32] This antibody did not cross-react with dystrophin because the antigen amino acid sequence had no homologous sequence to dystrophin except for two amino acids located at either end. When the methanol-fixed teased EDL were incubated with this antipeptide antibody, the reactivity of DRP was localized at the end-plate region with a staining profile quite similar to

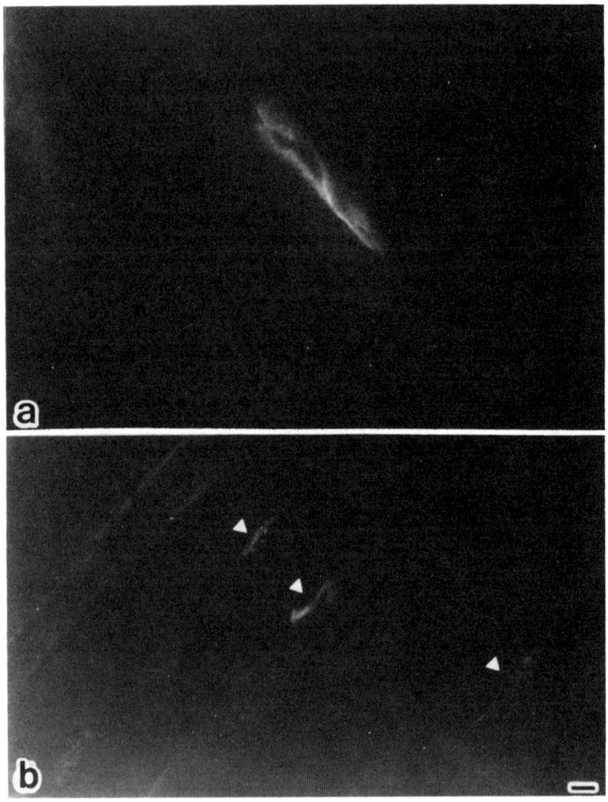

FIGURE 13. Dystrophin-related protein (DRP) in the mouse EOM end-plates. Methanol-fixed normal BALB/c mouse EOM fibers successively incubated with polyclonal rabbit antibodies to DRP and fluoresceinated anti-rabbit IgG. **a:** A single-terminal end-plate; **b:** multiterminal end-plates (bar = 5 μm).

that of α-BGT binding sites. We next examined the stainability of DRP in mouse EOM end-plates using the same polyclonal antibody (FIG. 13). We observed that DRP was present in both single and multiterminal end-plates but with a generally weaker staining intensity compared with that of the limb muscle end-plate. Relevant to this finding, it has been demonstrated ultrastructurally that the postsynaptic membranes of singly innervated fibers of mouse EOMs have sparse junctional folds when compared of twitch limb muscle fibers.[34] Inasmuch as DRP is thought to be localized on the postsynaptic membrane, the weak stainability of DRP might reflect the structural characteristics of the neuromuscular junctions of singly innervated EOM fibers. At present, as we have not yet quantified DRP, we cannot conclude that there is poor expression of DRP in EOM end-plates. However, if this finding is true, and furthermore, if this protein does play a role in stabilizing the AChR molecules in the postsynaptic membrane, we will extend the study to cover the structures constructing the microenvironment surrounding the AChR molecules in the EOM fibers.

SUMMARY

There are two types of motor nerve innervation patterns and AChR distributions in human EOM: single and multiple. The latter is further divided into two subgroups that are restricted to EOM and are not found in limb muscles. Epitopes that are unique to EOM end-plates exist. Some OMG patients have antibodies that are specifically targeted to those epitopes. These antibodies are functionally active and can cause AChR loss in EOM end-plates. In addition to AChRs, specific components constructing the microenvironment surrounding them may also be involved in the susceptibility of EOM in myasthenia gravis.

ACKNOWLEDGMENTS

The author is very grateful to Dr. M. Oono (Professor of the Department of Ophthalmology, Saga Medical School) for providing EOM materials and to Dr. M. Takemitu and other members of the laboratory of Dr. H. Sugita (National Institute of Neuroscience, NCNP) for the gift of the antibodies against DRP. I thank Dr. Y. Ito (Professor of the Department of Pharmacology, Faculty of Medicine, Kyushu University) for valuable discussion.

REFERENCES

1. Osserman, K. E. 1967. Ocular myasthenia gravis. Invest. Ophthalmol. **6:** 277–287.
2. Simpson, J. A. 1960. Myasthenia gravis: a new hypothesis. Scot. Med. J. **5:** 419–436.
3. Oosterhuis, H. J. G. H. 1982. The ocular signs and symptoms of myasthenia gravis. Doc. Ophthalmol. **52:** 363–378.
4. Osserman, K. E. & G. Genkins. 1971. Studies in myasthenia gravis: Review of a twenty-year experience in over 1200 patients. Mt. Sinai J. Med. **38:** 497–537.
5. Drachman, D. B., S. De Silva, D. Ramsay & A. Pestronk. 1987. Humoral pathogenesis of myasthenia gravis. Ann. N.Y. Acad. Sci. **505:** 90–105.
6. Oda, K. & Y. Ito. 1981. Myasthenia gravis. Antibodies to acetylcholine receptor in ocular myasthenia gravis. J. Neurol. **225:** 251–258.
7. Tindall, R. S. A. 1981. Humoral immunity in myasthenia gravis: biochemical characterization of acquired antireceptor antibodies and clinical correlations. Ann. Neurol. **10:** 437–447.
8. Oda, K. & H. Shibasaki. 1988. Antigenic difference of acetylcholine receptor between single and multiple form endplates of human extraocular muscle. Brain Res. **449:** 337–340.
9. Kaminski, H. J., E. Maas, P. Spiegel & R. L. Ruff. 1990. Why are eye muscles frequently involved in myasthenia gravis? Neurology **40:** 1663–1669.

10. CHIARANDINI, D. J. & J. DAVIDOWITZ. 1979. Structure and function of extraocular muscle fibers. Curr. Top. Eye. Res. **1:** 91–142.
11. KUPFER, C. 1960. Motor innervation of extraocular muscle. J. Physiol. (Lond.) **153:** 522–526.
12. NAMBA, T., T. NAKAMURA & D. GROB. 1968. Motor nerve endings in human extraocular muscle. Neurology **18:** 403–407.
13. SADEH, M. & L. Z. STERN. 1984. Observations on the innervation of human extraocular muscles. J. Neurol. Sci. **66:** 295–305.
14. ODA, K. 1986. Motor innervation and acetylcholine receptor distribution of human extraocular muscle fibres. J. Neurol. Sci. **74:** 125–133.
15. ANDERSON, M. J. & M. W. COHEN. 1974. Fluorescent staining of acetylcholine receptors in vertebrate skeletal muscle. J. Physiol. (Lond.) **237:** 385–400.
16. MORGAN, D. L. & U. PROSKE. 1984. Vertebrate slow muscle: its structure, pattern of innervation, and mechanical properties. Physiol. Rev. **64:** 103–169.
17. DIONNE, V. E. 1989. Two types of nicotinic acetylcholine receptor channels at slow fibre end-plates of the garter snake. J. Physiol. (Lond.) **409:** 313–331.
18. MISHINA, M., T. TAKAI, M. IMOTO, K. NODA, M. TAKAHASHI, S. NUMA, C. METHFESSEL & B. SAKMANN. 1986. Molecular distinction between fetal and adult forms of muscle acetylcholine receptor. Nature (Lond.). **321:** 406–411.
19. WITZEMANN, V., H.-R. BRENNER & B. SAKMANN. 1991. Neural factors regulate AChR subunit mRNAs at rat neuromuscular synapses. J. Cell Biol. **114:** 125–141.
20. HENDERSON, L. P. & P. BREHM. 1989. The single-channel basis for the slow kinetics of synaptic currents in vertebrate slow muscle fibers. Neuron **2:** 1399–1405.
21. RUFF, R. L. 1989. Tonic muscle fibers have two types of endplate acetylcholine receptor channels. Neurology **39:** 267.
22. COMPSTON, D. S. A., A. VINCENT, J. NEWSOM-DAVIS & J. R. BATCHELOR. 1980. Clinical, pathological, HLA antigen and immunological evidence for disease heterogeneity in myasthenia gravis. Brain **103:** 579–601.
23. MOSSMAN, S., A. VINCENT & J. NEWSOM-DAVIS. 1988. Passive transfer of myasthenia gravis by immunoglobulins: lack of correlation between AChR with antibody bound, acetylcholine receptor loss and transmission defect. J. Neurol. Sci. **84:** 15–28.
24. TOYKA, K. V., D. B. DRACHMAN, A. PESTRONK & I. KAO. 1975. Myasthenia gravis: passive transfer from man to mouse. Science **190:** 397–399.
25. ODA, K., S. KORENAGA & Y. ITO. 1981. Myasthenia gravis: passive transfer to mice of antibodies to human and mouse acetylcholine receptor. Neurology **31:** 282–287.
26. ODA, K. 1984. Age changes of motor innervation and acetylcholine receptor distribution on human skeletal muscle fibres. J. Neurol. Sci. **66:** 327–338.
27. DRACHMAN, D. B., S. DESILVA, D. RAMSAY & A. PESTRONK. 1987. "Sero-negative" myasthenia gravis: a humorally mediated variant of myasthenia. Neurology **37** (suppl. 1): 214.
28. LOVE, D. R., D. F. HILL, G. DICKSON, N. K. SPURR, B. C. BYTH, R. F. MARSDEN, F. S. WALSH, Y. H. EDWARDS & K. E. DAVIES. 1989. An autosomal transcript in skeletal muscle with homology to dystrophin. Nature (Lond.) **339:** 55–58.
29. KHURANA, T. S., S. C. WATKINS, P. CHAFEY, J. CHELLY, F. M. S. TOME, M. FARDEAU, J-C. KAPLAN & L. M. KUNKEL. 1991. Immunolocalization and developmental expression of dystrophin related protein in skeletal muscle. Neuromuscular Dis. **1:** 185–194.
30. OHLENDIECK, K., J. M. ERVASTI, K. MATSUMURA, S. D. KAHL, C. J. LEVEILLE & K. P. CAMPBELL. 1991. Dystrophin-related protein is localized to neuromuscular junctions of adult skeletal muscle. Neuron **7:** 499–508.
31. NGUYEN THI MAN, J. M. ELLIS, D. R. LOVE, K. E. DAVIES, K. C. GATTER, G. DIKSON & G. E. MORRIS. 1991. Localization of the DMDL gene-encoded dystrophin-related protein using a panel of nineteen monoclonal antibodies: presence at neuromuscular junctions, in the sarcolemma of dystrophic skeletal muscle, in vascular and other smooth muscles, and in proliferating brain cell lines. J. Cell Biol. **115:** 1695–1700.
32. TAKEMITSU, M., S. ISHIURA, R. KOGA, K. KAMAKURA, K. ARAHATA, I. NONAKA

& H. SUGITA. 1991. Dystrophin-related protein in the fetal and denervated skeletal muscles of normal and *mdx* mice. Biochem. Biophys. Res. Commun. **180:** 1179–1186.

33. FROEHNER, S. C. 1991. The submembrane machinery for nicotinic acetylcholine receptor clustering. J. Cell Biol. **114:** 1–7.
34. SALPETER, M. M., F. A. MCHENRY & H-H. FENG. 1974. Myoneural junctions in the extraocular muscles of the mouse. Anat. Rec. **179:** 201–224.

Phenotypic and Functional Abnormalities of B Cells in Myasthenia Gravis[a]

SAMIA RAGHEB AND ROBERT P. LISAK

Division of Neuroimmunology
Department of Neurology
Wayne State University
Detroit, Michigan 48201

Acquired autoimmune myasthenia gravis (MG) is a B-cell mediated disease in which antibodies are directed against the nicotinic acetylcholine receptor (AChR).[1,2] At present, little is known about the pathological mechanisms that lead to the breakdown of immune tolerance to the AChR. Anti-AChR antibodies interfere with normal neuromuscular transmission by blocking the ligand binding site on the AChR, increasing AChR turnover, and leading to AChR degradation through complement fixation.[1-3] Anti-AChR antibodies have been well characterized in terms of isotypes, allotypes, idiotypes, and antigenic specificities.[1-6] The focus of this paper is on the antibody-producing B lymphocytes in MG.

B CELLS IN PERIPHERAL BLOOD

The frequency of IgG-producing AChR-specific B cells in patients with MG has been estimated to be one in every 14,085 peripheral blood mononuclear cells (PBMC).[7] Even though MG is a B-cell mediated disease, the frequency of total B cells is not increased in the peripheral blood of patients with MG. Based on the expression of surface immunoglobulin (sIg) and CD20, the percentage of B cells in the peripheral blood of MG patients was the same as for normal, nonmyasthenic controls[8,9] (TABLE 1). Furthermore, as TABLE 1 shows, the expression of other B-cell surface markers did not differ between MG patients and controls. Of particular interest, the expression of cell-surface markers, which are found on activated B cells (CD25, CD71, and B5), was not increased in patients with MG. These data, however, do not tell us anything regarding the state of activation of AChR-specific B cells.

In vitro experiments have shown that mononuclear cell cultures from the peripheral blood of MG patients produce AChR-specific antibodies.[10-13] The amount of antibody production *in vitro* correlates with the serum anti-AChR antibody titer. PBMC from normal controls do not produce anti-AChR antibody *in vitro* even though they produce comparable amounts of total IgG and IgM as PBMC from patients with MG.[11,12] Even following the polyclonal stimulation of B cells with pokeweed mitogen (PWM), an antigen-presenting cell and T cell–dependent B-cell mitogen, PBMC from nonmyasthenic controls do not produce detectable anti-AChR antibody.[11,12]

The *in vitro* production of AChR-specific antibodies is T cell–dependent.[10,14-16] In an effort to assess the role of T cells in regulating anti-AChR antibody production, CD8^{+} (cytotoxic/suppressor) T cells were depleted from the PBMC popula-

[a] This project was supported by the Muscular Dystrophy Association of America.

tion.[12] Following such treatment, the *in vitro* production of total IgG, IgM, and AChR-specific antibody increased for PBMC from MG patients. For controls, there was an increase in total IgG and IgM production. Control PBMC, however, still failed to produce detectable levels of anti-AChR antibodies. These data suggest that $CD8^+$ lymphocytes are regulating anti-AChR antibody production in patients with MG. The lack of anti-AChR antibody production in nonmyasthenic individuals, however, is not simply due to $CD8^+$ T-cell regulation. The cumulative evidence suggests that there is a difference in frequency or state of activation of AChR-specific B cells in the peripheral blood of patients with MG versus normal, healthy individuals.

TABLE 1. Phenotypic Characterization of B Cells in the Peripheral Blood[a]

	MG patients	Normal controls
$CD20^+$	7.0 ± 4.0[b]	7.3 ± 4.6
$HLA\text{-}DR^+$ B cells	90.4 ± 7.5	90.3 ± 9.1
$CD25^+$ B cells	1.4 ± 3.5	3.9 ± 5.6
$CD71^+$ B cells	20.4 ± 12.6	15.5 ± 15.2
$B5^+$ B cells	47.0 ± 15.3	51.0 ± 15.9
$CD40^+$ B cells	72.6 ± 9.9	73.9 ± 11.2

[a] PBMC were labeled by immunofluorescence. The percentage of positively stained cells was determined by flow cytometry. For double-labeled cells, results are expressed as the percentage of B cells that bear a particular marker.

[b] Mean percent positive cells ± standard deviation.

B CELLS IN THE THYMUS

B lymphocytes in the myasthenic thymus have also been characterized phenotypically and functionally. About 60–80% of patients with MG have thymic hyperplasia, which primarily affects the medulla.[17,18] Germinal centers, which are immunohistologically identical to those found in a reactive lymph node, are present. The presence of these germinal centers is indicative of B-cell activation and proliferation occurring within the myasthenic thymus. Thymic B cells express CD20 on the cell surface as do B cells in peripheral blood. The expression of sIg, however, is poor on thymic B cells compared to peripheral blood B lymphocytes. Thymic B lymphocytes also express cell-surface activation markers such as interleukin-2 (IL-2) receptors (CD25) and transferrin receptors (CD71). Therefore, they appear to be highly activated.[19]

In vitro cultures of dispersed thymic cells have shown that the frequency of immunoglobulin-secreting cells is higher in thymic cultures than in PBMC cultures from the same MG patient.[20] Thymic cell cultures produce anti-AChR antibodies,[12,15,20–24] which are primarily of the IgG isotype.[25,26] By comparison, PBMC cultures produce both IgG and IgM antibodies. To determine whether thymic and peripheral blood T lymphocytes were providing different regulatory signals to their respective B cells, T-cell depleted peripheral blood B cells were cultured with T cells obtained from the thymus of the same MG patient. These cultures produced comparable levels of IgG and IgM, suggesting that thymic and peripheral blood B cells were somehow different.[26]

For patients with MG, anti-AChR antibody production by thymic cell cultures correlated with thymic histology.[21–24] Cultures from patients with thymic lymphoid hyperplasia produced more antibody than those from patients with thymoma or a normal (involuted) thymus. Anti-AChR antibody production, however, did not necessarily correlate with the percentage of B lymphocytes in these cultures. Despite the low frequency of B cells in the thymus compared to blood, antibody production by thymic cell cultures usually exceeded that by PBMC cultures from the same patient. This was especially true for nonstimulated cultures.[12,24]

CD5$^+$ B CELLS IN MG

The CD5 molecule is a 67 kDa cell-surface glycoprotein commonly referred to as Leu-1 in humans. Its physiologic role in lymphocyte activation remains largely unknown. CD5 was originally described as a pan–T-cell marker, but it is also expressed at a low density on some B lymphocytes.[27,28] On the B-cell surface CD5 is expressed in conjunction with other typical B-cell markers (sIg, CD20, or class II major histocompatibility antigens). CD5 was first discovered on B-cell lymphocytic leukemias,[29] but it is also present on normal B lymphocytes.[30,31] In humans, there is an increased frequency of B cells that express CD5 in the peripheral blood of patients with some, but not all, immunopathologic autoimmune diseases. CD5$^+$ B-cell frequency is increased in patients with rheumatoid arthritis,[32] Sjögren's syndrome,[33] Hashimoto's thyroiditis,[34] and insulin-dependent diabetes mellitus.[35] Patients with systemic lupus erythematosus do not have an increased frequency of CD5$^+$ B cells.[36] Patients with multiple sclerosis are reported to have an increased frequency of CD5$^+$ B cells in the cerebrospinal fluid.[37]

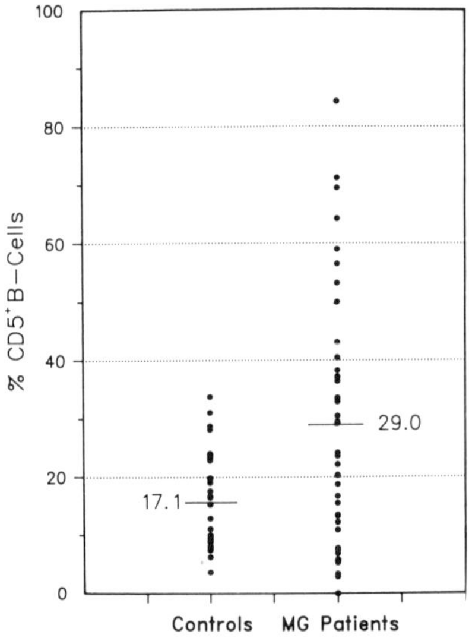

FIGURE 1. The frequency of CD5$^+$ B cells in the peripheral blood of normal controls and patients with MG. The frequency of CD5$^+$ B cells in MG patients (n = 41) and controls (n = 27) was determined by immunofluorescence. Bars represent the mean for each group. The difference between the two groups was statistically significant ($p < 0.05$, Mann-Whitney test).

TABLE 2. Comparison of the Frequency of Total B Cells and CD5$^+$ B Cells in the Thymus and Blood of Patients with MG

		Percent CD5 B Cells		Percent B Cells	
MG	Histology[a]	Thymus	Blood	Thymus	Blood
1. +	Involuted	5.3	27.9	2.5	6.8
2. +	Hyperplasia	13.3	11.3	7.6	9.0
3. +	Hyperplasia	37.1	ND[b]	3.2	ND
4. +	Hyperplasia	36.8	24.2	2.0	5.9
5. +	Thymoma	39.3	11.3	1.7	5.1
6. +	Hyperplasia	40.0	8.3	0.6	10.2
7. +	Involuted	25.0	ND	1.3	ND
8. +	Hyperplasia	41.6	33.1	1.3	12.3
9. +	Hyperplasia	8.0	5.0	30.8	4.1
10. +	Hyperplasia	45.0	2.0	3.1	16.7
11. +	Thymoma	65.8	50.0	1.1	2.5
12. +	Hyperplasia	58.9	6.8	1.2	5.7
13. +	Hyperplasia	17.9	ND	17.4	ND
14. +	Hyperplasia	9.1	14.2	40.5	9.0
15. +	Hyperplasia	27.3	12.8	2.1	3.9
16. +	Hyperplasia	82.3	37.2	3.0	8.2
17. +	Hyperplasia	23.5	ND	1.8	ND

[a] Histology of the thymus.
[b] ND = Not done.

In vitro, purified CD5$^+$ B cells have a high rate of spontaneous proliferation.[27,28] CD5$^+$ B cells of both murine and human origin produce polyreactive, low-affinity autoantibodies that are primarily of the IgM isotype, but they also produce some IgG autoantibodies. B cells that are CD5$^+$ rarely produce high-affinity antibodies to exogenous antigens.[27,28,38] The accumulated evidence strongly suggests that CD5$^+$ B cells are involved in the development of an autoimmune response.

We have found that in some patients with MG there is a high frequency of CD5$^+$ B cells in the peripheral blood[9] (FIG. 1). A high-frequency phenotype was defined as ≥30% of all B cells expressing the CD5 molecule. Based on the data summarized in FIGURE 1, 44% of MG patients have a high frequency of CD5$^+$ B cells compared to only 7% of normal, nonmyasthenic controls. Race, sex, age, or age of onset of disease had no effect on whether an individual had a high or low frequency of CD5$^+$ B cells. The presence of an associated autoimmune disease or other serum autoantibodies also did not seem to affect CD5$^+$ B-cell frequency.

CD5$^+$ B cells are also present in the myasthenic thymus. In TABLE 2, the frequency of CD5$^+$ B cells in the thymus and peripheral blood of the same patient is compared. In thirteen cases both thymus and peripheral blood samples were compared. In nine out of thirteen cases, the frequency of B cells that express CD5 was higher in the thymus than in the peripheral blood. We have also examined four cases of thymoma without MG (TABLE 3). Three out of the four cases had a high frequency of CD5$^+$ B cells in the thymus. Unfortunately, we were not able to obtain peripheral blood samples from these patients for comparison. The data from TABLES 2 and 3 are grouped by thymic histology and summarized in TABLE 4. What becomes apparent is that the frequency of total B lymphocytes (based on CD20 expression) is higher in thymic hyperplasia. This has been previously reported.[39] We have not as yet examined the frequency of CD5$^+$ B cells in normal

TABLE 3. Comparison of the Frequency of Total B Cells and $CD5^+$ B Cells in the Thymus and Blood of Patients without MG

		Percent CD5 B Cells		Percent B Cells	
MG	Histology[a]	Thymus	Blood	Thymus	Blood
1. –	Thymoma	44.4	ND[b]	1.3	ND
2. –	Thymoma	96.3	27.8	1.2	4.0
3. –	Thymoma	33.3	ND	0.3	ND
4. –	Thymoma	0	ND	0.8	ND

[a] Histology of the thymus.
[b] ND = Not done.

thymus. In the involuted myasthenic thymus, however, the frequency of $CD5^+$ B cells appears to be lower than in cases of thymic hyperplasia and thymoma (with or without MG). It has recently been reported that the majority of B cells in the normal murine thymus do express CD5.[40] Conflicting data have been reported for the normal human thymus.[41,42] Therefore, the presence of a high frequency of $CD5^+$ B cells in the thymus may not be unique to patients with MG.

Many questions concerning the origin and function of $CD5^+$ B cells still remain unanswered. It is unclear whether $CD5^+$ B cells represent a separate B-cell lineage from conventional B cells.[27,28] Alternatively, CD5 may be a maturational or activation marker for B lymphocytes. To address this question, PBMC were obtained from patients with MG who had a low frequency of B cells that expressed CD5. Following culture with PWM, the frequency of $CD5^+$ B cells increased in every case (TABLE 5). It is unclear whether this increase reflects proliferation of existing $CD5^+$ B cells or differentiation of $CD5^-$ B cells into $CD5^+$ cells. There is evidence to indicate that $CD5^-$ B cells can become $CD5^+$ in the presence of T-cell derived factors (lymphokines).[43] Interestingly, as TABLE 5 shows, the frequency of $CD5^+$ B cells also increased in unstimulated PBMC cultures, which may reflect a high rate of spontaneous proliferation for $CD5^+$ B cells and a very low rate of proliferation for $CD5^-$ B cells in the absence of stimulation.

Serial studies of PBMC from MG patients and normal, nonmyasthenic controls showed that although the frequency of $CD5^+$ B cells remained fairly stable for controls, there were larger fluctuations in the frequency of $CD5^+$ B cells for the

TABLE 4. The Frequency of Total B Cells and $CD5^+$ B Cells in the Thymus

	Thymic Histology-MG		
	Involuted (n = 2)	Hyperplasia (n = 13)	Thymoma (n = 2)
B cells	1.9 ± 0.9[a]	8.8 ± 12.9	1.4 ± 0.4
$CD5^+$ B cells	15.2 ± 13.9	33.9 ± 21.1	52.6 ± 18.7
	Thymoma, non-MG (n = 4)		
B cells	0.9 ± 0.5		
$CD5^+$ B cells	43.5 ± 39.9		

[a] Mean percent positive cells ± standard deviation.

patients.[44] Long-term therapy with immunosuppressive drugs did appear to lower the frequency of $CD5^+$ B cells in peripheral blood.[44] There was a trend for a higher frequency of $CD5^+$ B cells in the more clinically affected patients; however, this was not invariably the case.

The relevance of an increased frequency of $CD5^+$ B cells in the peripheral blood of some patients with MG is unclear. It is not likely that $CD5^+$ B cells are the primary producers of anti-AChR antibodies, the majority of which are IgG. Rather, $CD5^+$ B cells may participate in the defective immunoregulatory pathways that lead to anti-AChR antibody production and subsequent development of clinical MG. The ligand for CD5 has been identified as CD72.[45,46] The CD5 molecule is expressed on T and B lymphocytes, whereas CD72 is expressed on B cells. Therefore, CD5/CD72 interactions are likely to be important in T-cell/B-cell interactions and in $CD5^+/CD5^-$ B-cell interactions. Furthermore, it has been proposed that CD5/CD72 interactions on the same B lymphocyte may account for the self-renewing capacity of $CD5^+$ B cells.[46]

TABLE 5. PWM Stimulation Increases the Frequency of $CD5^+$ B Cells[a]

Freshly isolated PBMC	Post–8-Day Culture: Unstimulated	Post–8-Day Culture: PWM
7.7[b]	17.7	60.0
6.1	22.2	33.3
1.5	11.1	66.7
0.0	23.1	59.7

[a] The frequency of $CD5^+$ B cells was determined in freshly isolated PBMC and following culture in medium alone (unstimulated) or in medium plus PWM (0.5 μg/mL).

[b] Frequency of $CD5^+$ B cells.

REFERENCES

1. LISAK, R. P., A. I. LEVINSON & B. ZWEIMAN. 1985. Autoimmune aspects of myasthenia gravis. Concepts Immunopathol. **2:** 65–101.
2. LINDSTROM, J., D. SHELTON & Y. FUJII. 1988. Myasthenia gravis. Adv. Immunol. **42:** 233–284.
3. DRACHMAN, D. B., S. DE SILVA, D. RAMSAY & A. PESTRONK. 1987. Humoral pathogenesis of myasthenia gravis. Ann. N.Y. Acad. Sci. **505:** 90–105.
4. TZARTOS, S. J., M. E. SEYBOLD & J. M. LINDSTROM. 1982. Specificities of antibodies to acetylcholine receptors in sera from myasthenia gravis patients measured by monoclonal antibodies. Proc. Natl. Acad. Sci. USA **79:** 188–192.
5. RATNAM, M., P. B. SARGENT, V. SARIN, J. L. FOX, D. L. NGUYEN, J. RIVIER, M. CRIADO & J. LINDSTROM. 1986. Location of antigenic determinants on primary sequences of subunits of nicotinic acetylcholine receptor by peptide mapping. Biochemistry **25:** 2621–2632.
6. VINCENT, A., P. J. WHITING, M. SCHLUEP, F. HEIDENREICH, B. LANG, A. ROBERTS, N. WILLCOX & J. NEWSOM-DAVIS. 1987. Antibody heterogeneity and specificity in myasthenia gravis. Ann. N.Y. Acad. Sci. **505:** 106–120.
7. LINK, H., O. OLSSON, J. SUN, W.-Z. WANG, G. ANDERSSON, H.-P. EKRE, T. BRENNER,

O. Abramsky & T. Olsson. 1991. Acetylcholine receptor-reactive T and B cells in myasthenia gravis and controls. J. Clin. Invest. **87:** 2191–2196.

8. Lisak, R. P., B. Zweiman & S. M. Phillips. 1978. Thymic and peripheral blood T- and B-cell levels in myasthenia gravis. Neurology **28:** 1298–1301.
9. Ragheb, S. & R. P. Lisak. 1990. The frequency of CD5$^+$ B lymphocytes in the peripheral blood of patients with myasthenia gravis. Neurology **40:** 1120–1124.
10. Newsom-Davis, J., N. Willcox & L. Calder. 1981. Thymus cells in myasthenia gravis selectively enhance production of anti-acetylcholine receptor antibody in autologous blood lymphocytes. N. Engl. J. Med. **305:** 1313–1318.
11. Lisak, R. P., C. Laramore, B. Zweiman & A. Moskovitz. 1983. *In vitro* synthesis of antibodies to acetylcholine receptor by peripheral blood mononuclear cells of patients with myasthenia gravis. Neurology **33:** 604–608.
12. Lisak, R. P., A. I. Levinson, B. Zweiman & M. J. Kornstein. 1987. *In vitro* synthesis of IgG and antibodies to AChR by peripheral and thymic lymphocytes. Ann. N.Y. Acad. Sci. **505:** 39–49.
13. Brenner, T., Y. Timore, I. Wirguin, O. Abramsky & M. Steinitz. 1989. *In vitro* synthesis of antibodies to acetylcholine receptor by Epstein-Barr virus–stimulated B lymphocytes derived from patients with myasthenia gravis. J. Neuroimmunol. **24:** 217–222.
14. Shinomiya, N. & J. Yata. 1981. B and T cell involvement in anti-acetylcholine receptor antibody formation in myasthenia gravis. Clin. Exp. Immunol. **46:** 277–285.
15. Willcox, H. N. A., J. Newsom-Davis & L. R. Calder. 1984. Cell types required for anti-acetylcholine receptor antibody synthesis by cultured thymocytes and blood lymphocytes in myasthenia gravis. Clin. Exp. Immunol. **58:** 97–106.
16. Limburg, P. C., E. Hummel-Tappel, H. J. Oosterhuis & T. H. The. 1985. *In vitro* T-cell dependent B-cell activity in myasthenia gravis. Clin. Exp. Immunol. **61:** 31–38.
17. Castleman, B. 1966. The pathology of the thymus gland in myasthenia gravis. Ann. N.Y. Acad. Sci. **135:** 496–503.
18. Staber, F. G., V. Fink & W. Sack. 1975. B lymphocytes in the thymus of patients with myasthenia gravis. N. Engl. J. Med. **292:** 1032–1033.
19. Cohen-Kaminsky, S., C. Leprince, P. Galanaud, Y. Richards & S. Berrih-Aknin. 1989. B- and T-cell activation in the thymus of patients with myasthenia gravis. Thymus **14:** 187–193.
20. Levinson, A. I., B. Zweiman, R. P. Lisak, A. Dziarski & A. R. Moskovitz. 1984. Thymic B-cell activation in myasthenia gravis. Neurology **34:** 462–468.
21. Smiley, J. D., J. Bradley, D. Daly & M. Ziff. 1969. Immunoglobulin synthesis *in vitro* by human thymus: comparison of myasthenia gravis and normal thymus. Clin. Exp. Immunol. **4:** 387–399.
22. Vincent, A., G. K. Scadding, H. C. Thomas & J. Newsom-Davis. 1978. *In vitro* synthesis of anti-acetylcholine receptor antibody by thymic lymphocytes in myasthenia gravis. Lancet **1:** 778–779.
23. Scadding, G. K., A. Vincent, J. Newsom-Davis & K. Henry. 1981. Acetylcholine receptor antibody synthesis by thymic lymphocytes: correlation with thymic histology. Neurology **31:** 935–943.
24. Lisak, R. P., A. I. Levinson, B. Zweiman & M. J. Kornstein. 1986. Antibodies to acetylcholine receptor and tetanus toxoid: *in vitro* synthesis by thymic lymphocytes. J. Immunol. **137:** 1221–1225.
25. Levinson, A. I., B. Zweiman & R. P. Lisak. 1987. Isotype restriction of thymic B-cell immunoglobulin synthesis in myasthenia gravis. Ann. N.Y. Acad. Sci. **505:** 701–703.
26. Levinson, A. I., B. Zweiman & R. P. Lisak. 1990. Pokeweed mitogen-induced immunoglobulin secretory responses of thymic B cells in myasthenia gravis: selective secretion of IgG versus IgM cannot be explained by helper functions of thymic T cells. Clin. Immunol. Immunopathol. **57:** 211–217.
27. Gadol, N. & K. A. Ault. 1986. Phenotypic and functional characterization of human Leu 1 (CD5) B cells. Immunol. Rev. **93:** 23–34.
28. Hayakawa, K. & R. R. Hardy. 1988. Normal, autoimmune, and malignant CD5$^+$ B cells: the Ly-1 B lineage? Ann. Rev. Immunol. **6:** 197–218.

29. BOUMSELL, L., A. BERNARD, V. LEPAGE, L. DEGAS, J. LEMERLE & J. DAUSSET. 1978. Some chronic lymphocytic leukemia cells bearing surface immunoglobulins share determinants with T cells. Eur. J. Immunol. **8:** 900–904.
30. CALIGARIS-CAPPIO, F., M. GOBBI, M. BOFILL & G. JANOSSY. 1982. Infrequent normal B lymphocytes express features of B-chronic lymphocytic leukemia. J. Exp. Med. **155:** 623–628.
31. MANOHAR, V., E. BROWN, W. M. LEISERSON & T. M. CHUSED. 1982. Expression of Lyt-1 by a subset of B lymphocytes. J. Immunol. **129:** 532–538.
32. PLATER-ZYBERK, C., R. N. MAINI, K. LAM, T. D. KENNEDY & G. JANOSSY. 1985. A rheumatoid arthritis B cell subset expresses a phenotype similar to that in chronic lymphocytic leukemia. Arthritis Rheum. **28:** 971–976.
33. DAUPHINÉE, M., Z. TOVAR & N. TALAL. 1988. B cells expressing CD5 are increased in Sjögren's syndrome. Arthritis Rheum. **31:** 642–647.
34. SURANYI, P., G. SZEGEDI, S. DAMJANOVICH, F. JUHASZ, V. STENSZKY & N. R. FARID. 1989. B lymphocyte subsets in Hashimoto's thyroiditis. Immunol. Lett. **22:** 147–150.
35. NICOLETTI, F., P. L. MERONI, W. BARCELLINI, P. Ø. STEINSVÄG, M. DI MAURO, M. LUNETTA, M. PAGANO, U. CONSOLI & C. ZANUSSI. 1990. Enhanced percentage of $CD5^+$ B lymphocytes in newly diagnosed IDDM patients. Immunol. Lett. **23:** 211–216.
36. CASALI, P., S. E. BURASTERO, J. E. BALOW & A. L. NOTKINS. 1989. High-affinity antibodies to ssDNA are produced by $CD5^-$ B cells in systemic lupus erythematosus patients. J. Immunol. **143:** 3476–3483.
37. MIX, E., T. OLSSON, J. CORREALE, S. BAIG, V. KOSTULAS, O. OLSSON & H. LINK. 1990. B cells expressing CD5 are increased in cerebrospinal fluid of patients with multiple sclerosis. Clin. Exp. Immunol. **79:** 21–27.
38. CASALI, P. & A. L. NOTKINS. 1989. Probing the human B-cell repertoire with EBV: polyreactive antibodies and $CD5^+$ B lymphocytes. Ann. Rev. Immunol. **7:** 513–535.
39. DURELLI, L., M. F. FERRIO, U. MASSAZZA, R. CAVALLO, R. FERRI & L. BERGAMINI. 1990. Human T cell intrathymic ontogeny: Differences between myasthenic and normal thymus. Int. J. Neurosci. **51:** 321–323.
40. MIYAMA-INABA, M., S.-I. KUMA, K. INABA, H. OGATA, H. IWAI, R. YASUMIZU, S. MURAMATSU, R. M. STEINMAN & S. IKEHARA. 1988. Unusual phenotype of B cells in the thymus of normal mice. J. Exp. Med. **168:** 811–816.
41. KUMAMOTO, T., M. INABA, H. IMAMURA, K.-I. NANGO, Y. ADACHI, S. THAN, K. INABA, T. KAGAWA & S. IKEHARA. 1991. Characterization of B cells in human thymus. Immunobiology **183:** 88–93.
42. SPENCER, J., M. CHOY, T. HUSSELL, L. PAPADAKI, J. P. KINGTON & P. G. ISAACSON. 1992. Properties of human thymic B cells. Immunology **75:** 596–600.
43. WERNER-FAVRE, C., T. L. VISCHER, D. WOHLWEND & R. H. ZUBLER. 1989. Cell surface antigen CD5 is a marker for activated human B cells. Eur. J. Immunol. **19:** 1209–1213.
44. RAGHEB, S. & R. P. LISAK. 1992. Effect of clinical status and treatment on the frequency of $CD5^+$ B cells in patients with myasthenia gravis. Neurology **42:** 1076–1080.
45. VAN DE VELDE, H., I. VON HOEGEN, W. LUO, J. R. PARNES & K. THIELEMANS. 1991. The B-cell surface protein CD72/Lyb-2 is the ligand for CD5. Nature **351:** 662–665.
46. LUO, W., H. VAN DE VELDE, I. VON HOEGEN, J. R. PARNES & K. THIELEMANS. 1992. Ly-1 (CD5), a membrane glycoprotein of mouse T lymphocytes and a subset of B cells, is a natural ligand of the B cell surface protein Lyb-2 (CD72). J. Immunol. **148:** 1630–1634.

Effector Mechanisms of Myasthenic Antibodies[a]

DAVID P. RICHMAN, ROBERT L. WOLLMANN,[b] RICARDO A. MASELLI, CHRISTOPHER M. GOMEZ,[c] ALEXANDER L. COREY,[d] MARK A. AGIUS, AND ROBERT H. FAIRCLOUGH

Department of Neurology and the Center for Neuroscience University of California, Davis Davis, California 95616

[b]*Departments of Pathology and Neurology University of Chicago Chicago, Illinois 60637*

The serendipitous observation of Patrick and Lindstrom,[1] that an immune response to the nicotinic acetylcholine receptor (AChR) at the neuromuscular junction results in an experimental form of myasthenia gravis (MG) that in many ways is identical to the naturally occurring human disease, led to the identification of a similar immune response in MG.[2–4] Over the last 15 years information has accrued from extensive study of the animal model, now called experimental autoimmune myasthenia gravis (EAMG), and of the human disease demonstrating that the autoimmune response involves, primarily, T cell–controlled antibody production.[2–4] The anti-AChR antibodies produced in this autoimmune response cause the disordered neuromuscular transmission that is the hallmark of MG.

The histopathology of MG, especially at the ultrastructural level, has demonstrated an extensive effect on the postsynaptic membrane, with simplification of the postsynaptic folds and reduction in the amount of AChR in the end-plate membrane.[2,5] In spite of the presence of membrane debris, along with elements of the complement cascade, within the synaptic cleft in MG and EAMG, there has been the widespread belief that MG represents the major example of a neurologic disease, previously thought to be degenerative in nature, that is, in fact, mediated by noninflammatory autoimmune processes.[4–6] Hence, the histologic findings in MG have been used to support a number of hypotheses that various degenerative diseases of the nervous system, including amyotrophic lateral sclerosis, paraneoplastic cerebellar degeneration, stiff-man syndrome, and Huntington's disease, are the result of similar noninflammatory autoimmune processes.[7–11]

The most widely studied form of EAMG, induced by active immunization of Lewis rats with AChR purified from *Torpedo californica*, is biphasic.[2–4] About one

[a] This work was supported in part by Grants from the National Institutes of Health (NS15462, NS19779, NS24304, NS01226), the Muscular Dystrophy Society, and the Amyotrophic Lateral Sclerosis Society of America.

[c] Present address: Department of Neurology, University of Minnesota, Minneapolis, MN.

[d] Deceased.

week after immunization, animals develop a short-lived acute phase, occurring at the time the serum titer of anti-AChR antibody is just beginning to rise. This phase, which lasts only a few days, is characterized by weakness and muscle end-plate necrosis, with infiltration of macrophages into the neuromuscular junction. Two weeks later the second, or chronic, phase develops, with weakness progressing, in many cases, to death. During this phase, inflammatory cells are absent from the neuromuscular junction. Instead, ultrastructural analysis reveals simplification of the end-plate membrane folds and reduction in the amount of AChR normally found at the peaks of the folds, similar to the findings in MG.

Our studies of the pathogenesis of EAMG have made extensive use of anti-AChR monoclonal antibodies (MAbs) to analyze the mechanisms by which anti-AChR antibodies induce the structural and physiologic abnormalities at the neuromuscular junction. These MAbs are able to produce a passively induced form of EAMG that is amenable to precise analysis. Two days after injection of anti-AChR MAb into naive animals, an acute phase occurs that is identical to the acute phase in actively induced EAMG. However, the chronic phase of the passive-transfer disease, which occurs about three weeks after MAb injection, is not associated with clinical weakness but is characterized by the histological findings that are identical to the chronic phase of actively induced EAMG. The absence of clinical findings in the MAb-induced form of chronic EAMG represents a substantial discrepancy between the actively and passively induced models.

More significantly, there is a major difference between both these model diseases and the human disease. Although the histologic findings in the chronic phase of the actively and passively induced forms of EAMG are identical to those described above for MG, an acute phase of MG has not been reported. The absence of findings suggestive of an acute phase represents one of the major disparities between MG and its animal models.[2,4,6]

There has been, in fact, some evidence for a role of inflammation at the neuromuscular junction in MG. In 1905, Buzzard[12] observed mononuclear cell infiltrates in muscles of MG patients, which he called "lymphorrhages", although the cell types involved and the relationship of the cells to the neuromuscular junction were never determined. Most contemporary studies, which have primarily involved intercostal muscle biopsies obtained at the time of thymectomy, have failed to demonstrate inflammatory infiltrates.[13–16] However, one study[17] of a triceps muscle biopsy from a patient with a relatively recent onset of MG demonstrated collections of mononuclear cells overlying an end-plate region identified by cholinesterase staining. Unfortunately, the nature of the infiltrating cells was not determined.

We have made use of a combination of morphologic, immunologic, and electrophysiologic methods to analyze the pathogenic mechanisms by which anti-AChR antibodies induce the experimental and human forms of MG. The animal studies have taken advantage of the precision in analysis afforded by the single set of characteristics associated with each individual MAb. We have then applied this information to a study of the human disease.

METHODS

Rat anti-AChR MAbs were produced, purified, and analyzed as previously described.[18,19] Clinical assessment of EAMG was carried out on a five-point scale.[19] Histological analysis and determination of end-plate index (EI) as a measure of end-plate membrane folding was also previously described.[20] The electromyographic (EMG) techniques for studying both EAMG and MG were also described.[18,21,22] Microelectrode studies, which involved both animal and human muscle, used standard methods.[21,23] Determination of AChR content in muscle involved immunoprecipitation of [^{125}I]alpha bungarotoxin–labeled muscle extracts.[24]

RESULTS AND DISCUSSION

Acute Phase of MAb-induced Passive-Transfer EAMG

Injection of Lewis rat anti-AChR MAbs into naive Lewis rats induces disabling weakness involving the forelimbs more than the hind limbs, along with a low-slung, unbalanced gait.[25] The weakness begins at 12–24 h, is maximal at 48 h, and clears by the fourth or fifth day. Histologic analysis reveals phagocytosis of end-plate regions by infiltrating macrophages.[20] In some electron micrographs, macrophages appear to be pinching off the end-plate from the remainder of the muscle fiber (FIG. 1).

Chronic Phase of MAb-induced Passive-Transfer EAMG

A series of five animals, which had recovered from the acute phase of EAMG induced by a single injection of anti-AChR MAb, were analyzed at 21 days.[20] Each was clinically normal and had normal EMG studies. Ultrastructural analysis, however, revealed marked simplification of the end-plate membrane, resulting in reduced EI, 3.0 ± 0.73 compared with 5.0 ± 0.28 for control animals.

A second series of five animals received injections of anti-AChR MAb[20] twice weekly. These animals also recovered from the weakness induced by the first injection at 4–5 days. The subsequent MAb injections did not induce additional episodes of weakness. At the end of the period of injections, ultrastructural analysis revealed end-plate simplification (FIG. 2) with reduction in EI (1.55 ± 0.07 vs. 5.0 ± 0.28) compared with controls. The miniature end-plate potential (MEPP) amplitudes and the AChR content of muscle were also moderately reduced in these animals.[24]

The histological findings in these animals with the chronic phase of EAMG were identical to those that had been reported in MG.[20] None of those chronically injected animals, however, had clinical weakness. Also, the reduction in MEPP amplitudes was considerably less than in the chronic phase of EAMG induced by active immunization or in MG. It appears that a polyclonal attack is required to induce the marked effect on neuromuscular transmission seen in those other forms of myasthenia.

The results following a single injection of MAb in which the ultrastructural findings of end-plate simplification are noted at 21 days, along with the similar findings in chronically injected animals and in patients with MG, raise the possibility that these findings represent the "healed" or partially healed end-plate following the acute injury that occurs during the acute phase. The end-plate in acute EAMG, which is eventually phagocytosed, is probably reformed in a more simplified form. It is also possible that the continued presence of anti-AChR antibody in chronic actively induced EAMG, in MG, and in the chronic MAb injection model, plays a role in prolonging the period of end-plate simplification.

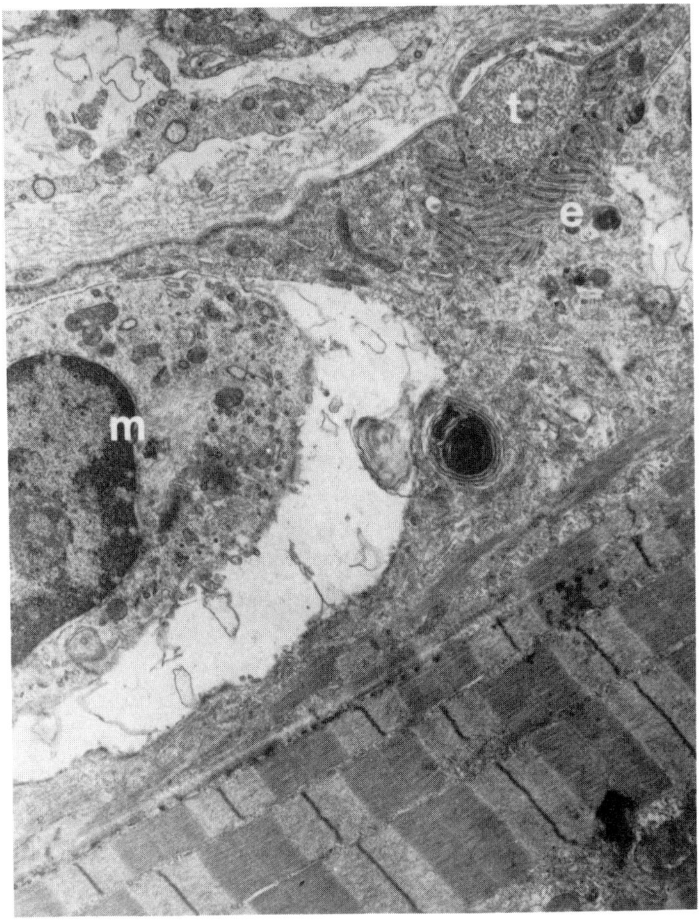

FIGURE 1. Neuromuscular junction from forelimb muscle of a rat, 48 h after injection of rat anti-AChR MAb 56I. A macrophage (m) is seen appearing to be separating the end-plate membrane (e) and axon terminus (t) from the underlying myofiber. The axon terminus is closely apposed to the end-plate, which is highly folded and appears morphologically intact. × 10,000. (Gomez *et al.*[20] With permission from *Acta Neuropathologica*).

Hyperacute EAMG Induced by Passive Transfer of Anti-AChR MAbs That Block AChR Function

Within our extensive library of anti-AChR MAbs, none has been demonstrated to both block binding of alpha-bungarotoxin and to bind to syngeneic rat AChR. On the other hand a number of MAbs block alpha-bungarotoxin and acetylcholine binding and also bind to AChR from lower vertebrates. We studied the effect of injection of some of these Lewis rat MAbs into chickens.[26] Inasmuch as rat immunoglobulin does not fix chicken complement, the MAbs that did not affect

AChR function produced no findings when they were injected into chickens. For the subgroup of MAbs that blocked alpha-bungarotoxin binding, however, injections into chickens produced a hyperacute form of EAMG characterized by generalized paralysis within one hour of intravenous injection, along with marked reduction in MEPP amplitudes.[26] Light microscopic and electron microscopic analysis of these animals revealed no abnormalities.

Hence, the hyperacute form of EAMG appears to result from pharmacologic blockade of AChRs. This observation has led to the hypothesis that the absence

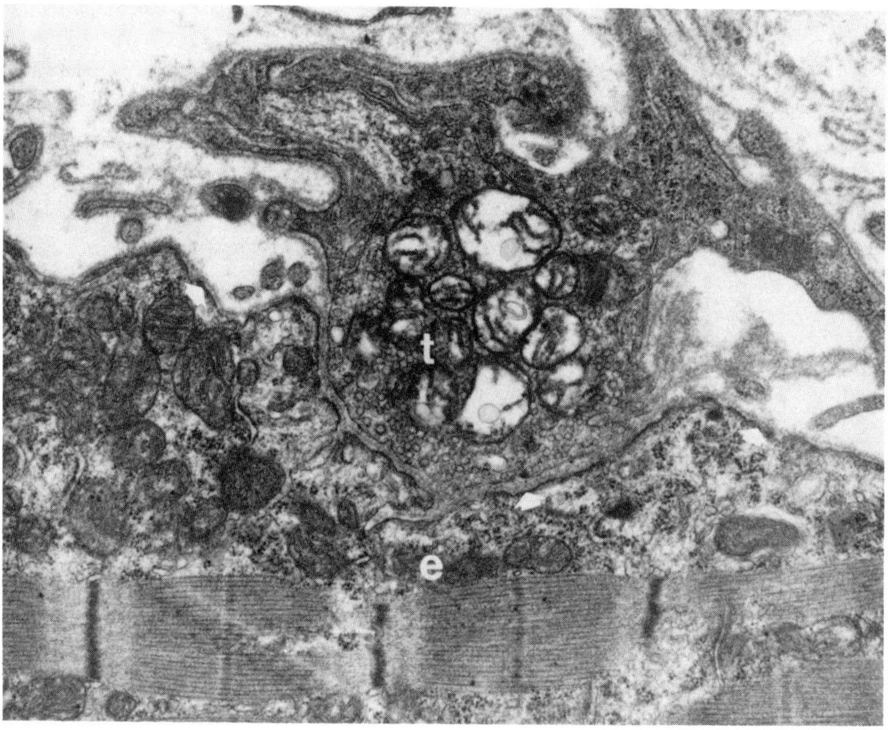

FIGURE 2. Highly simplified, but intact, end-plate (e) membrane (white arrow) from an animal having received multiple injections of anti-AChR MAb 132A. t = nerve terminus. × 25,000.[20] (Gomez *et al.*[20] With permission from *Acta Neuropathologica*).

of weakness in the MAb-induced chronic EAMG in rats is the consequence of the fact that none of the MAbs injected blocks acetylcholine binding in rats. Hence, the "complete" chronic EAMG syndrome or the complete MG picture requires both reduction in end-plate AChR that accompanies the end-plate simplification, presumably from antibody-induced complement-mediated membrane damage, increased AChR turnover, and inflammatory destruction[27] (see below), as well as pharmacologic blockade of the remaining AChRs by antibodies that block agonist binding. It is the latter antibodies that are lacking in our MAb-induced chronic phase model.

Refractoriness to a Second Episode of MAb-induced Acute EAMG

In the course of the study of the chronic phase of the MAb-induced disease, we observed that subsequent injection of MAb did not reproduce the acute phase seen after the first injection.[20,28] To study the nature of this refractoriness for a second episode of acute EAMG, we carried out an additional analysis. If the second injection involves an unrelated anti-AChR MAb or even polyclonal anti-AChR antiserum,[28] the refractoriness is still observed. Study of the time course of this refractory phase revealed that it lasts for up to 11 weeks after the initial injection.[29] During this time period, EI and AChR content remain low. The return of susceptibility for acute EAMG after 11 weeks correlates temporally with increases in EI and AChR content toward normal values. Our current hypothesis is that the refractoriness is the result of the initial reduction in end-plate AChR density resulting from the inflammation and necrosis that occurs during the preceding acute phase.[28,29] This decreased antigen (AChR) density at the end-plate could result in a lower concentration of bound antibody, specifically a lower concentration of antibody Fc portion and, hence, a lower concentration of complement component C1q bound by Fc. The C1q concentration could then be below the threshold needed for progression of the complement cascade to the generation of sufficient quantities of chemoattractant to induce macrophage infiltration.[28–30] We would further hypothesize that at 11 weeks following a single injection, the end-plate membrane is adequately reformed so that the AChR density is sufficient to induce inflammation after MAb injection.

Apparent Disparity between the Inflammation Seen in EAMG and the Histologic Findings in MG

Our initial hypothesis to explain the paucity of reports of inflammation at the neuromuscular function in MG was that most end-plates that had been analyzed were in the "refractory state" with respect to inflammation. That is, the equivalent of the acute phase does occur in MG, but very early in the course of the disease, perhaps before symptoms are sufficient to lead to the diagnosis or the undertaking of a muscle biopsy. Our recent study of neuromuscular junction histology and physiology in MG, however, has demonstrated that end-plate inflammation is present in MG biopsies.[22] This study made use of the newly developed anconeus muscle biopsy technique[21] to assess the histology and physiology of the neuromuscular junction in MG. Of eight patients studied, seven were found to have inflammatory cells at some end-plates (ranging from 6 to 33% of end-plates examined). In most of these patients, the inflammation was of a milder degree than in the acute phase of EAMG. In one patient, however, an end-plate was identified with inflammation as severe as that found in acute EAMG. In this patient macrophages were seen appearing to remove end-plates from the remainder of the muscle fiber (FIG. 3). Subsequently reexamination of biopsies from intercostal muscle has revealed mild inflammation as well (A. G. Engel, personal communication).

Hence, the apparent discrepancy between MG and EAMG appears not to hold up. Rather, the acute phase does occur not uncommonly in MG. The implication of this revised view of the pathogenesis in MG is the elimination of the reservations

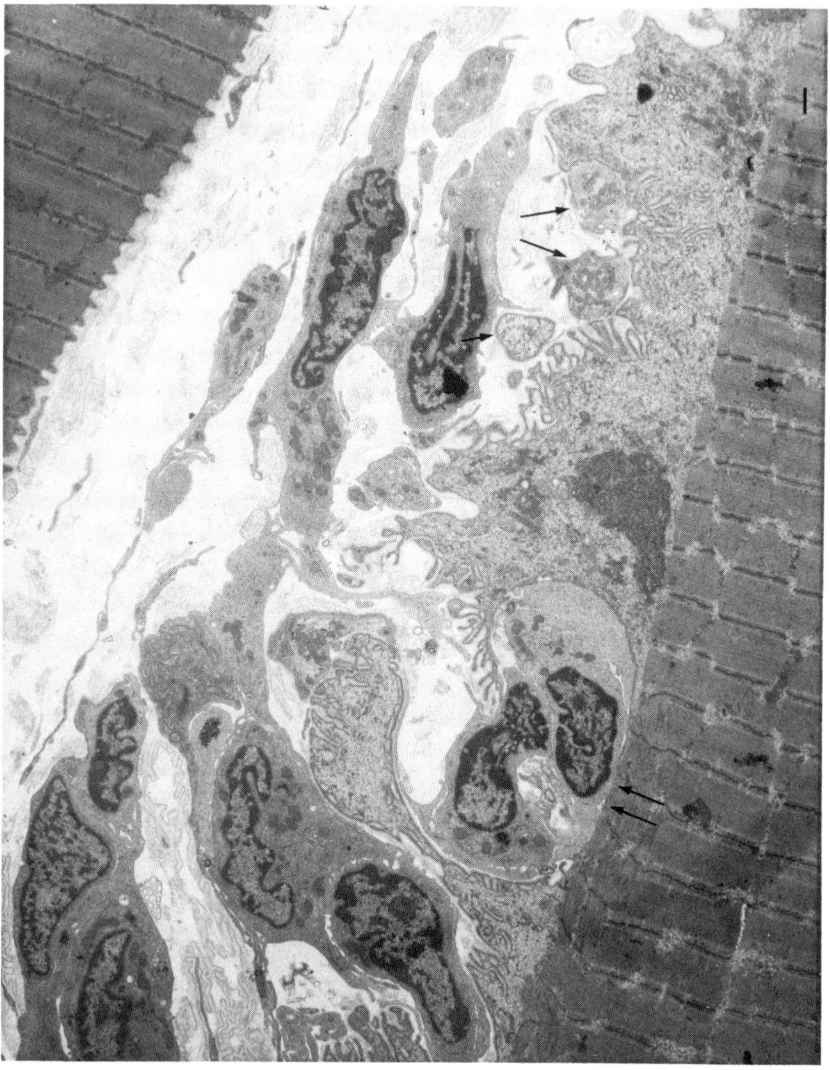

FIGURE 3A. A neuromuscular junction from a patient with myasthenia gravis. Macrophages appear to be removing postsynaptic sarcolemmal membrane from a myofiber. At one point, macrophages are seated directly adjacent to the sarcolemmal membrane with no intervening basal lamina (double arrows). Small segments of axon termini are opposed to simplified postsynaptic membrane (single arrows). Horizontal bar = 1 μm. (Maselli *et al.*[22] With permission from *Neurology*.)

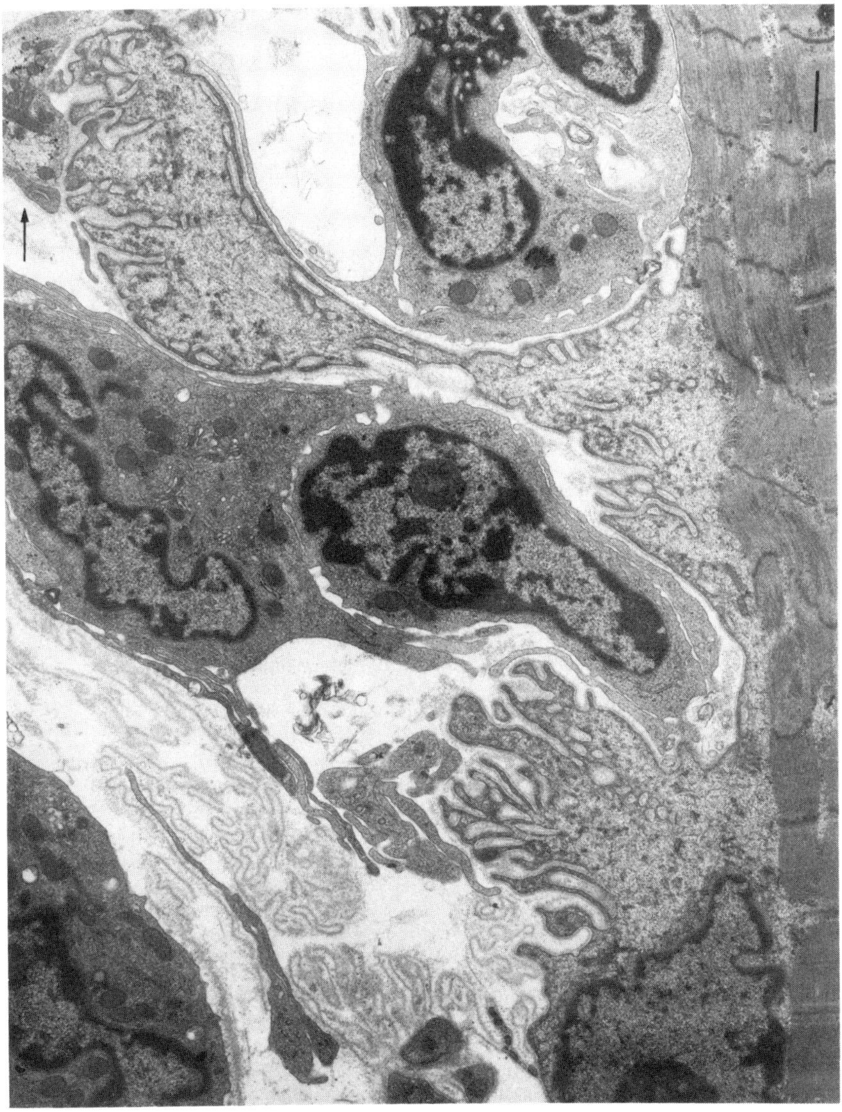

FIGURE 3B. Higher power view of the neuromuscular junction in FIG. 3A, demonstrating that the macrophages appear to be pinching off a segment of postsynaptic membrane, with axon terminus (arrow) still attached. Horizontal bar = 1 μm. (Maselli *et al.*[22] With permission from *Neurology*.)

concerning the application to MG of information gained from EAMG on the pathogenic mechanisms involved in the antibody-mediated effects. Perhaps more importantly, it is clear from these results that MG is an inflammatory autoimmune disease, and that it no longer can be considered as a model of a "degenerative disease" caused by noninflammatory autoimmune processes.

REFERENCES

1. PATRICK, J. & J. LINDSTROM. 1973. Autoimmune response to acetylcholine receptor. Science **180:** 871–872.
2. LINDSTROM, J., D. SHELTON & Y. FUJII. 1988. Myasthenia gravis. Adv. Immunol. **42:** 233–284.
3. LEVIN, K. H. & D. P. RICHMAN. 1989. Myasthenia gravis. Clin. Aspects Autoimmunity **4:** 23–31.
4. DRACHMAN, D. B. 1987. Myasthenia gravis: a model disorder of acetylcholine receptors. Res. Publ. Assoc. Res. Nerv. Ment. Dis. **65:** 65–82.
5. ENGEL, A. G. & B. BANKER, Eds. 1986. Myology. Vols. I–II. McGraw-Hill. New York.
6. WAKSMAN, B. H., Ed. 1990. Immunologic mechanisms in neurologic and psychiatric disease. Res. Publ. Assoc. Res. Nerv. Ment. Dis. **68:** 1–336.
7. DRACHMAN, D. B. 1989. Amyotrophic lateral sclerosis: an unconventional autoimmune disease? Ann. Neurol. **26:** 269–274.
8. ANDERSON, N. E., J. M. CUNNINGHAM & J. B. POSNER. 1987. Autoimmune pathogenesis of paraneoplastic neurological syndromes. Crit. Rev. Neurobiol. **3:** 245–299.
9. STECK, A. J. & E. NARDELLI. 1988. Paraneoplastic autoimmunity. Ann. N.Y. Acad. Sci. **540:** 91–98.
10. SANBERG, P. R. & J. T. COYLE. 1984. Scientific approaches to Huntington's disease. Crit. Rev. Neurobiol. **1:** 1–44.
11. SOLIMENA, N., F. FOLLI, R. APARISI, G. POZZA & P. DECAMILLI. 1990. Autoantibodies to GABA-ergic neurons and pancreatic beta cells in Stiffman syndrome. N. Engl. J. Med. **22:** 1555–1560.
12. BUZZARD, E. F. 1905. The clinical history and post-mortem examination of five cases of myasthenia gravis. Brain **28:** 438.
13. ENGEL, A. G. & T. SANTA. 1971. Histometric analysis of the ultrastructure of the neuromuscular junction in myasthenia gravis and in the myasthenic syndrome. Ann. N.Y. Acad. Sci. **183:** 46–63.
14. SANTA, T., A. G. ENGEL & E. H. LAMBERT. 1972. Histometric study of neuromuscular junction ultrastructure. 1. Myasthenia gravis. Neurology **22:** 71–82.
15. TSUJIHATA, M., R. HAZAMA, N. ISHII, Y. IDE, N. MORI & M. TAKAMORI. 1979. Limb muscle endplates in ocular myasthenia gravis: quantitative ultrastructural study. Neurology **29:** 654–661.
16. ENGEL, A. G., K. SAHASHI & G. FUMAGALLI. 1981. Part II. Effect of anti-acetylcholine receptor antibodies on acetylcholine receptor structure and function: the immunopathology of acquired myasthenia gravis. Ann. N.Y. Acad. Sci. **377:** 158–174.
17. PASCUZZI, R. M. & J. F. CAMPA. 1988. Lymphorrhage localized to the muscle endplate in myasthenia gravis. Arch. Pathol. Lab. Med. **112:** 934–937.
18. RICHMAN, D. P., C. M. GOMEZ, P. W. BERMAN, S. A. BURRES, F. W. FITCH & B. G. W. ARNASON. 1980. Monoclonal anti-acetylcholine receptor antibodies can cause experimental myasthenia. Nature **286:** 738–739.
19. GOMEZ, C. M. & D. P. RICHMAN. 1985. Monoclonal anti-acetylcholine receptor antibodies with differing capacities to induce experimental autoimmune myasthenia gravis. J. Immunol. **135:** 234–241.
20. GOMEZ, C. M., R. L. WOLLMANN & D. P. RICHMAN. 1984. Induction of the morphologic changes of both acute and chronic experimental myasthenia by monoclonal antibody directed against acetylcholine receptor. Acta. Neuropathol. (Berlin) **63:** 131–143.
21. MASELLI, R. A., D. P. MASS, B. J. DISTAD & D. P. RICHMAN. 1991. Anconeus muscle:

a human muscle preparation suitable for *in-vitro* microelectrode studies. Muscle & Nerve **14:** 1189–1192.
22. MASELLI, R. A., D. P. RICHMAN & R. L. WOLLMANN. 1991. Inflammation at the neuromuscular junction in myasthenia gravis. Neurology **41:** 1497–1504.
23. MASELLI, R. A., D. J. NELSON & D. P. RICHMAN. 1989. Effects of a monoclonal antiacetylcholine receptor antibody on the avian end-plate. J. Physiol. (Lond). **411:** 271–283.
24. GOMEZ, C. M. & D. P. RICHMAN. 1987. Chronic experimental autoimmune myasthenia gravis induced by monoclonal antibody to acetylcholine receptor: biochemical and electrophysiologic criteria. J. Immunol. **139:** 73–76.
25. BURRES, S. A., J. W. CRAYTON, C. M. GOMEZ & D. P. RICHMAN. 1981. Myasthenia induced by monoclonal anti-acetylcholine receptor antibodies: clinical and electrophysiological aspects. Ann. Neurol. **9:** 563–568.
26. GOMEZ, C. M. & D. P. RICHMAN. 1983. Anti-acetylcholine receptor antibodies directed against the alpha-bungarotoxin binding site induce a unique form of experimental myasthenia. Proc. Natl. Acad. Sci. USA **80:** 4089–4093.
27. ENGEL, A. G. & G. FUMAGALLI. 1982. Mechanisms of acetylcholine receptor loss from the neuromuscular junction. Ciba Found. Symp. **90:** 197–224.
28. COREY, A. L., D. P. RICHMAN, C. A. SCHUMAN, C. M. GOMEZ & B. G. W. ARNASON. 1985. Use of monoclonal antiacetylcholine receptor antibodies to investigate the macrophage inflammation of acute experimental myasthenia gravis: refractoriness to a second episode of acute disease. Neurology **35:** 1455–1460.
29. COREY, A. L., D. P. RICHMAN, R. L. WOLLMANN & M. A. AGIUS. 1987. Refractoriness to a second episode of experimental myasthenia gravis: correlation with acetylcholine receptor concentration and morphologic appearance of the post-synaptic membrane. J. Immunol. **138:** 3269–3275.
30. LENNON, V. A., M. E. SEYBOLD, J. M. LINDSTROM, C. COCHRANE & R. ULEVITCH. 1978. Role of complement in the pathogenesis of experimental autoimmune myasthenia gravis. J. Exp. Med. **147:** 973–983.

Molecular Structure of a Monoclonal Antiacetylcholine Receptor Antibody and of a Corresponding Monoclonal Antiidiotopic Antibody[a]

MARK A. AGIUS,[b] BHARATI SANYAL,[c] AND DAVID P. RICHMAN[b]

[b]*Department of Neurology and the Center for Neuroscience University of California, Davis Davis, California 95616*

[c]*Department of Neurology University of Chicago Chicago, Illinois 60637*

INTRODUCTION

The antiacetylcholine receptor monoclonal antibody (MAb) 132A (gamma 2a/kappa) is directed against the main immunogenic region of the acetylcholine receptor (AChR) and induces passive transfer experimental myasthenia (EAMG) in Lewis rats.[1] MAb HC4A (gamma 2a/kappa) is an isogeneic antiidiotopic (anti-Id) MAb that is directed against the variable region of MAb 132A[1] and inhibits the binding of MAb 132A to AChR. MAb HC4A protects animals actively immunized with AChR from developing EAMG and consequently appears to recognize a regulatory Id.[2] Although both HC4A and AChR have similar affinities for binding to 132A, the inhibition of 132A binding to AChR by HC4A is considerably greater than the inhibition of HC4A binding to 132A by AChR. The lack of reciprocity in these inhibition studies, despite similar affinities, suggests that the anti-Id–mediated inhibition is an allosteric effect.[1] To investigate such a molecular mechanism, we have cloned and sequenced the entire heavy and light chains of MAb 132A and MAb HC4A.

RESULTS AND DISCUSSION

The strategy employed involved the synthesis of complementary DNA libraries from messenger RNA derived from hybridoma cells producing these antibodies. Full-length clones were obtained for the heavy and light chains for both the anti-AChR MAb 132A and the anti-Id MAb HC4A. The sequences include the 5′ untranslated regions, coding regions as well as 3′untranslated regions with poly (A) signals and poly (A) tails. The two gamma 2a constant region sequences are identical to one another and differ in one base from the genomic sequence described

[a] This work was supported by NIH Grants # NS01226, NS15462, and NS19779.

for rat gamma 2a.[3] The difference is located in the 3′untranslated sequence. The two kappa sequences are also identical to each other and to the rat kappa sequence previously reported.[4]

The deduced amino acid sequences corresponding to the variable regions of MAbs 132A and HC4A will allow us to synthesize peptides corresponding to the hypervariable regions. These peptides will be used in inhibition experiments to map the residues contributing to AChR binding as well as to anti-Id binding.

REFERENCES

1. AGIUS, M. A., C. G. GEANNOPOULOS, R. H. FAIRCLOUGH & D. P. RICHMAN. 1988. Monoclonal anti-idiotopic antibodies against myasthenia-inducing anti-acetylcholine receptor monoclonal antibodies. J. Immunol. **140:** 62–68.
2. AGIUS, M. A. & D. P. RICHMAN. 1986. Suppression of development of experimental autoimmune myasthenia gravis with isogeneic monoclonal antiidiotopic antibody. J. Immunol. **137:** 2195–2198.
3. BRUGGEMANN, M. 1988. Evolution of the rat immunoglobulin gamma heavy-chain gene family. Gene **74:** 473–482.
4. SVASTI, J. & C. MILSTEIN. 1972. The complete amino acid sequence of a mouse kappa light chain. Biochem. J. **128:** 427–444.

HLA-A2–Restricted T-Cell Line Recognizing an Epitope of the Human Acetylcholine Receptor[a]

F. BAGGI, R. MANTEGAZZA,[b] A. VINCENT, AND J. NEWSOM-DAVIS

Neurosciences Group
Institute of Molecular Medicine
University of Oxford
Oxford OX3 9DU, United Kingdom

[b]*Neurological Institute "C. Besta"*
Milan, Italy

INTRODUCTION

In myasthenia gravis (MG), CD4$^+$ acetylcholine receptor (AChR)–specific T-helper lymphocytes are present and can be easily propagated *in vitro*. Interestingly, these autoreactive T cells can also be isolated from healthy subjects (see Willcox *et al.*, this volume), suggesting that they are part of the normal immune T-cell repertoire. We have looked for potential class I–restricted cytotoxic T cells (CTL) because, in northern Europeans at least, the stronger association is with HLA-B8 rather than HLA-DR3, and such cells might play a role in pathogenesis or possibly in regulation of the response to AChR. We had identified six peptides of the human AChR α-subunit able to bind to HLA-A2 (p17′-25′/59-65, p87-103, p108-129, and p157-188 in the extracellular domain, and p310-327 and p391-417 in the cytoplasmic loop). One of these, p391-417, contains a nine residue stretch (corresponding to p408-416) of homology with a well-defined HLA-A2–restricted epitope of the influenza A virus matrix protein, M58-66. These two peptides share the same "anchor" residues at position 2 and 9 (I and L), consistent with strong binding to HLA-A2. We have decided to use these two peptides in order to raise T-cell lines from MG and control PBL. Here we report preliminary results.

METHODS

The influenza-type A matrix peptide M58-66 (GILGFVFTL) and AChR α-subunit p408-416 (HILLGVFML) were synthesized on a solid-phase peptide synthesizer (Applied Biosystem). The purity was >95% by HPLC on a C_{18} reverse-phase column. CTL lines specific for the two peptides were generated as follows: PBL (2×10^6/mL) were cultured *in vitro* with 20 μM peptide for 6 days in RPMI 1640 and 10% fetal calf serum. The cells were maintained *in vitro* by weekly

[a] F. Baggi was supported by the International Association on MG (IMG Italy) and by Associazione per la Promozione delle Ricerche Neurologiche (Arin, Italy).

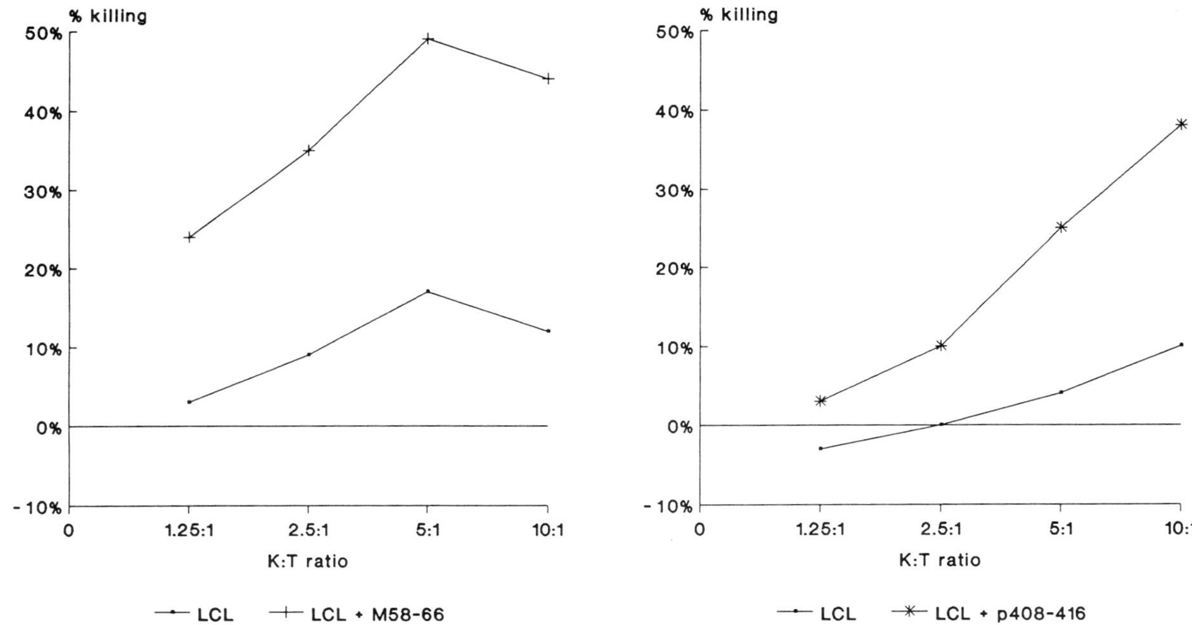

FIGURE 1. MR M58-66 (left panel) and MR p408-416 (right panel) CTLs were isolated from an MG patient (HLA type A2,26 B13,62 C9 DR4(w4),7 (DQ2,3). The cytotoxic activity was measured by the JAM test.[1] Briefly, autologous EBV-LCL target cells were labeled overnight with [^{3}H]thymidine (5–10 μCi/mL) and washed with RPMI medium. The target cells were plated at 1×10^4 cells/well in the presence of 5 μM peptide (alternatively, the target cells were prepulsed with the peptide) and the killer cells added at the indicated killer to target ratios. The killing was calculated as % killing (=DNA loss) = (S − E) × 100/S, where S = cpm, retained DNA in the absence of killers (spontaneous); and E = cpm, experimentally retained DNA in presence of killers.

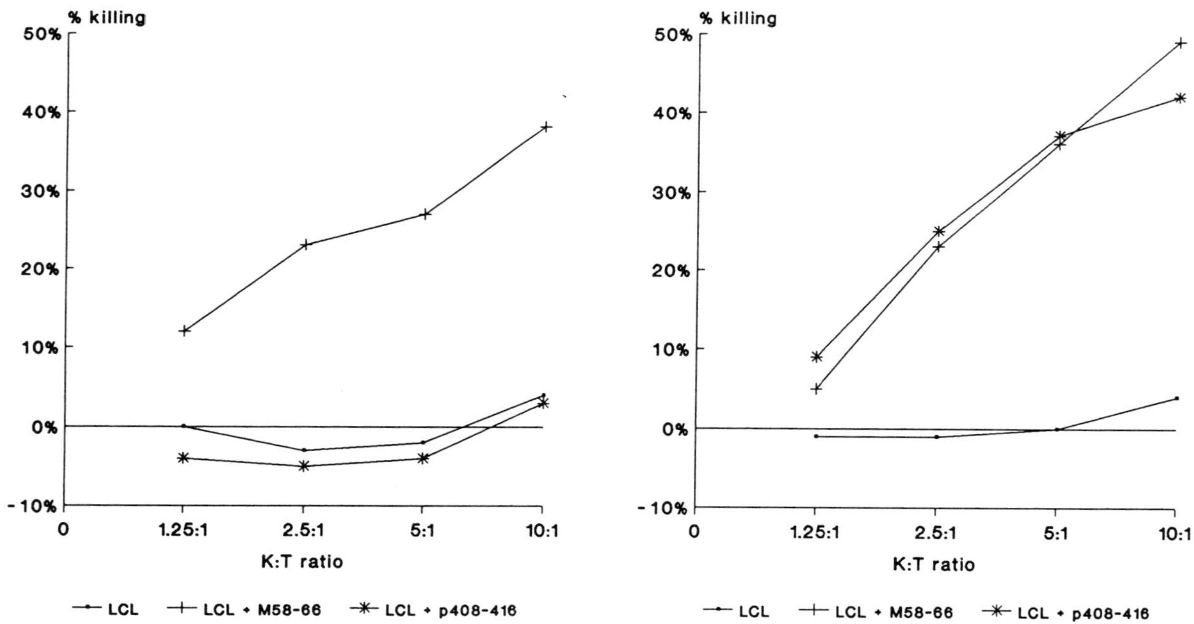

FIGURE 2. MN M58-66 (left panel) and MN p408-416 (right panel) CTLs were isolated from a healthy donor (HLA type A2,24 B35,60 C4,10 DR1,8 DQ4,5). Only MN p408-416 CTL (right panel) showed cross-reactivity against the heterologous peptide.

restimulation with irradiated autologous PBL (3000 rads) or HLA-A2 LCL pulsed with the appropriate peptide and IL-2 (10 U/mL); they retained their cytolytic activity for two months. The cytotoxic activity was evaluated by the JAM method.[1]

RESULTS

We have isolated four cytotoxic T-cell lines (2/4 MG patients and 2/3 controls) recognizing the influenza A matrix peptide M58-66 and two cytotoxic T-cell lines specific for the AChR α-subunit p408-416 (1/4 MG patients and 1/2 controls). These T-cell lines were able to kill HLA-A2$^+$ target cells (max killing 45%) in the presence of (FIG. 1), or pre-pulsed with, the homologous peptide. The p408-416–specific CTL from the healthy control MN also showed good heterologous cross-reactive killing with M58-66–sensitized targets (FIG. 2). One of the M58-66–specific CTL, which was raised against influenza A virus–infected B cells, also showed partial cross-reactivity (data not shown).

CONCLUSIONS

At least one (p408-416) of the six identified HLA-A2–binding AChR peptides can elicit class I–restricted CTL. The presence of these CTLs in both MG and controls suggests that the immune repertoire includes AChR-specific class I–restricted T lymphocytes. It is premature to claim any pathogenic or regulatory role for such T cells, particularly because we do not know whether endogenous antigen can be processed to generate this epitope.

REFERENCE

1. MATZINGER, P. 1991. The JAM test: A simple assay for DNA fragmentation and cell death. J. Immunol. Methods **145:** 185–192.

Differential Recognition by T Cells of the P3A$^+$ and P3A$^-$ Isoforms of the Human Acetylcholine Receptor Alpha-Subunit

A-P. BATOCCHI,[a] G. HARCOURT, D. BEESON,
N. PANTIC, S. HAWKE, N. WILLCOX, AND
J. NEWSOM-DAVIS

Neurosciences Group
Institute of Molecular Medicine
John Radcliffe Hospital
Oxford OX3 9DU, United Kingdom

The production of anti-AChR antibodies by B cells in myasthenia gravis (MG) is believed to be under the control of helper T cells, and the main target for both classes of lymphocyte is the AChR alpha-subunit.[1] In order to study the mechanisms of such T-cell help in myasthenia gravis, we isolated AChR-specific T cells from MG patients and healthy controls using recombinant human alpha-subunit (r1-437) as the stimulating antigen. The present paper describes the characterization of T cells from one MG patient and one healthy control.

From patient MG3 (FIG. 1), several T-cell lines and clones specific for the alpha-subunit were isolated from peripheral blood six months after onset of the disease. Using shorter recombinant alpha-subunit fragments and synthetic peptides, the recognition site of these cells was identified as part of an isoform of the human alpha-subunit first described by Beeson *et al.*[2] This contains an extra 25–amino acid insert (1′-25′) between positions 58 and 59 encoded by a newly identified exon, P3A. The actual T-cell epitope includes part of the insert plus some adjacent amino acids (α 17′-25′/59-65) (FIG. 1), and its recognition is restricted by HLA-DQ1w5. By contrast, a T-cell line raised from a healthy individual, C1 (FIG. 1), using a shorter recombinant polypeptide, r37-181, only recognizes alpha-subunit constructs lacking this insert (P3A$^-$), which would actually disrupt its epitope, α 51-65. Neither population of T cells is able to recognize the alternative isoform. Recognition of α 51-65 by C1 is restricted by HLA-DRw53.

The level of expression of the P3A$^+$ alpha-subunit in human muscle is unknown, but in the human rhabdomyosarcoma line, TE671, it is about 50% of the total alpha-subunit.[3] Attempts to show recognition of native AChR purified from human muscle by MG3 clones or the C1 T-cell line were initially unsuccessful when soluble AChR was used, possibly because of the low concentration of the antigen. When the AChR was bound onto magnetic beads, however,[4] by way of antibodies to a synthetic P3A peptide (55-58/1′-25′/59-65) for MG3 or to a pool of synthetic peptides covering AA 62-128 for C1, proliferative responses were observed (FIG. 1). When the AChR was deliberately denatured with cyanogen bromide, the re-

[a] Present address: Department of Neurology, University Cattolica, Rome, Italy.

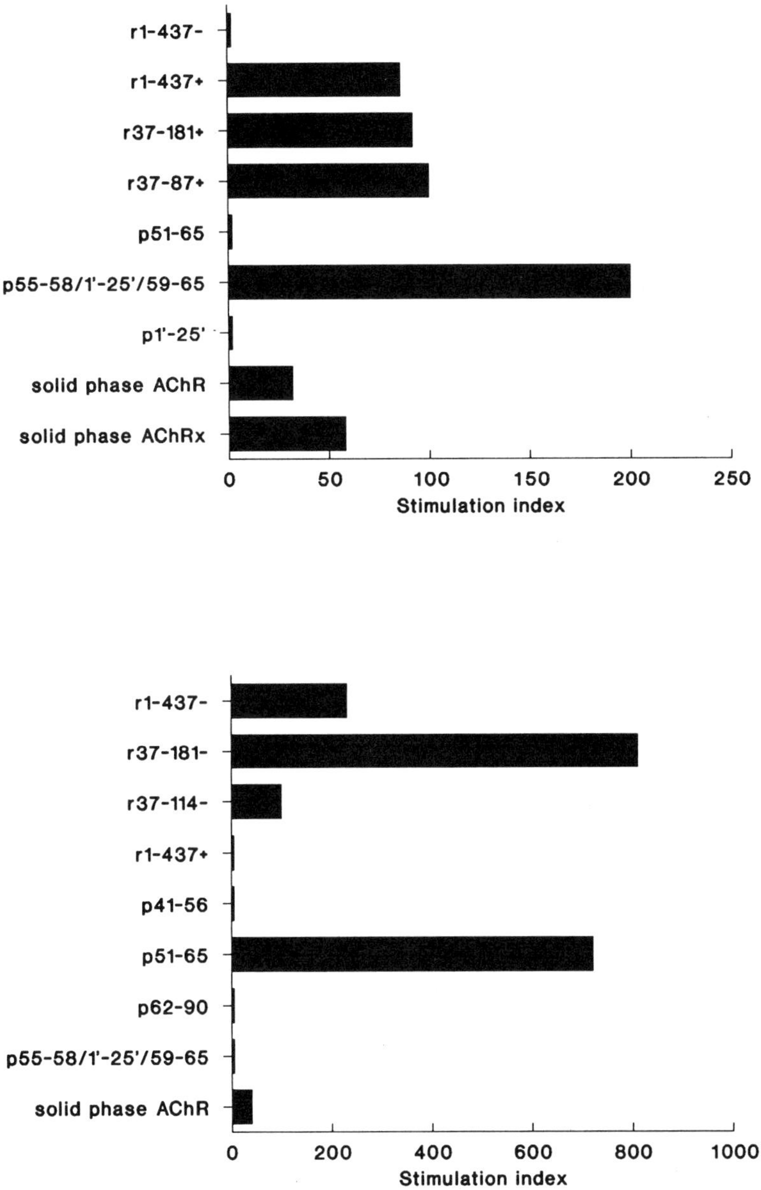

FIGURE 1. Proliferative responses of T cells from (top) patient MG3 and (bottom) C1, a healthy control raised against recombinant human AChR alpha-subunit. r1-437 etc + or − equals recombinant human alpha-subunit polypeptides with or without P3A insert. Prefix "r" denotes recombinant product; prefix "p" denotes synthetic peptide. Solid phase AChR = purified native human AChR bound to magnetic beads with specific antibodies. Solid phase AChRx = purified human AChR denatured with cyanogen bromide and bound to magnetic beads with specific antibodies (see text).

sponse of MG3 was enhanced (FIG. 1, AChRx), suggesting increased expression of this sequential epitope and improved capture by the beads.

Antigen-presenting cells from the healthy control, C1, were capable of processing and presenting antigen to either cell line, because this individual was both DQ1w5$^+$ and DRw53$^+$. This antigenicity of both forms of the alpha-subunit within an individual's repertoire suggests a common processing site downstream from position 65. In MG the close proximity of the main immunogenic region at α67-76 might facilitate presentation of these particular T-cell epitopes by MIR-specific B cells.

REFERENCES

1. HOHLFELD, R., K. TOYKA, S. TZARTOS, W. CARSON & B. CONTI-TRONCONI. 1987. Proc. Natl. Acad. Sci. USA **84:** 5379–5383.
2. BEESON, D., A. MORRIS, A. VINCENT & J. NEWSOM-DAVIS. 1990. EMBO J. **9:** 2101–2106.
3. MORRIS, A., D. BEESON, L. JACOBSON, F. BAGGI, A. VINCENT & J. NEWSOM-DAVIS. 1991. FEBS Lett. **295:** 116–118.
4. HAWKE, S., N. WILLCOX, G. HARCOURT, A. VINCENT & J. NEWSOM-DAVIS. 1992. J. Immunol. Methods **155:** 41–48.

Abnormal Immunoregulation Involving the IL-2/IL-2 Receptor Complex in Myasthenia Gravis

S. COHEN-KAMINSKY, J. M. DUQUENOY AND
S. BERRIH-AKNIN

CNRS URA-1159
Hôpital Marie-Lannelongue
92350 Le Plessis-Robinson, France

In myasthenia gravis (MG), anti-acetylcholine receptor (anti-AChR) antibody (Ab) production is dependent on T cells, and T-cell proliferation is regulated by interleukin-2 (IL-2). We have previously demonstrated abnormalities related to the IL-2/IL-2 receptor complex in MG patients that could account for T-lymphocyte hyperactivity in patients, namely a high sensitivity to recombinant IL-2 in MG thymic[1] and blood lymphocytes[2] without previous stimulation and a high expression of soluble IL-2 receptor (IL-2R) in the sera of MG patients.[3] No correlation was found between these abnormalities and IL-2R p55 chain expression in fresh MG lymphocytes, raising the hypothesis of a defect in expression and/or regulation of IL-2R in MG patients. Thus, we investigated the expression of the high-affinity IL-2R in >90% of enriched blast-cell fractions[4] from PHA-stimulated peripheral blood lymphocytes by binding techniques with iodinated IL-2[5] and by double color cytofluorographic analysis using Abs to the p55 and the p75 chains of the IL-2R.

The study included 12 MG patients and 10 control subjects. All patients were already thymectomized at the time of the study. Disease severity was graded according to Osserman's classification: I (1 patient), IIA (5 patients), IIB (6 patients). Anti-AChR Ab titer ranges were as follows: < 0.1 nM (2 patients), 0.1–1 nm (5 patients), 1–10 nM (3 patients), 10–100 nM (2 patients). The monoclonal Ab to the P75 chain of the IL-2R (Mikβ1) is a kind gift of Dr. Tsudo.[6] The stainings were performed essentially as previously described.[1,2]

These experiments allowed us to quantify the expression of high-affinity IL-2R sites by two independent techniques: calculation of the numbers of high-affinity IL-2 binding sites (after Scatchard analysis of the binding experiments with iodinated IL-2) and analysis of the percentages of cells expressing simultaneously the p55 and p75 chains of the IL-2R (by cytofluorographic studies). Using these two techniques, we did not find a significant difference between MG patients and controls. In addition, no phenotypic changes in the distribution of lymphocytes expressing one or two chains of the IL-2R were found in resting or PHA-activated lymphocytes. By contrast, Scatchard analysis of IL-2–binding experiments indicated a significant increase of the affinity for IL-2 in MG patients compared to controls ($p < 0.04$, Mann Whitney test) (TABLE 1). The affinity ratios between the control subjects and the MG patients studied in the same experiments were higher than 2 in 10 out 12 patients, and this ratio could reach the value of six.

This result brings the first argument in favor of a defect in IL-2 receptor expression and/or regulation in MG and confirms the role of the IL-2/IL-2 receptor complex in the abnormal immunoregulation found in MG. Indeed, the higher affinity of the IL-2R could explain the highest sensitivity to IL-2 previously demonstrated in T lymphocytes from MG patients compared to controls.[1,2] Moreover this defect in high-affinity IL-2R could act as a pathogenic cofactor in this plurifactorial disease, leading to aberrant immunoregulation in triggering and/or exacerbation of the autoimmune process. Such a hypothesis has been raised in a number of autoimmune diseases in humans and in experimental models (see ref. 7 for review). The molecular basis of this biochemical alteration of the IL-2R deserves to be investigated. It could be due either to the involvement of other IL-2R chains (known or unknown) participating in the IL-2R function, to differences in post-translational modifications of the IL-2R chains, or to IL-2R polymorphism. As the most important actors of the immune and the autoimmune response (T-cell receptors, major histocompatibility complex molecules, and antigens) are polymorphic, this last characteristic in the other molecules involved as the cytokines and their receptors could also contribute to the notion of disease susceptibility.

TABLE 1. Expression of the High-Affinity IL-2R in MG Patients Compared to Controls

	K_d Values for the High-Affinity IL-2R		Numbers of High-Affinity Sites for IL-2 per Cell	
	Median	Range	Median	Range
Controls	51 pM	25–91 pM	2325	285–4100
Patients	26 pM	13–55 pM	2750	100–3995

[a] Quantification of the numbers of high-affinity IL-2 binding sites and of their affinity by Scatchard analysis of iodinated IL-2–binding techniques. Receptor numbers and dissociation constants (K_d) were measured as previously described[5] with minor modifications. High-affinity IL-2R (K_d = 10–90 pM) were selectively detected, using iodinated IL-2 concentrations, ranging between 5 and 400 pM, too low to bind to intermediate-affinity (K_d = 1 nM) or low-affinity (K_d = 10 nM) receptors. Of note, the experimental site numbers were corrected, according to cytofluorographic results from double immunostaining with anti-p55 and anti-p75.

REFERENCES

1. COHEN-KAMINSKY, S., P. LEVASSEUR, J. P. BINET & S. BERRIH-AKNIN. 1989. J. Neuroimmunol. **24:** 75–85.
2. COHEN-KAMINSKY, S., C. GAUD, E. MOREL & S. BERRIH-AKNIN. 1989. J. Autoimmunity **2:** 241–258.
3. COHEN-KAMINSKY, S., Y. JACQUES, C. AIME, D. SAFAR, E. MOREL & S. BERRIH-AKNIN. 1992. Clin. Immunol. Immunopathol. **62:** 190–198.
4. KURNICK, J. T., L. OSTBERG, M. STEGAGNO, A. K. KIMURA, A. ORN & O. SJOBERG. 1979. Scand. J. Immunol. **10:** 563–573.
5. ROBB, R. J., W. C. GREENE & C. M. RUSK. 1984. J. Exp. Med. **160:** 1126–1146.
6. TSUDO, M., F. KITAMURA & M. MIYASADA. 1989. Proc. Natl. Acad. Sci. USA **86:** 1982–1986.
7. KROEHER G. & AND G. WICK. 1989. Immunol. Today **10:** 246–251.

V_H Gene Family Utilization of Antiacetylcholine Receptor Antibodies in Experimental Autoimmune Myasthenia Gravis

YVO GRAUS, JAN VERSCHUUREN, NICOLAAS BOS,[a]
PETER VAN BREDA VRIESMAN, AND MARC DE BAETS

University of Limburg
Maastricht, the Netherlands

[a]*University of Groningen*
Groningen, the Netherlands

In the mouse EAMG model we determined the utilization of immunoglobulin heavy chain variable region (V_H) gene families by anti-AChR MAbs obtained from two mouse strains differing in their susceptibility for induction of EAMG by immunization with *Torpedo* AChR. Previously, a preferential usage of D-proximal V_H gene families in autoimmune diseases was suggested. To investigate a possible bias in V_H gene family usage in the EAMG model, we determined the utilization of V_H gene families by 65 anti-AChR and 20 anti-KLH hybridomas. V_H gene family utilization was determined by slot-blot hybridization of hybridoma RNA to DNA probes representative for 9 of the 12 known mouse V_H gene families. Fifty-eight out of 65 (89%) anti-AChR hybridomas used V_H genes from 6 V_H gene families (J558, 3609, Vgam3.8, 3660, Q52, and PC7183). The V_H gene family usage of anti-AChR MAbs approached a stochastic distribution over the different V_H gene families except for the J606 family, which was significantly underrepresented ($p < 0.01$), and the Vgam3.8 family, which was overrepresented ($p < 0.05$). The V_H gene family utilization of anti-AChR hybridomas showed no bias in utilization towards D-proximal V_H gene families (FIG. 1A). The Vgam3.8 family was used by 3 out of 6 MAbs directed to the α-bungarotoxin binding site. The overrepresentation of V_H genes from the small Vgam3.8 family could be due to selection of related V_H genes for binding to a defined antigenic region. No correlation was found, however, between V_H gene family usage and antibody specificity for the main immunogenic region. Previously we identified cross-reactive idiotopes (CRI) on both rat and mouse anti-AChR MAbs.[1] In order to relate expression of CRI to V_H gene family usage, we determined the appearance of CRI on mouse anti-AChR MAbs encoded by 6 different V_H families. Expression of CRI was found to be irrespective of the V_H gene family used. The structural basis for expression of a CRI among anti-AChR MAbs may not be directly related to overall sequence homology. Although V_H genes originating from different families share less than 70% overall sequence homology, they could share small highly homologous DNA sequences encoding similar structural features, that is, CRI of anti-AChR antibodies.

C57bl/6 (H-2^b, Igh^d) or BALB/c (H-2^b, IgH^a) mice show high and low susceptibility for EAMG, respectively. We therefore investigated if disease susceptibility can be related to an aberrant V_H gene family utilization of anti-AChR MAbs. The V_H gene family utilization of anti-AChR MAbs in BALB/c mice showed a relative

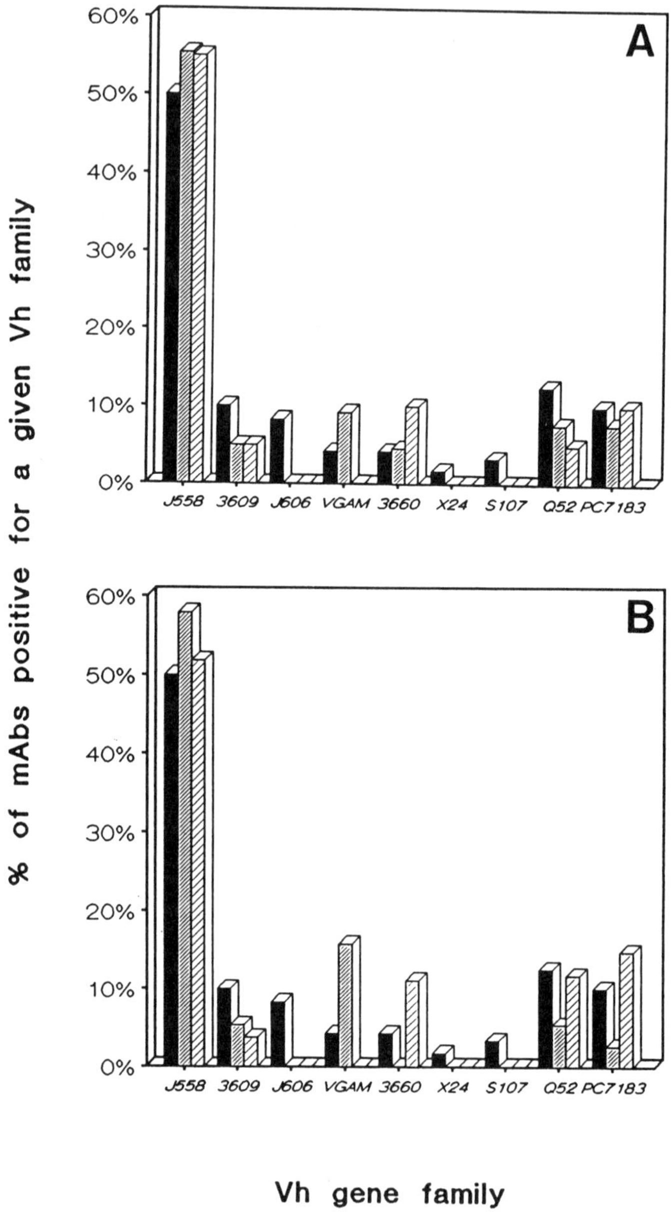

FIGURE 1. The V_H gene family distribution of anti-AChR and anti-KLH MAbs over nine V_H gene families. V_H family usage was expressed as the percentage of MAbs positive for a given V_H gene family and was compared to a stochastic distribution based on the estimated size of the family. Seven anti-AChR MAbs and 3 anti-KLH MAbs could not be assigned to one of the tested V_H gene families. **A:** V_H gene family usage of anti-AChR versus anti-KLH MAbs. ■, stochastic utilization; ▨, AChR MAbs, n = 65; ▧, KLH MAbs, n = 20. **B:** V_H gene family usage of anti-AChR MAbs from EAMG-susceptible C57bl/6 and EAMG-resistant BALB/c mice. ■, stochastic utilization; ▨, C57bl/6, n = 38; ▧, BALB/c, n = 27.

bias towards the D-region proximal families Q52 and PC7183 compared to C57bl/6 mice (FIG. 1B). The 3660 V_H gene family was exclusively used by BALB/c anti-AChR MAbs, whereas the Vgam3.8 family was only used by C57bl/6 anti-AChR MAbs (FIG. 1B). When V_H gene family utilization, however, was compared between BALB/c- and C57Bl/6-derived MAbs, no significant difference could be observed. The differences observed in V_H gene family utilization of C57bl/6- and BALB/c-derived anti-AChR MAbs were characteristic for the used mouse strains, and no association could be made between disease susceptibility and abnormalities in V_H gene family usage.

Taken together, these results indicate that the immune response against the AChR was encoded by many different V_H genes reflecting the polyclonal nature of anti-AChR antibodies in EAMG.

REFERENCE

1. VERSCHUUREN, J. J. G. M., Y. M. F. GRAUS, S. TZARTOS, P. J. C. VAN BREDA VRIESMAN & M. H. DE BAETS. 1991. J. Immunol. **146:** 941–948.

Stimulation of Specific T Cells by Human AChR Adsorbed to Immunomagnetic Particles

S. HAWKE, N. WILLCOX, G. HARCOURT, A. VINCENT, AND J. NEWSOM-DAVIS

Neurosciences Group
Institute of Molecular Medicine
University of Oxford
Oxford OX3 9DU, United Kingdom

INTRODUCTION

Cognate interactions between AChR-specific T and B cells are believed to be crucial to the induction of the autoantibodies in myasthenia gravis. These heterogeneous high-affinity antibodies recognize intact human AChR, implying that the T cells that control their production recognize peptides processed by B cells from intact autoantigen. Purified native human AChR would therefore be the ideal antigen to select and expand pathogenic T cells *in vitro,* but its scarcity is widely held to be prohibitive, and alternative strategies have evolved.

Our approach is to use recombinant α-subunit when characterizing T-cell lines and clones from patients and controls. One of these recognizes a new insert (see Batocchi *et al.* this volume) and another (PM-A) the α-subunit peptide p138-167.[1] Remarkably PM-A moderately responds to soluble AChR purified from human muscle. If AChR is adsorbed to the surface of biologically inert immunomagnetic particles (Dynabeads®), however, and targeted directly to antigen-presenting cells, strikingly efficient T-cell stimulation ensues. This system is versatile, and its specificity can be rigorously controlled. Antigen binding and washing is rapid and technically simple.[2]

RESULTS AND DISCUSSION

PM-A T cells respond well to recombinant AChR α-subunit and moderately also to AChR solubilized from muscle, although responses gradually decline (probably due to AChR degradation, even at $-70°C$). If AChR is concentrated onto Dynabeads® by way of specific antibodies, however, maximal responses are usually observed (FIG. 1), even with 10^3 times smaller amounts of AChR. Both the AChR and the antibodies are required; the latter can be either monoclonal-specific for extrajunctional AChR,[3] for example, or affinity-purified rabbit antibodies to cytoplasmic loop peptides. In stark contrast, 10^7 times higher molarities of synthetic peptides are required.

Both affinity-purified AChR preparations, that have long since lost their ability to stimulate, and small volumes of normal muscle extracts are satisfactory sources of AChR for bead adsorption (thus bypassing a purification step). Quantitation of AChR bound to Dynabeads® can be performed using [^{125}I]α-BuTx: about 10–20 times more AChR was adsorbed from affinity-purified preparations than from denervated or normal muscle extracts, perhaps because of aggregation or precipitation at low cholate concentrations in the former.

A second T-cell line specific for part of the 25 amino acid P3A insert and the adjacent sequence (AChR α 59-65) can be stimulated even when antibody to the insert is used to adsorb recombinant α-subunit to the beads (FIG. 2); evidently the epitope is not blocked in the process. The same antibody can also

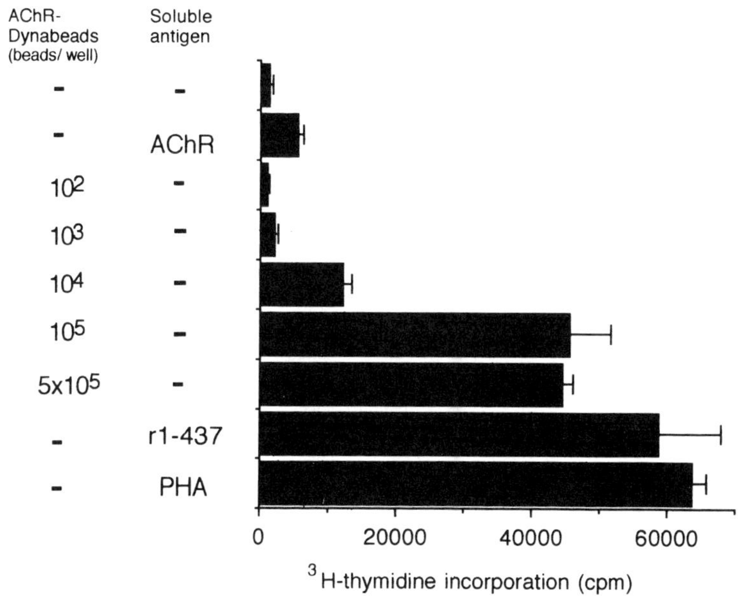

FIGURE 1. PM-A: response to soluble and solid-phase human AChR. Affinity-purified AChR (10 μL) was incubated with 10^7 anti-human AChR MAb-coated Dynabeads® for 30 minutes. They were washed twice and then resuspended in medium prior to addition to 200 μL wells containing 2×10^4 PM-A T cells and 2×10^5 APC (irradiated PBMC) in a standard proliferation assay.[1] Error bars indicate the standard deviation of triplicate wells.

capture antigen from muscle AChR preparations to stimulate this line, whereas antibodies to distant and/or conformational epitopes cannot, suggesting that the stimulatory antigen has been nicked/unfolded (Batocchi *et al.*, this volume).

Similar results have been obtained with hemocyanin (and polyclonal antisera), so its usefulness is not restricted to purified antibodies. The ability to wash bound antigens in low SDS concentrations and then stimulate T cells helped considerably in proving that certain "unmappable" T-cell lines raised with recombinant proteins were specific for *E. coli* contaminants (see Willcox *et al.*, this volume).

CONCLUSION

This technique will be particularly useful where cellular responses to scarce self or environmental antigens are sought. The ability to concentrate (and purify) AChR from toxic detergent extracts in amounts that maximally stimulate established T-cell lines allows rapid assessment of whether T cells selected using surrogate human AChR antigens *in vitro* might potentially react to processed native antigen. Early work suggests that antigen fragmentation, either artificial or spontaneous, may allow selection of T cells that are unable to respond to intact antigen. We are currently assessing polyclonal T-cell responses to bead-bound AChR from PBL and thymus and attempting to use this form of antigen to select and expand AChR-specific T cells.

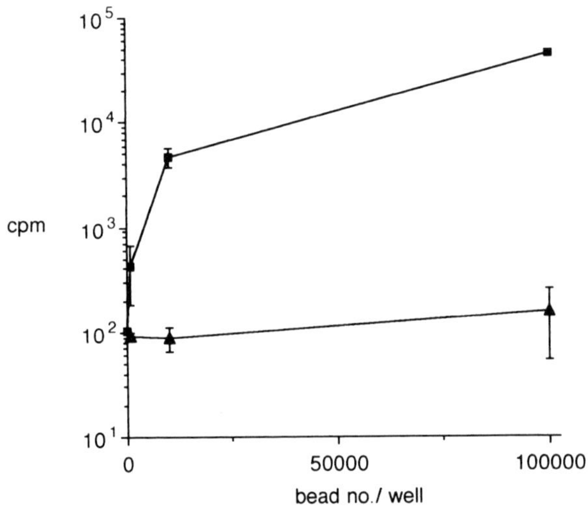

FIGURE 2. MG1DB4 response to solid-phase recombinant full-length α-subunit. Full-length recombinant human α-subunit (2 μg) was adsorbed to sheep anti-rabbit IgG Dynabeads® (2.8 μm), uncoupled (triangular symbols) or coupled (square symbols) to affinity-purified rabbit anti-peptide p51-58, 1′-25′, 59-65 antibodies prior to incubation with 2×10^4 clone DB4 and 2×10^5 irradiated heterologous PBL in standard proliferation assay. T cells + APC, 104 cpm. Soluble r1-437 2 μg/mL as a positive control gave 80346 cpm. Error bars indicate the standard deviations of triplicate wells.

REFERENCES

1. Ong, B., N. Willcox, P. Wordsworth, D. Beeson, A. Vincent, D. Altmann, J. Lanchbury, G. Harcourt, J. Bell & J. Newsom-Davis. 1991. Critical role for the Val/Gly^{86} HLA-DRβ dimorphism in autoantigen presentation to human T cells. Proc. Natl. Acad. Sci. USA **88:** 7343–7347.
2. Hawke, S., N. Willcox, G. Harcourt, A. Vincent & J. Newsom-Davis. 1992.

Stimulation of human T cells by sparse antigens captured on immunomagnetic particles. J. Immunol. Methods **155:** 41–48.
3. WHITING, P., A. VINCENT, M. SCHLUEP & J. NEWSOM-DAVIS. 1986. Monoclonal antibodies that distinguish between normal and denervated human acetylcholine receptor. J. Neuroimmunol. **11:** 223.

T-Cell Receptor Expression in Murine Myasthenia Gravis

A. J. INFANTE,[a,b,c] J. FALER,[b] S. CHANG,[a] P. CURRIER,[a]
P. THOMPSON,[c] K. KROLICK,[c] K. ZBOROWSKI,[d]
AND E. KRAIG[d]

Departments of [a]Pediatrics, [b]Biochemistry, [c]Microbiology, and [d]Cellular and Structural Biology
7703 Floyd Curl Drive
University of Texas Health Science Center
San Antonio, Texas 78284-7810

INTRODUCTION

The idea that lymphocytes involved in autoimmunity might be oligoclonal dates back to Burnet. This hypothesis has recently been supported by results from the analysis of T cells in certain animal models. This finding implies that antigen/MHC specificity and various tolerance mechanisms severely restrict the available T-cell receptor (TCR) repertoire for autoantigens. In addition to illuminating the ways by which susceptibility to autoimmunity is determined, study of oligoclonal T-cell responses may reveal useful targets of specific immunotherapy. Experimental autoimmune myasthenia gravis (EAMG) is an interesting animal model of human MG induced in susceptible strains of mice by immunization with *Torpedo* acetylcholine receptor (AChR). Susceptibility to EAMG involves recognition of a limited number of specific AChR peptides[1] and is influenced by mutation in the mouse major histocompatibility complex (MHC).[2] Because of the clonal dominance of T cells recognizing a particular peptide, residues 146-162 of the α-subunit in disease-susceptible C57BL/6 (B6) mice but not in disease-resistant MHC-mutant congenic bm12 mice,[3] we sought to determine whether the TCR used in this response were limited in expression of V, D, or J gene elements or had conserved residues encoded by N-diversity.

RESULTS

A panel of $CD4^+$ helper T-cell clones isolated from AChR-immunized B6 mice all recognized the immunodominant 146-162 peptide.[3] Seven of nine clones were $V\beta6^+$ by flow cytometry analysis.[4] The remaining two "clones" contained $V\beta6^+$ and $V\beta8^+$ cells. $V\beta8^+$ subclones, although reactive with AChR, did not recognize peptide 146-162. The predominance of $V\beta6^+$ clones in the response to AChR was confirmed by immunization and *in vitro* expansion of $CD4^+$ lymph node cells with AChR or peptide and analysis by two-color flow cytometry. Greater than 50% of AChR-expanded $CD4^+$ cells and greater than 75% of peptide-expanded cells from B6 mice were $V\beta6^+$.[4] By contrast, T cells from bm12 mice, although responsive to AChR, did not recognize peptide 146-162, nor were they $V\beta6^+$. B6 × bm12 F_1 mice demonstrated that peptide reactivity and $V\beta6$ expression were codominant; there was no evidence for negative selection.

Name	Source	Vß6	NDßN		Jß		Cß0
6.1	Clone	TGT GCC GGC AGT C A G S	ATT GAA CAG AAC GG I E Q N G	A	AGA TTA TTT TTC R L F F	[1.4]	+
6.9	Clone	TGT GCC GGC AGT C A G S	ATT GAA CAG I E Q	AAC N	GAA AGA TTA E R L	[1.4]	+
7.2	Clone	TGT GCC AGC AGT C A S S	ATA GAA CTG GGC GG I E L G G	C	ACC GGG CAG CTC T G Q L	[2.2]	-
7.3	Clone	TGT GCC AGC AGT C A S S	ATA GAG I E	TAT Y	GAA CAG TAC E Q Y	[2.6]	-
7.9	Clone	TGT GCC AGC AGT C A S S	ATT GGG GA I G E	A	ACG CTG TAT TTT T L Y F	[2.3]	-
CBR1	Line	TGT GCC AGC AGT C A S S	ATA GAG GGG AG I E G S	C	AAC GAA AGA TTA N E R L	[1.4]	+

FIGURE 1. Nucleotide and translated protein sequences of Vβ6$^+$ TCR β chains involved in response to AChR 146-162/I-A^b. T-cell populations were prepared as described in refs. 3 and 4. Total cellular RNA was extracted and cDNA synthesized using an oligo dT primer. PCR amplification was done using Vβ6 and Cβ primers incorporating 5′ restriction sites, and the amplified fragments were digested and ligated into pUC19. Plasmid DNA from transformed *E. coli* strain TB1 was directly sequenced in both directions using Sequenase (U.S. Biochemicals) and appropriate primers.

In order to examine other regions of AChR-specific TCR, particularly the V-D-J junctional regions proposed to make contact with the antigenic peptide fragment bound to MHC, we used a PCR strategy to amplify and sequence $V\beta6^+$ RNA-derived cDNA from T-cell clones, polyclonal T-cell lines, and immune lymph node T-cell populations (FIG. 1). Two findings were striking. (1) Despite differences in codon usage, the first residue predicted for the junctional protein sequence was always isoleucine (I) followed by glutamic acid (E) either directly or separated by one other residue. The presence of negative charge (E) in this region of the TCR may correlate with recognition of a critical lysine residue (positive charge) at position 155 in the 146-162 peptide (Infante and Wall, unpublished results). J segment usage did not appear restricted. (2) Several sequences included a seldom-used constant region exon, $C\beta0$, which is known to be enriched in thymic TCR messages compared to those in peripheral lymphoid tissue.[5]

DISCUSSION

TCR usage in the AChR response of EAMG-susceptible B6 mice is clearly restricted in $V\beta$ expression and appears to also involve limited heterogeneity in V-D-J junctional sequence. Circumstantial evidence suggests that these findings are due to constraints imposed by recognition of peptide/MHC. First, bm12 mice, which are genetically identical to B6 except for three amino acids in their MHC class II molecules, do not express $V\beta6^+$ TCR in response to AChR immunization,[4] nor do they recognize peptide 146-162.[3] Second, the predicted conserved junctional amino acids are encoded by different codons, suggesting selection of appropriate TCR β chains at the protein level. Third, the junctional sequences suggest a plausible charged pair interaction between TCR and peptide. Thus, the major determinant of restricted TCR gene usage in murine EAMG appears to be a structural requirement for recognition of the peptide/MHC complex. The additional finding of frequent use of $C\beta0$ in TCR from T cells involved in an autoimmune response is novel. Inasmuch as $C\beta0$ message has previously been found to be expressed at higher levels in thymus than in peripheral lymphoid tissue, this finding suggests that AChR reactive T cells in this model of EAMG may represent a subset of T cells that is usually prevented from leaving the thymus, perhaps by negative selection. Thus, $C\beta0$ may represent an independent marker of autoimmune T cells, unrelated to antigen/MHC specificity.

REFERENCES

1. ATASSI, M. Z., B. MULAC-JERICEVIC, T. YOKOI & T. MANSHOURI. 1987. Fed. Proc. Fed. Am. Soc. Exp. Biol. **46:** 538.
2. CHRISTADOSS, P., J. M. LINDSTROM, R. W. MELVOLD & N. TALAL. 1985. Immunogenetics **21:** 33.
3. INFANTE, A. J., P. A. THOMPSON, K. A. KROLICK & K. A. WALL. 1991. J. Immunol. **146:** 2977.
4. INFANTE, A. J., H. LEVCOVITZ, V. GORDON, K. A. WALL, P. A. THOMPSON & K. A. KROLICK. 1992. J. Immunol. **148:** 3385.
5. DENT, A. L., P. J. FINK & S. M. HEDRICK. 1989. J. Immunol. **143:** 322.

EAMG Induced in Rabbits by Immunization against Peptides Representing Human AChR α138-199

L. JACOBSON, A. VINCENT, P. SHILLITO, AND
J. NEWSOM-DAVIS

Neuroscience Group
Institute of Molecular Medicine
University of Oxford
Oxford OX3 9DU, United Kingdom

INTRODUCTION

The pathogenic nature of antibodies to the acetylcholine receptor (AChR) in myasthenia gravis (MG) has been well documented; however, antibodies do not bind well to isolated subunits or synthetic peptides and, conversely, antibodies raised against synthetic peptides do not bind well to the intact AChR. We have immunized NZW rabbits with synthetic peptides, either singly or in pools of two to three peptides, with sequences from the proposed extracellular domain of the AChR α-subunit α1-199 and a region of the cytoplasmic domain. Rabbits immunized concurrently with peptides α138-167, α156-187, and α185-199 developed acute muscle weakness, anti-AChR antibodies, and clear physiological evidence of experimental autoimmune myasthenia gravis (EAMG).

MATERIAL AND METHODS

Peptides were synthesized on a semiautomatic LKB Biolinx 4175 peptide synthesizer. NZW rabbits were immunized subcutaneously at 0 and 4 weeks with 500 μg of each peptide in Freund's complete adjuvant (FCA) and killed if definite signs of weakness were apparent. Serum antibodies were assessed by Western blot using the peptide dot-blot technique, detected by [^{125}I]protein A/protein G, and by immunoprecipitation of [^{125}I]α-bungarotoxin (α-BuTx)–labeled rabbit or human AChR (obtained from Triton X 100–extracted control leg muscles). Miniature end-plate potential (mepp) amplitudes were measured by microelectrode recordings in the rabbit phrenic nerve/diaphragm preparation using conventional techniques. AChR was measured in extracted leg muscle by immunoprecipitation of [^{125}I]α-BuTx-AChR.

RESULTS

All but one of the rabbit sera bound to the immunizing peptide (TABLE 1), but most of the sera did not bind appreciably to intact AChR. The exceptions were the three animals immunized against pool α138-199 and the one given the cytoplasmic sequence α309-368. The first (no. 6) immunized against pool α138-199 died unex-

TABLE 1. Anti-AChR Antibody Titers against Peptides, Recombinant α-Subunit (r701), and Rabbit AChR

Immunizing Peptide or Peptide Pool (rabbit)	Dot-blot Detection	Anti-[^{125}I]r701[a] (nM)	Anti-[^{125}I]α-BuTx AChR (nM)
(1) α1-56	+ + +[b]	27.5	—
(2) α45-65	−	—	—
(3) α51-P3A-65	+ + +	25.3	—
(4) α62-129	+ + +	8.1	—
(5) α124-143	+ +	—	—
(6) α138-199	died unexpectedly		
(7) α309-368	+ + +	414.7	807.6
(8) α138-199	+ + +	557.5	16.8
(9) α138-199	+ + +	ND	2.1

[a] Sera were tested against [^{125}I]recombinant α-subunit (r-701).
[b] + + +, 0.01 μg detected; + +, 1.0 μg detected; +, 10.0 μg detected; −, not detected; ND = not determined.

pectedly. The second and third (numbers 8 and 9) showed clear evidence of weakness within 10–15 days of the second injection and were moribund within a few days. Miniature end-plate potential amplitudes and [^{125}I]α-BuTx binding sites were significantly reduced (TABLE 2) in rabbits 8 and 9.

DISCUSSION

In recent years there have been several reports of EAMG, reduced mepp amplitudes and AChRs, induced in rats by immunization with α183-200[1] or α125-147.[2,3] In each case anti-AChR antibodies were detected by immunoprecipitation of [^{125}I]α-BuTx–rat AChR. In this study, we immunized rabbits with a pool of peptides, representing α138-199, of the human AChR α-subunit. Anti-rabbit AChR

TABLE 2. Miniature End-plate Potentials and AChRs in Rabbits

	Miniature End-plate Potential Amplitude in Rabbit Diaphragm (mV) ± SD	[^{125}I]α-BuTx Bound to Leg AChR Extracts (nM)
Control 1	0.44 ± 0.02 (n = 5)	0.21
Control 2	0.38 ± 0.01 (n = 6)	0.23
Control 3	0.47 ± 0.06 (n = 5)	0.33
	mean 0.43 ± 0.04	mean 0.26 ± 0.06
	died	died
Rabbit (6)	0.24 ± 0.05 (n = 6)	0.11
Rabbit (8)	0.32 ± 0.04 (n = 6)	0.10
Rabbit (9)	mean 0.28 ± 0.06[a]	mean 0.11 ± 0.007[a]

[a] $p = 0.05$.

titers were moderately high (>2 nM), and there was a significant reduction in mepp amplitude at diaphragm end-plates and a reduction in muscle AChR. More important, the rabbits became acutely weak. The results demonstrate that at least one of the peptides in this pool is very immunogenic and can induce highly pathogenic anti–AChR antibodies.

REFERENCES

1. Takamori, M., S. Okumura, M. Nagata & H. Yoshikawa. 1988. Myasthenogenic significance of synthetic α-subunit peptide 183-200 of *Torpedo californica* and human acetylcholine receptor. J. Neurol. Sci. **85:** 121–129.
2. Lennon, V. A., Z-X. Huang, D. J. McCormick, G. E. Griesmann, N. Fujii & E. H. Lambert. 1988. Synthetic peptide of human acetylcholine receptor α-subunit sequence 125-147 (methionine 144), a more potent autoantigen than its norleucine 144 analogue. Ann. N.Y. Acad. Sci. **540:** 516–519.
3. Matsuo, H., M. Tsujihata, A. Satoh, G. Takeo, T. Yoshimura & S. Nagataki. 1992. Myasthenogenicity of a human acetylcholine receptor α-subunit peptide: morphology and immunology. Muscle & Nerve **15:** 282–287.

A Novel Therapy for Myasthenia Gravis by Reducing the Endocytosis of Acetylcholine Receptors[a]

RALPH W. KUNCL, ILAN WITTSTEIN, ROBERT N. ADAMS, WINSTON W. WIGGINS, ORLANDO AVILA, ALAN PESTRONK, KEVIN McINTOSH, DONNA LUCAS, SHARI DeSILVA, MOHAMED LEHAR, AND DANIEL B. DRACHMAN

Department of Neurology
The Johns Hopkins University School of Medicine
Meyer 5-119, 600 North Wolfe Street
Baltimore, Maryland 21205

The number of available acetylcholine receptors (AChRs) at the neuromuscular junction is reduced in myasthenia gravis (MG). This is due in part to accelerated endocytosis of antibody-bound AChRs.[1] Methylation of phospholipids in skeletal muscle[2] may enhance the endocytosis of receptors by altering the polarity of phospholipids, with resultant changes in microfluidity of the membrane lipid bilayer.[3,4] We asked whether a potent inhibitor of phospholipid methylation, 3-deazaadenosine (3DZA),[5,6] could decrease endocytosis of AChRs, as a potential therapeutic strategy in MG. We tested the effect of 3DZA on AChR turnover *in vitro* and in two animals models: passive transfer MG in the mouse and experimental autoimmune MG (EAMG) in the rat.

METHODS

Skeletal Muscle Cultures, AChR Turnover, and Pharmacologic Agents

Primary cultures of rat skeletal muscle were used at the 8-9 day myotube stage. The phospholipid methylation inhibitor, 3DZA (Burroughs Wellcome Co.), was dissolved in Me_2SO and diluted in culture medium. L-Homocysteine thiolactone hydrochloride was used to potentiate the inhibition of methylation by 3DZA, and L-methionine was used to reverse inhibitor effects.

To measure total surface AChRs, parallel sets of cultures were saturated with 0.2 μg [^{125}I]α-bungarotoxin ([^{125}I]α-BuTx) per mL (specific activity, 2–5 $\times$ 10^4

[a] This work was supported by Grants from the Muscular Dystrophy Association, the National Institute of Neurological Diseases and Stroke (NS26455, NS23719), and by gifts from the Jay Slotkin Fund for Neuromuscular Research. R. W. Kuncl was the recipient of a Teacher Investigator Development Award (NINDS NS00734), an Andrew W. Mellon Foundation Clinician Scientist Career Development Award, and the Capitol Award for Scientific Achievement of the Myasthenia Gravis Foundation of Greater Washington, in partial support of this research.

Ci/mol) at 37°C for 30 min at the end of an experimental period. After washing, cultures were extracted with 2% Triton X-100 to solubilize the [^{125}I]α-BuTx–labeled AChRs for gamma counting. The rate of degradation of AChRs was measured by an indirect method that depends on the specific labeling of the receptors and the release into the culture medium of radioactive material derived from degraded [^{125}I]α-BuTx-AChR complexes, as described previously.[4,7] To induce antibody-accelerated AChR degradation, a myasthenic patient's serum that was known to produce marked acceleration of AChR degradation was added to the culture medium at a final concentration of 20 percent. To assay the rate of AChR synthesis and insertion, preexisting AChRs in skeletal muscle cultures were first blocked by incubation with unlabeled α-BuTx, and excess α-BuTx was removed by washing. After incubation for 6 h, [^{125}I]α-BuTx was added to label the new AChRs that had been subsequently inserted in the membrane. Cultures were extracted with 2% Triton X-100 and the radioactivity counted.

Passive Transfer Myasthenia Gravis

Female $B_6D_2F_1$ mice were given crude immunoglobulin (prepared by precipitation with 50% NH_4SO_4) from a myasthenic patient's serum by daily ip injections containing 10–11 mg IgG and 0.6–1.0 mg IgM. Control animals received immunoglobulin from pooled normal sera. AChRs were counted in junctional strips of hemidiaphragms and sternomastoid muscles by quantitative [^{125}I]α-BuTx binding. The injected myasthenic immunoglobulin was sufficient to reduce mouse junctional AChRs by 56% within one week. Myasthenic mice were treated with 3DZA (controls with saline) given ip in two divided doses daily, and a dose response curve was determined.

Experimental Autoimmune Myasthenia Gravis

Female Lewis rats were immunized with 400 pmol of *Torpedo* AChR in Freund's complete adjuvant. Control animals received saline. AChRs were counted in junctional strips of sternomastoid muscles and hemidiaphragms by quantitative [^{125}I]α-BuTx binding. Immunization was sufficient to reduce rat junctional AChRs in sternomastoid muscles by 57% within one month. Rats with EAMG were treated with 3DZA (controls with saline) given ip in two divided doses daily.

Statistical Methods

To compare the data from multiple experiments, results were normalized as a percent of the control group for each experiment. Comparisons could then be made between grand means of all cultures or animals treated in a particular way. All data are expressed as mean ± SEM. The significance of differences among experimental groups was calculated by one-way analysis of variance. When F values were significant, individual treatments were compared by Student's *t* test (two-tailed).

RESULTS

Normal Culture Conditions

3DZA inhibited the degradation of AChRs in culture by as much as 50%, and there was a broad dose response beginning as low as 2 μM ($p < 0.00001$). The effect was potentiated by L-homocysteine and reversed by L-methionine, indicating that the drug works by inhibiting a methyltransferase reaction. The inhibitory effect on degradation corresponded with inhibition of phospholipid methylation (data not shown). With 3DZA continuously in the culture medium, the inhibition of receptor degradation persisted for at least 72 h, the longest time tested.

3DZA had no acute effect on the synthesis of AChRs (at concentrations and incubation times that affected AChR degradation). A variety of other control measures were checked, and 3DZA did not affect total protein synthesis, cell morphology by phase microscopy, release of creatine kinase into the medium (a measure of cytotoxicity), or the creatine kinase content of cells (a measure of cell growth).

As a result of the decreased rate of degradation in the presence of a constant rate of synthesis, the number of AChRs increased on the surface of muscle cells as much as 24.5% above controls within 8 h ($p < 0.00001$). Concentrations of 3DZA ranging from 2 μM to 100 μM appeared to be optimal for decreasing the turnover of AChRs in normal, uninnervated skeletal muscle cells.

Simulated Myasthenic Conditions in Culture

Degradation Rate

3DZA inhibited the accelerated degradation and loss of AChRs produced by human myasthenic immunoglobulin. Serum from a patient with MG markedly increased the AChR degradation rate nearly threefold from the control rate of 4.5%/h to 12.2%/hour. Under these simulated myasthenic conditions during 8 h in culture, 3DZA reduced the accelerated rate of degradation by 5.6% to 24% ($p < 0.00001$) in a dose-responsive manner over the concentration range of 20–1000 μM. The effect plateaued by 24 h in culture, reaching a maximal inhibition of 50 percent.

Total Surface AChRs

In the presence of human anti-AChR antibody, rat skeletal muscle fibers in culture lost 60% of their AChRs in 8 hours. 3DZA at concentrations up to 100 μM significantly increased surface receptors by up to 35% under these conditions (TABLE 1). Again, the effect demonstrated dose-response and significant potentiation by L-homocysteine ($p < 0.00001$). The combination of 100 μM 3DZA and 100 μM L-homocysteine allowed a 50% protection of the receptors otherwise lost to antibody-accelerated degradation, for a net increase of 75% in the number of receptors compared with the myasthenic condition. This effect lasted at least 24 h in culture.

Experimental in Vivo *Models of Myasthenia Gravis*

In passive transfer MG, 3DZA significantly increased junctional AChRs in myasthenic mouse diaphragms by 35% ± 9% within 1 week of treatment ($p < 0.005$). This therapeutic effect was not due to any change in levels of circulating anti-AChR antibody. An *in vivo* dose response curve was determined in this model, and the optimally effective dose of 3DZA was 5 to 15 mg/kg/day (dosage divided b.i.d.).

In EAMG, 3DZA significantly increased junctional AChRs in the myasthenic rat sternomastoid muscles by 21% ± 10% within 4 weeks of treatment ($p < 0.005$). This therapeutic effect was not due to a change in anti-AChR antibody titer. Despite a stable or increasing antibody titer, 3DZA was able to decrease the loss of ACh receptors in EAMG, consistent with a mechanism by inhibition of receptor endocytosis.

TABLE 1. Effect of 3-Deazaadenosine on the Accelerated Loss of Surface AChRs Induced by Myasthenic Serum[a]

Culture Conditions	[^{125}I]α-BuTx Binding (CPM/dish ± SEM)	Percent Increase[b]	p[b]
8-h Experiments:			
Normal control	17,966 ± 741		
Myasthenic serum alone	7,166 ± 399	—	—
Myasthenic serum plus			
3DZA 20 μM	7,507 ± 137	4.8%	<0.05
3DZA 20 μM + L-HCT 100 μM	8,907[c] ± 222	24.3%	<0.00001
3DZA 100 μM	9,653 ± 469	34.7%	<0.00001
3DZA 100 μM + L-HCT 100 μM	12,530[d] ± 333	74.9%	<0.00001
24-h Experiments:			
Normal control	27,798 ± 888		
Myasthenic serum alone	8,363 ± 249	—	—
Myasthenic serum plus 3DZA 20 μM	9,252 ± 269	10.6%	<0.00001
Myasthenic serum plus 3DZA 100 μM	10,055 ± 381	20.2%	<0.00001

[a] Parallel cultures were treated with or without 3DZA. Data are presented as mean ± SEM.

[b] Percent increase and significance of difference compared with "myasthenic serum alone" condition.

[c,d] Significantly different than parallel treatment without L-homocysteine thiolactone (L-HCT); [c]$p < 0.001$, [d]$p < 0.00001$.

CONCLUSION

Because 3DZA allows accumulation of AChRs in the surface membrane of muscle cells by reducing endocytotic degradation, it provides a potential strategy for therapy in human myasthenia gravis. The potential advantages of such therapy include the following: (1) 3DZA targets a key pathogenetic mechanism: antibody-enhanced endocytosis of the AChR; (2) 3DZA has a rapid therapeutic effect, showing an increase in the number of AChRs within a few half-lives of receptor turnover (*i.e.*, weeks); (3) the mechanism of action of 3DZA is independent of that of other available therapies, making it potentially additive in effect; (4) 3DZA has been shown to be safe in phase I trials in humans; and (5) disease-modifying drugs could be of benefit either as sole therapy or as adjunct to other therapeutic agents. Given the safety margin of neuromuscular transmission and the fact that successful transmission is a threshold phenomenon, even a moderate increase in the number of available AChRs could prove clinically beneficial in myasthenia gravis.

ACKNOWLEDGMENT

The helpful discussion and cooperation of Dr. Steven H. Grossman (Burroughs Wellcome Co.) is gratefully acknowledged.

REFERENCES

1. Kao, I & D. B. Drachman. 1977. Myasthenic immunoglobulin accelerates acetylcholine receptor degradation. Science **196:** 527–529.
2. Kuncl, R. W., D. B. Drachman & Y. Kishimoto. 1985. Phospholipid methylation in skeletal muscle membranes. Muscle & Nerve **8:** 426–434.
3. Schroeder, F. 1981. Altered phospholipid composition affects endocytosis in cultured LM fibroblasts. Biochim. Biophys. Acta **649:** 162–174.
4. Kuncl, R. W., D. B. Drachman & R. Adams. 1988. Inhibition of methyltransferase reduces the turnover of acetylcholine receptors. Proc. Natl. Acad. Sci. USA **85:** 4032–4036.
5. Zimmerman, T. P., G. Wolberg, C. R. Stopford & G. S. Duncan. 1979. 3-Deazaadenosine as a tool for studying the relationship of cellular methylation reactions to various leukocyte functions. *In* Transmethylation. E. Usdin, R. T. Borchardt & C. R. Creveling, Eds.: 187–196. Elsevier/North-Holland. New York.
6. Schanche, J-S., T. Schanche & P. M. Ueland. 1981. Inhibition of phospholipid methyltransferase(s) from rat liver plasma membranes by analogues of S-adenosylhomocysteine. Mol. Pharmacol. **20:** 631–6.
7. Devreotes, P. N. & D. M. Fambrough. 1975. Acetylcholine receptor turnover in membranes of developing muscle fibers. J. Cell Biol. **65:** 335–358.

The hu-SCID Myasthenic Mouse

A New Tool for the Investigation of Seronegative Myasthenia Gravis[a]

GIANVITO MARTINO,[b,c] LUIGI M. E. GRIMALDI,[c,e]
ROBERT L. WOLLMANN,[b] PAOLO BONGIOANNI,[b]
JOSÈ QUINTANS,[d] AND BARRY G. W. ARNASON[b]

[b]*Department of Neurology and*
The Brain Research Institute
University of Chicago
Chicago, Illinois

[c]*Neuroimmunology Unit–Di.Bi.T.*
Clinica Neurologica IV
Università di Milano
IRCCS H. San Raffaele
20132 Milano, Italy

[d]*Department of Pathology*
University of Chicago
Chicago, Illinois

INTRODUCTION

Severe combined immunodeficient (SCID) mice lack functional B and T cells due to a deletion of the entire J region during gene rearrangement.[1] Thus, human peripheral blood lymphocytes (PBL) injected intraperitoneally (ip) into SCID mice survive within the peritoneum for at least 3 months after transplantation and produce human immunoglobulin (Ig).[2–4] Myasthenia gravis (MG) is a T cell–dependent and antibody-mediated autoimmune disorder manifested by weakness and fatigability of voluntary muscles.[5,6] Pathological changes at the muscular level follow an autoimmune attack directed against the acetylcholine receptor (AChR) on the postsynaptic end-plate. Anti-AChR antibodies can be detected in sera from 85–90% of MG patients, but antibody titers do not correlate with disease severity.[7] We attempted to reproduce the immunopathological findings of MG in SCID mice by transplantation into their peritoneal cavity of human PBL from patients with MG (MG-SCID) with or without detectable serum anti-AChR antibodies.

METHODS

Fifteen to 30 × 10[6] PBL from 6 MG patients (2 women, 4 men; mean age 47) were injected ip into 15 C.B-17 *scid*/*scid* mice (from 2 to 4 mice from each donor).

[a] This work was supported by the Telethon-Italy. We thank Bonita DuPont for her support in this project.

[e] To whom correspondence should be sent.

Eight SCID mice transplanted with the same number of PBL from 7 healthy subjects (3 men, 4 women; mean age 29) were also studied as the control group. The MG patients were assigned to group I (1 patient), IIA (2 patients), and IIB (3 patients) according to the Osserman classification of disease severity.[8] Single-fiber electromyography and response to intravenous edrophonium were positive in all five patients tested. Four of six patients were positive for antibodies against human AChR by conventional RIA. Three of these seropositive MG patients were also positive for anti-*Torpedo* AChR by ELISA, using membrane-bound *Torpedo* AChR as antigen (prepared as described by Klimkowsky[9]) and an Fc-specific peroxidase-conjugated goat antihuman IgG as detector antibody. Animals were bled every two weeks from the tail vein and sacrificed between 8 and 12 weeks after transplantation. Anti-AChR antibody was determined in SCID mouse serum samples by ELISA (see above). To localize human IgG at the muscular end-plate, cryostat sections from muscles of 3 MG-SCID (2 transplanted with PBL from 2 different seronegative MG patients; 1 with PBL from 1 seropositive MG patient) and 2 control SCID mice (transplanted with PBLs from 2 different healthy donors) were stained with a rabbit antihuman IgG antiserum using direct immunocytochemistry. Muscles from 3 MG-SCID mice (transplanted with PBL from 1 seronegative and 2 seropositive MG patients) and 2 control SCID mice (transplanted with PBL from healthy subjects) were processed for electron microscopy. To quantify end-plate damage we measured an end-plate index (EI) (determined as the length of postsynaptic membrane/length of apposed presynaptic membrane) for each neuromuscular junction identified by electron microscopy.

RESULTS AND CONCLUSION

All nine SCID mice transplanted with PBL from seropositive MG patients and five of six SCID mice transplanted with PBL from seronegative MG patients showed human anti-AChR IgG from 2 to 4 weeks after transplantation. Human anti-AChR IgG levels increased rapidly in MG-SCID mice between 2 and 4 weeks, remained high for another 4 weeks, and began to fall by 10 weeks. No differences were found between production and kinetics of anti-AChR antibody in SCID mice transplanted with PBL from anti-AChR antibody seropositive or seronegative MG patients. Human IgG was found, by immunocytochemistry, on the muscle side of end-plates of all MG-SCID mice transplanted with PBL from seropositive or seronegative MG patients. In addition, EI measurement by electron microscopy revealed clear simplification of postsynaptic membranes of neuromuscular junctions in 3 MG-SCID mice transplanted with PBL from 3 different MG patients (1 seronegative and 2 seropositive). The differences in EI between the 3 MG-SCID mice and the 2 control SCID mice were significant ($p < 0.0001$). Anti-AChR IgG was found in only one of the 8 control SCID mice. However, in this animal (as well as in all remaining control animals) electron microscopy showed normal neuromuscular junctions.

In conclusion, we have induced MG in an immunodeficient murine environment by transplanting human PBL from MG patients. In this model the production of antibodies by cells from seronegative MG patients indicates that B cells committed to AChR are present in "antibody-negative" patients and that a similar pathological mechanism (anti-AChR antibody-mediated) may account for their muscle damage. Why anti-AChR IgG are detectable in SCID mice transplanted with PBL from seronegative MG patients is unclear. We can speculate that SCID mice lack a suppressive mechanism that operates in the patients. Fluctuations within an idiotypic-antiidiotypic network[10] could also play a role in determining human antibody detection in the SCID mouse. The MG-SCID model may provide a useful tool for studies on anti-AChR antibody formation and detection in seronegative MG patients.

REFERENCES

1. Schuler, W., I. J. Weiler, A. Schuler, R. A. Philips, N. Rosemberg, T. W. Mak, J. F. Kearney, R. P. Perry & M. J. Bosma. 1986. Cell **46:** 963–972.
2. Mosier, D. E., R. J. Gulizia, S. M. Baird & D. B. Wilson. 1988. Nature **335:** 256–259.
3. McCune, J. M., R. Namikawa, H. Kaneshima, L. D. Shultz, M. Lieberman & I. L. Weissman. 1988. Science **241:** 1632–1639.
4. Martino, G., J. Anastasi, J. Feng, C. McShan, L. J. DeGroot, J. Quintans & L. M. E. Grimaldi. 1993. Eur. J. Immunol. In press.
5. Vincent, A. 1980. Physiol. Rev. **60:** 756–824.
6. Lindstrom, J., D. Shelton & Y. Fujii. 1988. Adv. Immunol. **42:** 233–284.
7. Lindstrom, J., M. E. Seybold, V. A. Lennon, S. Whittingham & D. D. Duane. 1976. Neurology **26:** 1054–1059.
8. Osserman, K. E. 1958. *In* Myasthenia Gravis, 79–86. Grune and Stratton. New York.
9. Klimkowsky, M. W., J. E. Heuser & R. M. Stroud. 1980. J. Cell Biol. **85:** 823–838.
10. Dwyer, D. S., R. J. Bradley, C. K. Urquhart & J. F. Kearney. 1983. Nature **301:** 611–614.

Tolerance to AChR Induced by AChR-Coupled Syngeneic Cells: Clonal Anergy?[a]

KEVIN R. McINTOSH, PAMELA C. ROEHM, AND DANIEL B. DRACHMAN

Department of Neurology
School of Medicine
The Johns Hopkins University
5-119 Meyer Building
600 North Wolfe Street
Baltimore, Maryland 21287

Immunotherapy of myasthenia gravis should be possible by inducing immunologic tolerance to AChR. This goal has remained elusive thus far, as AChR is a highly immunogenic (*i.e.*, nontolerogenic) molecule. Tolerance can be induced to a variety of antigens by intravenous administration of antigen-coupled syngeneic cells.[1–3] This form of tolerance has been attributed to induction of antigen-specific suppressor T cells and direct inactivation of antigen-specific T lymphocytes.[4,5]

The purpose of this study was to determine whether tolerance to AChR could be induced in Lewis rats by intravenous administration of AChR-coupled syngeneic spleen cells. As a first step in understanding the mechanism of tolerance induced by AChR-coupled cells, we investigated the role of Ia antigens (on the carrier spleen cells) and the possible role of suppressor cells in the induction of tolerance.

Our initial objective was to determine whether AChR-coupled syngeneic cells could induce antigen-specific tolerance of the primary antibody response to AChR in Lewis rats. In these experiments, AChR-coupled spleen cells (AChR-SC) or control mock-coupled spleen cells (MOCK-SC) were injected one week prior to immunization with AChR plus ovalbumin (the response to OVA was used as a specificity control). As shown in FIGURE 1, the antibody response to AChR was inhibited in rats treated with AChR-SC, compared to the MOCK-SC–treated control group (61% inhibition). The antibody response to OVA was unaffected by this treatment, demonstrating that tolerance was specific for AChR. In addition, lymphocytes from the draining lymph nodes of tolerant rats demonstrated reduced proliferative responses to AChR *in vitro,* assessed by [^{3}H]thymidine uptake (data not shown). This indicated that the T-cell compartment was tolerant, inasmuch as lymphoproliferation under these circumstances has been attributed primarily to T cells.[6]

To determine the role of Ia determinants on the carrier cells in tolerance induction, spleen cells were fractionated into Ia-positive and Ia-negative populations prior to coupling with AChR and injection into rats. Controls were injected with the untreated subpopulations. The results (FIG. 2) showed that only AChR-coupled Ia-positive cells were capable of inducing tolerance specific for AChR.

[a] This work was supported in part by Grants from the Myasthenia Gravis Foundation, the Muscular Dystrophy Association, and the NIH (1RO1 NS23719).

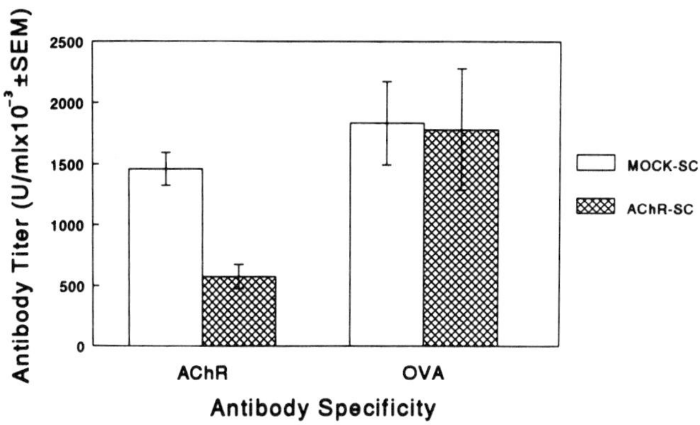

FIGURE 1. Specific inhibition of the primary antibody response by AChR-coupled cells. Purified *Torpedo* AChR was coupled to rat spleen cells (AChR-SC) using chromic chloride.[9] Control mock-coupled spleen cells (MOCK-SC) were prepared by treating spleen cells with chromic chloride alone. 20×10^6 AChR-SC or control MOCK-SC were injected intravenously into the animals. One week later, the rats were challenged intradermally with 50 μg AChR and 100 μg OVA in complete Freund's adjuvant (CFA). Eight weeks later, the rats were bled, and serum antibody titers to AChR and OVA were determined by kinetic ELISA as previously described.[9]

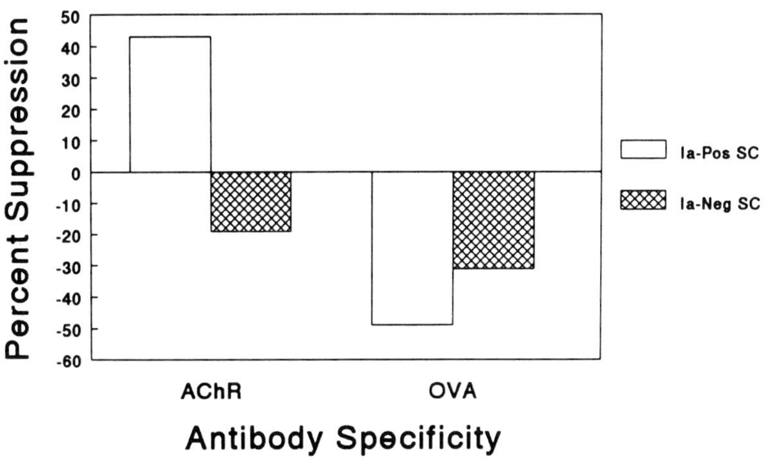

FIGURE 2. Requirement for MHC class II expression on carrier spleen cells. Spleen cells were fractionated into Ia-positive (Ia-Pos SC) and Ia-negative (Ia-Neg SC) populations, prior to AChR coupling and injection, by a combination of techniques. Ia-positive cells were prepared by collection of nylon wool adherent cells and subsequent removal of contaminating T cells using anti–T-cell monoclonal antibodies (W3/13 + OX19) and magnetic beads. Ia-negative cells were prepared from nylon wool nonadherent cells, treatment with anti-Ia monoclonal antibodies (OX3), and removal of antibody-coated cells with magnetic beads. Purities of these populations exceeded 95%. Tolerance was induced as described in the FIGURE 1 legend. Antibody titers were determined 4 weeks after immunization by kinetic ELISA. Percent suppression represents percent inhibition or enhancement (negative numbers) induced by AChR-coupled Ia-Pos or Ia-Neg spleen cells relative to the appropriate control group.

These results suggested that tolerance may occur through antigen presentation by AChR-coupled Ia-positive spleen cells. As suggested by others,[7] antigen could be presented to helper T cells or to suppressor T cells. To evaluate whether suppressor T cells played a role in tolerance, we performed adoptive transfer assays to determine whether unresponsiveness could be transferred to naive recipient rats by spleen cells from tolerant animals. As shown in FIGURE 3, spleen cells from tolerant rats (taken one week after tolerance induction) were not suppressive for the primary antibody response in the recipients, even when purified T cells were transferred. Furthermore, suppression could not be observed when spleen cells were transferred three weeks after tolerance induction, nor could suppression be demonstrated in a sensitive *in vitro* suppressor assay (data not shown).

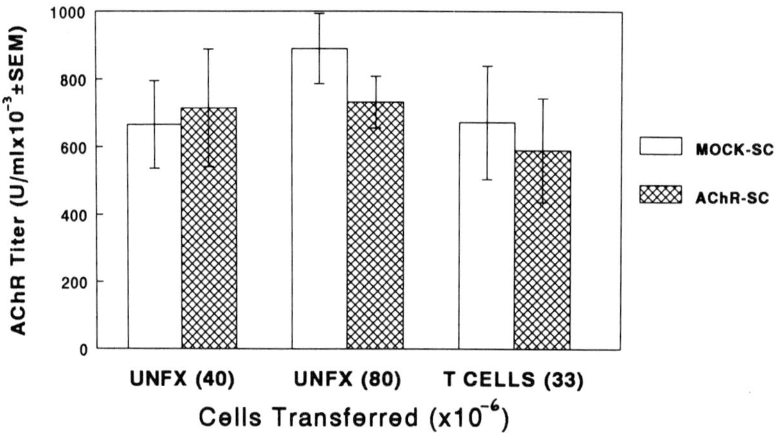

FIGURE 3. Failure to demonstrate suppressor cell activity in tolerant rats. Rats were tolerized with AChR-SC or injected with control MOCK-SC as described in the FIGURE 1 legend. One week later, the animals were killed, and various numbers of their unfractionated (UNFX) spleen cells were transferred intravenously into naive recipient Lewis rats. In one experiment, splenic T cells were purified (nylon wool nonadherent cells, treated with anti–B cell (OX33 + OX12) and anti-Ia (OX3) monoclonal antibodies and removed with magnetic beads) prior to transfer. One day later, the recipients were immunized intradermally with 50 μg AChR in CFA. Antibody titers to AChR were determined by kinetic ELISA 4 weeks after immunization.

We have demonstrated that intravenous administration of AChR-coupled syngeneic spleen cells induced tolerance of the primary antibody response in Lewis rats. These results demonstrate that AChR, which is known to be highly immunogenic, can be rendered tolerogenic. The mechanism of tolerance induction by AChR-SC appears to occur through antigen presentation. Inasmuch as suppressor cells were not activated in this system, and lymphoproliferative responses to AChR were specifically compromised, it is likely that AChR-specific helper T cells were directly tolerized by AChR-SC. We propose that AChR-coupled Ia-positive cells process the bound AChR and present it to T cells, inducing clonal anergy as described by R. H. Schwartz and colleagues.[8] Presumably, AChR-SC may not be able to provide adequate costimulation for T-cell activation, due to treatment with chromic chloride, resulting in clonal anergy. Studies are currently in progress to test this hypothesis.

REFERENCES

1. Claman, H. N. 1976. Tolerance and contact sensitivity to DNFB in mice. V. Induction of tolerance with DNP compounds and with free and membrane-associated DNFB. J. Immunol. **116:** 704–709.
2. Braley-Mullen, H. 1980. Suppression of the antibody response to type III pneumococcal polysaccharide with antigen coupled to syngeneic lymphoid cells. Cell Immunol. **52:** 132–139.
3. Kennedy, M. K., M. C. Dal Canto, J. L. Trotter & S. D. Miller. 1988. Specific immune regulation of chronic-relapsing experimental allergic encephalomyelitis in mice. J. Immunol. **141:** 2986–2993.
4. Miller, S. D., R. P. Wetzig & H. N. Claman. 1979. The induction of cell-mediated immunity and tolerance with protein antigens coupled to syngeneic lymphoid cells. J. Exp. Med. **149:** 758–773.
5. Sherr, D. H., M. Krikor, K. M. Heghinian, B. Benacerraf & M. E. Dorf. 1979. Immune suppression *in vivo* with antigen-modified syngeneic cells. III. Distinctions between T cell tolerance and T cell mediated suppression. J. Immunol. **123:** 2682–2688.
6. Corradin, G., H. M. Etlinger & J. M. Chiller. 1977. Lymphocyte specificity to protein antigens. I. Characterization of the antigen-induced *in vitro* T-cell dependent proliferation response with lymph node cells from primed mice. J. Immunol. **119:** 1048–1053.
7. Sherr, D. H., K. M. Heghinian, B. Benacerraf & M. E. Dorf. 1980. Immune suppression *in vivo* with antigen-modified syngeneic cells. IV. Requirement for Ia + adherent cells for induction. J. Immunol. **124:** 1389–1395.
8. Mueller, D. L., M. K. Jenkins & R. H. Schwartz. 1989. Clonal expansion versus functional clonal inactivation: A costimulatory signalling pathway determines the outcome of T cell antigen receptor occupancy. Ann. Rev. Immunol. **7:** 445–480.
9. McIntosh, K. R. & D. B. Drachman. 1992. Tolerance to acetylcholine receptor induced by AChR-coupled syngeneic cells. J. Neuroimmunol. **38:** 75–84.

Acetylcholine Receptor-specific T Cells Are Present in the Normal Immune Repertoire

A Study with Recombinant Polypeptides of the Human Acetylcholine Receptor α-Subunit

A. MELMS,[a] G. MALCHEREK,[a] R. SCHOEPFER,[c]
N. SOMMER,[a] H. KALBACHER,[b] AND J. LINDSTROM[d]

Departments of [a]Neurology and [b]Biochemistry
University of Tübingen
D7400 Tübingen, Germany

[c]Center for Molecular Biology
University of Heidelberg
D6900 Heidelberg, Germany

[d]David Mahoney Institute of Neurological Sciences
University of Pennsylvania
Philadelphia, Pennsylvania 19104-6074

Acetylcholine receptor (AChR)-specific T-helper cells are the prerequisite for the generation of the humoral autoimmune response against AChR that will eventually impair the neuromuscular transmission leading to the clinical symptoms of myasthenia gravis (MG). The α-subunit of the AChR has been demonstrated to be the major target for both T and B lymphocytes.[1] Here, we studied the proliferative response of T lymphocytes from MG patients and healthy volunteers after stimulation with recombinant polypeptides of the human AChR α-subunit (α1-437).

A complementary DNA clone of the α-subunit from the cell line TE671, which was found to express human muscle AChR, was used for the construction of bacterial expression plasmids.[2] Expression was based on the T7 RNA polymerase system from Studier, Rosenberg, and co-workers.[3] Polypeptides of the human AChR α-subunit were purified from inclusion bodies by extraction with 8M urea and nonionic detergents (FIG. 1). Antigen specificity of T-cell lines and clones was examined in microproliferation assays using purified polypeptides of the α-subunit with various truncations of the C-terminus. Control antigens included lysed and nontransformed bacteria, an irrelevant recombinant polypeptide, and tetanus toxoid.

Using polypeptides with the authentic sequence of the human AChR α-subunit, T cells from 20 of 28 MG patients (71%) and 7 of 19 healthy donors (37%) responded significantly in primary proliferation assays. Subsequently, AChR-specific T-cell lines and clones were raised from patients and controls that had features expected for T-helper lymphocytes. The majority of T-cell lines and clones from patients and controls recognized fragments corresponding to the extracellular and N-terminal part of the molecule (α1-103 and α1-209). T cells recognizing these fragments cross-reacted with a fusion protein encoding sequence α6-216 of the murine AChR α-subunit[4] (FIG. 2). Few lines recognized a fragment from close to the C-terminus (α327-398). In spite of more than 80% of amino acid sequence homology with human AChR, none of these lines recognized *Torpedo*-AChR that was previously used as a substitute for human muscle AChR, suggesting that T cells in this study had a bias for unique human α-sequences. Similar observations have been recently

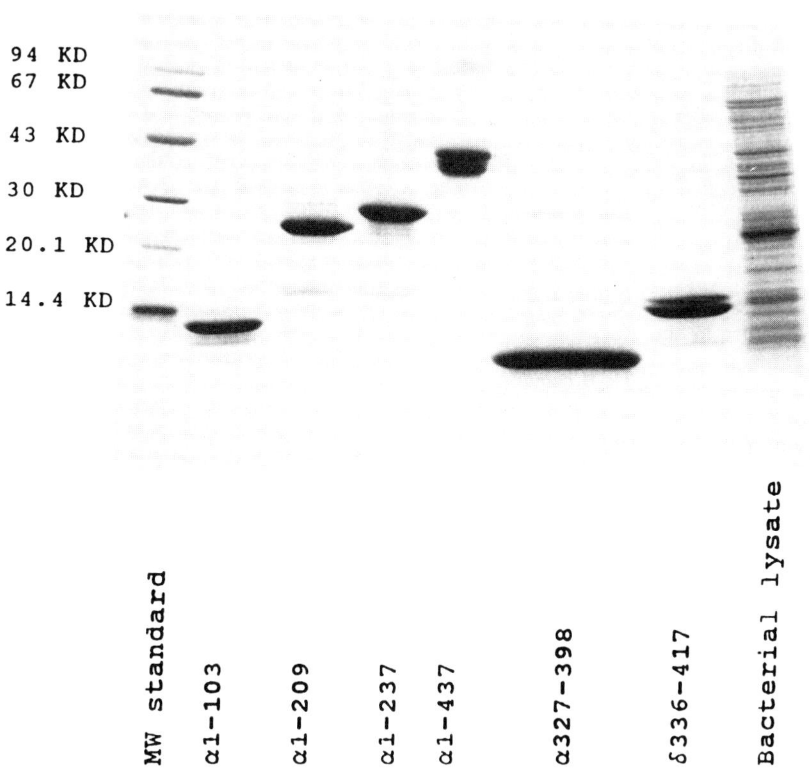

FIGURE 1. SDS-PAGE of purified polypeptides of the human AChR α-subunit; 5 to 20% polyacrylamide gradient and Coomassie brilliant blue staining. Lane 1: low molecular weight standards (Pharmacia); lane 2: α1-103; lane 3: α1-209; lane 4: α1-237; lane 5: α1-437; lane 6: α327-398; lane 7: δ336-417; lane 8: total lysate of nontransformed bacteria BL 21 DE3 (plysS).

reported by others.[5,6] There was no preference of certain MHC class II molecules such as HLA-DR3 to be used as a restriction element in antigen presentation. A given AChR fragment could be presented in the context of several MHC class II molecules, and various fragments could be presented in the context of the same MHC molecule. These results support earlier observations describing a rather heterogeneous situation in dealing with AChR as a multideterminant autoantigen both on the level of T cells and antigen-presenting cells. The demonstration of AChR-specific T cells in normal individuals indicated that the mere presence of

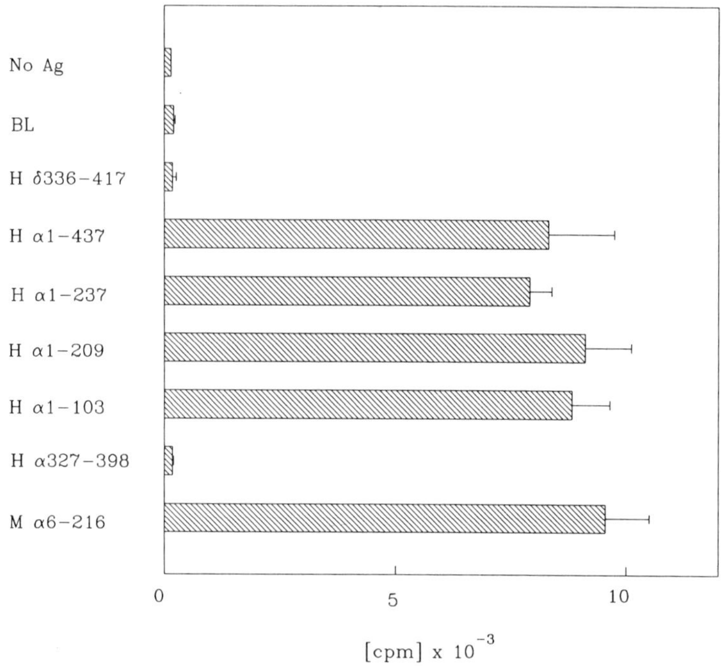

FIGURE 2. AChR-specific T-cell line from a healthy donor. Proliferation assay, [^{3}H]thymidine incorporation, mean of triplicates ±SD. T cells were stimulated with polypeptides of the human (H) or murine (M) AChR α-subunit; BL, bacterial lysate.

specific T cells is not sufficient to cause clinical signs of an (antibody-mediated) autoimmune disease and emphasized the role of immunoregulatory mechanisms for maintaining a state of tolerance.

REFERENCES

1. Hohlfeld, R., K. V. Toyka, S. J. Tzartos, W. Carson & B. M. Conti-Tronconi. 1987. Proc. Natl. Acad. Sci. USA **84:** 5379–5381.
2. Schoepfer, R., M. Luther & J. Lindstrom. 1988. FEBS Lett. **226:** 235–240.
3. Studier, F. W., A. H. Rosenberg, J. J. Dunn & J. W. Dubbendorff. 1990. Methods Enzymol. **185:** 60–89.
4. Melms, A., S. Chrestel, B. Schalke, H. Wekerle, A. Mauron, M. Ballivet & T. Barkas. 1989. J. Clin. Invest. **83:** 785–790.
5. Salvetti, M., S. Jung, S. F. Chang, H. Will, B. Schalke & H. Wekerle. 1991. Ann. Neurol. **29:** 508–516.
6. Newsom-Davis, J., G. Harcourt, N. Sommer, D. Beeson, N. Willcox & J. B. Rothbard. 1989. J. Autoimmunity **2**(Suppl.) 101–108.

T-Cell Receptor Gene Usage of Acetylcholine Receptor-specific T-Helper Cells

A. MELMS,[a] J. R. OKSENBERG,[c] G. MALCHEREK,[a]
R. SCHOEPFER,[d] C. A. MÜLLER,[b] J. LINDSTROM,[e] AND
L. STEINMAN[c]

[a]Neuroimmunology Laboratory
Department of Neurology
[b]Section of Transplantation Immunology
University of Tübingen
D 7400 Tübingen, Germany

[c]Departments of Neurology and Neurological Sciences
Stanford University Medical Center
Stanford, California 94305-5235

[d]Zentrum für Molekulare Biologie Heidelberg
University of Heidelberg
D 6900 Heidelberg, Germany

[e]David Mahoney Institute of Neurological Sciences
University of Pennsylvania
Philadelphia, Pennsylvania 19104-6074

There is experimental evidence suggesting that autoimmune T lymphocytes employ only a restricted heterogeneity of T-cell receptor genes for autoantigen recognition.[1] This might be of potential interest for selective immunointervention in human autoimmune diseases. We wanted to study the basis for such a strategy in a well-defined autoimmune disease like myasthenia gravis (MG). We examined ten CD4$^+$ T-cell lines and clones which had been raised from four patients with myasthenia gravis (MG) and one healthy volunteer after *in vitro* stimulation with a recombinant polypeptide of the human acetylcholine receptor α-subunit, which represents the major autoimmune target in MG. Previously, these T cells had been characterized in terms of antigen specificity and MHC restriction.[2]

RNA was extracted and reverse-transcribed from 1 to 2 million AChR-specific T cells and amplified by the polymerase chain reaction. Amplification was performed in single reactions employing appropriate sets of oligonucleotide primers specific for 18 Vα and 21 Vβ gene families of the human T-cell receptor.[3-5]

Even though some T cells from different donors shared the antigen specificity or the HLA restriction element, AChR-specific T cells used individually different sets of both Vα and Vβ gene segments of their T-cell receptor genes. Certain Vα or Vβ gene families (such as Vα8 or Vβ12) were detected more frequently, but there was no association between TCR genes and MHC class II molecules serving as restriction elements or AChR epitope specificity (TABLE 1).

These results demonstrated that there is a restricted but individual heterogeneity among TCR V$\alpha\beta$-gene transcripts employed in autoantigen recognition. These findings indicate that a highly selective immunotherapy directed at the T-cell

TABLE 1. T-Cell Receptor Gene Usage of AChR-specific T Cells

Subject Age/Sex	T-cell Line	AChR α-subunit Specificity	HLA Restriction	T-cell Receptor Family	
				Vα	Vβ
Patient 1 19/F	line 6	α1-103	DR3	α8	β7
	clone 5.2	α1-103	DR3	α8	β7
	clone 5.21	α1-103	DR3	α8	β7
Patient 2 20/M	line 1.5	α327-398	DRw11	α1	β6
Patient 3 49/M	line C2	α238-327 or α398-437[a]	DRw11	α8, (α5)[b]	β2, (β5.1)[b]
	line C4	α1-103	DRw11	α5, (α8)[b]	β5.1
Patient 4 65/M	line 1.5	not tested	DR2	α2, 3, 8	β3
Healthy control 34/M	line 1	α1-103	DRw8	α2, 16	β2, 12, 18
	line 6	α1-103	DRw8	α2, 7	β6, 12, 18
	clone 14	α1-103	DRw8	α3, 10	β12

[a] Subunit region determinated by exclusion.
[b] Weak signal.

receptor will probably be more complicated or less efficient than previously anticipated.

REFERENCES

1. ACHA-ORBEA, H., D. J. MITCHELL, L. TIMMERMANN, D. C. WRAITH, G. S. TAUSCH, M. K. WALDOR, S. S. ZAMVIL, H. O. MCDEVITT & L. STEINMAN. 1988. Cell **54:** 263–273.
2. MELMS, A., G. MALCHEREK, U. GERN, H. WIETHÖLTER, C. A. MÜLLER, R. SCHOEPFER & J. LINDSTROM. 1992. Ann. Neurol. **31:** 311–318.
3. CHOI, Y., B. KOTZIN, L. HERRON, J. CALLAHAN, P. MARRACK & J. KAPPLER. 1989. Proc. Natl. Acad. Sci. USA **86:** 8941–8945.
4. BRAGADO, R., P. LAUZURICA, D. LOPEZ & J. A. LOPEZ DE CASTRO. 1990. J. Exp. Med. **171:** 1189–1204.
5. WUCHERPFENNIG, K. W., K. OTA, N. ENDO, J. G. SEIDMAN, A. ROSENZWEIG, H. L. WEINER & D. A. HAFLER. 1990. Science **248:** 1016–1019.

Production of Interferon Gamma and Interleukin-2 by Peripheral Blood Lymphocytes of Patients with Myasthenia Gravis and Other Autoimmune Diseases[a]

F. MOKHTARIAN, D. SHIRAZIAN, AND D. GROB

Division of Immunology
Department of Medicine
Maimonides Medical Center and
State University of New York
Health Science Center
Brooklyn, New York 11219

INTRODUCTION

The production of interferon gamma (IFN-g) and interleukin-2 (IL-2) by T-cell lines established from the peripheral blood of patients with myasthenia gravis (MG), other autoimmune diseases (OAD), and normal controls, in the absence of AChR and in the presence of IL-2, was investigated. Production, after only one week in culture, of IFN-g and IL-2 by T cells of patients with MG was lower than by T cells of OAD and normal controls. Later, however, the production of IFN-g and IL-2 increased and was significantly higher by T cells of MG and OAD patients than by normal controls. An immunoregulatory defect may be operating in the T cells of MG and OAD patients that results in an outburst of activation and lymphokine production, which may play a role in the pathogenesis of these diseases.

Myasthenia gravis is thought to be an autoimmune disease primarily mediated by antibodies to the acetylcholine receptor (AChR).[1] The severity of disease, however, does not correlate well with serum antibody levels.[1] Many studies, including ours, have found that T cells from the peripheral blood of MG patients respond, by lymphoproliferation, to AChR or to AChR synthetic peptides, to a greater extent than T cells from normal subjects.[2–7] These lines, however, were all obtained by *in vitro* culture of T cells with antigens. Thus, the specific function of these T cells, and their involvement in the regulation of autoantibody production, has not been fully elucidated. The lack of correlation between antibody titer and disease severity suggests that other factors, such as cytokine production, may play an important role in determining the severity of the disease produced.

[a] This work was supported by NIH Grant NS24688 and by a Grant from Maimonides Research and Development Foundation.

METHODS

Studies were carried out in eight patients with MG, of whom one had mild weakness at the time of study, five had moderate weakness, and two were severely weak, with difficulty breathing, swallowing, and getting out of bed. The patients' ages ranged from 25 to 84 years, and the duration of disease ranged from 1 to 25 years. Serum antibodies to AChR were significantly elevated in all patients. Six patients manifested OAD: one with Guillain Barré syndrome, one with Sjögren's syndrome, one with lupus erythematosis, and three with chronic inflammatory myeloneuropathies. Eight healthy staff members served as normal controls. Age and sex distributions were comparable in patients with MG and OAD, and in normal controls. Mononuclear cells were isolated from PBL of patients with MG and OAD and from normal controls by Lymphoprep (Accurate Chemical, Westbury, NY) density gradient centrifugation, and 2×10^5 cells/well were then cultured for determination of lymphokine production in 0.2 mL of standard RPMI medium containing 10% FCS, HEPES buffer, and 15 units of IL-2 per milliliter. At weekly intervals, supernatants were harvested for determination of lymphokines (IFN-g and IL-2). IFN-g and IL-2 were measured by a solid-phase enzyme immunoassay (ELISA), employing the multiple-antibody sandwich principle (Genzyme).

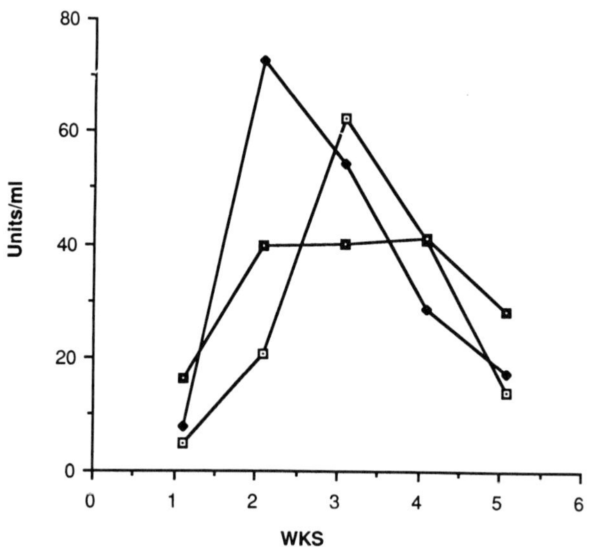

FIGURE 1. IFN-gamma production (units/mL) by IL-2–activated PBL lines from patients with myasthenia gravis (MG) and other autoimmune diseases (OAD), and from normal subjects (CTL) at 5 weekly intervals. Wk 1, MG < CTL ($p < 0.03$) and OAD < CTL ($p < 0.01$). Wk 2, MG < CTL ($p < 0.03$) and OAD > CTL ($p < 0.03$). Wk 3, MG and OAD > CTL ($p < 0.03$). Wk 5, MG and OAD < CTL ($p < 0.02$). —□—, MG; —■—, OAD; —■—, CTL.

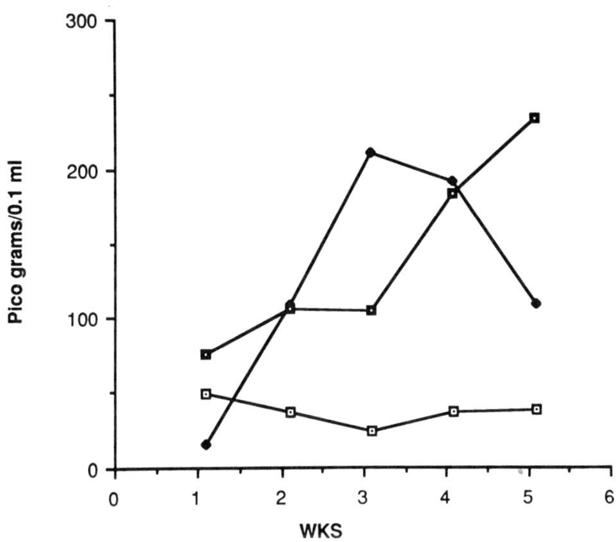

FIGURE 2. IL-2 production (pg/mL × 10^{-1}) by IL-2–activated PBL lines from patients with MG and OAD, and from normal subjects (CTL) at 5 weekly intervals. Wk 1, MG < CTL ($p < 0.05$). Wk 2, MG and OAD > CTL ($p < 0.05$). Wk 3, MG > CTL ($p < 0.01$) and OAD > CTL ($p < 0.03$). Wk 4, MG and OAD > CTL ($p < 0.01$). Wk 5, MG > CTL ($p < 0.02$) and OAD > CTL ($p < 0.001$). —☐—, CTL; —■—, MG; —■—, OAD.

RESULTS

IFN-g

Production, after one week in culture, of lymphocytes from MG and OAD patients was significantly lower than from normal controls ($p = 0.01$ for MG and $p = 0.03$ for OAD) (FIG. 1). This level of lymphokine production reached its peak, by lymphocytes from normal controls in 2 weeks, from patients with OAD in 2 weeks, and in MG patients in 3 weeks and gradually decreased in the fourth and fifth weeks. In cultures of lymphocytes from normal controls, IFN-g production increased after the first week, did not significantly change during weeks 2–4, and declined in the fifth week. IFN-g produced by lymphocytes of MG and OAD patients was substantially greater than from normal controls at the end of the third week of culture ($p < 0.03$ and $p = 0.03$, respectively) (FIG. 1). By the fifth week, production of IFN-g by lymphocytes of MG and OAD patients fell to significantly lower levels than that by lymphocytes of normal controls ($p < 0.05$ and $p = 0.02$, respectively).

IL-2

Production by lymphocytes of normal controls changed little during the 5-week culture, whereas that of MG and OAD patients increased significantly (FIG. 2).

Production by lymphocytes of MG patients reached a peak at the end of 3 weeks, gradually declined in weeks 4 and 5, and was significantly higher than production by lymphocytes of normal controls during the second through the fifth weeks of culture. IL-2 production by lymphocytes of OAD patients steadily increased during the 5 weeks tested and was also significantly higher than production by lymphocytes of normal controls from the second ($p < 0.05$) through the fifth weeks of culture ($p < 0.002$) (FIG. 2).

In conclusion, an immunoregulatory defect may be operating in the T cells of MG and OAD patients, which results in an outburst of activation and lymphokine production that may play a role in the pathogenesis of MG and OAD.

REFERENCES

1. LINDSTROM, J. M., A. G. ENGEL, M. E. SEYBOLD, V. A. LENNON & E. H. LAMBERT. 1976. J. Exp. Med. **144:** 739–753.
2. RICHMAN, D., J. PATRICK & B. G. W. ARNASON. 1976. N. Engl. J. Med. **294:** 694–698.
3. BROOKE, S., C. BRAUTBAR, L. STEINMAN, O. ABRAMSKY, J. ROTHBARD, D. NEUMANN, S. FUCHS & E. MOZES. 1988. J. Clin. Invest. **82:** 1894–1900.
4. MOKHTARIAN, F., M. PINO, W. OFOSSU-APPIAH & D. GROB. 1990. J. Clin. Invest. **86:** 2099–2108.
5. HOHLFELD, R., K. V. TOYKA, L. L. MINER, S. L. WALGRAVE & B. M. CONTI-TRONCONI. 1988. J. Clin. Invest. **81:** 657–660.
6. HARCOURT, G. C., N. SOMMER, J. ROTHBARD, H. N. A. WILLCOX & J. NEWSOM-DAVIS. 1988. J. Clin. Invest. **82:** 1295–1300.
7. MELMS, S., B. C. G. SCHALKE, T. KIRCHNER, H. K. MULLER-HERMELINK, E. ALBERT & H. WEKERLE. 1988. J. Clin. Invest. **81:** 902–908.

Cells from Mice Exposed Chronically to D-Penicillamine Show Proliferative Responses to D-Penicillamine–Treated Self (Macrophage/Dendritic Cells): A Graft-versus-Host Response?[a]

AUDREY S. PENN AND JEFFREY J. JACQUES[b]

Department of Neurology
College of Physicians and Surgeons of Columbia University
630 West 168th Street
New York, New York 10032

INTRODUCTION

There are still few clues to the mechanisms that trigger the autoimmune response to acetylcholine receptor (AChR) that culminates in myasthenia gravis. Inasmuch as deliberate immunization with AChR produces similar electrophysiological and serological phenomena in experimental animals, at times associated with overt weakness, tolerance to AChR itself or to similar epitopes on other tissues, must be broken. D-Penicillamine (β-mercaptovaline), which is used successfully to treat rheumatoid arthritis, produces MG in susceptible individuals of specific MHC backgrounds.[1] We have shown that this sulfhydryl-reducing agent will reduce and bind to isolated purified *Torpedo* AChR.[1] The bound D-penicillamine (D-Pen) molecules may serve as haptens and trigger an immune response to AChR epitopes directly. Observations, however, that T cells can be specifically sensitized to spleen cells modified by D-Pen suggests that this could represent the initiating step with subsequent cross-reaction with D-Pen–modified AChR.[2]

The syngeneic or autologous mixed lymphocyte reaction (MLR) is the proliferation that occurs when T cells are cultured with autologous or syngeneic non–T cells and respond to self Ia molecules.[3] This is similar to the graft-versus-host reaction in which grafted T cells attack allogeneic MHC antigens of the host. An autologous MLR also occurs after modification of monocytes with the trinitrophenyl group.[4] Among monocytes, the dendritic cell (DC), a strongly Ia-positive accessory cell, has been shown to be extremely effective in triggering an MLR.[5] We have shown previously that lymphocyte proliferative responses to AChR occur in mice of Ia^k background that have been exposed chronically to D-Pen.[6] In the experiments reported here, we tested the response of lymph node and spleen cells from C3H strain mice exposed to D-Pen HCl for 8–11 weeks to a dendritic cell–enriched fraction pretreated with D-Pen from C3H spleen. We found a significantly increased MLR, indicating that chronic D-Pen exposure sensitized these mice to D-Pen presented on a population enriched in Ia-bearing DC.

[a] This work was supported by NIH Grant NS-17904.

[b] J. J. Jacques is a Viets Fellow of the Myasthenia Gravis Foundation.

MATERIAL AND METHODS

Eight- to 12-week-old C3Heb/FeJ (Ia^k) mice were given intraperitoneal D-Pen HCl (40 mg/kg) for 8 to 11 weeks. Single cell suspensions were prepared from spleen and popliteal lymph nodes (PLN), removed 9 days after administration of CFA in saline into 1 hind foot-pad and intraperitoneal (ip), to serve as responders. Dendritic cells were enriched by isolation of 18-hour nonadherent cells from collagenase-treated, RBC-depleted C3H spleens. This cell fraction contained low-density ($p < 1.062$ g/mL), Ia^+, Mac-1 (antimurine and human macrophage) negative cells, which spread and elaborated long projections when allowed to re-adhere on glass (FIG. 1). These cells were treated with 1.35 mM D-Pen, during the 18-hour adherence step, to serve as stimulators. After collection at 350 g, they were washed twice before use. Stimulating and responding cell populations were co-cultured for 72 hours and were pulsed with 1 μC [^{3}H]thymidine 6 hours before harvesting onto glass fiber paper for scintillation counting. All cells were in Click's medium prepared without 2-mercaptoethanol to prevent loss of D-Pen bound by disulfide bond formation.

RESULTS

Our results (TABLE 1) indicate that both PLN and spleen cells from mice chronically exposed to D-Pen proliferated in culture when exposed to D-Pen–

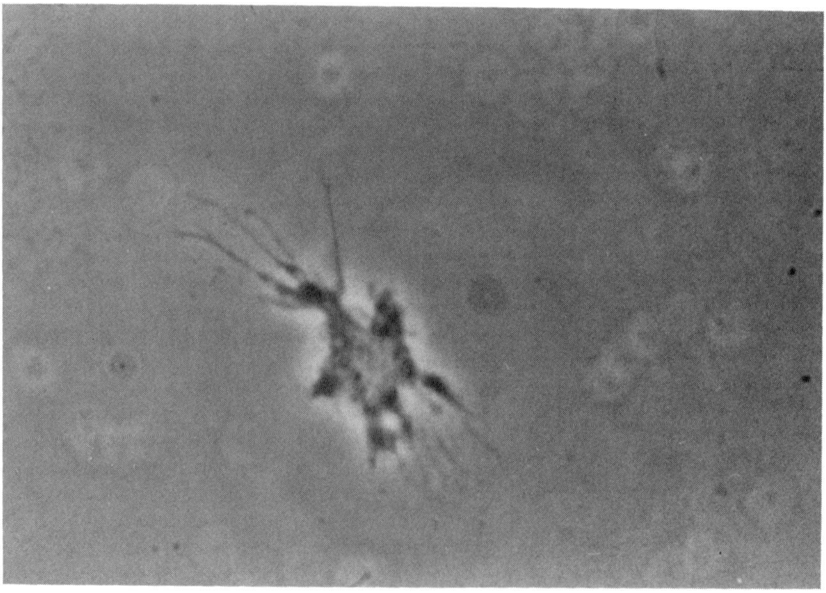

FIGURE 1. Dendritic cell in 18-hour nonadherent population of C3Heb/FeJ mouse spleen spreading and elaborating processes upon readherence to glass. Phase optics; ×1440.

TABLE 1. Mixed Lymphocyte Reaction: Proliferative Responses of Cells from D-Penicillamine–Treated C3H Mice

	[^{3}H]Thymidine incorporation: cpm × 10^{-3} (S.I.)[c] Responding cells from			
	D-Pen Rx mice		Saline Rx mice	
	S[a]	PLN[b]	S[a]	PLN[b]
Stimulation				
#1				
Medium	0.26 ± 0.05	0.18 ± 0.03	0.02 ± 0.03	0.2 ± 0.05
D-Pen DC[d]	1.14 ± 0.01(4.4)	0.84 ± 0.1(4.7)	0.75 ± 0.06	0.4 ± 0.07
Con A	138.09 ± 6.8	122.09 ± 5.4	99.07 ± 14.6	19.4 ± 9.01
#2				
Medium	1.01 ± 0.1	0.37 ± 0.05	0.09 ± 0.2	0.8 ± 0.01
D-Pen DC[d]	5.02 ± 1.2(5.1)	3.00 ± 0.3(8.1)	3.01 ± 0.2	1.8 ± 0.02
Con A	51.06 ± 4.4	96.08 ± 6.5	45.05 ± 0.5	92.7 ± 3.02
#3				
Medium	0.45 ± 0.02	0.2 ± 0.05	0.38 ± 0.07	0.37 ± 0.01
D-Pen DC[d]	1.2 ± 0.02(2.7)	1.9 ± 0.15(12.8)	0.86 ± 0.05	0.62 ± 0.05
Con A	60.8 ± 7.07	144.5 ± 18.05	70.05 ± 5.09	108.2 ± 4.03

[a] Spleen cells.
[b] Popliteal lymph node cells.
[c] Stimulation index (cpm, experimental/cpm control).
[d] 18-hour nonadherent fraction of D-Pen–treated (1.35 mM) syngeneic spleen cells in 2-mercaptoethanol–free medium.

treated DC-enriched syngeneic cells. The responses of cells from control mice reflects response to Ia antigens on the DC. The significantly greater response of D-Pen–exposed mice indicates that cells from these mice had been sensitized to D-Pen *in vivo,* and responded when reexposed to D-Pen. This amplified syngeneic MLR, with D-Pen–treated DC as stimulators, suggests that either D-Pen is recognized as a haptenic determinant in association with the Ia molecules to which the MLR is directed, or that D-Pen actually modifies Ia molecules. Class II MHC molecules have at least two readily reducible disulfide bonds that can be reduced and then bound by D-Pen in a manner similar to those of AChR.[7] D-Pen administration is associated with graft-versus-host–like autoimmune phenomena in certain individuals.[3] It is also associated with the development of myasthenia in individuals of certain HLA backgrounds, of augmented antibody titers to AChR in A/J and C3H mice, which share Ia^k, and of proliferative responses to AChR in Ia^k mice.[8] A contribution of an initial autologous MLR directed at D-Pen associated with Ia antigens to these autoimmune responses seems entirely possible. The presence of excessive numbers of HLA-DR positive macrophage/dendritic cells in myasthenic thymuses,[9] which may also be involved in the intrathymic immune activity of these glands, makes our current findings more intriguing.

REFERENCES

1. BEVER JR., C. T., H. W. CHANG, A. S. PENN, I. A. JAFFE & E. BOCK. 1982. Chemical alteration of acetylcholine receptor by penicillamine: A mechanism for induction of myasthenia gravis. Neurology **32:** 1077–1082.
2. GLIMCHER, L. H., D. L. LONGO, I. GREEN & R. H. SCHWARTZ. 1981. Murine syngeneic

mixed lymphocyte response. Target antigens are self Ia antigens. J. Exp. Med. **154:** 1652–1670.

3. NAGATA, N., U. HURTENBACH & E. GLEICHMANN. 1986. Specific sensitization of LYT-1+2− T cells to spleen cells modified by the drug D-penicillamine or a stereoisomer. J. Immunol. **136:** 136–142.
4. SIEGEL, R. L., E. J. YUNIS & R. S. GEHA. 1982. Nature of the immunogenic moiety recognized by the human T cell proliferating in response to trinitrophenyl-modified self-antigens. Hum. Immunol. **4:** 287–299.
5. NUSSENZWEIG, M. & R. M. STEINMAN. 1980. Contribution of dendritic cells to stimulation of the murine syngeneic mixed leukocyte reaction. J. Exp. Med. **151:** 1196–1212.
6. PENN, A. S., M. ASCIONE-LAMBERT & A. E. EGGERS. 1986. D-penicillamine initiates a T cell response to acetylcholine receptor. Ann. Neurol. **20:** 162.
7. SPRINGER, T. A., R. J. ROBB, C. TERHORST & J. L. STROMINGER. 1977. Subunit and disulfide structure of monomeric and dimeric forms of detergent-soluble HLA antigens. J. Biol. Chem. **252:** 4694–4700.
8. BEVER, JR., C. T., K. L. DRETCHEN, G. J. BLAKE, H. W. CHANG, A. S. PENN & R. ASOFSKY. 1984. Augmented anti-acetylcholine receptor responses following chronic penicillamine administration. Ann. Neurol. **16:** 9–13.
9. PENN, A. S., A. JARETZKI III & M. WOLFF. 1990. Excess Ia-positive macrophage/dendritic cells in myasthenic thymus glands. Neurology **40:** 349.

Response of Peripheral Blood Mononuclear Cells from Myasthenia Gravis Patients to Exogenous Interleukin-2

SAMIA RAGHEB,[a] MARIA FEDERICA FERRIO,[b]
AND ROBERT P. LISAK[a]

[a]*Department of Neurology*
Wayne State University
Detroit, Michigan 48201

[b]*Department of Neurology*
University of Turin
Turin, Italy

Acquired myasthenia gravis (MG) is an autoimmune disease in which antibodies are directed to self-acetylcholine receptor at the neuromuscular junction.[1] Both B and T lymphocytes participate in the immunopathogenesis of MG. Activated lymphocytes have an increased expression of cell-surface interleukin-2 (IL-2) receptor (CD25) and transferrin receptor (CD71).[2] To determine whether peripheral blood mononuclear cells (PBMC) from patients with MG were in an *in vivo* state of activation, CD25 and CD71 expressions were measured for B and T lymphocytes. In addition, the *in vitro* sensitivity of PBMC to exogenous recombinant human IL-2 was measured. PBMC from sex- and age-matched normal controls were also studied for comparison. TABLE 1 shows the profiles of the two populations studied.

PBMC were isolated from MG patients and controls. The phenotypic expression of cell-surface markers was determined by immunofluorescence; the percentage of positively stained lymphocytes was determined by FACS. The frequency of total T cells was based on CD2 expression, and the frequency of total B cells was based on CD20 expression. For patients with MG, the mean frequency ± SD of B cells that expressed CD25 was 2.0 ± 4.4; for B cells that expressed CD71, the mean frequency was 20.4 ± 12.6 (control B cells: CD25 = 3.6 ± 5.3, CD71 = 15.5 ± 15.2). For patients with MG, the frequency of T cells that expressed CD25 was 0.6 ± 0.6; for T cells that expressed CD71, the frequency was 4.7 ± 11.0 (control T cells: CD25 = 0.7 ± 0.7, CD71 = 2.6 ± 3.0). There were no significant differences in the expression of CD25 or CD71 on either B or T lymphocytes between MG patients and normal controls.

IL-2 sensitivity was determined by measuring the *in vitro* proliferative response of PBMC to exogenous recombinant human IL-2. During initial experiments, we determined that IL-2 concentrations greater than 100 U/mL were inhibitory for PBMC proliferation. Therefore, PBMC (2×10^5 cells) were cultured with IL-2 (1, 5, 10, 25, and 50 U/mL) or control medium for a total of 2, 4, or 7 days. During the last 18–24 hours of the culture period, cells were pulsed with tritiated thymidine. If PBMC from patients with MG were more activated *in vivo* than PBMC from controls, then a short-term culture with a suboptimal IL-2 concentration should reflect any differences between the two groups. Data from the

TABLE 1.

	MG Patients n = 23	Normal Controls n = 18
Males	11	8
Females	12	10
No Therapy	4	18
Therapy	19	0
Therapy for MG patients:		
Pyridostigmine bromide	18	
Plasmapheresis	3	
Thymectomy	7	
Prednisone	4	
Azathioprine	4	

IL-2–dependent proliferation assays were analyzed in terms of net cpm, Δ cpm, and stimulation indices. Patient PBMC had an increased background proliferation at each time point tested compared to PBMC from controls. More importantly, patient and control PBMC did not differ in their IL-2–dependent proliferative response at any IL-2 concentration or at any length of culture. When patients with MG were divided into two groups based on treatment, we found that PBMC from patients who were not treated with immunomodulatory therapy (n = 12, patients receiving pyridostigmine bromide or no therapy at all were included in this group) had an increased *in vitro* IL-2 sensitivity compared to treated patients (n = 11), but they were still not significantly different from controls.

In conclusion, MG patients do not appear to have an increased frequency of $CD25^+$ or $CD71^+$ T or B cells. This does not rule out the possibility that, between MG patients and controls, differences may exist in the density of these markers or in their affinity for their ligands. Finally, for individual patients or controls, the magnitude of the IL-2–dependent *in vitro* proliferative response did not correlate with CD25 expression on either B or T cells.

REFERENCES

1. LINDSTROM, J., D. SHELTON & Y. FUJII. 1988. Myasthenia gravis. Adv. Immunol. **42:** 233–284.
2. LEONARD, W. J. 1986. The human interleukin-2 receptor. Annu. Rev. Immunol. **4:** 69–95.

Acetylcholine Receptor Presentation by B Cells Using Heterobifunctional Antibody Conjugates[a]

JOHANNES REIM,[b] KEVIN R. McINTOSH,
AND DANIEL DRACHMAN

Department of Neurology
School of Medicine
The Johns Hopkins University
600 North Wolfe Street
Baltimore, Maryland 21205

INTRODUCTION

Helper T cells recognize processed peptides derived from antigens, presented in association with self-MHC class II antigens. B cells can serve as potent antigen-presenting cells (APC) when antigen is targeted to B-cell surface structures.[1-4] We have targeted acetylcholine receptor (AChR) to surface immunoglobulin (sIg) using heterobifunctional conjugates consisting of antibodies directed against B-cell sIg and monoclonal antibody directed against AChR.

In this study we have assessed the effectiveness of conjugate plus AChR-treated B cells (AChR-APC) in targeting AChR-specific T cells. Time-course experiments were performed to assess the difference in time of AChR uptake, processing, and presentation by B-cell preparations treated with conjugate followed by AChR, as compared with control cells treated with AChR alone.

METHODS

Preparation of Antibody Conjugate

The *Torpedo* AChR-specific MAb 35 was conjugated to affinity-purified goat anti-rat IgM using the heterobifunctional coupling reagent SPDP. The resulting dimeric antibody conjugate was separated from unconjugated antibodies by gel filtration. Testing of the conjugate for each of the antibody components by ELISA assays showed that the conjugate had both anti-AChR and anti-IgM activities that were quantitatively nearly equivalent to those of the native (unconjugated) antibody components (data not shown).

[a] This work was supported in part by Grants from the Dana Foundation and the Muscular Dystrophy Association.

[b] Address for correspondence: Johannes Reim, Department of Biophysics and Biophysical Chemistry, The Johns Hopkins University, School of Medicine, 711 Hunterian Building, 725 N. Wolfe Street, Baltimore, Maryland 21205.

TABLE 1. Conjugate Enhances Antigen Presentation to AChR-primed T Cells[a]

	Lymphoproliferative response (Δ cpm)			
	[AChR] μg/mL × 2.5			
	10^{-5}	10^{-4}	10^{-3}	10^{-2}
APC with conjugate	250	200	1350	14200
APC without conjugate	0	0	200	250

[a] B cell–enriched APC treated with conjugate were subsequently incubated with AChR at the concentrations indicated and used to stimulate either AChR-primed T cells (TABLE 1) or a cycled AChR-specific T-cell line (TABLE 2). Control APC were not treated with conjugate. Data points are means of delta cpm in triplicate cultures.

Preparation of AChR-presenting B Cells (AChR-APC)

B cell–enriched, macrophage-depleted, lymph node cells (LNC) from Lewis rats were used as APC. The B cell–enriched LNC were adjusted to 5 million cells/mL, incubated with 4 μg/mL conjugate in culture medium for 1 hour at 4°C, and washed. The conjugate-treated cells were then incubated with AChR (concentrations up to 2.5 μg/mL) at 4°C for 1 hour. In some experiments, the AChR-APC were incubated at 37°C for the times indicated to allow antigen processing.

For stimulation experiments, conjugate-treated cells were distributed in 96-well plates at 250,000 cells/well, and the indicated concentrations of AChR in culture medium were added and incubated for 1 hour at 4°C. Responder cells (see below) were added to the wells and co-cultured as described.

Preparation of AChR-responsive T-Cell Populations

Two types of T-cell populations were used. (1) Fresh T cells: four to six weeks after immunization with AChR, draining inguinal lymph nodes from Lewis rats were removed and enriched for T cells by passage through a nylon wool column, treatment with a cocktail of mouse monoclonal antibodies directed against rat B cells and Ia-positive cells, and subsequent removal of these cells with magnetic beads. The resulting populations had <1% B cells, as determined by fluorescence microscopy. (2) "Cycled" T-cell lines: Lymphocytes prepared from AChR-primed lymph nodes from Lewis rats were stimulated in bulk cultures with AChR (1.25 μg/mL) and subsequently expanded in IL-2–containing medium. After two cycles of stimulation and expansion, immunofluorescence using OX3 MAb showed less

TABLE 2. Conjugate Enhances Antigen Presentation to a Cycled AChR-specific T-Cell Line[a]

	Lymphoproliferative response (Δ cpm × 10^3)			
	[AChR] μg/mL × 2.5			
	10^{-6}	10^{-4}	10^{-3}	10^{-2}
APC with conjugate	16	23	60	105
APC without conjugate	0	16	14	17

[a] See footnote to TABLE 1.

than 1% Ia-positive cells. These cycled cells proliferated in response to AChR only in the presence of added APC.

Lymphoproliferation Assays

AChR-specific T cells (125,000/well) were stimulated by culture with 250,000 AChR-APC, using concentrations of AChR as described for each experiment. For the final 18 hours, [^{3}H]thymidine (1 μCi/well) was added. [^{3}H]thymidine

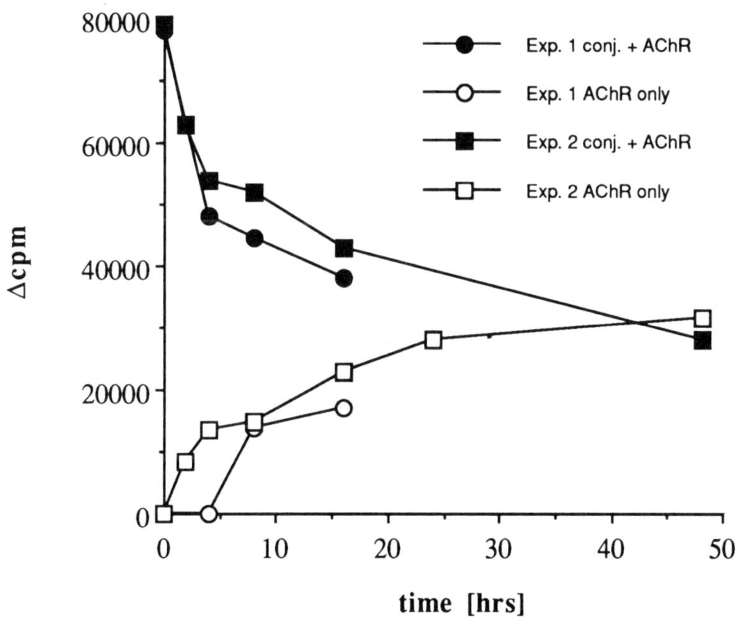

FIGURE 1. Conjugate accelerates AChR presentation to AChR-specific T cells. The results of two experiments are shown. B cell–enriched APC treated with conjugate at 4°C were subsequently incubated with AChR at 37°C for the times indicated to allow antigen processing. Control APC were incubated with AChR only for the times indicated. The conjugate-treated or control APC were then used to stimulate cycled, AChR-specific T cells. Data points are means of delta cpm in triplicate cultures.

incorporation was determined by liquid scintillation counting. Results are expressed as Δ cpm, after subtraction of background responses by control (unstimulated) cultures.

RESULTS AND DISCUSSION

Our results show that B cell–enriched cell preparations treated with conjugate plus AChR present AChR 100- to 1000-fold more effectively than preparations treated with AChR alone, as evidenced by lymphoproliferative responses of both AChR-primed T cells (TABLE 1) and AChR-specific T-cell lines (TABLE 2). Time-course experiments (FIG. 1) indicate a marked acceleration of AChR uptake, processing, and presentation by the B cell–enriched APC treated with conjugate followed by AChR, as compared with control cells treated with AChR alone.

The pathogenesis of myasthenia gravis involves a T-cell dependent, antibody-mediated autoimmune response directed against AChRs. Our results demonstrate that AChR-specific T cells are targeted effectively by conjugate plus antigen-treated, syngeneic B cells serving as AChR antigen-presenting cells. Our goal, however, is to eliminate the anti-AChR response specifically. We have induced antigen-specific "anergy" of AChR-specific T cells by targeting them with fixed APC.[5,6] Inactivation of AChR-specific T cells should down-regulate the anti-AChR immune response, resulting in therapeutic benefit. In experiments described in detail elsewhere,[7] unresponsiveness of AChR-specific T cells was induced *in vitro* by culturing AChR-specific T cells with paraformaldehyde-fixed AChR-APC.

REFERENCES

1. SNIDER, D. P. & D. M. SEGAL. 1987. Targeted antigen presentation using crosslinked antibody heteroaggregates. J. Immunol. **139:** 1609–1616.
2. SNIDER, D. P. & D. M. SEGAL. 1989. Efficiency of antigen presentation after antigen targeting to surface IgD, IgM, FcRII, and B220 molecules on murine splenic B cells. J. Immunol. **143:** 59–65.
3. SNIDER, D. P., A. KAUBISCH & D. M. SEGAL. 1990. Enhanced immunogenicity induced by bispecific antibodies. J. Exp. Med. **171:** 1957–1963.
4. CASTEN, L. A. & S. K. PIERCE. 1988. Receptor mediated B cell antigen processing. J. Immunol. **140:** 404–410.
5. JENKINS, M. K. & R. H. SCHWARTZ. 1987. Antigen presentation by chemically modified splenocytes induces antigen-specific T cell unresponsiveness *in vitro* and *in vivo*. J. Exp. Med. **165:** 302–319.
6. GILBERT, K. M., K. D. HOANG & W. O. WEIGLE. 1990. Th1 and Th2 clones differ in their response to a tolerogenic signal. J. Immunol. **144:** 2063–2071.
7. REIM, J., K. MCINTOSH, S. MARTIN & D. DRACHMAN. 1992. Specific immunotherapeutic strategy for myasthenia gravis. J. Neuroimmunol. **41:** 61–70.

Molecular Immunopathogenesis of Myasthenia Gravis Using MHC Class II Mutant and Transgenic Mice[a]

MOHAN SHENOY,[b] CHELLA DAVID,[c]
MINAKO OSHIMA,[d] M. ZOUHAIR ATASSI,[d]
AND PREMKUMAR CHRISTADOSS[b,e]

[b]*The Department of Microbiology*
University of Texas Medical Branch
Galveston, Texas 77555-1019

[c]*The Department of Immunology*
Mayo Clinic
Rochester, Minnesota

[d]*The Department of Biochemistry*
Baylor College of Medicine
Houston, Texas 77030

Experimental autoimmune myasthenia gravis (EAMG) susceptibility has been linked to the MHC, and recombinant analysis has mapped the gene to the I-A subregion.[1] In order to study the intricate association of MHC class II molecule with acetylcholine receptor (AChR) epitopes and their recognition by the T-cell receptor (TCR), we have been studying the trimolecular complex (MHC + AChR + TCR) interaction in the development of EAMG in mice.

The Effect of bm12 Mutation

The I-Aβ chain mutant bm12 was screened to study the role of I-Aβ chain in EAMG pathogenesis. A gene conversion event between I-Eβ^b and I-Aβ^b in the bm12 strain altered three amino acids in the C-terminal half of the first domain of I-Aβ^b of B6 mice (Ile-67 $\rightarrow$ Phe; Arg-70 $\rightarrow$ Gln; Thr-71 $\rightarrow$ Lys). B6 mice (I-A^b) immunized with AChR in complete Freund's adjuvant (CFA) developed autoimmunity to AChR and muscle weakness. By contrast, bm12 mutant mice showed a reduced autoimmune response to AChR and failed to develop muscle weakness. Therefore, a mutation in the bm12 I-Aβ chain at positions 67, 70, and 71 prevented the disease.[2] Following this study, Bell *et al.* reported a linkage of HLA-DQβ chain (mouse analogue of I-Aβ chain) polymorphism in human MG.[3]

[a] This work was supported by the MDA.

[e] Address correspondence to Premkumar Christadoss, MD, Department of Microbiology, 3.108 Medical Research Building, Rt. J19, University of Texas Medical Branch, Galveston, TX 77555-1019.

The Effect of $A\alpha^k$ Transgene and $A\beta^b : A\alpha^k$ Hybrid Molecule

To study the contribution of I-Aα^k gene (EAMG resistant gene) and Aβ^b : Aα^k transpair in EAMG pathogenesis, we screened B10.Aα^k transgenic mice.[4] The I-Aα^k chain paired with the endogenous I-Aβ^b chain, and the Aβ^b : Aα^k hybrid molecules, were expressed on the cell surface of B10.Aα^k transgenic mice. The I-Aα^k expression in I-Aα^{k+} mice have also positively selected Vβ11$^+$ TCR.[4] The expression of Aβ^b : Aα^k transpair in B10.Aα^k transgenic mice suppressed both the cellular and humoral autoimmune responses to AChR and reduced the incidence of clinical MG.[4]

The Effect of I-E Molecule in $E\alpha^k$ Transgenic Mice

Mice of the H-2^b haplotype have a genomic deletion of the I-Eα gene and, therefore, fail to express the I-E molecule on their cell surface. To test the hypothesis of whether failure to express I-E molecule on the cell surface contributed to EAMG pathogenesis, I-Eα^k transgenic B10 mice expressing cell surface I-E molecule were screened. The expression of the I-E molecule on the cell surface of B10 mice significantly reduced the incidence of muscle weakness and suppressed antibody production.[5] The resistance to EAMG observed in bm12,B10.Aα^k and B10.E$^+$ transgenic mice may be due to (1) clonal deletion and/or anergy, (2) effective suppression/regulation, and (3) inefficient binding and/or presentation of epitopes to T cells.

Interaction of I-Aβ Subunit with AChR-α Chain T-cell Epitopes

To study the effect of gene conversion at I-Aβ positions 67, 70, and 71 on the T-cell responses to epitopes of T-AChR-α subunit, B6 and bm12 mice were primed with *Torpedo* AChR (T-AChR) in CFA, and the profile of T-lymphocyte proliferation was determined for the 18 synthetic overlapping peptides encompassing the entire extracellular portion of amino acids of the T-AChR-α subunit.[6] A marked reduction in the proliferative response of T-AChR–primed bm12 lymphocytes to peptides α146-162 and α182-198 was observed, when compared with the corresponding response of B6 T-AChR–primed lymphocytes. The proliferative response to the other dominant AChR-α subunit, namely α111-126, was well preserved in bm12. Therefore, I-Aβ residues encompassing the region 67-71 determine the immunogenicity to at least two of the T-AChR-α subunit T-cell epitopes. Because bm12 resistance to EAMG correlates with reduced proliferative response to the otherwise immunodominant epitopes within α146-162 and α182-198 on the T-AChR-α subunit, the data implicated the importance of these two epitopes in EAMG pathogenesis. The region α146-162 within the AChR molecule may be involved in the generation of specific helper T cells that could provide help for the production of pathogenic antibodies against the same regions, or against other regions of the AChR molecule through T-T or T-B cell intersite influences (epitope linkages). In a recent study, we could significantly reduce the incidence of EAMG by neonatally tolerizing B6 mice with peptide α146-162, implicating the pathogenic role of this epitope in EAMG. The TCR utilization of α146-162 peptide–specific T cells is being evaluated. We believe that these studies should culminate in the development of peptide- or clone-specific therapy for EAMG and eventually MG.

REFERENCES

1. Christadoss, P. 1989. CRC Crit. Rev. Immunol. **9:** 247–278.
2. Christadoss, P., J. M. Lindstrom, R. W. Melvold & N. Talal. 1985. Immunogenetics **21:** 33.
3. Bell, J., S. Smoot, C. Newly, K. Toyka, L. Rassenti, K. Smith, R. Hohlfeld, H. McDevitt & L. Steinman. 1988. Lancet **i:** 1058–1060.
4. Christadoss, P., C. S. David & S. Keve. 1992. Clin. Immunol. Immunopathol. **62:** 235–239.
5. Christadoss, P., C. S. David, M. Shenoy & S. Keve. 1990. Immunogenetics **31:** 241–244.
6. Mulac-Jericevic, B. & M. Z. Atassi. 1987. J. Protein Chem. **6:** 365–373.

Regulation of Experimental Autoimmune Myasthenia Gravis by Synthetic Peptides of the Acetylcholine Receptor

MIRIAM C. SOUROUJON,[a,b] SARA CARMON,[a]
AND SARA FUCHS[a]

[a]*Department of Chemical Immunology*
The Weizmann Institute of Science
Rehovot 76100, Israel

[b]*The Open University of Israel*
Tel-Aviv 61392, Israel

A series of synthetic peptides corresponding to sequences of the acetylcholine receptor (AChR) were employed to identify possible antigenic determinants within the receptor that can modulate the anti-AChR response and experimental autoimmune myasthenia gravis (EAMG). Three *Torpedo* peptides (Tα73-89, Tα351-368, and Tδ354-367) were chosen because they seem to represent immunopotent regions of AChR.[1-3] The two peptides corresponding to residues 351-368 from the α-subunits of *Torpedo* and of human muscle receptors[4] were employed to compare the effects of foreign and mammalian peptides.

RESULTS

Rabbits preimmunized with peptides corresponding to sequences of the AChR, prior to injection with AChR, were protected against EAMG. Namely, six out of eight rabbits survived three inoculations of AChR and lived at least five months after the third injection with AChR, whereas all four control rabbits got sick and died shortly following one or two injections of AChR. Most of the protected rabbits exhibited transient EAMG that did not develop into the fully expressed, lethal form of the disease (FIG. 1).

The ability of specific synthetic peptides to protect against EAMG did not correlate with their ability to elicit a high anti-AChR antibody titer or to induce the production of antibodies reacting with the native receptor. On the other hand, there seemed to be a correlation between the protection against the lethal effect of AChR inoculation and a delayed reactivity with all four subunits of the receptor, as tested by Western blot analysis. Control rabbits that died shortly after the second injection of AChR developed antibodies to all AChR subunits as early as 13-17 days following AChR inoculation, whereas peptide-preimmunized protected rabbits exhibited the reactivity towards all subunits at day 24 or later (TABLE 1).

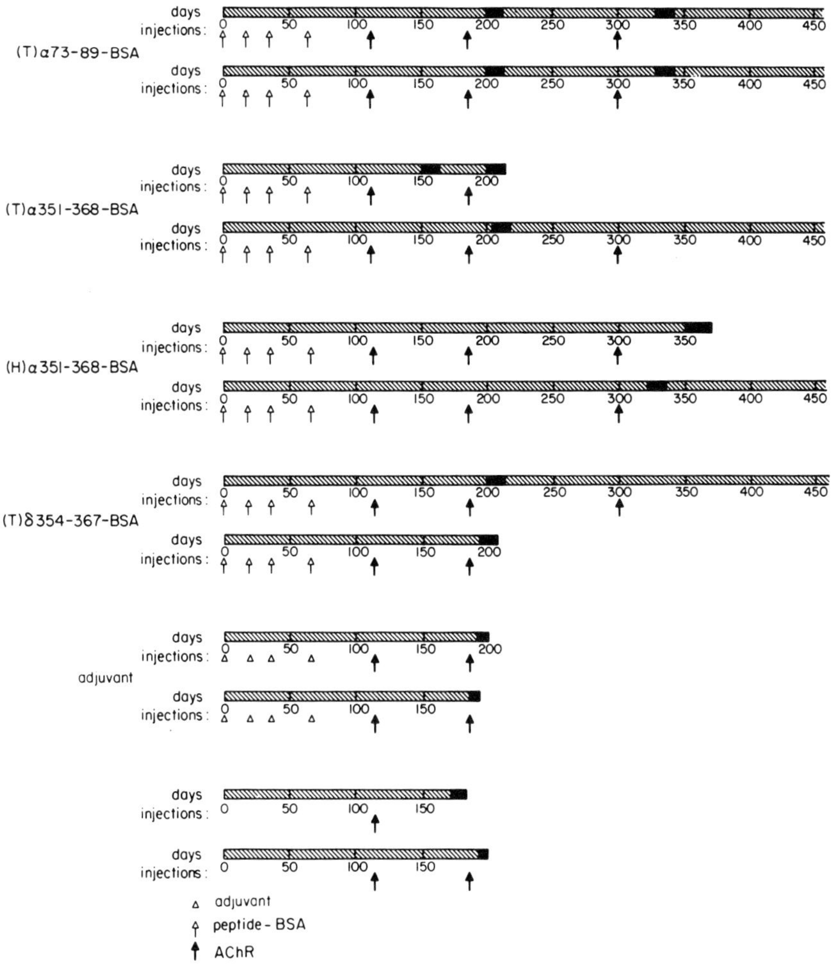

FIGURE 1. Effect of peptide preimmunization on EAMG. The FIGURE describes the immunization schedule and development of EAMG in rabbits preimmunized with peptide conjugates (open arrows) or adjuvant (open arrow heads), and challenged later with *Torpedo* AChR (solid arrows). Black areas indicate the manifestation of clinical signs of EAMG. The top and bottom lines for each immunogen represent two rabbits (designated i and ii, respectively, in TABLE 1).

CONCLUSION

Our results indicate that the reactivity with the entire *Torpedo* receptor molecule may be related to the loss of tolerance towards self-AChR that induces the autoimmune attack in EAMG. Peptide-preimmunization can possibly divert the anti-AChR response towards the nonpathogenic determinant represented by the

TABLE 1. Correlation between EAMG and Reactivity with AChR in Protection Experiment

Immunogen		Survival time[a] (days)	Reactivity with other subunits[b] (days)
Tα73-89-BSA	(i)[c]	>350	24–42
	(ii)	>350	24–42
Tα351-368-BSA	(i)	95	13–17
	(ii)	>350	42–62
Hα351-368-BSA	(i)	225	24–42
	(ii)	>350	42–64
Tδ354-367-BSA	(i)	>350	24–42
	(ii)	95	17–24
Adjuvant	(i)	90	13–17
	(ii)	85	13–17
—	(i)	65	13–17
	(ii)	90	13–17

[a] The immunization schedule is as described in FIG. 1. Survival time after the first AChR inoculation is recorded.

[b] Time of appearance of reactivity in Western blots with subunits other than the one including the immunizing peptide sequence.

[c] (i) and (ii) represent two different rabbits for each immunogen.

peptide, thereby delaying the pathological autoimmune process. Although the mechanism is unknown, we believe that these results are encouraging, in view of the possible application of AChR-specific synthetic peptides for the regulation of human myasthenia.

REFERENCES

1. SOUROUJON, M. C., D. NEUMANN, S. PIZZIGHELLA, A. SAFRAN & S. FUCHS. 1986. Biochem. Biophys. Res. Commun. **135:** 82–89.
2. SOUROUJON, M. C., D. NEUMANN, S. PIZZIGHELLA, M. FRIDKIN & S. FUCHS. 1986. EMBO J. **5:** 543–546.
3. BARKAS, T., A. MAURON, B. ROTH, C. ALLIOD, S. J. TZARTOS & M. BALLIVET. 1987. Science **235:** 77–80.
4. FUCHS, S., D. NEUMANN, A. SAFRAN, S. PIZZIGHELLA, R. MANTEGAZZA, M. P. DANIELS & Z. VOGEL. 1987. Biochemistry **26:** 4611–4616.

Effects of D-Penicillamine on Multiple Immunogenic Epitopes of the Acetylcholine Receptor[a]

QUN XU, ROBERT H. FAIRCLOUGH, AND
DAVID P. RICHMAN

Department of Neurology and Center for Neuroscience
University of California, Davis
Davis, California 95616

Some patients treated with D-penicillamine (D-PA) for rheumatoid arthritis develop myasthenia gravis (MG) that clears after discontinuation of the drug.[1] D-PA, an amino acid with a free sulfhydryl (FIG. 1), can potentially covalently react with protein disulfide bonds. We here describe a study of the effects of D-PA and two chemical analogues upon the binding of several antiacetylcholine receptor monoclonal antibodies (MAbs) to the acetylcholine receptor (AChR). The results suggest a "haptenization" of the AChR by D-PA, which Penn *et al.*[2] have shown directly inhibits the binding of AChR agonists, and which, we propose, alters the immunologic identity of the AChR. It is possible that a similar mechanism may be operative in the onset of idiopathic myasthenia gravis.

METHODS

Inhibition of Binding of Anti-AChR MAbs by D-PA

Polyvinyl 96-well microtiter plates were coated with 100 μL of 10^{-8} M toxin sites of membrane-bound *Torpedo* AChR. The AChR coating was treated with 10 mM D-PA, modified D-PA, serine (Ser), or L-cysteine (Cys) for 2 hours. Treated membranes were then titrated with anti-AChR MAbs. Bound MAbs were detected with goat anti-rat IgG conjugated to horseradish peroxidase, (HRP), and the assay was developed by adding HRP substrates H_2O_2 and orthophenylenediamine, followed by measurement of the optical density at 450 nm. D-PA was modified by treating with iodoacetamide ($IAcNH_2$) or *N*-ethylmaleimide (NEM) prior to its addition to the AChR. Both $IAcNH_2$ and NEM can alkylate -SH groups.

Binding Study of Cysteine and Iodoacetamide to the AChR

One millimolar [^{35}S]Cys at 9 cpm/pmole was incubated with membrane-bound AChR for three hours. Unreacted Cys was separated from covalently bound Cys

[a] This work was supported in part by Grants from the National Institutes of Health (NS15462, NS19779, NS24304, NS01226), the Muscular Dystrophy Society, and the Amyotrophic Lateral Sclerosis Society of America.

by Airfuge sedimentation of the membranes at 20 psi for five minutes followed by three washes with phosphate-buffered saline. [^{35}S]Cys pretreated with $IAcNH_2$ served as a control to determine ^{35}S non-specifically sedimenting with the membrane vesicles. ^{35}S covalently incorporated was counted in EcoScint scintillation solution with a Beckman LS5000TD scintillation counter. Covalent modification of the AChR with [^{14}C]$IAcNH_2$ was attempted following D-PA or L-Cys treatment. ^{14}C incorporation was determined as for [^{35}S]Cys.

RESULTS AND DISCUSSION

Reaction of AChR with D-PA enhances the binding of delta subunit–directed MAb 85E[3], partially inhibits the binding of main immunogenic region (MIR) directed MAbs 35 and 132,[3] and completely blocks the binding of agonist-blocking MAbs 370A[3] and 387D (FIG. 2a).

D-PA is a D-amino acid with an -SH group side chain similar to that of Cys (FIG. 1). To test the hypothesis that the -SH group of D-PA is the chemically

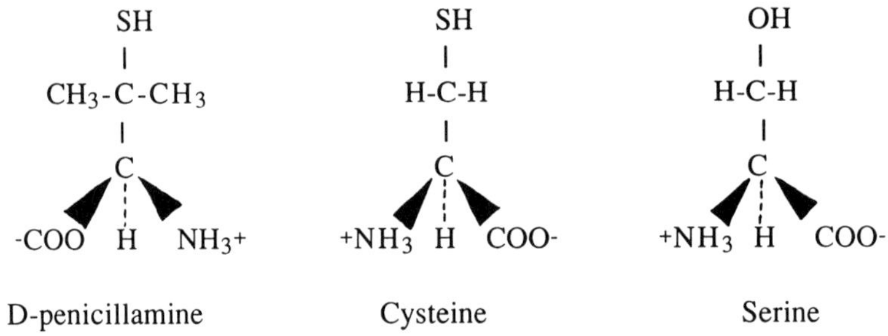

FIGURE 1. Structure of D-PA, Cys, and Ser. Apart from the enantiomeric difference, Cys lacks the two methyl groups of D-PA on the beta carbon atom. Like D-PA, Cys has a free sulfhydryl group. Ser differs from Cys in having an -OH group in place of the -SH.

active agent that covalently modifies anti-AChR MAb epitopes, we treated the AChR with D-PA, chemically modified D-PA, as well as with two analogues: Cys and Ser. When the free -SH groups of D-PA are specifically alkylated with $IAcNH_2$ or NEM, the modified D-PA does not affect MAb binding (FIG. 2b). Cys is the L-enantiomeric analogue of D-PA lacking the methyl groups on the beta carbon atom (FIG. 1). Treatment of the AChR with Cys inhibits MAb binding in a manner similar to that of D-PA (FIG. 2c). Treatment of AChR with Ser, which differs from Cys in having an -OH group in place of the -SH group, has no effect on the binding of MAbs (FIG. 2d). The latter result suggests that the amino and carboxyl groups of D-PA are not involved in inhibition of MAb binding. Taken together, these data suggest that the free -SH group of D-PA reacts with certain AChR disulfides and affects several anti-AChR MAb epitopes. The most dramatically affected epitopes are those of the agonist blocking MAbs, like 387D. This result suggests that the alpha-subunit Cys 192-193 disulfide near the agonist bonding site is a possible target of D-PA modification. To see whether 1 mM Cys acts simply to reduce

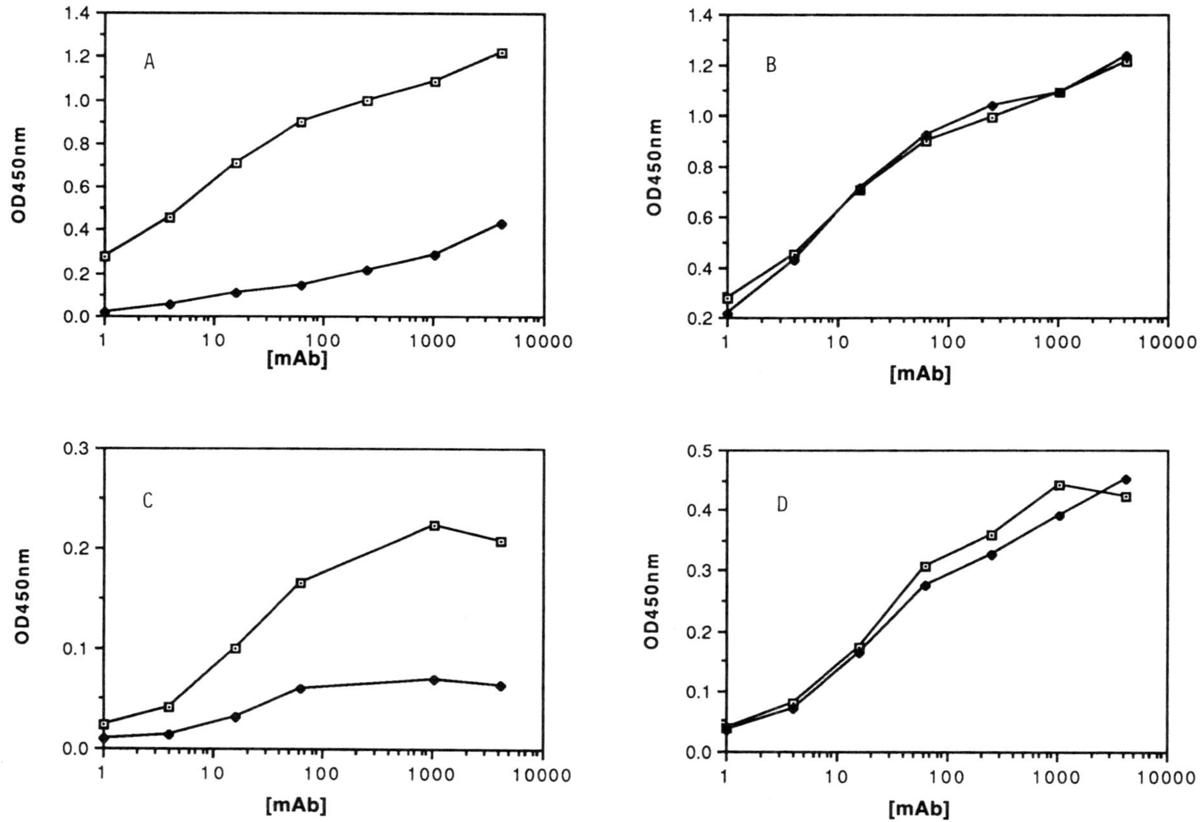

FIGURE 2. A: Titration of AChR-enriched membranes with MAb 387D in the absence (⊡) and presence (◆) of D-penicillamine (D-PA). B: Titration of AChR-enriched membranes with MAb 387D in the absence (⊡) and presence (◆) of D-PA inactivated with *N*-ethylmaleimide. C: Titration with MAb 387D in the absence (⊡) and presence (◆) of cysteine. D: Titration with MAb 387D in the absence (⊡) and presence (◆) of serine. [mAb] = relative antibody concentration.

disulfide bonds to -SH groups, we reacted AChR that had been treated with Cys with 0.1 mM [^{14}C]IAcNH$_2$. No additional ^{14}C was incorporated into Cys-treated membranes compared to untreated membranes, suggesting no new -SH groups were produced by Cys treatment (TABLE 1). This result suggests that these agents are not merely reducing AChR disulfides but are derivitizing the AChR at these sites. To test this hypothesis we measured the stoichiometry of [^{35}S]Cys incorporation into the AChR. The incorporation of one equivalent of [^{35}S]Cys/toxin site with no new free -SH groups suggests incorporation at each of Cys 192 and Cys 193 of only one of the two alpha-subunits (TABLE 1).

The view that emerges from these studies is that D-PA covalently incorporates into the AChR through disulfide exchange at alpha-Cys 192 Cys 193 and thereby severely inhibits the binding of anti-AChR MAbs that block agonist binding. The binding of other anti-AChR MAbs that do not affect agonist binding is less affected

TABLE 1. Incorporation of Isotope into AChR-Enriched Membranes[a]

Treatment	pmole Toxin Sites	Total cpm Added	cpm (pellet)	Specific Incorporation (pmole Agent)/ (pmole Toxin Site)
[^{35}S]Cys + AChR	20	1,789,108	506 ± 7	0.9
([^{35}S]Cys + IAcNH$_2$) + AChR	20	1,789,108	341 ± 36	—
(AChR + Cys) + [^{14}C]IAcNH$_2$	20	412,200	465 ± 10	1.1
AChR + [^{14}C]IAcNH$_2$	20	412,200	404 ± 16	1.0

[a] [^{14}C]IAcNH$_2$-specific activity: 21.0 cpm/pmole. [^{35}S]Cys-specific activity: 9.0 cpm/pmole.

by D-PA treatment. It is this effect on the antigenic structure of the native AChR at the neuromuscular junction that may be responsible for its change in antigenicity and the establishment of the autoimmune response to AChR in D-PA–induced myasthenia gravis.

REFERENCES

1. TORRES, C. F., R. C. GRIGGS, J. BAUM & A. S. PENN. 1980. Arthritis Rheum. **23:** 505–508.
2. CHRISTOPHER, JR., T. B., H. W. CHANG, A. S. PENN, I. A. JAFFE & E. BOCK. 1982. Neurology **32:** 1077–82.
3. XU, Q., R. H. FAIRCLOUGH & D. P. RICHMAN. 1989. Ann. Neurol. **26:** 145.

Human Muscle Acetylcholine Receptor Reactive T and B Lymphocytes in Myasthenia Gravis

QING YI AND ANN KARI LEFVERT

Immunological Research Laboratory and Department of Medicine Karolinska Hospital, S-104 01 Stockholm, Sweden

INTRODUCTION

Myasthenia gravis (MG) is an autoimmune disease caused by an autoimmune attack against the acetylcholine receptors (AChR). Antibodies against the AChR are present in >90% of patients with MG.[1] The synthesis of these antibodies is regulated by AChR-specific T-helper cells.[2]

The use of human AChR for the investigation of the relevant T- and B-cell reactivities in MG has been hampered by the difficulty of obtaining sufficient quantities of the purified receptor. Most studies have used *Torpedo* AChR or synthetic peptides covering part of or the whole sequence of the α-subunit of human AChR.

In this study, we analyzed the prevalence and number of T and B cells in peripheral blood mononuclear cells (PBMC) from 33 patients with MG and 27 healthy controls on the single cell levels with regard to their reactivities to affinity-purified human skeletal muscle AChR.

METHODS

PBMC were prepared from heparinized blood and resuspended in culture medium (RPMI-1640) supplemented with 10% human serum.

Enumeration of human muscle AChR–reactive T and B cells was done by using solid-phase enzyme-linked immunospot (ELISPOT) assays[3] for detection of IFN-γ or IL-2 secretion by stimulated T cells or anti-AChR IgG antibody production by B cells. AChR from human skeletal muscle preparation were purified by its binding to α-bungarotoxin and used as the antigen for T and B cells. Controls including medium or toxin only and Con A (20 μg/mL) were tested in parallel.

RESULTS

Human AChR-reactive T Cells

The stimulated cells, obtained by subtraction of the numbers of spots secreted spontaneously without antigen, are used (FIG. 1). Using a cut-off level of the mean $\pm$ 2 SD of the mean of the values obtained with cells from the healthy controls,

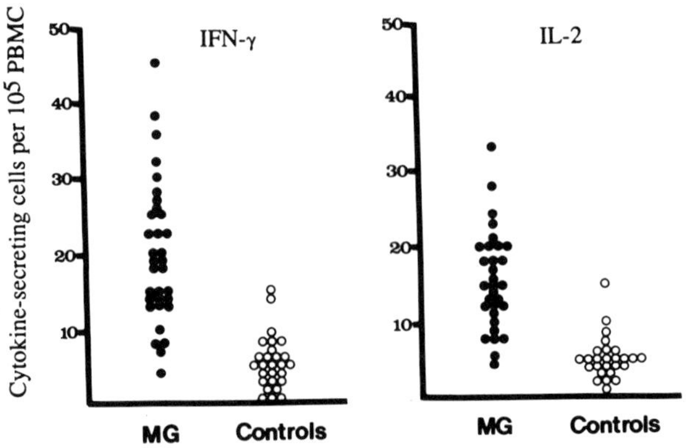

FIGURE 1. Number of IFN-γ– and IL-2–secreting cells/10^5 PBMC after incubation with 1/20 dilution of human muscle AChR in patients with myasthenia gravis (MG) and in healthy controls (controls).

29/33 (87.7%) of the patients and 2/27 (7.4%) of the controls responded with T-cell activation ($p < 0.001$). The mean value of stimulated AChR-reactive IFN-γ–secreting T cells in patients was 19.6/10^5 PBMC, corresponding to 1/5102 mononuclear cells. Comparable results were also obtained with the numbers of IL-2–secreting cells. A correlation was found between the AChR-stimulated numbers of IFN-γ– and IL-2–secreting cells ($r = 0.68$, $p < 0.001$).

Human AChR-reactive B Cells

Anti-AChR IgG-secreting cells were detected in peripheral blood from 30/33 (91%) MG patients (FIG. 2). The mean value of the antibody-secreting cells was 11.7 cells/10^6 PBMC (range: 0–30), corresponding to 1/70,422 mononuclear cells,

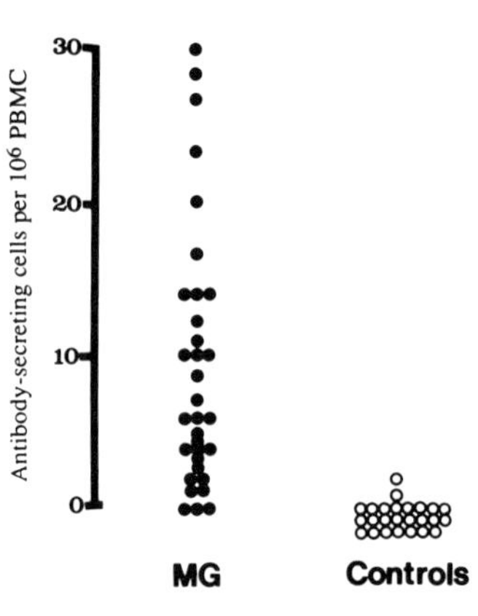

FIGURE 2. Number of antihuman AChR IgG antibody-secreting cells/10^6 PBMC in patients with myasthenia gravis (MG) and healthy controls (controls).

whereas in the controls, 2/25 (8%) had the antibody-secreting cells, with a mean value of 0.16 cell/10^6 PBMC (range: 0–2). The difference between the two groups was significant ($p < 0.001$). Correlations were found between the AChR-stimulated numbers of IFN-γ– or IL-2–secreting cells and the numbers of anti-AChR IgG antibody-secreting cells ($r = 0.527$, $p < 0.0001$; and $r = 0.442$, $p < 0.003$).

CONCLUSIONS

This study has shown that PBMC from 90% of patients with MG contain T and B cells reactive with the human AChR. The prevalence of such cells was about 1/5000 PBMC for T cells and 1/70,000 PBMC for B cells. The approach to detect human AChR-reactive T and B cells by using the sensitive cell assays will be of importance for the mapping of the repertoire and the elucidation of the role of T and B cells in the disease.

REFERENCES

1. Lefvert, A. K., K. Bergström, G. Matell, P. O. Osterman & R. Pirskanen. 1978. J. Neurol. Neurosurg. Psychiatry **41:** 394–403.
2. Hohlfeld, R., K. V. Toyka, K. Heininger, H. Gross-Wilde & I. Kalies. 1984. Nature **310:** 244–246.
3. Czerkinsky, C., G. Andersson, H. P. Ekre, L. Å. Nilsson, L. Klareskog & Ö. Ouchterlony. 1988. J. Immunol. Methods **110:** 29–36.

Molecular Properties of Calcium Channels in Skeletal Muscle and Neurons

WILLIAM A. CATTERALL, KAREN DE JONGH, ERIC ROTMAN, JOHANNES HELL, RUTH WESTENBROEK, STEFAN J. DUBEL,[a] AND TERRY P. SNUTCH[a]

Department of Pharmacology, SJ-30
University of Washington
Seattle, Washington 98195

[a]*Biotechnology Laboratory*
University of British Columbia
Vancouver, British Columbia, V6T 1Z3 Canada

MOLECULAR PROPERTIES OF L-TYPE CALCIUM CHANNELS

Four classes of voltage-gated calcium channels have been defined based on electrophysiological and pharmacological properties.[1] L-type, or dihydropyridine-sensitive, calcium channels mediate long-lasting calcium currents and are most abundant in skeletal muscle where they are localized to the transverse tubule membrane. This tissue contains 50- to 100-fold more dihydropyridine receptor sites than other tissues and hence has been the principal tissue for molecular studies of the L-type channel. The most abundant form of the rabbit skeletal muscle L-type calcium channel is a complex of 5 subunits (reviewed in refs. 2–5): $\alpha 1$ (175 kDa), $\alpha 2$ (143 kDa), β (55 kDa), γ (30 kDa), and δ (24-27 kDa). The $\alpha 1$-, β-, and γ-subunits have been isolated and are products of distinct genes.[6–8] Protein chemistry studies indicate that the $\alpha 2$- and δ-subunits are encoded by the same gene, whose protein product is proteolytically processed and disulfide-bonded to form the $\alpha 2\delta$ complex.[9]

The $\alpha 1$-subunit alone can function as a voltage-gated ion channel when expressed in *Xenopus* oocytes[10] or mammalian cells.[11] The cDNA sequence of the $\alpha 1$-subunit predicts a protein of 1873 amino acids, whose structure is similar to the sodium channel α-subunit, with four internally homologous domains, each of which contains six transmembrane helices that have been named S1 to S6 (FIG. 1A[6]). By analogy with the potassium channels, whose primary structure predicts a model corresponding to one of these four domains, we infer that the extracellular loops between transmembrane segments S5 and S6 in each domain fold across the membrane as a β sheet structure.[12] The calcium flux through the channel is modulated by several classes of compounds, including the dihydropyridines, phenylalkylamines, and benzothiazepines, which bind to the channel at distinct receptor sites (reviewed in ref. 13). It is the $\alpha 1$-subunit that contains the receptor sites for these various channel modulators. The $\alpha 1$-subunit, as well as the β-subunit, is also the target for phosphorylation by a number of kinases, and it is well established that the calcium flux through the channel is regulated by phosphorylation-dependent events (reviewed in refs. 3, 14).

Thus the sites at which the $\alpha 1$-subunit is phosphorylated, as well as the sites that form the binding domains for pharmacological compounds, are functionally important regions in the α1-subunit. Understanding where these regions occur will enhance our understanding of the channel structure and function and may prove useful for the design of novel therapeutic compounds. To this end we have used site-directed antibodies to study functionally important regions in the α1-subunit.

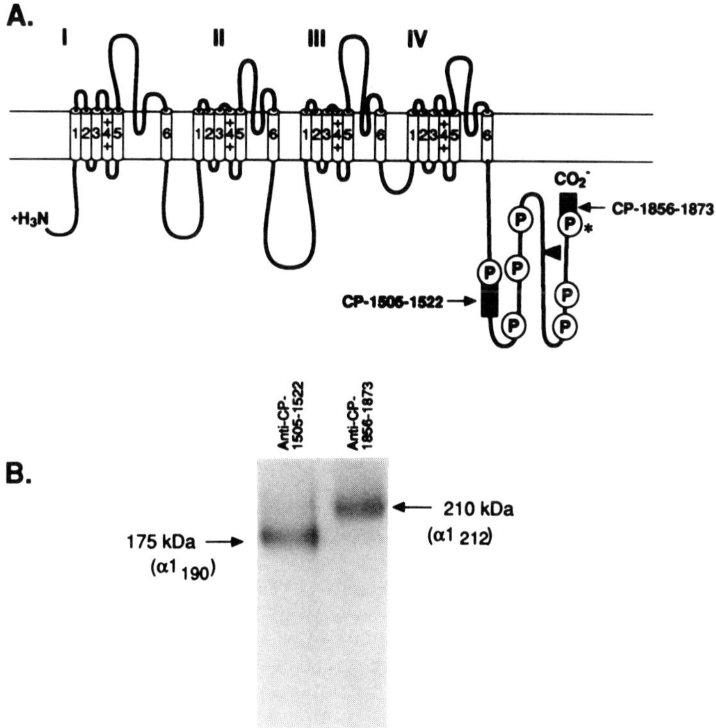

FIGURE 1. A: The proposed folding pattern of the α1-subunit of the rabbit skeletal muscle L-type calcium channel predicted from the amino acid sequence. The recognition sites of antibodies directed against residues 1505-1522 (anti–CP-1505-1522) and 1856-1873 (anti–CP-1856-1873) are shown. The arrowhead indicates the region in which the C-terminus of $\alpha 1_{190}$ occurs. Ⓟ indicates cAMP-dependent phosphorylation sites in the C-terminal region, with * indicating the most rapidly phosphorylated site (serine 1854). **B:** Identification of full-length and truncated forms of the calcium channel α1-subunit. Immunoprecipitation of purified ^{32}P-labeled calcium channels and analysis by SDS-PAGE were carried out using the antibodies indicated above according to the procedure outlined in De Jongh *et al.*[16]

Molecular Properties of Two Size Forms of the α1-Subunit

We have demonstrated the existence of two size forms of the α1-subunit in purified preparations of skeletal muscle calcium channels, T-tubule membranes, and intact skeletal muscle cells in culture.[5,15] Antibodies directed against a peptide corresponding to residues 1505-1522 of the α1-subunit recognize a form of the α1-subunit with an apparent molecular mass of 175 kDa (FIG. 1B). Antibodies directed against a peptide that represents the C-terminal 18 amino acids, predicted by cloning and sequence analysis of the α1-subunit, recognize a 212 kDa form of the α1-subunit (FIG. 1B), which is present at a tenfold lower level than the 175 kDa form. The minor 212-kDa form ($\alpha 1_{212}$) contains the complete amino acid sequence encoded by the α1 mRNA, whereas the major form, corresponding to the 175-kDa protein observed in purified preparations, is truncated at its C terminus. A more accurate estimate of its molecular mass by Ferguson analysis, which compensates for anomalous effects of acrylamide concentration in SDS-PAGE, indicated this truncated subunit was approximately 190 kDa, and we refer to this form of the α1-subunit as $\alpha 1_{190}$. Antibody mapping of the C-terminal region of $\alpha 1_{190}$ placed the C-terminus between residues 1685 and 1699, in agreement with the size predicted by Ferguson analysis (FIG. 1A[16]).

The skeletal muscle L-type calcium channel has been proposed to serve dual roles as a voltage-gated ion channel mediating slow calcium currents and as a voltage sensor for initiation of excitation-contraction coupling. Microinjection of cDNA encoding the α1-subunit into muscle cells cultured from mice with the muscular dysgenesis mutation indicates that a single α1-subunit gene is responsible for both these functions in skeletal muscle.[17,18] We have proposed that calcium channels in skeletal muscle containing the two forms of α1 may fulfill these two roles; channels containing $\alpha 1_{212}$ may be specialized for ion conductance, whereas those containing $\alpha 1_{190}$ may be specialized as the voltage sensor for excitation-contraction coupling.[5]

Differential Phosphorylation and Functional Significance of the Two Forms of the α1-Subunit

It is well established that calcium flux through L-type calcium channels is regulated by cAMP-dependent protein phosphorylation. Phosphorylation of purified calcium channels increases the number that are active in ion conductance.[19,20] Agents that elevate intracellular cAMP levels increase activation of L-type calcium currents in skeletal muscle cells, concomitant with an increase in contractile force,[21] whereas both forms of the α1-subunit are phosphorylated in response to physiological stimuli that activate cAMP-dependent protein phosphorylation in intact muscle cells.[15] There are a number of sites in the α1-subunit that are consensus sequences for phosphorylation by cAMP-dependent protein kinase, most of which occur in the C-terminal region. The full-length α1-subunit contains several potential cAMP-dependent phosphorylation sites that are missing in $\alpha 1_{190}$ (FIG. 1A), suggesting a possible mechanism for differential regulation of the two forms of α1. Because phosphorylation of these sites in the C terminus of α1 may be important for regulation of ion conductance activity, we have examined the phosphorylation of the two forms of the α1-subunit by cAMP-dependent protein kinase.

Two-dimensional tryptic phosphopeptide mapping of denatured α1-subunits that had been incubated with cAMP-dependent protein kinase for 30 minutes indicated $\alpha 1_{190}$ was phosphorylated on three major and three minor peptides (FIG. 2A); $\alpha 1_{212}$ was phosphorylated on all six of these phosphopeptides as well as two additional peptides (T4 and T5; FIG. 2B). Thus the phosphopeptides derived from $\alpha 1_{190}$ are present in $\alpha 1_{212}$, which contains two additional tryptic phosphopeptides. In order to assess the relative importance of these phosphorylation sites, tryptic phosphopeptide maps were generated from $\alpha 1$-subunits that had been incubated with cAMP-dependent protein kinase for brief time periods. Time course experiments indicated that a single site on the C-terminal portion of $\alpha 1_{212}$, corresponding to phosphopeptide T5, is the most intensely and rapidly phosphorylated and that phosphorylation of this peptide was maximal after 10 seconds. Further studies carried out under denaturing conditions indicated phosphorylation occurs almost exclusively on this site unless the channel is first denatured to expose other phosphorylation sites within the remainder of $\alpha 1$ (FIG. 2C). Immunoprecipitation

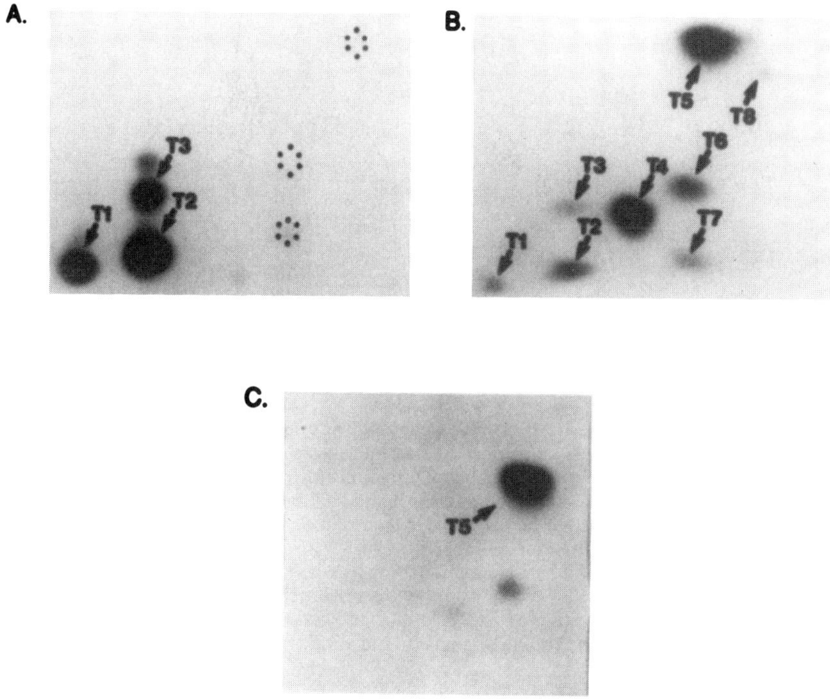

FIGURE 2. Two-dimensional tryptic phosphopeptide mapping of purified calcium channel $\alpha 1$-subunits. Purified calcium channels were phosphorylated for 30 minutes in the presence **(A** and **B)** or absence **(C)** of 1% Triton X-100 before isolation of $\alpha 1_{190}$ **(A)** or $\alpha 1_{212}$ **(B** and **C)**. Phosphopeptide mapping then was carried out as outlined in Lai *et al.*[15]

with an antipeptide antibody enabled us to identify this phosphorylation site as serine 1854.

This is the C-terminal–most cAMP-dependent phosphorylation consensus sequence in the $\alpha 1$-subunit (FIG. 1A). The two nearby consensus sequences (serine residues 1757 and 1772) are only phosphorylated after denaturation or proteolytic cleavage of $\alpha 1$. Studies with the cardiac calcium channel have suggested an important role for the C-terminus in regulation of the channel by phosphorylation.[22]

Phosphorylation of Ser 1854 in the C-terminus of the full-length $\alpha 1$ may be important in this modulation of the ion conductance of the channel. This is in agreement with the proposal that calcium channels containing the two forms of $\alpha 1$ serve dual functions;[5] channels containing the full-length form of the $\alpha 1$-subunit may serve as cAMP-regulated calcium channels, whereas those containing the truncated $\alpha 1$ may serve as the voltage sensor for coupling excitation of muscle to contraction.

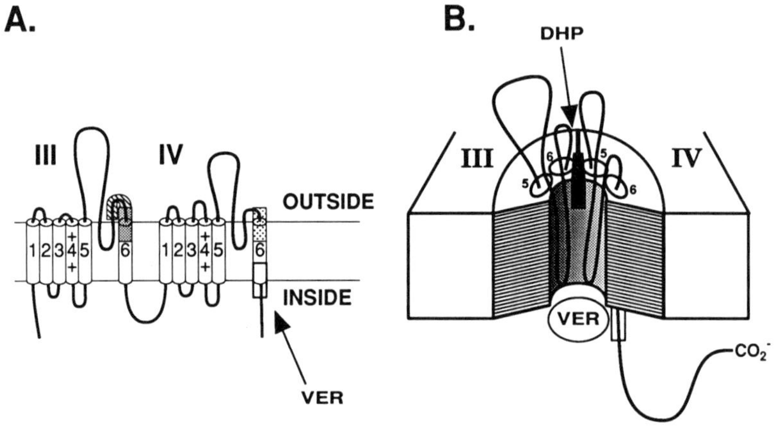

FIGURE 3. Model for phenylalkylamine and dihydropyridine binding. **A:** Transmembrane folding model of domains III and IV of the $\alpha 1$-subunit showing the core of the dihydropyridine binding site that is located at the extracellular end of transmembrane helix IIIS6 (filled gray rectangle). The subsite labeled by compounds such as azidopine, whose photoreactive group is located some distance from the dihydropyridine ring, is indicated by the hatched curved bar. The secondary site of photolabeling in domain IV is indicated by the dotted rectangle. The site of covalent labeling by verapamil derivatives is indicated by the rectangle labeled VER. **B:** Cross-sectional view of the $\alpha 1$-subunit showing the proposed binding cleft for the dihydropyridines located at the extracellular surface between domains III and IV and the proposed intracellular channel-blocking position of verapamil and other phenylalkylamines.

Identification of the Phenylalkylamine- and Dihydropyridine-Binding Regions within the α1-Subunit

The phenylalkylamines are a diverse group of compounds that exert their effects by binding to the intracellular side of the calcium channel.[23] These drugs behave as open-channel blockers,[24,25] suggesting they may bind in or near the intracellular side of the transmembrane pore of the calcium channel. The photoreactive phenylalkylamine ludopamil (LU 49888), which is a derivative of verapamil, covalently labels the $\alpha 1$-subunit of the calcium channel.[26] Mapping of labeled peptide fragments with a panel of site-directed antipeptide antibodies restricted the site of covalent labeling to a fragment extending between Glu-1349 and Trp-1391 (FIG. 3). This region includes the sixth transmembrane segment of domain IV and several extracellular and intracellular residues. It is known that phenylalkylamines bind to the intracellular side of the channel, and thus the verapamil binding site must be formed in part by the intracellular end of transmembrane segment IVS6 and the adjacent intracellular residues. By analogy with recent studies of potassium channels, we postulate that the intracellular ends of the four S6 regions together with the adjacent residues may form part of the intracellular opening of the transmembrane pore and that verapamil and related phenylalkylamines may bind there and block the calcium channel from the intracellular side.

A number of dihydropyridine compounds have been shown to modulate the voltage-dependent gating of the L-type calcium channel. These compounds can act as either activators or inhibitors of the channel and bind with high affinity to inactivated calcium channels.[27] Dihydropyridines bind to their receptor site, which has been localized to the α1-subunit,[3,13] from the extracellular surface of the channel,[28] suggesting that the dihydropyridine binding domain may form part of the extracellular surface of the α1-subunit. The antibody mapping approach has been used to localize this binding domain to the extracellular ends of the sixth transmembrane segments in domains III and IV (FIG. 3 refs. 29, 30). Because the dihydropyridines modulate the function of the calcium channel by modifying gating functions, the interface between domains III and IV may be important in activation gating.

N-TYPE CALCIUM CHANNELS

Electrophysiological and pharmacological criteria have defined four main types of Ca channel in vertebrate neurons (T, L, N, and P; for reviews see refs. 31, 32). Of these, N-type Ca channels are distinct in that they have only been described in neurons and are blocked by the neuropeptide toxin, ω-conotoxin GVIA (ω-CgTx).[31–37] The correlation between ω-CgTx–sensitive N-type Ca channels and the sensitivity of neurotransmitter release to ω-CgTx[38–41] has led to the proposal that N-type Ca channels are concentrated as presynaptic nerve termini and that they play a major role in mediating chemical synaptic transmission.[42] Furthermore, ω-CgTx–sensitive N-type Ca channels are a target for modulation by neurotransmitters and neuropeptides[43–46] and have also been implicated as targets of autoantibodies in small cell lung carcinoma associated with Lambert-Eaton myasthenic syndrome (B. Lang *et al.*, this volume).

Recent studies show that the ω-CgTx receptor from brain is a multisubunit complex that includes components homologous to the skeletal muscle L-type Ca channel.[47–49] Molecular cloning experiments show that rat brain expresses at least four major classes of Ca channel α1-subunit (designated rbA, rbB, rbC, and rbD).[50] The rbC and rbD isoforms are most closely related to DHP-sensitive Ca channels from various tissues,[6,8,10,51–53] suggesting that the brain expresses two distinct forms of L-type Ca channel. The rbA and rbB isoforms, however, are only moderately related to L-type Ca channels and are, therefore, of particular interest with respect to identifying Ca channels that mediate specific functions in neurons.

cDNA Cloning and Primary Structure of Rat Brain Class B Ca Channel

To isolate cDNAs encoding the full-length rbB Ca channel α1-subunit, synthetic oligonucleotides corresponding to the previously determined rbB-10 cDNA sequence[50] were radiolabeled and used to probe a size-selected rat brain cDNA library. Five overlapping cDNAs totaling >28 kb were sequenced in their entirety to define a complete open reading frame of 7008 bp encoding a 2336 amino acid protein. The calculated molecular mass of the rbB-I protein (262 kDa) is larger than any of the previously cloned rat brain Ca channel α1-subunits.[54,55]

The deduced amino acid sequence of rbB-I is similar in overall structure to other Ca channel α1-subunits.[6,8,10,50–56] The rbB-I protein is composed of four predominantly hydrophobic homologous domains (I-IV) that are linked by hydrophilic segments of various lengths, as for the α1-subunit of skeletal muscle calcium channels (FIG. 1). Hydrophobicity analysis reveals that each domain contains five mainly hydrophobic segments (S1, S2, S3, S5, and S6), each of which can potentially cross the membrane once. Similar to other voltage-gated ion channels, each domain of rbB-I also contains an S4 segment, in which every third amino acid is a positively charged arginine or lysine. This segment probably represents the voltage sensor of the channel. There are 41 potential phosphorylation sites for cAMP-dependent protein kinase, cGMP-dependent protein kinase, and protein kinase C; and calmodulin kinase II in the rbB-I protein according to established consensus sequences.[57] Thirty-seven of these consensus sites are predicted to be in cytoplasmic regions of the molecule. Three of the 13 consensus sites for N-linked glycosylation found are predicted to be on the extracellular side of the rbB-I protein.

The rbB protein shows only moderate identity to DHP-sensitive Ca channels (≈34% amino acid identity to cardiac, lung, and skeletal muscle L-type Ca channels). Of Ca channels cloned to date, rbB-I is most closely related to the rat brain class A (rbA-I) Ca channel (58% amino acid identity overall).[54] Comparison of the rbA-I and rbB-I proteins shows that they are most identical in the amino terminus and in the four repeated domains (79–87% identity). Unlike cloned L-type Ca channels, both rbA-I and rbB-I contain a large hydrophilic segment (>400 residues) linking domains II and III. Other than the region immediately flanking the carboxyl end of the domain II S6 segment, however, there is little identity between rbA-I and rbB-I in this region. Also of note is that both the rbA-I and rbB-I Ca channels possess a large hydrophilic carboxyl region after domain IV but show only limited amino acid identity in this region. Although the functional significance of these two major differences between the rbA-I and rbB-I Ca channels is unknown, they may reflect distinct cytoplasmic interactions involving channel modulation.

Immunoprecipitation of Brain ω-CgTx–Binding Sites

To examine the relationship between rbB-I and the other cloned rat brain Ca channels, polyclonal antisera were generated against synthetic peptides corresponding to the amino acid sequences of the rbA-I, rbB-I, and rbC-I proteins (designated CNA 1, CNB 1, and CNC 1, respectively). The peptides comprised sequences specific for the variable domain II-domain III cytoplasmic segment of each of the cloned brain Ca channel α1-subunits, with the rbC peptide common to both rbC-I and rbC-II splicing variants.[55] Ca channels in rat forebrain membranes were labeled with [^{125}I]ω-CgTx (fraction GVIA), and solubilized and immunoprecipitated with the specific antisera. Immunoprecipitation of forebrain membranes labeled with [^{125}I]ω-CgTx shows that the CNA 1 and CNC 1 antisera precipitate only trace amounts of toxin-binding sites (FIG. 4). By contrast, the CNB 1 antiserum specifically precipitates forebrain ω-CgTx–binding sites. Immunoprecipitation is effectively blocked by preincubation of the antiserum with purified rbB peptide

(FIG. 4), indicating the specificity of the CNB 1 antiserum for a single class of Ca channel. The forebrain membranes were labeled under conditions where only high-affinity, irreversible ω-CgTx sites are occupied. Because ω-CgTx blocks N-type Ca currents irreversibly and with high affinity,[35–37] immunoprecipitation with the CNB 1 antiserum suggests that the rbB-I cDNA encodes the α1-subunit of an ω-CgTx–sensitive N-type Ca channel.

Distribution of rbB-I Expression

Autoradiographic studies of [^{125}I]ω-CgTx–binding sites in the rat brain show a wide distribution of binding, with the highest densities in the cerebral cortex, hippocampus, olfactory bulb, and cerebellar cortex.[58] To examine the spatial pattern of rbB-I expression, total cellular RNA was isolated from various regions

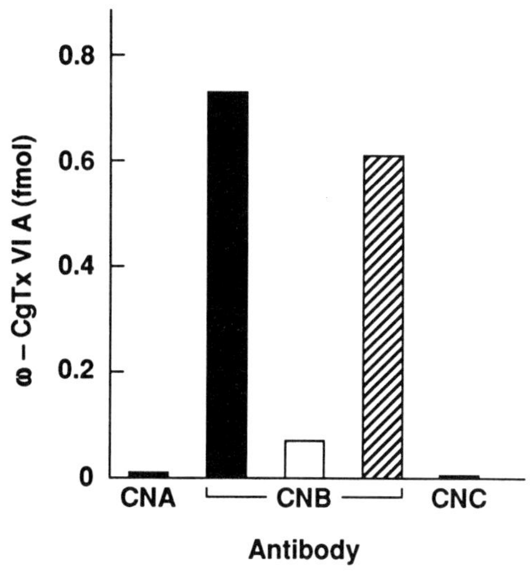

FIGURE 4. Immunoprecipitation of forebrain [^{125}I]ω-CgTx–binding sites by the CNA, CNB, and CNC antisera (filled bars). Peptide block of immuno-precipitation of ω-CgTx–binding sites. Antiserum CNB-1 was incubated with the rbB (open bar) or rbC (hatched bar) peptides as indicated before immunoprecipitation. Nonspecific precipitation of receptors as determined with a nonspecific antiserum was subtracted (0.1 fmol for [^{125}I]ω-CgTx).

of adult rat brain and probed with a 2.5-kb cDNA fragment specific for the rbB-I Ca channel. FIGURE 5 shows that the probe detects a single major transcript of ≈10 kb. Significantly, the rbB-I transcript is present in all regions of the rat central nervous system, including the spinal cord. The relative levels of rbB-I expression vary slightly between regions and are quite distinct from the expression pattern of the rbA-I Ca channel, which is predominantly expressed in the cerebellum.[54] The wide distribution of rbB-I transcripts agrees with autoradiographic studies of [^{125}I]ω-CgTx–binding sites in rat brain and is consistent with the putative role of ω-CgTx–sensitive Ca channels in neurotransmitter release.

The distribution of rbB-I expression in other rat tissues was examined by using the highly sensitive polymerase chain reaction assay. First-strand cDNA was synthesized from total RNAs by using a downstream primer specific for the 3′ nontranslated region of rbB-I. Subsequently, the cDNA was amplified by using the downstream primer and a second rbB-I–specific upstream primer. Examination of the PCR products by agarose gel electrophoresis showed that a single 367-bp fragment resulted from brain RNA. Hybridization of a radiolabeled oligonucleotide specific for the 3′ nontranslated region of rbB-I confirmed that rbB-I transcripts are detected only in brain and not in other rat tissues. The rbB-I tissue distribution is distinct from that of rbA-I, which is expressed in rat brain, heart, and pituitary.[54] Similarly, the rbB-I expression pattern is distinct from the rbC-I and rbC-II patterns, which are expressed in all tissues examined.[55] Thus, rbB-I is the only cloned Ca channel demonstrated to be specifically localized to the nervous system. To probe the relationship between the rbB-I protein and N-type Ca channels further, we examined the expression of rbB-I in several rat cell lines that have been shown to possess various voltage-gated Ca channels. We found an exact correspondence between expression of rbB-I and the presence of N-type Ca channels in these clonal cell lines, and there is no correlation with L- or T-type Ca channels.[59] Thus, several lines of evidence support the identification of the rbB cDNA as encoding an N-type calcium channel.

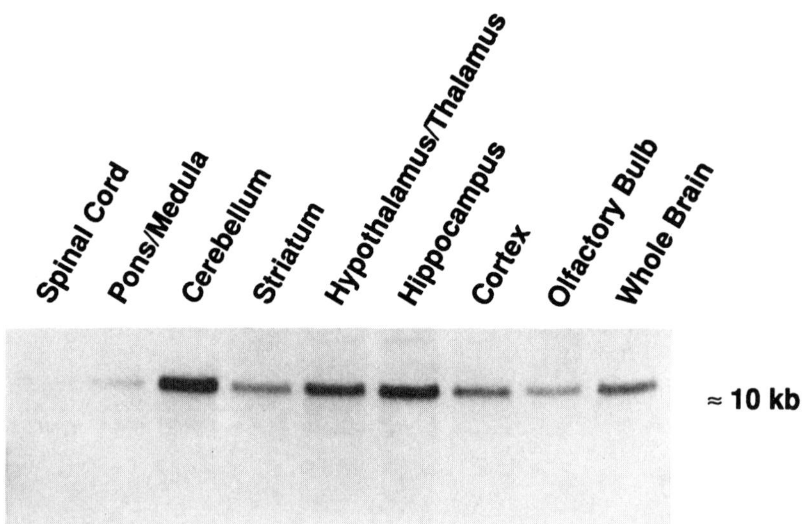

FIGURE 5. Northern blot analysis of rbB-I expression in the rat central nervous system. Autoradiograph of blot hybridization of a 2.5-kb cDNA probe specific for the 3′ noncoding region of rbB-I to 30 μg of various rat brain total RNAs. Size of the rbB-I transcript was determined by using RNA standards (BRL). Autoradiography was for 7 days with an intensifying screen.

Biochemical Properties and Subcellular Localization of N-type Calcium Channels Encoded by rbB cDNA

CNB 1 antibodies immunoprecipitated up to 50% of the total N-type calcium channels labeled with [^{125}I]ω-CgTx (FIG. 6A). These results suggest that there may be multiple subtypes of ω-conotoxin–sensitive N-type calcium channels that are differentially recognized by this antibody. Immunoblotting with CNB 1 re-

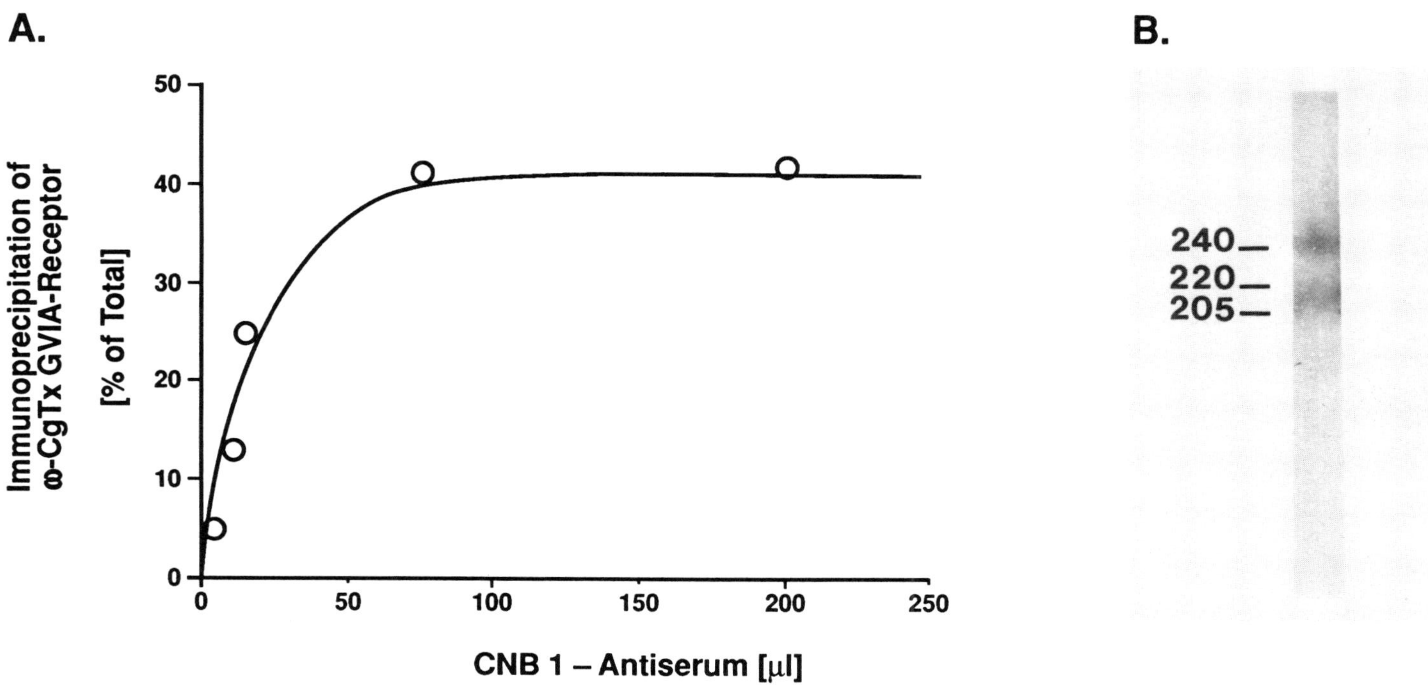

FIGURE 6. Immunoprecipitation and immunoblotting of an ω-CgTx GVIA–sensitive calcium channel. **A:** Calcium channels in brain synaptosomal membranes were labeled by binding of [^{125}I]ω-CgTx GVIA, solubilized, and partially purified.[59] The purified calcium channels were incubated with increasing concentrations of antiserum CNB 1 as indicated, and the antigen-antibody complexes were adsorbed to protein A Sepharose, collected by centrifugation, and quantitated by gamma counting.[59] **B:** A partially purified preparation of ω-CgTx GVIA–sensitive calcium channels was analyzed by immunoblotting with antibodies purified from antiserum CNB 1 by antigen-affinity chromatography.

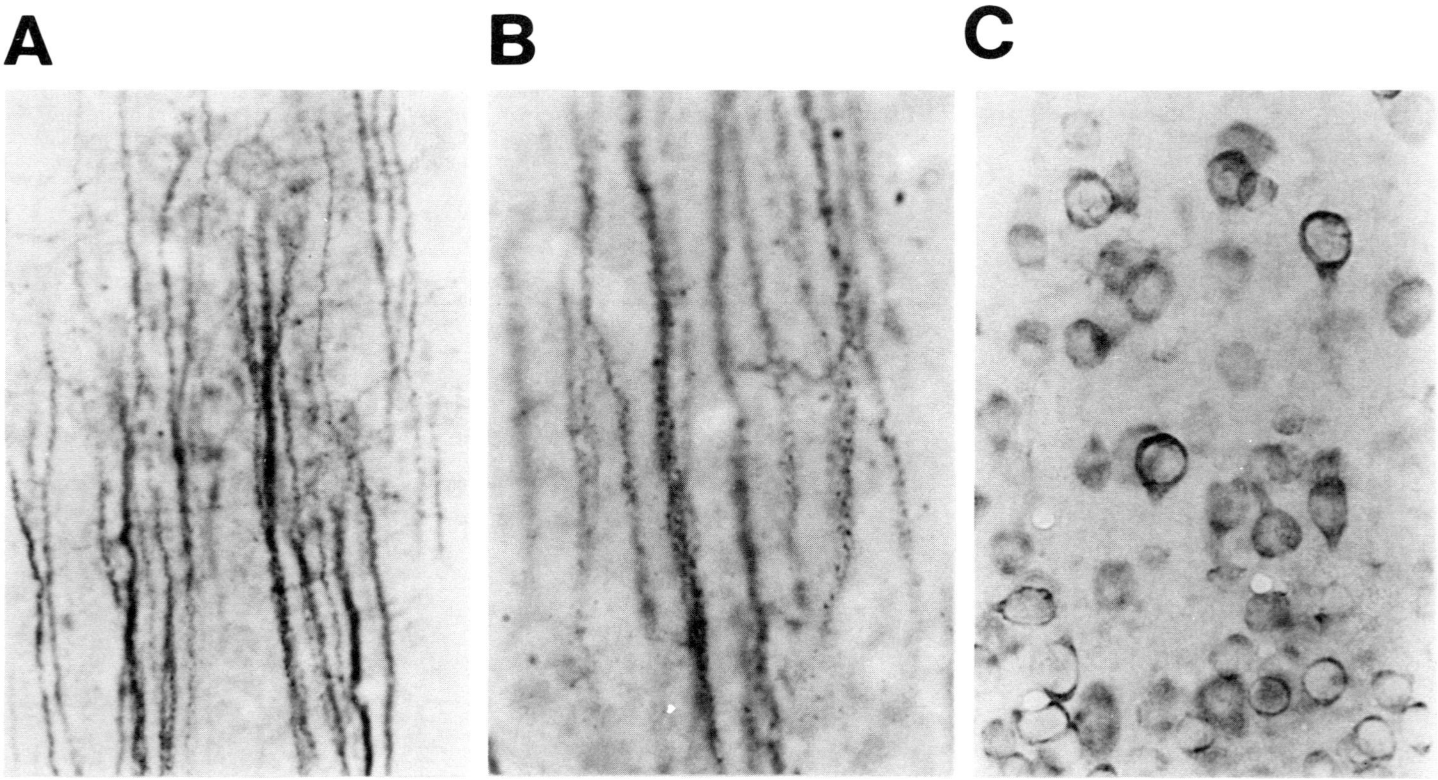

FIGURE 7. Complementary localization of N- and L-type calcium channels on cerebral cortical neurons. **A:** Sagittal sections of rat brain cerebral cortex were incubated with affinity-purified CNB 1 antibodies, and the bound antibodies were visualized by the peroxidase-antiperoxidase method. Faintly labeled neuronal cell bodies and densely labeled dendrites are observed. **B:** Higher resolution view of dendrites of cortical pyramidal cells. Punctate staining of nerve terminals is observed superimposed on the general staining of the dendrites. **C:** Sagittal sections were stained with monoclonal antibody MANC-1, which labels L-type calcium channels.[60,61] Only cell bodies and proximal dendrites are labeled.

vealed proteins of 210 and 240 kDa (FIG. 6B) that precisely comigrated with ω-conotoxin–binding activity on sucrose gradient sedimentation. These two size forms of the α1-subunit of an N-type calcium channel may reflect differential posttranscriptional or posttranslational processing of the carboxyl terminal region as has been described for the α1-subunit of skeletal muscle calcium channels above.

CNB 1 was used in combination with the immunofluorescence and indirect peroxidase-anti-peroxidase techniques to investigate the distribution of N-type Ca channels in rat brain. Ca channels recognized by CNB 1 are localized along dendritic processes and on occasional somata. For example, CNB 1 immunoreactivity is localized along the full length of apical dendrites of hippocampal pyramidal neurons with little labeling of the somata, whereas somata of pyramidal cells in the dorsal cortex are frequently labeled, and the staining of apical dendrites decreases as distance from the cell body increases (FIG. 7A). In addition, there is intense punctate staining of structures (probably presynaptic terminals) along the dendrites and on occasional cell bodies in the dorsal cortex (FIG. 7B), subiculum, and cerebellar Purkinje cells. These observations contrast with studies of L-type calcium channels, which have been shown to be localized in a narrow zone around the periphery of neuronal somata and in clusters in proximal dendrities of most major projection neurons such as hippocampal and cortical pyramidal cells, Purkinje cells, and spinal motor neurons (FIG. 7C).[60,61] These results indicate that the ω-CgTx–sensitive Ca channels recognized by CNB 1 have a distinct distribution from L-type Ca channels in central neurons that may correlate with distinct functional roles for these channel subtypes. Specific staining of presynaptic terminals with antibodies against N-type, but not L-type, calcium channels is consistent with the physiological evidence for a major role of ω-conotoxin–sensitive N-type calcium channels in transmitter release.

REFERENCES

1. TSIEN, R. W., P. T. ELLINOR & W. A. HORNE. 1991. Trends Pharmacol. Sci. **12:** 349–354.
2. CAMPBELL, K. P., A. T. LEUNG & A. H. SHARP. 1988. Trends Neurosci. **11:** 425–430.
3. CATTERALL, W. A., M. J. SEAGAR & M. TAKAHASHI. 1988. J. Biol. Chem. **263:** 3535–3538.
4. CATTERALL, W. A. 1991. Science **253:** 1499–1500.
5. DE JONGH, K. S., D. K. MERRICK & W. A. CATTERALL. 1989. Proc. Natl. Acad. Sci. USA **86:** 8585–8589.
6. TANABE, T., H. TAKESHIMA, A. MIKAMI, V. FLOCKERZI, H. TAKAHASHI, K. KANGAWA, M. KOJIMA, H. MATSUO, T. HIROSE & S. NUMA. 1987. Nature **328:** 313–318.
7. RUTH, P., A. RÖHRKASTEN, M. BIEL, E. BOSSE, S. REGULLA, H. E. MEYER, V. FLOCKERZI & F. HOFMANN. 1989. Science **245:** 1115–1118.
8. ELLIS, S. B., M. E. WILLIAMS, N. R. WAYS, R. BRENNER, A. H. SHARP, A. T. LEUNG, K. P. CAMPBELL, E. MCKENNA, W. J. KOCH, A. HUI, A. SCHWARTZ & M. M. HARPOLD. 1988. Science **241:** 1661–1664.
9. DE JONGH, K. S., C. WARNER & W. A. CATTERALL. 1990. J. Biol. Chem. **265:** 14738–14741.
10. MIKAMI, A., K. IMOTO, T. TANABE, T. NIIDOME, Y. MORI, H. TAKESHIMA, S. NARUMIYA & S. NUMA. 1989. Nature **340:** 230–233.
11. PEREZ-REYES, E., H. S. KIM, A. E. LACERDA, W. HORNE, X. WEI, D. RAMPE, K. P. CAMPBELL, A. M. BROWN & L. BIRNBAUMER. 1989. Nature **340:** 233–236.
12. YELLEN, G., M. E. JURMAN, T. ABRAMSON & R. MACKINNON. 1991. Science **251:** 939–942.

13. GLOSSMANN, H. & J. STRIESSNIG. 1990. Rev. Physiol. Biochem. Pharmacol. **114:** 1–105.
14. HOSEY, M. M. & M. LAZDUNSKI. 1988. J. Membr. Biol. **104:** 81–105.
15. LAI, Y., M. J. SEAGAR, M. TAKAHASHI & W. A. CATTERALL. 1990. J. Biol. Chem. **265:** 20839–20848.
16. DE JONGH, K. S., C. WARNER, A. A. COLVIN & W. A. CATTERALL. 1991. Proc. Natl. Acad. Sci. USA **88:** 10778–10782.
17. TANABE, T., K. G. BEAM, J. A. POWELL & S. NUMA. 1988. Nature **336:** 134–139.
18. ADAMS, B. A., T. TANABE, A. MIKAMI, S. NUMA & K. G. BEAM. 1990. Nature **346:** 569–572.
19. FLOCKERZI, V., H.-J. OEKEN, F. HOFMANN, D. PELZER, A. CAVALIE & W. TRAUTWEIN. 1986. Nature **323:** 66–68.
20. NUNOKI, K., V. FLORIO & W. A. CATTERALL. 1989. Proc. Natl. Acad. Sci. USA **86:** 6816–6820.
21. ARREOLA, J., J. CALVO, M. C. GARCIA & J. A. SÁNCHEZ. 1987. J. Physiol. **393:** 307–330.
22. HESCHELER, J. & W. TRAUTWEIN. 1988. J. Physiol. (Lond.) **404:** 259–274.
23. HESCHELER, J., D. PELZER, G. TRUBE & W. TRAUTWEIN. 1982. Pflügers Arch. **393:** 287–291.
24. HONDEGHEM, L. M. & B. G. KATZUNG. 1984. Annu. Rev. Pharmacol. Toxicol. **24:** 387–423.
25. LEE, K. S. & R. W. TSIEN. 1983. Nature **302:** 790–794.
26. STRIESSNIG, J., H. GLOSSMANN & W. A. CATTERALL. 1990. Proc. Natl. Acad. Sci. USA **87:** 9108–9112.
27. BEAN, B. P. 1984. Proc. Natl. Acad. Sci. USA **81:** 6388–6392.
28. KASS, R. S. & J. P. ARENA. 1989. J. Gen. Physiol. **93:** 1109–1127.
29. NAKAYAMA, H., M. TAKI, J. STRIESSNIG, H. GLOSSMANN, W. A. CATTERALL & Y. KANAOKA. 1991. Proc. Natl. Acad. Sci. USA **88:** 9203–9207.
30. STRIESSNIG, J., B. J. MURPHY & W. A. CATTERALL. 1991. Proc. Natl. Acad. Sci. USA **88:** 10769–10773.
31. BEAN, B. P. 1989. Annu. Rev. Physiol. **51:** 367–384.
32. HESS, P. 1990. Annu. Rev. Neurosci. **13:** 337–356.
33. OLIVERA, B. M., J. M. MCINTOSH, L. J. CRUZ, F. A. LUQUE & W. R. GRAY. 1984. Biochemistry **23:** 5087–5090.
34. NOWYCKY, M. C., A. P. FOX & R. W. TSIEN. 1985. Nature **316:** 440–443.
35. MCCLESKEY, E. W., A. P. FOX, D. H. FELDMAN, L. J. CRUZ, B. M. OLIVERA, R. W. TSIEN & D. YOSHIKAMI. 1987. Proc. Natl. Acad. Sci. USA **84:** 4327–4331.
36. PLUMMER, M. R., D. E. LOGOTHETIS & P. HESS. 1989. Neuron **2:** 1453–1463.
37. REGAN, L. J., D. W. Y. SAH & B. P. BEAN. 1991. Neuron **6:** 269–280.
38. REYNOLDS, I. J., J. A. WAGNER, S. H. SNYDER, S. A. THAYER, B. M. OLIVERA & R. J. MILLER. 1986. Proc. Natl. Acad. Sci. USA **83:** 8804–8807.
39. HIRNING, L. D., A. P. FOX, E. W. MCCLESKEY, B. M. OLIVERA, S. A. THAYER, R. J. MILLER & R. W. TSIEN. 1988. Science **239:** 57–61.
40. YEAGER, R. E., D. YOSHIKAMI, J. RIVER, L. J. CRUZ & G. P. MILJANICH. 1987. J. Neurosci. **7:** 2390–2396.
41. STANLEY, E. F. & G. GOPING. 1991. J. Neurosci. **11:** 985–993.
42. MILLER, R. J. 1987. Science **235:** 46–52.
43. WANKE, E., A. FERRONI, A. MALGAROLI, A. AMBROSINI, T. POZZAN & J. MELDOLESI. 1987. Proc. Natl. Acad. Sci. USA **84:** 4313–4317.
44. GROSS, R. E. & R. L. MACDONALD. 1987. Proc. Natl. Acad. Sci. USA **84:** 5469–5473.
45. LIPSCOMBE, D., S. KONGSAMUT & R. W. TSIEN. 1989. Nature **340:** 639–641.
46. BLEY, K. R. & R. W. TSIEN. 1990. Neuron **2:** 379–391.
47. MCENERY, M. W., A. M. SNOWMAN, A. H. SHARP, M. E. ADAMS & S. H. SNYDER. 1991. Proc. Natl. Acad. Sci. USA **88:** 11095–11099.
48. SAKAMOTO, J. & K. P. CAMPBELL. 1991. J. Biol. Chem. **266:** 18914–18919.
49. AHLIJANIAN, M. K., J. STRIESSNIG & W. A. CATTERALL. 1991. J. Biol. Chem. **266:** 20192–20197.
50. SNUTCH, T. P., J. P. LEONARD, M. M. GILBERT, H. A. LESTER & N. DAVIDSON. 1990. Proc. Natl. Acad. Sci. USA **87:** 3391–3395.

51. BIEL, M., P. RUTH, R. HULLIN, W. STUHMER, V. FLOCKERZI & F. HOFMANN. 1990. FEBS Lett. **269:** 409–412.
52. KOCH, W. J., P. T. ELLINOR & A. SCHWARTZ. 1990. J. Biol. Chem. **265:** 17786–17791.
53. WILLIAMS, M. E., D. H. FELDMAN, A. F. MCCUE, R. BRENNER, G. VELICELEBI, S. B. ELLIS & M. M. HARPOLD. 1992. Neuron **8:** 71–84.
54. STARR, T. V., W. PRYSTAY & T. P. SNUTCH. 1991. Proc. Natl. Acad. Sci. USA **88:** 5621–5625.
55. SNUTCH, T. P., W. J. TOMLINSON, J. P. LEONARD & M. M. GILBERT. 1991. Neuron **7:** 45–57.
56. MORI, Y., T. FRIEDRICH, M.-S. KIM, A. MIKAMI, J. NAKAI, P. RUTH, E. BOSSE, F. HOFMANN, V. FLOCKERZI, T. FURUICHI, K. MIKOSHIBA, K. IMOTO, T. TANABE & S. NUMA. 1991. Nature **350:** 398–402.
57. KENNELLY, P. J. & E. G. KREBS. 1991. J. Biol. Chem. **266:** 15555–15558.
58. KERR, L. M., F. FILLOUX, B. M. OLIVERA, H. JACKSON & J. K. WAMSLEY. 1988. Eur. J. Pharmacol. **146:** 181–183.
59. DUBEL, S. J., T. V. B. STARR, J. HELL, M. K. AHLIJANIAN, J. J. ENYEART, W. A. CATTERALL & T. P. SNUTCH. 1992. Proc. Natl. Acad. Sci. USA **89:** 5058–5062.
60. AHLIJANIAN, M. K., R. E. WESTENBROEK & W. A. CATTERALL. 1990. Neuron **4:** 819–832.
61. WESTENBROEK, R. E., M. K. AHLIJANIAN & W. A. CATTERALL. 1990. Nature **347:** 281–284.

Calcium Channel Types at the Neuromuscular Junction

D. WRAY AND V. PORTER

Pharmacology Department
Leeds University
Leeds LS2 9JT, United Kingdom

INTRODUCTION

Calcium channels at the neuromuscular junction are of key importance in neuromuscular transmission. They are also of great interest in the Lambert-Eaton myasthenic syndrome (LEMS),[1,2] where they are functionally impaired by autoantibodies. There is very little known, however, about the detailed nature of these voltage-operated channels at the presynaptic membrane. This is mainly because the nerve terminal is too small to make either intracellular or patch-clamp recordings.

Calcium channels have been characterized in other tissues on the basis of their electrophysiological and pharmacological properties.[3,4] The main classification has been into L-, T-, and N-type channels. The T-type channel is low threshold-activated, and currents are transient in nature. Both L- and N-type channels are high-threshold activated, while L-type currents are sustained and N-type currents are often (slowly) inactivating. The L-type channel is blocked by dihydropyridines, whereas the N-type channel is blocked by ω-conotoxin. More recently, a further channel type has been proposed[5,6] (P-type), which is a high-threshold current not blocked by either dihydropyridines or ω-conotoxin, but blocked by toxin from funnel web spider venom. Evidence is available[4–6] suggesting that the location of N- and P-type channels is predominantly neuronal.

The object of this paper is to review earlier evidence and present new data on the immunological and pharmacological characterization of the type of calcium channel present at the neuromuscular junction.

LAMBERT-EATON MYASTHENIC SYNDROME ANTIBODY

It is now quite clear that LEMS is an autoimmune syndrome;[1,2] this was shown by our passive transfer experiments and supported by ample clinical evidence. We further showed that the presynaptic calcium channel is the target for the autoantibodies.[7] It was of interest, therefore, to determine the type of calcium channel that the autoantibodies bind to. We showed that when these antibodies were applied to neuroblastoma (NG 108 15) cells for 24 h they produce a block of L-type but not T-type channels.[8] This might appear to suggest that the calcium channels at the neuromuscular junction are also L-type; however, dihydropyridine agonists and antagonists are ineffective on calcium channels at the neuromuscular junction.[9] These data, taken together, suggest that the channels at the neuromuscular junction are not L-type, but that there are some antigenic similarities, and hence structural similarities, with some region at least of the L-type channel.

Biochemical evidence indicates that LEMS antibody binds to the ω-conotoxin receptor in various cell lines (*i.e.* the N-type channel) in immunoprecipitation assays,[10,11] and this is the basis of a diagnostic assay for LEMS. Other electrophysiological data have confirmed the picture that LEMS antibody acts on N- (as well as L-, but not T-) type currents in a variety of cells.[12] ω-Conotoxin, however, has

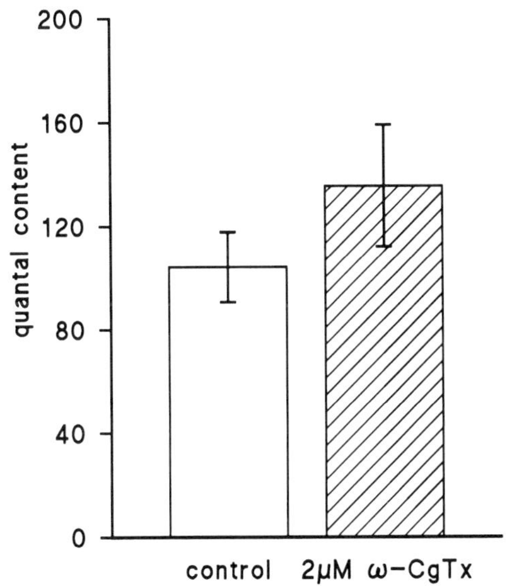

FIGURE 1. Effect of 2 μM ω-conotoxin (ωCgTx) on quantal content. ω-Conotoxin was applied to the bath; the mean quantal content (±SEM) is shown for 17 control and 10 treated end-plates from 5 muscles. There was no significant difference between test and control (Student's *t* test). Methods for this and all subsequent FIGURES were as follows. Isolated mouse phrenic nerve diaphragm preparations were bathed in oxygenated (95% O_2, 5% CO_2) Krebs solution, at room temperature (21–24°C). The composition of the Krebs solution was (mM) NaCl, 118; KCl, 4.7; $MgSO_4$, 1.2; KH_2PO_4, 1.2; $NaHCO_3$, 25.0; glucose, 11.1; $CaCl_2$, 2.52, and tubocurarine (3–5 μM), sufficient to prevent muscle contraction. Intracellular voltage recordings were made using glass microelectrodes as previously described.[29] The nerve was stimulated at 0.5 Hz (pulse width 0.1 msec, supramaximal voltage 5–15 V). All data were stored on magnetic tape and subsequently analyzed by computer after digitization at 5 kHz. The quantal content was calculated from the end-plate potential amplitude using the variance method.

no effect on electrically evoked release at the mammalian neuromuscular junction itself[13,14] (at least in the mouse, although it does act at the frog neuromuscular junction[14,15]). Our own data confirm this (FIG. 1). Thus the calcium channel involved in electrically stimulated release at the mammalian neuromuscular junction does not appear to be N-type, but (because of antibody cross-reactivity with N-type channels) it does seem to have a region of antigenic and hence structural similarity with the N-type channel (and also with the L-type channel; see above).

ANTIBODIES RAISED AGAINST L-TYPE CALCIUM CHANNELS FROM SKELETAL MUSCLE T TUBULES

The L-type calcium channel consists of several subunits: a large pore–containing α_1-subunit, a large α_2-subunit, together with smaller subunits (β, γ, δ).[16] The function of the subunits other than α_1 is not known, but they may be involved in the regulation of channel function (*e.g.* refs. 17, 18). Monoclonal antibodies were raised against the α_2-subunit of the skeletal muscle L-type channel, and we tested them[19] on L-type currents in neuroblastoma (NG 108 15) cells by the whole-cell patch clamp technique. One monoclonal antibody caused reductions in L-type currents after 24 h incubations, so it was of interest to look for effects of such antibodies at the neuromuscular junction. For this, mice were immunized with the α_2-subunit (two or more ip injections of 5–10 μg α_2-subunit protein). There was no significant effect of this antibody on quantal content in diaphragm muscles.[20] This suggests that there are differences between the L-type channel in neuroblastoma cells and the calcium channel at the neuromuscular junction (which might not possess an α_2-subunit). This result is weakened somewhat, however, by the fact that this is a polyclonal antibody, whereas a monoclonal antibody was applied to the neuroblastoma cells.

As the amino acid sequence of the L-type α_1-subunit from skeletal muscle t-tubules is known,[21] we raised peptide-specific polyclonal antibodies against various regions of the sequence. One such antibody was raised against residues 143–158, which is thought to correspond to a portion of the putative extracellular loop between the membrane spanning regions S3 and S4 of the first domain. This antibody, raised against L-type channels, was tested on calcium channel currents in patch-clamped neuroblastoma NG 108 15 cells after 24 h applications and found to reduce T and N currents.[19,22] Thus this antibody, raised against L-type channels, also cross-reacted with T- and N-type channels, indicating that this antibody probes a region of L-, T-, and N-type channels, which are structurally similar in NG 108 15 cells. Such an antibody would perhaps be expected to be a suitable probe that might also cross-react at the neuromuscular junction. When mice were injected, however, with the peptide (conjugated with keyhole limpet hemocyanin, 100 μg, one sc injection), quantal content in the diaphragm was not reduced (FIG. 2). Antibody titers were low in these experiments, so further experiments were carried out applying the same antibody (as was used on the neuroblastoma cells) directly to mouse muscle *in vitro*. As can be seen in FIGURE 2, there was again no effect of this antibody on quantal content in the mouse diaphragm. Thus this antibody did not seem to affect the calcium channels at the neuromuscular junction, even though it recognised N-, T-, and L-type channels in neuroblastoma cells, providing further evidence that the neuromuscular calcium channel is different from N-, T-, and L-type channels.

METAL IONS

In further attempts to characterize the calcium channel types present at the neuromuscular junction, we have investigated the effects of metal ions selective for certain types of calcium channel.

The nickel ion is known to be a selective blocker of T-type channels at low concentrations (20 μM for NG 108 15 cells).[23] Therefore we have tested this concentration of nickel at the mouse diaphragm neuromuscular junction. FIGURE 3 shows that there was no effect on end-plate potential (epp) amplitude and, by implication, no effect on the nerve terminal calcium channels concerned with electrically evoked release. This suggests that the channel at the neuromuscular junction is not T-type. A higher concentration of nickel ion (1 mM) did affect epp amplitude at the neuromuscular junction (FIG. 4); however, such a higher concentration is expected to be relatively nonselective between channel types anyway.[24]

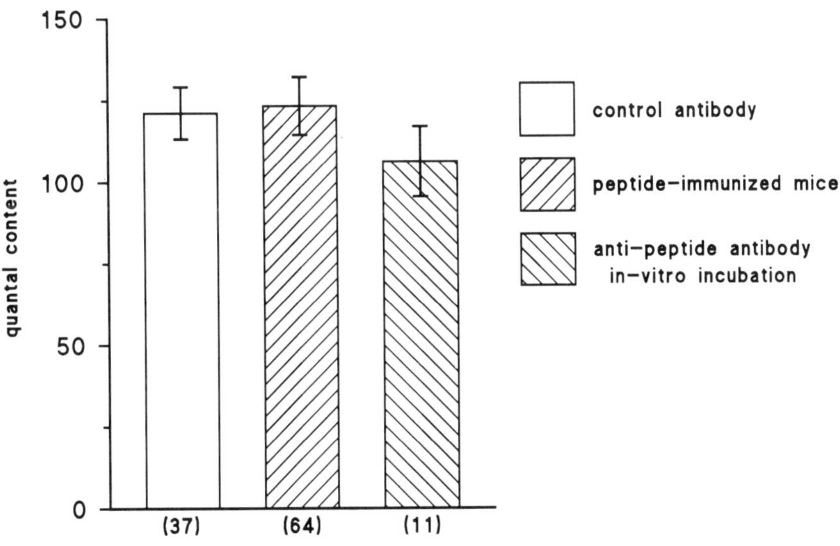

FIGURE 2. The effect of a peptide-specific anti-α_1-antibody on quantal content. Mice were immunized with an injection sc of 100 μg of the synthesized peptide (corresponding to residues 143–158 of the α_1-subunit of the L-type Ca^{2+} channel) conjugated to keyhold limpet hemocyanin in complete Freund's adjuvant; control mice were given only the adjuvant. Six to 14 weeks later the quantal content was measured from the isolated perfused (5–10 mL/min) diaphragm of each mouse. Also shown is the effect of 15 h *in vitro* incubation of mouse diaphragm muscles with the peptide-specific antibody at the same concentration (25 μg/mL) known to be effective on N and T currents in NG 108 15 cells. The mean quantal content ($\pm$SEM) is shown; the number of end-plates in each case is given in parenthesis. There were no significant effects (Student's t test).

The gadolinium ion is thought to give selective block of N-type channels at concentrations $\leq$5 μM in NG 108 15 cells[25] (in fact, at 5 μM for NG 108 15 cells, in our hands, not only N- but also T- and L-type channels were blocked; data not shown). We have therefore tested out the effect of this blocker at the mouse neuromuscular junction. FIGURE 5 shows that gadolinium (5 μM) had no effect on epp amplitudes, indicating an absence of N-type channels (and probably also T- and L-type channels) concerned with electrically evoked release

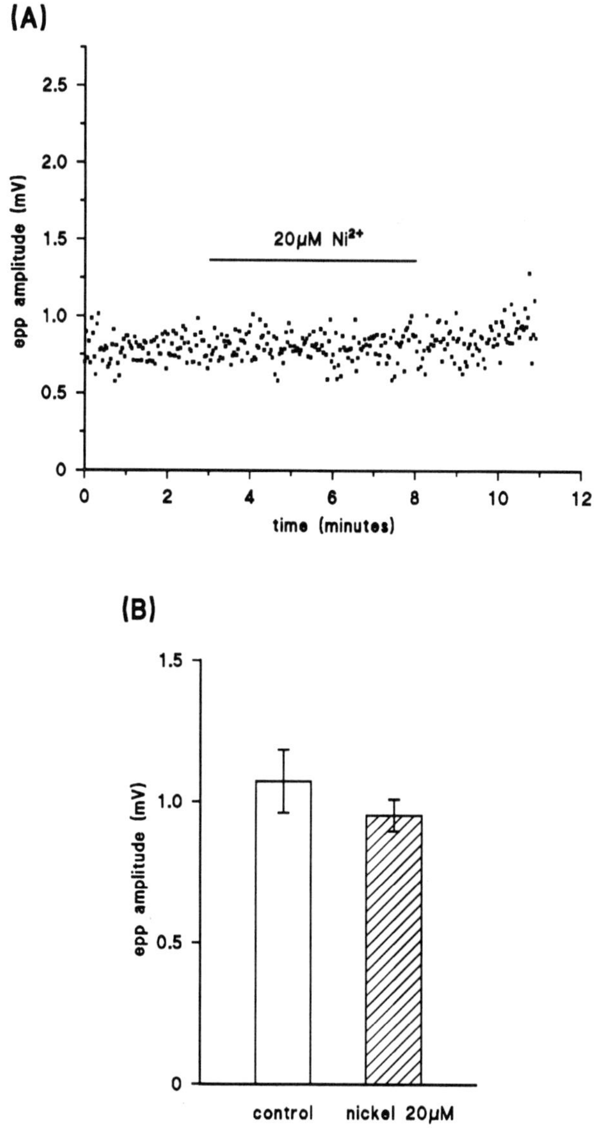

FIGURE 3. The effect of nickel chloride (20 μM) on end-plate potential. **A:** This shows the amplitude of each epp plotted against time from a single end-plate before, during, and after perfusion with nickel chloride (20 μM). **B:** This shows the mean ($\pm$SEM) end-plate potential amplitude, measured at five such end-plates. There was no significant effect (Student's t test).

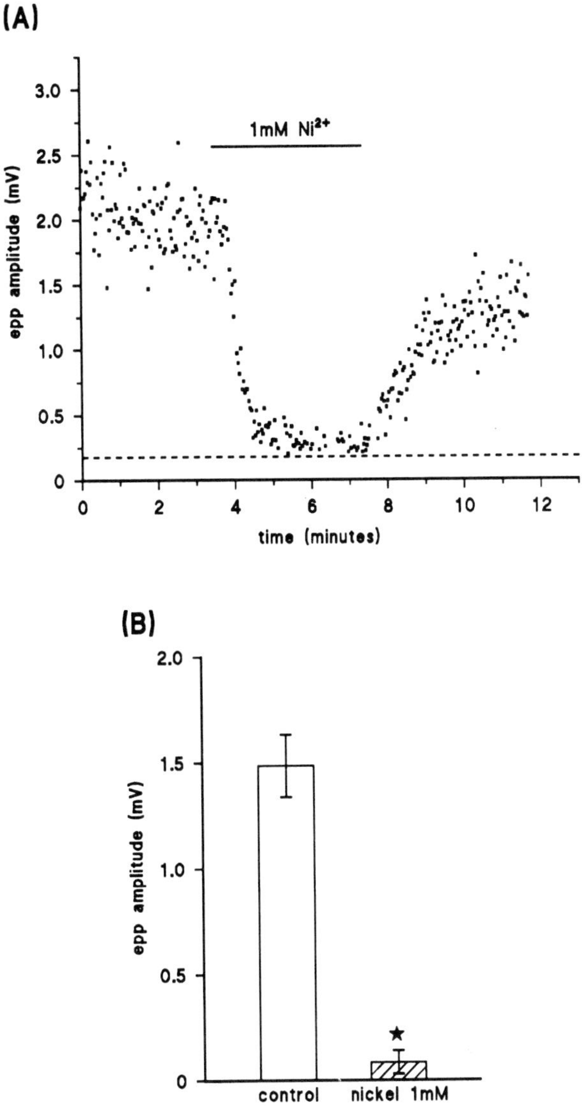

FIGURE 4. The effect of nickel chloride (1 mM) on end-plate potential. **A:** This shows the amplitude of each epp plotted against time from a single end-plate before, during, and after perfusion with nickel chloride (1 mM). The dotted line indicates the level of resolution of the epps within the noise. **B:** This shows the mean (±SEM) epp amplitude, measured at five such end-plates. *, significant difference from control (Student's *t* test, $p < 0.001$).

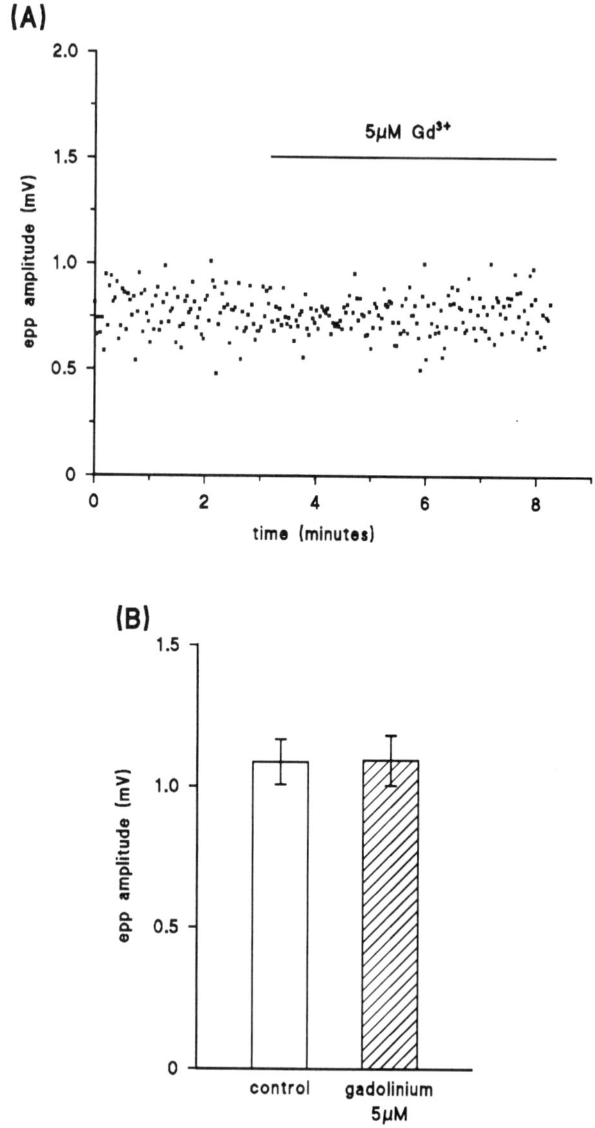

FIGURE 5. The effect of gadolinium chloride (5 μM) on end-plate potential. **A:** This shows the amplitude of each epp plotted against time from a single end-plate before, during, and after perfusion with gadolinium chloride (5 μM). **B:** This shows the mean ($\pm$SEM) epp amplitude, measured at five such end-plates. There was no significant effect (Student's *t* test).

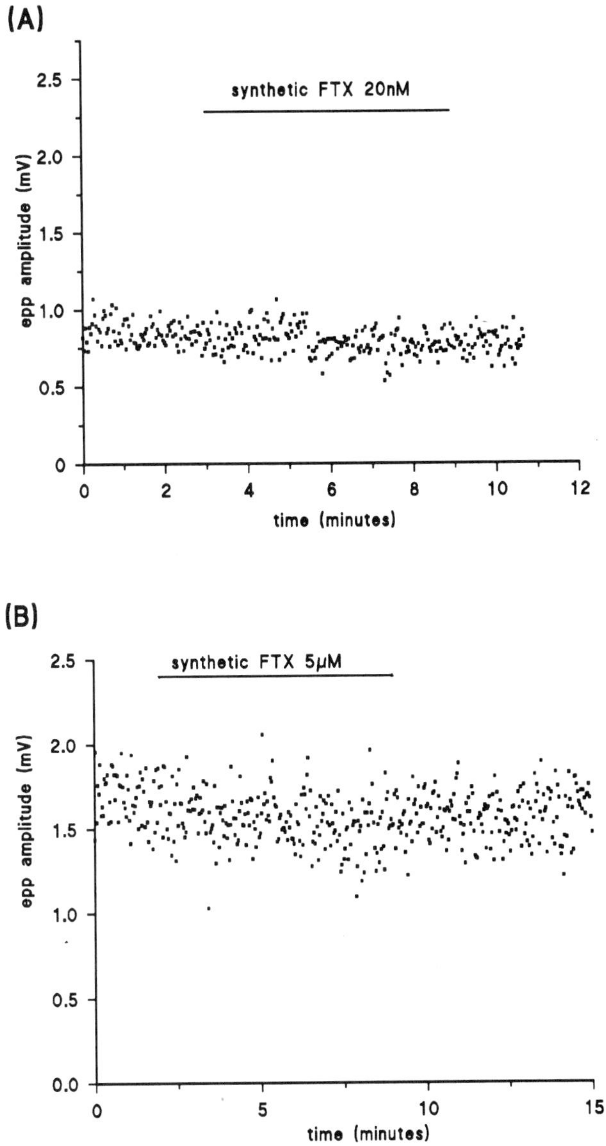

FIGURE 6. The effect of synthetic FTX on end-plate potential. The FIGURES both show the amplitude of each epp plotted against time from a single end-plate before, during, and after perfusion with synthetic FTX. **A:** 20 nM FTX; **B:** 5 μM FTX.

at mouse nerve terminals. In contrast to this result with mammalian muscle, gadolinium does seem to have effects at frog nerve terminals,[26] particularly at much higher doses.

SYNTHETIC FUNNEL WEB SPIDER TOXIN

As mentioned above, P-type channels are characterized by their ability to be blocked by certain components of funnel web spider toxin.[5,6] One such component of the toxin is a polyamine with a structure close to arginine polyamine ("synthetic FTX"). We have therefore investigated the effect of synthetic FTX (more precisely, FTX 3:3 see ref. 6) on the mouse neuromuscular junction. FIGURES 6 and 7 show that at concentrations of 20 nM, 5 μM, and 0.1 mM, synthetic FTX had no effects on epp amplitude and quantal content. Thus even high concentrations had no effect, suggesting the absence of P-type calcium channels at the mouse neuromuscular junction. Uchitel *et al.*,[27] however, have reported blocking effects of synthetic FTX at higher concentrations (EC_{50} around 1 mM), but such effects are likely to be nonspecific, with actions of synthetic FTX reported on both low and high voltage–activated channels at >0.01 mM in dorsal root ganglia.[28] Perhaps more significantly, applying the spider toxin itself (rather than synthetic FTX) to the neuromuscular junction[27] did lead to block of release at the motor nerve terminal, so that possibly calcium channels at the neuromuscular junction may, after all, have some characteristics of P-type channels.

CONCLUSIONS

The above pharmacological and immunological experiments were carried out in attempts to characterize the type of channel present at the mouse skeletal neuromuscular junction. One might take the mouse as "typical" of a mammalian neuromuscular junction, and of course it would be interesting to confirm this with other mammalian species. Results for mouse muscle are, in certain cases, different from frog muscle, which may, in its turn, also be different from other amphibia.

Applications of antibody, toxins, and metal ions specific for different calcium channel types indicated that the motor nerve terminal calcium channel cannot be simply defined as either L-, T-, N-, or P-type. Cross-reactivity, however, with LEMS antibody indicates that there are some regions of antigenic and hence structural similarity between nerve terminal calcium channels and L- and N-type channels. There may also be some similarity with P-type channels, because, although synthetic FTX was without effect at concentrations up to 0.1 mM, the spider toxin itself did have an effect.

The main subunits (α_1) of the L- and P-type channels have each been cloned and sequenced.[18,21] They are both proposed to have a membrane topography consisting of four domains, each comprising six membrane-spanning regions. It is tempting to speculate that the nerve-terminal calcium channels at the neuromuscular junction also have this topography. Our data could be explained if parts of the sequence of these channels are different from those for N-, T-, L-, and P-type channels in other tissues, but certain regions are highly homologous to L-, N-, and P-type channels.

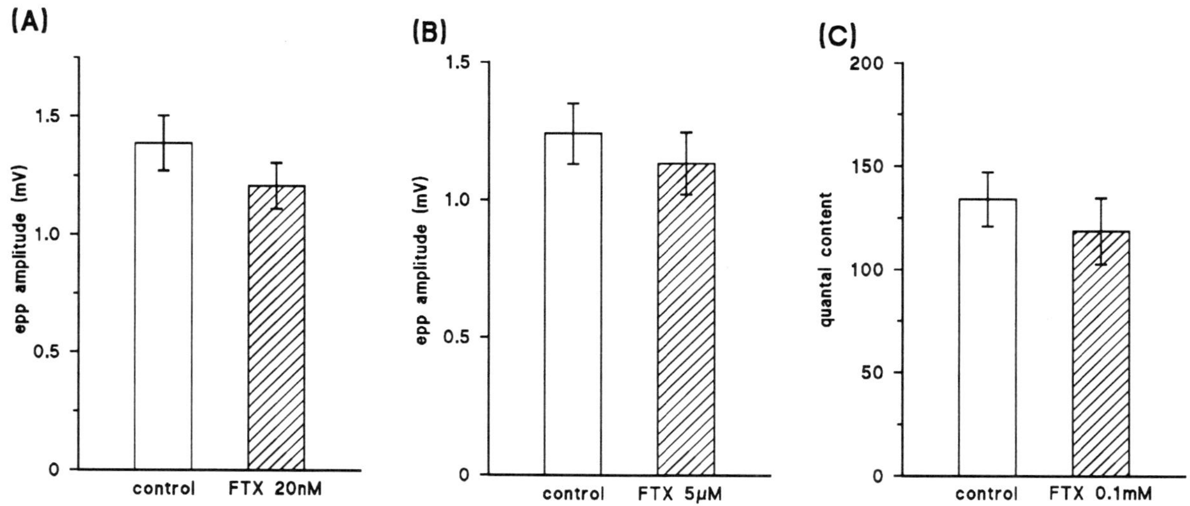

FIGURE 7. The effect of synthetic FTX on end-plate potential and quantal content. **A:** This shows the mean (±SEM) epp amplitude (eight end-plates) before and after perfusion with synthetic FTX, 20 nM. **B:** As in **A,** but with 5 μM FTX (five end-plates). **C:** This shows mean quantal content (±SEM), sampled from 13 control end-plates and 13 different end-plates, in the presence of synthetic FTX (0.1 mM) (5 muscles). There were no significant effects (Student's *t* test).

ACKNOWLEDGMENTS

We thank Dr. I. A. Pullar for a donation of synthetic FTX.

REFERENCES

1. WRAY, D. 1990. Electrophysiology of neuromuscular transmission. *In* Neuromuscular transmission: basic and applied aspects. A. Vincent & D. Wray, Eds.: 27–40. Manchester University Press. Manchester.
2. WRAY, D. W.-, B. LANG, J. NEWSOM-DAVIS & C. PEERS. 1989. Antibodies against calcium channels in the Lambert-Eaton myasthenic syndrome. Ann. N.Y. Acad. Sci. **560:** 269–277.
3. NOWYCKY, M. C., A. P. FOX & R. W. TSIEN. 1985. Three types of neuronal calcium channel with different calcium channel agonist sensitivities. Nature **316:** 440–444.
4. DASCAL, N. 1990. Analysis and functional characteristics of dihydropyridine-sensitive and -insensitive calcium channel proteins. Biochem. Pharmacol. **40:** 2, 1171–1178.
5. LLINÁS, R., M. SUGIMORI, J.-W. LIN & B. CHERKSEY. 1989. Blocking and isolation of a calcium channel from neurons in mammals and cephalopods utilizing a toxin fraction (FTX) from funnel-web spider poison. Proc. Natl. Acad. Sci. USA **86:** 1689–1693.
6. CHERKSEY, B. D., M. SUKIMORI & R. R. LLINÁS. 1991. Properties of calcium channels isolated with spider toxin, FTX. Ann. N.Y. Acad. Sci. **635:** 80–89.
7. LANG, B., J. NEWSOM-DAVIS, C. PEERS, C. PRIOR & D. W. WRAY. 1983. The effect of myasthenic syndrome antibody on presynaptic calcium channels in the mouse. J. Physiol. **390:** 257–270.
8. PEERS, C., B. LANG, J. NEWSOM-DAVIS & D. WRAY. 1990. Selective action of Lambert-Eaton myasthenic syndrome antibodies on Ca^{2+} channels in the neuroblastoma-glioma hybrid cell line NG 108 15. J. Physiol. **421:** 293–308.
9. BURGES, J. & D. W.-WRAY. 1989. Effect of the calcium-channel agonist CGP 28392 on transmitter release at mouse neuromuscular junctions. Ann. N.Y. Acad. Sci. **560:** 297–300.
10. SHER, E., C. GOTTI, N. CANAL, C. SCOPETTA, G. PICCOLO, A. EVOLI & F. CLEMENTI. 1989. Specificity of calcium channel autoantibodies in Lambert-Eaton myasthenic syndrome. Lancet **ii:** 640–643.
11. LANG, B., I. JOHNSTON, K. KEYS, G. ELRINGTON, B. MARQUEZE, C. LEVEQUE, N. MARTIN-MOUTOT, M. SEAGAR, T. HOSHINO, M. TAKAHASHI, M. SUGIMORI, B. D. CHERKSEY, R. LLINÁS & J. NEWSOM-DAVIS. 1993. Autoantibody specificities in Lambert-Eaton myasthenic syndrome. Ann. N.Y. Acad. Sci. This volume.
12. KIM, Y. I., J. K. W. BLANDINO & T. J. O'SHAUGHNESSY. 1993. Inhibitory action of Lambert-Eaton syndrome IgG on calcium currents in a thyroid C-cell line. Ann. N.Y. Acad. Sci. This volume.
13. PROTTI, D. A., L. SZCZUPAK, F. S. SCORNIK & O. D. UCHITEL. 1991. Effect of ω-conotoxin GVIA on neurotransmitter release at the mouse neuromuscular junction. Brain Res. **557:** 336–339.
14. SANO, K., K. ENOMOTO & T. MAENO. 1987. Effects of synthetic ω-conotoxin, a new type Ca^{2+} antagonist on frog and mouse neuromuscular transmission. Eur. J. Pharmacol. **141:** 235–241.
15. KERR, L. M. & D. YOSHIKAMI. 1984. A venom peptide with a novel presynaptic blocking action. Nature **308:** 282–284.
16. CATTERALL, W. A., M. J. SEAGAR, M. TAKAHASHI & K. NUNOKI. 1989. Molecular properties of dihydropyridine-sensitive calcium channels. Ann. N.Y. Acad. Sci. **560:** 1–14.
17. VARADI, G., P. LORY, D. SCHULTZ, M. VARADI & A. SCHWARTZ. 1991. Acceleration of activation and inactivation by the β subunit of the skeletal muscle calcium channel. Nature **352:** 159–162.
18. MORI, Y., T. FREIDRICH, M. S. KIM, A. MIKAMI, J. NAKAI, P. RUTH, E. BOSSE,

F. HOFMANN, V. FLOCKERZI, T. FURUICHI, K. MIKOSHIBA, K. IMOTO, T. TANABE & S. NUMA. 1991. Primary structure and functional expression from complementary DNA of a brain calcium channel. Nature **350:** 398–402.

19. PHILLIPS, H. J., G. G. WILSON, S. BALDWIN, R. NORMAN & D. WRAY. 1991. Effect of antibodies on calcium channels. Ann. N.Y. Acad. Sci. **635:** 489–491.
20. WRAY, D. W., C. PEERS & R. NORMAN. 1987. Effect of antibodies raised against calcium channels on neuromuscular transmission in the mouse. International Union of Pharmacology, 10th Int. Congr. Pharmacol. P172.
21. TANABE, T., H. TAKESHIMA, A. MIKAMI, V. FLOCKERZI, H. TAKAHASHI, K. KANGAWA, M. KOJIMA, H. MATSUO, T. HIROSE & S. NUMA. 1987. Primary structure of the receptor for calcium channel blockers from skeletal muscle. Nature **328:** 313–318.
22. WILSON, G. G., S. BALDWIN, R. I. NORMAN & D. WRAY. 1991. Possible similarity of structure between different types of calcium channels. Eur. J. Neurosci. (Suppl) **4:** 203.
23. CAULFIELD, M. P., J. ROBBINS & D. A. BROWN. 1992. Neurotransmitters inhibit the ω-conotoxin–sensitive component of Ca current in neuroblastoma × glioma hybrid (NG 108-15) cells, not the nifedipine-sensitive component. Pflugers Arch. **420:** 486–492.
24. KASAI, H. & E. NEHER. 1992. Dihydropyridine-sensitive and ω-conotoxin–sensitive calcium channels in a mammalian neuroblastoma-glioma cell line. J. Physiol. **448:** 161–188.
25. DOCHERTY, R. J. 1988. Gadolinium selectively blocks a component of calcium current in rodent neuroblastoma × glioma hybrid (NG108-15) cells. J. Physiol. **398:** 33–47.
26. MOLGO, J., E. DEL POZO, J. E. BANOS & D. ANGAUT-PETIT. 1991. Changes of quantal transmitter release caused by gadolinium ions at the frog neuromuscular junction. Br. J. Pharmacol. **104:** 133–138.
27. UCHITEL, O. D., D. A. PROTTI, V. SANCHEZ, B. D. CHERKSEY, M. SUGIMORI & R. LLINÁS. 1992. P-type voltage-dependent calcium channel mediates presynaptic calcium influx and transmitter release in mammalian synapses. Neurobiology **89:** 3330–3333.
28. SCOTT, R. H., M. I. SWEENEY, E. M. KOBRINSKY, H. A. PEARSON, G. H. TIMMS, I. A. PULLAR, S. WEDLEY & A. C. DOLPHIN. 1992. Actions of arginine polyamine on voltage and ligand-activated whole cell currents recorded from cultured neurones. Br. J. Pharmacol. **106:** 199–207.
29. WRAY, D. 1981. Prolonged exposure to acetylcholine, noise analysis and channel inactivation in cat tenuissimus muscle. J. Physiol. **310:** 37–56.

Presynaptic Calcium Channels and the Transmitter Release Mechanism

ELIS F. STANLEY

National Institute on Neurological Disorders and Stroke
Section on Synaptic Mechanisms
Building 9, Room 1E124
National Institutes of Health
Bethesda, Maryland 20892

INTRODUCTION

The mechanism whereby the influx of calcium ions triggers the exocytosis of classical neurotransmitters remains one of the key unsolved issues in neurobiology. Despite much effort, only three of the components involved in this process have been well established, the voltage-sensitive calcium channel, the calcium ions themselves, and the transmitter-filled secretory vesicle. Thus, membrane depolarization activates the channels, and the calcium ions that enter are believed to activate a synaptic vesicle release mechanism. Several studies have examined the characteristics of calcium ion influx, measured as an inward current through many simultaneously activated channels, and its relation to the secretory process, detected as a postsynaptic response, to further elucidate the secretory process. These studies have been limited, however, by an inability to examine secretion at the molecular level.

PRESYNAPTIC CALCIUM CHANNELS

We have recently developed a preparation that allows direct recording from the nerve terminal transmitter release face at the molecular level. This preparation is based on the unusual calyx-type synapse of the chick ciliary ganglion. The presynaptic nerve terminal at this synapse does not form small, bouton-type, contacts but instead spreads out to envelope the postsynaptic ciliary neuron in a thin sheet (see ref. 1). We isolated this calyx-type nerve terminal by enzymatic dissociation and applied patch clamp techniques in the whole-cell[2,3] and on-cell[4] configurations. With these techniques it was possible to characterize the properties of this vertebrate presynaptic calcium channel in some detail. The channels activated at about -40 mV and reached maximum activation at about 0 mV (FIG. 1). Pharmacological characterization demonstrated a resistance to dihydropyridine blockers,[5] but the inward current was blocked by ω-conotoxin.[3] The on-cell recording revealed one predominant calcium channel type with a 12-14 pS conductance (in 110 Ba). These channels could be classified, therefore, as within the N-type[6] group but were distinguished by a relatively low activation threshold and the near absence of voltage-dependent inactivation. The calcium channels were found to be located only on the release-face side of the nerve terminal and not on the opposite Schwann cell face[4] (FIG. 2), consistent with a role in transmitter secretion.[7]

The voltage range over which these presynaptic calcium currents are activated is remarkably similar to that of two other presynaptic structures: the nerve terminal in the squid giant synapse[8,9] and the hair cell of the inner ear.[10] The simplest explanation for this commonality may be the requirement to open in response to a typical axon-like action potential that only has a duration of 1 to 2 milliseconds. Within this short period the calcium influx must be sufficient to trigger the secretion of enough transmitter quanta to activate the postsynaptic neuron. The lower the threshold and the steeper the activation curve, the more channels will be recruited during a brief depolarization. On the other hand, the voltage threshold for activation must be significantly above the resting potential to avoid indiscriminate calcium channel activation and continuous transmitter release. Thus, in a synapse where one-to-one transmission of impulses is the norm, the activation kinetics of an "ideal" presynaptic calcium channel combines a low-voltage threshold with a steep dependence of activation on voltage (cf. also refs. 8 and 10).

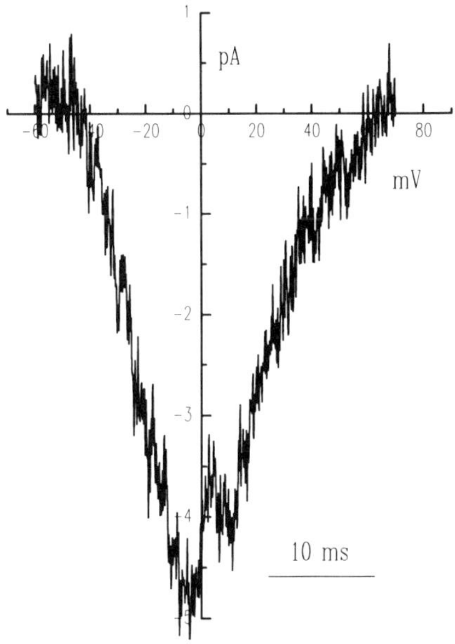

FIGURE 1. Current-to-voltage relation of nerve terminal calcium channels at low Ba concentration (6 mM Ba) with the electrode positioned on the release face of the nerve terminal. The patch included many calcium channels, probably several hundred (see ref. 4), and the current reflects average, rather than single-channel, behavior (see FIG. 2). The membrane potential was held at −70 mV and was ramped to +70 mV; the depicted current is the mean of two trials and was corrected for leak.

CORRELATION OF CHANNEL ACTIVITY AND QUANTAL SECRETION

The next question we addressed was, How close are the calcium channels to the vesicle release mechanism calcium binding sites? We have adapted a sensitive luminescent assay[11] for the detection of single quanta of acetylcholine (ACh) and have included the components of this assay in the patch pipette. This has enabled us to correlate single calcium channel openings with the evoked release of ACh.[12] We found that very brief openings of one or, at most, a very few calcium channels that admit several hundred calcium ions were sufficient to gate the release of a transmitter quantum (manuscript submitted). The affinity of the calcium binding site had been reported to be relatively low, requiring calcium ion concentrations in excess of 10^{-5} M to trigger secretion.[13–15] Inasmuch as such high concentrations are only generated within a few tens of nanometers from the calcium channel pore,[16–18] these results are consistent with the hypothesis that the vesicle release mechanism is located within a very small radius of the calcium channel.

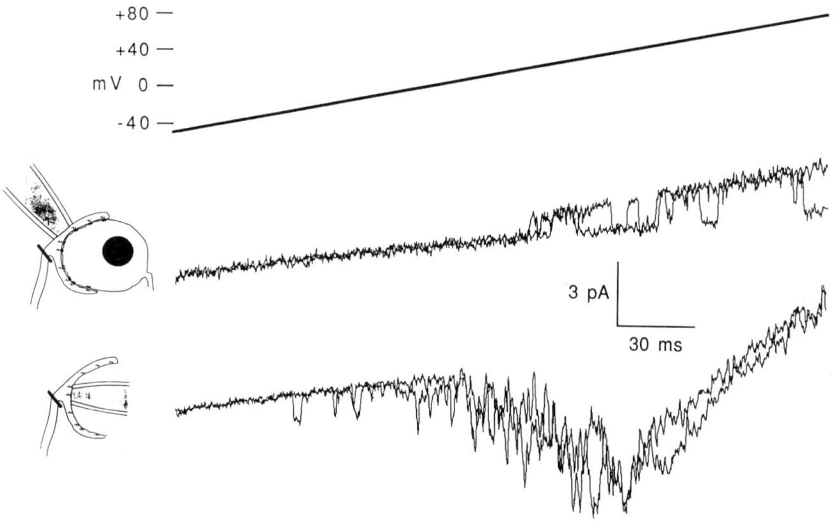

FIGURE 2. On-cell patch clamp recording of single ion channels from the outer and inner aspects of the presynaptic calyx nerve terminal during voltage ramps from a holding potential of −80 mV (the sloping line denotes membrane potential; only the segment from −50 to +80 is shown). Only outwardly directed currents were detected on the external aspect, whereas inward currents through single-calcium channels (in 110 Ba) were only recorded on the internal aspect. Single-channel fluctuations are seen at threshold voltages that were higher than in FIG. 1 (−20 to 0 mV) due to the high external Ba. (Modified from Stanley[4]).

ORGANIZATION OF THE NERVE TERMINAL RELEASE FACE

The main conclusion of this work is that the main components of gated exocytosis of secretory vesicles in fast-transmitting nerve terminals are located within a very small area of the transmitter release face. It is more than likely, therefore, that these structures comprise a multimolecular complex, a concept that is intuitively attractive because it would minimize the required calcium ion influx and any inherent delays due to ion diffusion. The complex must include a calcium channel, a vesicle docking site, a release mechanism with calcium binding sites, and, presumably, a vesicle reloading mechanism. Other proteins may be involved in holding these components together, as has recently been suggested,[19,20] and, inasmuch as release sites are usually clustered, additional crossbridges may link the release site complexes.

Perhaps the most intriguing questions relate to the nature of the calcium binding sites and the vesicle fusion mechanism. Very little is known about the latter, whereas any calcium binding protein in the region must be considered as a possible candidate. Proteins that have been suggested to play such a role include synaptophysin[21] and synaptotagmin.[22] It might be expected that the calcium-binding protein is an integral membrane protein because such a protein would be favorably located to translate the calcium signal into membrane fusion. It has been generally assumed, however, that this binding site is a part of the secretory vesicle, but it could equally well be a component of the nerve-terminal transmitter release face.

CONCLUSION

The calyx-type synapse of the chick ciliary ganglion was used to record single-calcium channels on the release face of a vertebrate presynaptic nerve terminal, and the activity of these channels was related to the transmitter release mechanism. The calcium channels had a threshold of about -40 mV (in 6 mM Ba) and were fully activated at about 0 mV. These kinetics are consistent with activation by a 1 to 2 ms action potential. A model of release-face organization is presented that involves a functional link between individual calcium channels and the transmitter release mechanism.

REFERENCES

1. Landmesser, L. & G. Pilar. 1972. The onset and development of transmission in the chick ciliary ganglion. J. Physiol. (Lond.) **222:** 691–713.
2. Stanley, E. F. 1989. Calcium currents in a vertebrate presynaptic nerve terminal: the chick ciliary ganglion calyx. Brain. Res. **505:** 341–345.
3. Stanley, E. F. & G. Goping. 1991. Characterization of a calcium current in a vertebrate cholinergic presynaptic nerve terminal. J. Neurosci. **11:** 985–993.
4. Stanley, E. F. 1991. Single calcium channels on a cholinergic presynaptic nerve terminal. Neuron **7:** 585–591.
5. Stanley, E. F. & A. H. Atrakchi. 1990. The calcium current in the presynaptic nerve terminal of the chick giant synapse is insensitive to the dihydropyridine nifedipine. Proc. Natl. Acad. Sci. USA **87:** 9683–9687.
6. Fox, A. P., M. C. Nowycky & R. W. Tsien. 1987. Kinetic and pharmacological properties distinguishing three types of calcium currents in chick sensory neurons. J. Physiol. (Lond.) **394:** 149–172.

7. PUMPLIN, D. W., T. S. REESE & R. R. LLINÁS. 1981. Are the presynaptic membrane particles the calcium channels? Proc. Natl. Acad. Sci. USA **78:** 7210–7213.
8. LLINÁS, R. R., I. Z. STEINBERG & K. WALTON. 1981. Presynaptic calcium currents in squid giant synapse. Biophys. J. **33:** 289–321.
9. AUGUSTINE, G. J., M. P. CHARLTON & S. SMITH. 1985. Calcium entry into voltage-clamped presynaptic terminals of squid. J. Physiol. (Lond.) **367:** 143–162.
10. HUDSPETH, A. J. & R. S. LEWIS. 1988. Kinetic analysis of voltage- and ion-dependent conductances in saccular hair cells of the bull-frog, *Rana catesbeiana*. J. Physiol. (Lond.) **400:** 237–274.
11. ISRAEL, M. & B. LESBATS. 1981. Chemiluminescent determination of acetylcholine, and continuous detection of its release from *Torpedo* electric organ synapses and synaptosomes. Neurochem. Int. **3:** 81–90.
12. STANLEY, E. F. 1992. Correlation of single calcium channel openings and acetylcholine release from the release face of a cholinergic nerve terminal. Soc. Neurosci. Abstr. **18:** 575. (Abstract).
13. ADLER, E. M., G. J. AUGUSTINE, S. N. DUFFY & M. P. CHARLTON. 1991. Alien intracellular calcium chelators attenuate neurotransmitter release at the squid giant synapse. J. Neurosci. **11:** 1496–1507.
14. YAMADA, W. M. & R. S. ZUCKER. 1992. Time course of transmitter release calculated from simulations of a calcium diffusion model. Biophys. J. **61:** 671–682.
15. LLINÁS, R. R., M. SUGIMORI & R. B. SILVER. 1992. Microdomains of high calcium concentration in a presynaptic terminal. Science **256:** 677–679.
16. CHAD, J. E. & R. ECKERT. 1984. Calcium domains associated with individual channels can account for anomalous voltage relations of Ca-dependent responses. Biophys. J. **45:** 993–999.
17. SIMON, S. M. & R. R. LLINÁS. 1985. Compartmentalization of the submembrane calcium activity during calcium influx and its significance in transmitter release. Biophys. J. **48:** 485–498.
18. FOGELSON, A. L. & R. S. ZUCKER. 1985. Presynaptic calcium diffusion from various arrays of single channels. Biophys. J. **48:** 1003–1007.
19. LEVEQUE, C., T. HOSHINO, P. DAVID *et al.* 1992. The synaptic vesicle protein synaptotagmin associates with calcium channels and is a putative Lambert-Eaton myasthenic syndrome antigen. Proc. Natl. Acad. Sci. USA **89:** 3625–3629.
20. BENNETT, M. K., N. CALAKOS & R. H. SCHELLER. 1992. Syntaxin: A synaptic protein implicated in docking of synaptic vesicles at presynaptic active zones. Science **257:** 255–259.
21. RHEM, H., B. WIEDENMANN & H. BETZ. 1986. Molecular characterization of synaptophysin, a major calcium-binding protein of the synaptic vesicle membrane. EMBO J. **5:** 535–541.
22. BROSE, N., A. G. PETRENKO, T. C. SÜDHOF & R. JAHN. 1992. Synaptotagmin: A calcium sensor on the synaptic vesicle surface. Science **256:** 1021–1025.

Neuronal Calcium Channels as Target for Lambert-Eaton Myasthenic Syndrome Autoantibodies[a]

E. SHER, E. CARBONE,[b] AND F. CLEMENTI

CNR Center of Cytopharmacology
Department of Medical Pharmacology
University of Milan
20129 Milan, Italy

[b]*Department of Anatomy and Human Physiology*
University of Turin
10125 Turin, Italy

INTRODUCTION

Lambert-Eaton myasthenic syndrome (LEMS) is an autoimmune synaptic disease characterized by a widespread deficiency in depolarization-induced neurotransmitter release from nerve terminals (see ref. 1 for a recent review). Over the last decade a lot of evidence has accumulated, suggesting that voltage-operated calcium channels (VOCC) are the molecular targets of this autoimmune syndrome.[2,3] As neuronal VOCCs form a large family of heterogeneous proteins,[4–9] the question is posed, On which VOCC subtypes do LEMS autoantibodies act?

Identification and characterization of neuronal VOCCs is not an easy task. Old criteria[10] based on kinetic differences among channel subtypes have been proved mostly inappropriate.[7,11] Presently, many laboratories are convinced that a more pharmacological approach based on the different sensitivity of the channels to drugs and toxins leads to more reliable results (TABLE 1).

Following these criteria, we have recently demonstrated that neurosecretory cells, previously believed to possess a single VOCC, may indeed express multiple VOCC subtypes[12–14] and that LEMS autoantibodies recognize a VOCC subtype specifically labeled by ω-conotoxin (ω-Ctx).

We will describe here our present knowledge of VOCC subtypes in neurosecretory cells, their pharmacological and functional relevance, and their role as autoantigens in LEMS.

VOCC SUBTYPES IN MAMMALIAN NEURONS AND SECRETORY CELLS

Using radioligand binding, fluorimetric, and electrophysiological methods, we screened several neuronal and secretory cells for their expression of VOCC sub-

[a] Part of the work described here was supported by CNR Grants (Special Project "Control of Neurotransmitter Release" and Oriented Project "Clinical Applications of Oncological Research") and by a Telethon-Italy Grant.

TABLE 1. Pharmacological Identification of VOCC Subtypes

Drugs and Toxins[a]	VOCC Subtypes			
	T	L	ω	P
DHP	+[b]	+ + +	–	–
ω-CTX (GVIA)	+[b]	–	+ + +	–
ω-Agatx[c] (IIIA)	–	+ +	+ +	–
ω-Agatx (IVA)	–	–	–	+ + +

[a] Effects of the agents: +, slight; + +, medium; + + +, strong; –, none.
[b] Both DHPs and ω-CTX have been shown to affect reversibly T-type VOCC in central and sensory neurons.
[c] Agatoxin.

types. Particularly, we analyzed several human neuroblastoma (NB) cell lines originating from the peripheral nervous system, rat pheochromocytoma PC12 cells, rat insulinoma RINm5F cells, and several human small-cell lung carcinoma (SCLC) cell lines. The latter were of particular interest because of the strong association between SCLC and LEMS.

Human NB IMR32 cells were shown to possess both low-voltage activated (LVA, T-type) and high-voltage activated (HVA, ω- and L-type) VOCCs.[15] The transient T-type VOCCs could be easily distinguished on the basis of their activation-inactivation kinetics, whereas ω- and L-type channels were identified primarily by their different sensitivity to ω-Ctx and dihydropyridine (DHP) derivatives. Kinetic differences between ω- and L-type were subtle and in disagreement with previously held views on VOCC properties.[10] Similar conclusions have been recently reinforced by other laboratories using murine NB cells[16] and primary cultures of rat[17] and chick[18] sensory neurons. In PC12,[12] SCLC,[13] and RINm5F[14] cells we also found evidence for heterogeneous VOCC subtypes. Besides the previously identified L-type channel, these cells also express a significant fraction of ω-type channels. In RINm5F cells there was also an unexpected large density of "P-like" channels defined as ω-Ctx- and DHP-insensitive HVA VOCCs. These findings are summarized in TABLE 2.

The relative proportion of VOCC subtypes varies among different cell lines and also, within the same clonal line, depends on cell differentiation. For instance, ω-type channels represent the majority (80–90%) of the HVA VOCCs in differenti-

TABLE 2. VOCC Subtypes in Different Neuronal and Secretory Cell Lines

Cell line	VOCC Subtypes[a]			
	T	L	ω	P[b]
IMR-32 (human neuroblastoma)	+	+	+ + +	–
PC12 (rat pheochromocytoma)	–	+ +	+ +	–[c]
RINm5F (rat insulinoma)	+	+ + +	+	+ +
SCLC (human small-cell lung carcinoma)	–	+ +	+ +	–[c]

[a] Fraction of HVA channels expressed: +, 10–20%; + +, 30–50%; + + +, 60–90%.
[b] We here refer, loosely, to P-type channels as high-voltage activated VOCC unaffected by either DHPs or ω-CTX.
[c] There is evidence that both PC12 and SCLC cells might also express a small fraction of P-type VOCC.

ated, secretory-competent,[19] human NB IMR32 cells,[15] but only a minority (10–20%) of HVA VOCCs in RINm5F cells.[14] On the other hand, undifferentiated IMR32 cells express mainly T- and L-type channels and a minority of ω-Ctx–sensitive VOCCs (10–20%). Similar differentiation-induced changes in VOCC-subtype expression have been also reported for PC12 cells.[20]

FUNCTIONAL ROLE OF VOCC SUBTYPES

Having proved the existence of multiple VOCC subtypes, the next step was to understand where they were located on the plasmamembrane, and what their functional role and their involvement was as targets of specific modulatory events. Recent studies have shown that L-type VOCCs are preferentially located in neuronal cell bodies and at the bases of major dendrites.[21] On the other hand, both we[22] and others[23,24] have found ω-type VOCCs to be clustered at the level of presynaptic active zones. P-type channels appear more ubiquitous, being located at the level of the soma, dendrites, axons, and nerve terminals in a variety of neurons.[25]

These localization studies, together with a number of functional data (reviewed in ref. 7), suggest that both ω- and P-type VOCCs could be involved in the control of neurotransmitter release, whereas neuronal L-type VOCCs (and P-type VOCCs) may be primarily implicated in integration processes occurring at the cell body level.

The picture appears quite different in small endocrine cells or in undifferentiated neurons. In RINm5F cells[14] and in human pancreatic β cells (Sher *et al.*, manuscript submitted), L-type VOCCs play a crucial role in the control of glucose-induced insulin release. Other VOCCs, however, are likely to be involved in this process. In RINm5F cells ω-Ctx blocks a component of insulin release,[14] and in human pancreatic β cells, P-like VOCCs may also be involved inasmuch as insulin secretion in these cells is not fully prevented by high doses of DHPs.

A specific physiological role for each VOCC subtype is also indirectly suggested by their selective modulation by hormones and neurotransmitters.[6] For instance, in both IMR32[26] and RINm5F cells,[27] we found that adrenergic α2 agonists potently inhibit the activation kinetics of the ω-type channel in a strict voltage-dependent manner, suggesting a dominant role of this channel in the control of neurotransmitter and hormone release and its modulation. The effect is dose-dependent, membrane limited, and mediated by a pertussis toxin–sensitive G protein. On the contrary, in RINm5F cells, L-type channels are partially inhibited by noradrenaline in a voltage-independent manner, and P-like channels are largely unaffected by the amine.[27a] Unspecific activation of G proteins achieved by intracellular injections of GTP-γ-S leads, however, to an overall marked inhibition of all three VOCCs expressed in these cells.

LEMS AUTOANTIBODIES SPECIFICITY

We have recently shown that autoantibodies obtained from LEMS patients are able to immunoprecipitate [^{125}I]ω-Ctx ω-type VOCCs,[28] a finding then confirmed by two other groups.[29,30] LEMS autoantibodies immunoprecipitate ω- type VOCCs extracted from both human NB,[28-30] human SCLC[13,29,30] rat insulinoma[14] cells, and rat brain membranes.[31] The titer of anti–ω-type VOCC antibodies determined using antigen extracted from the various sources is very similar, suggesting a strong immunogenic relationship between ω-type VOCCs expressed in different tissues and animal species. These human "polyclonal" antibodies were indeed the first antibodies able to recognize ω-type VOCCs.[28] More recently two specific antibodies directed respectively to the α2-[32] and the β-[33] subunits of the skeletal muscle L-type VOCC, have been shown to recognize not only neuronal L-type VOCCs, but to cross-react partially[32] or more extensively[33] with corresponding determinants of ω-type VOCC as well. Although there are little doubts concerning the recognition of ω-type VOCCs by LEMS autoantibodies, the issue of their possible cross-reactivity with other VOCC subtypes is still controversial.

Passive transfer experiments in the mouse, for instance, have demonstrated that LEMS autoantibodies react with presynaptic VOCCs, but not with cardiac (L-type) VOCCs.[34] Presynaptic VOCCs in the mouse seem to be insensitive to both ω-Ctx and DHPs[35,36] (but see also ref. 37), suggesting a possible recognition of both ω-type and "P-like" VOCCs by LEMS antibodies.

On the other hand, patch-clamp experiments performed on both bovine chromaffin[38] and murine NB[39] cells led the authors to propose that LEMS autoantibodies inhibit specifically L-type VOCCs rather than ω-type VOCCs. Most of the conclusions, however, were based on biophysical criteria (kinetic differences and holding potential sensitivities), which may result inappropriately, if not supported by pharmacological assays. Because both cell types are proved to express ω-Ctx–sensitive channels[16,40] with kinetic properties hardly distinguishable from L-type channels, these studies appear inconclusive concerning the issue of which VOCC subtype is recognized by LEMS autoantibodies. A combined use of toxins and LEMS antibodies in patch-clamp experiments might help in elucidating this point.

VOCCs are multimeric membrane proteins made up of the combination of different α1- (pore-forming), α2δ-, and β-subunits. In skeletal muscle, a γ-subunit is also present.[8] The role of α2δ- and β-subunits is not clear, but there is evidence for a modulatory effect of both subunits on VOCC expression and, for the β-subunit in particular, on VOCC kinetics.[41–43]

It is not yet known against which subunit(s) pathogenetic LEMS autoantibodies are directed. The fact that they affect VOCC function[38,44,45] and number[28,30,46,47] when added to intact cells suggests that they very likely recognize extracellular epitopes (the α1- and α2δ-subunits are transmembrane proteins, whereas the β-subunit is intracellular). Surprisingly enough, however, LEMS sera have been found to recognize VOCC's β-subunits (see M. Rosenfeld *et al.*, this volume). There is no evidence so far for a specific recognition of other VOCC subunits. Anti–β-subunit antibodies are found only in LEMS patients' sera and might represent a valuable marker for the disease, although no pathogenetic role for β-subunit–specific autoantibodies has yet been demonstrated.

In conclusion, LEMS autoantibodies, besides recognizing human ω-type VOCCs, might also cross-react with other VOCC subtypes. This would not be particularly surprising in light of the strong molecular homology observed among the different VOCC subunits cloned up to now.[8,9]

DIAGNOSTIC VALUE OF ANTI–ω-TYPE VOCC ANTIBODY ASSAY

We have already reported[28] that the detection of anti–ω-type VOCC antibodies was highly diagnostic for LEMS, as revealed also by double-blind experiments.

In that first series of experiments, we found 90% of LEMS patients (out of 22) positive in our assay; furthermore, 42% of patients with SCLC without LEMS (out of 7) were also positive. We have now expanded our case studies and found that out of 52 LEMS patients tested, 78.8% were positive for anti-VOCC antibodies, and that out of 32 patients with SCLC without LEMS, 40.6% were also positive. We can conclude that having expanded the population under study, the first results were essentially confirmed. The diagnostic value of the assay extends to those clinical cases in which the neurophysiological examination is not conclusive or in those cases in which more than one neurological disorder might overlap.

For instance "seropositive" LEMS can be associated with seropositive subacute sensory neuropathy (E. Sher *et al.*, unpublished results); in another case seropositive LEMS was associated with a peripheral neuropathy with antineuronal antibodies;[48] in a third case seropositive LEMS was associated with both a peripheral sensory neuropathy and the nonimmunological paraneoplastic syndrome SIADH.[49] The association of seropositive LEMS with seropositive myasthenia gravis has been also reported.[28,50] As in myasthenia gravis, the absolute anti–ω-type VOCC antibody titer does not strictly correlate with the clinical status of the patients. In the single patient, however, there is a good correlation between changes in anti–ω-type VOCC antibody titers and clinical condition.[30]

In conclusion, the anti–ω-type VOCC antibody assay could well become a useful aid in the differential diagnosis of neuromuscular diseases and in the follow-up of LEMS patients.

ANTI-VOCC ANTIBODIES IN LEMS AND LUNG CANCERS

LEMS was first recognized in association with a SCLC.[51] It is now known that 60–70% of LEMS cases are indeed associated with SCLC.[52] LEMS is, however, sporadically reported in association with other tumors as well.[52]

SCLC is composed of secretory cells that express a number of neuronal markers.[53] SCLC cells show calcium-dependent action potentials[54] and depolarization-induced calcium influx,[44,45] suggesting the presence of VOCCs.

Calcium influx in SCLC cells is reduced after treatment with LEMS antibodies,[43,44] and LEMS antibodies immunoprecipitate ω-type VOCCs from SCLC cell lines.[13,29] Most of these studies have been performed on *in vitro* growing SCLC cell lines. We have recently demonstrated, however, that an SCLC obtained at surgery from a patient with seropositive LEMS also contained [^{125}I]ω-Ctx–binding sites (TABLE 3), which were recognized by both the patient's antibodies and by other LEMS sera (TABLE 4). These results strengthened the hypothesis that ω-type VOCC expressed by SCLC cells could represent the primary trigger for the autoimmune anti-VOCC response leading to LEMS.[3] SCLC cells, however, also express other VOCC subtypes[13,44] (and Lang *et al.*, this volume), and it is still unclear whether they also play a role in inducing the autoimmune attack.

Contradictory results have been reported on the prevalence (and/or titer) of anti-VOCC antibodies in LEMS patients with or without SCLC.[28–30] These differences might be related to technical differences in the assay (which is new and might need further refinements) or to the still relatively small number of patients examined.

We have mentioned above that we found 30–40% of SCLC patients without LEMS having positive anti-VOCC antibody titers. At least three possibilities have to be considered: first, a patient might have LEMS that had not been recognized;

TABLE 3. $[^{125}I]\omega$-CTX Binding to Surgically Removed SCLC[a]

$[^{125}I]\omega$-CTX (pM)	Tumor 1	Tumor 2
	(fmoles/mg of protein)	
18	2.8	2.3
36	3.5	7.2
45	5.9	7.5
81	9.7	12.3
100	13.2	11.1
172	12.3	8.9
210	16.3	16.7

[a] $[^{125}I]\omega$-CTX binding was linearly dependent on tissue concentration (not shown). Each point was performed in triplicate. Nonspecific binding was calculated in parallel tubes in the presence of an excess (100 nM) of unlabeled ω-CTX and then subtracted from total binding. The data shown represent only specific binding.

second, a patient might have a subclinical form of LEMS; third, the antibodies in some SCLC patients might have no pathogenetic effect with regard to LEMS, but might induce some biological effect on tumor growth and/or could be important serological markers of SCLC. More extensive studies will clarify these points.

In a few cases, seropositive LEMS has been described in association with lung tumors other than SCLC. We have described the case of a patient who developed an epidermoid lung carcinoma one year after the clinical and serological diagnosis of LEMS.[55] Also Lennon and Lambert[29] found two seropositive LEMS patients who had non-SCLC lung tumors. Although some tumors might be associated with LEMS just by chance,[52] our idea is that other tumors besides SCLC might also express neuronal antigens (VOCC in particular) that might trigger autoimmunity. We are referring, in particular, to the recently recognized small percentage ($\approx$10%) of non-SCLC lung tumors that have been shown to express neuroendocrine markers.[56]

The recognition of a paraneoplastic syndrome can lead to the identification of a clinically unsuspected underlying tumor. A diagnosis of LEMS, possibly with the aid of the new serological test, is therefore a warning for the clinician and for

TABLE 4. Anti-VOCC Antibody Titers Determined Using as Antigen $[^{125}I]\omega$-CTX–VOCC Complexes Extracted from Surgically Removed SCLC[a]

Serum	Antibody titer (pM)
Control	17
Autologous LEMS	101
Heterologous LEMS	118

[a] Increasing amounts (from 1 to 5 μL) of control or LEMS serum were added to a fixed amount of antigen. Antihuman IgG was added, and the pellets were washed and counted for radioactivity. The antibody titer is expressed as pmoles of $[^{125}I]\omega$-CTX binding sites precipitated per liter of serum.[28]

the patient who might develop a lung cancer soon. Could the newly described serological test be really useful for making diagnoses of lung cancer independently of LEMS and, if so, with what degree of specificity?

We are still far off from giving a definitive answer to this question; however, some preliminary results are encouraging. We have tested the sera of 97 patients with lung diseases without clinically determined LEMS: 32 with SCLC, 31 with non-SCLC, and 34 with inflammatory lung diseases. All the patients with non-SCLC or inflammatory lung diseases (ILD) were negative in our assay. On the other hand, 40% of SCLC patients were positive.[56a] Although some of these SCLC patients might have an undiagnosed LEMS (see above), it is, in any case, worth noting the specificity of positive results for the SCLC population. Further studies will possibly confirm the validity of this assay in the diagnosis and follow-up of lung cancer patients.

CONCLUSION

We have briefly summarized our current knowledge of the physiological and pharmacological properties of voltage-operated calcium channels in neuronal and neurosecretory cells. At least four VOCC subtypes have been characterized, and the identification of newly cloned neuronal VOCC subtypes will certainly increase their number.

In our view, the only VOCC subtypes clearly proved to be recognized by LEMS autoantibodies are the human ω-type channels of human peripheral neurons, rat brain, and neuroendocrine cells and, possibly, the P-like channels of mouse neuromuscular junctions. The future availability of cloned calcium channel subunits will help to elucidate autoantibody specificity in LEMS and their mechanism of action. At present the anti–ω-type VOCC antibody assay appears to be the only valid and routinely useful serological test availabe for the detection of LEMS antibodies. The identification (and definition of the functional role) of anti-VOCC antibodies in lung cancer patients will bring more information on the immunity of cancer and might suggest new specific therapeutic approaches.

ACKNOWLEDGMENTS

We are indebted to Drs. M. Passafaro, E. Biancardi, A. Pollo, and M. Lovallo for their valuable work and continuous assistance. We also thank the many neurologists and pneumologists that made this work possible.

REFERENCES

1. Lambert, E. H. & D. Elmqvist. 1971. Ann. N.Y. Acad. Sci. **183:** 183–199.
2. Vincent A., B. Lang & J. Newsom-Davis. 1989. Trends Neurosci. **12:** 496–502.
3. Sher, E., E. Biancardi, M. Passafaro & F. Clementi. 1991. FASEB J. **5:** 2677–2683.
4. Hosey, M. M. & M. Lazdunski. 1988. J. Membr. Biol. **104:** 81–105.
5. Bean, B. P. 1989. Annu. Rev. Physiol. **51:** 367–384.
6. Carbone, E. & D. Swandulla. 1989. Prog. Biophys. Mol. Biol. **54:** 31–58.
7. Sher, E. & F. Clementi. 1991. Neurosci. **42:** 301–307.
8. Miller, R. J. 1992. J. Biol. Chem. **267:** 1403–1406.

9. Tsien, R. W., P. T. Ellinor & W. A. Horne. 1991. Trends Pharmacol. Sci. **12:** 349–353.
10. Tsien, R. W., D. Lipscombe, D. V. Madison, K. R. Bley & A. P. Fox. 1988. Trends Pharmacol. Sci. **11:** 431–437.
11. Swandulla, D., E. Carbone & H. D. Lux. 1991. Trends Neurosci. **14:** 46–51.
12. Sher, E., A. Pandiella & F. Clementi. 1988. FEBS Lett. **235:** 178–182.
13. Sher, E., A. Pandiella & F. Clementi. 1990. Cancer Res. **50:** 3892–3896.
14. Sher, E., E. Biancardi, A. Pollo, E. Carbone, G. Li, C. B. Wollheim & F. Clementi. 1992. Eur. J. Pharmacol. **216:** 407–414.
15. Carbone, E., E. Sher & F. Clementi. 1990. Pflügers Arch. **416:** 170–179.
16. Kasai, H. & E. Neher. 1992. J. Physiol. (Lond.) **448:** 161–188.
17. Regan, L. J., D. W. Y. Sah & B. P. Bean. 1991. Neuron **6:** 269–280.
18. Cox, D. H. & K. Dunlap. 1992. J. Neurosci. **12**(3): 906–914.
19. Sher, E., S. Denis-Donini, A. Zanini, C. Bisiani & F. Clementi. 1989. J. Cell Biol. **108:** 2291–2300.
20. Usowicz, M. M., H. Porzig, C. Becker & H. Reuter. 1990. J. Physiol. (Lond.) **426:** 95–116.
21. Ahlidanian, M. K., R. E. Westenbroek & W. A. Catterall. 1990. Neuron **3:** 819–832.
22. Torri Tarelli, F., M. Passafaro, F. Clementi & E. Sher. 1991. Brain Res. **547:** 331–334.
23. Robitaille, R., E. M. Adler & M. P. Charlton. 1990. Neuron **5:** 773–779.
24. Cohen, M. W., O. T. Jones, K. J. Angelides. 1991. J. Neurosci. **11:** 1032–1039.
25. Hillman, D., S. Chen, T. T. Aung, B. Cherksey, M. Sugimori & R. R. Llinás. 1991. Proc. Natl. Acad. Sci. USA **88:** 7076–7080.
26. Pollo, A., M. Lovallo, E. Sher & E. Carbone. 1992. Eur. J. Physiol. **422:** 75–83.
27. Aicardi, G., A. Pollo, E. Sher, E. Carbone. 1991. FEBS Lett. **281:** 201–204.
27a. Polla, A., M. Lavallo, E. Sher & E. Carbone. 1993. Eur. J. Physiol. In press.
28. Sher, E., C. Gotti, N. Canal, C. Scoppetta, G. Piccoli, A. Evoli & F. Clementi. 1989. Lancet **ii:** 640–643.
29. Lennon, V. A. & E. M. Lambert. 1989. Mayo Clin. Proc. **64:** 1498–1504.
30. Leys, K., B. Lang, I. Johnston & J. Newsom-Davis. 1991. Ann. Neurol. **29:** 307–314.
31. Leveque, C., T. Hoshino, P. David, Y. Shoji-Kasai, K. Leys, A. Omori, B. Lang, O. El Far, K. Sato, N. Martin-Moutot, J. Newsom-Davis, M. Takahashi & M. J. Seagar. 1992. Proc. Natl. Acad. Sci. USA **89:** 3625–3629.
32. Ahlijanian, M. K., J. Striessnig & W. A. Catterall. 1991. J. Biol. Chem. **266**(30): 20192–20197.
33. Sakamoto, J. & K. P. Campbell. 1991. J. Biol. Chem. **266**(28): 18914–18919.
34. Lang, B., J. Newsom-Davis & D. W. Wray. 1988. Proc. R. Soc. Lond. Biol. Sci. **235**(1278): 103–110.
35. Sano, K., K. Enomoto & T. Maeno. 1987. Eur. J. Pharmacol. **141:** 235–241.
36. De Luca, A., M. J. Rand, J. J. Reid & D. F. Story. 1991. Toxicon **29**(3): 311–320.
37. Protti, D. A., I. Szczupak, F. S. Scornik & D. D. Uchitel. 1991. Brain Res. **557:** 336–339.
38. Kim, Y. I. & E. Neher. 1988. Science **239:** 405–408.
39. Peers C., B. Lang, J. W. Newsom-Davis & D. Wray. 1990. J. Physiol. (Lond.) **421:** 293–308.
40. Artalejo, C. R., R. L. Perlman & A. P. Fox. 1992. Neuron **8:** 85–95.
41. Varadi, G., Ph. Lory, D. Schultz, M. Varadi & A. Schwartz. 1991. Nature **352:** 159–162.
42. Lacerda, A. E., H. S. Kim, P. Ruth, E. Perez-Reyes, V. Flockerzi, F. Hoffmann, L. Birnbaumer & A. M. Brown. 1991. Nature **352:** 527–530.
43. Williams, M. E., P. F. Brust, D. H. Feldman, S. Patthi, S. Simerson, A. Maroufi, A. F. McCue, G. Velicelebi, S. B. Ellis & M. M. Harpold. 1992. Science **257:** 389–395.
44. Roberts, A., S. Perera, B. Lang, A. Vincent, J. Newsom-Davis. 1985. Nature **317:** 737–739.

45. De Aizpurua, H. J., E. H. Lambert, G. E. Griesmann, B. M. Olivera & V. A. Lennon. 1988. Cancer Res. **48:** 4719–4724.
46. Nagel, A., A. G. Engel, B. Lang, J. Newsom-Davis & T. Fukuoka. 1988. Ann. Neurol. **24:** 552–558.
47. Passafaro, M., F. Clementi & E. Sher. 1992. J. Neurosci. **12:** 3372–3379.
48. Comola, M., R. Nemni, E. Sher, A. Quattrini, A. Faravelli, G. Comi, M. Corbo, F. Clementi & N. Canal. 1992. Eur. Neurol. In press.
49. Riva, M., E. Sher, A. Brioschi, L. Bucci, D. D'Urso, V. Soresi, G. Palmieri & F. Clementi. 1990. Farmaci & Terapia **7:** 226–230.
50. Newsom-Davis J., K. Leys, A. Vincent, I. Ferguson, G. Modi & K. Mills. 1991. J. Neurol. Neurosurg. Psychiatry **54:** 452–453.
51. Lambert, E. H., L. M. Eaton & E. D. Rooke. 1956. Am. J. Physiol. **187:** 612–613.
52. O'Neill, J. H., N. M. F. Murray & J. Newsom-Davis. 1988. Brain **111:** 577–596.
53. Gazdar, A. F., D. N. Carney, J. G. Guccion *et al.* 1981. *In* Small Cell Lung Cancer. 145–175. Grune & Stratton. New York.
54. McCann, F. V., O. S. Pettengill, J. J. Cole *et al.* 1981. Science **212:** 1155–1157.
55. Sher, E., M. Comola, R. Nemni, N. Canal & F. Clementi. 1990. Lancet **i:** 413.
56. Jensen, S. M., A. F. Gazdar, F. Cuttitta, E. K. Russell & R. I. Linnoila. 1990. Cancer Res. **50:** 6068–6074.
56a. Pelucchi, A. *et al.* 1993. Am. Rev. Respir. Dis. In press.

Autoantibody Specificities in Lambert-Eaton Myasthenic Syndrome

BETHAN LANG,[a] IRENE JOHNSTON,[a]
KATHERINE LEYS,[a] GILES ELRINGTON,[a]
BEATRICE MARQUEZE,[a,b] CHRISTIAN LEVEQUE,[b]
NICOLE MARTIN-MOUTOT,[b] MICHAEL SEAGAR,[b]
TOSHIMITSU HOSHINO,[c] MASAMI TAKAHASHI,[c]
MUTSUYUKI SUGIMORI,[d] BRUCE D. CHERKSEY,[d]
RODOLFO LLINÁS,[d] AND JOHN NEWSOM-DAVIS[a]

[a]*Neurosciences Group*
Institute of Molecular Medicine
University of Oxford
John Radcliffe Hospital
Oxford OX3 9DU, United Kingdom

[b]*Neurobiologie des Canaux Ioniques*
INSERM CJF 9016
Faculté de Médecine Nord
Bd. Pierre Dramard
13326 Marseille Cedex 15, France

[c]*Mitsubishi-Kasei Institute of Life Sciences*
11 Minamiooya
Machida-shi
Tokyo, Japan

[d]*Department of Physiology*
New York University Medical Center
New York, New York 10016

A myasthenic syndrome was first described in association with lung cancer by Anderson in 1953;[1] however, subsequent careful physiological studies of similar patients by Lambert and his colleagues clearly distinguished the syndrome from myasthenia gravis.[2]

Patients with the Lambert-Eaton myasthenic syndrome (LEMS) present with proximal muscle weakness and depressed or absent tendon reflexes. Autonomic symptoms such as dry mouth, constipation, and sexual impotence are also common. In approximately 60% of cases there is an associated small cell carcinoma of the lung (SCLC),[3] a tumor that is thought to be derived from neuroectodermal tissue. There is also an increased frequency of autoimmune diseases in these LEMS patients, first detailed by Guttmann.[4] Further reports have shown a higher than expected incidence of thyroid disease, vitiligo, and celiac disease.[3] The incidence of organ-specific autoantibodies (thyroid, gastric parietal cell, and skeletal muscle) also appears elevated in the noncancer-associated LEMS cases.[5] The inference that LEMS might be an autoimmune disease is strengthened by immunogenetic associations with HLA-B8 and the particular immunoglobulin heavy chain marker Glm(2).[6]

Direct evidence that LEMS might be antibody-mediated was provided by the improvement of the patients following a course of plasma exchange.[7] Subsequent physiological, morphological, and biochemical studies lead to the identification of the antigenic target. Injection of IgG prepared from the plasma of LEMS patients into mice reproduced the physiological abnormalities. There was a significant reduction in the mean quantal content of the end-plate potential (*i.e.* the number of acetylcholine (ACh) packages released by each nerve impulse).[8,9] Pharmacological studies further demonstrated a reduction in the amount of ACh released in response to a K^+-induced depolarization in the injected mice.[10] Freeze-fracture electromicroscopy of the diaphragms of these mice showed a disorganization in the normal orderly double parallel row array of the active-zone particles (AZP), thought to be the morphological representations of voltage-gated calcium channels (VGCC) on the presynaptic nerve terminal. In contrast with the orderly arrays of AZPs in control-injected animals, they were clustered and reduced in numbers in the test mice,[11] similar to the findings previously reported in LEMS patients.[12] Functional effects of VGCC were directly investigated by measuring the *in vitro* effects of LEMS IgG on the K^+-stimulated $^{45}Ca^{2+}$ flux in cultured human SCLC lines. Incubating the cells in LEMS IgG caused a significant reduction in $^{45}Ca^{2+}$ influx, compared with control cells, which was time- and concentration-dependent.[13,14]

The voltage-gated calcium channels have recently been classified into four main groups (T, L, N, and P) depending on their pharmacological and physiological characteristics.[15,16] In this paper we investigate the heterogeneity of the autoantibodies in LEMS and how they interact with the different VGCC and associated presynaptic proteins.

L- AND T-TYPE VGCC

The most widely investigated type of calcium channel is the L-type VGCC, which mediates long-lasting calcium currents following strong depolarization. Pharmacologically these channels interact with dihydropyridines, such as the agonist BAY K 8644 and the antagonist nitrendipine. Several authors have reported LEMS antibodies interacting with these channels. Kim and Neher[17] demonstrated that IgG fractions prepared from the plasma of LEMS patients reduced the VGCC currents by 40% in adrenal chromaffin cells, whereas similar effects were observed on the calcium-dependent hormone secretion in rat anterior pituitary cells.[18] Exocytosis in these secretory cells is considered to be mediated predominantly through L-type VGCC. Peers *et al.* also showed that LEMS IgG selectively reduced the L-type Ca^{2+} channel current in undifferentiated NG108-15 cells, although having no effect on the transient (T-type) calcium channels.[19]

N-TYPE CALCIUM CHANNELS

The purification of the polypeptide ω-conotoxin (ω-CgTx) from the venom of the marine fish-eating snail *Conus geographus* has permitted the investigation of another subgroup of VGCC (ω-type) to which it binds. This 27 amino acid peptide selectively blocks neuronal VGCC from a variety of preparations[20,21] and also from SCLC lines.[22,23] Using radiolabeled toxin, an immunoprecipitation assay has been set up for the detection of anti–ω-CgTx receptor antibodies in the serum of patients with LEMS.[24–26] This assay is extremely sensitive and is capable of detecting picomolar amounts of the antibodies.

In a previous study[27] we investigated 36 patients with LEMS for serum antibodies to ^{125}I-labeled ω-CgTx-VGCC (^{125}I-ω-CgTx-VGCC) complexes extracted from a human neuroblastoma cell line SKN-SH. Forty-four percent of these patients had significant levels of antibodies (range 30–1466 pM) compared to healthy controls. The incidence of positivity was greater in LEMS patients without a detectable carcinoma (NCD-LEMS) than in LEMS patients with SCLC (SCLC-LEMS). Low antibody titers (30–90 pM) were observed, however, in some SCLC patients without neurological disease and in 67% of patients with rheumatoid arthritis or systemic lupus erythematosus.

We have now further extended the study to include 74 LEMS patients, 55 SCLC patients without LEMS, 26 patients with other neurological and/or autoimmune diseases, and 20 healthy controls. False positives in this sensitive assay were frequent in disorders characterized by high levels of circulating immunoglobulins (*e.g.* rheumatoid arthritis and SLE). Therefore the concentration of total IgG was measured in each sample, and the presence of antibody titers among the control sera were found to correlate closely with elevated IgG levels in these patients. After correction for IgG concentration, only 2 out of 81 diseased controls (2.5%), both SCLC patients with no symptoms of LEMS, had significant levels of ^{125}I-ω-CgTx-VGCC antibodies, whereas 22/74 (30%) of all LEMS patients tested had titers greater than 3 SD above the mean of the healthy controls. Between patients, the titer does not relate to disease severity.[27] Within individual patients, however, longitudinal studies have shown that as the anti–ω-CgTx antibody titer decreases the patient gets stronger.[27]

Previous attempts to characterize the LEMS antigen using biochemical techniques, such as Western blotting of crude membranes, have not been very successful.[28,29] Here we describe further experiments using a partially purified ω-CgTx–sensitive calcium channel preparation.[30]

Rat brain synaptic membranes were labeled with ^{125}I-ω-CgTx and extracted in Triton X100. ω-CgTx receptors were purified approximately 250-fold by two-step affinity chromatography. LEMS IgG or serum was used to probe Western blots of this partially purified ω-CgTx receptor (FIG. 1). Five of the 19 samples initially tested labeled a single major protein of molecular mass 58 kDa. These samples represented IgG and/or serum taken from three LEMS patients, two of which had a high titer of anti–^{125}I-ω-CgTx-VGCC antibodies. Studies with a larger group of LEMS patients (TABLE 1) showed a high correlation between the antibody titer and reactivities with the 58 kDa protein in Western blotting. This band was not observed in any of the control preparations.

IgG from one of the LEMS patients was immunopurified by incubation with 58 kDa protein immobilized on nitrocellulose. The subsequently eluted IgG was still able to precipitate the solubilized ω-CgTx receptor in a concentration-dependent manner (FIG. 2A). The 58 kDa protein itself does not appear to be the ω-CgTx receptor, described as a polypeptide of molecular mass 200–300 kDa.[31] Sucrose gradient sedimentation of rat brain synaptosomes shows that the 58 kDa protein exists in two forms, the majority being separate from the ω-CgTx receptor (FIG. 2B). As Western blotting techniques only detect antibodies capable of binding to denatured proteins, the presence of antibodies directed against native ω-CgTx receptors cannot be excluded.

The 58 kDa protein is also recognized by a mouse monoclonal antibody originally raised against chick brain synaptic membranes (1D12). This monoclonal antibody immunoprecipitates ω-CgTx receptor complexes from rat brain synaptic membranes and reacts with the 58 kDa protein on Western blots. Immunopurifica-

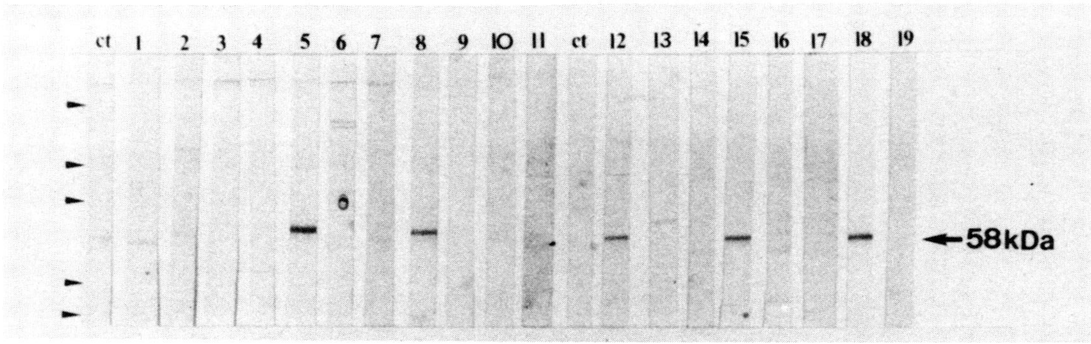

FIGURE 1. Western blots of partially purified ω-CgTx receptor were probed with plasma or IgG fractions from LEMS patients; patients with other neurological or autoimmune diseases; or healthy controls. Lanes 1–11, plasma. Lanes 12–19, IgG each preceded by a control (ct). Arrowheads (from top to bottom) indicate molecular mass standards 206, 110, 70, 43, and 28 kDa. Arrows indicate the 58 kDa immunoreactive band. Lanes: (1) LEMS (SCLC); (2) healthy; (3) neuromyotonia; (4) LEMS (SCLC); (5) LEMS (NCD); (6) SCLC; (7) Guillain Barré syndrome; (8) LEMS (NCD); (9) rheumatoid arthritis; (10) healthy; (11) LEMS (NCD); (12) LEMS (SCLC); (13) chronic inflammatory demyelinating neuropathy; (14) LEMS (SCLC); (15) LEMS; (16) healthy; (17) myasthenia gravis; (18) LEMS; and (19) LEMS (SCLC). (Leveque *et al.*[30] With permission from the *Proceedings of the National Academy of Sciences of the United States of America*.)

tion on an 1D12-Sepharose column resulted in a protein that was strongly recognized by LEMS IgG in Western blotting, indicating that the antigen recognized by 1D12 is the same as that recognized by LEMS IgG.

1D12 has been used to screen a λgt11 rat brain cDNA library, and four overlapping clones were obtained, the sequences of which were virtually homologous to the previously published sequence of the synaptic vesicle protein synaptotagmin.[32] These results therefore imply that 1D12 and the LEMS sera that can bind to the 58 kDa protein both interact with synaptotagmin associated with calcium channels.

Synaptotagmin, previously known as p65, is an abundant membrane protein existing in multiple isoforms. It spans the synaptic vesicle membrane. The C terminus of the molecule extends into the cytoplasm of the nerve terminal and contains two copies of an internal repeat homologous to the C_2 domain of the regulatory region of protein kinase C.[33] It also contains a phospholipid binding activity and is therefore thought to play an important role in the docking and fusion of vesicles during exocytosis. The N terminus of synaptotagmin normally

TABLE 1. Anti–ω-CgTx Receptor Antibody Titer and Anti-58 kDa Immunoreactivity in a Group of LEMS Patients

Patient	Anti–^{125}I-ω-CgTx-VGCC Titer (pM) Using Human Neuroblastoma Cell Line SKN-SH as a Source of Antigen	58 kDa Immunoreactivity in Western Blots of Purified ω-CgTx Receptor from Rat Brain
1	2000	+
2	1400	+
3	1400	+
4	733	+
5	602	±
6	65	±
7	58	±
8	47	−
9	45	−
10	<40	−
11	4	−
12	4	−
13	2	±

lies within the synaptic vesicle. During exocytosis, however, it would be exposed extracellularly in the synaptic cleft, potentially allowing interaction with circulating antibodies.

The pathological relevance of these antibodies is uncertain as they seem to be present in only a minority of patients. Nor is it yet known whether synaptotagmin is expressed in SCLC tumors, but one might speculate that abnormal expression at this site could trigger the observed antibody response. The mechanism triggering these anti–ω-CgTx-VGCC antibodies in non-SCLC cases is unknown.

P-TYPE CALCIUM CHANNELS

In a previous study,[14] we showed that IgG from all LEMS patients tested was able to inhibit K^+-stimulated $^{45}Ca^{2+}$ influx in a human SCLC line derived from a

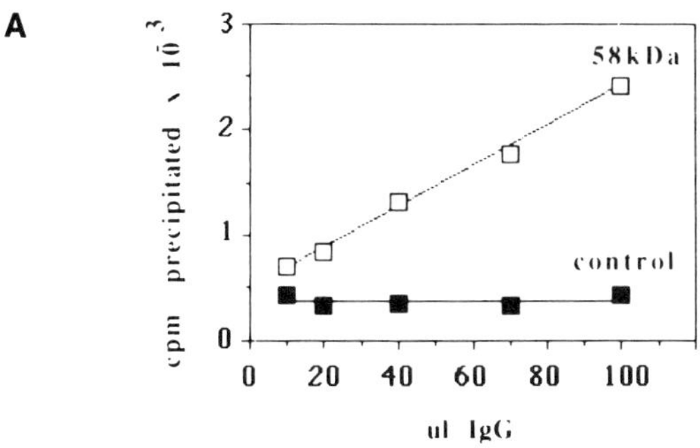

FIGURE 2. Relationship between 58 kDa LEMS antigen and the ω-CgTx receptor. **A:** LEMS IgG was affinity-purified on a nitrocellulose strip containing immunoreactive 58 kDa synaptosomal proteins (□) or an identically sized control strip (■) cut from the 70 kDa region of the same Western blot. Eluted IgG was tested for its ability to immunoprecipitate solubilized ^{125}I-ω-CgTx receptors. **B:** Rat brain synaptosomes were solubilized in Triton X100, and the relative rates of sedimentation of ^{125}I-ω-CgTx calcium channels (■) and the 58 kDa antigen (□) through a 5–20% sucrose gradient were compared. (Leveque *et al.*[30] With permission from the *Proceedings of the National Academy of Sciences of the United States of America.*)

patient without neurological symptoms. In order to relate the results of this functional assay to the titer of the anti–ω-CgTx-VGCC antibodies, we studied the effects of IgG prepared from 12 LEMS patients on a single SCLC line derived from a patient with LEMS who had undergone pneumonectomy. The line (MB) has now been in continuous culture in our laboratory for 18 months. Membranes derived from this line were used as a source of VGCC. High-affinity binding sites for ^{125}I-ω-CgTx were detected (9 fmoles/mg protein K_d = 10 pM). Using ^{125}I-ω-CgTx-VGCC complexes extracted from these cells, we were able to demonstrate by radioimmunoassay antibody binding in LEMS sera. To standardize the assays

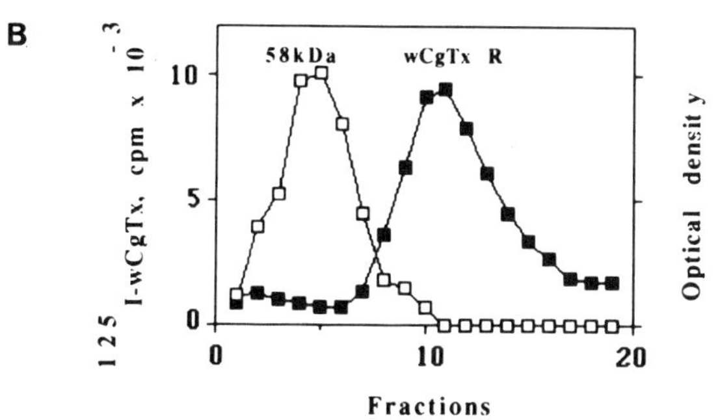

FIGURE 2B. See legend above.

further we prepared IgG from all test and control samples. Twelve LEMS patients, from whom we had a large quantity of plasma, were chosen. They were representative of the general population of LEMS patients: five patients with biopsy-proven SCLC (SCLC-LEMS), six patients followed for at least five years without detecting cancer (NCD-LEMS), and two patients whose status was uncertain (SU-LEMS). Four of the 12 were previously shown to have high anti–ω-CgTx-VGCC antibody titers (range 435–1466 pM) using human neuroblastoma cell line SKN-SH as a source of antigen.[27] Using the cell line MB, similar results were obtained. The same four were positive (96–1393 pM), whereas none of the controls showed titers greater than 25 pM. It is interesting that IgG derived from patient MB was unable to immunoprecipitate ω-CgTx receptors derived from her own cell line (see FIG. 3).

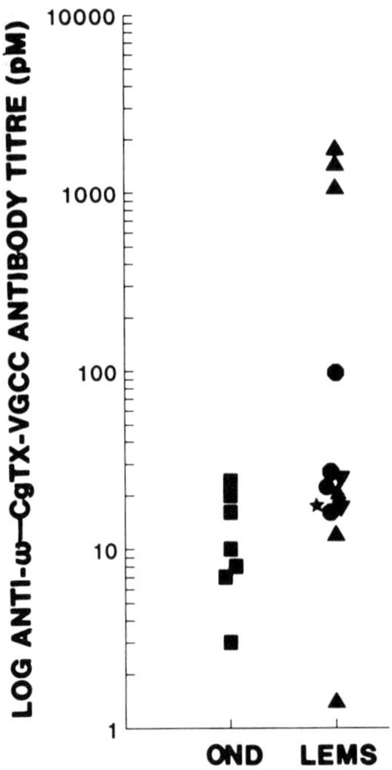

FIGURE 3. Antibody titers in human SCLC line MB. Anti–ω-CgTx-VGCC antibody titers of IgG prepared from LEMS or OND patients measured by immunoprecipitation of VGCC extracted from human SCLC line (MB). Standard preparations of IgG (10 mg/mL) were used in each case. ★ Indicates antibody titer of patient MB. ■, OND (other neurological disease); ○, SCLC-LEMS; ▼, SU-LEMS; ▲, NCD-LEMS.

The MB line showed a K^+-stimulated $^{45}Ca^{2+}$ influx of 2.73 ± 0.14 nmoles/mg protein. This flux was inhibited by high concentrations (10^{-5} M) of the VGCC antagonist methoxyverapamil but was only partially blocked by the dihydropyridines, nitrendipine and nifedipine, or by ω-CgTx (TABLE 2), implying that the majority (greater than 70%) of the calcium influx is through ω-CgTx–insensitive, dihydropyridine-insensitive channels.

Preincubation of the cells with IgG (2 mg/mL) prepared from LEMS patients produced a mean reduction in K^+-stimulated $^{45}Ca^{2+}$ flux of 64% (range 49–76%), as compared to cells grown in medium alone. IgG from control patients with other neurological diseases caused a mean reduction of 6% (FIG. 4). All LEMS IgG tested, including those negative in the immunoprecipitation assay, produced a

TABLE 2. Pharmacological Characteristics of Human SCLC Line (MB)

A. ^{125}I-ω-CgTx Binding

B_{max} = 9 fmoles/mg protein
K_d = 10 pM

B. K^+-stimulated $^{45}Ca^{2+}$ influx

= 2.73 ± 0.14 nmoles/mg protein (n = 21)

C. Effect of calcium channel antagonists on K^+-stimulated $^{45}Ca^{2+}$ flux

Antagonist	Concentration (M)	Inhibition K^+ Stimulated Ca^{2+} Flux (Percent)
Methoxyverapamil	10^{-5}	95
Nitrendipine	10^{-5}	17
Nifedipine	10^{-5}	22
ω-CgTx	10^{-6}	2

significant reduction in the K^+-stimulated flux. There was no correlation between the inhibition of flux caused by incubation in LEMS IgG and the anti–ω-CgTx VGCC antibody titer of the same sample (FIG. 5). These results thus imply that the inhibiting effect of the LEMS IgG on the K^+-stimulated $^{45}Ca^{2+}$ flux must be acting at least partially on ω-CgTx-insensitive channels.

Voltage-gated calcium conductance unaffected by dihydropyridines or by ω-CgTx was not included in the original classification of VGCC proposed by Tsien

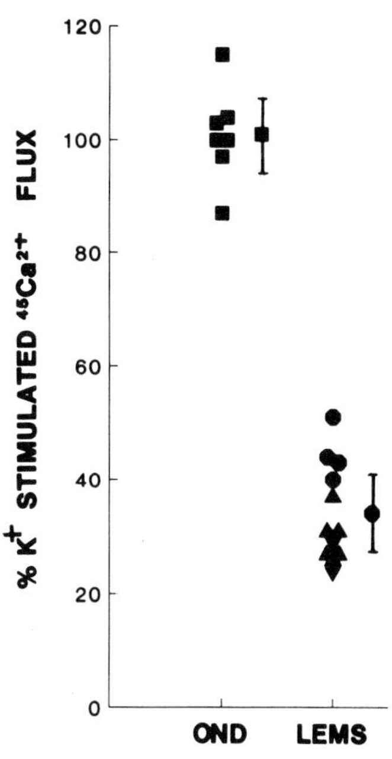

FIGURE 4. Effect of IgG on the K^+-stimulated $^{45}Ca^{2+}$ flux in MB cells. Cells were incubated for 3 days in IgG prepared from patients with either LEMS or other neurological disease (OND) at a final concentration of 2 mg/mL. At the end of incubation time the cells were harvested and assayed as previously described.[31] Each value is expressed as a percentage of the K^+-stimulated $^{45}Ca^{2+}$ flux of cells grown concurrently in medium alone. The mean ± standard deviation is shown for each group. For symbol definitions, see legend to FIG. 3.

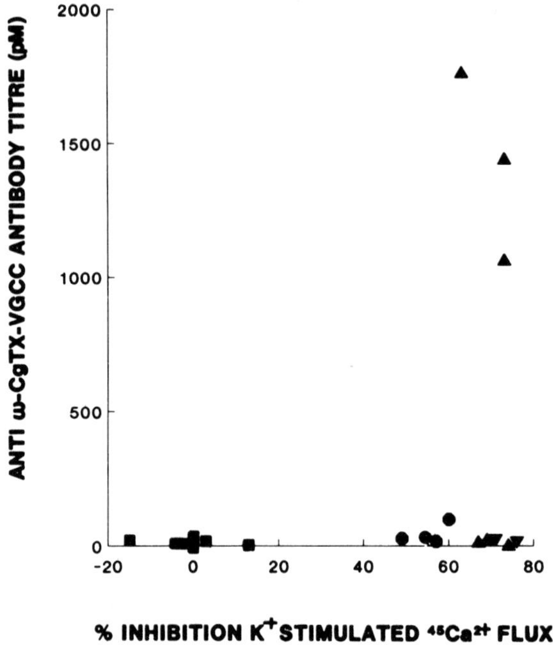

FIGURE 5. Relationship between the effect of LEMS IgG on K^+-stimulated $^{45}Ca^{2+}$ flux and the anti–ω-CgTx VGCC antibody titer. No correlation is observed between the inhibition of the K^+-stimulated $^{45}Ca^{2+}$ flux caused by LEMS IgG and the anti–ω-CgTx antibody titer in human SCLC line MB. For symbol definitions, see legend to FIG. 3.

et al.[15] Recent studies using a new family of calcium channel blockers have begun to define a further subtype of VGCC. A toxin (FTX) purified from the venom of the American funnel web spider (*Agelenopsis aperta*) has been shown to block dendritic spiking in mammalian Purkinje cells,[16] and neurotransmission in the squid giant synapse[16] and in the mammalian neuromuscular junction.[34] These VGCC, termed P-channels, have been shown to be insensitive to block by both dihydropyridines and ω-CgTx. We have tested the unpurified spider venom and also the partially purified FTX toxin on the K^+-stimulated $^{45}Ca^{2+}$ flux in the MB cell line. Both venom and toxin produced an inhibition of stimulated flux that exceeded 70% (FIG. 6), indicating the likely presence of P-type channels in this SCLC line. Inasmuch as LEMS IgG blocks $^{45}Ca^{2+}$ flux in MB to a similar extent, we conclude that some LEMS antibodies may also be acting on the P-type VGCC.

CONCLUSIONS

Our results indicate that the specific autoantibodies in LEMS are heterogeneous. Such a finding is not necessarily surprising, inasmuch as in myasthenia gravis, for example, patients are often reported with antibodies to other muscle antigens besides the acetylcholine receptor. In LEMS the target antigen is the VGCC, which is now known to comprise at least four subtypes (T, L, N, and P). We have discussed the evidence for LEMS autoantibodies to the L-channel and present new evidence for antibodies to P-type channels. Furthermore, some patients appear to have IgG antibodies to synaptotagmin, a synaptic vesicle protein that interacts with ω-CgTx receptors and that may be important in the process of exocytosis. Our data have not excluded the presence of LEMS autoantibodies directed against the ω-CgTx receptor itself. The relative contributions of these different autoantibodies to the functional disorder of transmitter release in LEMS remains to be determined.

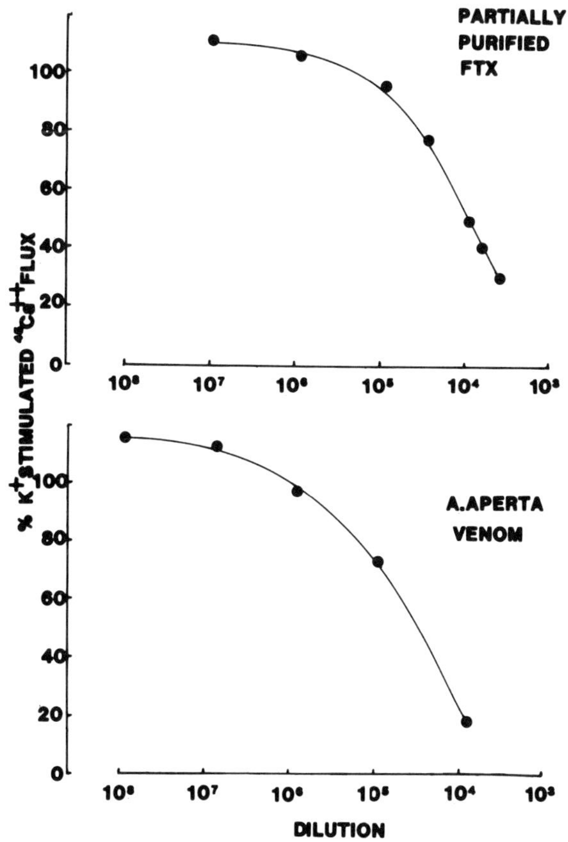

FIGURE 6. Effect of funnel web spider venom and toxin on K^+-stimulated $^{45}Ca^{2+}$ flux in human SCLC line (MB). MB cells were grown in culture until confluent and then incubated for 10 min in increasing concentrations of the venom or FTX. The cells were then assayed as previously described. The results are expressed as a percentage of the K^+-stimulated $^{45}Ca^{2+}$ flux of cells treated concurrently with buffer alone.

REFERENCES

1. Anderson, H. J., H. C. Churchill-Davidson & A. T. Richardson. 1953. Bronchial neoplasm with myasthenia. Lancet **ii:** 1291–1293.
2. Lambert, E. H., L. M. Eaton & E. D. Rooke. 1956. Defect of neuromuscular conduction associated with malignant neoplasma. Am. J. Physiol. **187:** 612–613.
3. O'Neill, J. H., N. M. Murray & J. Newsom-Davis. 1988. The Lambert-Eaton myasthenic syndrome. A review of 50 cases. Brain **111:** 577–596.
4. Guttmann, L., T. W. Crosby, M. Takamori & J. D. Martin. 1972. The Eaton-Lambert syndrome and autoimmune disorders. Am. J. Med. **53:** 354–356.
5. Lennon, V. A., E. H. Lambert, S. Whittingham & V. Fairbanks. 1982. Autoimmunity in the Lambert-Eaton myasthenic syndrome. Muscle & Nerve **5:** S21–S25.
6. Willcox, N., A. G. Demaine, J. Newsom-Davis, K. I. Welsh, S. A. Robb & S. G. Spiro. 1985. Increased frequency of IgG heavy chain marker Glm(2) and of HLA-B8 in Lambert-Eaton myasthenic syndrome with and without associated lung carcinoma. Hum. Immunol. **14:** 29–36.
7. Newsom-Davis, J. & N. M. F. Murray. 1984. Plasma exchange and immunosuppressive drug treatment in the Lambert-Eaton myasthenic syndrome. Neurology **34:** 480–485.
8. Lang, B., J. Newsom-Davis, D. Wray, A. Vincent & N. M. F. Murray. 1981. Autoimmune aetiology for myasthenic (Eaton-Lambert) syndrome. Lancet **ii:** 224–226.
9. Lang, B., J. Newsom-Davis, C. Peers, C. Prior & D. W. Wray. 1987. The effect of myasthenic syndrome antibody on presynaptic calcium channels in the mouse. J. Physiol. (Lond.) **390:** 257–270.
10. Lang, B., P. C. Molenaar, J. Newsom-Davis & A. Vincent. 1984. Passive transfer of Lambert-Eaton myasthenic syndrome in mice: decreased rates of resting and evoked release of acetylcholine from skeletal muscle. J. Neurochem. **42:** 658–662.
11. Fukunaga, H., A. G. Engel, B. Lang, J. Newsom-Davis & A. Vincent. 1983. Passive transfer of Lambert-Eaton myasthenic syndrome with IgG from man to mouse depletes the presynaptic membrane active zones. Proc. Natl. Acad. Sci. USA **80:** 7636–7640.
12. Fukunaga, H., A. G. Engel, M. Osame & E. H. Lambert. 1982. Paucity and disorganisation of presynaptic membrane active zones in the Lambert-Eaton myasthenic syndrome. Muscle & Nerve **5:** 686–697.
13. Roberts, A., S. Perera, B. Lang, A. Vincent & J. Newsom-Davis. 1985. Paraneoplastic myasthenic syndrome IgG inhibits $^{45}Ca^{2+}$ flux in a human small cell carcinoma line. Nature **317:** 737–739.
14. Lang, B., A. Vincent, N. M. Murray & J. Newsom-Davis. 1989. Lambert-Eaton myasthenic syndrome: immunoglobulin G inhibition of Ca^{2+} flux in tumor cells correlates with disease severity. Ann. Neurol. **25:** 265–271.
15. Tsien, R. W., D. Lipscombe, D. V. Madison, K. R. Bley & A. P. Fox. 1988. Multiple types of neuronal calcium channels and their selective modulation. TINS **II**(10): 431–438.
16. Llinás, R., M. Sugimori, J.-W. Lin & B. Cherskey. 1989. Blocking and isolation of a calcium channel from neurons in mammals and cephalopods utilising a toxin fraction (FTX) from funnel-web spider poison. Proc. Natl. Acad. Sci. USA **86:** 1689–1693.
17. Kim, Y. I. & E. Neher. 1988. IgG from patients with Lambert-Eaton syndrome blocks voltage dependent calcium channels. Science **239:** 405–408.
18. Login, I. S., Y. I. Kim, A. M. Judd, B. L. Spangelo & R. M. MacLeod. 1987. Immunoglobulins of Lambert-Eaton myasthenic syndrome inhibit rat pituitary hormone release. Ann. Neurol. **22:** 610–614.
19. Peers, C., B. Lang, J. Newsom-Davis & D. W. Wray. 1990. Selective action of myasthenic syndrome antibodies on calcium channel in a rodent neuroblastoma × glioma cell line. J. Physiol. (Lond.) **421:** 293–308.
20. Rivier, J., R. Galyean, W. R. Gray, A. Azimi-Zonooz, J. M. McIntosh, L. J.

CRUS & B. M. OLIVIERA. 1987. Neuronal calcium channels inhibitors. J. Biol. Chem. **262:** 1194–1198.
21. SHER, E., A. PANDIELLA & F. CLEMENTI. 1988 Ω-conotoxin binding and effects on calcium channel function in human neuroblastoma and rat pheochromocytoma cell lines. FEBS Lett. **235:** 178–182.
22. DE AIZPURUA, H. J., E. H. LAMBERT, G. E. GRIESMANN, B. O. OLIVIERA & V. A. LENNON. 1988. Antagonism of voltage-gated calcium channels in small cell carcinomas of patients with and without Lambert-Eaton myasthenic syndrome by autoantibodies, Ω-conotoxin and adenosine. Cancer Res. **48:** 4719–4724.
23. SHER, E., A. PANDIELLA & F. CLEMENTI. 1990. Voltage-operated calcium channels in small cell lung carcinoma cell lines: Pharmacological, functional, and immunological properties. Cancer Res. **50:** 3892–3896.
24. SHER, E., N. CANAL, G. PICCOLO, C. GOTTI, C. SCOPPETTA, A. EVOLI & F. CLEMENTI. 1989. Specificity of calcium channel autoantibodies in Lambert-Eaton myasthenic syndrome. Lancet **ii:** 640–643.
25. KEYS, K., B. LANG, A. VINCENT & J. NEWSOM-DAVIS. 1989. Calcium channel autoantibodies in Lambert-Eaton myasthenic syndrome. Lancet **ii:** 1107.
26. LENNON, V. A. & E. H. LAMBERT. 1989. Autoantibodies bind solubilized calcium channel–omega-conotoxin complexes from small cell lung carcinoma: a diagnostic aid for Lambert-Eaton myasthenic syndrome. Mayo Clin. Proc. **64:** 1498–1504.
27. LEYS, K., B. LANG, I. JOHNSTON & J. NEWSOM-DAVIS. 1991. Calcium channel autoantibodies in the Lambert-Eaton myasthenic syndrome. Ann. Neurol. **29:** 307–314.
28. CHESTER, K. A., B. LANG, J. GILL, A. VINCENT & J. NEWSOM-DAVIS. 1988. Lambert-Eaton syndrome antibodies: reaction with membranes from a small cell lung cancer xenograft. J. Neuroimmunol. **18:** 97–104.
29. CHAPMAN, J., R. RABINOWITZ, A. D. KORCZYN & D. M. MICHAELSON. 1990. Rats immunised with cholinergic synaptosomes: a model for Lambert-Eaton syndrome. Muscle & Nerve **13:** 726–733.
30. LEVEQUE, C., T. HOSHINO, P. DAVID, Y. KASAI, K. LEYS, A. OMORI, B. LANG, O. EL FAR, K. SATO, N. MARTIN-MOUTOT, J. NEWSOM-DAVIS, M. TAKAHASHI & M. J. SEAGAR. 1992. The synaptic vesicle protein synaptotagmin associates with calcium channels and is a putative Lambert-Eaton myasthenic syndrome antigen. Proc. Natl. Acad. Sci. USA. **89:** 3625-3629.
31. MARQUEZE, B., N. MARTIN-MOUTOT, C. LEVEQUE & F. COURAUD. 1988. Characterisation of the Ω-conotoxin binding molecule in rat brain synaptosome and cultured neurones. Mol. Pharmacol. **34:** 87–90.
32. PERIN, M. S., V. A. FRIED, G. MIGNERY, R. JAHN & T. C. SUDHOF. 1990. Nature **345:** 260–263.
33. SUDHOF, T. C. & R. JAHN. 1991. Proteins of synaptic vesicles involved in exocytosis and membrane recycling. Neuron **6:** 665–677.
34. UCHITEL, O. D., D. A. PROTTI, V. SANCHEZ, B. D. CHERKSEY, M. SUGIMORI & R. LLINÁS. 1992. P-type voltage-dependent calcium channel mediates presynaptic calcium influx and transmitter release in mammalian synapses. Proc. Natl. Acad. Sci. USA **89:** 3330–3333.

Lambert-Eaton Syndrome IgG Inhibits Dihydropyridine-sensitive, Slowly Inactivating Calcium Channels in Bovine Adrenal Chromaffin Cells[a]

JOYCE K. W. BLANDINO[b] AND YONG I. KIM[b,c]

Departments of [b]Biomedical Engineering and [c]Neurology
University of Virginia Health Sciences Center
Charlottesville, Virginia 22908

Lambert-Eaton syndrome (LES) is an autoimmune disorder characterized by an insufficient release of cholinergic neurotransmitters at the neuromuscular junction.[1,2] Nearly 60% of LES patients have concurrent small-cell lung cancer. The LES animal passive transfer studies,[1] freeze-fracture analysis of LES motor nerve terminal,[3] and patch-clamp studies of the pathogenic action of the autoantibodies[4,5] provided convincing evidence that the autoantibodies from LES cross-react with voltage-dependent calcium channels in the presynaptic membrane. Bovine adrenal chromaffin (BAC) cells are widely used as a model of sympathetic neurons and have recently been shown to possess two types of Ca^{2+} channels.[6] The aim of the present work has been to investigate the specificity of LES IgG on the two types of Ca^{2+} channels in BAC cells.

BAC cells were maintained in Dulbecco's modified Eagle medium with 15% heat-inactivated horse serum at 5% CO_2 and 95% air. The dihydropyridine (DHP) antagonist nicardipine was first dissolved in 100% ethanol, and the solution was diluted with distilled water before adding to the cells. Lyophilized ω-conotoxin GVIA (ω-CgTx) was dissolved in 1 mg/mL bovine serum albumin to make a 10 μM stock. Utilizing patch-clamp techniques, whole-cell calcium current (I_{Ca}) was elicited from a holding potential (V_h) of −90 mV. The pipette contained (in mM) $CaCl_2$ 3, $MgCl_2$ 2, CsOH/HEPES 100, and EGTA 30 (pH = 7.2); and the external solution was composed of (mM) $CaCl_2$ 5, $MgCl_2$ 2, TEA-Cl 10, HEPES 10, and Tris-trichloroacetate 110 (pH = 7.4).

First we identified the two types of Ca^{2+} channels in BAC cells based on inactivation kinetics and pharmacology. Results are summarized as follows. (1) I_{Ca} inactivated with an initial transient described by a time constant (τ_f) of 169 ± 28 msec (n = 14 cells) followed by a long inactivating component, τ_s = 1393 ± 129 ms (n = 14). FIGURE 1A shows a Ca^{2+} current trace and its components (solid lines) based on τ_s and τ_f for a depolarization to +20 mV. I_{Ca} obtained at various external Ca^{2+} concentrations (5, 2, 0.8, and 0.4 mM) had similar τ_s and τ_f. (2) Ten micromolar nicardipine reduced steady-state I_{Ca} (measured at the end of 2 s) by 81.5 ± 4.7% (n = 11), and the remaining component decayed with a single exponential, τ = 280 ± 22 ms (n = 11), similar to the τ_f of normal untreated cells (FIGURES 1B and 2). After nicardipine, I_{Ca} in most cells could

[a] This work was supported by National Institutes of Health Grant NS-18607 and a Research Grant from the Muscular Dystrophy Association.

not be fit with a double exponential. Thus the slow component, which is DHP-sensitive, exhibits characteristics consistent with L-type Ca^{2+} channels. (3) The τ (1090 ± 126 ms, n = 9) obtained with 1 μM ω-CgTx was similar to τ_s in normal untreated cells (FIGURES 1C and 2). ω-CgTx significantly reduced both peak (55.8 ± 5.2%) and steady-state (45.2 ± 8.3%, n = 9) I_{Ca}. Thus, the fast component exhibits characteristics consistent with N-type Ca^{2+} channels.

After BAC cells were exposed to LES IgG (1 mg/mL) for 24 hours, the peak and steady-state I_{Ca} fell by 70.6 ± 7.2% (n = 10) and 69.9 ± 7.1% (n = 10),

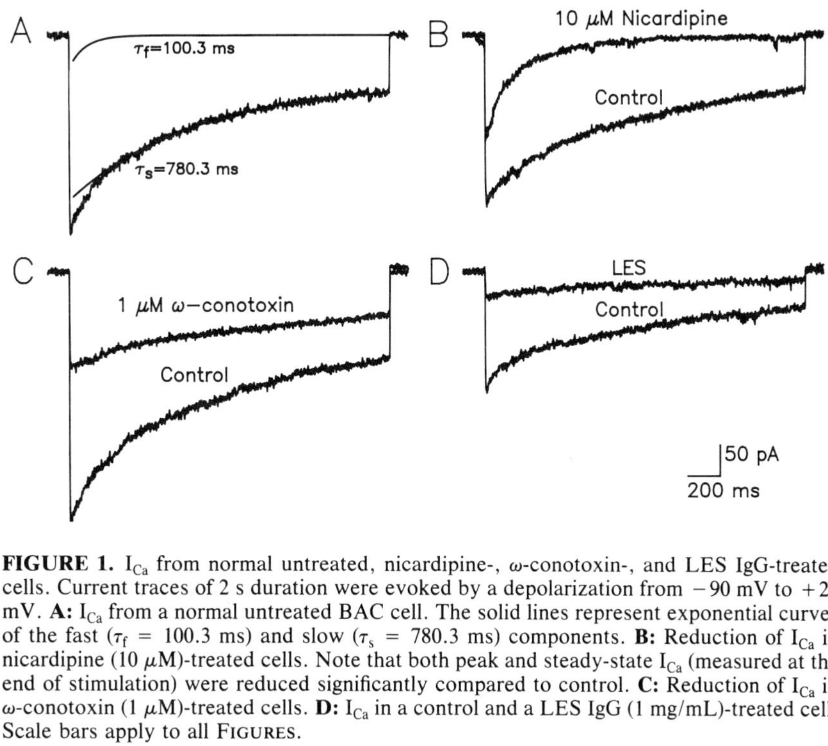

FIGURE 1. I_{Ca} from normal untreated, nicardipine-, ω-conotoxin-, and LES IgG-treated cells. Current traces of 2 s duration were evoked by a depolarization from −90 mV to +20 mV. **A:** I_{Ca} from a normal untreated BAC cell. The solid lines represent exponential curves of the fast (τ_f = 100.3 ms) and slow (τ_s = 780.3 ms) components. **B:** Reduction of I_{Ca} in nicardipine (10 μM)-treated cells. Note that both peak and steady-state I_{Ca} (measured at the end of stimulation) were reduced significantly compared to control. **C:** Reduction of I_{Ca} in ω-conotoxin (1 μM)-treated cells. **D:** I_{Ca} in a control and a LES IgG (1 mg/mL)-treated cell. Scale bars apply to all FIGURES.

respectively, compared to control IgG-treated cells. FIGURE 1D depicts the I_{Ca} from a control- and an LES IgG-treated cell. The ensemble variance analysis[7] of I_{Ca} showed a reduction of 48.4 ± 7.4% (n = 28) in the number of functional Ca^{2+} channels with no change in single Ca^{2+} channel current, consistent with the "all-or-nothing" inhibition of voltage-gated Ca^{2+} channels by LES IgG.[4] Inasmuch as τ for the fast component is 170 ms, the current amplitude for this component is negligible at 2 seconds. Hence, the steady-state current component is composed of L-type Ca^{2+} channels only. LES IgG is found to markedly inhibit this component. FIGURE 2 summarizes τ's obtained from normal untreated, nicardipine-, ω-CgTx-, and LES IgG-treated cells. Because the peak I_{Ca} declined with both ω-CgTx and LES IgG, this also suggests that the ω-CgTx–sensitive channels may be inhibited by LES IgG. I_{Ca} after LES IgG treatment could be fit with either a single or a double exponential; the single exponential fit, however, yielded a τ of 718 ± 103 ms (n = 6), which was between the τ's obtained after nicardipine (280 ms) and ω-CgTx (1090 ms) treatment.

We conclude that LES IgG reacts with dihydropyridine-sensitive, slowly inactivating (L-type) Ca^{2+} channels. These observations also support the possibility that LES IgG inhibits ω-conotoxin (N-type) Ca^{2+} channels.

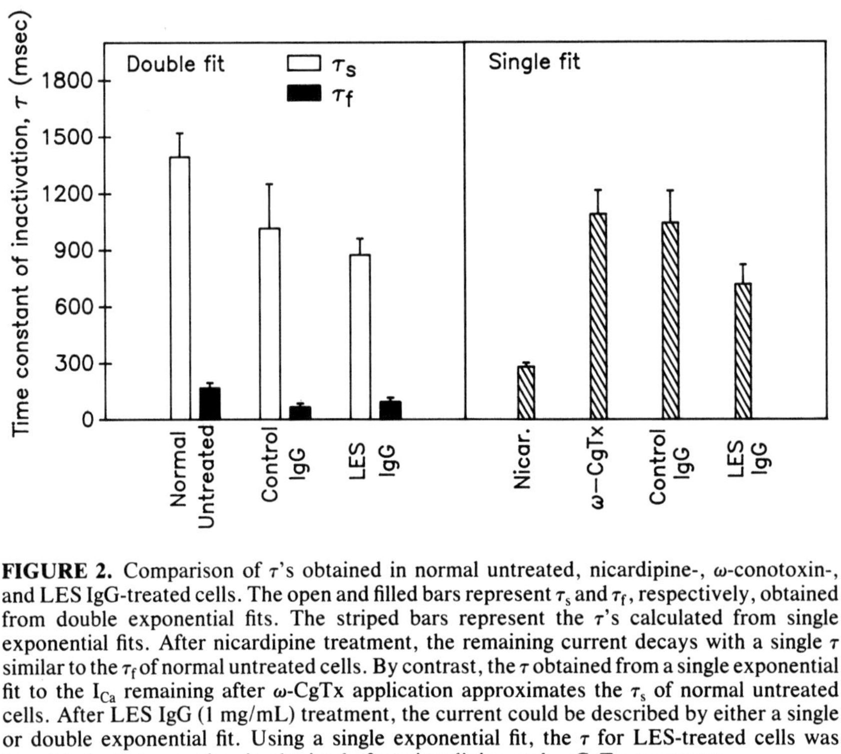

FIGURE 2. Comparison of τ's obtained in normal untreated, nicardipine-, ω-conotoxin-, and LES IgG-treated cells. The open and filled bars represent τ_s and τ_f, respectively, obtained from double exponential fits. The striped bars represent the τ's calculated from single exponential fits. After nicardipine treatment, the remaining current decays with a single τ similar to the τ_f of normal untreated cells. By contrast, the τ obtained from a single exponential fit to the I_{Ca} remaining after ω-CgTx application approximates the τ_s of normal untreated cells. After LES IgG (1 mg/mL) treatment, the current could be described by either a single or double exponential fit. Using a single exponential fit, the τ for LES-treated cells was found to be between the τ's obtained after nicardipine and ω-CgTx treatment.

REFERENCES

1. LANG, B., J. NEWSOM-DAVIS, D. WRAY, A. VINCENT & N. MURRAY. 1981. Autoimmune aetiology for myasthenic (Eaton-Lambert) syndrome. Lancet **ii:** 224–226.
2. ELMQVIST, D. & E. H. LAMBERT. 1968. Detailed analysis of neuromuscular transmission in a patient with the myasthenic syndrome sometimes associated with bronchogenic carcinoma. Mayo Clin. Proc. **43:** 689–713.
3. NAGEL, A., A. G. ENGEL, B. LANG, J. NEWSOM-DAVIS & T. FUKUOKA. 1988. Lambert-

Eaton myasthenic syndrome IgG depletes presynaptic membrane active zone particles by antigenic modulation. Ann. Neurol. **24:** 552–558.

4. Kim, Y. I. & E. Neher. 1988. IgG from patients with Lambert-Eaton syndrome blocks voltage-dependent calcium channels. Science **239:** 405–408.
5. Peers, C., B. Lang, J. Newsom-Davis & D. W. Wray. 1990. Selective action of myasthenic syndrome antibodies on calcium channels in a rodent neuroblastoma × glioma cell line. J. Physiol. (Lond.) **421:** 293–308.
6. Bossu, J. L., M. D. Waad & A. Feltz. 1991. Inactivation characteristics reveal two calcium currents in adult bovine chromaffin cells. J. Physiol. (Lond.) **437:** 603–620.
7. Sigworth, F. J. 1980. The variance of sodium current fluctuation at the node of ranvier. J. Physiol. (Lond.) **307:** 97–129.

Inhibitory Action of Lambert-Eaton Syndrome IgG on Calcium Currents in a Thyroid C-Cell Line[a]

YONG I. KIM,[b,c] JOYCE K. W. BLANDINO,[b] AND THOMAS J. O'SHAUGHNESSY[b]

Departments of [b]Biomedical Engineering and [c]Neurology
University of Virginia Health Sciences Center
Charlottesville, Virginia 22908

Lambert-Eaton syndrome (LES) is an autoimmune disorder of the presynaptic motor nerve terminal in which the release of cholinergic transmitters is reduced.[1,2] The disorder is often associated with small-cell lung cancer (SCLC).[3] Studies including passive transfer of LES to mice,[4,5] freeze-fracture analysis of LES motor nerve terminals,[6] and patch-clamp studies of the pathogenic action of the autoantibodies[7,8] support the notion that LES autoantibodies cross-react with voltage-dependent calcium channels. The objective of this study has been to investigate the action of LES IgG on the Ca^{2+} channels in a rat medullary thyroid C-cell line, 6-23 (clone 6). These cells have a neural crest origin and possess T-, N-, and L-type Ca^{2+} channels.[9]

We first studied the biophysical and pharmacological properties of the three types of Ca^{2+} currents in thyroid C cells. FIGURE 1A displays current traces at -20 and $+20$ mV depolarizations from holding potentials (V_h) of -80 and -40 mV. The corresponding current-voltage (I-V) relationship of I_{Ca} from the same cell is presented in FIGURE 1B. The T-type Ca^{2+} currents activated at -50 mV, and the current amplitude was maximum at -30 and -20 mV. Increasing the depolarizing voltage beyond -20 mV increased the current amplitude and activated a steady-state current component. The steady-state I_{Ca} (measured at the end of 350 ms stimulation) was maximum at $+20$ mV and was partially comprised of L-type Ca^{2+} currents. At V_h of -40 mV, the T-type Ca^{2+} currents were inactivated. By subtracting the current measured at V_h of -40 mV from that measured at V_h of -80 mV from the same cell, the N-type Ca^{2+} current was isolated as the peak current amplitude elicited with depolarizations > 0 mV. The N-type Ca^{2+} currents peaked at $+20$ mV.

Ten micromolar nicardipine reduced the peak, steady-state, and T-type I_{Ca} by $53.9 \pm 7.6\%$ (n = 13), $38.7 \pm 10.0\%$ (n = 13), and $81.2 \pm 13.3\%$ (n = 11), respectively. Application of 1 μM ω-conotoxin did not alter T-type I_{Ca} but inhibited peak ($-32.9 \pm 6.5\%$, n = 13) and steady-state ($-35.8 \pm 6.6\%$, n = 15) I_{Ca} significantly. Hence, the thyroid C cells possess both dihydropyridine-sensitive and ω-conotoxin–sensitive Ca^{2+} channels.

We also studied the exocytotic secretory process of these cells with the use of membrane capacitance (C_m) measurement techniques.[10] Applying 500 ms depo-

[a] This work was supported by National Institutes of Health Grant NS-18607 and a Research Grant from the Muscular Dystrophy Association.

larizing pulses resulted in an incremental increase in C_m (FIG. 1C), which appeared to correlate with the amplitude of I_{Ca}. When 5.8 mM Ca^{2+}/6 mM EGTA was applied through the pipette, C_m rose steadily over a 5 minute period. These observations indicate that thyroid C cells undergo Ca^{2+}-dependent exocytosis.

The effect of LES IgG on the Ca^{2+} currents of thyroid C cells was then investigated. When cells were exposed to LES serum (1 mg/mL IgG) for 12 hours, peak and steady-state I_{Ca} fell 51.5 ± 8.9% and 30.8 ± 14% (n = 11), respectively. A longer incubation period, however, was necessary for LES IgG (1 mg/mL) to

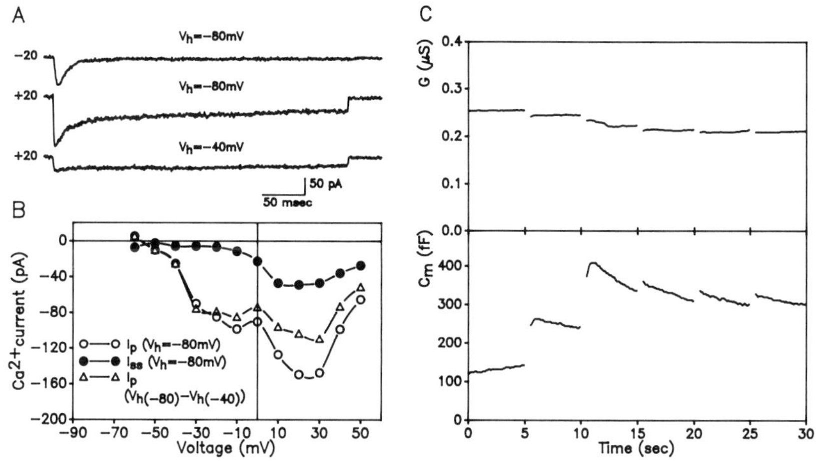

FIGURE 1. Multiple calcium currents and exocytosis in thyroid C cells. **A:** Current traces of 350 ms duration were evoked from V_h of −80 and −40 mV. **B:** I-V curves for peak (I_p, open circle), steady-state (I_{ss}, filled circle), and peak I_{Ca} (open triangle) calculated from the difference between I_{Ca} measured at V_h of −80 and −40 mV from the same cell as used in part A. External solution contained (mM) TEA-Cl 117; $CaCl_2$ 10; $MgCl_2$ 2; CsCl 5; HEPES 5; and TTX 0.005 (pH 7.4). Internal pipette solution was composed of (mM) CsCl 120; $CaCl_2$ 1; $MgCl_2$ 2; BAPTA 11; HEPES 10; and MgATP 1 (pH 7.2). **C:** Ca^{2+}-dependent exocytosis in response to voltage stimulation from a different cell. A 500 ms voltage step from V_h of −80 mV to +20 mV was applied to the cell at 5, 10, 15, 20, and 25 s from the start of the whole-cell recording, and the resulting changes in C_m were measured. The pipette was filled with (mM) CsCl 120; EGTA 0.5; HEPES 10; $MgCl_2$ 2; Na_2ATP 2; and TEA-Cl 20 (pH 7.2). The external solution contained (mM) $CaCl_2$ 10; HEPES 10; KCl 2; $MgCl_2$ 2; and NaCl 130 (pH 7.2).

inhibit I_{Ca} in these cells. In three experiments in which cells were incubated with LES IgG for 48 hours, we obtained about 30% (n = 17 − 24) reduction of peak and steady-state I_{Ca} at both V_h of −80 and −40 mV. This reduction was significant ($p < 0.05$) compared to the control. The T-type I_{Ca} was not affected by either LES serum or IgG. FIGURE 2 illustrates the I-V relationship of control and LES IgG-treated cells from three experiments. Both N- (FIG. 2B) and L-type (FIG. 2A, triangle) Ca^{2+} currents, not T-type (FIG. 2A, circle, at depolarizations <0 mV), appear to be inhibited by LES IgG.

We found, however, that in three other experiments, LES IgG was without significant effect. The reason for this discrepancy is not known, although there was clear tendency toward reduction. More experiments are necessary to determine the specific action of LES IgG on the T, N, and L-type Ca^{2+} channels in thyroid C cells.

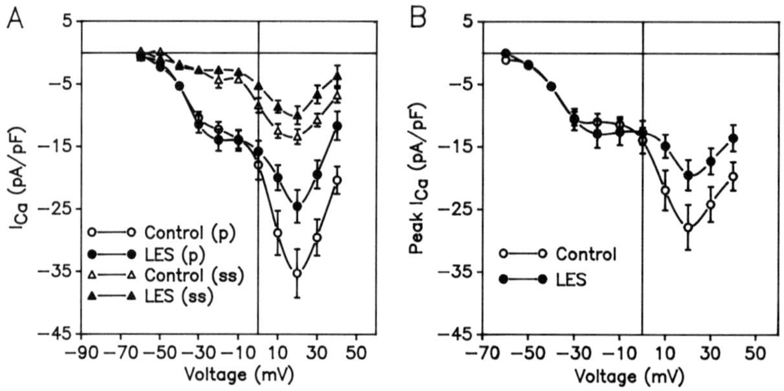

FIGURE 2. I_{Ca} in control and LES IgG-treated thyroid C cells. **A:** Peak (p, circles) and steady-state (ss, triangles) I_{Ca}, plotted against depolarizing voltage, ranged from −60 to +40 mV, in control (open symbol) and LES (filled symbol) IgG-treated cells (V_h = −80 mV). **B:** Peak I_{Ca} was calculated from the difference of I_{Ca} measured with V_h of −80 and −40 mV in control and LES IgG-treated cells. All currents are normalized by dividing them with C_m. Each data point is an average of at least 17 cells. The error bar represents standard error of the mean.

REFERENCES

1. Lang, B., J. Newsom-Davis, D. Wray, A. Vincent & N. Murray. 1981. Autoimmune aetiology for myasthenic (Eaton-Lambert) syndrome. Lancet **ii:** 224–226.
2. Elmqvist, D. & E. H. Lambert. 1968. Detailed analysis of neuromuscular transmission in a patient with the myasthenic syndrome sometimes associated with bronchogenic carcinoma. Mayo Clin. Proc. **43:** 689–713.
3. Lambert, E. H., E. D. Rooke, L. M. Eaton & C. H. Hodgson. 1961. Myasthenic syndrome occasionally associated with bronchial neoplasm: Neurophysiologic studies. *In* Myasthenia Gravis. H. R. Viets, Ed.: 362–410. Charles C. Thomas. Springfield, IL.
4. Kim, Y. I. 1987. Lambert-Eaton myasthenic syndrome: evidence for calcium channel blockade. Ann. N.Y. Acad. Sci. **505:** 377–379.
5. Lang, B., J. Newsom-Davis, J. Peers, C. Prior & D. W. Wray. 1987. The effect of myasthenic syndrome antibody on presynaptic calcium channels in the mouse. J. Physiol. (Lond.) **360:** 257–270.
6. Nagel, A., A. G. Engel, B. Lang, J. Newsom-Davis & T. Fukuoka. 1988. Lambert-Eaton myasthenic syndrome IgG depletes presynaptic membrane active zone particles by antigenic modulation. Ann. Neurol. **24:** 552–558.
7. Kim, Y. I. & E. Neher. 1988. IgG from patients with Lambert-Eaton syndrome blocks voltage-dependent calcium channels. Science **239:** 405–408.
8. Peers, C., B. Lang, J. Newsom-Davis & D. W. Wray. 1990. Selective action of

myasthenic syndrome antibodies on calcium channels in a rodent neuroblastoma × glioma cell line. J. Physiol. (Lond.) **421:** 293–308.

9. Biagi, B. A. & J. J. Enyeart. 1991. Multiple calcium currents in a thyroid C-cell line: biophysical properties and pharmacology. Am. J. Physiol. **260:** C1253–C1263.
10. Neher, E. & A. Marty. 1982. Discrete changes of cell membrane capacitance observed under conditions of enhanced secretion in bovine adrenal chromaffin cells. Proc. Natl. Acad. Sci. USA **79:** 6712–6716.

Analysis of Neuromuscular Transmission Failure Induced by Anticholinesterases

RICARDO A. MASELLI AND CYNTHIA LEUNG

Section of Neurosciences
Department of Neurology
University of California, Davis
Davis, California 95616

The effects of anticholinesterases on the neuromuscular junction have been extensively investigated both clinically and experimentally. The mechanism responsible for the failure of neuromuscular transmission induced by these agents, however, is still incompletely understood. We recently described two patterns of stimulation-induced end-plate depolarization in rats intoxicated with diisopropylfluorophosphate (DFP) that may have accounted for the impairment of neuromuscular transmission observed.[1] One pattern exhibited progressive staircase summation of long-decay end-plate potentials (EPP), resulting in maximal depolarization at the end of the stimulation period. The other type was characterized by maximal end-plate depolarization at the beginning of the stimulation period followed by a progressive return of the resting membrane potential to the prestimulation level. This latter pattern was proposed to be the underlying mechanism responsible for the decremental response followed by recuperation of compound muscle action potential (CMAP) amplitudes seen in humans at early stages of intoxication with organophosphate compounds.[2] The former type was thought to account for the straight decrement of CMAP amplitudes observed in humans intoxicated with higher doses of anticholinesterases.[3] In both cases the blockade of neuromuscular transmission appeared to result from inactivation of perijunctional sodium channels responsible for the generation and propagation of action potentials in the muscle fibers. In an effort to obtain further validation of these hypotheses, we have developed a preparation which permits simultaneous recordings of CMAPs and EPP during repetitive nerve stimulation (RNS).

RESULTS AND DISCUSSION

All experiments were performed using the "sector cut fiber" preparation. This preparation consisted of the left hemidiaphragm muscle of the rat in which the anterior portion was cut approximately 3 mm away from the insertion of the muscle fibers into the rib cage. This resulted in depolarization of the muscle fibers that prevented the generation of action potentials and muscle twitch, making it possible to record EPP in fibers from the anterior part of the muscle. Because the branches to the anterior and posterior sector of the diaphragm were left intact, electrical stimulation of the phrenic nerve resulted in simultaneous CMAPs in the

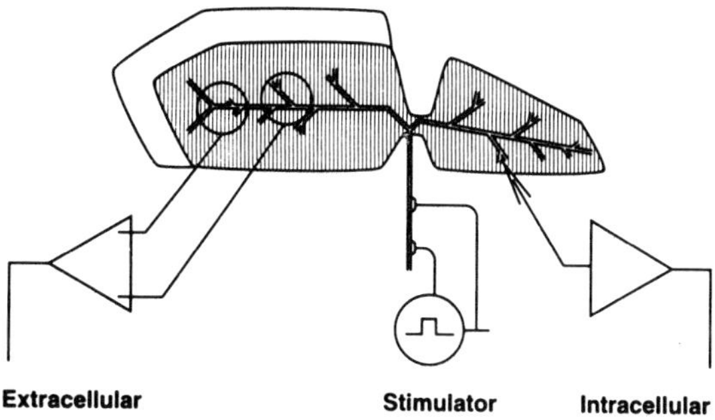

FIGURE 1. Schematic representation of the sector cut fiber preparation showing stimulation and recording sites.

posterior sector and EPP in muscle fibers of the anterior sector (FIG. 1). The intracellular recordings were performed using previously described techniques.[4] The preparations were exposed to several different concentrations of neostigmine methylsulfate.

In control muscles, RNS (2–20 Hz) resulted in no decrement of extracellularly recorded CMAPs (FIG. 2A). By contrast, significant amplitude decrement of intracellularly recorded EPP was always observed (FIG. 2B). This disparity may probably be accounted for by the large safety margin that characterizes the rat neuromuscular junction. At low concentrations of neostigmine (0.1–0.2 μM), RNS resulted in decrement with recuperation of CMAP amplitudes in more than 90% of the trials (FIG. 2C and E). The intracellular recordings showed a pattern of initial

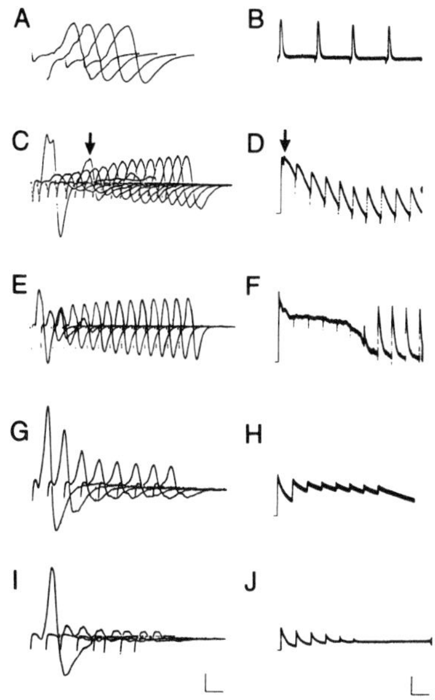

FIGURE 2. Examples of simultaneously evoked CMAPs (left) and EPP (right) recorded from preparations exposed to normal physiologic solution (A and B), 0.2 μM neostigmine methylsulfate (C, D, E, and F), 2.0 μM neostigmine methylsulfate (G and H), and 20 μM neostigmine methylsulfate (I and J). Stimulation frequency was 20 Hz in A–D; 5 Hz in E, F, I, and J; and 10 Hz in G and H. Calibration bars are 100 μV, 2 ms in A; 200 μV, 5 ms in C and E; 100 μV, 5 ms in G; 50 μV, 5 ms in I; 5 mV, 20 ms in B; 5 mV, 50 ms in D; 5 mV, 200 ms in F; 5 mV, 100 ms in H; and 2 mV, 200 ms in J.

depolarization of the end-plate followed by recuperation of the prestimulation resting membrane potential in more than 90% of the trials (FIG. 2D), often preventing impulse conduction through the motor nerve terminal (FIG. 2F). The first stimulus of the train elicited repetitive CMAPs (FIG. 2C, arrow) and repetitive EPP (FIG. 2D, arrow). At higher concentrations of neostigmine (1–10 μM), RNS induced straight decremental responses of CMAP amplitudes in more than 75% of the trials (FIG. 2G). This was usually associated with staircase summation of long decay EPP resulting in maximal depolarization at the end of the stimulation period (FIG. 2H). When very high concentrations of neostigmine were used (20 μM), the progressive decline of CMAP and EPP amplitudes was markedly pronounced, suggesting an additional component of end-plate receptor desensitization (FIG. 2I and J).

The mean miniature end-plate potential (MEPP) amplitude recorded from muscle fibers exposed to low concentrations of neostigmine was increased compared to that of MEPPs recorded from control preparations. By contrast, preparations exposed to high concentrations of neostigmine had MEPP amplitudes that were reduced compared to control values.

The mean decay time constants of miniature end-plate currents (MEPCs) recorded in the presence of neostigmine (all concentrations) were increased compared with control values. A single exponential component adequately fits the decays of control MEPCs. The decays of MEPCs recorded in the presence of high concentrations of neostigmine, however, often required a double exponential fitting. These results indicate that different patterns of transient end-plate depolarization account for the failure of neuromuscular transmission induced by low concentrations of anticholinesterases. These effects may mainly be explained by the inhibition of the cholinesterase. High concentrations of anticholinesterases induce additional desensitization of the end-plate receptor, which then becomes the dominant mechanism of failure of neuromuscular transmission. The double exponential decay of MEPCs recorded at high concentrations of neostigmine suggests an underlying blockade of the end-plate receptor channel in its open conformation.

REFERENCES

1. MASELLI, R. A. & B. C. SOLIVEN. 1991. Analysis of the organophosphate-induced electromyographic response to repetitive nerve stimulation: paradoxical response to edrophonium and D-tubocurarine. Muscle & Nerve **14:** 1182–1188.
2. MASELLI, R. A., J. H. JACOBSEN & J. P. SPIRE. 1986. Edrophonium: An aid in the diagnosis of acute organophosphate poisoning. Ann. Neurol. **19:** 508–510.
3. BESSER, R., L. GUTMANN, U. DILLMANN, L. S. WEILEMANN & H. C. HOPF. 1989. End-plate disfunction in acute organophosphate intoxication. Neurology **39:** 561–566.
4. MASELLI, R. A., D. P. RICHMAN & R. L. WOLLMANN. 1991. Inflammation at the neuromuscular junction in myasthenia gravis. Neurology **41:** 1497–1504.

Mammalian Neuromuscular Transmission Blocked by Funnel Web Toxin[a]

D. A. PROTTI,[b] V. A. SANCHEZ,[b] B. D. CHERKSEY,[c]
M. SUGIMORI,[c] R. LLINÁS,[c] AND O. D. UCHITEL[b]

[b]*Instituto de Biologia Celular*
Facultad de Medicina
Universidad de Buenos Aires
Paraguay 2155
Buenos Aires 1121, Argentina

[c]*Department of Physiology and Biophysics*
New York University Medical Center
550 First Avenue
New York, New York 10016

Different types of voltage-dependent calcium channels (VDCC) known as L-, N- and T-type have been described.[1] Most recently, a new VDCC, the P-type has been found in cerebellar Purkinje cells.[2] The Ca^{2+} channel responsible for transmitter release at the neuromuscular junction has not been characterized, however. The use of the polyamine blocking agent from funnel-web spider toxin (FTX) allowed Llinás *et al.* to isolate the P-channel protein and to further characterize its channel properties.[3] The P-type Ca^{2+} channel is blocked by divalent cations and by FTX but is not affected by ω-conotoxin (ω-Cgtx) and dihydropyridines (DHP), which are known to block N- and L-type Ca^{2+} channel.[1]

At the mammalian neuromuscular junction, ω-CgTx and DHP are also ineffective in blocking neurotransmitter release, indicating that Ca^{2+} channels involved in evoked transmitter release are insensitive to these drugs.[4,5]

We have characterized the Ca^{2+} channels at the mammalian neuromuscular junction by studying the effect of FTX and its synthetic analogue (sFTX) on transmitter release and on presynaptic currents at the mouse neuromuscular junction.[6] Muscle contraction was abolished a few minutes after adding FTX (1 μL/mL) to the incubating bath containing normal saline. The end-plate potentials (epp) recorded with intracellular microelectrodes were of small amplitude in the presence of FTX.

Presynaptic Ca^{2+} currents were recorded with extracellular microelectrodes inserted in the perineural sheath near the end-plate areas in the presence of K^+ channel blockers (tetraethylammonium and 3,4-diaminopyridine).[7] The addition of FTX (0.5 μL/mL) to the bath diminished the calcium wave, indicating that FTX action on neuromuscular transmission is due to a reduction in the influx of Ca^{2+} into the nerve terminal during the action potential.

We further studied synaptic transmission in a curarized nerve-muscle prepara-

[a] This work was supported by CONICET-Argentina, Fundación Roemmers, and Muscular Dystrophy of America Grants, NINCDS13742, AGO9480, and AFOSR-89-0270.

tion in order to analyze the effect of the inorganic Ca^{2+} channel blocker Cd^{2+} and FTX on transmitter release at low and high frequency stimulation. As expected, quantal content of the transmitter release was diminished by FTX and Cd^{2+}. In the absence of these agents, high-frequency stimulation produces a decline in epp amplitudes (FIG. 1A). In the presence of either Cd^{2+} or FTX, an early facilitation of the epp is seen (FIG. 1B and C). FIGURE 2 compares the mean values of m for each epp evoked by a train of stimuli at 40 Hz in muscles incubated with or without a Ca^{2+} channel blocker. The initial value for m, strikingly reduced during the train of stimuli in the control experiment, contrasts with the early

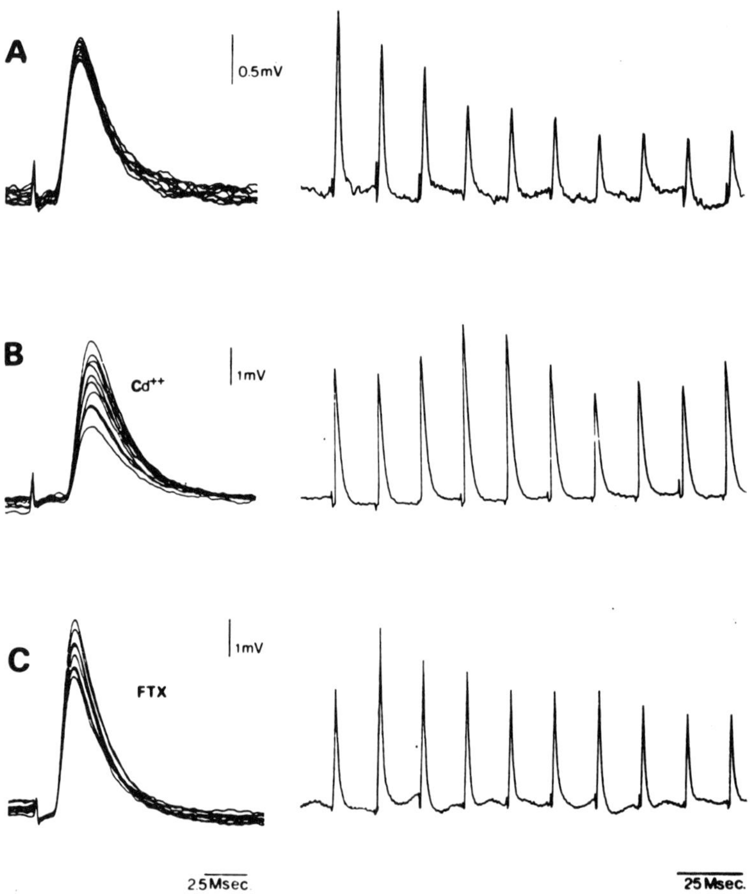

FIGURE 1. End-plate potentials recorded with intracellular microelectrodes from a mouse diaphragm muscle fiber in normal Ringer solution at 0.5 Hz (left) and 40 Hz (right) frequency stimulation. A: *d*-Tubocurine 1.8 μg/mL. B: *d*-tubocurarine 0.75 μg/mL and $CdCl_2$ 8 μM. C: *d*-tubocurarine 0.6 μg/mL and FTX 0.1 μl/mL. Ringer solution contained NaCl 137 mM, KCl 5 mM, $NaHCO_3$ 12 mM, NaH_2PO_4 1 mM, $MgSO_4$ 1 mM, $CaCl_2$ 2 mM, glucose 2 g/L, pH 7.4 bubbled with 95% O_2 and 5% CO_2. The phrenic nerve was stimulated by way of a suction electrode with 3 volts, 0.06 ms pulses.

facilitation and subsequent less-pronounced decline in the m value in the preparations treated with Cd^{2+} or FTX.

This small facilitation in transmitter release was also reported in animals treated with IgG from Lambert-Eaton human patients,[8] a treatment known to alter the presynaptic Ca^{2+} channels.

We conclude that mammalian neuromuscular transmission is very sensitive to the blocking effect of FTX and that a P-type of Ca^{2+} channel mediates transmitter release at the mammalian neuromuscular junction.

FIGURE 2. The quantal content (m) of the epp recorded during a train stimuli of 250 ms at 40 Hz was estimated as follows: (epp amplitude 40 Hz) × (quantal content 0.5 Hz)/mean epp amplitude 0.5 Hz). The quantal content at 0.5 Hz was estimated based on the variance of the amplitude of 50 to 100 epp recorded after 1 minute of stimulation.

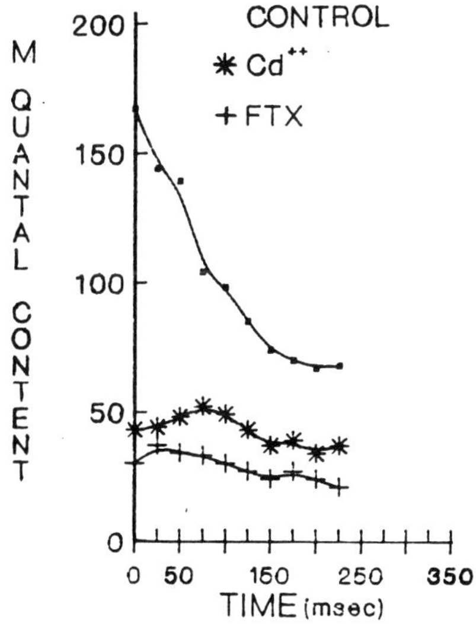

REFERENCES

1. Tsien, R. W., D. Lipscombe, D. V. Madison, K. R. Bley & P. Fox. 1988. Trends Neurosci. **11:** 431–438.
2. Llinás, R., M. Sugimori, J. W. Lin & B. Cherksey. 1989. Proc. Natl. Acad. Sci. USA **86:** 1689–1693.
3. Cherksey, B. D., M. Sugimori & R. R. Llinás. 1991. Ann. N.Y. Acad. Sci. **635:** 80–89.
4. Protti, D. A., L. Szczupak, F. S. Scornick & O. D. Uchitel. 1991. Brain Res. **557:** 336–339.
5. Atchison, W. D. 1989. J. Pharmacol. Exp. Ther. **251:** 672–678.
6. Uchitel, O. D., D. A. Protti, V. A. Sanchez, B. D. Cherksey, M. Sugimori & R. Llinás. 1992. Proc. Natl. Acad. Sci. USA. **89:** 3330–3333.
7. Penner, R. & F. Dreyer. 1986. Pflugers Arch **406:** 190–197.
8. Newsom-Davis, J., N. Murray, D. Wray, B. Lang, C. Prior, B. Gwilt & A. Vincent. 1982. Muscle & Nerve **5:** S17–S20.

Sera from Patients with Lambert-Eaton Myasthenic Syndrome Recognize the β-Subunit of Ca^{2+} Channel Complexes

MYRNA R. ROSENFELD,[a,b] ELLEN WONG,[a]
JOSEP DALMAU,[a,b] GEOFF MANLEY,[a] DAVID EGAN,[a]
JEROME B. POSNER,[b] EMANUELE SHER,[c]
AND HENRY M. FURNEAUX[a,d]

[a]*Laboratory of Molecular Neuro-oncology*
[b]*Department of Neurology*
Memorial Sloan-Kettering Cancer Center
New York, New York 10021

[c]*Consiglio Nazionale delle Richerche*
Centro per lo Studio della Farmacologia delle
Infrastructture Cellulari
Milano, Italy

Lambert-Eaton myasthenic syndrome (LEMS) is a paraneoplastic disorder of the neuromuscular junction characterized by a defect in the presynaptic quantal release of acetylcholine.[1] At least seventy percent of the patients with LEMS have small cell carcinomas of the lung.[2] LEMS patients develop an immune response against component(s) of a voltage-gated Ca^{2+} channel expressed in small cell carcinomas of the lung.[3] It has been suggested that this immune response cross-reacts with Ca^{2+} channels in the presynaptic neuromuscular junction and is responsible for the neurological dysfunction. To date there is no direct biochemical analysis of the subunits that interact with the IgG. The Ca^{2+} channel consists of 5 subunits ($\alpha 1$, $\alpha 2$, β, γ, δ).[4] The $\alpha 1$-subunit is the central ion-channel component. The β-subunit has been shown to accelerate both activation and inactivation of the calcium channel current and to increase the number of calcium antagonist binding sites upon transfection into LCa.11 cells and so appears to have an important regulatory function.[5]

In order to isolate and characterize LEMS antigens, we screened a human fetal brain λ ZAP II expression library using the serum of a patient with LEMS that contained an unusually high concentration (24, 188 pM/mL, normal < 30 pM/mL) of antibodies against Ca^{2+} channels, as determined by the ω-conotoxin assay.[6] A cDNA clone encoding a fusion protein of 35 kDa was isolated that was found to react with 3/7 Eaton-Lambert sera and 0/34 other irrelevant sera. Overlapping cDNAs were isolated, and the sequence of the gene product, which we call Mys B, was completed. The predicted amino-acid sequence was used in a homology search of Genpept and Swiss-Prot data bases and revealed striking homology (76.6% identity over 461 amino acids) with the β-subunits of Ca^{2+}

[d] Address correspondence to Henry M. Furneaux, Ph.D., Laboratory of Molecular Neuro-oncology, Memorial Sloan-Kettering Cancer Center, 1275 York Avenue, New York, NY 10021.

channel complexes. FIGURE 1 shows the homology alignment of Mys B with β-subunits cloned from rabbit skeletal muscle,[7] rat brain,[8] and human hippocampus.[9] Analysis of functional domains using the Blast program revealed the presence of an SH3 (Src homology 3) domain (residues 114–174) in the first highly conserved segment. FIGURE 2 shows that the sequence of the Mys B SH3 domain agrees well with a consensus SH3 segment. The function of SH3 domains is not well

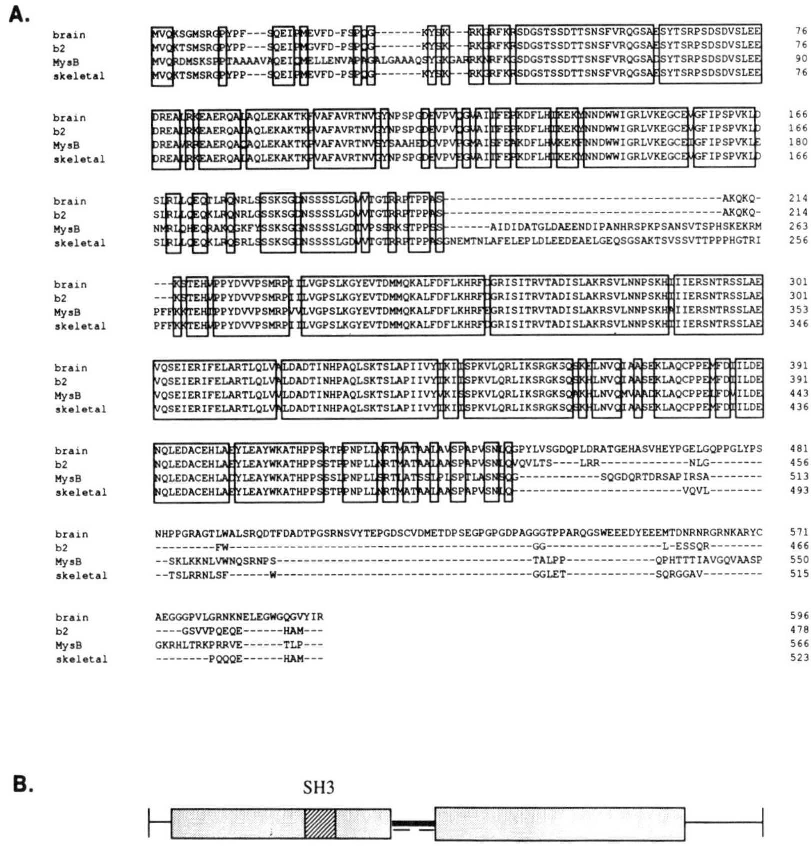

FIGURE 1. A: Homology alignment of Mys B peptide sequence with β2, brain, and skeletal β-subunits. B: Consensus β-subunit structure–shaded regions indicate domains of high homology between β-subunits.

understood but may imply that the β-subunit is involved in a cascade of proteins regulated by phosphorylation.[10]

From the data reported here we believe that LEMS patients harbor antibodies against β-subunits of Ca^{2+} channels. There is evidence that there may be a large family of β-subunit proteins associated with the different pharmacological types

```
Consensus     --ALYDY.-......LSF.KGD...-...-.....WW.........-..G.IPSNYV------

MysB          YSAAHEDDVPVPGMAISFEAKDFLH-VKE-KFNNDWWIGRLVKEGCEIGFIPSPVKLE----
p45-a CGD     AIADYEK---TSGSEMALSTGDVVE-VVE-KSESGWWFCQM-KA--KRGWIPASFL------
p47-b CGD     AIKAYTA---VEGDEVSLLEGEAVE-VIQ-KLLDGWWVIR--KDD-VTGYFPSMYL------
GAP           AILPYTK--VPDTDEISFLKGDMFI-VHN-ELEDGWMWVTNLRTD-EQGLIVEDLV------
HS-1          -VALYDYQ-GEGSDELSFDPDDVIT-DIE-MVDEGWWRG-RCHGH--FGLFPANYY------
vav-c         --ARYDFC-ARDRSELSLKEGDIIK-ILNKKGQQGWWRG-EIYGR--VGWFPANYVE-EDYS
Abl           FVALYDFV-ASGDNTLSITKGEKLR-VLGYNHNGEWCEA-QTKNG--QGWVPSNY-------
Plc-gam-1     --ALFDYK-AQREDELTFTKSAIIQN-VE-KQEGGWWRG-DYGGKK-QLWFPSNYVE-EMVS
Myosin-1B     --ALYDFA-AENPDELTFNEGAVVT-VIN-KSNPDWWEGEL--NG-QRGVFPASYV------
v-crk         --ALFDFK-GNDDGDLPFKKGDILK-IRD-KPEEQWWNAEDM-DG-KRGMIPVPYVEKCRPS
Fodrin-a      --ALYDYQ-EKSPREVTMKKGDIIT-IIN-STNKDWWKV-EV-ND-RQGFVPAAYVKKIDP-
Spectrin      --ALYDYQ-EKSPREVTMKKGDILT-LLN-STNKDWWKV-EV-ND-RQGFVPAAYV------
c-Fgr         --ALYDYE-ARTEDDLTFTKGEKFH-ILN-NTEGDWWEARSLSSG-KTGCIPSNYV-APVDS
Syn           V-ALYDYE-ARTEDDLSFHKGEKFQ-ILN-SSEGDWWEARSLTTG-ETGYIPSNYV------
v-YES         --ALYDYE-ARTTDDLSFKKGERFQ-IIN-NTEGDWWEARSIATG-KTGYIPSNYV------
c-src         --ALYDYE-SRTETDLSFKKGERLQ-IVM-NTEGDWWLAHSLTTG-QTGYIPSNYV-APSDS
Hck           V-ALYDYE-AIHHEDLSFQKGDQMV-VLE-ESDGEWWKARSLATR-KEGYIPSNYV------
Lyn           V-ALYPYD-GIHPDDLSFQKGDQMV-VLE-ESDGEWWKARSLATR-KEGYIPSNYV------
Lck           I-ALHSYE-PSHDGDLGFEKGEQLR-ILE-QSDGEWWKAQSLTTG-QEGFIPFNFV------
```

FIGURE 2. Consensus analysis of SH3 domains.

of Ca^{2+} channels.[11] The human antigen reported here, Mys B, appears to be a novel member of this gene family. The results suggest that the inhibition of Ca^{2+} channel activity by LEMS patients' sera may be due to binding of IgG to the β-subunit. Many questions remain with respect to the way in which IgG binding to the β-subunit may adversely affect Ca^{2+} channel activity.

REFERENCES

1. LAMBERT, E. H., L. M. EATON & E. D. ROOKE. 1956. Am. J. Physiol. **187:** 612–613.
2. O'NEILL, J. M., N. M. F. MURRAY & J. NEWSOM-DAVIS. 1988. Brain **111:** 577–596.
3. ROBERTS, A., S. PERERA, B. LANG, A. VINCENT & J. NEWSOM-DAVIS. 1985. Nature **317:** 737–739.
4. CATTERALL, W. A., K. NUNOKI, Y. LAI, K. DE JONGH, W. THOMSEN & S. ROSSIE. 1990. The Biology and Medicine of Signal Transduction. Raven Press. New York.
5. VARADI, G., P. LORY, D. SCHULTZ, M. VARADI & A. SCHWARTZ. 1991. Nature **352:** 159–162.
6. SHER, E., N. CANAL, G. PICCOLO, C. GOTTI, C. SCOPPETTA, A. EVOLI & F. CLEMENTI. 1989. Lancet **i:** 640–643.
7. RUTH, P., A. ROHRKASTEN, M. BIEL *et al.* 1989. Science **245:** 1115–1118.
8. PRAGNELL, M., J. SAKAMOTO, S. D. JAY & K. P. CAMPBELL. 1991. Fed. Eur. Biochem. Soc. **291:** 253–258.
9. WILLIAMS, M. E., D. H. FELDMAN, A. F. MCCUE, R. BRENNER, G. VELICELEBI, S. B. ELLIS & M. M. HARPOLD. 1992. Neuron **8:** 1–20.
10. KOCH, C. A., D. ANDERSON, M. F. MORAN, C. ELLIS & T. PAWSON. 1991. Science **252:** 668–674.
11. SAKAMOTO, J. & K. P. CAMPBELL. 1991. J. Biol. Chem. **266:** 18914–18919.

Na^+ Channel Distribution and Inactivation Properties of Human Type 1 and 2 Muscle Fibers[a]

R. L. RUFF[b,c,e] AND D. WHITTLESEY[d]

Departments of [b]Neurology, [c]Neurosciences, and [d]Surgery
Department of Veterans Affairs Medical Center
Case Western Reserve University School of Medicine
Cleveland, Ohio 44106

Na^+ channels on rat and rabbit skeletal muscle fibers have two inactivation processes with different kinetics and voltage dependence.[1-4] Fast inactivation is sufficiently rapid to terminate the action potential.[1] Slow inactivation is too slow to affect action potential termination. Slow inactivation, however, operates at more negative potentials than fast inactivation, so that the distribution of channels between closed and slow inactivated states may regulate the number of excitable Na^+ channels.[3]

In rat fast twitch muscle fibers, I_{Na} is about 5–10-fold higher immediately adjacent to the end-plate than at regions away from the end-plate.[5] I_{Na} declines rapidly with distance away from the end-plate, reaching the background level at 100–200 μm from the end-plate.

We studied voltage-clamp Na^+ currents (I_{Na}) in human intercostal muscle fibers from four men, 50 to 61 years old, who did not have muscle disease using the loose-patch-clamp technique. The fibers could be divided into two groups based upon the properties of I_{Na} (TABLE 1). The two groups of fibers were called type 1 and type 2. Histochemical staining indicated that type 1 fibers were slow twitch and that type 2 fibers were fast twitch. The Na^+ currents on type 1 and 2 fibers had different activation and inactivation properties and different densities. The membrane depolarization that produced a peak inward I_{Na}, $V_{I,Na,max}$, was more positive for type 1 fibers (TABLE 1). Both type 1 and type 2 fibers demonstrated fast and slow inactivation of I_{Na}. Na^+ currents on type 2 fibers were reduced by fast and slow inactivation at more negative potentials than type 1 fibers. The membrane potentials at which I_{Na} was reduced to 50% of the maximum value that could be obtained from a patch of membrane, $I_{Na,max}$, due to the fast inactivation process, $V_{h,1/2}$, and the slow inactivation process, $V_{s,1/2}$, for type 1 and type 2 fibers are shown in TABLE 1. At the resting potential about half of the maximal Na^+ current of type 2 fibers was not available due to slow inactivation, whereas, at the resting potential, inactivation did not appreciably reduce the Na^+ currents of type 1 fibers. The current densities on extrajunctional membrane (>200 μm from the end-plate) for type 2 fibers were >200% larger than type 1 fibers. Type 1 fibers had lower I_{Na}

[a] This work was supported by Merit Reviewed Funding from the Department of Veterans Affairs, and NIH Grant NS26661.

[e] Address inquiries to Dr. R. L. Ruff, Chief, Neurology Service 127(W), Cleveland VAMC, 10701 East Boulevard, Cleveland, OH 44106.

TABLE 1. Resting Membrane Potentials (RP), $I_{Na,max}$, $V_{I,Na,max}$, $V_{h,1/2}$, and A_h of Type 1 and Type 2 Human Intercostal Muscle Fibers

	Type 1 (n = 15)	Type 2 (n = 16)
RP (mV)	−85.9 ± 1.0[a]	−93.3 ± 1.0
$I_{Na,max}$ (mA/cm^2)[b]	7.33 ± 0.41[a]	18.0 ± 1.1
$V_{I,Na,max}$ (mV)	−19.8 ± 0.8[a]	−28.2 ± 0.7
$V_{h,1/2}$ (mV)	−61.1 ± 0.9[a]	−70.2 ± 0.9
$V_{s,1/2}$ (mV)[c]	−71.7 ± 3.5	−93.1 ± 2.6

[a] Type 1 and type 2 fibers were significantly different for RP ($p < 0.01$), $I_{Na,max}$ ($p < 0.001$), $V_{I,Na,max}$ ($p < 0.001$), $V_{h,1/2}$ ($p < 0.001$), and $V_{s,1/2}$ ($p < 0.001$).
[b] Measured >200 μm from the end-plate border.
[c] The number of fibers is n. For these studies n = 11 for type 1 fibers, and n = 12 for type 2 fibers.

density on the extrajunctional membrane (>200 μm from the end-plate) than type 2 fibers and required larger membrane depolarizations to inactivate Na^+ channels by fast or slow inactivation of I_{Na}.

Both type 1 and 2 fibers had an increase in Na^+ current density near the end-plate (TABLE 2). Measurement of membrane capacity suggested that the increase in I_{Na} at the end-plate border was due to increased Na^+ channel density rather than increased membrane folding; however, the increase in current density was greater for type 2 fibers. Type 1 fibers had a 3.2-fold increase and type 2 fibers had a 5.5-fold increase in Na^+ current density near the end-plate compared to extrajunctional membrane. The current density on the end-plate border for type 2 fibers was 400% larger than on type 1 fibers.

Differences in the properties of Na^+ channels between fast and slow twitch fibers may contribute to the ability of fast twitch fibers to operate at high firing frequencies and slow twitch fibers to be tonically active. The reduced susceptibility of type 1 fibers to inactivation may enable type 1 fibers to fire for prolonged times. Within the secondary synaptic folds the space constant is probably much smaller than along the surface of the membrane. Concentration of Na^+ channels near the end-plate may increase the transmission safety factor. Type 2 fibers may need a large safety factor because they operate at higher firing rates than type 1 fibers.

TABLE 2. Maximum I_{Na} Normalized to Membrane Capacitance on the End-plate Border and >200 μm Away from the End-plate for Type 1 and Type 2 Human Muscle Fibers

	Type 1 (n = 7)[a]	Type 2 (n = 7)[a]
End-plate Border		
$I_{Na,max}$ (mA/μF)	5.94 ± 0.27	25.2 ± 2.0
Extrajunctional		
$I_{Na,max}$ (mA/μF)	2.05 ± 0.62	4.87 ± 1.4
Ratio $I_{Na,max}$ (mA/μF)		
End-plate Border: extrajunctional	2.93 ± 0.14	5.24 ± 0.45

[a] The number of fibers is n. All of the values for type 2 fibers were larger than for type 1 ($p < 0.001$).

REFERENCES

1. ALMERS, W., W. M. ROBERTS & R. L. RUFF. 1984. Voltage clamp of rat and human skeletal muscle: measurements with an improved loose-patch clamp technique. J. Physiol. (Lond.) **347:** 751–768.
2. KIRSCH, G. E. & M. F. ANDERSON. 1986. Sodium channel kinetics in normal and denervated rabbit muscle membrane. Muscle & Nerve **9:** 738–747.
3. RUFF, R. L., L. SIMONCINI & W. STÜHMER. 1987. Comparison between slow sodium channel inactivation in rat slow and fast twitch muscle. J. Physiol. (Lond.) **383:** 339–348.
4. SIMONCINI, L. & W. STÜHMER. 1987. Slow sodium channel inactivation in rat fast-twitch muscle. J. Physiol. (London) **383:** 327–337.
5. CALDWELL, J. H., D. T. CAMPBELL & K. G. BEAM. 1986. Sodium channel distribution in vertebrate skeletal muscle. J. Gen. Physiol. **87:** 907–932.

Muscarinic Receptor-Mediated Modulation of Acetylcholine Release in Mice with Passively Transferred Lambert-Eaton Myasthenic Syndrome[a]

MASAHARU TAKAMORI

Department of Neurology
Kanazawa University School of Medicine
13-1, Takara-machi
Kanazawa, Japan 920

With regard to nicotinic (positive)- and muscarinic (negative)-feedback mechanisms in the release of acetylcholine (ACh) at the neuromuscular junction, presynaptic modulation by muscarinic acetylcholine receptor (AChR) was studied in mice with passively transferred Lambert-Eaton myasthenic syndrome (LEMS) in which the voltage-gated calcium channel (VGCC) function was deficient.[1] This occurs because the motor nerve terminal contains not only VGCC but also muscarinic AChR;[2] the latter is functionally coupled to the former in conjunction with inositol polyphosphate generation.[3]

Serum samples were obtained from a 74-year-old woman with small-cell lung carcinoma (SCLC) in whom the diagnosis of LEMS was made on the basis of characteristic electrophysiologic finding and elevated titer of antibody reactive with solubilized neuroblastoma calcium channels complexed with [^{125}I]ω-conotoxin, 405 pmol/L (normal value, $\leq$31 pmol/L).[4]

The disease was passively transferred to 4 mice by 6 daily intraperitoneal injections of 1.5 mL LEMS patient's serum, as confirmed by the reduction in EPP quantum content and immediately releasable stores at sacrifice on day 8 (cyclophosphamide 300 μg/g, injected on day 1) as compared to 3 mice injected with control sera (FIG. 1).

Atropine sulfate (1 μM), used as a blocker of presynaptic muscarinic AChR, which is involved in the negative feedback of ACh release,[5] induced a greater enhancement of ACh release and stores in LEMS mice as compared to control mice during which time the positive feedback mediated by presynaptic nicotinic AChR was suppressed by *d*-tubocurarine (FIG. 1).

Results are consistent with the contention that the release from muscarinic negative feedback control in cholinergic transmission is likely expressed in association with low intraneural Ca^{2+} and/or restricted ability of the nerve terminal to use external Ca^{2+}.[6]

In the present LEMS mice, whether the muscarinic AChR is expressed on SCLC cells[7] and what role, if any, antibody mechanisms play at the motor nerve terminal were not clarified by the present pharmacophysiologic study.

[a] This work was supported by Grants from the Japanese Ministries of Education, Science, and Culture (03454239), and Health and Welfare.

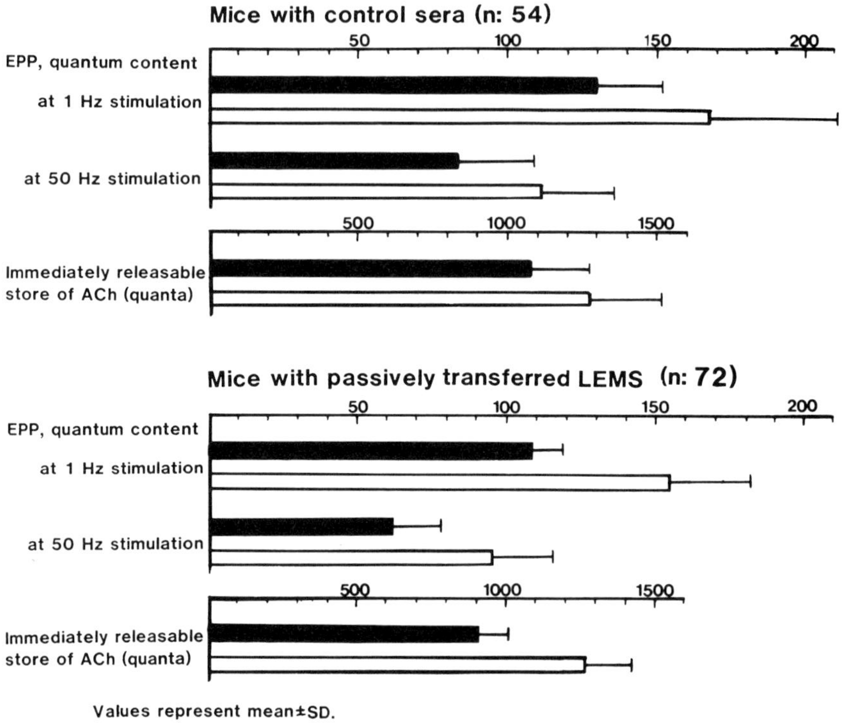

FIGURE 1. Microelectrode study of presynaptic function. The column and bar represent the mean and SD, respectively, in each electrophysiological index. n indicates the number of end-plates tested: 18 in each of 3 mice with control sera, and 18 in each of 4 mice with passively transferred LEMS. Significant reduction of electrophysiological indices was noted in mouse diaphragms treated with LEMS patient serum as compared with those treated with control sera ($p < 0.001$). Atropine (1 μM) enhanced them in both tested and control mice ($p < 0.001$). Enhancements were greater in tested mice than in control mice. (Percent increases of the mean values of ACh quantal release at 1 and 50 Hz and releasable ACh store were 42, 55, and 38%, respectively, in tested mice, whereas they were 29, 33, and 18%, respectively, in control mice.) ■, without atropine; □, with atropine.

REFERENCES

1. Lang, B., J. Newsom-Davis, C. Prior & D. Wray. 1983. Antibodies to motor nerve terminals: an electrophysiological study of a human myasthenic syndrome transferred to mouse. J. Physiol. (Lond.) **344:** 335–345.
2. Bowman, W. C., C. Prior & I. G. Marshall. 1990. Presynaptic receptors in the neuromuscular junction. Ann. N.Y. Acad. Sci. **604:** 69–81.
3. Williams, C. L. & V. A. Lennon. 1990. Activation of M_3 muscarinic acetylcholine receptors inhibits voltage-dependent calcium influx in small cell lung carcinoma. J. Biol. Chem. **265:** 1443–1447.
4. Lennon, V. A. & E. H. Lambert. 1989. Autoantibodies bind solubilized calcium channel-ω-conotoxin complexes from small cell lung carcinoma: a diagnostic aid for Lambert-Eaton myasthenic syndrome. Mayo Clin. Proc. **64:** 1498–1504.

5. Vizi, E. S. & G. T. Somogyi. 1989. Prejunctional modulation of acetylcholine release from the skeletal neuromuscular junction: link between positive (nicotinic)- and negative (muscarinic)-feedback modulation. Br. J. Pharmacol. **97:** 65–70.
6. Alberts, P. &. L. Stjärne. 1982. Role of calcium in muscarinic autoinhibition of ^{3}H-acetylcholine secretion in guinea-pig ileum myenteric plexus. Acta Physiol. Scand. **115:** 487–491.
7. Cunningham, J. M., V. A. Lennon, E. H. Lambert & B. Scheithauer. 1985. Acetylcholine receptors in small cell carcinomas. J. Neurochem. **45:** 159–167.

Effects of Lambert-Eaton Syndrome Serum and IgG on Calcium and Sodium Currents in Small-Cell Lung Cancer Cells[a]

MICHAEL P. VIGLIONE,[b] JOYCE K. W. BLANDINO,[c]
SANG J. KIM,[c] AND YONG I. KIM[c,d]

Departments of [c]Biomedical Engineering
[b]Neuroscience and [d]Neurology
University of Virginia Health Sciences Center
Charlottesville, Virginia 22908

The Lambert-Eaton myasthenic syndrome (LES) is an autoimmune disorder frequently associated with small-cell lung cancer (SCLC). Several lines of evidence indicate that the autoantibodies are directed against presynaptic calcium channels and that the putative stimulus for the production of the pathogenic antibodies is the tumor.[1] It has been demonstrated that LES IgG can inhibit voltage-dependent Ca^{2+} channels in bovine adrenal chromaffin (BAC) cells[2] and neuroblastoma cells[3] as well as reduce the influx of $^{45}Ca^{2+}$ into SCLC cells.[4,5] We have been studying the electrophysiologic properties of a human small-cell lung cancer cell line, NCI-H146, and the effects of LES serum and IgG on the voltage-dependent ionic currents expressed by these cells. Protocols for cell culture and whole-cell recording were carried out as in Pancrazio *et al.*[6] Values are presented as mean ± SEM; number of cells (n) are given for control and LES-treated cells. p values are for Student's t test; $p < 0.05$ is considered to be significant.

In response to step depolarizations from a holding potential of −80 mV, H-146 cells display inward currents that peak at +10 to +30 mV and are carried by Na^+ and Ca^{2+}. Calcium currents (I_{Ca}) recorded with 10 mM $[Ca^{2+}]_0$ decay with a single time constant (τ). In approximately half of the cells studied, I_{Ca} inactivated with both a fast (τ = 60 ms) and a slow (τ = 1100 ms) time constant. We also assessed the ability of ω-conotoxin (ω-CgTx GVIA, Peninsula Labs, Inc.) and the dihydropyridine (DHP) nicardipine to inhibit peak and steady-state I_{Ca} elicited with two-second depolarizations. Conotoxin (1 μM) was equally effective at reducing peak and plateau currents (−34 ± 10% and −33 ± 9%, respectively; n = 15, $p < 0.025$), whereas nicardipine reduced the plateau current (−84 ± 3%; n = 26, $p < 0.0005$) to a greater extent than it did the peak (−41 ± 13%; n = 26, $p < 0.01$). Additionally, increasing the holding potential from −80 mV to −40 mV decreased the peak current by 48 ± 9% (n = 17, $p < 0.005$) while reducing the plateau current by only 29 ± 12% (n = 17, $p < 0.1$) (see FIG. 1B).

The time-constant observations, pharmacologic sensitivities, and response to holding potential are all consistent with H-146 expressing multiple types of

[a] This work was supported by National Institutes of Health Grant NS-18607 and a Research Grant from the Muscular Dystrophy Association.

calcium channels. Based on this information, one of our objectives has been to determine whether LES serum and/or IgG inhibit specific types of calcium channels. Collectively, the following observations strongly suggest that LES IgG can inhibit DHP-sensitive channels (L-type). (1) Although inactivation profiles may not be diagnostic for Ca^{2+} channel types when used without other information, the fact that both LES IgG and nicardipine reduce the τ of inactivation from 1200 ms to approximately 600 ms support that both agents affect the same channels (FIG. 1A). (2) LES IgG is able to inhibit I_{Ca} evoked from either $V_h = -80$ mV or $V_h = -40$ mV (FIG. 1C). (3) LES IgG reduces both the peak and steady state currents; at least for one patient, the greatest

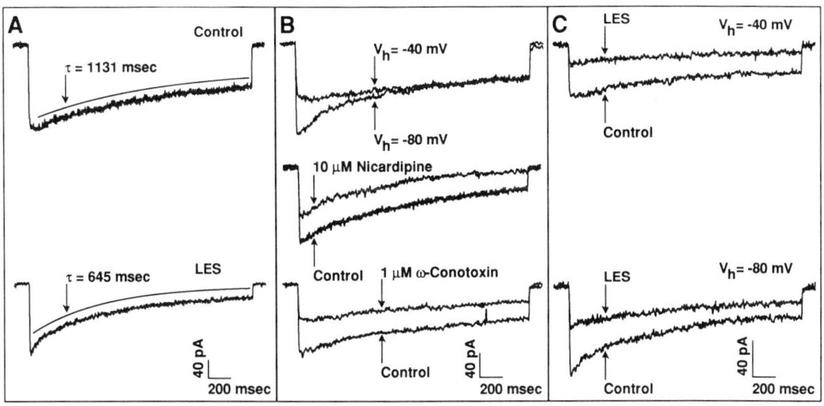

FIGURE 1. A: Peak I_{Ca} evoked from $V_h = -80$ mV was fit with the equation $y = A_0 + A_1 exp(-t/\tau)$ to obtain a value for the inactivation time constant τ. The top trace is from a control SCLC cell and the bottom from a cell incubated for 24 hours in LES serum. I_{Ca} recorded from LES cells and cells treated with nicardipine decay with a faster τ compared to controls. Exponential curves have been offset for visualization. **B:** (top) Effect of holding potential on I_{Ca} in SCLC cells. Shown is a cell in which raising V_h from -80 mV to -40 mV significantly reduces the peak current. (middle) 10 μM nicardipine reduces both peak and steady state but has a more profound effect on plateau I_{Ca} ($V_h = -80$ mV). (bottom) 1 μM ω-conotoxin inhibits both peak and steady state I_{Ca} equally ($V_h = -80$ mV). **C:** LES serum inhibits I_{Ca} evoked from -80 mV and from -40 mV. Depicted are representative cells exposed to either control or LES serum (1 mg/mL IgG) for 24 hours.

reduction observed was for the plateau current elicited from a depolarized holding potential. Nicardipine also maximally inhibited this same current.

In a previous study using BAC cells, we found that 30% of our LES patients had produced IgG capable of inhibiting I_{Na}.[7] An interesting observation of this study is that serum and IgG from one patient also significantly reduced I_{Na} in SCLC cells (FIG. 2B and C). Serum reduced I_{Na} by 26 ± 6% (n = 36, $p < 0.005$) after one day of exposure; the reduction, however, did not increase with an additional two days of incubation. IgG on the other hand was able to reduce I_{Na} by only 15% (n = 24) after one day of incubation, its efficacy increasing to 30% after three days of exposure (n = 16). For the serum and IgG from this same patient, the inhibition of I_{Ca} followed a time course similar to that observed for I_{Na} inhibition. Whereas serum was equally effective in decreasing I_{Ca} at 24, 48, and 72 hours of incubation (approximately 30% decrease, n = 22–38 cells for both control and LES at each time point), the IgG was maximally effective after 48 hours ($-49 \pm 6\%$, n = 17, $p < 0.01$) (see FIG. 2A, D, and E). The blockade of SCLC Ca^{2+} channels by LES IgG appears to be an "all-or-nothing" phenomenon. Estimates of single-channel properties using ensemble variance analysis[8] show that LES-treated cells (serum, 24 hours) have $27 \pm 8\%$ (n = 34, $p < 0.05$) fewer functional Ca^{2+} channels. The calcium channels that remain, however, have a single-channel current of 1.29 ± 0.08 pA (n = 34), similar to that of the control value of 1.16 ± 0.06 pA (n = 34, $p < 0.1$).

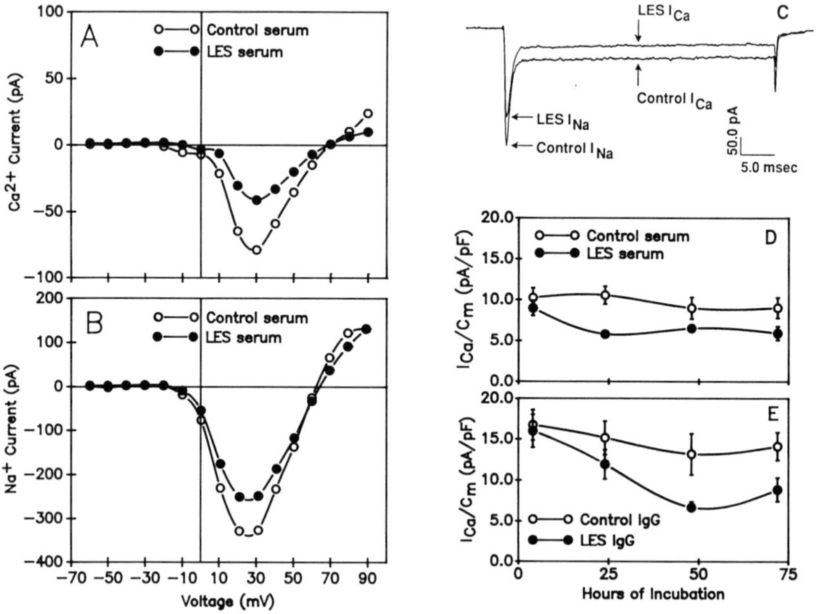

FIGURE 2. **A:** Calcium current-voltage relationships from a SCLC cell exposed to control serum and a cell incubated with LES serum (1 mg/mL IgG, 24 hours). **B:** Sodium current-voltage relationships from the same cell depicted in **A**. **C:** Current tracings from the cells described in **A** and **B** ($V_h = -80$ mV). **D:** Time course of action of LES serum on SCLC cells. I_{Ca} is expressed normalized for cell membrane capacitance. **E:** Time course of action of LES IgG on I_{Ca} from SCLC cells. Currents normalized as in **D**.

Even though it is clear that SCLC cells express more than one type of calcium channel, discerning which types are inhibited by the LES IgG has proven difficult. Nicardipine and holding potential differentially affect the peak and plateau phases of I_{Ca}. If the channels not affected by the DHP and/or holding potential are assumed to be different from those that are sensitive, then approximately half of the peak and 30% of the plateau currents are DHP- and V_h-insensitive, possibly ω-CgTx–sensitive channels. LES IgG or serum, on the other hand, can inhibit I_{Ca} by as much as 50%, reducing both peak and steady state I_{Ca} evoked from either −80 mV or a depolarized holding potential. These observations suggest that LES antibodies may interact with both DHP- and ω-CgTx–sensitive channels.

REFERENCES

1. VINCENT, A., B. LANG & J. NEWSOM-DAVIS. 1989. Autoimmunity to the voltage-gated calcium channels underlies the Lambert-Eaton myasthenic syndrome, a paraneoplastic disorder. Trends Neurosci. **12:** 496–502.
2. KIM, Y. I. & E. NEHER. 1988. IgG from patients with Lambert-Eaton syndrome blocks voltage-dependent calcium channels. Science **239:** 405–408.
3. PEERS, C., B. LANG, J. NEWSOM-DAVIS & D. WRAY. 1990. Selective action of myasthenic syndrome antibodies on calcium channels in a rodent neuroblastoma × glioma cell line. J. Physiol. (Lond.) **421:** 293–308.
4. ROBERTS, A., S. PERERA, B. LANG, A. VINCENT & J. NEWSOM-DAVIS. 1985. Paraneoplastic myasthenic syndrome IgG inhibits $^{45}Ca^{2+}$ influx in a human small cell carcinoma line. Nature **317:** 737–739.
5. DE AIZPURUA, H. J., E. H. LAMBERT, G. E. GRIESMANN, B. M. OLIVERA & V. A. LENNON. 1988. Antagonism of voltage-gated calcium channels in small cell carcinomas of patients with and without Lambert-Eaton myasthenic syndrome by autoantibodies, ω-conotoxin and adenosine. Cancer Res. **48:** 4719–4724.
6. PANCRAZIO, J. J., M. P. VIGLIONE, A. TABBARA & Y. I. KIM. 1989. Voltage-dependent ion channels in small-cell lung cancer cells. Cancer Res. **49:** 5901–5906.
7. VIGLIONE, M. P., C. E. CREUTZ & Y. I. KIM. 1992. Lambert-Eaton syndrome: Antigen-antibody interaction and calcium current inhibition in chromaffin cells. Muscle & Nerve. **15:** 1325–1333.
8. SIGWORTH, F. J. 1980. The variance of sodium current fluctuation at the node of ranvier. J. Physiol. (Lond.) **307:** 97–129.

The Tissue Distribution of $\alpha 2$-δ Subunit Genes of the Voltage-Gated Calcium Channel in Rodent Tissues and NG108-15 Cells[a]

A. S. WIERZBICKI,[b] D. M. W. BEESON, B. LANG, AND J. NEWSOM-DAVIS

Neurosciences Group
Institute of Molecular Medicine
John Radcliffe Hospital
Oxford OX3 9DU, United Kingdom

INTRODUCTION

Lambert-Eaton myasthenic syndrome (LEMS) is an autoimmune disease in which autoantibodies directed against the presynaptic voltage-gated calcium channels (VGCCs) cause a reduction in neurotransmitter release. In 60% of LEMS patients there is an associated small-cell lung carcinoma.[1] The syndrome can be passively transferred to mice by injection with IgG,[2] and this IgG can also inhibit potassium-stimulated $^{45}C^{2+}$ flux into cells of neuroectodermal origin, including NG108-15 cells.[3] The VGCC consists in skeletal muscle of five subunits: $\alpha 1$, $\alpha 2$, β, γ, and δ.[4] These are encoded by four genes, as the $\alpha 2$ and δ are products of the same gene.[5] This study investigated the tissue diversity of mRNA expression for the $\alpha 2$-δ gene in rodent tissues and in the rat glioma × mouse neuroblastoma line NG108-15.

METHODS

Poly(A)$^+$ RNA was prepared by the method of Chomcynski and Sacchi[6] followed by oligo-dT chromatography from 0.5 g of tissue from mouse and rat brain, lung, skeletal muscle, cardiac muscle, liver, and kidney. First strand cDNA was prepared using the reverse transcriptase Superscript[tm] (BRL Ltd), and polymerase chain reaction (PCR) amplification was carried out using 100 ng of cDNA from each source. Oligonucleotide primers were obtained to the rabbit skeletal muscle $\alpha 2$-δ gene[7] at nucleotides 2532-2554 and complementary to 3132-3155 (British Biotechnology Ltd). Amplification was conducted at 94°C for 1 minute, 55°C for 1 minute, and 72°C for 2 minutes in a volume of 100 μL containing 10 mM Tris-Cl, pH 8.4; 50 mM KCl; 3.5 mM $MgCl_2$; 5% dimethyl-sulphoxide; 0.01% Triton; 0.01% NP-40; and 0.5 U of *Taq* polymerase (Promega Corp.). The PCR products were

[a] A. S. Wierzbicki is supported by a Medical Research Council (UK) training fellowship.
[b] Present address: Department of Medical Biochemistry, University Hospital of Wales, Heath Park, Cardiff CF4 4XW, UK.

```
Rabbit skeletal muscle   KRNSDVDMDCVILDDGGFLLMANHDDYTNQIGF
Mouse brain              -----------------------------Y--S
Mouse lung               -----------------------------Y--S
Rat skeletal muscle      ----------------------I---------T
NG108-15                 FHTWIMF...---.....---SIY---RLHVRA

FGEIDPSLMRHLVNISVYAFNKSYDYQSVCEPGAAPKQGAGHRSAYVPSIADILQIG
--------------------------------------------------T------
--------------------------------------------------T------
--------------------------------------------------T------
-TV-GHY-VERPLRL---------------D------R-G-----------------

WWATAAAWSILQQFLLSLTFPRLLEAADMEDNDFTASMSLQSCITQQTQYFFDNDSL
----------------QFS-S-----VE--EDA------S-----------------
----------------QFS-S-----VE--EDA------S-----------------
------------------LL----------ED-------K-----E----------A
-------------L-CD-Q-------VE--EDA----L--------------K---I

SFSGVLDCGNCSRIFHVEKLMNTNLIFIMVESKGTCPCDTRLLIQAEQTSDGPDPCD
----------------------RQ-------R--H----------------------
----------------------RQ-------R--H----------------------
-------R-K------L--R-S---------R--------------G----------
----I-A----.QDL---T----T-GS-------D--------M----A-------S

MVK
---
---
---
---
```

FIGURE 1. Comparison of the amino acid sequences of VGCC α2-δ subunit PCR products from rabbit skeletal muscle (top line) with those from mouse brain, mouse lung, and rat skeletal muscle and with the homologous portion of the sequence from α2-δ subunit gene of NG108-15 cells isolated from a λgt10 cDNA library. PCR primers are shown in bold. The putative second membrane spanning region is underlined. Gaps are denoted by points and unchanged amino acids by dashes.

subcloned into pUC8 and sequenced. A portion of an $\alpha 2$-δ gene was obtained from NG108-15 cells by construction of a specifically primed cDNA library in λgt10[8] using a primer complementary to bp 3297-3312 of the rabbit skeletal muscle $\alpha 2$-δ gene. The library was screened at low stringency with a rabbit skeletal muscle $\alpha 2$-δ PCR product probe (bp 2532-3155) and clones overlapping the other PCR products isolated, subcloned into pUC8, and sequenced on both strands.

RESULTS

The primers used in this study were based on the rabbit skeletal muscle VGCC $\alpha 2$-δ gene sequence, and of the tissue sources examined, amplified products were only obtained from rat skeletal muscle, mouse brain, and mouse lung. A homologous sequence was obtained from positive clones isolated from a cDNA library from NG108-15 cells. The alignment of the derived amino acid sequences is shown in FIG. 1. The PCR products were 81% (rat), 87% (mouse), and 69% (NG108-15) homologous to the rabbit skeletal muscle gene. Nucleotide homology to the rabbit gene was 86%, 93%, and 75%, respectively.

DISCUSSION

Diversity of tissue expression at the protein and genetic level has been shown to occur in the $\alpha 1$-,[9] β-,[10] and γ-subunits[11] and has been suspected to occur in the $\alpha 2$- and δ-subunits.[11,12] This study shows that tissue-specific diversity does occur in the $\alpha 2$-δ gene expression and that in mice a portion of an $\alpha 2$-δ gene is identical in lung and brain.

REFERENCES

1. O'NEILL, J. H., N. M. F. MURRAY & J. NEWSOM-DAVIS. 1986. Brain **111:** 577–596.
2. LANG, B., J. NEWSOM-DAVIS, D. WRAY *et al.* 1981. Lancet **ii:** 224–226.
3. LANG, B., K. LEYS, A. VINCENT & J. NEWSOM-DAVIS. 1989. Ann. N.Y. Acad. Sci. **560:** 294–296.
4. CURTIS, B. & W. A. CATTERALL. 1984. Biochemistry **23:** 2113–2118.
5. JAY, S. D., A. H. SHARP, S. D. KAHL *et al.* 1991. J. Biol. Chem. **266:** 3287–3293.
6. CHOMCYNSKI, P. A. & G. S. SACCHI. 1989. Anal. Biochem. **162:** 156–162.
7. ELLIS, S. B., M. E. WILLIAMS, N. R. WAYS & K. P. CAMPBELL. 1989. Science **241:** 1661–1664.
8. BERGER, R. L. & A. E. KIMMEL. 1989. Guide to molecular cloning techniques. Academic Press. Philadelphia.
9. TSIEN, R. W., P. A. ELLINOR & W. T. HORNE. 1991. Trends. Pharmacol. Sci. **12:** 349–355.
10. PRAGNELL, W. A., J. SAKAMOTO, S. D. KAY & K. P. CAMPBELL. 1991. FEBS Lett. **291:** 253–258.
11. BIEL, M., R. HULLIN, S. FREUNDER *et al.* 1991. Eur. J. Biochem. **200:** 81–88.
12. WILLIAMS, M. E., D. H. FELDMAN, A. F. MCCUNE *et al.* 1992. Neuron **8:** 71–84.

The Investigation of Congenital Myasthenic Syndromes[a]

ANDREW G. ENGEL

Department of Neurology
Mayo Clinic
Rochester, Minnesota 55905

INTRODUCTION

Congenital myasthenic syndromes are genetically determined disorders in which the safety margin of neuromuscular transmission is affected by one or more pathologic mechanisms. The safety margin of transmission is defined as the difference between the actual end-plate potential (EPP) amplitude and that required to trigger a muscle fiber action potential. TABLE 1 presents an overview of the factors that can affect the safety margin, and TABLE 2 provides a current classification of congenital myasthenic syndromes. The classification is tentative because some syndromes are still under investigation, and new syndromes are likely to be discovered.

Clear understanding of the different congenital myasthenic syndromes is important for precise diagnosis and rational therapy. For example, patients with end-plate acetylcholinesterase (AChE) deficiency will not respond to AChE inhibitors because the enzyme is absent, and may not benefit from agents facilitating quantal release because this is restricted by the smallness of the nerve terminals and their encasement by Schwann cells.[1,2] Cholinergic agonists may be initially beneficial in patients with a prolonged open time of the acetylcholine receptor (AChR) channel, but can eventually worsen the end-plate myopathy, the AChR deficiency, and the clinical state.[3,4] In familial infantile myasthenia,[5] where the resynthesis or vesicular packaging of ACh is impaired, AChE inhibitors can be lifesaving during crises, and can prevent abnormal fatigability between crises. Finally, patients with AChR deficiency not complicated by a prolonged open time of the AChR channel are likely to benefit from AChE inhibitors or 3,4-diaminopyridine.

For the researcher, the congenital myasthenic syndromes are experiments of nature that shed light on the neurobiology of synaptic transmission. For example, the combined electrophysiological and ultrastructural findings in the syndrome of synaptic vesicle deficiency and reduced quantal release[6] provide direct support for the basic equation: $m = np$, where m is the number of quanta released by nerve impulse, n the number of synaptic vesicles readily available for release, and p the probability of quantal release.[7,8] The slow-channel syndrome, now clearly defined by patch-clamp recordings, can be considered an excitotoxic disease, in which the excessive ingress of electrolytes, and especially of calcium, through a ligand-gated ion channel results in cellular injury.[3,9] Further, it is likely that molecular genetic studies of the congenital myasthenic syndromes will uncover the mutational basis of the kinetic abnormalities of AChR demonstrated by electrophysiol-

[a] Work in the author's laboratory was supported by Grant NS 6277 from the National Institues of Health and by a Research Grant from the Muscular Dystrophy Association.

TABLE 1. Factors Affecting the Safety Margin of Neuromuscular Transmission

Quantal release mechanisms ($m = np$)[a]
Quantal size: the number of ACh molecules per synaptic vesicle
Quantal effects
End-plate geometry
AChE[c]—density in synaptic space and functional state of enzyme
AChR[d]—density on junctional folds, affinity for ACh, channel properties

[a] m = number of transmitter quanta released by nerve impulse; n = number of readily releasable transmitter quanta; p = probability of quantal release.
[b] ACh = acetylcholine.
[c] AChE = acetylcholinesterase.
[d] AChR = acetylcholine receptor.

ogy. This, in turn, will increase our understanding of the mechanisms that regulate the AChR channel.

THE INVESTIGATION OF CONGENITAL MYASTHENIC SYNDROMES

The correlation of clinical, morphologic, and *in vitro* electrophysiological data is essential for the complete investigation of a congenital myasthenic syndrome (TABLE 3). The clinical evaluation must include detailed electromyographic and serologic studies. The morphologic examination should include careful examination of the fine structure of the end-plate. The *in vitro* electrophysiologic studies must be sufficiently complete so that they provide information on parameters of quantal release and on the factors that effect the efficiency of the released quanta. It is also important to bear in mind that some congenital syndromes can result from or are associated with a kinetic abnormality of the AChR channel (TABLE

TABLE 2. Classification of Congenital Myasthenic Syndromes

Congenital, well characterized
Familial infantile myasthenia (defect in ACh resynthesis/packaging)[a]
Paucity of synaptic vesicles and reduced quantal release[b]
End-plate AChE deficiency[a]
Kinetic abnormality and deficiency of AChR
Classical slow-channel syndrome[c]
AChR deficiency, prolonged open time, low conductance, ε-subunit mutation[b]
AChR deficiency with short-channel open time[b]
Kinetic abnormality of AChR without AChR deficiency
High-conductance fast-channel syndrome[a]
Abnormal ACh-AChR interaction[b]
Congenital, partially characterized
AChR deficiency with paucity of secondary synaptic clefts[a]
Other AChR deficiencies[b]
Familial limb girdle myasthenia[a]

[a] Autosomal recessive inheritance.
[b] Autosomal recessive inheritance suspected.
[c] Autosomal dominant inheritance.

2), and that this may go undetected unless noise analysis or single-channel patch-clamp recordings are performed.

With the exception of end-plate AChE deficiency,[2] the molecular genetic basis of the congenital syndromes has not been investigated. This type of approach requires either positional cloning, or a candidate molecule (*e.g.*, a synaptic vesicle-related protein, a component of the basal lamina, an AChR subunit, a component of the postsynaptic cytoskeleton) and knowledge of the nucleotide sequence of the wild-type gene encoding the molecule.

TABLE 3. Investigation of Myasthenic Syndromes

- Clinical
  - History, examination, response to AChE inhibitor
  - EMG[a]: conventional, stimulation studies, SFEMG[b]
  - Serologic tests (AChR antibodies, tests for botulism)
- Morphologic studies
  - Routine histochemical studies
  - Cytochemical and immunocytochemical localizations of AChE, AChR, AChR subunits, IgG, C3, C9, MAC[c] at the neuromuscular junction
  - Estimate of the size, shape, and two-dimensional profile of AChE-reactive end-plates or end-plate regions on teased muscle fibers
  - Quantitative electron microscopy and electron cytochemistry
- $[^{125}\text{I}]\alpha$-bungarotoxin binding sites/end-plate
- *In vitro* electrophysiology studies
  - Conventional microelectrode studies: MEPP,[d] MEPC,[e] evoked quantal release (m, n, p)
  - Noise analysis: channel kinetics
  - Singel-channel patch-clamp recordings
- Molecular genetic analysis

[a] EMG, electromyography.
[b] SFEMG, single fiber EMG.
[c] MAC, C5b-9 complement membrane attack complex.
[d] MEPP, miniature end-plate potential.
[e] MEPC, miniature end-plate current.

CLINICAL OBSERVATIONS

History and Examination

A typical clinical history for a congenital myasthenic syndrome would include ocular, bulbar, or respiratory symptoms worsened by crying or activity in the neonatal period; fluctuating ocular palsies and abnormal fatigability on exertion during infancy and childhood; normal or delayed motor milestones; sometimes progression of symptoms during adolescence or adult life; and negative tests for anti-AChR antibodies. However, some syndromes (*e.g.* the slow channel syndrome[3] and familial limb-girdle myasthenia) may not present until the second or third decade of life; and in familial infantile myasthenia the symptoms can be episodic, with severe weakness and respiratory insufficiency appearing with fever, excitement, or without known cause.[5] A positive family history is consistent with the diagnosis. A negative family history does not exclude autosomal recessive inheritance, an incompletely penetrant autosomal dominant gene in one parent, or a new mutation.

On examination, the most important clue to a neuromuscular transmission defect is increasing weakness on sustained exertion. This can be documented by observing increasing ptosis during sustained upward gaze, measuring the arm abduction time, and counting the number of deep-knee bends the patient can perform. Patients with severe involvement of the truncal muscles, as in end-plate AChE deficiency, rapidly develop postural scoliosis and shift their weight from one foot to another on standing.[1,2] Selectively severe weakness of cervical and of wrist and finger extensor muscles is found in older patients with end-plate AChE deficiency[2] and in the slow-channel syndrome.[3] The pupillary light reflexes are delayed in patients with end-plate AChE deficiency.[2] Ocular muscle involvement can be absent or mild in some cases of end-plate AChE deficiency,[2] the slow-channel syndrome,[3] and familial limb-girdle myasthenia. The tendon reflexes are usually preserved but were hypoactive in one of five patients with end-plate AChE deficiency[2] and in a severely weak patient with a mutation affecting the ε-subunit of AChR.[4]

The Intravenous Edrophonium Test

The test is negative in patients with end-plate AChE deficiency, and in familial infantile myasthenia between attacks of weakness. It is inconsistently positive in the slow-channel syndrome and in the high-conductance fast-channel syndrome. A negative edrophonium test does not exclude the diagnosis of a congenital myasthenic syndrome; a positive test can be consistent with the diagnosis but does not differentiate it from acquired autoimmune myasthenia gravis (MG).

Electromyographic (EMG) Studies

The diagnosis must be supported by a decremental EMG response at low-frequency (2 Hz) stimulation in at least one muscle or by abnormal jitter and blocking on single-fiber EMG. The decremental response may be absent in familial infantile myasthenia between attacks; here a decremental response can be elicited by prolonged 10 Hz stimulation or by exercise for several minutes before 2 Hz stimulation.[5]

Single-nerve stimuli evoke a repetitive compound muscle action potential (CMAP) in patients taking AChE inhibitors, in end-plate AChE deficiency[1,2] and in the slow-channel syndrome.[3] The interval between the first and subsequent potentials is 5–10 ms, and the second and subsequent potentials decrease more rapidly than the first. Therefore, the test must be done in patients not exposed to AChE inhibitors, after a period of rest, and with single nerve stimuli.

The amplitude of the first evoked CMAP is usually normal, but a low-amplitude CMAP that facilitated on exercise and after neostigmine injection was found in a severely weak patient with a mutation affecting the ε-subunit of AChR.[4]

Observations in the EMG laboratory can provide an objective estimate of responsiveness to AChE inhibitors or other cholinergic agents. For example, one can compare the decrement observed in a given muscle before and 30 minutes after a subcutaneous dose of neostigmine methylsulfate.

Serologic Tests

A positive AChR antibody test excludes the diagnosis of a congenital syndrome, but a negative test in a sporadic case does not, because a proportion of patients with autoimmune MG are seronegative. Seronegative MG can sometimes be excluded by the clinical or EMG findings. The absence of immune deposits (IgG and complement) from the end-plates, or an *in vitro* electrophysiologic abnormality different from that in MG, is strong evidence against seronegative MG.

MORPHOLOGIC STUDIES

Routine Histochemical Studies

These may show no abnormality or only type-2 fiber atrophy. More frequently, there is type-1 fiber preponderance. This finding is not specific, but it should raise the question of a congenital myasthenic syndrome.

In the slow-channel syndrome, small groups of atrophic fibers and various myopathic alterations, including vacuolar change near the end-plate, occur in severely affected muscles.[3]

The mean muscle fiber diameter must be determined in all muscle specimens used for miniature end-plate potential (MEPP) measurements, as the MEPP amplitude is related by an inverse exponential function to the fiber diameter.[10]

Cytochemical Localization of AChE

End-plate AChE is associated with the basal lamina of the primary and secondary synaptic clefts.[11] Therefore, the light microscopic localization of AChE reveals the position and configuration of the synaptic gutter but provides no information on the pre- or postsynaptic components of the end-plate. When AChE is localized on fixed and teased fibers and the end-plate viewed face-on, the normal synaptic gutter has an oval outline and branches like arms, resembling a pretzel. In some, but by no means all, congenital syndromes,[3,4,12,13] and also in autoimmune MG,[14] the end-plate is remodeled due to sprouting of terminal or ultraterminal axons and formation of new end-plate regions. The AChE reaction now reveals multiple small regions, dispersed over an extended length of the muscle fiber surface. Rather inaccurately, this is described as "elongation" of the end-plate,[12,14] and the axial length of the fiber surface on which AChE-positive zones are dispersed is designated as "the length of the end-plate".

In congenital end-plate AChE deficiency, there is no reaction for AChE in either fresh or fixed muscle.[1,2] When this diagnosis is suspected, AChE should be localized in the presence of 10^{-5} M iso-OMPA to inhibit any butyrylcholinesterase activity at or near the end-plate.[1,2]

Immunocytochemical Studies

These can (1) confirm the absence of immunoreactive AChE from the end-plate in end-plate AChE deficiency;[2] (2) prove or disprove the diagnosis of autoimmune MG by showing the presence or absence of immune deposits at the end-plate; (3) provide a qualitative estimate of the abundance of end-plate AChR;[3] and (4) demonstrate the presence or absence of subunit-specific AChR epitopes.[4]

Electron Microscopy and Electron Cytochemistry

Electron microscopy of the end-plate is essential to the investigation of a congenital myasthenic syndrome. The size and density of the synaptic vesicles, the size of the nerve terminal and its relation to the Schwann cell, and the structure of the junctional folds cannot be observed any other way. Quantitative electron microscopy allows morphometric reconstruction of the end-plate,[15] yields data for structure-function correlations, and, in some disorders, suggests the mechanism by which neuromuscular transmission is impaired. For example, (1) electron microscopy was required to demonstrate synaptic vesicle deficiency in a syndrome with reduced quantal release.[6] Without visualizing the synaptic vesicles, this disorder could be confused with the Lambert-Eaton syndrome. (2) The small nerve terminals and their encasement by Schwann cells provides an explanation for the reduced quantal release in end-plate AChE deficiency.[1,2] (3) In the slow-channel syndrome, the degeneration of the junctional folds explains the loss of AChR and the reduced MEPP amplitude.[3]

Electron cytochemical localization of AChR with peroxidase-labeled α-bungarotoxin or with anti-AChR antibodies displays the density and distribution of AChR over the postsynaptic membrane.[1,3,5,6,16–18] This information, together with the MEPP amplitude and the number of α-bungarotoxin binding sites per end-plate (see below), is used to assess the extent, significance, and possible mechanism of end-plate AChR deficiency.

[^{125}I] α-Bungarotoxin Binding Sites Per End-Plate

Two toxin molecules bind to each AChR. The number of toxin-binding sites per end-plate, or N, is the product of the $A_t \times D$, where A_t is the total postsynaptic membrane area that binds toxin, and D the average density (number per unit area) of binding sites. In infants and young children, N is smaller than in adults because the end-plates, and A_t, are smaller than in adults. In adults with end-plates of normal shape and size, a decrease in N argues for a decrease in D. In diverse congenital myasthenic syndromes, a decrease of N is associated with remodeling of the end-plate.[3,4,12,13] Here the decrease must involve A_t or D. (When N is reduced, D could be reduced and A_t reduced, normal, or increased; or D could remain constant while A_t is reduced.) Ultrastructural localization of peroxidase-labeled α-bungarotoxin at such end-plates shows either a diffuse decrease of AChR over intact junctional folds or a focal loss of AChR from degenerating folds.[3,4]

IN VITRO ELECTROPHYSIOLOGICAL STUDIES

Conventional Microelectrode Studies

A strip of muscle intact from origin to insertion, usually from an intercostal muscle[8] or from anconeus,[19] is used. The amplitude and frequency of the MEPP, the miniature end-plate current (MEPC), and parameters of evoked quantal release are routinely determined.

MEPP measurements are made in rested muscle with focally placed microelectrodes, and the amplitude is normalized for a resting membrane potential of −80 mV. The MEPP amplitude is affected by the cable properties of the sarcolemma,[10] the number of ACh molecules per quantum, end-plate geometry, and the density and kinetic properties of AChR.[20,21] With AChE intact, the decay phase of the MEPP is related to the open time of the AChR channel[22,23] and the cable properties of the sarcolemma.[10,24]

In familial infantile myasthenia there is a defect in the resynthesis or vesicular packaging of ACh; in routine measurements, the MEPP amplitude is normal but decreases significantly after prolonged 10 Hz stimulation. When this diagnosis is suspected, MEPP measurements are made both before and after 10 Hz stimulation for 5 to 10 minutes.[5]

The MEPC is recorded from the voltage-clamped muscle fiber. It is independent of the cable properties of the sarcolemma but otherwise is affected by the same factors as the MEPP.[25] The amplitude and duration of the MEPC are the macroscopic expression of the bursts of AChR channel activity occurring in response to one quantum of ACh.[26] With end-plate AChE intact, the decay time constant of the MEPC reflects, but is usually somewhat longer than, the mean burst duration.

Parameters of quantal release are evaluated in curarized muscle. The number of quanta released by nerve impulse (m) is determined at 1 Hz stimulation by the variance method.[27] The value of m is affected by the probability of quantal release (p) and the number of readily releasable quanta (n).[7,8] Therefore, p and n are also determined using brief trains of high-frequency stimuli.[28] The value of p is related to the calcium concentration within the nerve terminal. The value of n is affected by the density of the synaptic vesicles, the total nerve terminal volume at a given end-plate, and the total presynaptic membrane area not covered by Schwann cell.

Analysis of the ACh-Induced Current Noise

At the voltage clamped end-plate, the ACh-induced fluctuations in end-plate current correspond to the fluctuations in the number of channels open at a given moment. Analysis of these fluctuations, or noise, allows estimation of the open time and conductance of the AChR channel.[23] The mean channel open time, derived from the noise power spectrum, corresponds to the mean duration of opening bursts recorded from single channels by patch-clamp analysis. The mean channel conductance is estimated either from the noise power spectrum or from regression of the variance of the conductance fluctuations on the mean increase in channel conductance.[23] In this laboratory, noise analysis has identified a number of kinetic abnormalities of AChR that otherwise would not have been detected.[4,16,17,29,30]

Single-Channel Patch-Clamp Recordings

High resolution patch-clamp recordings of currents flowing through single AChR channels provide precise information on channel conductance and on the kinetics of single openings and bursts of openings.[31] Single channel recordings from the end-plate are difficult, however, because meticulous digestion and dissection of muscle are required before tight seals can be obtained on the postsynaptic membrane. Also, patch-clamp recordings in any one case reflect the activity of only a limited number of channels. Recent patch-clamp recordings in this laboratory from end-plates in the slow-channel myasthenic syndrome have revealed a population of bursts whose mean duration was normal, and a population whose mean duration was 8-fold longer than normal. All openings had a normal conductance. These findings suggest two populations of AChR, encoded by a mutant gene and by a wild-type gene, respectively.[9] Patch-clamp analysis will enhance the detection of kinetic defects of AChR in other congenital myasthenic syndromes.

APPLICATION OF THE INVESTIGATIVE APPROACH: RECOGNITION OF THE HIGH-CONDUCTANCE FAST-CHANNEL SYNDROME[32]

A 9-year-old girl had mild myasthenic symptoms since birth. At age 9, she had significant weakness only in the frontalis, levator palpebrae and neck flexor muscles. Electromyography showed no decrement in limb muscles, but single fiber examination of the facial muscles was consistent with a defect of neuromuscular transmission. The symptoms responded only partially to pyridostigmine. Tests for anti-AChR antibodies were negative. A younger sister had elements of the same disease. An intercostal muscle specimen was obtained from the older sister for end-plate studies. The quantal content of the end-plate potential was normal. Miniature end-plate currents were abnormally large and their decay time constant was abnormally short. AChR channel properties were studied by analysis of acetylcholine-induced current noise. The mean single-channel conductance was increased 1.7-fold, and the mean channel open time was 30 percent shorter than normal. The number of AChR per end-plate was normal. Electron microscopy revealed only a few degenerating or simplified end-plates, but the mean postsynaptic area of folds and cleft was significantly reduced. The channel abnormality in this disorder may stem from a point mutation in an AChR subunit affecting a single amino acid residue lining the pore of the AChR channel.[33–35] The mechanism by which the physiologic abnormality produces clinical symptoms is still unclear. However, we postulate that, as in the slow-channel syndrome,[3] in the clinically affected muscles the cholinergic overactivity causes an end-plate myopathy with loss of AChR and alteration of end-plate geometry.

REFERENCES

1. ENGEL, A. G., E. H. LAMBERT & M. R. GOMEZ. 1977. A new myasthenic syndrome with end-plate acetylcholinesterase deficiency, small nerve terminals, and reduced acetylcholine release. Ann. Neurol. **1:** 315–330.
2. HUTCHINSON, D. O., A. G. ENGEL, T. J. WALLS, S. NAKANO, S. CAMP, P. TAYLOR, C. M. HARPER & J. M. BRENGMAN. 1993. The spectrum of congenital end-plate acetylcholinesterase deficiency. Ann. N.Y. Acad. Sci. This volume.
3. ENGEL, A. G., E. H. LAMBERT, D. M. MULDER, C. F. TORRES, K. SAHASHI, T. E. BERTORINI & J. N. WHITAKER. 1982. A newly recognized congenital myasthenic syndrome attributed to a prolonged open time of the acetylcholine-induced ion channel. Ann. Neurol. **11:** 553–569.
4. ENGEL, A. G., D. O. HUTCHINSON, S. NAKANO, L. MURPHY, R. C. GRIGGS, Y. GU, Z. W. HALL & J. LINDSTROM. 1993. Myasthenic syndromes attributed to mutations affecting the epsilon subunit of the acetylcholine receptor. Ann. N.Y. Acad. Sci. This volume.
5. MORA, M., E. H. LAMBERT & A. G. ENGEL. 1987. Synaptic vesicle abnormality in familial infantile myasthenia. Neurology **37:** 206–214.

6. WALLS, T. J., A. G. ENGEL, A. S. NAGEL, M. C. HARPER & V. F. TRASTEK. 1993. Congenital myasthenic syndromes associated with paucity of synaptic vesicles and reduced quantal release. Ann. N.Y. Acad. Sci. This volume.
7. DEL CASTILLO, J. & B. KATZ. 1954. Quantal components of the end-plate potential. J. Physiol. (Lond.) **124:** 560–573.
8. ELMQVIST, D. & D. M. J. QUASTEL. 1965. A quantitative study of end-plate potentials in isolated human muscle. J. Physiol. (Lond.). **178:** 505–529.
9. HUTCHINSON, D. O., S. NAKANO, J. P. SIEB & A. G. ENGEL. 1993. Patch clamp analysis of the slow channel myasthenic syndrome. Neurology. In press. (Abstract)
10. KATZ, B. & S. THESLEFF. 1957. On the factors which determine the amplitude of the 'miniature end-plate potential'. J. Physiol. (Lond.). **137:** 267–278.
11. MCMAHAN, U. J., J. S. SANES & L. M. MARSHALL. 1978. Cholinesterase is associated with the basal lamina of the neuromuscular junction. Nature **271:** 172–174.
12. VINCENT, A., S. G. CULL-CANDY, J. NEWSOM-DAVIS, A. TRAUTMANN, P. C. MOLENAAR & R. L. POLAK. 1981. Congenital myasthenia: End-plate acetylcholine receptors and electrophysiology in five cases. Muscle & Nerve **4:** 306–318.
13. WOKKE, J. H. J., F. G. I. JENNEKENS, P. C. MOLENAAR, C. J. M. VAN DEN ORD, B. S. OEN & H. F. M. BUSCH. 1989. Congenital paucity of secondary synaptic clefts (CPSC) syndrome in 2 adult sibs. Neurology **39:** 648–654.
14. COËRS, C. & N. TELEMAN-TOPPET. 1976. Morphological and histochemical changes of motor units in myasthenia. Ann. N.Y. Acad. Sci. **274:** 6–19.
15. ENGEL, A. G. 1986. Quantitative morphological studies of muscle. *In* Myology. A. G. Engel & B. Q. Banker, Eds.: 1045–1079. McGraw-Hill. New York.
16. UCHITEL, O., A. G. ENGEL, T. J. WALLS, A. NAGEL, V. BRILL & V. F. TRASTEK. 1993. Congenital myasthenic syndrome attributed to an abnormal interaction of acetylcholine with its receptor. Ann. N.Y. Acad. Sci. This volume.
17. ENGEL, A. G. 1986. Quantitative morphologic studies of muscle. *In* Myology. A. G. Engel & B. Q. Banker, Eds.: 1045–1078. McGraw-Hill. New York.
18. ENGEL, A. G., J. LINDSTROM, E. H. LAMBERT & V. A. LENNON. 1977. Ultrastructural localization of the acetylcholine receptor in myasthenia gravis and in its experimental autoimmune model. Neurology **27:** 307–315.
19. MASELLI, R., D. P. MASS, B. J. DISTAD & D. RICHMAN. 1991. Anconeus muscle: A human muscle preparation suitable for *in-vitro* microelectrode studies. Muscle & Nerve **14:** 1189–1192.
20. ENGEL, A. G. 1987. The molecular biology of end-plate diseases. *In* The Vertebrate Neuromuscular Junction. M. M. Salpeter, Ed.: 361–424. Alan Liss. New York.
21. M. M. SALPETER. 1987. Vertebrate neuromuscular junctions: General morphology, molecular organization, and functional consequences. *In* The Vertebrate Neuromuscular Junction. M. M. Salpeter, Ed.: 1–54. Alan Liss. New York.
22. KATZ, B. & R. MILEDI. 1972. The statistical nature of the acetylcholine potential and its molecular components. J. Physiol. (Lond.) **224:** 665–699.
23. ANDERSON, C. R. & C. F. STEVENS. 1973. Voltage clamp analysis of acetylcholine produced end-plate current fluctuations at frog neuromuscular junction. J. Physiol. (Lond.) **235:** 655–691.
24. FATT, P. & B. KATZ. 1952. Spontaneous subthreshold activity at motor nerve endings. J. Physiol. (Lond.) **117:** 109–128.
25. MAGLEBY, K. L. 1986. Neuromuscular transmission. *In* Myology. A. G. Engel & B. Q. Banker, Eds.: 393–418. McGraw-Hill. New York.
26. COLQUHOUN, D. & A. G. HAWKES. 1983. The principles of the stochastic interpretation of ion channel mechanisms. *In* Single-Channel Recording. B. Sakmann & E. Neher, Eds.: 135–175. Plenum. New York.
27. MARTIN, A. R. 1955. A further study of the statistical composition of the end-plate potential. J. Physiol. (Lond.) **130:** 114–122.
28. KAMENSKAYA, M. A., D. ELQVIST & S. THESLEFF. 1975. Guanidine and neuromuscular transmission. II. Effect on transmitter release in response to repetitive stimulation. Arch. Neurol. **32:** 510–518.
29. NAGEL, A., A. G. ENGEL, T. J. WALLS, C. M. HARPER & H. A. WAISBURG. 1993.

Congenital end-plate acetylcholine receptor deficiency and short channel open time. Ann. N.Y. Acad. Sci. This volume.

30. ENGEL, A. G., T. J. WALLS, A. NAGEL & O. UCHITEL. 1990. Newly recognized congenital myasthenic syndromes. I. Congenital paucity of synaptic vesicles and reduced quantal release. II. High-conductance fast-channel syndrome. III. Abnormal acetylcholine receptor (AChR) interaction with acetylcholine. IV. AChR deficiency and short channel open-time. Prog. Brain. Res. **84:** 125–137.
31. HAMMILL, O. P., A. MARTY, E. NEHER, B. SAKMANN & F. J. SIGWORTH. 1981. Improved patch-clamp techniques for high-resolution current recording from cell-free membrane patches. Pflügers Arch. **391:** 85–100.
32. ENGEL, A. G., O. UCHITEL, T. J. WALLS, A. NAGEL & J. BODENSTEINER. 1990. Congenital myasthenic syndrome with high conductance and fast closure of the acetylcholine-induced ion channel. Neurology **40**(Suppl. 1): 277. (Abstract.)
33. LEONARD, R. J., C. J. LABARCA, P. CHARNET, N. DAVIDSON *et al.* 1988. Evidence that the M2 membrane spanning region lines the ion channel pore of the nicotinic receptor. Science **242:** 1578–1581.
34. IMOTO, K., T. KONNO, J. NAKAI, F. WANG *et al.* 1991. A ring of uncharged polar amino acids as a component of channel constriction in the nicotinic acetylcholine receptor. FEBS Lett. **289:** 193–200.
35. VILLARROEL, A., S. HERLITZE, V. WITZEMANN, M. KOENEN *et al.* 1992. Asymmetry of rat acetylcholine receptor subunits in the narrow region of the pore. Proc. R. Soc. Lond. B. Biol. Sci. **249:** 317–324.

Insights into Possible Skeletal Muscle Nicotinic Acetylcholine Receptor (AChR) Changes in Some Congenital Myasthenias from Physiological Studies, Point Mutations, and Subunit Substitutions of the AChR[a]

HENRY J. KAMINSKI AND ROBERT L. RUFF[b]

Departments of Neurology and Neuroscience
Department of Veterans Affairs Medical Center
Case Western Reserve University Medical School
Cleveland, Ohio 44106

This article considers the possible basis for some of the congenital disorders of neuromuscular transmission (CDNT),[1,2] also called congenital myasthenias, in which there appears to be an alteration in the function of the skeletal muscle nicotinic acetylcholine receptor (AChR). We will first review some of the information that is known about how alterations in microstructure or subunit composition of the AChR can affect the character of the ionic currents that transit through the channel. Then we will relate alterations in the currents produced by experimental alterations in AChR structure with the AChR currents described in some of the known CDNT.

SUBUNIT COMPOSITION OF THE AChR

The AChR is a membrane-bound allosteric glycoprotein. The AChR is a pentamer composed of four subunits.[3–6] In mammals the channel is know to exist in at least two isoforms. The mature or "innervated" form of the AChR is composed of two α-subunits and one copy of each of the β-, δ-, and ε-subunits.[5,7–9] The fetal or "denervated" form of the AChR has a γ-subunit in place of the ε-subunit. Studies in which cDNA associated with specific subunits was injected into *Xenopus* oocytes indicated that few if any functional channels were produced if the α-, β-, or δ-subunit was omitted, which suggests that α-, β-, δ-, and γ- or ε-subunits are necessary to form functional channels efficiently.[8,10] Muscle fibers express AChRs before there is any contact between motor axons and the muscle fibers.[11,12] AChRs collect into clusters prior to neuronal contact. Contact between a neurite growth cone and a muscle fiber appears to enhance AChR clustering in the vicinity of

[a] This work was supported by Merit Reviewed Funding from the Department of Veterans Affairs, and National Institutes of Health Grants NS26661 and EY09186.

[b] Address inquiries to Robert L. Ruff, M.D., Ph.D., Department of Neurology 127(W), Department of Veterans Affairs Medical Center, 10701 East Boulevard, Cleveland, OH 44106.

the contact. Before and after the initial contacts of nerve fibers on a muscle fiber, both the fetal- and adult-type AChRs are expressed with the fetal-type receptor predominating. During early development muscle fibers receive synaptic contacts from many axons at several points along a fiber. Later in development, at about the stage of synaptic rejection, when a twitch fiber supports only one synapse per fiber, the fiber suppresses expression of the γ-subunit and up-regulates expression of the ε-subunit.[5,7–9,13] Mature twitch fibers express the ε-subunit in much greater abundance than the γ-subunit, so that >97% of the AChR are the mature isoform.[12] In addition, in adult innervated twitch muscle fibers, the nuclei near the end-plate, transcribe AChR subunits more actively than other nuclei.[14,15] After denervation the γ-subunit is again expressed and non–end-plate nuclei express AChR subunits. Consequently, fetal-type receptors reappear on denervated muscle fibers. In an innervated muscle fiber the AChR are clustered at the top of the junctional folds on the postsynaptic membrane, and few AChR are present on the extrajunctional membrane. After denervation, AChR appear diffusely over the muscle surface but remain in relative abundance at the end-plate.[12]

Each α-subunit contains one binding site for acetylcholine (ACh). AChR channel opening usually requires that two ACh molecules bind to the AChR.[16] There is positive cooperactivity for ACh binding.[17] The dissociation constant is about 50 μM with one ACh molecule bound and about 10 μM for a receptor in the closed state with two ACh molecules bound. In mammals there appears to be two isoforms of the α-subunit.[18] Although both isoforms of the α-subunit are expressed in rhabdomyosarcoma cells in tissue culture, the significance of these isoforms is not presently understood. The difference in the binding affinity for the first and second ACh molecule could result from a conformation change induced by binding of the first ACh molecule, a difference in the subunits surrounding each α-subunit, or perhaps the slight difference between the two α-subunit isoforms.[17,19–21]

DIFFERENCES BETWEEN THE ADULT- AND FETAL-TYPE AChRs

Electrophysiologically, adult- and fetal-type AChRs can be distinguished by their single-channel conductances and channel open times. Adult-type channels in mammals and amphibians have shorter mean open times and a conductance that is about 50% larger than is found with fetal-type channels.[11,12] Mishina *et al.*[8] demonstrated that differences between adult- and fetal-type AChR channels in mammals could largely be accounted for by the replacement of the γ- with the ε-subunit. The AChR has kinase A, kinase C, and tyrosine kinase phosphorylation sites.[22,23] Phosphorylation and other forms of posttranslational modification can alter properties of the AChR.[11,24] In particular, subunit phosphorylation appears to regulate agonist-induced desensitization.[22,23] Both subunit substitution and post-translational modification may contribute to developmental and denervation-induced changes in AChR properties in amphibians and reptiles. Of note, avian and probably fish species do not show a developmental or denervation-induced change in AChR properties, and both the avian and fish adult AChRs contain the γ-subunit, not the ε-subunit.[12] It is not known if both the γ- and ε-subunits are expressed by amphibians or reptiles.[11,12,25–27]

Electrophysiological studies suggest that mature tonic fiber synapses in reptiles and amphibians appear to have both adult- and fetal-type AChR.[25–27] With enzymatic treatment, the nerve terminals of snake muscle fibers can be separated from the muscle fibers to expose bare postsynaptic membrane that is suitable for patch-clamp recording of single-channel currents (FIG. 1).[27] The postsynaptic membrane on tonic fibers contained two types of AChR channels (FIG. 2A). One channel had a conductance that was about 50% larger and a shorter open time compared to the other channel. The larger conductance tonic fiber AChR channel resembled that found on twitch fiber end-plates (FIG. 2B). The smaller conductance tonic fiber AChR channel resembled the channel found on the extrajunctional membrane of denervated twitch fibers (FIG. 3). The open time distribution of the larger conductance tonic fiber channels was similar to that of the innervated twitch fiber channels (FIG. 4), and the open time distribution of the smaller conductance tonic fiber channels resembled that found for AChR channels on denervated twitch fibers (FIG. 5).

The similarity of the relationship between the large and small conductance tonic fiber AChR channels and adult-type and fetal-type mammalian AChR channels led to the suggestion that mature tonic fibers, which are innervated at multiple sites by more than one axon, may express both γ- and ε-subunits.[25-27] If mature tonic fibers in reptiles and amphibians express both γ- and ε-subunits, then perhaps multiinnervated mammalian extraocular muscle fibers, which resemble tonic fibers, also express γ- and ε-subunits.[28,29]

MECHANISM OF ION PASSAGE

The AChR forms an ion channel that is selective for cations but is relatively nonselective among cations compared to other voltage-gated cation-selective channels, such as sodium, calcium, and potassium channels.[3,6,30-32] The selectivity among channels is based partly on the size of the pore. Based on the size of permeant cations, the size of the AChR pore at its narrowest point is approximately 6.5 angstroms, which is considerably larger than that of the voltage-gated ion channels.[30] Ions do not pass through the channel freely in a bulk flow fashion; instead ions bind to specific sites within the pore and traverse the channel by moving from one site to the next. To traverse the narrowest portion of the channel, called the selectivity filter, an ion must be stripped of accompanying water molecules, which requires energy.[31] In order for an ion to traverse the AChR channel it must pass through the selectivity filter, and the free energy of the ion bound to a binding site within the channel must be less than or similar to the free energy of the hydrated ion.

The structure of the binding sites within the pore is critical to ion passage.[32] If the interaction between a binding site and ion is too weak, the ion will not pass through the channel. If the interaction is too strong then the ion will remain fixed at the binding site and effectively plug the channel. Several noncompetitive AChR inhibitors appear to block current flow by binding to and remaining at sites within the pore.[16] The charge of the amino acid side chains that line the pore and that form the inner and outer vestibule contribute to the ion selectivity of the channel.[4,6,31,32] The binding of ACh to the channel likely leads to a change in pore structure that allows ions to traverse by opening the channel.[33,34] In addition, ACh binding may also trigger a change in the properties of ion binding sites within the pore.

MICROSTRUCTURE OF AChR SUBUNITS

There is homology in the amino acid sequences among the subunits of the AChR.[7,8,10,12] Each of the subunits appears to have its N-terminal and C-terminal

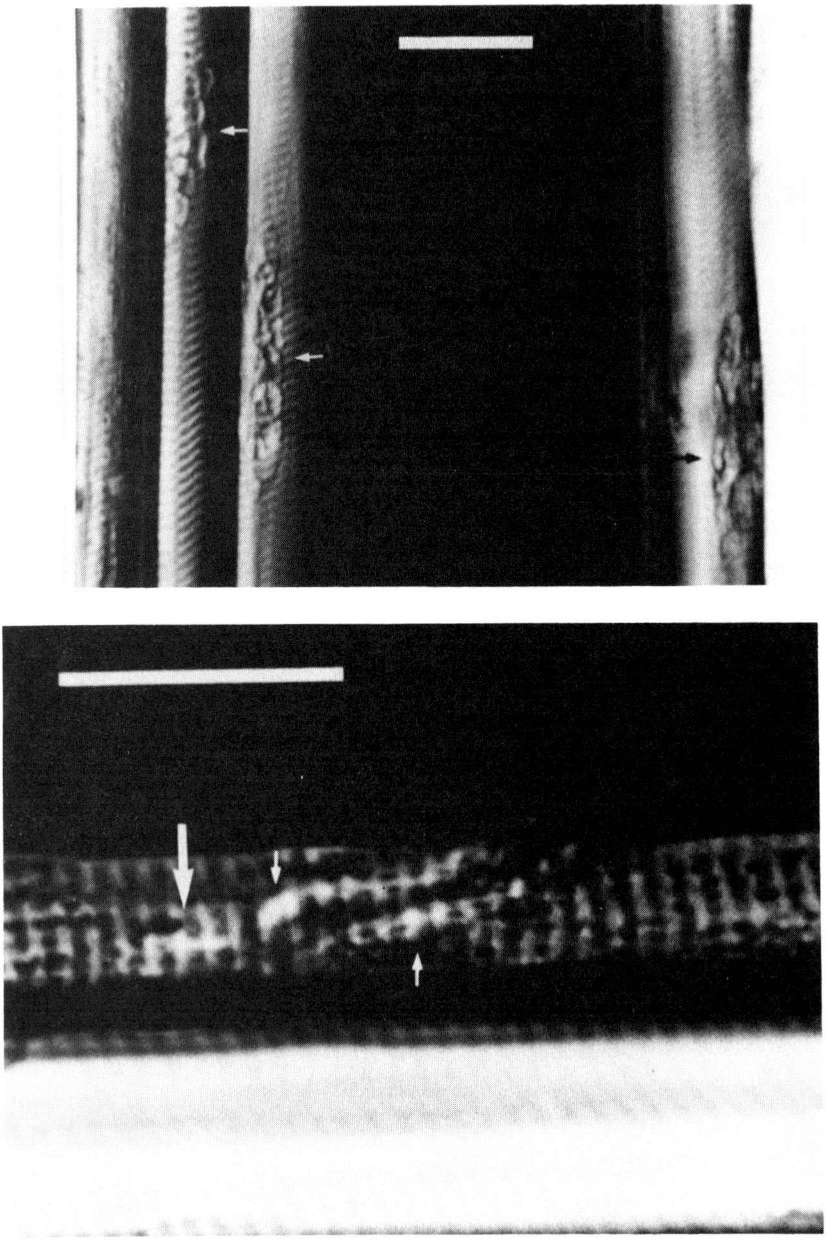

FIGURE 1. Appearance of postsynaptic membrane from twitch (top) and tonic (bottom) garter snake muscle fibers after enzymatic treatment to remove the nerve terminals. Top: The bare postsynaptic membrane of several en plaque synapses are indicated with arrows. Bottom: Partial enzymatic removal of two en grappe synapses (small arrows) from a garter snake tonic fiber to expose a small region of postsynaptic membrane, which appears as a relatively brighter spot on the sarcolemma (large arrow). The scale bars indicated 50 μm. (Ruff & Spiegel.[27] With permission of the American Physiological Society.)

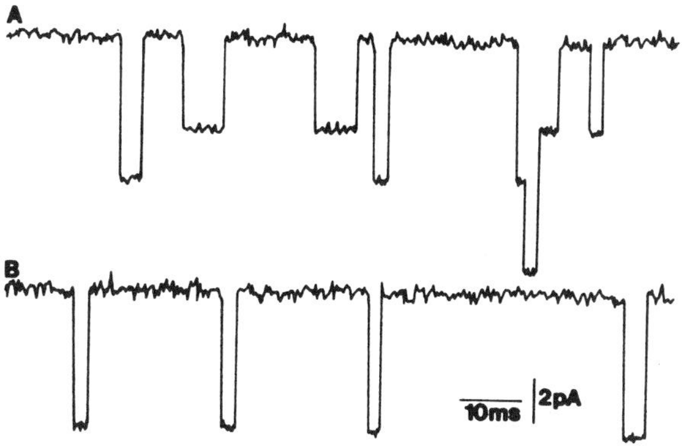

FIGURE 2. Currents evoked from the bare postsynaptic membranes of a tonic fiber (A) and a twitch fiber (B). The effective membrane potentials were −126 mV for the tonic fiber and −127 mV for the twitch fiber. Note that the tonic fiber had two sizes of single-channel currents, whereas the twitch fiber had one class of single-channel currents. Mean amplitude of the larger conductance events for the tonic fiber were 6.27 pA. The mean amplitude of the twitch fiber currents was 6.22 pA. The mean amplitude of the smaller conductance tonic fiber currents was 4.30 pA. The pipette contained 0.5 μM ACh. (Ruff & Spiegel.[27] With permission of the American Physiological Society.)

regions in the extracellular space. In addition, each subunit contain 4 α-helices, called M1 to M4, that probably span the membrane.[3,35] The extracellular portions of the subunits, consisting of the N- and C-terminal regions and the region between M2 and M3, form a large extracellular vestibule that surrounds the transmembrane ion channel. The regions between M1 and M2 and between M3 and M4 form a smaller vestibule around the intracellular orifice of the ion channel.[3,33,34]

ACh Binding Site

The ACh binding site on each α-subunit is located on the long N-terminal segment. The ACh binding site is located close to a reducible disulfide bond that is formed between cysteines at positions 192 and 193. Reduction of the disulfide bond greatly alters ACh binding affinity.[16] The binding site for α-bungarotoxin is also located on the N-terminal region, close to, but distinct from, the ACh binding site. Amino acid substitutions in the N-terminal region could therefore alter the

FIGURE 3. Extrajunctional AChR channel currents recorded from a denervated twitch fiber. The resting potential of the cell was −86 mV, and there was an additional −40 mV holding potential. The average current amplitude was 4.32 pA.

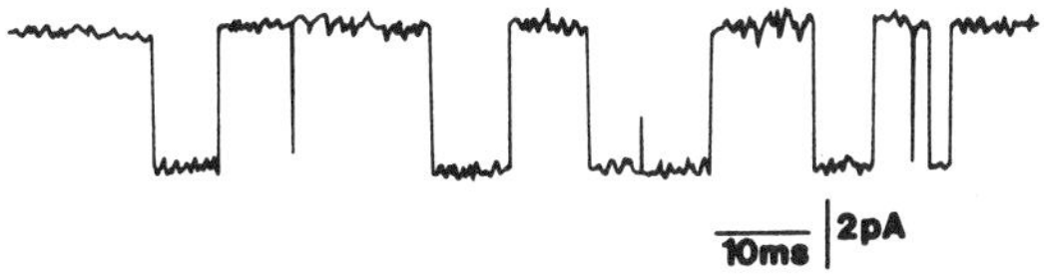

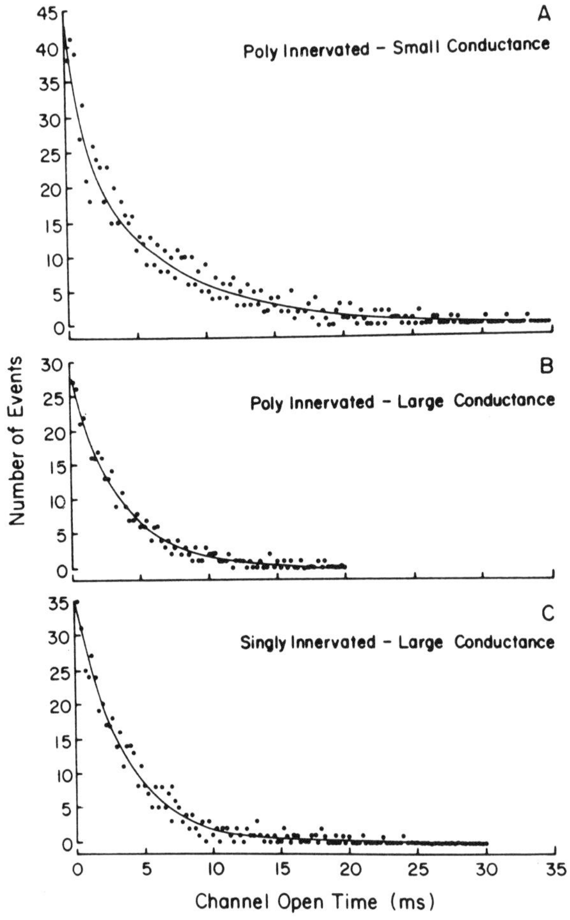

FIGURE 4. Open time distributions for 811 small conductance channel openings on polyinnervated tonic fibers (A), 360 larger conductance openings on polyinnervated tonic fibers (B), and 492 openings on singly innervated twitch fibers. The smooth curves are one (B, C) or two (A) component exponential curves that are least-squares fits to the data. The time constants for the curves fitted to the larger conductance events were 3.54 ms for tonic fibers (B) and 3.43 ms for twitch fibers (C). The time constants for the two-component curve fitted to the smaller conductance events were 1.18 and 7.51 ms. (Ruff & Spiegel.[27] With permission of the American Physiological Society.)

binding affinity of ACh, other nicotinic agonists, or competitive inhibitors such as α-bungarotoxin or *d*-tubocurarine.[36]

The Ionic Pore Is Probably Formed by the M2 Segments

Several experiments suggest that the M2 segment of the subunits line the ion channel. Giraudat *et al.*[37] examined the structure of the high-affinity binding site

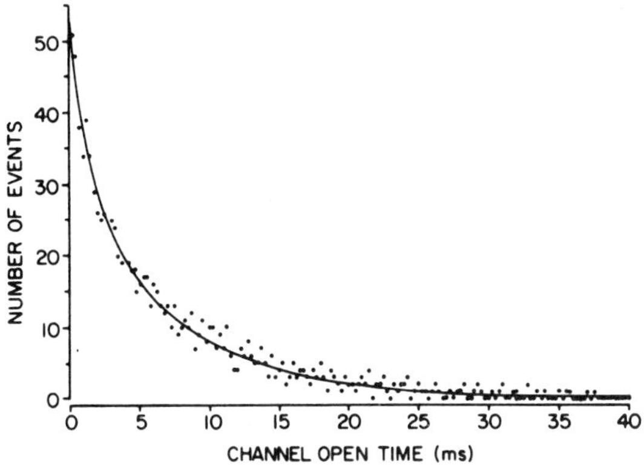

FIGURE 5. Open time distribution for 1015 channel openings on the denervated twitch fiber shown in FIG. 4. The open times were best fitted by two exponentials with time constants of 7.33 ms and 1.23 ms. (Ruff & Spiegel.[27] With permission of the American Physiological Society.)

for the chlorpromazine class of noncompetitive channel blockers by photolabeling the subunits of the *Torpedo* AChR channel using [^{3}H]chlorpromazine. They found that incorporation of radioactivity into all subunits was reduced by the addition of phencyclidine, which competes for the high-affinity binding site for noncompetitive blockers. They isolated the δ-subunit and found that the serine residue at δ-262, which is located in the M2 segment, was labeled by [^{3}H]chlorpromazine (see FIG. 6; TABLE 1 lists the amino acid abbreviations and the charge or polarity of selected amino acids). Labeling of this residue was prevented by phencyclidine. Hucho *et al.*[38] studied the high-affinity binding site for noncompetitive AChR channel blockers using [^{3}H]triphenmethylphosphonium. They found a binding site in the α-, β- and δ-subunits. The photolabel was located in the M2 region of each of the subunits at α-248, β-254, and δ-262.

Permanently charged local anesthetic analogues such as QX222 block AChR channels by entering the extracellular orifice of the channel and binding to a site within the channel that is close to the intracellular orifice.[39,40] Based upon the

		M2 SEGMENT
Mouse	α	MTLSISVLLSLTVFLLVIV
	β	MGLSIFALLTLTVFLLLLA
	δ	TSVAISVLLAQSVFLLLIS
		⇑
		S⇒A
		F⇒S

FIGURE 6. Amino acid sequences of the M2 segments of the mouse α-, β- and δ-subunits. The arrow indicates the site of amino acid substitutions. Note that the arrow points to the serine at δ-262, which corresponds to the site labeled by [^{3}H]chlorpromazine in studies of the *Torpedo* δ-subunit.

work with noncompetitive inhibitors, the serine at δ-262 and amino acids on comparable positions of the other subunits appeared to be good candidates for the local anesthetic binding site. Leonard *et al.*[41] used site-directed mutagenesis of mouse AChR subunits to study the effects of interchanging polar and nonpolar amino acids on the binding affinity of QX222 for the AChR channel. The amino acid substitutions they studied were exchanging the serine at δ-262 and the analogous serine on the α-subunit with a nonpolar alanine, and replacing a nonpolar phenylalanine on the β-subunit with a polar serine (FIG. 6). Inasmuch as QX222 is positively charged, the affinity of QX222 for its channel binding site should be related to the number of polar and nonpolar amino acids near the binding site. As the number of serines decreased there was a decrease in the residence time for QX222 in the channel and consequently a decrease in the equilibrium binding affinity of QX222. Receptors with three serines replaced with alanines displayed a decrease in outward single channel currents, which would be consistent with the hypothesis that reducing the complement of polar amino acids near the intracellular orifice of the pore would inhibit intracellular cations from entering and traversing through the channel. The effects of the amino acid substitutions in M2 on the binding affinity of QX222 for the channel and on the direction of the current rectification support the hypothesis that the M2 segments contribute to the lining of the ion channel.

Altering the Fixed Charges in the Segments Surrounding M2 Change the Size of Single Channel Currents

Experiments using site-directed mutagenesis demonstrated that the amount of net negative charge in the intra- and extracellular vestibules surrounding the AChR ion pore strongly influences the size of the single channel currents. Negative charges in the vestibules attract cations, thereby increasing the concentration of current-carrying ions at the pore orifices. Imoto *et al.*[42] observed that in an extracellular solution with low divalent cation concentrations, the bovine AChR channel (formed with the γ-subunit) had a smaller conductance for inward currents than the *Torpedo* AChR channel. Replacement of the δ-subunit of *Torpedo* with the bovine δ-subunit produced a chimeric channel with conductance similar to the channel composed entirely of bovine subunits. The experimenters then constructed chimeras consisting of sections of the *Torpedo* and bovine δ-subunit. They found that a region comprising the M2 segment and the extracellular segment

TABLE 1. Classification of Amino Acids Based upon Charge and Polarity

Amino Acid	[charge]
Alanine (A)	[Nonpolar]
Phenylalanine (F)	[Nonpolar]
Valine (V)	[Nonpolar]
Isoleucine (I)	[Nonpolar]
Glutamine (Q)	[Polar]
Serine (S)	[Polar]
Aspartate (D)	[−]
Glutamate (E)	[−]
Lysine (K)	[+]
Arginine (R)	[+]

			INNER VESTIBULE		OUTER VESTIBULE	
Torpedo	α	M1⇒	DSG,EK	⇐M2⇒	ELIPSTSSAVPLIGK	⇐M3
	β	M1⇒	DAG,EK	⇐M2⇒	DKVPETSLSVPIIIR	⇐M3
	γ	M1⇒	QAGGQK	⇐M2⇒	QKVPETSLNVPLIGK	⇐M3
	δ	M1⇒	ESG,EK	⇐M2⇒	QRLPETALAVPLIGK	⇐M3
Bovine	γ	M1⇒	KAGGQK	⇐M2⇒	KKVPETSQAVPLISK	⇐M3
	ε	M1⇒	QAGGQK	⇐M2⇒	QKTPETSLSVPLLGG	⇐M3
	δ	M1⇒	DCG,EK	⇐M2⇒	KRLPATSMAIPLIGK	⇐M3
Mouse	α	M1⇒	DSG,EK	⇐M2⇒	ELIPSTSSAVPLIGK	⇐M3
	β	M1⇒	DAG,EK	⇐M2⇒	IKVPETSLAVPIIIK	⇐M3
	δ	M1⇒	DCG,EK	⇐M2⇒	KRLPATSMAIPLVGK	⇐M3
			1 23		45 6	

FIGURE 7. Amino acid sequences of the inner and outer vestibules surrounding the M2 segment for several *Torpedo*, bovine, and mouse AChR subunits. The positively charged amino acids are double underlined. The negatively charged amino acids are singly underlined. The numbers 1-6 indicate the locations of amino acid substitutions to change the charge in the inner or outer vestibule.

between M2 and M3 was the key factor in determining the conductance of channels made with a chimeric δ-subunit. The segment between M2 and M3 is part of the extracellular, or outer, vestibule surrounding the ion pore. FIGURE 7 shows that the outer vestibule of the *Torpedo* δ-subunit contains one net positive charge, whereas the outer vestibule of the bovine δ-subunit contains three net positive charges.

Imoto *et al.*[43] went on to make a series of site-directed mutations of *Torpedo* subunits in the inner and outer vestibules surrounding M2. FIGURE 7 indicates the locations (1-6) of the mutations. They found that changing the charge in the inner vestibule altered the conductance for outward flowing currents, and changing the charge of the outer vestibule altered the channel conductance for inward flowing currents. Increasing the net negative charge in a vestibule increased the single channel current.

As mentioned above the adult form of the bovine AChR with the ε-subunit has a larger single channel conductance than the bovine channel with the γ-subunit. FIGURE 7 shows that the inner vestibule of the γ-subunit has two net positive charges compared to one net positive charge for the ε-subunit. The outer vestibule of the bovine γ-subunit has two net positive charges compared to no net charge for the ε-subunit (FIG. 7).

Possible Effect of the Amino Acid Sequence in M2 on Channel Open Times

Two lines of evidence suggest that the amino acid sequence in M2 may play a role in regulating the stability of the open state and hence the open time of the channel. One of the prime differences in mammalian AChR channels composed with the ε-subunit (adult-type) compared to the channels composed with the

γ-subunit (fetal-type) is that the channel with the ε-subunit has a shorter open time.[8] Although the difference in open times could be attributed to several differences between the γ-subunit and ε-subunit, let us consider the differences in the M2 segments between the γ-subunit and ε-subunit (FIG. 8).[7] There are two amino acid differences. The first is that an alanine in the γ-subunit is replaced by a serine in the ε-subunit. The second difference is that a valine in the γ-subunit is replaced by an isoleucine. As indicated in TABLE 1, alanine, valine, and isoleucine are nonpolar amino acids; serine is a polar amino acid. Hence the switch of alanine for serine changes the polarity of the M2 segment, whereas the exchange of valine for isoleucine does not change the polarity of the M2 segment. We suggest that replacing a nonpolar amino acid with a polar amino acid in the M2 segment in a critical region near the cytoplasmic orifice of the channel may destabilize the open state, resulting in shorter open times. Support for this suggestion comes from the work of Leonard *et al.*,[41] who found that when a nonpolar phenylalanine in the β-subunit was replaced by a polar serine, the open time of the channel was reduced (FIG. 6). When a polar serine in the α-subunit or δ-subunit was replaced by a nonpolar alanine, the channel open time increased (FIG. 6).[41] Though circumstantial, the evidence supports the hypothesis that the presence of polar versus nonpolar amino acids near the cytoplasmic end of M2 may regulate the channel open time.

Possible Etiology of Two Component Miniature End-plate Currents (MEPCs) and Current Fluctuations

In normal end-plates, the MEPCs decay with a single exponential time course, the channel open time distributions are characterized by a single exponential, and the power spectrum relationship of the amplitude of ACh-induced current fluctuations plotted against frequency can be described by a single component Lorentzian curve.[4,32,39] Detailed analysis of the open time of single AChR channels indicates that there is an excess number of very brief openings.[17,19,20,21,44,45] In addition, successive openings of the same AChR channel have correlated open times.[45] The excess number of brief open times and correlation in open times for successive openings can be explained by kinetic schemes in which the open state has access to at least two unique closed states.[27] Most of the single channel and current fluctuation observations, however, are consistent with a simple kinetic model in which two ACh molecules sequentially and rapidly bind to the AChR to form the closed RA_2 state (FIG. 9), and then the AChR undergoes a relatively slow conformation change from the closed state to the opened state, RA_2^*.[4,17,32] As will be discussed below, CDNTs may be associated with complex changes in AChR kinetics, resulting in two-component MEPCs and current fluctuations that are described by two Lorentzian curves. The two component MEPCs and current fluctuations can be produced by having two populations of channels with different open times or by a kinetic scheme in which the conformation changes between RA_2 and RA_2^* are no longer the only rate-limiting steps.[39,40]

		M2 SEGMENT
Bovine	γ	CTVAINVLLAQTVFLFLVA
	ϵ	CTVSINVLLAQTVFLFLIA
		⇑ ⇑

FIGURE 8. Amino acid sequences of the M2 segments of the bovine γ- and ε-subunits. The arrows indicate the site of amino acid differences.

Two Populations of Receptors

$$R_F + A \rightleftharpoons R_FA + A \rightleftharpoons R_FA_2 \underset{\alpha_F}{\overset{\beta_F}{\rightleftharpoons}} R_FA_2^{open}$$

$$R_S + A \rightleftharpoons R_SA + A \rightleftharpoons R_SA_2 \underset{\alpha_S}{\overset{\beta_S}{\rightleftharpoons}} R_SA_2^{open}$$

Channel Blockade

$$R + A \rightleftharpoons RA + A \rightleftharpoons RA_2 \underset{\alpha}{\overset{\beta}{\rightleftharpoons}} RA_2^{open} \underset{K_U}{\overset{K_B}{\rightleftharpoons}} RA_2^{blocked}$$

Slower Agonist-Receptor Interaction

$$R + A \rightleftharpoons RA + A \underset{K_1}{\overset{K_2}{\rightleftharpoons}} RA_2 \underset{\alpha}{\overset{\beta}{\rightleftharpoons}} RA_2^{open}$$

FIGURE 9. Three different kinetic schemes that could produce two component MEPCs, current fluctuation power spectra, and single-channel current open time distributions.

FIGURE 9 illustrates three possible kinetic schemes that could produce two component MEPCs, current fluctuations, and open time distributions.[39,40] In the first scheme there are two populations of receptors: one with faster transitions between the open and closed states and one with slower transitions. This kinetic scheme would predict that the single channel conductance measured by current fluctuations would be an average of the single channel conductances of the two classes of receptors. There would be a fixed relationship between the decay rate of the MEPC components, which would be α_F and α_S, and the characteristic frequencies of the components of the power spectrum. The amplitude of the MEPC could be increased or decreased depending upon the conductances of the two classes of receptors.

In the second model the channel is blocked after it opens. In this model the decay rates of the components of the MEPCs and the characteristic frequencies of the Lorentzian components of the current fluctuations would be determined by the rate constants α, β, K_B, and K_U. The blocking agent could be the agonist. Sine and Steinbach[46] observed that ACh blocked AChRs in the chick muscle cell line BC3H1 when [ACh] > 300 μM. In usual experimental conditions, the agonist concentration experienced by AChRs during an MEPC is much higher than during stimulation for current fluctuation studies. If channel block was affected by an agonist molecule, then K_B would depend on the agonist concentration, and there would not be a fixed relationship between the MEPC decay rate constants and the characteristic frequencies of the components of the power spectrum. Unless the open channel block is very fast, the mean channel conductance would be the conductance of the unblocked channel and the initial peak amplitude of the MEPC would be normal. If channel block was very fast it might be difficult to resolve the rapid channel block, and reopenings and the channel would appear to have a lower single channel conductance then the actual open state conductance.[46]

In the third model the binding and unbinding of the second ACh molecule are slow. Such a model might correspond to mutation in which there was negative rather than positive cooperativity for ACh binding.[17] The two components of the MEPC result from closure of the open and then reopening of the closed channel because the first unbinding step is relatively slow. In this model there would be a fixed relationship between the decay rates of the components of the MEPC and the components of the power spectrum. The single channel conductance would be the conductance of the open state. The mean end-plate current would be reduced because fewer receptors reach the RA_2^* state.

EXAMPLES OF CDNT IN WHICH AChR FUNCTION IS DISTURBED

High-conductance Fast-Channel Syndrome

Engel *et al.*[2] described a 9 year old with a poor suck and cry in the neonatal period. Intermittent weakness was provoked or worsened by exertion and by exposure to heat. She had mild weakness of the facial and cervical limb muscles which did not improve with anticholinesterase drugs. Her EMG showed a mild decrement in the compound muscle action potential (CMAP) after three minutes of exercise. An intercostal muscle biopsy revealed normal quantal content of the end-plate potential (EPP). The miniature end-plate potential (MEPP) and MEPC amplitudes were large, and their decay times were short. Current fluctuation analysis indicated that the single-channel conductance was 1.64- to 1.67-fold larger than normal. The mean channel open time determined from the decay constant of the MEPC was 39% shorter than normal. The channel open time deduced from the power spectrum of ACh-induced current fluctuations was 29% shorter than normal. The number of α-bungarotoxin binding sites were normal. Some fibers had degenerative changes in the junctional sarcoplasm. The increased single channel conductance could have resulted from a mutation that increases the amount of negative charge in the external vestibule of the AChR channel.[42,43] The reduced open time could have resulted from a mutation that destabilizes the open state, as described above when phenylalanine was replaced by a serine in the M2 segment of the mouse β-subunit.[41] An interesting possibility is that the degeneration seen may have been due, in part, to increased calcium conductance of the channel so that there was local injury due to calcium-activated protease activity.[2]

Abnormal ACh-AChR Interaction

Engel *et al.*[2] described a 21-year-old woman with generalized weakness and abnormal fatigability of all muscles since birth. Her symptoms were partially improved by anticholinesterase drugs. No other family members were affected. The EMG showed a decremental response at 2 Hz stimulation. Analysis of a muscle biopsy revealed normal quantal content of the EPP. The MEPP and MEPC amplitudes were reduced. The number of α-bungarotoxin binding sites per end-plate were in the high normal range, and AChR were located in a normal pattern at the top of the junctional folds. The mean single channel conductance was normal. The ACh-induced current fluctuation power spectrum was abnormal and was best fitted by the sum of two Lorentzian functions. The recorded MEPCs were too small to determine whether the decay phase was biexponential. The small MEPP and MEPCs in the presence of a normal number of receptors could be explained by a decrease in ACh content of the vesicles or an abnormality of the AChR. That the synaptic vesicles were not reduced in size points away from reduced vesicle ACh content.[2] In addition, the abnormal AChR kinetics could be explained by an abnormal interaction of ACh with its receptor. As suggested above, three kinetic schemes could explain the findings: (1) two populations of receptors, (2) blocking of the ion channel, and (3) a decrease in the rate constant for agonist dissociation or another abnormality so that closure of the channel is no longer the only rate-limiting step (FIG. 9). Two populations of receptors is unlikely because at least one class of AChR must have a low conductance to explain the small MEPC size, which would be inconsistent with the finding of normal single-channel conductance by noise analysis. Blocking of the channel by an agonist molecule does not seem consistent with the clinical improvement seen with cholinesterase medication and with an increase in a MEPP amplitude with neostigmine. The reduced affinity of the receptor for ACh could explain the small MEPC, normal channel conductance, response to anticholinesterase, and the two-component channel kinetics. A mutation in the N-terminal region of the α-subunit could alter agonist binding. In addition, a mutation that results in negative rather than positive cooperativity for ACh binding would also compromise agonist binding.

AChR Deficiency and Short-Channel Open-time

Engel *et al.*[2] reported a 27-month-old girl who required respiratory assistance for the first three weeks of life. She was hypotonic with facial diplegia and ophthalmoplegia. Her symptoms were improved by anticholinesterase drugs, and the EMG showed a decremental response at 2 Hz stimulation. An intercostal muscle specimen revealed normal quantal content of the EPP. The MEPP amplitude was small but increased with neostigmine. The number of α-bungarotoxin binding sites was only 7% of the mean control value. The mean single-channel conductance was normal but the mean channel open time was short: 1.61 ms compared to the control value of 2.27 milliseconds. The short open time could have resulted from a mutation of one of the subunits that destabilized the open state. The reduced number of receptors may have been due to reduced synthesis, impaired membrane incorporation, or decreased stability of the AChR with a mutant subunit.

AChR with Low-conductance, Increased Open Time, and Reduced Density

Engel *et al.*[47] described three cases of a familial congenital myasthenic syndrome in which all the patients had ophthalmoparesis from infancy. Electrophysiologic tests revealed that the duration and half decay times of the MEPP and EPP were markedly prolonged, and these values were increased further by prostigmine. The mean MEPP amplitude was reduced. The MEPP frequency and quantal content were normal in each instant. Single-channel conductance was reduced by 20 percent. The findings are qualitatively similar to those produced by substitution of the γ-subunit for the ε-subunit. The channel open time, however, was much longer than is seen comparing fetal-type and adult-type AChRs. Perhaps there is a mutation that reduces the net negative charge in the vestibule surrounding the pore and that increases the lifetime of the opened state.

Slow Channel Syndrome

The patients described with this syndrome have presented from infancy to adulthood with intermittent weakness on the backdrop of a slowly progressive myopathy.[1] The upper extremities and trunk were weaker than the lower limbs. Weakness did not improve with anticholinesterase treatment. EMG showed decremental responses to repetitive stimulation, and single electrical stimulation produced repetitive CMAPs. MEPPs were prolonged with decreased amplitude. The findings resembled those seen with acetylcholinesterase (AChE) deficiency;[2,48] however, AChE stains reveal the presence of functional AChE. The prolongation of the MEPP most likely resulted from a prolongation of the open time of the AChR due to increased stability of the open state. The increased open time would prolong the depolarization of the membrane beyond the refractory period for action potential generation, producing the observed repetitive CMAPs.

The only anatomical abnormality identified was degeneration of the junctional folds. The AChR density was normal. The membrane damage may have resulted from the prolonged depolarization of the end-plate, stimulating increased influx of calcium through the AChR or specific calcium channels. The increased calcium influx could stimulate calcium-activated proteases.[2]

AChR with Altered d*-Tubocurarine Binding*

Morgan-Hughes *et al.*[49] described a 32-year-old man with a 14-month history of fluctuating generalized and bulbar weakness. An EMG showed a slight decrement in the CMAP after 5 Hz stimulation. Light microscopic analysis of a triceps biopsy revealed tubular aggregates. Electron microscopy of some end-plates revealed reduced synaptic vesicles and short, irregular junctional folds. Microelectrode recordings were not performed. The AChR density was reduced as measured by α-bungarotoxin binding. The AChR affinity for the competitive inhibitor *d*-tubocurarine was increased tenfold. Perhaps an alteration of the N-terminal region of the α-subunit increased the binding affinity for *d*-tubocurarine and impaired insertion of AChRs into the end-plate membrane.

REFERENCES

1. Engel, A. G., E. H. Lambert, D. M. Mulder, C. F. Torres, K. Sahashi, T. E. Bertorini & J. N. Whitaker. 1982. A newly recognized congenital myasthenic syndrome attributed to a prolonged open time of the acetylcholine-induced ion channel. Ann. Neurol. **11:** 553–569.
2. Engel, A. G., T. J. Walls, A. Nagel & O. Uchitel. 1990. Newly recognized congenital myasthenic syndromes: I. Congenital paucity of synaptic vesicles and reduced quantal release. II. High conductance fast-channel syndrome. III. Abnormal acetylcholine receptor (AChR) interaction with acetylcholine. IV. AChR deficiency and short channel open time. Prog. Brain Res. **84:** 125–137.
3. Guy, H. R. & F. Hucho. 1987. The ion channel of nicotinic acetylcholine receptor. Trends Neurosci. **10:** 318–322.
4. Ruff, R. L. 1986. Ionic channels II. Voltage- and agonist-gated and agonist-modified channel properties and structure. Muscle & Nerve **9:** 767–786.
5. Witzemann, V., B. Barg, M. Criado, E. Stein & B. Sakmann. 1989. Developmental regulation of five subunit specific mRNAs encoding acetylcholine receptor subtypes in rat muscle. FEBS Lett. **242:** 419–424.

6. Dani, J. A. 1989. Site-directed mutagenesis and single-channel currents define the ionic channel of the nicotinic acetylcholine receptor. Trends Neurosci. **12:** 125–130.
7. Takai, T., M. Noda, M. Mishina, S. Shimizu, Y. Furutani, T. Kayano, T. Ikeda, K. Tai, H. Takahashi, T. Takahashi, M. Kuno & S. Numa. 1985. Cloning, sequencing and expression of cDNA for a novel subunit of acetylcholine receptor from calf muscle. Nature **315:** 761–764.
8. Mishina, M., T. Takai, K. Imoto, M. Noda, T. Takahashi, S. Numa, C. Methfessl & B. Sakmann. 1986. Molecular distinction between fetal and adult forms of muscle acetylcholine receptor. Nature **321:** 406–411.
9. Witzemann, V., B. Barg, Y. Nishikawa, B. Sakmann & S. Numa. 1987. Differential regulation of muscle acetylcholine receptor γ- and ε-subunit mRNAs. FEBS Lett. **223:** 104–112.
10. Mishina, M., T. Kurosaki, T. Tobimatsu, Y. Morimoto, M. Noda, T. Yamamoto, M. Terao, J. Lindstrom, T. Takahashi, M. Kuno & S. Numa. 1984. Expression of functional acetylcholine receptor from cloned cDNAs. Nature **307:** 604–608.
11. Brehm, P. 1989. Resolving the structural basis for developmental changes in muscle ACh receptor function: it takes nerve. Trends Neurosci. **12:** 174–177.
12. Schuetze, S. M. & L. W. Role. 1987. Developmental regulation of nicotinic acetylcholine receptors. Annu. Rev. Neurosci. **10:** 403–457.
13. Noda, M., Y. Furutani, H. Takahashi, M. Toyosato, T. Tanabe, S. Shimizu, S. Kikyotani, T. Kayano, T. Hirose, S. Inayama & S. Numa. 1983. Cloning and sequence analysis of calf cDNA and human genomic DNA encoding alpha-subunit precursor of muscle acetylcholine receptor. Nature **305:** 818–823.
14. Merlie, J. P. & J. R. Sanes. 1985. Concentration of acetylcholine receptor mRNA in synaptic regions of adult muscle fibers. Nature **317:** 66–68.
15. Sanes, J. R., Y. R. Johnson, P. T. Kotzbauer, J. Mudd, T. Hanley, J.-C. Martinou & J. P. Merlie. 1991. Selective expression of an acetylcholine receptor-lacZ transgene in synaptic nuclei of adult muscle fibers. Development **113:** 1181–1191.
16. Karlin, A., P. N. Kao & M. DiPaola. 1986. Molecular pharmacology of the nicotinic acetylcholine receptor. Trends Pharmacol. Sci. **4:** 304–308.
17. Sine, S. M. & J. H. Steinbach. 1987. Activation of acetylcholine receptors on clonal mammalian BC3H-1 cells by high concentrations of agonist. J. Physiol. (Lond.) **385:** 325–359.
18. Morris, A., D. Beeson, L. Jacobson, F. Baggi, A. Vincent & J. Newsom-Davis. 1991. Two isoforms of the muscle acetylcholine receptor of alpha-subunit are translated in the human cell line TE671. FEBS Lett. **295:** 116–118.
19. Sine, S. M. & J. H. Steinbach. 1984. Activation of a nicotinic acetylcholine receptor. Biophys. J. **45:** 175–185.
20. Sigworth, F. J. & S. M. Sine. 1987. Data transformations for improved display and fitting of single-channel dwell time histograms. Biophys. J. **52:** 1047–1054.
21. Auerbach, A. B. & F. Sachs. 1984. Single channel currents from acetylcholine receptors in embryonic chick muscle. Kinetic and conductance properties of gaps within bursts. Biophys. J. **45:** 187–198.
22. Qu, Z., E. Moritz & R. L. Huganir. 1990. Regulation of tyrosine phosphorylation of the nicotinic acetylcholine receptor at the rat neuromuscular junction. Neuron **2:** 367–378.
23. Safran, A., C. Provenzano, R. Sagi-Eisenberg & S. Fuchs. 1990. Phosphorylation of membrane-bound acetylcholine receptor by protein kinase C: characterization and subunit specificity. Biochemistry **29:** 6730–6734.
24. Rozental, R. 1991. *In vitro* denervation of frog skeletal muscle: expression of several conductance classes of nicotinic receptors. Neurosci. Lett. **133:** 65–67.
25. Dionne, V. E. 1989. Two types of nicotinic acetylcholine receptor channels at slow fibre end-plates of the garter snake. J. Physiol. (Lond.) **409:** 313–331.
26. Henderson, L. P. & P. Brehm. 1989. The single-channel basis for the slow kinetics of synaptic currents in vertebrate slow muscle fibers. Neuron **2:** 1399–1405.
27. Ruff, R. L. & P. Spiegel. 1990. Ca sensitivity and AChR currents of twitch and tonic snake muscle fibers. Am. J. Physiol. **259** (Cellular Physiology): C911–C919.

28. RUFF, R. L., H. J. KAMINSKI, E. MASS & P. SPIEGEL. 1989. Ocular muscles: physiology and structure-function correlations. Bull. Soc. Belge Ophtalmol. **237:** 321–352.
29. KAMINSKI, H. J., E. MAAS, P. SPIEGEL & R. L. RUFF. 1990. Why are eye muscles frequently involved by myasthenia gravis? Neurology **40:** 1663–1669.
30. DWYER, T. M., D. J. ADAMS & B. HILLE. 1980. The permeability of the endplate channel to organic cations in frog muscle. J. Gen. Physiol. **75:** 469–492.
31. DANI, J. A. & G. EISENMAN. 1987. Monovalent and divalent cation permeation in acetylcholine receptors channel. J. Gen. Physiol. **89:** 959–983.
32. RUFF, R. L. 1986. Ionic Channels: I. The biophysical basis for ion passage and channel gating. Muscle & Nerve **9:** 675–699.
33. TOYOSHIMA, C. & N. UNWIN. 1988. Ion channel of acetylcholine receptor reconstructed from images of postsynaptic membranes. Nature **336:** 247–250.
34. UNWIN, N., C. TOYOSHIMA & E. KUBALEK. 1988. Arrangement of the acetylcholine receptor subunits in the resting and desensitized states, determined by cryoelectron microscopy of crystallized *Torpedo* postsynaptic membranes. J. Cell Biol. **107:** 1123–1138.
35. CHAVEZ, R. A. & Z. W. HALL. 1992. Expression of fusion proteins of the nicotinic acetylcholine receptor from mammalian muscle identifies the membrane-spanning regions in the α and δ subunits. J. Cell Biol. **116:** 385–393.
36. NEUMANN, D., D. BARCHAN, M. HOROWITZ, E. KOCHVA & S. FUCHS. 1989. Snake acetylcholine receptor: Cloning of the domain containing the four extracellular cysteines of the α-subunit. Proc. Natl. Acad. Sci. USA **86:** 7255–7259.
37. GIRAUDAT, J., M. DENNIS, T. HEIDMANN, J.-Y. CHANG & J.-P. CHANGEUX. 1986. Structure of the high-affinity binding site for noncompetitive blockers of the acetylcholine receptor: Serine-262 of the δ subunit is labeled [^{3}H]chlorpromazine. Proc. Natl. Acad. Sci. USA **83:** 2719–2723.
38. HUCHO, F., W. OBERTHÜR & F. LOTTSPEICH. 1986. The ion channel of the nicotinic acetylcholine receptor is formed by the homologous helices M II of the receptor subunits. FEBS Lett. **205:** 137–142.
39. RUFF, R. 1977. A quantitative analysis of local anesthetic alteration of miniature endplate current fluctuations. J. Physiol. (Lond.) **264:** 89–124.
40. RUFF, R. L. 1982. The kinetics of local anesthetic blockade of endplate channels. Biophys. J. **37:** 625–631.
41. LEONARD, R. J., C. G. LABARCA, P. CHARNET, N. DAVIDSON & H. A. LESTER. 1988. Evidence that the M2 membrane-spanning region lines the ion channel pore of the nicotinic receptor. Science **242:** 1578–1581.
42. IMOTO, K., C. METHFESSEL, B. SAKMANN, M. MISHINA, Y. MORI, T. KONNO, K. FUKUDA, M. KURASAKI, H. BUJO, Y. FUJITA & S. NUMA. 1986. Location of a δ subunit region determining ion transport through the acetylcholine receptor channel. Nature **324:** 670–674.
43. IMOTO, K., C. BUSCH, B. SAKMANN, M. MISHINA, T. KONNO, J. NAKAI, H. BUJO, Y. MORI, K. FUKUDA & S. NUMA. 1988. Rings of negatively charged amino acids determine the acetylcholine receptor channel conductance. Nature **305:** 645–648.
44. SINE, S. M. & J. H. STEINBACH. 1986. Activation of acetylcholine receptors on clonal mammalian BC3H-1 cells by low concentrations of agonist. J. Physiol. (Lond.) **373:** (129–162).
45. JACKSON, M. B., B. S. WONG, C. E. MORRIS, H. LECAR & C. N. CHRISTIAN. 1983. Successive openings of the same acetylcholine receptor channel are correlated in open time. Biophys. J. **42:** 109–114.
46. SINE, S. M. & J. H. STEINBACH. 1984. Agonists block currents through acetylcholine receptor channels. Biophys. J. **46:** 277–284.
47. ENGEL, A. G., E. H. LAMBERT, D. M. MULDER, C. F. TORRES, K. SAHASHI, T. E. BERTORINI & J. N. WHITAKER. 1979. Investigations of 3 cases of a newly recognized familial, congenital myasthenic syndrome. Trans. Am. Neurol. Assoc. **104:** 8–11.
48. ENGEL, A. 1984. Myasthenia gravis and myasthenic syndrome. Ann. Neurol. **16:** 519–534.
49. MORGAN-HUGHES, J. A., B. R. F. LECKY, D. N. LANDON & N. M. F. MURRAY. 1981. Alterations in the number and affinity of junctional acetylcholine receptors in a myopathy with tubular aggregates. A newly recognized receptor defect. Brain **104:** 279–295.

Clinical and Experimental Observations in Patients with Congenital Myasthenic Syndromes[a]

A. VINCENT,[b,g] J. NEWSOM-DAVIS,[b,c] D. WRAY,[d,e]
P. SHILLITO,[b] J. HARRISON,[b] M. BETTY,[b] D. BEESON,[b]
K. MILLS,[b] J. PALACE,[b] P. MOLENAAR,[f]
AND N. MURRAY[c]

[b]*Department of Clinical Neurology*
University of Oxford
Oxford, United Kingdom

[c]*The National Hospital for Nervous Diseases*
London, United Kingdom

[d]*Department of Pharmacology*
University of Leeds
Leeds, United Kingdom

[e]*Department of Pharmacology*
Royal Free Hospital School of Medicine and
[f]*Department of Physiology*
University of Leiden
Leiden, the Netherlands

Over the last 15 years we have performed muscle biopsies on over 30 patients with myasthenic symptoms. About one-third of these were diagnosed as seronegative myasthenia gravis, largely on the history of acquired disease and a clear clinical response to plasma exchange and immunosuppression. Some of the results are given elsewhere (Vincent *et al.*, this volume). The remaining 22 patients are thought to have congenital or hereditary disorders of neuromuscular transmission[1] (reviewed in ref. 2). In this review we summarize our clinical and experimental findings.

CLINICAL OBSERVATIONS

TABLE 1 summarizes the clinical observations. Fourteen of the patients were European, five were from the Middle East, and three were from Central Asia. They were aged from 9 to 44 at the time of the biopsy. The great majority (19/22) presented at or before the age of two, mainly during the perinatal period (15/22). There was a strong bias towards males (M : F ratio 16 : 6). Five had a family history of muscle weakness or of perinatal respiratory failure in siblings, and six were

[a] We are very grateful to the Myasthenia Gravis Association and the Muscular Dystrophy Group of Great Britain for support.

[g] Address for correspondence: Neurosciences Group, Institute of Molecular Medicine, John Radcliffe Hospital, Oxford, OX3 9DU, UK.

TABLE 1. Clinical Observations in 22 Patients[a] with Congenital Myasthenic Syndromes

	Yes	No
Male	16	6
Age of onset <2 years	19	3
Parental consanguinity	6	16
Family history	5	17
Generalized symptoms	22	0
Ocular symptoms	16	6
Diagnosis		
Decrement (>10%)	14	6
Jitter (mean > 53 μs)	15	2
Repetitive responses to single stimuli	3	17
Good response to treatment		
Anti-AChE	17	5
3,4-DAP[b]	13	6

[a] 14 from Europe, 5 from the Middle East, 3 from Asia.
[b] 3,4-DAP = 3,4-diaminopyridine.[13]

the products of consanguineous parents. All patients had generalized weakness, but six had no ocular symptoms, and a further three had ptosis only, with no external opthalmoplegia.

INVESTIGATIONS

On electromyography, abnormal decrement of the compound muscle action potential, during 3 Hz stimulation, and increased jitter were present in the majority (TABLE 1). Many also showed a positive Tensilon test. Of particular interest was a double or repetitive response to single stimuli of the ulnar nerve. This was only found in three individuals (see below).

MUSCLE BIOPSIES

Intercostal muscle biopsies were performed to investigate further the nature of the abnormality in neuromuscular transmission. Muscle was taken under general anesthesia and kept in oxygenated Krebs solution throughout. The muscle was dissected into bundles and investigated by recording with microelectrodes (as in ref. 3), incubating in [^{125}I]α-bungarotoxin (α-BuTx) to measure end-plate acetylcholine receptors (AChRs), and staining for end-plate acetylcholine esterase (AChE) (see Vincent *et al.*[1] for details). Control muscles were obtained mainly from patients undergoing thoracic surgery and gave values similar to those reported previously[1] (TABLE 2).

In biopsies from patients with congenital myasthenic syndromes (CMS), miniature end-plate potential (mepp) amplitudes were reduced (<0.7 mV) in 14 out of 17 cases, and [^{125}I]α-Butx binding to the muscle end-plates was reduced ($<0.8 \times 10^7$ α-BuTx binding sites per end-plate) in 17 of 21. In 14 biopsies the cholinesterase-stained end-plates were abnormally large; in 2 of these there was very poor staining even after prolonged incubation. In 8 cases they were of normal or small size but with patchy or weak staining in some instances. All the results are summarized in TABLE 2, where they are grouped according to the categories indicated below. There was a significant negative correlation between end-plate length and α-BuTx binding ($r = -0.59$, $n = 21$, $p < 0.01$), but mepp amplitude and α-BuTx binding did not correlate (FIG. 1a and b).

Small mepp Amplitudes, Reduced [^{125}I]α-BuTx Binding, and Elongated End-plates

This was the largest category (nine patients), eight of whom were males. All had ocular symptoms, and most responded well to anti-AChE and/or 3,4-DAP

TABLE 2. Results of Intercostal Muscle Biopsies in Controls and 22 CMS Patients Studied since 1976

	Oc[a]	DAP	ChE	Mepp Amplitude mV	AChRs α-BuTx sites × 10^{-7}	End-plate Size Sarcomeres
1.[b]	+	+	+	0.10	0.18	16.6
2.	+	ND[f]	+	0.15[c]	0.14	15.4
3.	+	+	+	0.33[c]	0.15	11.8
4.	+	−	+	0.37	0.55	>15
5.	+	+	+	0.20	0.69	15.7
6.	+	−	−	0.28	0.17	>12
7.	+	+	+	0.29	0.10	>12
8.	+	−	−	0.20	0.32	11.0
9.	+	+	+	0.36[c]	0.39	10.8
10.	−	+	+	0.56	0.60	7.1
11.	+	−	+	0.31	0.55	8.8
12.	−	−	−	0.75	0.68	8.7[d]
13.	−	+	+	0.07	1.6	7.7
14.	−	−	−	0.32[c]	0.84	5.5
15.	+	ND	−	0.80	0.69	13.4[d]
				prolonged		
16.	−	+	+	0.80	1.48	7.0[e]
17.	+	+	+	0.22[c]	ND	7.0
18.	p	+	ND	ND	0.38	>12
19.	p	+	ND	ND	0.26	>12
20.	−	+	+	ND	0.50	11.7[d]
21.	+	+	+	ND	0.18	11.5
22.	p	ND	+	ND	0.97	6.3
Total				0.36 ± 0.23 (17)**	0.54 ± 0.42 (21)**	10.8 ± 3.3 (21)*
Controls				0.8 ± 0.1 (4)	1.2 ± 0.3 (7)	7.5 ± 0.9 (3)

*$p < 0.02$; **$p < 0.001$.

[a] Oc = ocular symptoms present (+) or absent (−); p = ptosis only; DAP = response (+) or no response (−) to 3,4-DAP treatment; ChE = response (+) or no response (−) to antiacetylcholinesterase drug treatment.

[b] Patients.

[c] Measured in the presence of eserine.

[d] Repetitive response to single stimulus.

[e] Reduced quantal content.

[f] ND = Not determined. All values given are means or means ± SD (no.).

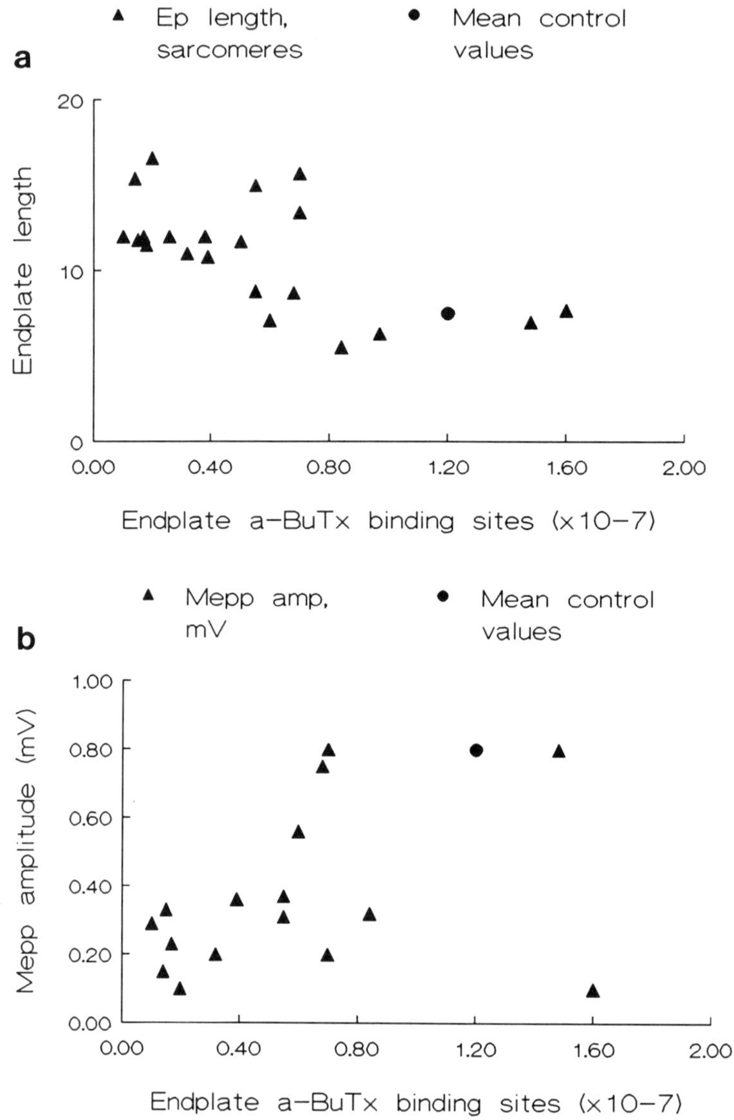

FIGURE 1. Relationship between end-plate length **(a)** and mepp amplitude **(b)** and end-plate [^{125}I]α-BuTx binding sites in patients with CMS. Based on values from TABLE 2.

(TABLE 2). The reduced number of AChRs would probably be sufficient to account for the reduced mepp amplitudes and consequent weakness. The primary defect in these cases was likely either to have been in the AChR itself or in structural proteins that might have been concerned in determining end-plate size and the insertion or stabilization of AChRs and of other functional proteins. In favor of structural defects in some individuals were the abnormal appearances of the cholinesterase-stained end-plates (FIG. 2). Some end-plates were elongated compared to controls (FIG. 2a) but otherwise appeared normal (FIG. 2b), whereas in one individual (no. 9), the end-plates were "spiky" (FIG. 2c). In another patient (no. 8) the cholinesterase staining was very weak, and end-plates could only be seen after prolonged exposure to the stain (FIG. 2d).

Normal-sized End-plates with Slightly Reduced mepp Amplitude and [^{125}I]α-BuTx Binding

In three patients, mepp amplitudes were normal or somewhat reduced and α-BuTx binding was lower than normal. The end-plates were of normal size, but cholinesterase staining was weak. In one patient (no. 12) there was diffuse staining around the end-plate area. This female patient had no ocular symptoms and had shown a repetitive response to single stimuli, but she did not have the grossly prolonged mepp decay times typical of the slow-channel syndrome. The defect in her case may have been in the end-plate form of AChE or in a structural component of the basal lamina involved in anchoring AChE. She failed to respond to cholinesterase inhibitors or to 3,4-DAP.

Small mepp Amplitudes with Normal [^{125}I]α-BuTx Binding and AChE Staining

Two patients (13 and 14) had normal end-plate staining and α-BuTx binding, but mepp amplitudes were reduced dramatically in patient 13. They were both male and did not have ocular symptoms. The sister of one (patient 13, previously described in ref. 1) had complained of fatigability, and to investigate a possible presynaptic abnormality, muscle biopsies were obtained from her and, subsequently, from their father (who had no evident weakness). Results are given in TABLE 3. Both had normal α-BuTx binding and AChE staining of the end-plates. Whereas mepp amplitudes were normal in the sister, they were very small, however, in the father. The fresh ACh content was not abnormal in either the sister or father. The amount of ACh released from the sister's muscle during K^+-evoked depolarization of the nerve terminal was drastically reduced compared to control values. Curiously, the amount of ACh synthesized during the incubation was not so low (437 pmols as compared with 1000 ± 290 in controls), suggesting that the synthesis and storage of ACh were not affected in this individual. It is possible that in the sister there is a defect in the vesicular uptake of ACh, which only becomes manifest during the prolonged KCl stimulation, or a defect in vesicular release. In the father the K^+-evoked release was much higher, perhaps within normal limits. In him and the propositus, therefore, the reduced mepp amplitudes may be the result of a further abnormality, possibly in the response of the AChR itself.

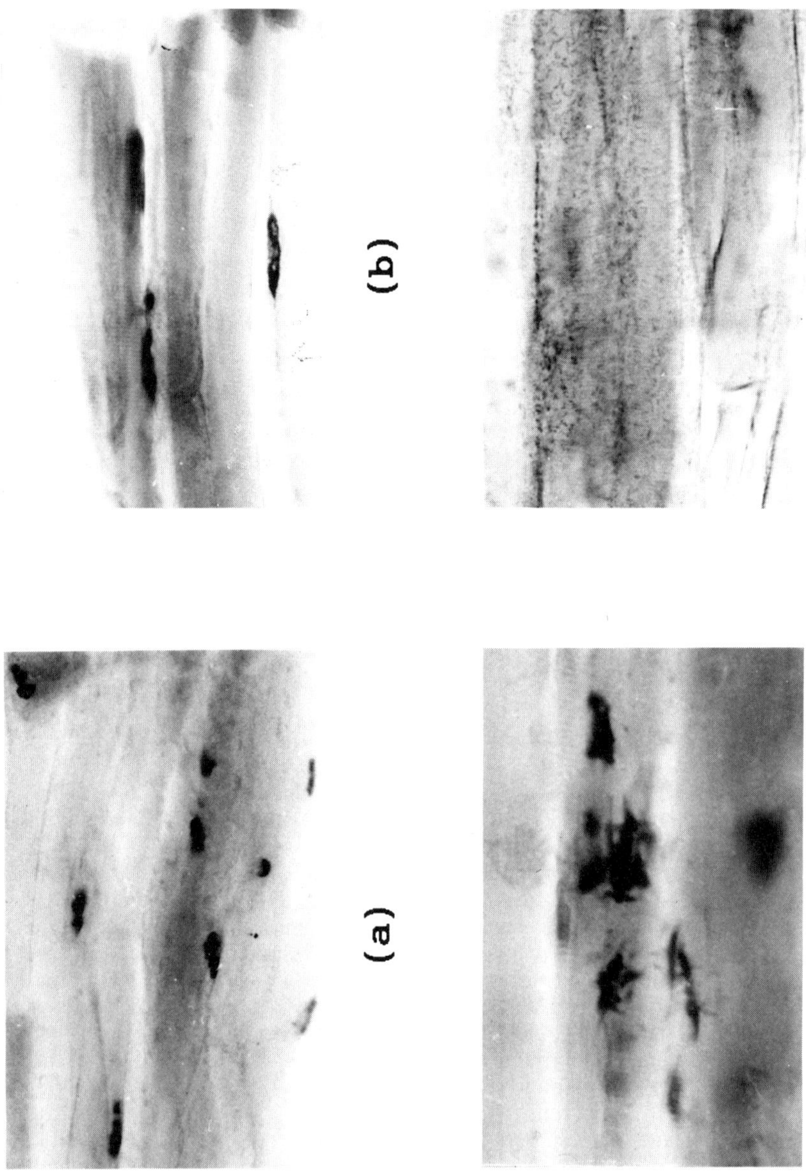
(a)
(b)
(c)
(d)

TABLE 3. Investigations in Intercostal Muscle Biopsies in Controls, in Patient 13, and in His Sister and Father[a]

	mepp Amplitude mV	Fresh ACh Content (pmol g^{-1})[b]	K^+-evoked ACh Release (pmol)[b] 0–60 min	K^+-evoked ACh Release (pmol)[b] 60–150 min	ACh Synthesized during the Incubation (pmol)[b]
Controls	0.8	220 ± 31 (11)	100 ± 37 (6)	95 ± 20 (2)	1000 ± 290 (5)[c]
Patient 13	0.07	160	nd	nd	nd
Sister	0.7	239	13	5	437[c]
Father	0.2	259	56	nd	nd

[a] In all three individuals [^{125}I]α-BuTx binding and AChE staining of end-plates were normal.

[b] Cholinesterase was inhibited by DFP or Soman. ACh content was determined from freshly dissected muscle bundles as described earlier.[1] ACh release during a period in normal Ringer was subtracted from that during the subsequent incubation in the presence of 50 mM KCl. During the last 90 min incubation, the medium contained 10 μM hemicholinium-3 to suppress ACh synthesis.

[c] In both the sister and in the controls there was much more ACh synthesized than released during the incubation. This surplus store of ACh, from which ACh cannot be released, has been described previously.[4]

The Slow-Channel Syndrome

One patient (15), described in detail elsewhere,[5] had the slow-channel syndrome (see ref. 2). She showed a double response to a single ulnar nerve stimulus. The mepp amplitudes were normal but the decay of the mepps and of miniature end-plate currents was prolonged. [^{125}I]α-BuTx binding was slightly reduced and end-plates were elongated.

Congenital Lambert-Eaton Myasthenic Syndrome

One 15-year-old boy (no. 16) had normal mepp amplitudes, α-BuTx binding, and AChE staining of end-plates. However, the quantal content was reduced (24 ± 10, n = 3 compared with values of 79 ± 9, n = 15 in controls). This suggests a congenital disorder analogous to acquired Lambert-Eaton syndrome.

Remaining Patients

In the remaining six patients the data were incomplete or unclear. In all of these except one (no. 22), however, a reduction in α-BuTx binding or in mepp amplitude was found, confirming the diagnosis of CMS (TABLE 2). The elongated

FIGURE 2. AChE-stained end-plates in CMS muscle. **a:** Control muscle shows well-stained end-plates of normal size. **b:** Well-stained but elongated end-plates typical of many patients with CMS. **c:** "Spiky" end-plates found in patient 9. **d:** Very poor and diffuse staining after prolonged incubation in patient 8.

TABLE 4. Restriction Fragment Length Polymorphisms (RFLPs) Found in 16 Healthy Individuals

Subunit	Enzymes	Fragments kb	Allele Frequency
α	BglII	5.5, 4.1	94%
		9.6	6%
	SstI	7.0, 6.0	94%
		13.0	6%
β	PstI	1.7	94%
		1.6	6%
	BglII	9.4	96%
		6.4	4%
γ	PvuII	1.5	72%
		1.7	28%
δ	PstI	1.2	62%
		1.1	38%
	PvuII	1.7	81%
		1.9	19%
ε	PvuII	3.8	48%
		2.0	52%

end-plates with reduced [^{125}I]α-BuTx binding, found in patients 18–21, suggests that these would probably fall into category A.

Restriction Fragment Length Polymorphisms

We have cloned and sequenced each of the genes for the five subunits of the human acetylcholine receptor (see Beeson *et al.*, this volume), enabling cDNAs to be used as probes to look for deletions or insertions in genomic DNA. Abnormalities of the AChR might be found in congenital myasthenic syndromes, particularly in those patients with the slow-channel syndrome or in whom mepp amplitudes or end-plate α-BuTx binding were reduced, although reduction in the number of end-plate AChRs may well be due to defects in other components of the neuromuscular junction (see discussion). We have performed restriction fragment length polymorphisms (RFLPs) in 25 patients (12 from whom muscle biopsies were studied) and 16 healthy controls using 13 restriction enzymes. Several polymorphisms were found in DNA from healthy (and CMS individuals) (TABLE 4), including some of those previously reported.[6] There were no obvious insertions or deletions in the patients, but in three individuals, RFLPs were found that were different from those observed in the controls (TABLE 5).[7,8] The presence of these two polymorphisms, however, in one healthy parent of the affected (Asian) individuals (not shown) suggests that they are not related to the defect in neuromuscular transmission.

DISCUSSION

In the first four patients with congenital myasthenic syndromes that we examined there were at least two different forms of the disorder.[1] Our results in 18 further biopsies have confirmed the heterogeneity of this condition. The majority of patients were male, and in the populations that we examined (Europe, the Middle East, and Central Asia), the age of onset was mainly at or near birth. All had generalized weakness, but the six patients without ocular symptoms included two with repetitive responses. Moreover, 5/6 patients without ocular symptoms had normal or small-sized end-plates (mean 7.9 ± 2.1; n = 6), whereas 14/16 of those with ocular symptoms had elongated end-plates (mean 11.9 ± 2.9; n = 16, see TABLE 2). This difference was highly significant ($p < 0.002$) and indicates that measurement of end-plate size alone may be helpful in suspected cases of CMS with ocular weakness. In other respects, it would appear difficult to predict the pattern of experimental findings on the basis of the clinical observations. Equally, a good response to anti-AChE or 3,4-DAP treatment was common but not easy to foretell (TABLE 2). The great majority of the patients had either reduced mepp amplitude, reduced numbers of AChRs, or abnormal cholinesterase staining. In over half the patients in whom data were complete, all three were abnormal. It was difficult on the basis of these relatively simple observations, however, to determine the exact defect, though few, if any, would fit into the categories described by Engel and his colleagues (see refs. 2 and 9, and this volume). Patient 12 may have an abnormality of AChE similar to those described by Hutchinson *et al.* (this volume); and patients 13 and 14 are somewhat similar to one patient described by Lecky *et al.*,[10] and the one with abnormal ACh/AChR interaction described by Uchitel *et al.* (this volume), though that patient had ocular weakness. The cases of AChR deficiency reviewed by Nagel *et al.* (this volume) differed from our cases 1 to 9 in having normal sized end-plates. The only patients in which elongated end-plates were found were the ones with a defect in the epsilon subunit (Engel *et al.*, this volume), which was associated with prolonged AChR open time. We only found prolonged open times in the one patient with slow-channel syndrome.[5] On the other hand, patient 16, a boy with normal mepp amplitudes, AChRs, and AChE-staining, had reduced quantal contents, suggesting a congenital form of Lambert-Eaton myasthenic syndrome. A clinical study of such a patient has been reported.[11]

Jennekens *et al.*[12] have recently reported a case of AChE deficiency with associated reduction in AChR. In discussion they suggest that a defect in a component of the basal lamina, such as Agrin, which may affect the anchoring of AChRs and of AChE, could be involved. The 22 patients in our study are heterogeneous and almost certainly involve a number of abnormalities. The neuromuscular junction is a highly complex structure in which many different proteins, in the cytoplasm and in the basal lamina, interact to maintain the close apposition of pre-

TABLE 5. RFLPs Found Only in CMS Patients

Subunit	Enzymes	Fragments kb	Allele Frequency within 25 CMS Patients[a]
α	KpnI	13.8	2% n = 1
		9.4	98% n = 24
β	PvuII	2.4	4% n = 2
		3.0	96% n = 23

[a] The common alleles were found in healthy controls and in most CMS patients. The uncommon alleles, found in three Asian CMS individuals (one per patient), were also found in one healthy parent.

and postsynaptic functional elements. The possibility of defects in all of these proteins needs to be addressed.

REFERENCES

1. Vincent, A., S. Cull-Candy, J. Newsom-Davis, A. Trautmann, P. C. Molenaar & R. Polak. 1981. Congenital myasthenia: endplate acetylcholine receptors and electrophysiology in five cases. Muscle & Nerve **4:** 306–318.
2. Engel, A. G. 1990. Congenital myasthenic syndromes. *In* Neuromuscular Transmission: Basic and Applied Aspects. A. Vincent & D. Wray, Eds.: 200–225. Manchester University Press. Manchester.
3. Wray, D. 1981. Prolonged exposure to acetylcholine: noise analysis and channel inactivation in cat tenuissimus muscle. J. Physiol. (Lond) **310:** 37–56.
4. Molenaar, P. C., B. Oen, R. L. Polak. 1987. Surplus acetylcholine and acetylcholine release in the rat diaphragm. J Physiol. (Lond.) **385:** 147–167.
5. Oosterhuis, H. J., J. Newsom-Davis, J. H. Wokke, P. C. Molenaar, T. V. Weerden, B. S. Oen, F. G. Jennekens, H. Veldman, A. Vincent, D. W. Wray *et al.* 1987. The slow channel syndrome. Two new cases. Brain **110:** 1061–1079.
6. Lobos, E., C. H. Rudnick, M. S. Watson & K. E. Isenberg. 1989. Linkage disequilibrium study of RFLPs detected at the human muscle nicotinic acetylcholine receptor subunit genes. Am. J. Hum. Genet. **44:** 522–533.
7. Harrison, J. P., D. Beeson, A. Vincent & J. Newsom-Davis. 1990. Hereditary (congenital) myasthenia: an investigation of possible genetic defects of muscle acetylcholine receptor (AChR). J. Neurol. Sci. **98**(Supplement): 190.
8. Betty, M., D. Beeson, J. Harrison, A. Vincent & J. Newsom-Davis. 1991. Analysing the AChR genes in hereditary myasthenics. J. Autoimmunity **4,** no. 6: v. (Abstr.)
9. Engel, A. G., T. J. Walls, A. Nagel & O. Uchitel. 1990. Newly recognized congenital myasthenic syndromes: I. Congenital paucity of synaptic vesicles and reduced quantal release. II. High-conductance fast-channel syndrome. III. Abnormal acetylcholine receptor (AChR) interaction with acetylcholine. IV. AChR deficiency and short channel-open time. Prog. Brain Res. **84:** 125–137.
10. Lecky, B. R. F., J. A. Morgan-Hughes, N. Murray, D. Landon, D. Wray & C. Prior. 1986. Congenital myasthenia: further evidence of disease heterogeneity. Muscle & Nerve **9:** 233–242.
11. Bady, B., G. Chauplannaz & H. Carrier. 1987. Congenital Lambert-Eaton myasthenic syndrome. J. Neurol. Neurosurg. Psychiatry **50:** 476–478.
12. Jennekens, F. G. I., L. F. G. M. Hesselmans, H. Veldman, E. N. H. Jansen, F. Spaans & P. C. Molenaar. 1992. Deficiency of acetylcholine receptors in a case of end-plate acetylcholinesterase deficiency: A histochemical investigation. Muscle & Nerve **15:** 63–72.
13. Palace, J., C. M. Wiles & J. Newsom-Davis. 1991. 3,4-Diaminopyridine in the treatment of congenital (hereditary) myasthenia. J. Neurol. Neurosurg. Psychiatry **54:** 1069–1072.

Congenital Myasthenic Syndrome Associated with Paucity of Synaptic Vesicles and Reduced Quantal Release[a]

T. J. WALLS,[b] A. G. ENGEL,[c,f] A. S. NAGEL,[e]
C. M. HARPER,[c] AND V. F. TRASTEK[d]

[b]*The Department of Neurology*
Newcastle General Hospital
Newcastle upon Tyne, NE4 6BE, United Kingdom

[c]*Department of Neurology and*
Neuromuscular Research Laboratory
[d]*Department of Surgery*
Mayo Clinic
Rochester, Minnesota 55905

[e]*Department of Neurology*
University of Würzburg
Würzburg, Germany

INTRODUCTION

In this communication we describe a congenital myasthenic syndrome caused by a presynaptic defect. The onset of symptoms was in the neonatal period, and antiacetylcholine receptor antibodies were absent. The safety margin of neuromuscular transmission was compromised by a reduced quantal content of the end-plate potential. This was due to a decrease in the number of readily releasable quanta and was accompanied by a proportionate decrease in the number of synaptic vesicles in otherwise normal nerve terminals. The probability of quantal release was slightly increased. No significant postsynaptic defect was documented. The present syndrome superficially resembles the Lambert-Eaton syndrome in that the quantal content of the end-plate potential is reduced in both. In the Lambert-Eaton syndrome, however, it is the probability of release and not the store of readily releasable quanta that is decreased, and the synaptic vesicles are normally abundant. Detailed *in vitro* electrophysiological studies and quantitative electron microscopy were essential to characterize the new disorder.

CLINICAL DATA

A female infant was born to unrelated South American parents in 1966. In the neonatal period she sucked poorly, choked on feeding, and had a weak cry. During

[a] This work was supported by Grant NS 6277 from the National Institutes of Health and by a Research Grant from the Muscular Dystrophy Association. Dr. T. J. Walls was the recipient of a Fulbright Fellowship from the U.S. Government.

[f] Address correspondence and reprint requests to A. G. Engel, M.D., Mayo Clinic, Rochester, MN 55905.

childhood she had fluctuating muscle weakness with abnormal fatigability, variable ptosis, and recurrent episodes of bulbar dysfunction. Motor development was delayed, but intellectual development was normal. Her symptoms were partially improved by small doses of anticholinesterase drugs. Tests for acetylcholine receptor antibodies were repeatedly negative. She was referred for detailed assessment at the age of 23 years.

The patient had a small stature, slender muscle bulk, hypernasal voice, fluctuating ophthalmoparesis with pronounced ptosis, and moderately severe weakness of the facial, tongue, neck flexor, and extremity muscles. The weakness and especially the ptosis increased rapidly with sustained exertion. The arm abduction time was only 12 seconds, and the patient could not rise from squatting without help. The tendon reflexes were preserved.

Clinical electromyographic studies revealed normal motor and sensory nerve conduction velocities. The compound muscle action potential (CMAP) was of normal amplitude in rested muscle, and single nerve stimuli did not evoke a repetitive CMAP. Stimulation of the ulnar nerve at 2 Hz produced a 30% decrement of the CMAP in the abductor digiti minimi, and a similar decrement was seen in all muscles studied. Stimulation at 50 Hz did not increase the amplitude of the CMAP.

INVESTIGATIONS

An intercostal muscle specimen was removed under general anesthesia. Control intercostal muscle specimens were obtained from subjects undergoing thoracotomy for other medical reasons. None of the control subjects had any evidence neuromuscular disease.

Light Microscopy

Portions of the specimen were processed for routine histochemical studies. These revealed no abnormalities. There was abundant reactivity for acetylcholinesterase at all end-plates. The mean end-plate length, measured from the histochemical reaction product for acetylcholinesterase in teased single muscle fibers was 22.7 μm (control range 20–29 μm).

Acetylcholine Receptors

The mean number of acetylcholine (ACh) binding sites per end-plate was estimated using radioiodinated α-bungarotoxin[1] to be 11.58×10^6 (mean $\pm$ SEM in 11 controls: $13.23 \pm 0.85 \times 10^6$). The distribution of acetylcholine receptors at the end-plates was examined with the electron microscope by the use of peroxidase conjugated α-bungarotoxin.[2] FIGURE 1 demonstrates a typical neuromuscular junction showing a normal distribution of reaction product over the terminal expansions of the junctional folds. The acetylcholine receptor index,[2] measured from such sections, was 3.43. In five controls this index ranged from 3.13 to 3.31.

In Vitro *Microelectrode Studies*

Isolated nerve/muscle preparations with the fibers intact from tendon to tendon were dissected from the intercostal muscle specimen and perfused in Tyrode's solution. Standard intracellular recording techniques were used to measure the amplitude, frequency, and time course of the miniature end-plate potential (MEPP). The number of transmitter quanta released by nerve impulse (m) at 1 Hz stimulation was measured by the variance method[3] after the addition of *d*-tubocurarine to the perfusing fluid. The curarizing dose was 0.5 μg/mL. Estimates of the number of quanta available for release (n) and of the probability of release (p) were obtained from the rundown of the end-plate potential amplitude during short trains of stimuli at 40 Hz.[4,5]

Miniature end-plate currents (MEPC) were recorded using a two-electrode voltage clamp. Properties of single acetylcholine-induced ion channels were studied using analysis of current noise recorded using a two-electrode voltage clamp

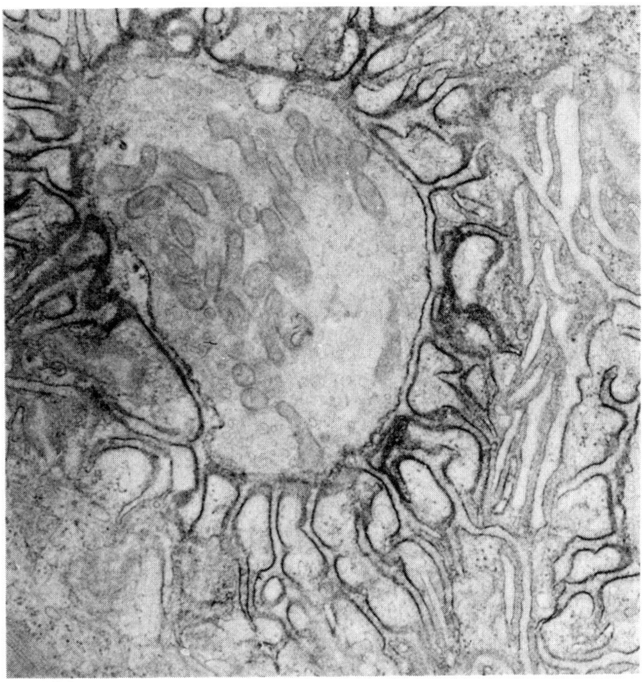

FIGURE 1. Ultrastructural localization of AChR with peroxidase-labeled α-bungarotoxin shows a normal distribution and density of AChR on intact junctional folds; $\times$18,100.

during the iontophoresis of acetylcholine.[6,7] Single-channel conductance was estimated both from the variance of the acetylcholine-induced noise and from its power spectrum, examined using Fourier analysis. The mean channel open time was derived from the power spectrum.

Miniature End-Plate Potentials and Currents (TABLE *1*)

The mean MEPP and MEPC amplitudes were not significantly different from the corresponding control values. The resting MEPP frequency was normal. When

TABLE 1. *In Vitro* Physiology—Miniature End-Plate Potentials and Currents

Parameter	Patient[a]	Controls[b]
MEPP amplitude (mV)[c]	0.79 ± 0.05 (15)	0.99 ± 0.06 (15)
MEPP amplitude after 10 min stimulation at 10 Hz (mV)[c]	0.77 ± 0.11 (5)	
Resting MEPP frequency (Hz)[c]	0.08 ± 0.01 (15)	0.12 ± 0.03 (12)
Rise in MEPP frequency in $[K^+]$ = 20 mM	× 39 (18)	× 21 (3)
MEPC amplitude (nA)[d]	3.09 ± 0.13 (12)	3.92 ± 0.14 (10)

[a] Mean ± SE (number of fibers).
[b] Mean ± SE (number of subjects).
[c] Corrected for a resting membrane potential of −80 mV; 28.5 ± 0.5°C.
[d] −80 mV; 22.5 ± 0.5°C.

the nerve terminals were depolarized by increasing the $[K^+]$ in the perfusing solution to 20 mM, the increase in the MEPP frequency was comparable to the increase obtained in the controls. The mean MEPP amplitude was not significantly reduced after 10 minutes of indirect stimulation at 10 Hz.

Single-Channel Properties (TABLE 2)

The time course of the decay phase of MEPC, the measured single-channel conductance, and the mean channel open time were all normal.

Evoked Transmitter Release (TABLE 3)

The mean number of transmitter quanta released at 1 Hz stimulation (m) was reduced to about 20% of normal. When the $[Ca^{2+}]$ in the perfusing fluid was raised to 5 mM, m rose by an amount comparable to that observed in the controls. The reduction of m in the patient was due to a reduction in the number of quanta immediately available for release (n) to about 20% of normal; the probability that an individual quantum would be released was at least as high as normal.

TABLE 2. *In Vitro* Physiology—Kinetic Properties of the Acetylcholine Receptor Ion Channel

Parameter[a]	Patient[b]	Controls[c]
MEPC decay time constant (ms)	3.66 ± 0.13 (12)	3.19 ± 0.14 (10)
Mean channel open time (ms) (power spectrum)	2.33 ± 0.10 (5)	2.37 ± 0.09 (9)
Single-channel conductance (pS) (noise variance)	33.13 ± 1.77 (6)	32.23 ± 0.75 (9)
Single-channel conductance (pS) (power spectrum)	30.86 ± 2.46 (5)	31.41 ± 0.96 (9)

[a] 22.5 ± 0.5°; −80 mV.
[b] Mean ± SE (number of fibers).
[c] Mean ± SE (number of subjects).

Ultrastructural Studies

Motor end-plates were localized for electron microscopy[2] and quantitatively analyzed[8] by established methods. Simple inspection of the end-plates revealed a paucity of synaptic vesicles in otherwise normal nerve terminals and no postsynaptic abnormality.

On quantitative analysis, the mean nerve terminal area and presynaptic membrane length were normal, but synaptic vesicle density (number/μm^2) was reduced to 20% of normal (FIG. 2 and TABLE 4). In some nerve terminals the remaining synaptic vesicles were concentrated around the periphery of the nerve terminal.

Quantitative analysis of the postsynaptic regions (TABLE 4) demonstrated a 16% reduction in the postsynaptic membrane density when compared with controls, but with a slightly increased acetylcholine receptor index this is unlikely to be of functional significance. There were no other significant postsynaptic abnormalities.

TABLE 3. *In Vitro* Physiology—Parameters of Quantal Release by Nerve Impulse

Parameter[a]	Patient[b] (no. of fibers)	Controls[b] (no. of subjects)
Quantal content (m) (1 Hz)	8.05 ± 0.85 (17)	51.1 ± 4.5 (15)
Rise in m in [Ca^{2+}] = 5.0 mM	× 2.26	× 1.62, × 1.84 (2)
Number of quanta available for release (n)	77 (12) 1[c]	360 ± 28 (12)
Probability of release (p)	0.235 (12)[c]	0.175 ± 0.013 (12)[d]

[a] Amplitude of evoked potentials was corrected for a resting membrane potential of −80 mV and for nonlinear summation; 28.5 ± 0.5°C.

[b] Mean ± SE.

[c] Pooled values for 12 end-plates.

[d] 8–12 end-plates per subject.

DISCUSSION

The Findings Define a New Syndrome

The clinical features and electromyographic findings in this patient were consistent with a defect of neuromuscular transmission. The age of onset and lack of acetylcholine receptor antibodies, the absence of end-plate AChR deficiency, and the character of the electrophysiological defect indicate that this is a congenital disorder different from acquired autoimmune myasthenia gravis. The normal number, distribution, and ion channel properties of the acetylcholine receptor at the end-plates, the normal MEPP and MEPC amplitudes, and the ultrastructural analysis of the end-plates exclude any postsynaptic abnormality in the present syndrome. The unchanged MEPP amplitude after prolonged 10 Hz stimulation is evidence against as abnormality in acetylcholine resynthesis or repackaging, the putative cause of familial infantile myasthenia.[9]

The number of quanta released by nerve impulse at 1 Hz stimulation (m) was only 20% of the control mean. The value of m depends on the number of readily releasable quanta (n) and the probability of quantal release (p) according to the relationship $m = np$. The repetitive stimulation studies *in vitro* suggested that the decrease in m was due to a reduction in n and that p was normal or even higher than normal. This assumption was supported by three additional observations: (1) Increasing [Ca^{2+}] in the bath, which increases m by increasing p, produced at least as large a proportionate rise in m as in controls; (2) increasing [K^+] in the bath, which depolarizes the nerve terminal and enhances Ca^{2+} influx into the nerve terminal through voltage-sensitive calcium channels, increased the MEPP frequency normally; (3) there was no facilitation of the CMAP during trains of high-frequency stimuli as occurs in the Lambert-Eaton syndrome.[10]

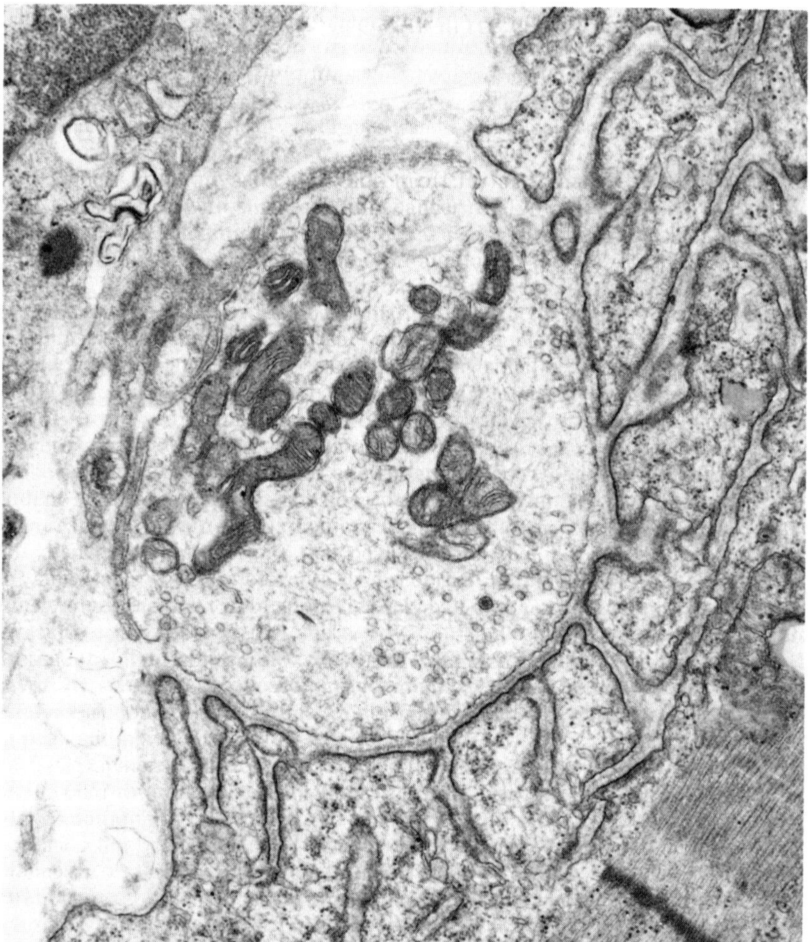

FIGURE 2. End-plate region in the patient. Note paucity of synaptic vesicles in the nerve terminal. The postsynaptic region is normal; ×23,200.

TABLE 4. Morphometric Analysis of End-Plate Regions

Parameter	Patient[a]	Controls[b]
Nerve terminal area (μm^2)	4.36 ± 0.45	3.88 ± 0.39
Presynaptic membrane length (μm)	5.12 ± 0.44	6.11 ± 0.40
Synaptic vesicle density (No./μm^2)	12.2 ± 0.84	50.3 ± 3.60
Postsynaptic area (μm^2)	9.05 ± 0.51	10.6 ± 0.79
Postsynaptic membrane density ($\mu m/\mu m^2$)	4.88 ± 0.13	5.83 ± 0.25

[a] Mean ± SE; 72–80 regions, 36 end-plates; more than one region can occur at an end-plate.
[b] Mean ± SE; 47–64 regions, 13 subjects.

The reduced density of synaptic vesicles in the nerve terminals, the fact that the reduction in m was associated with a decrease in n, and the normal response of the MEPP frequency to increased [K^+] exclude the possibility that our patient suffered from a form of the Lambert-Eaton syndrome. In that syndrome the synaptic vesicle density is consistently normal,[11] the response of the MEPP frequency to increased [K^+] is impaired,[10,12] and n is normal while p is abnormally small.[12]

Pathologic Mechanisms

Two processes could reduce m by reducing n: (1) A reduction in the size of the nerve terminals or a reduction in the presynaptic membrane area due to encasement of the nerve terminal by Schwann cells may have this result. Both of these findings can be observed in patients with congenital end-plate acetylcholinesterase deficiency.[12,13] (2) A reduction in the number of synaptic vesicles within the nerve terminals may have the same effect. Quantitative ultrastructural analysis of the end-plate demonstrated no decreases in the nerve terminal or presynaptic membrane areas. Synaptic vesicle density, however, was significantly reduced and to a degree that was comparable to the decrease in m and n.

Synaptic vesicles are produced in the perikaryon of the anterior horn cell and are carried distally along motor axons to the nerve terminal by fast axonal transport.[14–16] Packaging of the vesicles with ACh occurs within the nerve terminal. After ACh has been released by exocytosis, the vesicle membranes are recycled and are repacked with ACh.[17] In the present syndrome, a reduction in synaptic vesicle density could arise from (1) a defect of synaptic vesicle formation by the anterior horn cell, (2) a defect in axonal vesicle transport, or (3) impaired recycling of the synaptic vesicles in the nerve terminal. The fact that synaptic vesicle density was reduced even in unstimulated nerve terminals would favor a defect in synaptic vesicle formation or transport.

REFERENCES

1. ITO, Y., R. MILEDI, A. VINCENT & J. NEWSOM-DAVIS. 1978. Acetylcholine receptors and end-plate electrophysiology in myasthenia gravis. Brain **101:** 345–368.
2. ENGEL, A. G., J. LINDSTROM, E. H. LAMBERT & V. A. LENNON. 1977. Ultrastructural localization of the acetylcholine receptor in myasthenia gravis and in its experimental autoimmune model. Neurology **27:** 307–315.

3. Martin, A. R. 1955. A further study of the statistical composition of the end-plate potential. J. Physiol. (Lond.) **130:** 114–122.
4. Elmqvist, D. & D. M. J. Quastel. 1965. A quantitative study of end-plate potentials in isolated human muscle. J. Physiol. (Lond.) **178:** 505–529.
5. Kamenskaya, M. A., D. Elmqvist & S. Thesleff. 1975. Guanidine and neuromuscular transmission. II. Effect on transmitter release in response to repetitive stimulation. Arch. Neurol. **32:** 510–518.
6. Anderson, C. R. & C. F. Stevens. 1973. Voltage clamp analysis of acetylcholine produced end-plate current fluctuations at frog neuromuscular junction. J. Physiol. (Lond.) **235:** 655–691.
7. Cull-Candy, S. G., R. Miledi & O. D. Uchitel. 1982. Properties of junctional and extrajunctional acetylcholine receptor channels in organ cultured human muscle fibres. J. Physiol. (Lond.) **333:** 251–267.
8. Engel, A. G., M. Tsujihata, J. M. Lindstrom & V. A. Lennon. 1976. The motor end plate in myasthenia gravis and in experimental autoimmune myasthenia gravis. A quantitative ultrastructural study. Ann. N.Y. Acad. Sci. **274:** 60–79.
9. Mora, M., E. H. Lambert & A. G. Engel. 1987. Synaptic vesicle abnormality in familial infantile myasthenia. Neurology **37:** 206–214.
10. Lambert, E. H. & D. Elmqvist. 1971. Quantal components of end-plate potentials in the myasthenic syndrome. Ann. N.Y. Acad. Sci. **183:** 183–199.
11. Engel, A. G. & T. Santa. 1971. Histometric analysis of the ultrastructure of the neuromuscular junction in myasthenia gravis and the myasthenic syndrome. Ann. N.Y. Acad. Sci. **183:** 46–63.
12. Engel, A. G., E. H. Lambert & M. R. Gomez. 1977. A new myasthenic syndrome with end-plate acetylcholinesterase deficiency, small nerve terminals and reduced acetylcholine release. Ann. Neurol. **1:** 315–330.
13. Hutchinson, D. O., A. G. Engel, T. J. Walls, S. Nakano, S. Camp, P. Taylor, C. M. Harper & J. M. Brengman. 1993. The spectrum of congenital end-plate acetylcholinesterase deficiency. Ann. N.Y. Acad. Sci. This volume.
14. Bööj, S., P.-A. Larsson, A.-G. Dahllöf & A. Dahlström. 1986. Axonal transport of synapsin 1 and cholinergic synaptic vesicle-like material; further immunohistochemical evidence for transport of axonal cholinergic transmitter vesicles in motor neurons. Acta Physiol. Scand. **128:** 155–165.
15. Kiene, M. L. & H. Stadler. 1987. Synaptic vesicles in electromotoneurons. I. Axonal transport, site of transmitter uptake and processing of a core proteoglycan during maturation. EMBO J. **6:** 2209–2215.
16. Llinás, R., M. Sugimori, J. W. Lin, P. L. Leopold & S. T. Brady. 1989. Dependent directional movement of rat synaptic vesicles injected into presynaptic terminal of squid giant synapse. Proc. Natl. Acad. Sci. USA **86:** 5656–5660.
17. Südhof, T. C. & R. Jahn. 1991. Proteins of synaptic vesicles involved in exocytosis and membrane recycling. Cell **6:** 665–677.

The Spectrum of Congenital End-plate Acetylcholinesterase Deficiency[a]

D. O. HUTCHINSON,[b] A. G. ENGEL,[b,e] T. J. WALLS,[c] S. NAKANO,[b] S. CAMP,[d] P. TAYLOR,[d] C. M. HARPER,[b] AND J. M. BRENGMAN[b]

[b]*Department of Neurology*
Mayo Clinic
Rochester, Minnesota 55905

[c]*Department of Neurology*
Newcastle General Hospital
Newcastle Upon Tyne, England

[d]*Department of Pharmacology*
University of California
San Diego, California

Acetylcholinesterase (AChE) is the enzyme responsible for terminating the action of acetylcholine (ACh) at the neuromuscular junction (NMJ). In 1977, Engel, Lambert, and Gomez described a 14-year-old boy (patient 1) with a congenital myasthenic syndrome secondary to end-plate AChE deficiency.[1] We here describe our findings in four additional patients (patients 2–5) studied in the Mayo Muscle Laboratory between 1988 and 1991 and present evidence for genetic and phenotypic heterogeneity of the syndrome.

CLINICAL STUDIES

Patients 2 and 3 were sisters; the other cases were sporadic. The ages of the patients ranged from 4 months to 43 years at the time of study (see TABLE 1). Onset of symptoms was in the neonatal period or shortly thereafter. All had weakness increased by exertion (FIGURES 1 and 2), involving the facial, bulbar, cervical, torso, and limb muscles. Ptosis and extraocular muscle weakness were absent in the two sisters but were prominent in the others (FIG. 2). All patients had slow pupillary responses to light. Of the three who were beyond infancy, two had scoliosis, and the third had exaggerated lumbar lordosis (FIGURES 1 and 2). The tendon reflexes were normal. None of the patients responded favorably to cholinesterase inhibitors.

The most severely affected was patient 4. This infant boy fed poorly and had

[a] This work was supported by Grant NS 6277 from the National Institues of Health and by Research Grants from the Muscular Dystrophy Association. Dr. T. J. Walls was the recipient of a Fulbright Fellowship from the U.S. Government.

[e] To whom correspondence should be addressed.

multiple hospital admissions with respiratory failure. Eight months after our study, he had a tracheostomy and needed nearly continuous assisted ventilation.

Patient 5 had relatively mild weakness of ocular, cervical, and upper limb muscles until his fourth decade; he then became gradually weaker and developed bulbar symptoms. At age 43 he had moderate ptosis, marked weakness of the cervical, wrist, and finger extensor muscles but only mild weakness of the leg muscles (FIG. 2, right).

A repetitive compound muscle action potential (CMAP) occurred 6-10 ms after the first response (FIG. 3) in all patients except the infant. The second CMAP disappeared rapidly at stimulation rates greater than 0.2 Hz.

With repetitive stimulation at 2 Hz, a decremental response was found in all muscles tested (FIG. 3). The decrement was slightly diminished immediately after exercise (postactivation facilitation) in patients 2 and 5. Parenteral administration of edrophonium (patient 4) or neostigmine (patient 5) had no effect on the decrement. Motor unit potentials (MUP) were of short duration and low amplitude in all patients. Some MUP varied in amplitude from moment to moment, and an increased proportion of the MUP were polyphasic. Motor and sensory nerve conduction velocities were normal.

Tests of Autonomic Function

Heart rate and blood pressure responses to posture and Valsalva maneuver, and quantitative sudomotor axon reflex tests were performed in the two sisters (patients 2 and 3). These studies were normal except for equivocal, patchy reduction of sudomotor function in the younger sister (patient 3). Formal pupillography in patients 1, 2, and 3 confirmed the slow pupillary responses; the amplitudes of the responses were at the lower limits of normal. The responses to 0.125% pilocarpine did not suggest denervation of the iris.

SPECIAL STUDIES

Intercostal muscle specimens were obtained from patients 2, 3, and 5, and a deltoid muscle specimen from patient 4. In patients 2, 3, and 4 routine

TABLE 1. Summary of the Clinical Data

	Patient			
	2	3	4	5
Age when studied/sex	12 yr/F	8 yr/F	4 mo/M	43 yr/M
Age at onset symptoms	neonate	neonate	neonate	6 mo
Diffuse weakness	+	+	+	+
Ophthalmoparesis	−	−	+	+
Slowed pupil responses	+	+	+	+
Spinal deformity	+	+	−	+
Tendon reflexes	normal	normal	normal	normal
Episodes of respiratory distress	+	−	+	−
Response to antiesterase treatment	−	−	−	−
EMG[a] decrement at 2 Hz	+	+	+	+
Repetitive CMAP[b]	+	+	−	+

[a] EMG = electromyogram.
[b] CMAP = compound muscle action potential.

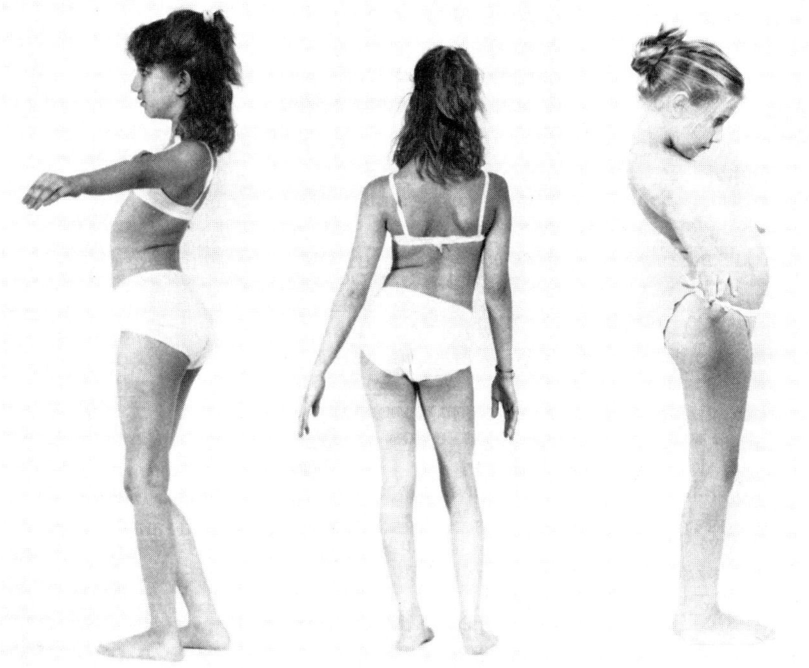

FIGURE 1. Two sisters (patients 2 and 3) with end-plate AChE deficiency. The lordosis and scoliosis in the older sister (left and center panels) and the lordosis in the younger sister (right panel) became worse during standing for 15 seconds. Lateral views show maximal arm abduction time after 15 seconds. At that time, the younger sister was unable to hold her head erect (right panel). (A. G. Engel. With permission from *Seminars in Neurology* **10:** 12–26. 1990.)

histochemical studies were either normal or showed minimal nonspecific abnormalities. Patient 5 had rare necrotic fibers and a mild increase of internal nuclei. End-plate AChE activity, determined with α-naphthyl acetate[2] and with dithio-bis-acetic acid[3] as substrates, was absent in patients 3, 4, and 5. In patient 2, traces of enzyme activity were noted at the end-plate after incubation with α-naphthyl acetate for 10 minutes (normal incubation time 5 minutes) and with dithio-bis-acetic acid for 30 minutes (normal incubation time 15 minutes). Even this activity, however, was abolished by 10^{-5} M iso-OMPA, a specific pseudocholinesterase inhibitor.

The number of acetylcholine receptors (AChR) per end-plate was determined in three patients using $[^{125}I]\alpha$-bungarotoxin.[4] In patients 2 and 3, the toxin-binding sites per end-plate were reduced to 6.6 and 7.2 $\times$ 10^6, respectively; in patient 5, the count (9.6 $\times$ 10^6) fell in the low-normal range. In 11 nonweak control subjects, the values ranged from 9.3 to 18.7 $\times$ 10^6 toxin-binding sites per end-plate.

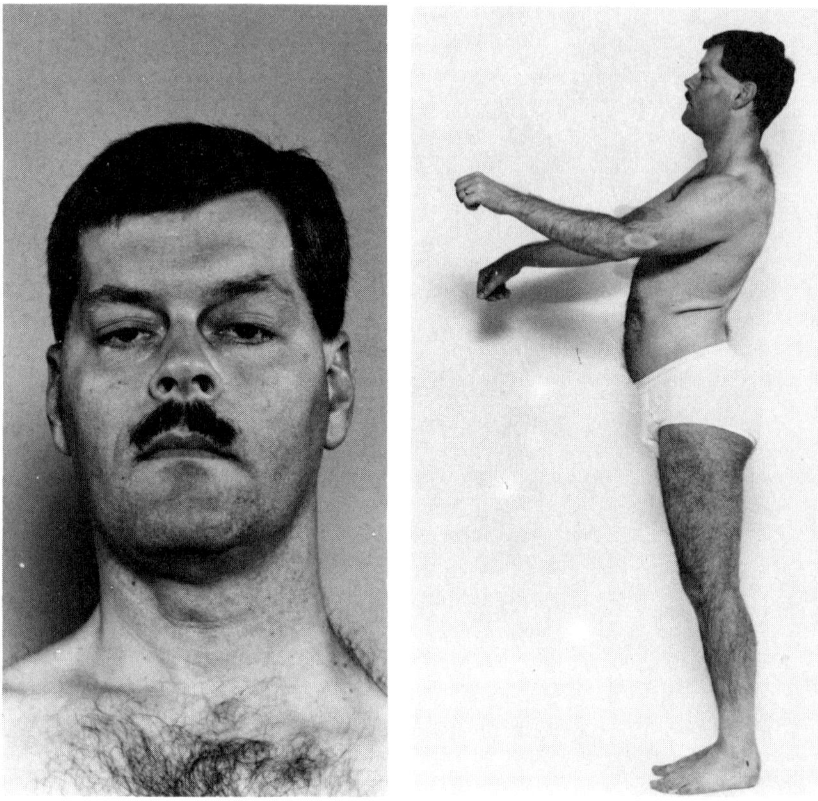

FIGURE 2. Patient 5 with AChE deficiency. Note ptosis and lordosis. Right panel shows maximal arm elevation after 15 seconds. Patient could not extend his fingers even at the beginning of the test.

In vitro *Microelectrode Studies*

Miniature end-plate potential (MEPP) and current (MEPC) amplitudes were reduced in patient 5 and were normal in the others. The MEPP (FIG. 4) and MEPC decay-time constants were uniformly prolonged. The addition of 1 μg/mL neostigmine methylsulfate to the bath had no significant effect on the amplitude or duration of the patient's MEPPs (TABLE 2).

The quantal content of the end-plate potential (EPP) on 1 hz stimulation was abnormally low in patients 2, 3, and 4 and was normal in patient 5. In all three, the reduction of quantal content was secondary to a decrease in the number of quanta readily available for release (*n*); the probability of release (*p*) was high-normal, or higher than normal (TABLE 3).

Noise analysis was done on ACh-induced fluctuations of the end-plate current in patients 3 and 5. The values derived for single-channel conductance and mean channel open time were normal (TABLE 4). This excluded the possibility that abnormal gating of the AChR ion channel was responsible for the prolonged MEPP and MEPC.

Immunocytochemistry

AChR was localized with rhodamine-conjugated α-bungarotoxin. AChE was immunolocalized in the same sections using mouse primary antibodies (one polyclonal antibody and five monoclonal antibodies recognizing different epitopes on the catalytic subunit of AChE) in combination with a fluorescein isothiocyanate-labeled second antibody.[5] With each antibody, AChE was absent from all endplates of all patients.

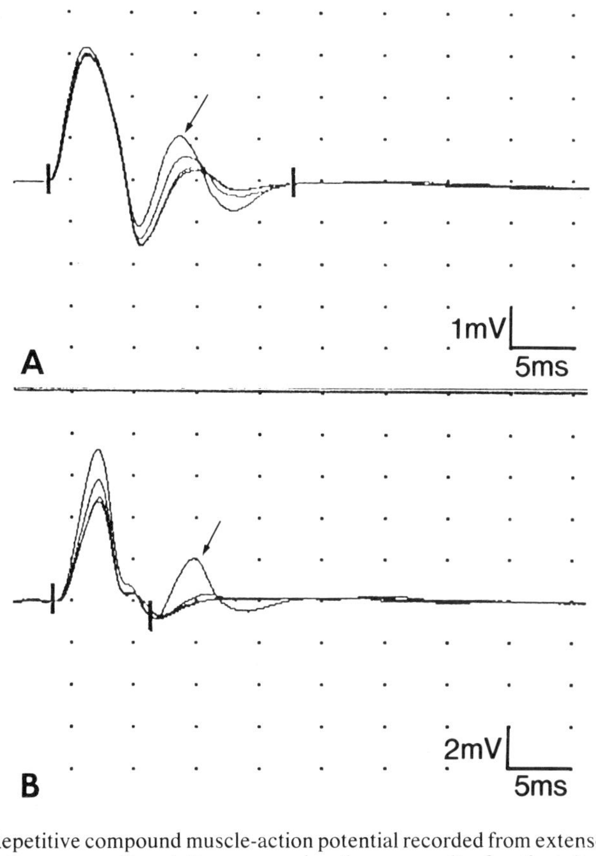

FIGURE 3. Repetitive compound muscle-action potential recorded from extensor digitorum brevis muscle (upper panel) and thenar muscles (lower panel) of patient 2 during 2 Hz stimulation of peroneal and median nerves. The second response (arrow) decreases more rapidly than the first. (A. G. Engel. With permission from *Seminars in Neurology* **10:** 12–26. 1990.)

Electron Microscopy

End-plate ultrastructure was studied quantitatively (TABLE 5) using previously established methods.[6] In all patients, a proportion of the nerve terminals was abnormally small, and the mean nerve terminal area was significantly reduced (FIGURES 5–7). At many end-plates, the small nerve terminals covered only a fraction of the underlying postsynaptic region (FIGURES 5 and 6). Furthermore, the presynaptic membrane of many nerve terminals was partially or completely encased by Schwann cell processes, resulting in further reduction in the area of the presynaptic membrane available for ACh release (FIGURES 6A, B and 7).

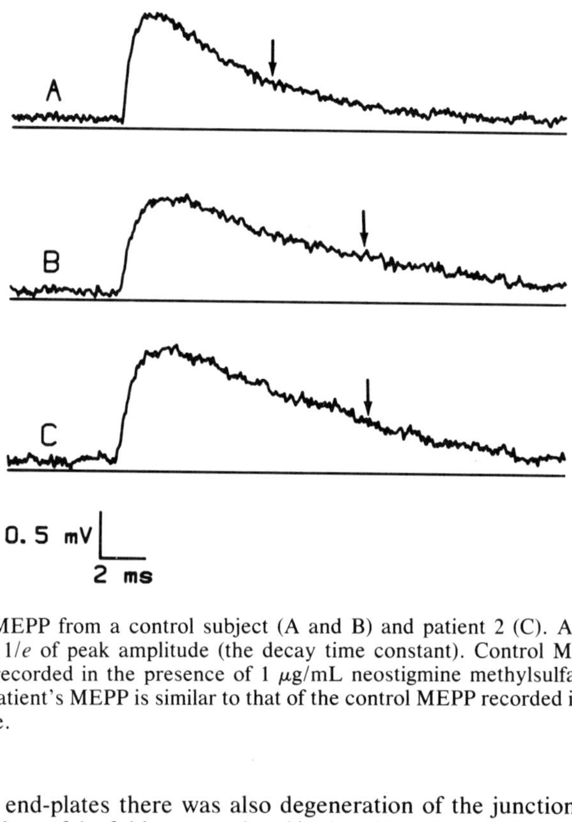

FIGURE 4. MEPP from a control subject (A and B) and patient 2 (C). Arrows indicate time to reach $1/e$ of peak amplitude (the decay time constant). Control MEPP shown in panel B was recorded in the presence of 1 μg/mL neostigmine methylsulfate. The decay phase of the patient's MEPP is similar to that of the control MEPP recorded in the presence of neostigmine.

At many end-plates there was also degeneration of the junctional folds, and globular residues of the folds accumulated in the widened synaptic spaces (FIGURES 6A, B and 8A). This resulted in simplification of many postsynaptic regions and in a reduction of the postsynaptic membrane densities (TABLE 5). At some end-plates the junctional folds were honeycombed by labyrinthine membranous networks.

A number of abnormalities were observed in the junctional sarcoplasm and in nearby muscle fiber regions. These included degeneration of membranous organelles, the appearance of small membrane-bound vesicles, focal loss and disappear-

TABLE 2. Miniature End-plate Potentials and Currents[a]

	Patients (no. of fibers)				Controls (no. of subjects)
	2	3	4	5	
MEPP[b] amplitude (mV)	0.90 ± 0.05 (10)	0.91 ± 0.07 (13)	0.92 ± 0.11 (12)	0.62 ± 0.04 (10)	0.99 ± 0.06 (15)
MEPP amplitude with neostigmine	—	0.80 ± 0.05 (20)	0.78 ± 0.09 (13)	0.70 ± 0.07 (6)	1.69 ± 0.23 (3)
MEPP τ[c] (ms)	6.57 ± 0.49 (9)	5.68 ± 0.37 (13)	6.76 ± 0.40 (12)	6.44 ± 0.23 (10)	4.65 ± 0.19 (14)
MEPP τ with neostigmine	6.31 ± 0.62 (3)	—	6.00 ± 0.47 (7)	5.54 ± 0.49 (6)	7.54 ± 0.34 (3)
MEPC[d] amplitude (nA)	—	3.85 ± 0.31 (9)	—	2.57 ± 0.12 (21)	3.92 ± 0.14 (10)
MEPC τ (ms)	—	5.14 ± 0.35 (9)	—	4.50 ± 0.12 (21)	3.19 ± 0.14 (10)

[a] Mean ± SE are shown. MEPPs were recorded at 28.5 ± 0.5°C and amplitudes were corrected to a resting membrane potential of −80 mV. For patients 2, 3, and 4, the MEPP amplitudes were multiplied by $[d/50]^{3/2}$ to correct for small fiber diameter, where d = mean fiber diameter in μm.[14] MEPCs were recorded at 22.5 ± 0.5°C by two-electrode voltage clamp at a holding potential of −80 mV. MEPP and MEPC τ values were determined by computerized least squares fit of the curve $y_t = y_o.e^{-t/\tau}$ to the decay phase of each potential, where y_o = peak amplitude and t = time. Neostigmine methylsulfate concentration was 1 μg/mL.

[b] MEPP = miniature end-plate potential.

[c] τ = decay time constant.

[d] MEPC = miniature end-plate current.

TABLE 3. Parameters of Quantal Release by Nerve Impulse[a]

	Patients (no. of fibers)				Controls (no. of subjects)
	2	3	4	5	
Quantal content (m) at 1 Hz	19 ± 4.0 (12)	8 ± 1.8 (14)	8 ± 0.6 (15)	62 ± 11.4 (22)	51 ± 4.5 (15)
Readily releasable quanta (n)	85 (12)	40 (10)	42 (4)	251 (21)	360 ± 28 (12)
Probability of release (p)	0.29 (12)	0.34 (10)	0.28 (4)	0.42 (21)	0.18 ± 0.01 (12)

[a] EPPs were recorded at 28.5 ± 0.5°C in the presence of tubocurarine and were corrected to a resting membrane potential of −80 mV[15] and for nonlinear summation.[16] Quantal content was calculated from 64 consecutive EPPs by the variance method.[17] Calculation of n and p was from a plot of quantal content of the first four EPPs at 40 Hz stimulation against the sum of quanta released by the preceding stimuli.[17]

ance of mitochondria (FIG. 8A), focal myofibrillar degeneration with streaming of Z disks (FIG. 8A), and degenerating nuclei (FIG. 8B).

Electron cytochemical localization of AChE with dithio-bis acetic acid as the substrate[3] in glutaraldehyde-fixed muscle specimens of patient 2 confirmed the end-plate AChE deficiency (FIG. 9).

Biochemical Studies

In vertebrates, AChE exists either as multimers of catalytic subunits (globular forms) or as one, two, or three tetramers of catalytic subunits attached to a collagen-like tail (asymmetric forms) (see FIG. 10).[7] The asymmetric collagen-tailed forms, particularly A12, are highly concentrated at the NMJ.[8,9]

Total muscle AChE was assayed spectrophotometrically using acetylthiocholine iodide as the substrate.[10] In patients 2, 3, and 4, as in patient 1,[1] AChE activity was less than in 12 controls, but in patient 5 the activity fell at the lower limit of

TABLE 4. Kinetic Properties of the AChR[a]

	Patients (no. of fibers)		Controls (no. of subjects)
	3	5	
Single-channel conductance from conductance plot (pS)[b]	29.4 ± 3.2 (3)	32.3 ± 1.8 (11)	32.2 ± 0.7 (9)
Single-channel conductance from power spectrum (pS)	30.1 ± 1.6 (3)	33.6 ± 1.3 (11)	31.4 ± 1.0 (9)
Single-channel open-time from power spectrum (ms)	2.25 ± 0.04 (3)	2.71 ± 0.2 (11)	2.37 ± 0.09 (9)

[a] The results were derived from analysis of ACh-induced current fluctuations.[18] Values indicate Mean ± SE. To compensate for loss of noise variance above the cut-off frequency of the filter, single-channel conductance derived from the conductance plot was divided by $(2/\pi)\tan^{-1}(f_{lp}/f_c)$, where f_{lp} = low pass filter setting and f_c = half-power frequency of the power spectrum.[19]

[b] pS = picosiemens.

TABLE 5. Morphometric Analysis of End-plate Regions[a]

	Patient				
	2	3	4	5	13 Controls
Nerve terminal area (μm^2)	$1.93^b \pm 0.28$	$1.22^b \pm 0.12$	$1.14^b \pm 0.17$	$1.71^b \pm 0.17$	3.88 ± 0.39
Presynaptic length (μm)	$1.90^b \pm 0.20$	$1.65^b \pm 0.09$	$0.97^b \pm 0.13$	$1.87^b \pm 0.22$	6.11 ± 0.40
Vesicle density (no./μm^2)	46 ± 7.8	60 ± 4.7	$132^b \pm 11.6$	60 ± 5.7	50 ± 3.6
Postsynaptic membrane density ($\mu m/\mu m^2$)	$3.86^b \pm 0.16$	$3.91^b \pm 0.11$	$3.41^b \pm 0.17$	$4.39^b \pm 0.15$	5.83 ± 0.25
End-plate regions analyzed	35	65	28	35–56	47–64

[a] Results are mean ± SE for the number of end-plate regions analyzed. More than one region can occur at an end-plate.

[b] Significantly different from control value ($p < 0.001$).

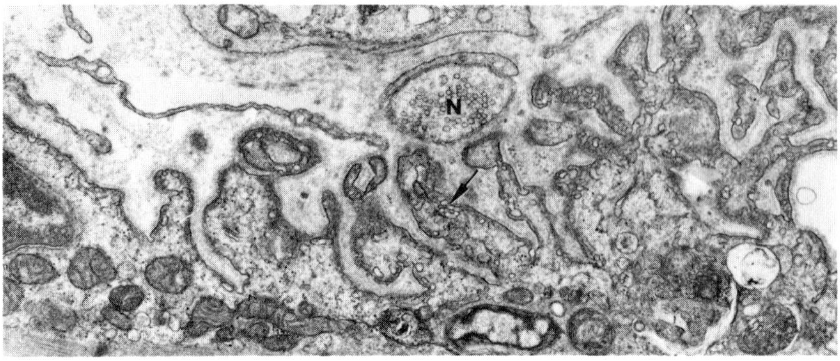

FIGURE 5. End-plate region in patient 2. Small nerve terminal (N) covers only a fraction of the postsynaptic region. Note the numerous endocytotic vesicles in the junctional folds (arrow); ×11,500. (A. G. Engel. With permission from *Seminars in Neurology* **10:** 12–26. 1990.)

normal. The K_m values of the muscle enzyme, determined in patients 2, 4, and 5, were normal (TABLE 6), and these values were also normal in patient 1.[1] The erythrocyte AChE activity and K_m values, determined in patients 1, 2, and 5, also fell in the normal range.

We used density-gradient centrifugation to study the molecular forms of AChE present in muscle from patients 3, 4, and 5. In patients 3 and 4, as in patient 1,[1] no asymmetric AChE was detected (see FIG. 11), but in patient 5 the pattern was similar to that found in controls.

Molecular Genetic Studies in Patients 3 and 4

The gene that encodes the catalytic subunit of human AChE has been cloned.[11] The exon organization of the human gene is identical to that of the mouse gene. The first 535 amino acids of the subunit are identical in the various forms of human AChE; differences at the carboxyl terminus arise from alternative mRNA processing of the exons at the 3′ end of the open reading frame. The tail portion of asymmetric AChE, which anchors the enzyme in the basal lamina of the synaptic space, is encoded by a different gene. This gene has been cloned for *Torpedo marmorata* but not for higher vertebrates.[21]

FIGURE 6. End-plate regions in patients 2 **(A)** and 4 **(B)**. The nerve terminals are small and cover only a fraction of the postsynaptic region; the presynaptic membranes are partially covered by Schwann cell processes, further reducing the area of membrane able to release ACh (arrows). In **A,** there is extensive degeneration of the tips of the junctional folds, and the widened synaptic space is filled with globular residues of the preexisting folds (asterisk). In **B,** the junctional folds are degenerating to the left of the nerve terminal (X). **A,** ×23,000; **B,** ×22,100.

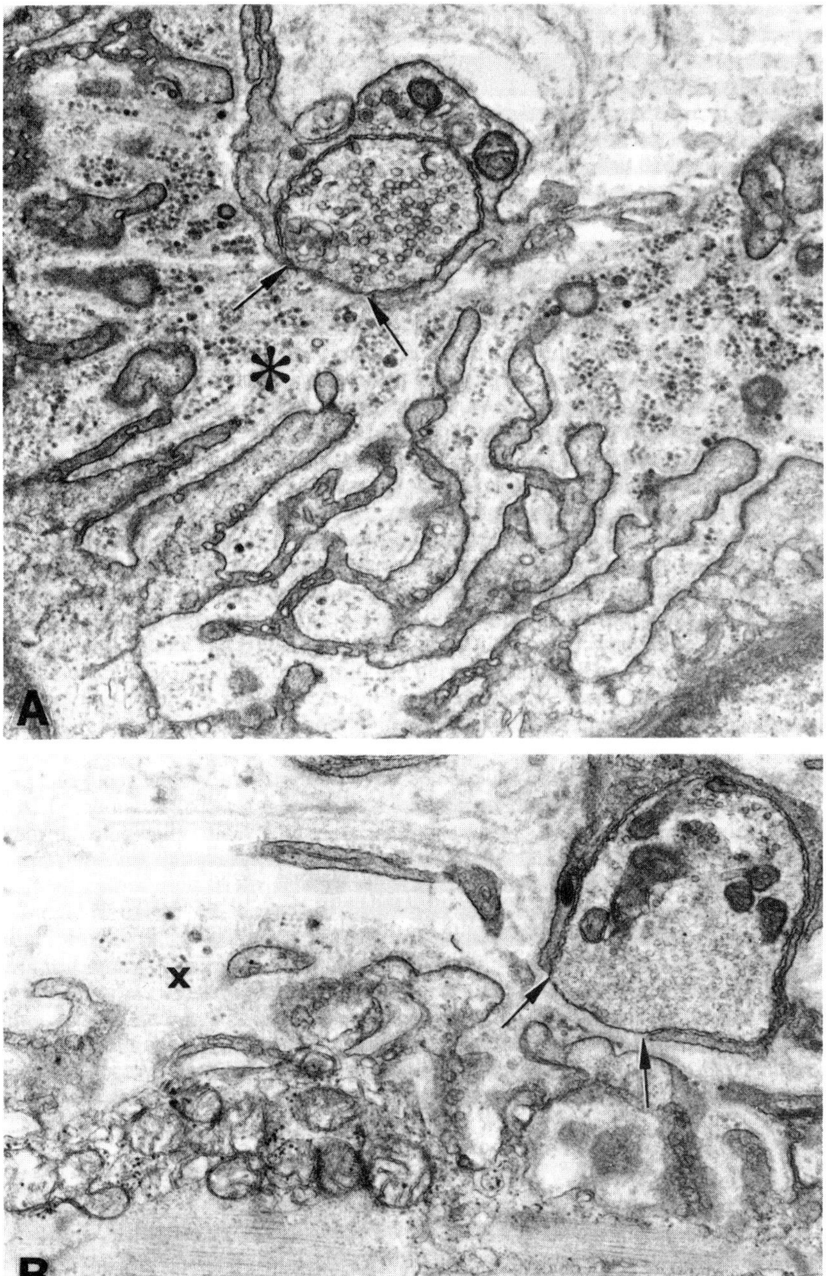
A
B
x

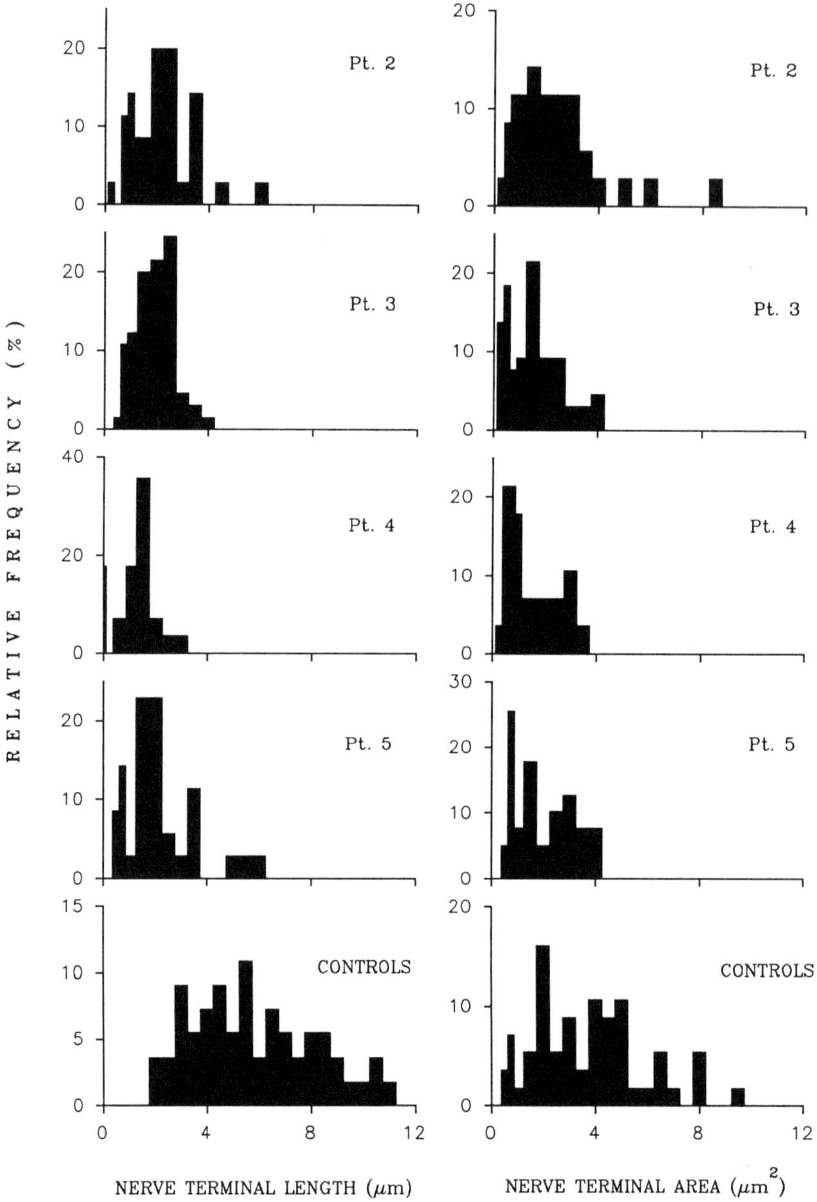

FIGURE 7. Frequency distribution of the presynaptic membrane length (left panels) and nerve terminal areas (right panels) in patients 2 to 5 and in 13 control subjects. The distributions are shifted to the left in the patients.

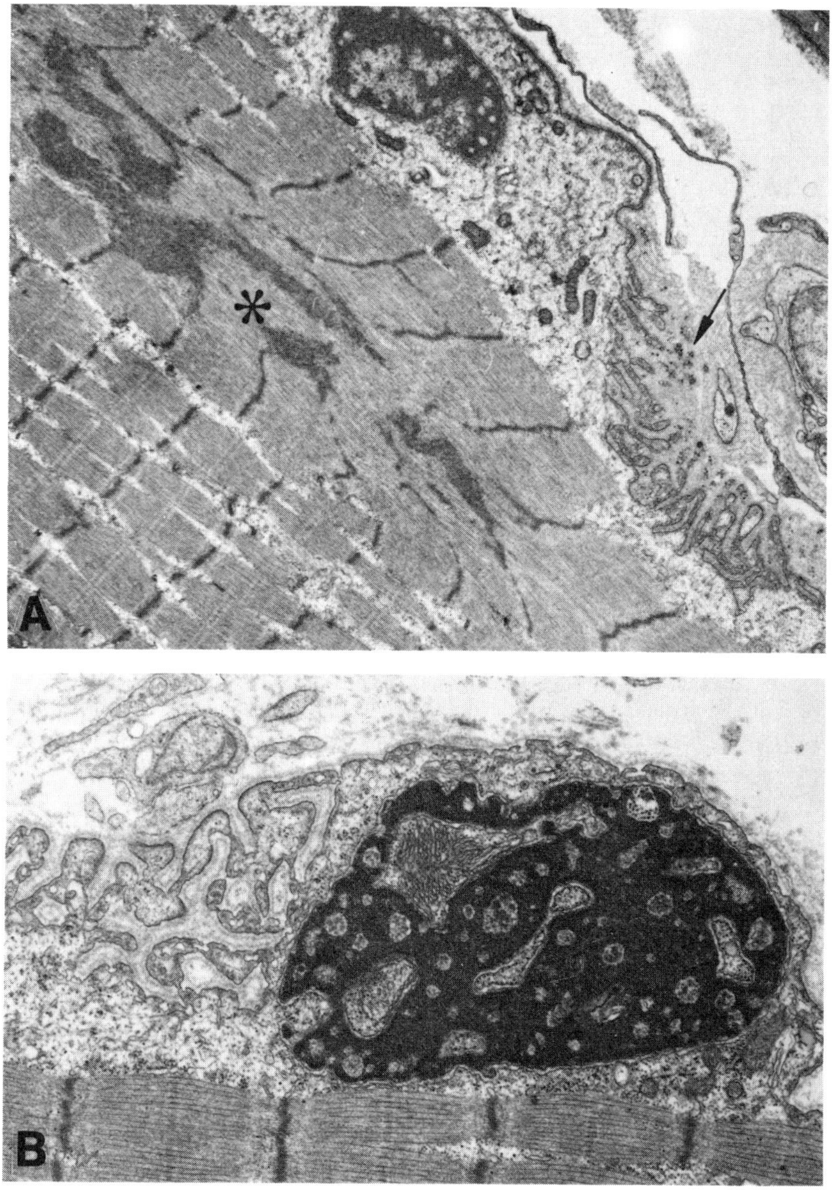

FIGURE 8. End-plate myopathy in patient 5. In **A,** the end-plate region at the right shows degeneration of junctional folds (arrow), and these folds are covered only by a small Schwann cell process; the underlying muscle fiber region is devoid of mitochondria and shows focal myofibrillar degeneration (asterisk). In **B,** the degenerating nucleus in the junctional sarcoplasm is filled by electron dense material and invaginated sarcoplasmic organelles. **A,** ×8,100; **B,** ×16,200.

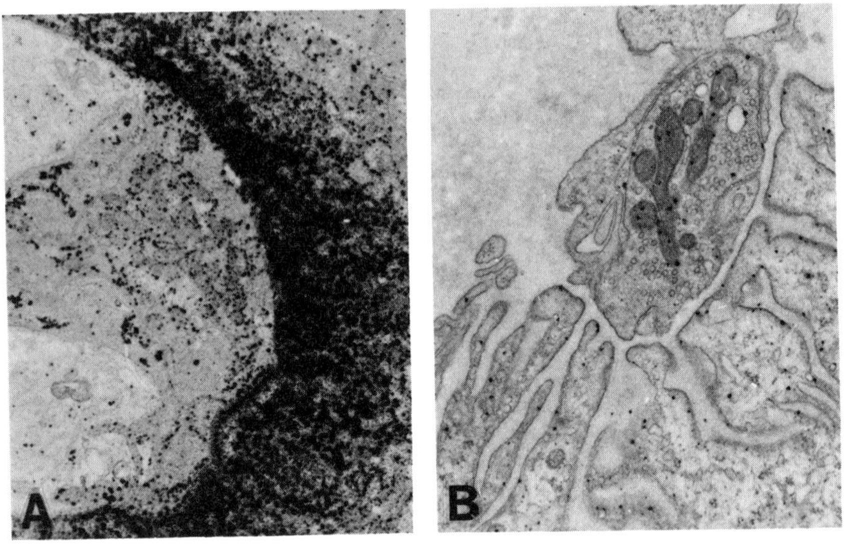

FIGURE 9. Electron cytochemical localization of AChE at a control end-plate **(A)** and at an end-plate from patient 2 with dithio-bis-acetic acid as substrate. The control end-plate was incubated for 30 minutes at room temperature. Heavy reaction product completely fills the synaptic space and extends into the adjacent regions. End-plate from patient 2 was incubated for 45 minutes at room temperature. There is no reaction product in the synaptic space. Sparse dark granules over the nerve terminal and junctional folds represent background staining. **A,** ×13,200; **B,** ×16,200.

In patients 2 and 3, restriction fragment analysis of genomic DNA encoding the catalytic subunits of AChE revealed no abnormalities to suggest a deletion in the gene. In these patients, the regions between exons 4 and 6 of the gene were amplified by the polymerase chain reaction. The amplified fragments were cloned and partially sequenced by dideoxy methods, and their sequence was compared with that of an amplified control sample and with the published sequence. No differences were found in the region encoding the cysteine, which is positioned four residues from the carboxyl terminus and is required for linking the catalytic to the tail subunit, or in the exon 4 donor site, or in the exon 6 acceptor site.

COMMENTS

Clinical Aspects

The clinical syndrome comprises (1) generalized weakness, which is present from a young age and increased by exertion; (2) slow pupillary light responses; (3) postural and then fixed spinal column deformity; (4) no response to AChE inhibitors; (5) an EMG decrement at 2 Hz stimulation; (6) a repetitive CMAP following single nerve stimuli (except in the youngest and most severely affected patient); and (7) a tendency for slow progression with time. The clinical features in patients 2–5 described in this paper were very similar to those observed in the first patient.[1] Ophthalmoparesis was not observed in patients 2 and 3 (the two sisters), however, and patient 5 had selectively severe involvement of cervical and dorsal forearm muscles but only slight weakness of the leg muscles.

The Safety Margin of Neuromuscular Transmission

The safety margin of neuromuscular transmission can be defined as the difference between the EPP amplitude and the amplitude necessary to initiate a muscle fiber action potential. In this syndrome, the safety margin is likely to be impaired by more than one mechanism: (1) In patients 2, 3, and 4 (as well as in patient 1), the quantal content of the EPP (*m*) was reduced. This can be explained by smallness of the nerve terminals, plus Schwann cell encasement of nerve terminals, which further reduced the presynaptic membrane

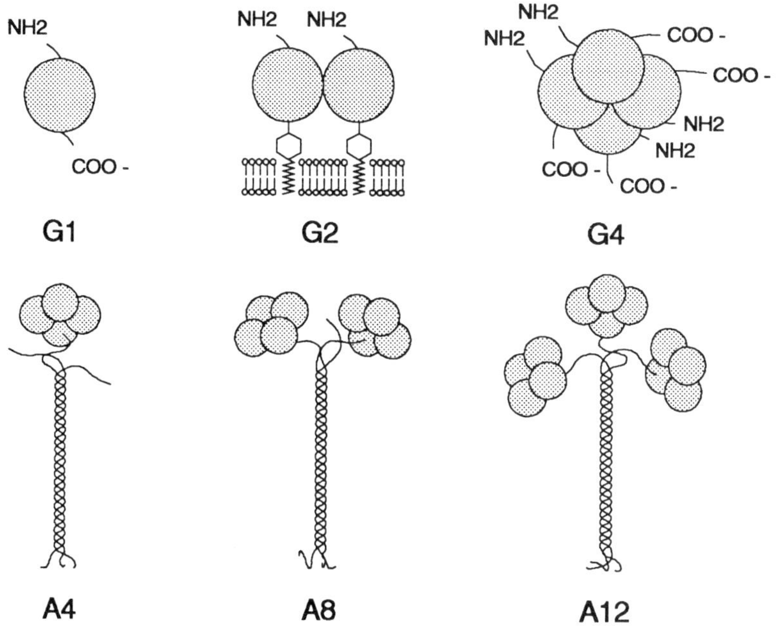

FIGURE 10. Molecular forms of AChE. Upper row shows globular forms; these can exist as monomers or oligomers either in a free hydrophilic form (as shown for G_1 and G_4) or in a glycophospholipid-linked amphiphilic form (as shown for G_2). Lower row shows asymmetric forms in which one, two, or three tetramers of catalytic subunits are disulfide-bonded to one end of a collagen-like tail (A_4, A_8, and A_{12} in diagram). The opposite end of the tail is attached to the basal lamina of the end-plate. Tetrameric catalytic subunits have also been found to be disulfide-bonded to a lipid-linked 20 kDa subunit. This species associates with the plasma membrane.

area available for ACh release. (2) In patients 2 and 3, end-plate AChR was deficient, probably secondary to degeneration of junctional folds. (3) End-plate geometry was disturbed in all patients, altering the usual close relationship of the nerve terminal to the junctional folds. This could allow dilution of ACh in the synaptic space, or diffusion of ACh from there, before it could reach the postsynaptic AChR. (4) In all patients, the prolonged exposure of AChR to ACh on sustained activity could induce desensitization of the receptors.[12] (5) Finally, the prolonged action of ACh at the end-plate might also produce sustained end-plate depolarizations at fast rates of stimulation. This has been described in experimental organophosphate poisoning[13] and could contribute to the failure of neuromuscular transmission by inactivating voltage-sensitive sodium channels in the vicinity of the end-plate.

Possible Causes of the End-plate AChE Deficiency

In patients 1, 3, and 4, asymmetric AChE was totally absent from muscle, but the globular form of the enzyme was preserved in muscle and erythrocytes. (Density gradient centrifugation was not performed in patient 2, but it is likely that she had the same defect as her sister.) These findings, plus the molecular genetic studies described above suggests that the defect in these patients resides in (1) the gene encoding the collagen tail of AChE, or (2) a protein necessary for the assembly of the catalytic and tail subunits. Less likely, the mutation still resides in the catalytic subunit, perhaps at a locus influencing splicing patterns or conformation of the gene product.

TABLE 6. Biochemical Data[a]

		Total AChE (nmol/min/mg protein)	K_m ($\times$ 10^{-4})
Patients	2	9.4	0.8
	3	2.1	
	4	11.8	1.7
	5	17.5	1.3
12 controls	Mean	24.3	1.0
	Range	17.5–32.6	0.6–1.4

[a] AChE was extracted from muscle by homogenization in a high-ionic strength buffer.[8] Total AChE was measured spectrophotometrically[10] in the presence of 10^{-5} M iso-OMPA.

Patient 5 differed phenotypically from the other patients. He was the least severely affected, had normal quantal release of ACh, and was able to assemble asymmetric forms of AChE within muscle. This indicates that he had a different genetic defect from the other patients. Plausible explanations for the absence of AChE at the neuromuscular junction in this patient include (1) a defect involving the N-terminal end of the tail subunit, which prevents its binding to the basal lamina;[21] (2) a defect in a basal lamina binding site for the tail subunit; or (3) absence of a signal peptide necessary for transporting asymmetric AChE from the Golgi apparatus to the cell surface.

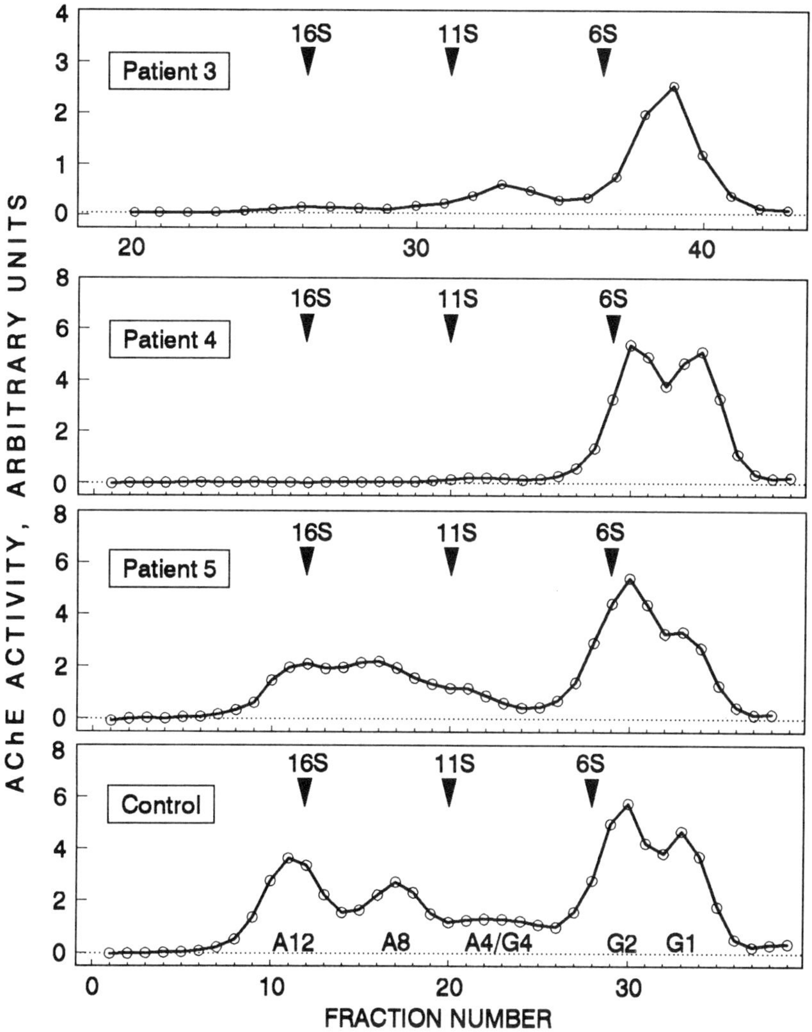

FIGURE 11. Sucrose density gradient ultracentrifugation of muscle extracts from patients 3, 4, and 5 and a control subject. Each gradient contained 1 mg of extract protein, marker enzymes with known sedimentation constants, and protease inhibitors. Arrows and sedimentation constants in each panel indicate positions of the marker enzymes. AChE was assayed in each fraction by Potter's method,[20] using 1.25 mM [^{14}C]acetylcholine as substrate in the presence of 10^{-5} M iso-OMPA. The species of AChE corresponding to each major peak are shown by a letter and number in the control panel: A and G denote asymmetric and globular species, and the numbers indicate the number of catalytic subunits present in the species. The asymmetric forms of AChE are completely absent in patients 3 and 4 but are detectable in patient 5.

REFERENCES

1. ENGEL, A. G., E. H. LAMBERT & M. R. GOMEZ. 1977. A new myasthenic syndrome with end-plate acetylcholinesterase deficiency, small nerve terminals, and reduced acetylcholine release. Ann. Neurol. **1:** 315–330.
2. BANCROFT, J. D. 1967. An introduction to histochemical technique. Butterworth and Company. London.
3. GAUTRON, J. 1974. Cytochimie ultrastructurale des acétylcholinestérases. Microscopie. **21:** 259–264.
4. ITO, Y., R. MILEDI, A. VINCENT & J. NEWSON-DAVIS. 1978. Acetylcholine receptors and end-plate electrophysiology in myasthenia gravis. Brain **101:** 345–368.
5. FAMBROUGH, D. M., A. G. ENGEL & T. L. ROSENBERRY. 1982. Acetylcholinesterase of human erythrocytes and neuromuscular junctions: Homologies revealed by monoclonal antibodies. Proc. Natl. Acad. Sci. USA **79:** 1078–1082.
6. ENGEL, A. G., M. TSUJIHATA, J. M. LINDSTROM & V. A. LENNON. 1976. The motor end plate in myasthenia gravis and in experimental autoimmune myasthenia gravis. A quantitative ultrastructural study. Ann. N.Y. Acad. Sci. **274:** 60–79.
7. TAYLOR, P. 1991. The cholinesterases. J. Cell Biol. **266:** 4025–4028.
8. HALL, Z. W. 1973. Multiple forms of acetylcholinesterase and their distribution in endplate and non-endplate regions of rat diaphragm muscle. J. Neurobiol. **4:** 343–361.
9. ABRAMSON, S. N., M. H. ELLISMAN, T. J. DEERINCK, Y. MAULET, M. K. GENTRY, B. P. DOCTOR & P. TAYLOR. 1989. Differences in structure and distribution of the molecular forms of acetylcholinesterase. J. Cell Biol. **108:** 2301–2311.
10. ELLMAN, G. L., K. D. COURTNEY, V. ANDRES & R. M. FEATHERSTONE. 1961. A new and rapid colorimetric determination of acetylcholinesterase activity. Biochem. Pharmacol. **7:** 88–95.
11. LI, Y., S. CHAMP, T. L. RACHINSKY, D. GETMAN & P. TAYLOR. 1991. Gene structure of mammalian acetylcholinesterase. J. Biol Chem. **266:** 23083–23090.
12. MAGLEBY, K. L. & B. S. PALLOTTA. 1981. A study of desensitization of acetylcholine receptors using nerve-released transmitter in the frog. J. Physiol. (Lond.) **316:** 225–250.
13. MASELLI, R. A. & B. C. SOLIVEN. 1991. Analysis of the organophosphate-induced electromyographic response to repetitive nerve stimulation: paradoxical response to edrophonium and *d*-tubocurarine. Muscle & Nerve. **14:** 1182–1188.
14. KATZ, B. & S. THESLEFF. 1957. On the factors which determine the amplitude of the 'miniature end-plate potential'. J. Physiol. (Lond.) **137:** 267–278.
15. CULL-CANDY, S. G., R. MILEDI & A. TRAUTMANN. 1979. End-plate currents and acetylcholine noise at normal and myasthenic human end-plates. J. Physiol. (Lond.) **287:** 247–265.
16. MARTIN, A. R. 1955. A further study of the statistical composition of the end-plate potential. J. Physiol. (Lond.) **130:** 114–122.
17. ELMQVIST, D. & D. M. J. QUASTEL. 1965. A quantitative study of end-plate potentials in isolated human muscle. J. Physiol. (Lond.) **178:** 505–529.
18. ANDERSON, C. R. & C. F. STEVENS. 1973. Voltage clamp analysis of acetylcholine produced end-plate current fluctuations at frog neuromuscular junction. J. Physiol. (Lond.) **235:** 655–691.
19. COLQUHOUN, D., W. A. LARGE & H. P. RANG. 1977. An analysis of the action of a false transmitter at the neuromuscular junction. J. Physiol. (Lond.) **266:** 361–395.
20. POTTER, L. T. 1967. A radiometric microassay of acetylcholinesterase. J. Pharmacol. Exp. Ther. **156:** 500–506.
21. KREJCI, E., F. COUSSEN, N. DUVAL, J. M. CHATEL, C. LEGAY & M. PUYPE *et al.* 1991. Primary structure of a collagenic tail peptide of *Torpedo* acetylcholinesterase: coexpression with catalytic subunit induces the production of collagen-tailed forms in transfected cells. EMBO J. **10:** 1285–1293.

Congenital Myasthenic Syndrome Attributed to an Abnormal Interaction of Acetylcholine with Its Receptor[a]

O. UCHITEL,[b] A. G. ENGEL,[c,h] T. J. WALLS,[d] A. NAGEL,[e]
V. BRIL,[f] AND V. F. TRASTEK[g]

[b]*Medical Faculty*
University of Buenos Aires
Buenos Aires, Argentina

[c]*Department of Neurology and*
Neuromuscular Research Laboratory
[g]*Department of Surgery*
Mayo Clinic
Rochester, Minnesota 55905

[d]*Department of Neurology*
Newcastle General Hospital
Newcastle Upon Tyne, England

[e]*Department of Neurology*
University of Würzburg
Würzburg, Germany

[f]*Department of Neurology*
University of Toronto
Toronto, Ontario, Canada

INTRODUCTION

We here describe a myasthenic syndrome presenting at birth and responding partially to anticholinesterase therapy. There were no immune deposits at the end-plates, and tests for acetylcholine receptor (AChR) antibodies were negative. The quantal content of the end-plate potential was normal. Miniature end-plate potentials and currents were very small, but the number of AChR per end-plate was normal. Electron microscopy revealed no pre- or postsynaptic abnormality; synaptic vesicles were of normal size, junctional folds were intact, and the density and distribution of AChR on the folds, visualized with peroxidase-labeled α-bungarotoxin, were normal. AChR channel properties were studied by analysis of acetylcholine (ACh)-induced current noise. The mean single-channel conductance was normal. The noise power spectrum was abnormal, containing two components of different time course. This could result from two populations of AChR or from abnormal ACh-AChR interaction. The former is less likely because both types of

[a] This work was supported by Grant NS 6277 from the National Institues of Health and by Research Grants from the Muscular Dystrophy Association. T. J. Walls was the recipient of a Fulbright Fellowship from the U.S. Government.

[h] Send reprint requests to A. G. Engel, Mayo Clinic, Rochester, Minnesota 55905.

AChR would have to have low conductance to explain the small miniature end-plate current, but the effective conductance of the channels that did open was normal. Ultrastructural and *in vitro* electrophysiologic studies, including noise analysis, were essential to characterize this disorder.

CLINICAL DATA

The patient was investigated at the Mayo Clinic at age 21, in 1989. She had been weak since birth with poor suck, stridor, ptosis, and feeble cry in the neonatal period. In the first year of life, she had intermittent choking spells and poor head control. She sat up at 7 months and walked at 10 months but had to hold onto the furniture for support. EMG studies and an edrophonium test at age 13 months suggested myasthenia gravis.

Between the ages of 2 and 7 the patient continued to fatigue easily but developed no new symptoms. Between the ages of 7 and 19 she developed ptosis, ocular palsies, and had difficulty in chewing and swallowing. Her symptoms became worse during the latter part of the day, before menses, and on exposure to heat. The patient's parents, two younger sisters, and other family members were unaffected.

Treatment with neostigmine bromide from the second year of life gave incomplete relief of symptoms. Plasmapheresis and ACTH therapy were ineffective. Prednisone treatment may have been of some benefit but had to be discontinued at age 9 because she became cushingoid and hyperglycemic. Ambenonium chloride, 60 to 70 mg daily since age 12, and prednisone, 25 mg on alternate days since age 20, were of limited benefit. Repeated tests for anti-AChR antibodies, including blocking and modulating antibodies, were negative.

Examination at age 21 showed moderately severe to severe weakness involving all voluntary muscles. The patient could not abduct her arms against gravity and could do only three deep-knee bends with assistance. The tendon reflexes were normal. EMG studies now showed decrements of 7, 24, and 40 percent on 2 Hz stimulation of the ulnar, facial, and musculocutaneous nerves, with posttetanic potentiation followed by postactivation exhaustion.

SPECIAL STUDIES

An intercostal muscle specimen was obtained at age 21 for morphological and *in vitro* electrophysiological studies.

Light Microscopy

Routine histochemical studies demonstrated no abnormality. The muscle fibers were of normal diameter, and there was a normal distribution of the histochemical fiber types. Acetylcholinesterase (AChE) activity was preserved at all end-plates. The mean length of the patient's end-plates, evaluated from the cytochemical reaction product for AChE on teased single muscle fibers, was 20.5 μm. In five adult control subjects the corresponding values ranged from 20–29 μm.

Specific cytochemical probes were employed to localize IgG,[1] complement component C3,[1] and the complement membrane attack complex (MAC)[2] at motor end-plates in cryostat sections. No reaction for IgG, C3, or MAC was found at any of the end-plates.

Acetylcholine Receptors

The reaction for AChR at the patient's end-plates, detected with rhodamine-labeled α-bungarotoxin, was like that observed at control end-plates.

The mean number of ACh binding sites per end-plate, estimated with radioiodinated α-bungarotoxin,[3] was 17.49×10^6. The corresponding value in 11 adult controls was $13.23 \pm 0.85 \times 10^6$ (mean ± SE). Because the length of the patient's end-plates fell in the low to normal range, the findings suggested a normal or higher than normal density of AChR at the patient's end-plates.

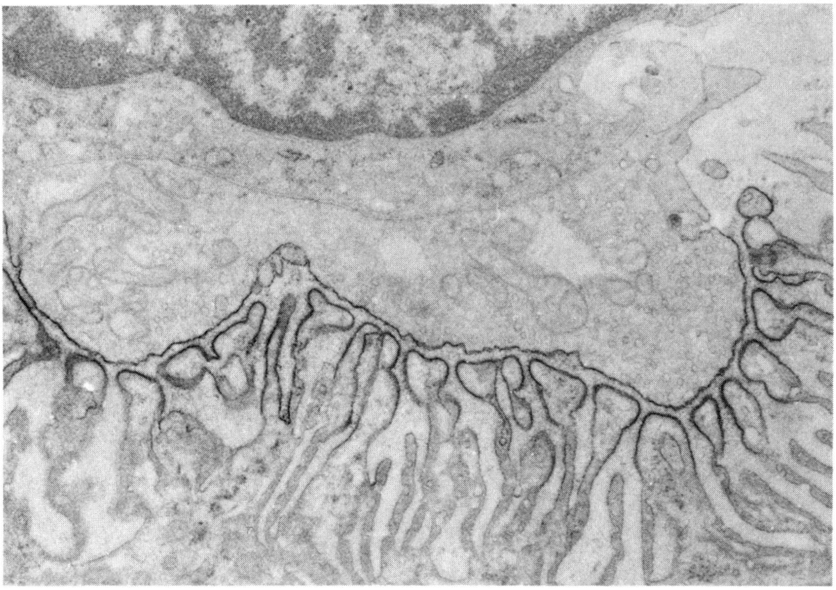

FIGURE 1. End-plate reacted for AChR with peroxidase-labeled α-bungarotoxin; unstained section. The density and distribution of AChR on the junctional folds is normal; ×15,700. (From A. G. Engel *et al.*, Progress in Brain Research **84:** 125–137, 1990, by permission.)

Electron microscopy of end-plates labeled with peroxidase-conjugated α-bungarotoxin[4] demonstrated a normal density and distribution of AChR on the junctional folds (FIG. 1). The mean AChR index (length of the postsynaptic membrane-binding peroxidase-conjugated α-bungarotoxin normalized for the length of the primary synaptic cleft[4]) for 40 end-plate regions was 3.21; in 5 control subjects the mean AChR indices ranged from 3.13 to 3.31.

Morphometric Reconstruction of the End-Plate

No pre- or postsynaptic abnormalities were detected on simple inspection (FIG. 2). Morphometric reconstruction[5,6] of the end-plate showed nerve terminals of normal size. The synaptic vesicles were normally abundant and of normal diameter. The mean values for the nerve terminal area, postsynaptic area, and postsynaptic membrane density were similar to the control means (TABLE 1).

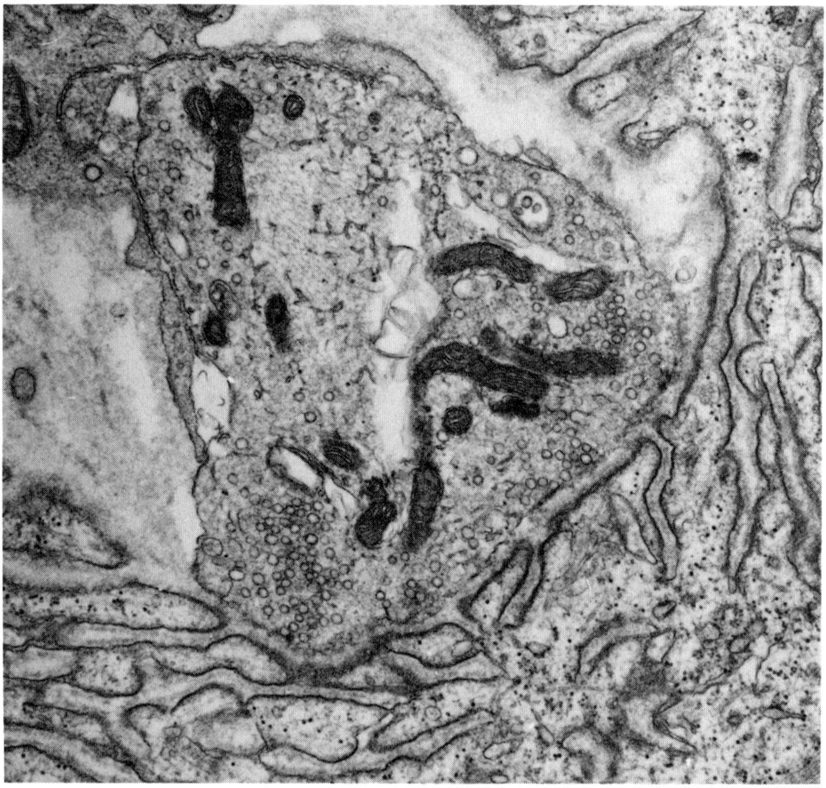

FIGURE 2. Electron micrograph of a representative end-plate region. The density and dimensions of the synaptic vesicles are normal, and the junctional folds are intact; ×21,300. (From A. G. Engel *et al.*, Progress in Brain Research **84:** 125–137, 1990, by permission.)

In Vitro *Microelectrode Studies*

Standard intracellular recording techniques were used to measure the amplitude, frequency, and time course of the miniature end-plate potential (MEPP).

TABLE 1. Morphometric Analysis of End-Plate Regions

Parameter	Patient[a]	Controls[b]
Nerve terminal area (μm^2)	3.28 ± 0.30	3.88 ± 0.39
Synaptic vesicle density (no./μm^2)	45.2 ± 2.19	50.3 ± 3.60
Synaptic vesicle diameter (nm)	61.6 ± 0.29	58.6 − 63.6[c]
Postsynaptic area (μm^2)	10.2 ± 0.60	10.6 ± 0.79
Postsynaptic membrane density ($\mu m/\mu m^2$)	5.55 ± 0.12	5.83 ± 0.25

[a] Mean ± SE; 73 regions observed at 37 end-plates were analyzed. N = 73 for all entries except for synaptic vesicle diameter, for which N = 500. Diameters of randomly selected synaptic vesicles were measured.

[b] Mean ± SE; 47–64 regions, 13 subjects.

[c] Range of 3 control means; 500 randomly selected synaptic vesicles were measured in each case.

The number of transmitter quanta released by nerve impulse (m) at 1 Hz stimulation was measured by the variance method[7] after the addition of *d*-tubocurarine to the perfusing fluid. Miniature end-plate currents (MEPC) were recorded using a two-electrode voltage clamp. The kinetic properties of the AChR channel were studied by analysis of the acetylcholine-induced current noise recorded by a two-electrode voltage clamp.[8,9] Single-channel conductance was estimated from regression of the variance of the ACh-induced conductance fluctuations on the mean increase in the ACh-induced end-plate conductance. The mean channel open time was derived from the noise power spectrum.[8,9]

The MEPPs were of low amplitude, and many were lost in the background noise. For this reason the MEPP amplitude was estimated by dividing the mean end-plate potential (EPP) amplitude by m, the quantal content of EPP. The mean MEPP amplitude for the patient was only 17% of the control mean. The addition of 1 μg/mL neostigmine methylsulfate to the bath increased the MEPP amplitude only slightly. Miniature end-plate currents were also abnormally small and could be recorded only in the presence of neostigmine methylsulfate. The quantal content of the EPP on 1 Hz stimulation was normal (TABLE 2).

The mean single-channel conductance fell in the control range. The noise power

TABLE 2. *In Vitro* Physiology—Miniature End-Plate Potentials and Currents and Quantal Release by Nerve Impulse

Parameter	Patient[a]	Controls[b]
MEPP amplitude (mV)[c]	0.17 ± 0.02 (17)[d]	0.99 ± 0.06 (15)
	0.28 ± 0.08 (4)[e]	1.69 ± 0.23 (3)[e]
MEPC amplitude (nA)[f]	<1.5 (5)[e]	3.92 ± 0.14 (10)
EPP quantal content (m) at 1 Hz[g]	34.4 ± 3.5 (11)	51.1 ± 4.5 (15)

[a] Mean ± SE (number of fibers).

[b] Mean ± SE (number of subjects).

[c] Corrected for a resting membrane potential of −80 mV; 28.5 ± 0.5°C.

[d] Estimated from m/(mean EPP amplitude)

[e] Measured in the presence of 1 μg/mL neostigmine methylsulfate.

[f] −80 mV; 22.5 ± 0.5°C.

[g] EPP amplitude corrected for a resting membrane potential of −80 mV and for nonlinear summation; 28.5 ± 0.5°C; m determined by the variance method.

spectrum was abnormal; it was best fitted by the sum of two Lorentzian functions, consistent with biexponential channel kinetics (FIG. 3 and TABLE 3).

DISCUSSION

The hallmarks of the present syndrome are a small amplitude of the MEPP and MEPC without AChR deficiency, abnormal AChR channel kinetics, normal end-plate structure, and a limited response to AChE inhibitors *in vivo* and *in vitro*.

The age of onset, the repeatedly negative tests for anti-AChR antibodies, the normal abundance and distribution of end-plate AChR, and the lack of immune deposits (IgG, C3, and MAC) and of any degenerative changes at the end-plates clearly distinguish this disorder from acquired autoimmune myasthenia gravis.

Possible Causes of Low MEPP and MEPC Amplitude without AChR Deficiency

The small MEPP and MEPC in the presence of normal amounts of end-plate AChR and normal end-plate geometry could be explained by (1) a decrease in the ACh content of the synaptic vesicles, (2) AChE overactivity from either an increased density of AChE in the synaptic basal lamina or an increased affinity of AChE for ACh, or (3) an abnormality in AChR. The fact that the size of the synaptic vesicles was normal points away from a reduced synaptic vesicle ACh content.[10,11] Overactivity of AChE is unlikely because in that case cholinesterase inhibitors should have removed the patient's symptoms and neostigmine should have restored the MEPP amplitude to normal. In fact, however, the patient responded poorly to anticholinesterase therapy, and there was only a slight increase of the MEPP amplitude in the presence of neostigmine (TABLE 2). On the other hand, the abnormal AChR channel kinetics point to an abnormality in AChR.

Possible Causes of Biexponential AChR Channel Kinetics

Biexponential AChR channel kinetics could result from (1) a dual population of AChR, as at the immature end-plate;[12] (2) blocking of the AChR channel;[13,14] (3) a decrease in the rate constant for agonist dissociation (k_{-2}) from diliganded

FIGURE 3. Power density spectrum of ACh-induced current noise at an end-plate in a control (upper panel) and in the patient (lower panel). The current signals were filtered to produce a band width of 2–600 Hz and digitized at 2000 Hz. The results shown are the average of three applications in the patient and four in the control. At the patient's end-plate, the mean induced current was 13.57 nA. This spectrum was best fitted by two Lorentzian curves with half-power frequencies of approximately 25.5 and 396 Hz; apparent channel open times 6.24 and 0.4 ms. At the control end-plate, the induced current was 9.87 nA. The spectrum was fitted with a single Lorentzian curve: half-power frequency 59 Hz; channel open time 2.7 ms. (From A. G. Engel *et al.*, Progress in Brain Research **84:** 125–137, 1990, by permission.)

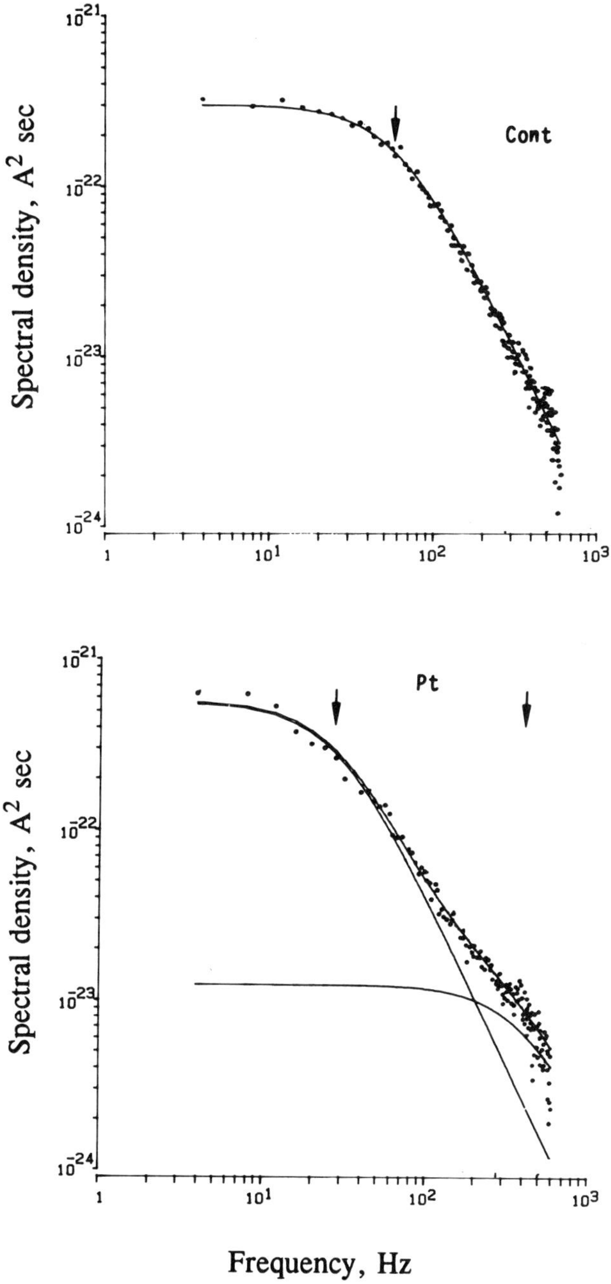
Spectral density, A² sec
Cont
Pt
Frequency, Hz
10⁻²¹
10⁻²²
10⁻²³
10⁻²⁴
1
10¹
10²
10³

closed AChR;[15] or (4) another abnormality of AChR, so that closure of the diliganded channel is no longer rate-limiting.[16]

A dual population of AChR channels with different rate constants of closure is an unlikely explanation, because both types of receptor would have to have a low conductance to explain the small MEPC. The effective conductance of the population of channels that did open, however, was normal (TABLE 3).

Blocking of the AChR channel by ACh would have to occur before channel opening, to also account for the small MEPC. If this were the case, however, then it would be difficult to explain even the small increase in the MEPP amplitude induced by neostigmine *in vitro*.

From the available data, we cannot prove or disprove the possibility that the biexponential channel kinetics were due to decrease in k_{-2} (the rate constant for agonist dissociation), or to another abnormality of AChR, for example, a reduced affinity of ACh for AChR, or to impaired transduction of the agonist-induced signal for isomerization of AChR. Further characterization of the AChR abnormality in this syndrome will require single channel recordings.

TABLE 3. *In Vitro* Physiology—Noise Analysis of the Kinetic Properties of the Acetylcholine Receptor Ion Channel[a]

Parameter	Patient[b]	Controls[c]
Channel open time (ms)[d]	4.97 ± 0.49 (6) and 0.44 ± 0.02 (6)	2.37 ± 0.09 (9)
Single-channel conductance (pS)[e]	33.69 ± 1.92 (7)	32.23 ± 0.75 (9)

[a] ACh-induced current noise was recorded at 22.5 ± 0.5° at a holding potential of −80 mV.
[b] Mean ± SE (number of fibers).
[c] Mean ± SE (number of subjects).
[d] From noise power spectrum.
[e] From regression of the variance of the ACh-induced conductance fluctuation on the ACh-induced conductance.

Possible Other Instances of this Disorder

Other less completely investigated patients may have had the same disorder. In a patient described by Vincent *et al.* (case 1), the MEPP amplitude was low, but the number of α-bungarotoxin binding sites per end-plate was normal.[17] This could have been due to a decreased conductance of the AChR channel, decreased number of ACh per quantum, or an abnormal ACh-AChR interaction. Specific studies to distinguish between these possibilities were not carried out. In a patient investigated by Lecky *et al.* (case 1), a reduced MEPP amplitude was associated with a normal number of α-bungarotoxin binding sites per end-plate, normal end-plate structure, and normal *d*-tubocurarine affinity. Limited observations on a single end-plate suggested that the effectiveness of ACh to open ion channels was reduced.[18]

REFERENCES

1. ENGEL, A. G., E. H. LAMBERT & F. M. HOWARD. 1977. Immune complexes (IgG and C3) at the motor end-plate in myasthenia gravis. Ultrastructural and light microscopic localization and electrophysiologic correlations. Mayo Clin. Proc. **52:** 267–280.
2. ENGEL, A. G. & K. ARAHATA. 1987. The membrane attack complex of complement at the endplate in myasthenia gravis. Ann. N.Y. Acad. Sci. **505:** 326–332.
3. ITO, Y., R. MILEDI, A. VINCENT & J. NEWSOM-DAVIS. 1978. Acetylcholine receptors and end-plate electrophysiology in myasthenia gravis. Brain **101:** 345–368.
4. ENGEL, A. G., J. LINDSTROM, E. H. LAMBERT & V. A. LENNON. 1977. Ultrastructural localization of the acetylcholine receptor in myasthenia gravis and in its experimental autoimmune model. Neurology **27:** 307–315.
5. ENGEL, A. G. & T. SANTA. 1971. Histometric analysis of the ultrastructure of the neuromuscular junction in myasthenia gravis and in the myasthenic syndrome. Ann. N.Y. Acad. Sci. **183:** 46–63.
6. ENGEL, A. G., M. TSUJIHATA, J. M. LINDSTROM & V. A. LENNON. 1976. The motor end plate in myasthenia gravis and in experimental autoimmune myasthenia gravis. A quantitative ultrastructural study. Ann. N.Y. Acad. Sci. **274:** 60–79.
7. MARTIN, A. R. 1955. A further study of the statistical composition of the end-plate potential. J. Physiol. (Lond.) **130:** 114–122.
8. ANDERSON, C. R. & C. F. STEVENS. 1973. Voltage clamp analysis of acetylcholine produced end-plate current fluctuations at frog neuromuscular junction. J. Physiol. (Lond.) **235:** 655–691.
9. CULL-CANDY, S. G., R. MILEDI & O. D. UCHITEL. 1982. Properties of junctional and extrajunctional acetylcholine receptor channels in organ cultured human muscle fibers. J. Physiol. (Lond.) **333:** 251–267.
10. JONES, S. F. & S. KWANBUNBUMPEN. 1970. The effects of nerve stimulation and hemicholinium on synaptic vesicles at the mammalian neuromuscular junction. J. Physiol. (Lond) **207:** 31–50.
11. NAGEL, A. & A. G. ENGEL. 1989. Vesamicol decreases both quantal and synaptic vesicle size at the neuromuscular junction. Soc. Neurosci. (Abstracts) **15:** 814.
12. FISCHBACH, G. D. & S. M. SCHUETZE. 1980. A post-natal decrease in acetylcholine channel open time at rat end-plates. J. Physiol. (Lond.) **303:** 125–137.
13. RUFF, R. L. 1977. A quantitative analysis of local anaesthetic alteration of miniature end-plate currents and end-plate current fluctuations. J. Physiol. (Lond.) **264:** 89–124.
14. OGDEN, D. C. & D. COLQUHOUN. 1985. Ion channel block by acetylcholine, carbachol and suberidylcholine at the frog neuromuscular junction. Proc. R. Soc. Lond. [Biol.] **225:** 329–355.
15. COLQUHOUN, D. & A. G. HAWKES. 1977. Relaxation and fluctuations of membrane currents that flow through drug-operated ion channels. Proc. R. Soc. Lond. [Biol.] **199:** 231–262.
16. ADAMS, P. R. 1987. Transmitter action at endplate membrane. *In* The Vertebrate Neuromuscular Junction. M. M. Salpeter, Ed.: 317–359. Allan Liss. New York.
17. VINCENT, A., S. G. CULL-CANDY, J. NEWSOM-DAVIS, A. TRAUTMANN, P. C. MOLENAAR & R. L. POLAK. 1981. Congenital myasthenia: End-plate acetylcholine receptors and electrophysiology in five cases. Muscle & Nerve **4:** 306–318.
18. LECKY, B. R. F., J. A. MORGAN-HUGHES, N. M. F. MURRAY, D. N. LANDON & D. WRAY. 1986. Congenital myasthenia. Further evidence of disease heterogeneity. Muscle & Nerve **9:** 233–242.

Myasthenic Syndromes Attributed to Mutations Affecting the Epsilon Subunit of the Acetylcholine Receptor[a]

A. G. ENGEL,[b] D. O. HUTCHINSON,[b] S. NAKANO,[b]
L. MURPHY,[b] R. C. GRIGGS,[c] Y. GU,[d] Z. W. HALL,[d]
AND J. LINDSTROM[e]

[b]*Department of Neurology*
Mayo Clinic
Rochester, Minnesota 55905

[c]*Department of Neurology*
University of Rochester School of Medicine
Rochester, New York

[d]*Department of Physiology*
University of California School of Medicine
San Francisco, California

[e]*Institute of Neuroscience*
University of Pennsylvania
Philadelphia, Pennsylvania

INTRODUCTION

Two unrelated women had myasthenic symptoms since birth, negative tests for acetylcholine receptor (AChR) antibodies, and end-plate AChR deficiency. Analysis of acetylcholine (ACh)-induced current noise revealed AChR channels with a reduced conductance and a prolonged open time in both, and biexponential channel kinetics in patient 2. Patient 1, who was less severely affected, had a less severe kinetic abnormality of AChR and better preserved end-plate architecture than patient 2. The electrophysiologic findings suggested a persistent fetal γ-subunit or mutation of another subunit of AChR. Immunocytochemical analysis of end-plate AChR with subunit-specific antibodies gave no evidence for a persistent γ-subunit in either patient. Three antibodies in patient 1 and one antibody in patient 2 directed against epitopes in the long cytoplasmic loop of the ε-subunit failed to bind to AChR. The findings argue for two distinct mutations, each affecting the long cytoplasmic loop of ε, and provide clues for molecular genetic studies designed to identify the mutations.

CLINICAL DATA

Patient 1. This patient was a 48-year-old woman. She had mild myasthenic symptoms since age 18 months. At age 35, exposure to a muscle relaxant during surgery

[a] This work was supported by Grant NS 6277 from the National Institutes of Health and by Research Grants from the Muscular Dystrophy Association.

caused ventilatory failure for 8 hours. After recovering from this episode, a positive edrophonium test and a decremental electromyographic (EMG) response were noted. A myasthenic disorder was suspected, and she was treated with pyridostigmine, 120-180 mg daily. Initially this was beneficial, but after a few years it became ineffective and was discontinued. Repeated tests for anti-AChR were negative. Other family members were unaffected.

At age 48, the patient had moderate bilateral ptosis and ophthalmoparesis, and mild weakness of facial, neck flexor, and selected limb muscles. She could rise from squatting only 6 times. The tendon reflexes were intact.

Patient 2. This patient was a 16-year-old girl. She had myasthenic symptoms since the neonatal period. As a child she ran awkwardly and could not keep up with her peers in sports. During early school years she fatigued easily on exertion. Her symptoms increased at about age 12. At age 13 she became wheelchair-dependent outside her home. Treatment with guanidine, 500 mg daily, was started at age 14; this improved the patient's symptoms only temporarily. Repeated tests for anti-AChR antibodies were negative. Other family members were unaffected.

At age 16, the patient had moderately severe to severe diffuse weakness of all limb and truncal muscles, but only mild weakness of ocular, facial, and bulbar muscles. Arm elevation time was less than 30 seconds, and she could only stand for one minute without assistance. The tendon reflexes were hypoactive.

Electromyographic Studies

Patient 1 had normal nerve conduction velocities, but the peroneal sensory nerve action potential was absent. Single-nerve stimuli evoked single compound muscle action potentials (CMAP) of normal amplitude. Two hertz stimulation of the ulnar, peroneal, and facial nerves resulted in a 10%, 20%, and 48% decremental response of the evoked CMAP, respectively. The decrement was partially repaired by exercise or injection of 1 mg neostigmine methylsulfate. The motor unit potentials varied in amplitude from moment to moment.

Patient 2 had normal nerve conduction velocities and normal sensory nerve action potentials. The CMAP recorded from the hypothenar and anterior tibial muscles were 3.4 and 0.5 mV, respectively (normal >6 mV and >5.1 mV). There were no repetitive CMAP after single stimuli to motor nerves. Two hertz stimulation of the ulnar and peroneal nerves resulted in decrements of 60% and 66 percent. Immediately after brief exercise, the CMAP evoked by ulnar and peroneal nerve stimulation increased 1.6-fold and 3.4-fold, respectively, with repair of the decrement. Either 1 mg neostigmine methylsulfate sc or 15 mg 3,4-diaminopyridine, orally, increased the CMAP amplitude 2- to 3-fold and repaired the decrement partially. Voluntary activity produced small motor unit potentials that varied in amplitude from moment to moment.

SPECIAL STUDIES

Intercostal muscle specimens were obtained from each patient for morphological and *in vitro* electrophysiological studies.

Light Microscopy

Both patients had type I fiber preponderance. In addition, patient 1 had type II fiber atrophy, and patient 2 had an increased variation of muscle fiber diameter. Acetylcholinesterase (AChE) activity was preserved at all end-plates in both. The configuration of the end-plates, evaluated from the cytochemical reaction product for AChE on teased single muscle fibers, was abnormal in both, with 3 to 12 small

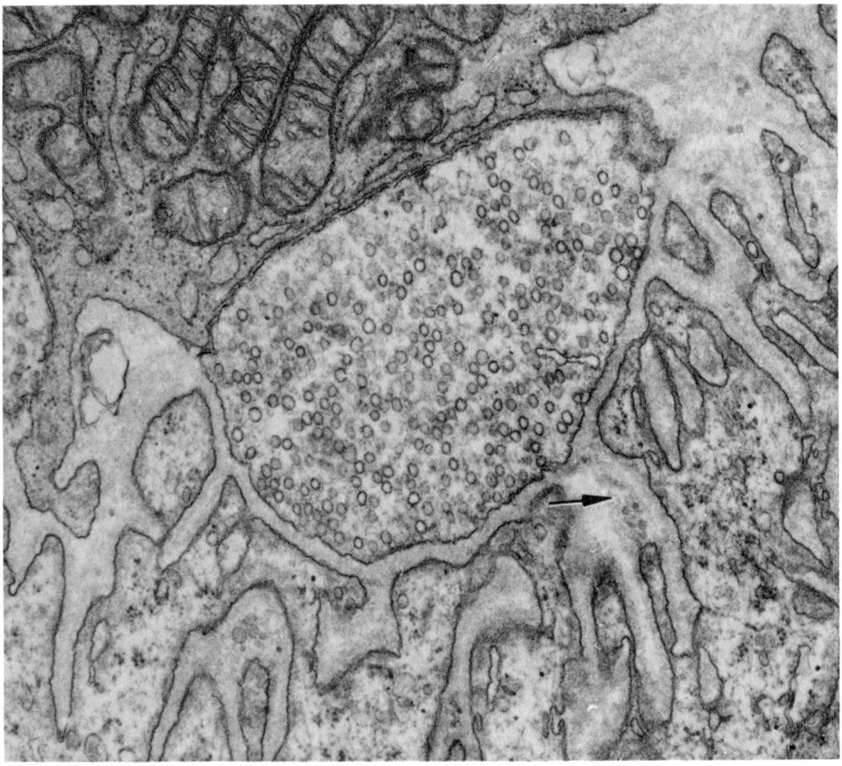

FIGURE 1. End-plate region in patient 1. The region appears normal except for degeneration of the tip of one junctional fold (arrow). Note numerous synaptic vesicles in the nerve terminal; ×28,000.

AChE-reactive regions distributed over an extended length of the muscle fiber surface membrane.

End-Plate Ultrastructure

In patient 1, most end-plates appeared normal on simple inspection (FIG. 1). A few end-plates had simplified postsynaptic regions or degenerating junctional folds. At some end-plates the junctional sarcoplasm contained dilated vesicles, myeloid structures, and degenerating mitochondria.

In patient 2, most end-plates appeared abnormal, with many degenerating junctional folds and many junctional folds studded with endocytotic vesicles. The synaptic spaces were frequently widened and filled with degenerate remnants of the junctional folds (FIG. 2). The junctional sarcoplasm often contained dilated vesicles, degenerating mitochondria, or necrotic nuclei. Muscle fiber regions near end-plates frequently showed loss of mitochondria and focal myofibrillar degeneration, and sometimes harbored small autophagic vacuoles.

On morphometric reconstruction of the end-plate,[1,2] the mean area of individual nerve terminals and the mean postsynaptic area were significantly reduced in both patients. The mean postsynaptic membrane density was significantly reduced in patient 2 only (TABLE 1).

α-Bungarotoxin Binding Studies

The mean number of ACh binding sites per end-plate, estimated using radioiodinated α-bungarotoxin,[3] was reduced to 1.79×10^6 in patient 1 and to 5.22×10^6 in patient 2. The corresponding values in 11 adult controls ranged from 9.32 to 18.72×10^6 (mean ± SE: $13.23 \pm 0.85 \times 10^6$).

AChR was demonstrated ultrastructurally with peroxidase-labeled α-bungarotoxin.[4] In patient 1, there was a diffuse decrease of AChR over the crests of all junctional folds (FIG. 3). In patient 2, there was total loss of AChR where the junctional folds had degenerated; where the junctional folds were intact, the density and distribution of AChR appeared normal (FIG. 4).

In Vitro *Microelectrode Studies*

Standard intracellular techniques were used to record miniature end-plate potentials (MEPP) and end-plate potentials (EPP). Amplitudes were corrected for nonlinearity of the response and to a resting membrane potential of −80 mV. The number of transmitter quanta released by nerve impulse (m) at 1 Hz stimulation was measured by the variance method in the presence of *d*-tubocurarine.[5]

Miniature end-plate currents (MEPC) were recorded with a two-electrode voltage clamp. The decay time constant of the MEPC was derived by computerized best fit of one or two exponential functions to the decay phase of the MEPC.

The kinetic properties of the AChR channel were evaluated by analysis of the decay phase of the MEPC and by analysis of the ACh-induced current noise.[6,7] The mean channel open time was derived from the noise power spectrum.[6,7] Single-channel conductance was estimated from regression of the variance of conductance fluctuations on the mean increase in end-plate conductance induced by ACh (conductance plot) and from the noise power spectrum.

The quantal content of the EPP at 1 Hz stimulation was slightly higher than normal in patient 1 and normal in patient 2 (TABLE 2).

The MEPP and MEPC amplitudes were significantly reduced in both patients (TABLE 2). In both, 1 μg/mL neostigmine methylsulfate increased the mean MEPP amplitude 1.4-fold.

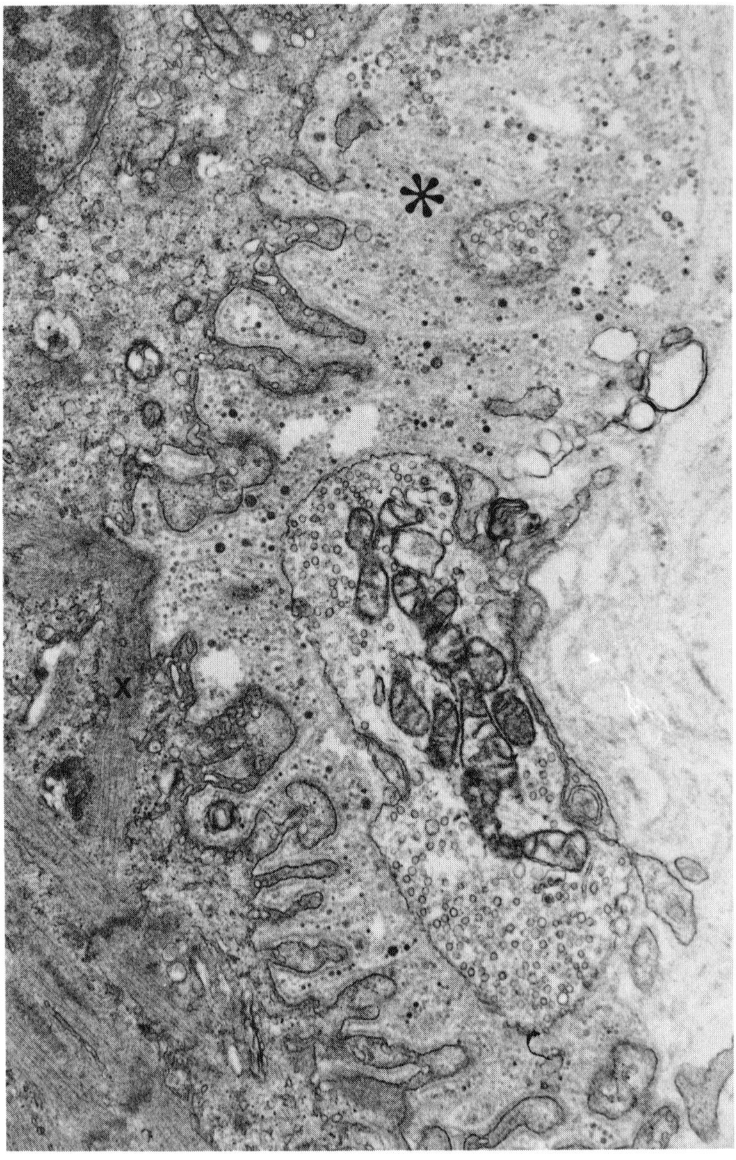

FIGURE 2. End-plate region in patient 2. All junctional folds are degenerating, and globular residues of the degenerating folds are accumulating in the widened synaptic space (asterisk). The junctional sarcoplasm contains many small vesicles and few degenerating mitochondria and is traversed by an aberrant myofibril (×). At end-plates like this, quantal efficiency would be compromised by (a) dilution, diffusion, and destruction by AChE of ACh molecules before they can reach AChR, (b) loss of AChR from the tips of the junctional folds, and (c) the reduced conductance of the AChR channel; ×17,400.

TABLE 1. Morphometric Analysis of End-Plate Regions

Parameter	Patient 1[a]	Patient 2[b]	Controls[c]
Nerve terminal area (μm^2)	2.16 ± 0.18[d]	2.53 ± 0.21[e]	3.88 ± 0.39
Synaptic vesicle density (no./μm^2)	50.4 ± 3.06	54.7 ± 4.33	50.3 ± 3.60
Postsynaptic area (μm^2)	5.09 ± 0.37[d]	8.26 ± 0.70[f]	10.60 ± 0.79
Postsynaptic membrane density ($\mu m/\mu m^2$)	5.43 ± 0.13	4.16 ± 0.14[d]	5.83 ± 0.25

[a] Mean ± SE; 76 regions observed at 28 end-plates were analyzed.
[b] Mean ± SE; 67 regions observed at 45 end-plates were analyzed.
[c] Mean ± SE; 47–64 regions, 13 subjects.
[d] Significantly different from control: $p < 0.001$.
[e] Significantly different from control: $p < 0.005$.
[f] Significantly different from control: $p < 0.05$.

Kinetic Properties of the AChR Channel

The decay phase of the MEPP, MEPC, and EPP were prolonged in both patients. In patient 1, the decay phases were adequately defined by single exponential functions, but in patient 2 the sums of two exponential functions were required for an adequate fit. In patient 1, detailed analysis of the MEPC decay time constant, which is closely related to the mean open time of the AChR channel,[6] showed it

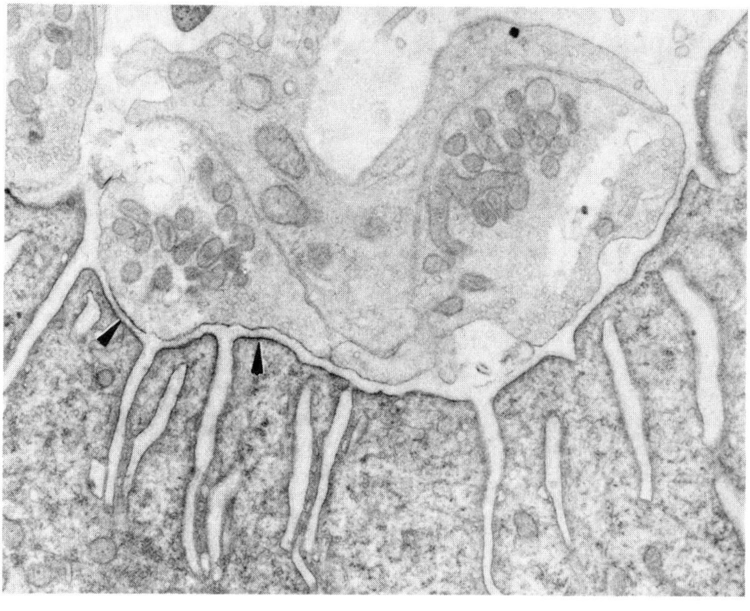

FIGURE 3. Localization of AChR with peroxidase-labeled α-bungarotoxin in patient 1. Note marked, diffuse decrease AChR over the junctional folds. Only short segments of the folds appear to react for AChR (arrowheads); ×15,500.

to be 1.3-fold longer than the corresponding control value. In patient 2, the two components of the MEPC decay curve yielded two time constants, one shorter and one much longer than normal (FIG. 5 and TABLE 3).

Analysis of the ACh-induced current noise confirmed that the prolonged decay phase of the MEPC was due to an increased open time of the AChR ion channel. In patient 1, the noise power spectrum could be fitted with a single Lorentzian function, and the channel open time was 1.8-fold longer than normal. In patient 2, the power spectrum was best fitted by two Lorentzian functions; these yielded two channel open times, one not significantly different from normal and one much longer than normal (TABLE 3).

The mean channel conductance was reduced to about 80% of normal in patient 1 and to about 60% of normal in patient 2 (TABLE 3).

Immunocytochemical Analysis of AChR

A kinetic abnormality of the AChR channel could be due to a mutation involving an AChR subunit. Alternatively, the prolonged open time and reduced conductance of the channel in patient 1 could stem from a persistent fetal-type γ-subunit in AChR, and the double-exponential channel kinetics in patient 2 could be caused by two populations of AChR, one of which could contain the fetal γ-subunit.[8–10]

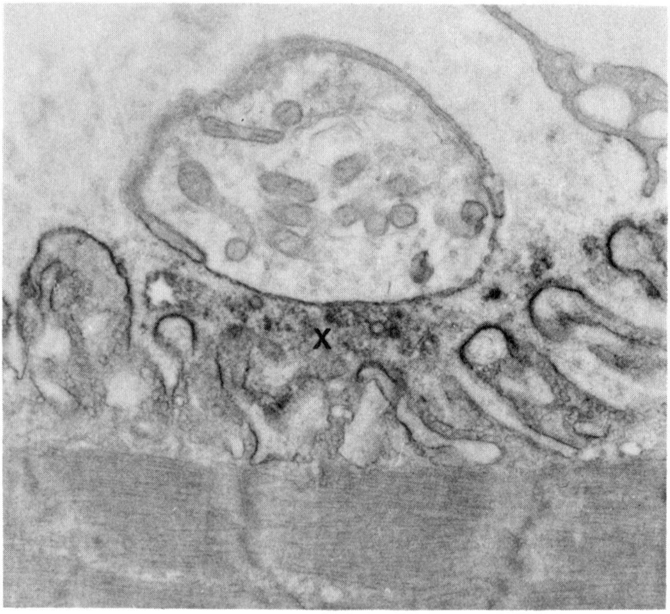

FIGURE 4. Localization of AChR with peroxidase-labeled α-bungarotoxin in patient 2. AChR is lost from the junctional folds where the folds are degenerating. Some degenerated globular residues of the folds still react for AChR (x). Where the folds are preserved, the density and distribution of AChR are normal; $\times$18,000.

TABLE 2. *In Vitro* Physiology—End-Plate Potentials and Currents

Parameter	Patient 1[a]	Patient 2[b]	Controls[c]
EPP quantal content at 1 Hz[d]	111 ± 9.0[h]	47 ± 6.1	51.1 ± 4.5 (range 23 – 91)[e]
MEPP ampltidue (mV)[d]	0.46 ± 0.03[g]	0.61 ± 0.06[g]	1.00 ± 0.03
MEPC amplitude (nA)[f]	1.85 ± 0.17[g]	2.12 ± 0.160[h]	3.92 ± 0.14

[a] Mean ± SE (8–12 fibers).
[b] Mean ± SE (8–12 fibers).
[c] Mean ± SE (78–190 fibers from 10–15 controls)
[d] Corrected to a resting membrane potential of −80 mV; 28.5 ± 0.5°C.
[e] Range of means in 15 controls.
[f] −80 mV; 22.5 ± 0.5°C.
[g] Significantly different from control: $p < 0.001$.
[h] Significantly different from control: $p < 0.01$.

We therefore probed the structure of AChR with subunit-specific antibodies. The antibodies recognized cytoplasmic or extracellular epitopes of the subunits (TABLE 4). A two-color immunofluorescence system was used in which the binding of a given anti-subunit antibody and of α-bungarotoxin were compared at the same end-plates.

The fetal γ-subunit was not detected by Polyab 485 at the patients' end-plates or in control end-plates of adult muscle but was detected in control end-plates in fetal muscle. All remaining antibodies bound to control end-plates in adult muscle and the intensity of the fluorescent signal obtained with each was proportionate to that obtained with α-bungarotoxin. In the two patients, the fluorescent signals obtained with antibodies recognizing epitopes of the α-, β- and δ-subunits, and the carboxy-terminal epitopes of the ε-subunit, were proportionate to that obtained with α-bungarotoxin. By contrast, the antiserum raised against ε360-369 failed to bind to the end-plates in either patient, and the two monoclonal antibodies recognizing epitopes within ε364-373 failed to bind to the end-plates in patient 1 (TABLE 4). The antibodies that failed to bind were directed against epitopes in the long cytoplasmic loop of ε, which is between the M3 and M4 membrane-spanning regions of the subunit.[10,13,20] Therefore, the findings imply ablation or alteration of epitopes in the long cytoplasmic loop of ε. The affected regions in the two patients are not identical, but both include ε364-369 (TABLE 4).

DISCUSSION

Phenotypic Differences between Patients 1 and 2

The two patients described here differed clinically, morphologically and, to a lesser extent, electrophysiologically. Patient 1 had a mild clinical course, a mild-to-modest EMG decrement, a mild end-plate myopathy, but severe and diffuse end-plate AChR deficiency. Patient 2 had severe disability by age 16, a striking EMG decrement, a severe end-plate myopathy, and focal loss of AChR from those junctional folds that had degenerated. In both patients, the end-plate potentials and currents were prolonged due to a prolonged open time of the AChR channel, and their amplitude was reduced due to deficiency of AChR and the reduced single-channel conductance. In patient 2, however, the kinetic abnormalities of the AChR channel were more marked than in patient 1, and analysis of the MEPC and of the noise power spectrum indicated biexponential channel kinetics.

Pathologic Mechanisms

In both patients, the efficiency of the released transmitter quanta was compromised by the end-plate AChR deficiency and the reduced conductance of the AChR ion channel. In patient 2, quantal efficiency was further impaired by the altered end-plate geometry, which included widening of the synaptic space and accumulation of degenerated remnants of the junctional folds in that space (FIG.

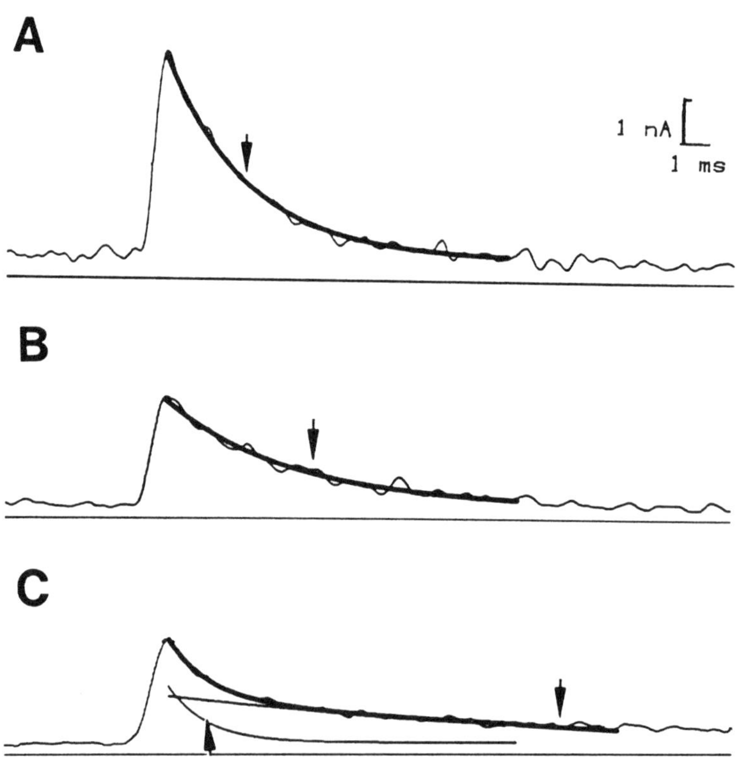

FIGURE 5. Miniature end-plate currents recorded from a control subject (A), patient 1 (B), and patient 2 (C). Heavy line superimposed on the decay phase of the MEPC represents the exponential function fitted by computer. Vertical arrows indicate decay time constants. The amplitude of the MEPC is reduced in both patients. In patient 1, the decay phase of the MEPC is monoexponential and longer than normal. In patient 2, the decay phase is best fitted by the sum of two exponential functions (shown by two separate lines); these yield two time constants of decay, one shorter and one much longer than normal.

TABLE 3. *In Vitro* Physiology—Kinetic Properties of the Acetylcholine Receptor Channel

Parameter[a]	Patient 1[b]	Patient 2[c]	Controls
MEPC decay time constant (ms)	4.11 ± 0.13[d]	1.50 ± 0.14[e] 16.5 ± 0.97[e]	3.23 ± 0.06[f]
Single-channel open time (ms) (from noise power spectrum)	4.09 ± 0.17[d]	1.9 ± 0.23[e] 12.4 ± 0.75[d,e]	2.30 ± 0.043[g]
Single-channel conductance (pS) (from noise power spectrum)	26.1 ± 1.69[d]		32.1 ± 0.72[g]
Single-channel conductance (pS) (from conductance plot)[h]	24.6 ± 1.30[d]	19.3 ± 2.4[d]	32.50 ± 0.81[g]

[a] 23°; −80 mV.
[b] Mean ± SE; 11 fibers.
[c] Mean ± SE; 14 fibers.
[d] Significantly different from control ($p < 0.005$).
[e] Biexponential channel kinetics.
[f] Mean ± SE; 78 fibers from 10 controls.
[g] Mean ± SE; 52 fibers from 9 controls.
[h] Corrected for fraction of noise variance lying beyond low-pass filter frequency.

2). The end-plate myopathy in this patient is attributed to an increased calcium influx into the junctional folds and sarcoplasm secondary to the markedly prolonged end-plate current.[21–24]

In patient 1, there was a diffuse decrease of AChR over junctional folds that were mostly intact. This suggests decreased synthesis or accelerated destruction of AChR. A mutation in an AChR subunit could account for this; for example, experimentally mutated α-subunits expressed in fibroblasts are rapidly degraded whether or not they associate with another subunit.[25] In patient 2, the end-plate AChR deficiency was less marked than in patient 1, and the density and distribution of AChR on those folds that were not degenerating appeared normal. This suggests that in patient 2 the end-plate AChR deficiency was predominantly due to the focal destruction of the junctional folds.

Implications of the Immunocytochemical Analysis

In both patients, immunocytochemical studies of AChR subunits revealed an ablation or alteration of epitopes in the long cytoplasmic loop of the ε-subunit. The defects in the two patients, however, are likely to be different because Polyab 360 did not bind to AChR in either patient, whereas MAb 168 and 154 did not bind in patient 1 but did bind in patient 2 (TABLE 4).

The exact sequence of amino acids comprised by the ablated epitopes remains unknown for two reasons: (1) the sequence of the human ε-subunit is still unpublished; and (2) polyab 360 was raised against calf ε360-369, and MAb 168 and 154 were raised against the *Torpedo* γ-subunit (the equivalent of human ε). The ε360-369 sequence in the calf, however, is identical with that in the rat[26] and the mouse[27] and is likely to be highly homologous with the corresponding ε sequence in humans.

We postulate that the immunocytochemical observations in the two patients are accounted for by two different mutations, each involving the long cytoplasmic loop of ε and each affecting ε364-369. Alternatively, the mutations may not directly involve the long cytoplasmic loop of ε but deform its tertiary structure so that

TABLE 4. Immunocytochemical Analysis of AChR Subunits

Antibody	Antigen	Specificity	Patient 1	Patient 2	Adult Controls
MAb[a] 35[11]	Eel AChR	includes α 68-71, MIR[c,f,12,13]	+[d]	+	+
MAb 198[14]	Human AChR	includes α 68-71, MIR[f,12]	+	+	+
MAb 210	Calf AChR	includes α 68-71, MIR[f,12,15]	+	+	+
MAb 61[11]	Eel AChR	includes α 375-385[g,13,16]	+	+	+
MAb 336	Human recombinant β-subunit	within β 326-425[g]	+	+	+
MAb 148[17]	Torpedo AChR	within β 360-410[g,13,18]	+	+	+
MAb 7[19]	Torpedo AChR	δ 377-388[g,13,20]	+	+	+
PAb[b] 486[10]	Mouse δ-subunit peptide	δ 486-496[f,10]	+	+	+
PAb 485[10]	Mouse γ-subunit peptide	γ 485-497[f,h,10]	−[e]	−	−
PAb 360[10]	Calf ε-subunit peptide	ε 360-369[g,10]	−	−	+
MAb 168[19]	Torpedo γ-subunit	within ε 364-373[g,13,20]	−	+	+
MAb 154[19]	Torpedo γ-subunit	within ε 364-373[g,13,20]	−	+	+
PAb 459[10]	Calf ε-subunit peptide	ε 459-471[f,10]	+	+	+

[a] Monoclonal antibody.
[b] Polyclonal antibody.
[c] MIR = main immunogenic region.
[d] + = immunoreactivity present.
[e] − = immunoreactivity absent.
[f] Extracellular AChR epitope.
[g] Cytoplasmic AChR epitope.
[h] Recognizes AChR in human fetal muscle.

segments of it lose their immunoreactivity. Precise identification of the mutations will require knowledge of the sequence of the human ε-subunit and molecular genetic analysis.

If the mutations alter the long cytoplasmic loop of ε directly, then it also follows that (1) different mutations involving nearby and overlapping regions of the deep cytoplasmic loop of ε can produce divergent clinical phenotypes, and (2) a mutation in a nonmembrane spanning region of a subunit can affect the conductance and open time of the AChR channel. That the functional properties of the AChR channel can be allosterically affected is consistent with studies involving mutagenesis of AChR subunits.[28]

REFERENCES

1. ENGEL, A. G. & T. SANTA. 1971. Histometric analysis of the ultrastructure of the neuromuscular junction in myasthenia gravis and in the myasthenic syndrome. Ann. N.Y. Acad. Sci. **183:** 46–63.
2. ENGEL, A. G., M. TSUJIHATA, J. M. LINDSTROM & V. A. LENNON. 1976. The motor end plate in myasthenia gravis and in experimental autoimmune myasthenia gravis. A quantitative ultrastructural study. Ann. N.Y. Acad. Sci. **274:** 60–79.
3. ITO, Y., R. MILEDI, A. VINCENT & J. NEWSOM-DAVIS. 1978. Acetylcholine receptors and end-plate electrophysiology in myasthenia gravis. Brain **101:** 345–368.
4. ENGEL, A. G., J. LINDSTROM, E. H. LAMBERT & V. A. LENNON. 1977. Ultrastructural localization of the acetylcholine receptor in myasthenia gravis and in its experimental autoimmune model. Neurology **27:** 307–315.
5. MARTIN, A. R. 1955. A further study of the statistical composition of the end-plate potential. J. Physiol. (Lond.) **130:** 114–122.
6. ANDERSON, C. R. & C. F. STEVENS. 1973. Voltage clamp analysis of acetylcholine produced end-plate current fluctuations at frog neuromuscular junction. J. Physiol. (Lond.) **235:** 655–691.
7. CULL-CANDY, S. G., R. MILEDI & O. D. UCHITEL. 1982. Properties of junctional and extrajunctional acetylcholine receptor channels in organ cultured human muscle fibres. J. Physiol. (Lond.) **333:** 251–267.
8. FISCHBACH, G. G. & S. M. SCHUETZE. 1980. A postnatal decrease in acetylcholine channel open time at rat end-plates. J. Physiol. (Lond.) **303:** 125–137.
9. SCHUETZE, S. M. 1987. Developmental regulation of nicotinic acetylcholine receptors. Annu. Rev. Neurosci. **10:** 403–457.
10. GU, Y. & Z. W. HALL. 1988. The immunological evidence for a change in subunits of the acetylcholine receptor in developing and denervated rat muscle. Neuron **1:** 117–125.
11. TZARTOS, S. J., D. E. RAND, B. E. EINARSON & J. LINDSTROM. 1981. Mapping of surface structures of electrophorus acetylcholine receptor using monoclonal antibodies. J. Biol. Chem. **256:** 8635–8645.
12. SAEDI, M. S., R. ARNAND, W. G. CONROY & J. LINDSTROM. 1990. Determination of amino acids critical to the main immunogenic region of intact acetylcholine receptors by *in vitro* mutagenesis. FEBS Lett. **267:** 55–59.
13. SARGENT, P. B., B. E. HEDGES, L. TSAVALER, S. CLEMMONS, S. TZARTOS & J. LINDSTROM. 1984. Structure and transmembrane nature of the acetylcholine receptor in amphibian skeletal muscle as revealed by crossreacting monoclonal antibodies. J. Cell Biol. **98:** 609–618.
14. TZARTOS, S., L. LANGEBERG, S. HOCHSCHWENDER & J. LINDSTROM. 1983. Demonstration of a main immunogenic region on acetylcholine receptors from human muscle using monoclonal antibodies to human receptor. FEBS Lett. **267:** 55–59.
15. DAS, M. & J. LINDSTROM. 1989. The main immunogenic region of the nicotinic acetylcholine receptor. Interaction of monoclonal antibodies with synthetic peptides. Biochem. Biophys. Res. Commun. **165:** 865–871.

16. RATNAM, M., D. LENGUYEN, J. RIVIER, P. B. SARGEANT & J. LINDSTROM. 1986. Transmembrane topography of the nicotinic acetylcholine receptor: Immunochemical tests contradict theoretical predictions based on hydrophobicity profiles. Biochemistry **25:** 2633–2643.
17. TZARTOS, S., L. LANGEBERG, S. HOCHSCHWENDER, L. W. SWANSON & J. LINDSTROM. 1986. Characteristics of monoclonal antibodies to denatured *Torpedo* and to native calf acetylcholine receptors: Species, subunit, and region specificity. J. Neuroimmunol. **10:** 235–253.
18. RATNAM, M., P. B. SARGEANT, V. SARIN, J. L. FOX, DE LENGUYEN, J. RIVIER, M. CRIADO & J. LINDSTROM. 1986. Location of antigenic determinants on primary sequences of the subunits of the nicotinic acetylcholine receptor by peptide mapping. Biochemistry **25:** 2621–1632.
19. TZARTOS, S. & J. LINDSTROM. 1980. Monoclonal antibodies used to probe acetylcholine receptor structure: Localization of the main immunogenic region and detection of similarities between subunits. Proc. Natl. Acad. Sci. USA **77:** 755–759.
20. NELSON, S., G. D. SHELDON, S. LEI, J. LINDSTROM & B. M. CONTI-TRONCONI. 1992. Epitopoe mapping of monoclonal antibodies to *Torpedo* acetylcholine receptor γ subunits which specifically recognize the ε subunit of mammalian muscle acetylcholine receptor. J. Neuroimmunol. **36:** 13–27.
21. EVANS, R. H. 1974. The entry of labelled calcium into innervated region of the mouse diaphragm muscle. J. Physiol. (Lond.) **240:** 517–533.
22. MILEDI, R., I. PARKER & G. SCHALOW. 1980. Transmitter induced calcium entry across the postsynaptic membrane at frog end-plates measured using arsenazo III. J. Physiol. (Lond.) **300:** 197–212.
23. LEONARD, J. P. & M. M. SALPETER. 1979. Agonist induced myopathy at the neuromuscular junction. J. Cell Biol. **82:** 811–819.
24. ENGEL, A. G., E. H. LAMBERT, D. M. MULDER, C. F. TORRES, K. SAHASHI, T. E. BERTORINI & J. N. WHITAKER. 1982. A newly recognized congenital myasthenic syndrome attributed to a prolonged open time of the acetylcholine-induced ion channel. Ann. Neurol. **11:** 553–569.
25. BLOUNT, P. & J. P. MERLIE. 1990. Mutational analysis of muscle nicotinic acetylcholine receptor subunit assembly. J. Cell Biol. **111:** 2613–2622.
26. CRIADO, M., V. WITZEMANN, M. KOENEN & B. SAKMANN. 1988. Nucleotide sequence of the rat muscle acetylcholine receptor epsilon subunit. Nucleic Acids Res. **16:** 10920.
27. BUONANNO, A., J. MUDD & J. P. MERLIE. 1988. Isolation and characterization of the β and ε subunit genes of mouse muscle acetylcholine receptor. J. Biol. Chem. **264:** 7611–7616.
28. CHANGEUX, J.-P., A. DEVILLERS-THIÉRY, J.-L. GALZI & D. BERTRAND. 1992. New mutants to explore nicotinic receptor functions. Trends Pharmacol. Sci. **131:** 299–301.

Congenital End-plate Acetylcholine Receptor Deficiency and Short Channel Open Time[a]

A. NAGEL,[b] A. G. ENGEL,[c,f] T. J. WALLS,[d]
C. M. HARPER,[c] AND H. A. WAISBURG[e]

[b]*Department of Neurology*
University of Würzburg
Würzburg, Germany

[c]*Department of Neurology*
Mayo Clinic
Rochester, Minnesota

[d]*Department of Neurology*
Newcastle General Hospital
Newcastle Upon Tyne, England

[e]*Larrea 1474*
Buenos Aires, Argentina

INTRODUCTION

We here describe a congenital myasthenic syndrome associated with a deficiency and kinetic abnormality of the acetylcholine receptor (AChR). The disease presented at birth. Immune deposits were absent from the end-plates, and tests for anti-AChR antibodies were negative. On electron microscopy, most postsynaptic regions appeared normal, but the density of AChR on the junctional folds was diffusely reduced. The number of transmitter quanta released by nerve impulse was normal. The amplitude of miniature end-plate potentials and currents was abnormally low. Noise analysis of the kinetic properties of AChR revealed a significant decrease in mean channel open time; mean channel conductance was normal. The safety margin of neuromuscular transmission in this disorder is compromised by the deficiency and abnormal kinetic properties of AChR. The diffuse decrease in AChR density on the junctional folds could be due to reduced synthesis, accelerated destruction, or decreased membrane insertion of the mutant AChR. Electron microscopy and a study of the kinetic properties of AChR by noise analysis were essential to characterize this disorder. A similar kinetic abnormality of AChR has not been described to date.

[a] This work was supported by Grant NS 6277 from the National Institues of Health and by Research Grants from the Muscular Dystrophy Association. T. J. Walls was the recipient of a Fulbright Fellowship from the U.S. Government.

[f] Send reprint requests to A. G. Engel.

CLINICAL DATA

The patient was the firstborn child of unrelated healthy parents. There was no history of muscle disease in the family. The pregnancy was normal. Delivery was complicated by premature rupture of the membranes and low-forceps delivery. Her respiratory efforts were insufficient immediately after birth and for the next 24 days she required ventilatory support. She was unable to suck and was fed by a nasogastric tube for 6 months. During the first year of life she was hypotonic, diffusely weak, had fluctuating ptosis, fatigued rapidly during any activity, and had frequent bouts of respiratory insufficiency. She attained head control at age 5 months and began walking at age 16 months, but fell frequently after beginning to walk. Subsequently, she had intermittent laryngeal stridor worsened by exertion, she had difficulty closing her mouth and chewing tough foods, and her speech became slurred and hypernasal when she was tired. Her mental development was normal.

At age 5 5/12 years, the patient was dolichocephalic with a high-arched palate and mild malocclusion. She had moderately severe ptosis but no limitation of the ocular ductions. There was severe weakness of facial, tongue, laryngeal, and masticatory muscles, and mild to moderately severe weakness of cervical, truncal, and limb muscles. The tendon reflexes were hypoactive. Arm abduction time was limited to 30 seconds, and the patient could only do two deep-knee bends.

Pyridostigmine, 5 mg/kg daily, started at age 11 months, improved the patient's symptoms. Prednisone, 1 mg/kd/day, started at age 15 months, appeared to be of additional benefit. Subsequently, pyridostigmine was increased to 10 mg/kg daily, and prednisone was tapered so that by age of 5 she took only 1 mg on alternate days.

Tests for anti-AChR antibodies were negative. Repeated electromyographic studies revealed a normal amplitude of the evoked compound muscle action potential (CMAP) in rested muscles. There was no repetitive CMAP response to single nerve stimuli. At age 10 months, 2 Hz stimulation of the ulnar nerve produced a 10% decrement of the evoked CMAP. At age 27 months, ulnar nerve stimulation did not evoke a decremental response, but facial nerve stimulation revealed a 25% decrement. At age 44 months, stimulation of the ulnar, peroneal or femoral nerves produced a 5% decrement; the decremental response in the facial muscles was unchanged.

SPECIAL STUDIES

An intercostal muscle specimen was obtained at age 27 months for morphological and *in vitro* electrophysiological studies. At that time the patient was treated with 5 mg/kg pyridostigmine daily and 1 mg/kg prednisone on alternate days.

Light Microscopy

Routine histochemical studies demonstrated type I fiber preponderance. The muscle fiber diameter, 22 ± 0.7 μm (mean $\pm$ SE), was normal for the patient's age.

Acetylcholinesterase (AChE) activity was preserved at all end-plates by enzyme cytochemical and immunocytochemical criteria. The mean length of the patient's end-plates, evaluated from the cytochemical reaction product for AChE on teased single muscle fibers, was 25.2 μm. The mean end-plate length in an intercostal muscle of a 3½-year-old girl without muscle weakness was 17.1 μm; in 5 adult controls the corresponding values ranged from 20–29 μm. Thus, the patient's end-plates were abnormally large for her age and muscle fiber size.[1]

No immune deposits (IgG, C3, C9, or the C5b-9 membrane attack complex) were detected at the patient's end-plates. Examination of end-plates for AChR with rhodamine-labeled α-bungarotoxin revealed a marked decrease in AChR. A comparable decrease in AChR was noted on immunostaining with a polyclonal anti-AChR antibody, indicating that the decreased toxin binding could not be accounted for by a selective failure of the α-subunit of AChR to bind toxin, or by a selective absence of the α-subunit from AChR.

Number of Acetylcholine Receptors per End-Plate

The mean number of acetylcholine (ACh) binding sites per end-plate, estimated using radioiodinated α-bungarotoxin,[2] was 0.97×10^6. The corresponding value in 11 adult controls was $13.23 \pm 0.85 \times 10^6$ (mean ± SE). Because the size of the patient's end-plates was comparable to that of adult controls, the low number of ACh binding sites per end-plate in the patient suggested a reduced density of end-plate AChR.

Electron Microscopy

Electron microscopy of the end-plates labeled with peroxidase-conjugated α-bungarotoxin[3] demonstrated a marked decrease in the density of AChR on the junctional folds (FIG. 1). At most end-plates the junctional folds were abundant and well developed, but some postsynaptic regions appeared simplified. Morphometric analysis[4] of the end-plates showed a normal postsynaptic membrane density. The mean nerve terminal area and postsynaptic area were significantly smaller than in control end-plates (TABLE 1).

In Vitro *Microelectrode Studies*

The miniature end-plate potential (MEPP) amplitude, normalized[5] for a muscle fiber diameter of 60 μm, was abnormally small. Miniature end-plate currents (MEPC) were detected only after the addition of 1 μg/mL neostigmine methylsulphate to the bath, but even then they were abnormally small. The quantal content of the end-plate potential on 1 Hz stimulation fell in the normal range (TABLE 2).

The properties of the AChR ion channel were investigated by analysis of the ACh-induced current noise[6,7] (TABLE 3). Mean single-channel conductance was derived from the noise power spectrum and from regression of the variance of the ACh-induced conductance on the induced ACh conductance. Both methods of analysis yielded normal values for mean single-channel conductance. The mean open time of the AChR ion channel was determined from the noise power spectrum. The mean channel open time for the patient, 1.61 ms, was smaller than the corresponding value in any of 9 adult controls and was significantly smaller than the lowest open time, 2.14 ms, observed in the controls ($p < 0.001$). The channel open time was also significantly shorter than at end-plates of a 27-month-old patient without muscle weakness.

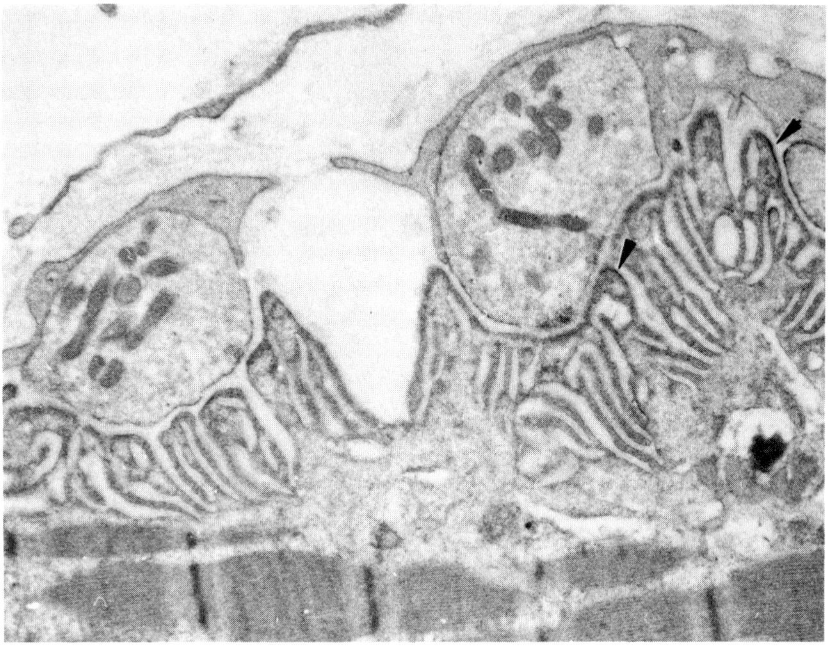

FIGURE 1. Electron micrograph of end-plate reacted for the localization at AChR with peroxidase-labeled α-bungarotoxin. The section is unstained. The junctional folds are numerous and well developed but barely show any reaction for AChR. Arrowheads indicate where faint reaction for AChR can be discerned; ×12,200. (From A. G. Engel *et al.*, *Progress in Brain Research* **84:** 125–137, 1990; reproduced with permission.)

DISCUSSION

The onset of the illness in the neonatal period, the negative tests for anti-AChR antibodies, the absence of immune deposits from the end-plates, the generally good preservation of the junctional folds, and the abnormal kinetic properties of AChR distinguish this disorder from autoimmune myasthenia gravis.

The findings in our patient appear to be unique among those patients with

TABLE 1. Morphometric Analysis of End-Plate Regions

Parameter	Patient[a]	Controls[b]
Nerve terminal area (μm^2)	2.85 ± 0.28	3.88 ± 0.39
Synaptic vesicle density (no./μm^2)	57.8 ± 3.90	50.3 ± 3.60
Postsynaptic area (μm^2)	5.13 ± 0.27	10.6 ± 0.79
Postsynaptic membrane density ($\mu m/\mu m^2$)	5.87 ± 0.13	5.83 ± 0.25

[a] Mean ± SE; 52–55 regions, 32 end-plates; more than one region can occur at an end-plate.

[b] Mean ± SE; 47–64 regions, 13 subjects.

TABLE 2. *In Vitro* Physiology—Miniature End-Plate Potentials, and Currents and Quantal Release by Nerve Impulse

Parameter	Patient[a]	Controls[b]
MEPP amplitude (mV)[c]	0.22 ± 0.09 (15)	0.99 ± 0.06 (15)
MEPC amplitude (nA)[d]	3.09 ± 0.13 (12)[e]	3.92 ± 0.14 (10)
EPP quantal content at 1 Hz[f]	41.1 ± 3.5 (11)	51.1 ± 4.5 (15)

[a] Mean ± SE (number of fibers).
[b] Mean ± SE (number of subjects).
[c] Normalized for a fiber diameter of 60 μm and corrected for a resting membrane potential of −80 mV; 28.5 ± 0.5°C.
[d] −80 mV; 22.5 ± 0.5°C.
[e] Measured in the presence of 1 μg/mL neostigmine methylsulfate.
[f] EPP amplitude corrected to a resting membrane potential of −80 mV and for nonlinear summation; 28.5 ± 0.5°C; variance method.

congenital AChR deficiency described to date. In 1981, Vincent *et al.*[8] described three patients with end-plate AChR deficiency (patients 2, 4, and 5). The kinetic properties of the AChR ion channel were studied in patients 4 and 5, who were siblings: normal channel conductance was observed in a total of three muscle fibers, and a normal channel open time, inferred from the decay time constant of MEPC, was found in a total of four muscle fibers.

In another patient with end-plate AChR deficiency that was investigated by Lecky *et al.*[9] in 1986 (patient 2), the end-plates appeared elongated on light microscopy, but their ultrastructure was normal. No electrophysiological studies were done in this case.

In four other patients, end-plate AChR deficiency was associated with poorly developed junctional folds and a paucity of secondary synaptic clefts.[10–12] In two patients the disorder presented at birth, and both had severe symptoms in early childhood.[10,11] A milder form of the syndrome was recognized in a 64-year-old man and his 58-year-old sister.[12] Both had ptosis and exercise intolerance since childhood and gradual worsening of symptoms in adult life. The kinetic properties of the AChR ion channel were not investigated in these patients.

The observations in our patient indicate that AChR deficiency can be associated with a kinetic abnormality of the AChR ion channel. The kinetic abnormality is likely to stem from a mutation in an AChR subunit. The AChR deficiency may be

TABLE 3. *In Vitro* Physiology—Noise Analysis of the Kinetic Properties of the Acetylcholine Receptor Ion Channel[a]

Parameter	Patient[b]	Controls[c]
Channel open time (ms)[d]	1.61 ± 0.04 (6)	2.37 ± 0.09 (9)
Single channel conductance (pS)[e]	36.33 ± 4.07 (6)	32.23 ± 0.75 (9)
Single channel conductance (pS)[d]	34.69 ± 4.75 (6)	31.41 ± 0.96 (9)

[a] ACh-induced current noise was recorded at 22.5 ± 0.5°C at a holding potential of −80 mV.
[b] Mean ± SE (number of fibers).
[c] Mean ± SE (number of subjects).
[d] From noise power spectrum.
[e] From regression of the variance of the ACh-induced conductance fluctuation on the ACh-induced conductance.

due to reduced synthesis, decreased membrane incorporation, or altered metabolic stability of the mutant AChR. It now appears likely that diverse mutations of AChR may result in end-plate AChR deficiency. It also follows that an adequate investigation of any case of end-plate AChR deficiency should include a study of the kinetic properties of the AChR channel.

REFERENCES

1. HARRIS, J. B. & R. R. RIBCHESER. 1979. The relationship between end-plate size and transmitter release in normal and dystrophic muscles of the mouse. J. Physiol. (Lond.) **296:** 245–265.
2. ITO, Y., R. MILEDI, A. VINCENT & J. NEWSOM-DAVIS. 1978. Acetylcholine receptors and end-plate electrophysiology in myasthenia gravis. Brain **101:** 345–368.
3. ENGEL, A. G., J. LINDSTROM, E. H. LAMBERT & V. A. LENNON. 1977. Ultrastructural localization of the acetylcholine receptor in myasthenia gravis and in its experimental autoimmune model. Neurology **27:** 307–315.
4. ENGEL, A. G., M. TSUJIHATA, J. M. LINDSTROM & V. A. LENNON. 1976. The motor end plate in myasthenia gravis and in experimental autoimmune myasthenia gravis. A quantitative ultrastructural study. Ann. N.Y. Acad. Sci. **274:** 60–79.
5. KATZ, B. & S. THESLEFF. 1957. On the factors which determine the amplitude of the 'miniature end-plate potential'. J. Physiol. (Lond.) **137:** 267–278.
6. ANDERSON, C. R. & C. F. STEVENS. 1973. Voltage clamp analysis of acetylcholine produced end-plate current fluctuations at frog neuromuscular junction. J. Physiol. (Lond.) **235:** 655–691.
7. CULL-CANDY, S. G., R. MILEDI & O. D. UCHITEL. 1982. Properties of junctional and extrajunctional acetylcholine receptor channels in organ cultured human muscle fibres. J. Physiol. (Lond.) **333:** 251–267.
8. VINCENT, A., S. G. CULL-CANDY, J. NEWSOM-DAVIS, A. TRAUTMANN, P. C. MOLENAAR & R. L. POLAK. 1981. Congenital myasthenia: End-plate acetylcholine receptors and electrophysiology in five cases. Muscle & Nerve **4:** 306–318.
9. LECKY, B. R. F., J. A. MORGAN-HUGHES, N. M. F. MURRAY, D. N. LANDON & D. WRAY. 1986. Congenital myasthenia. Further evidence of disease heterogeneity. Muscle & Nerve **9:** 233–242.
10. SMIT, S. M. E., F. G. I. JENNEKENS, H. VELDMAN & P. G. BARTH. 1984. Paucity of secondary synaptic clefts in a case of congenital myasthenia with multiple contractures: ultrastructural morphology of a developmental disorder. J. Neurol. Neurosurg. Psychiatry **47:** 1091–1097.
11. SMIT, L. M. E., G. HAGEMAN, H. VELDMAN, P. C. MOLENAAR, B. S. OEN & F. G. I. JENNEKENS. 1988. A myasthenic syndrome with congenital paucity of secondary synaptic clefts: CPSC syndrome. Muscle & Nerve **11:** 337–348.
12. WOKKE, J. H. J., F. G. I. JENNEKENS, P. C. MOLENAAR, C. J. M. VAN DEN ORD, B. S. OEN & H. F. M. BUSCH. 1989. Congenital paucity of secondary synaptic clefts (CPSC) syndrome in 2 adult sibs. Neurology **39:** 648–654.

Strategies for Treatment of Myasthenia Gravis[a]

DANIEL B. DRACHMAN, KEVIN R. McINTOSH, JOHANNES REIM, AND LAURA BALCER

Department of Neurology
Johns Hopkins University
600 North Wolfe Street
5-119 Meyer Building
Baltimore, Maryland 21287-7519

Sixty years have passed since the introduction of anticholinesterase (anti-ChE) drugs for the treatment of myasthenia gravis (MG) by Mary Walker.[1] This discovery not only proved the first effective therapy for MG, but also provided important evidence for the neuromuscular junction as the site of the defect in this disease. Although treatment with anti-ChE agents produced some benefit in most myasthenic patients, the prognosis remained poor. In 1958, when anti-ChE drugs were widely used, the mortality rate for generalized MG was still 30%, and more than 60% of patients either failed to improve or actually deteriorated.[2] More recently, with the use of immunosuppressive therapy, the outlook for MG has improved dramatically.[3] With proper treatment, the great majority of patients can be restored to fully productive lives. Nevertheless, currently available immunotherapeutic agents themselves present considerable problems. Generalized immunosuppression compromises the immune system as a whole, with increased susceptibility to infection or neoplasia. The available agents may have adverse side effects unrelated to immune suppression, such as hypertension, osteoporosis, and renal or hepatic toxicity. Lifelong treatment is often required; the majority of patients undergo relapse after discontinuation of therapy.[3,4]

As yet, the goal of a "cure" has remained elusive, despite our extensive knowledge of the pathophysiology, immunology, biochemistry, and molecular biology involved in MG. Ideally, treatment of MG should (a) specifically inhibit the autoimmune response to acetylcholine receptor (AChR), without otherwise interfering with the immune system; (b) be nontoxic; and (c) be long-lasting or permanent. Indeed, the goal of achieving discrete control of a specific immune response has been considered the "holy grail" of immunology. Given the detailed knowledge now available, it should be possible to design rational and specific therapy for MG; a strategy designed for MG could lead the way to treatment for a wide variety of other immune disorders as well.

SPECIAL CONSIDERATIONS IN MG

The pathogenetic mechanisms of MG have been thoroughly described[5–8] and will not be reviewed in detail here. There are several features, however, that

[a] The original work in this paper was carried out with partial support from the NIH #1RO1 NS 23719, the Muscular Dystrophy Association, and the Myasthenia Gravis Foundation.

are important in the design of therapeutic strategies and are therefore worth emphasizing.

(1) The loss of AChRs is due to an antibody-mediated process. The mechanisms by which AChR-specific antibodies reduce the number of receptors at neuromuscular junctions include (a) accelerated endocytosis of cross-linked AChRs;[9–11] (b) blockade of receptor sites of AChRs;[12,13] and (c) damage to the postsynaptic membrane, attributed to the effects of antibodies acting in conjunction with complement.[14,15]

In contrast to MG, several other autoimmune conditions for which novel therapeutic methods have recently been described are thought to be cell-mediated, including thyroiditis,[16] experimental autoimmune encephalitis (EAE),[17] and adjuvant arthritis.[18] As will be discussed later, approaches that produce specific inhibition of cellular responses do not necessarily inhibit antibody responses.

(2) The antibody response to AChR in MG is T cell–dependent. The requirement for T cells has been formally demonstrated in experimental autoimmune MG (EAMG) in rats.[19] Further, T cells from animals with EAMG and humans with MG respond specifically to AChR stimulation.[20–29]

(3) AChR is a highly immunogenic antigen. Immunization with AChR induces vigorous responses of both T cells and B cells. Perhaps because of this factor, immune responses to AChR may not be as amenable to certain therapeutic maneuvers that are successful for less powerful antigens.

(4) Immune responses to AChR are highly heterogeneous. In both MG and EAMG, antibodies directed to a wide variety of different AChR epitopes are found (see ref. 5 for review). Similarly, T cells in both human MG and animal models of EAMG respond to multiple different AChR epitopes.[23–29]

SITES FOR THERAPEUTIC INTERVENTION

In considering the entire arc of pathogenesis of MG, from antigen presentation to impairment of neuromuscular transmission, there is an astonishingly large number of sites at which therapeutic intervention can be applied. TABLE 1 outlines many possible approaches. They are arranged in more or less reverse order, starting at the neuromuscular junction and working backwards. Each of these approaches has either been used clinically or tested experimentally in MG or other autoimmune conditions (see references). The very length of the list implies that an ideal therapeutic strategy has yet to be devised.

The main foci of this presentation are the strategies marked by asterisks in TABLE 1, in which B cells or T cells involved in the autoimmune response are targeted in various ways.

Targeting B Cells: "Hot Antigen Suicide"

The goal of this strategy is to eliminate B cells that specifically produce anti-AChR antibodies. Because these B cells bear antibodies that act as surface receptors for epitopes of the AChR molecule, they are capable of binding these epitopes. If the AChR molecule has been armed with a lethal "warhead," the relevant B cells will bind and take up the lethal antigen, and will be killed. In principle, this should produce specific killing of the population of B cells responsible for production of the anti-AChR antibodies of MG. Experimentally, immunotoxins

TABLE 1. Therapeutic Interventions in Myasthenia Gravis

Neuromuscular junction:	
Anticholinesterases	Prolong ACh-AChR interaction.
3,4-diaminopyridine	Increase ACh release.[30]
3-deazaadenosine (3DZA)	Decrease AChR turnover.[31,32]
Antibody depletion:	
Plasmapheresis	
Semiselective pheresis	Preferential removal of AChR Ab; (polyvinyl alcohol with tryptophane).[38]
General immune treatment:	
Thymectomy	
Adrenal corticosteroids	
Azathioprine	
Cyclosporine A[34,35]	
Cyclophosphamide	
Total lymphoid irradiation[36]	
Intravenous immunoglobulin	
Unusual immunosuppressive agents tried:	
Dimethylsulfoxide	Suppressive in EAMG.[37]
Thalidomide	Used in bone marrow transplantation; no benefit in EAMG.[38]
"Pristine immune system":	
Cyclophosphamide + TBI;[a] bone marrow "rescue"	Abolishes immune response to AChR, in EAMG.[39]
Deplete AChR-specific B cells	
[b]* "Hot antigen suicide"	Effective in naive animals, culture. See text.
a) AChR + ricin A[40]	
b) AChR + ^{125}I[41]	
Deplete T-helper cells	
Anti-CD4 Ab[42]	
* IL2 toxin	See text.
Target antigen-specific T cells	
* T-cell vaccination	Effective in EAE; not EAMG; see text.
* Immunize with TCR peptide(s)	EAE; see text.
* APC used to target AChR-specific T cells	See text.
Specific suppression	
T-suppressor cells?[43,44]	
Large suppressor cells	Antigen-specific macrophages.[45–47]
Antigen-presenting cells	
Ab vs. MHC class II[48]	
Ab vs. MHC class II + peptide	EAE[49]
MHC-blocking peptide[50]	
Antiidiotypic approach	Mixed results in EAMG.[51–54]
Tolerance induction	
Deaggregated antigen[55]	
Route of administration	Oral, inhalation.[56]
TLI + antigen	Effective for some antigens; not AChR.[57]
Anti-CD4 + antigen	Effective for some antigens.[58]
Antigen-coupled cells	Effective in EAMG.[59]
? Tolerogenic fragment[60]	

[a] Total body irradiation.
[b] Asterisks mark main foci.

comprised of AChR, and either the toxic A chain of ricin[40] or radioiodine,[41] have been used successfully *in vitro* or in naive animals. There are serious potential problems, however, in applying this method to ongoing autoimmune diseases in animals, or eventually patients. First, the surface "address" of AChR-specific B cells consists of the antibodies they produce. These antibodies are highly heterogeneous[5] and bind to a wide variety of AChR epitopes, some of which are determined by their three-dimensional conformations, rather than by linear sequences.[61] Therefore, to target the entire spectrum of AChR-specific B cells, conformationally intact AChR must be used. In ongoing MG, precisely the same antibodies are present in the circulation. The AChR immunotoxin must therefore "run the gauntlet" of circulating antibodies. The immunotoxic molecules may be complexed by the antibodies, and the complexes are precipitated in the lungs, liver, or kidneys without ever reaching their goal, but with the risk of producing damage to these organs. Second, if some memory B cells remain, they may subsequently proliferate, resulting in recurrence of the autoimmune response. Even if elimination of the AChR-specific B cells is complete, the ability of B-cell genes to mutate and rearrange allows for the appearance of new AChR-specific B cells, which can be up-regulated in response to antigenic stimulation, and may result in recurrence of MG. These problems apply to virtually any strategy designed to target antigen-specific B cells.

T Cell–Directed Approaches

As noted above, T cells play a critical role in the autoimmune response in MG, and T cells have certain characteristics that make them amenable to targeting approaches. T-cell receptors "see" linear epitopes, in association with MHC class II, rather than conformational epitopes. As will be discussed later, this permits targeting of the T cells, while reducing the possible problem of interception by circulating antibodies. Unlike B cells, T cell–receptor genes do not undergo mutation, eliminating a source of resupply of potentially pathogenic cells. Further, various surface markers on T cells can serve as semispecific "addresses" for targeting (see TABLE 1 and below.) Finally, as will be discussed in another section, T cells can be inactivated, or rendered "anergic," which presents a powerful approach to immunotherapy.

Targeting Activated T Cells: IL2 Toxin

This strategy is based on the fact that T cells involved in an active immune response express receptors for the cytokine growth factor interleukin 2 (IL2), in contrast to resting and memory T cells, which do not express IL2 receptors.[62,63] This property can be exploited to target activated T cells, including those involved in the immune response to AChR. For this purpose, we have carried out experiments using a recently developed IL2 toxin, produced by recombinant DNA technology.[64] This fusion protein (designated DAB_{486}-IL2) is composed of the receptor-binding moiety of IL2 and the cytotoxic portion of diphtheria toxin. It binds to cells bearing IL2 receptors, is internalized by endocytosis, and irreversibly inhibits protein synthesis, resulting in death of the target T cell.[65] A single molecule of endocytosed diphtheria toxin is lethal to the cell.[66]

We examined the effects of IL2 toxin both in C57BL/6 mice with EAMG and in lymphocyte cultures prepared from Lewis rats with EAMG.[67] For the *in vivo* experiments, mice were immunized with 60 μg of purified *Torpedo* AChR absorbed on bentonite, administered intraperitoneally (ip). They were treated with IL2 toxin (1 mg/kg/d) given twice daily by the intraperitoneal route. Weekly serum samples were obtained, and anti-AChR antibody levels were measured by a kinetic ELISA method.[67] When IL2 toxin treatment was begun at the time of AChR immunization, it significantly suppressed the primary antibody response to AChR by approximately 50%, as compared to controls ($p < 0.02$). IL2 toxin, however, had no suppressive effect on the AChR antibody levels in ongoing EAMG, nor did IL2 toxin suppress the secondary response to a booster immunization in AChR in mice with established EAMG.

In order to study the effects of IL2 toxin on the immune responses to AChR more directly, we carried out experiments *in vitro*. Primed lymph node cells (LNC) were prepared from rats previously immunized with AChR and were restimulated in culture with AChR. Various doses of IL2 toxin were added, and the cultures were assayed for lymphoproliferation by [^{3}H]thymidine uptake and for AChR antibody production by kinetic ELISA. Both lymphoproliferation and antibody production were inhibited by more than 90%, at an IL2-toxin concentration of 1 μg/mL (well below the cytotoxic level of 10 μg/mL). Even at 0.1 μg/mL, the responses were suppressed by 50–60 percent. Thus, IL2 toxin had a profound inhibitory effect on these secondary immune responses. Similarly, when IL2 toxin was added to ongoing cultures of stimulated AChR-specific LNC, it inhibited lymphoproliferation profoundly (82 ± 10%).

We wondered whether the disparity between the striking effectiveness of IL2 toxin *in vitro* and the modest effect *in vivo* might be due to poor bioavailability in the intact animal and therefore measured the circulating levels of IL2 toxin at 15 minute intervals after ip injection of 1 mg/kg into mice. The results showed no measurable IL2 toxin at any time, with the limit of sensitivity of the assay being $50–100 \times 10^{-9}$ g/mL.[68]

Our results show that IL2 toxin can powerfully inhibit immune responses to AChR, provided that it gains access to the relevant target cell population. The fact that it has a very short half-life *in vivo*, estimated to be 5–10 minutes in rats,[68] suggests that it may have difficulty gaining access to the relevant lymphocyte population, which is relatively protected by the complex architecture of the lymph nodes within which they presumably reside. By contrast, previous studies using IL2 toxin to inhibit graft rejection[69] and delayed-type hypersensitivity responses[70] have been effective *in vivo*, which probably reflects the greater accessibility of the T cells in these situations.

These studies indicate that a targeting strategy aimed at IL2 receptors of T cells can provide semiselective (although not antigen-specific) suppression of an active autoimmune response. Its effectiveness in treating autoimmune diseases, however, depends critically on prolonging the activity of the agent *in vivo*, thus allowing it to gain access to the relevant T cells.

"T-Cell Vaccination": Immunization with T-Cell Receptor

Previous reports, notably by Cohen and his colleagues, have shown that specific immunological unresponsiveness to a given antigen can be induced by immunizing animals with T cells activated to that antigen.[71] This method termed T-cell vaccination has been shown to suppress immune responses to allografts[72] and to suppress several experimental autoimmune disorders.[73–76] In all of these disorders, the immune response is mediated by effector T cells. We carried out studies to determine whether a similar strategy could be used in EAMG, which is humorally mediated.[77] Enriched populations of AChR-sensitized T cells were prepared by stimulating primed LNC with AChR in culture and then expanding them with IL2. These cells were terminally stimulated with concanavalin A and subjected to extremely high pressure in a French press, as described for the vaccination protocol.[78] Lewis rats received three vaccinations (20×10^6 cells each) ip at two-week intervals. Control rats received similar injections of cells from nonimmunized animals. One week after the last stimulation, the vaccinated rats were challenged with AChR. The rats were bled before challenge, and at weekly intervals thereafter, for measurement of AChR antibody titers.

In addition, we used separate groups of vaccinated and control rats for *in vitro* studies of suppression by spleen cells. They were treated exactly as above, but were not challenged with AChR; instead, spleen cells were harvested one week after the third vaccination and used for the culture studies.

The results showed that the vaccinated rats had low but significant AChR antibody levels (mean = 8.4×10^{-9} M) even before challenge. That is, treatment with the AChR-sensitized cells evoked a minimal but definite immune response against AChR, which we believe was mediated through an antiidiotypic route. When challenged with AChR, the vaccinated rats developed significantly higher titers of antibodies to AChR than did control rats, which was the opposite of the effect we had hoped to produce. The apparent AChR priming of the vaccinated rats could not be attributed to carryover of the antigen by the cells used for vaccination; detailed studies demonstrated that these cells had no detectable AChR, as demonstrated by their lack of binding of α-bungarotoxin and their failure to stimulate AChR-primed LNC *in vitro*. We concluded that the functional priming of vaccinated rats was likely due to an antiidiotypic mechanism similar to that described by Lider *et al.*[79]

In contrast to the increased response to AChR *in vivo,* we found that splenic lymphocytes from the vaccinated rats produced antigen-specific suppression *in vitro*. When spleen cells from vaccinated rats were co-cultured with LNC primed to both AChR and an unrelated antigen (KLH), they specifically suppressed the AChR antibody response but not the response to KLH. Spleen cells from control rats did not show a specific suppressive response in these co-cultures. Thus, our observations indicated that vaccination with antigen-specific T cells can elicit both a positive antibody response and a suppressive cellular response in the same animal. Indeed, these findings are consistent with the results of cell vaccination in autoimmune thyroiditis in mice.[75] In that study, vaccination with T-cell lines specific for thyroglobulin inhibited the development of thyroiditis, which is a cell-mediated condition but did not affect the antithyroglobulin antibody response. Other studies have shown that T-cell vaccination has an inhibitory effect in several cell-mediated autoimmune animal models, including EAE,[73] interstitial nephritis,[74] and adjuvant-induced arthritis.[76]

The inhibitory effect induced by T-cell vaccination is believed to result from an immune response directed against the specific antigen receptors of the T cells used in the vaccination procedure.[79] If immunization with whole T cells can induce an inhibitory response against T-cell receptors, then immunization with the relevant T-cell receptors in pure form might also be effective. The key question is, How many T-cell receptors, and which ones, are relevant in a given autoimmune disease? This issue has been approached in studies of EAE, where the antigen used to produce the disease is myelin basic protein (MBP). Interest in EAE stems from its resemblance to multiple sclerosis (MS) in humans and the hypothesis that MBP may play a role in the pathogenesis of MS. In certain strains of rats and mice, the number of T-cell receptors used in EAE appears to be relatively restricted.[80,81] Under these circumstances, treatment with antibodies directed against the predominant T-cell receptors has been reported to prevent the development of EAE. Recently, active immunization with peptide moieties of the relevant T-cell receptors has had a similar beneficial effect in preventing or modifying experimental EAE in rats.[81,82] It should be emphasized, however, that such an approach directed against identified T-cell receptors requires that the receptor usage in the particular autoimmune disease must be markedly restricted. This requirement applies not only to the T-cell repertoire of each individual but also to the potential variation between unrelated individuals. At present, it appears that T-cell responses in human autoimmune diseases are likely to be more heterogeneous than those in inbred rodents, though they are only beginning to be studied in detail (Theofilopoulos *et al.*, this volume).

Antigen-Presenting Cells Used to Target AChR-Specific T Cells

In the case of MG in humans and EAMG in rats, there is now abundant evidence that the epitopes of AChR to which T cells respond are highly heterogeneous.[23–29,83,84] Not only do T cells from individual myasthenic patients respond to multiple epitopes, but the patterns of responsiveness also differ in different patients.[23,25–29,83,84] Using a battery of synthetic peptides that encompass the sequences of human AChR subunits, Conti-Tronconi and her colleagues (this volume) have found that T cells from myasthenic patients respond to over 30 different peptides! Similarly, in rats with EAMG, there is heterogeneity of responses to AChR peptides within each strain, as well as differences in the patterns of peptides to which T cells from different strains of rats respond.[24] Further, T cells recognize epitopes only in association with the appropriate self–MHC class II, which entails yet another level of heterogeneity.[85,86] Thus, in order to target the repertoire of AChR-specific T cells of any individual patient, it will require presentation of the entire spectrum of processed AChR peptides appropriate to that individual, in association with the proper MHC class II. Unless some restrictive principle limits the spectrum of pathogenic peptides and MHC antigens, this is likely to be an extremely complex undertaking.

Fortunately, the immune system itself provides a means of targeting all the relevant T cells: that is the job of antigen-presenting cells (APC). For any immune response, the APC naturally process the antigen correctly for that individual and present the processed peptides in association with the individual's own appropriate MHC class II. The individual's own APC can thus serve as "guided missiles" to target the entire repertoire of T cells specific for AChR.

How to obtain enough APC poses another problem: "Professional" APC consist mainly of macrophages and dendritic cells,[85,86] which are neither readily obtainable nor sufficiently numerous to use in a therapeutic strategy. B cells, however, are also known to serve as effective APC.[87–89] Yet B cells are normally effective presenters only for the antigen recognized by, and subsequently bound to, their surface antibody.[90–93] Even in a patient or experimental animal with MG, only a small proportion of B cells are specific for AChR, and therefore the vast majority would be ineffective as presenters of AChR epitopes to T cells. A recently described approach circumvents this problem: B cells can present any antigen if it is bound to their surface immunoglobulin.[93–95] We have used this principle in designing a novel strategy for the treatment of myasthenia, using the rat model of EAMG. To induce B cells to present AChR, we have constructed a conjugate consisting of antibody directed against the B-cell's surface IgM (goat anti-rat IgM) linked to a monoclonal antibody directed against *Torpedo* AChR (MAb 35[96]). The conjugate was prepared by the use of SPDP, a heterobifunctional coupling reagent that introduces a disulfide bond linking the two antibodies,[97] and was purified by column chromatography.[94,98] The coupling method produces heteroconjugates, not homoconjugates. ELISA and radioimmunoassays confirmed that conjugate binds to both rat IgM and *Torpedo* AChR, and that its binding activity was about 90% that of equimolar amounts of each of its component antibodies. The anti-IgM : anti-AChR conjugate links AChR to surface IgM of B cells, regardless of their inherent antigen specificity. The B cells internalize the bound AChR, process it, and present in association with the appropriate MHC class II. When conjugate plus AChR-treated B cells (AChR-APC) are used to present AChR to AChR-primed rat T cells *in vitro,* they stimulate them efficiently to proliferate and produce IL2. The conjugate increased the efficiency of presentation markedly, by two to three orders of magnitude.[98]

The goal of this strategy, however, is to induce inactivation, rather than stimulation, of the targeted AChR-specific T cells. Recent work has demonstrated that fixation of APC, using cross-linking reagents (paraformaldehyde, ECDI), renders them incapable of fully stimulating T cells. Interaction between the fixed APC and the specific T cells induces long-lasting inability of the T cells to respond to subsequent antigen stimulation, termed anergy.[99,100] It is believed that the fixed APC are incapable of delivering a costimulatory "second signal" required for full stimulation of the T cell's activation program.[100] The incomplete activation produced by fixed APC results in inability of the T cells to undergo further proliferation and/or to provide help to B cells.[100,101]

To evaluate the potential usefulness of this strategy for the treatment of MG, we tested the ability of fixed, AChR-presenting B cells to induce anergy in a population of AChR-specific T cells, in culture.[98] The AChR-responsive T-cell populations were prepared from lymph node cells of AChR-primed Lewis rats. For some experiments they were directly depleted of APC, by passage through a nylon wool column, followed by treatment with a cocktail of mouse monoclonal antibodies directed against rat B cells and Ia positive cells and removal with paramagnetic beads coated with rat anti-mouse antibodies.[102] In other experiments, they were stimulated with AChR and expanded with IL2. After two cycles of stimulation and IL2 treatment, the expanded T-cell populations contained less than 1% Ia positive cells. Both types of T-cell preparations were capable of responding to AChR stimulation only in the presence of added APC. For antigen presentation, we prepared B cell–enriched populations from lymph node cells of naive rats, by a combination of nylon wool fractionation to deplete T cells and adherence to plastic dishes to deplete macrophages. To bind AChR, they were first incubated with conjugate at 4°C, followed by incubation at 4°C with purified *Torpedo* AChR. The washed cells with bound conjugate plus AChR were then incubated at 37°C for 4 to 8 hours, to allow antigen processing and presentation. They were then briefly fixed in dilute paraformaldehyde (0.15% for 1 min at 37°C) and thoroughly washed. Control fixed APC consisted of an enriched B-cell population that was not treated with conjugate or AChR, but was fixed with paraformaldehyde as above. The fixed AChR-APC or control-fixed APC were precultured with responder T cells overnight at a ratio of 5 : 1. After the preculture period, the T cells were recovered by centrifugation over Ficoll-Hypaque and restimulated with fresh conjugate-treated APC in the presence of AChR.

The results showed that preculture with fixed AChR-APC induced markedly decreased responsiveness to specific restimulation with AChR, as compared with controls. Preculture with fixed AChR-APC virtually eliminated IL2 production by restimulated AChR-specific T cells. After restimulating with fresh AChR-APC, the T cells produced no detectable IL2, as determined by a bioassay using an IL2-dependent cell line (CTLL cells[103]). By contrast, AChR-specific T cells that were precultured with control fixed APC produced abundant IL2. AChR-specific T-cell proliferation was also markedly reduced after preculture with fixed AChR-APC. Preculture of purified primed T cells with fixed AChR-APC resulted in a 54% decrease in the T cells' proliferative response ([^{3}H]thymidine incorporation), as compared with the proliferative response of cells cultured with control-fixed APC. Preculture of the expanded T-cell line with fixed AChR-APC resulted in a 59% reduction of [^{3}H]thymidine incorporation in response to AChR stimulation, as compared with controls.

In order to evaluate the antigen specificity of unresponsiveness induced by fixed AChR-APC, we examined their effects on T cells specific for an unrelated antigen, ovalbumin (OVA). Preculture with fixed AChR-APC had no significant effect on the ability of OVA-specific T cells to proliferate in response to antigen stimulation.

This method of using conjugate plus antigen-treated B cells as APC to "target" and induce anergy in antigen-specific T cells has many potential advantages for eventual application to a variety of immune disorders: (1) The use of antibody conjugates allows virtually all mature B cells to be used as APC. B cells are abundant and readily obtained from humans or experimental animals. (2) Because the individual's own APC can be used, they naturally process the antigen (AChR) appropriately, in association with the correct MHC class II for that individual. They should therefore be capable of targeting the entire repertoire of T cells responsive to the antigen, for that individual. (3) Inasmuch as the fixed AChR-APC bear processed antigen fragments, they should evade circulating antibodies, which are generally directed against conformationally determined epitopes. This is an important consideration in ongoing autoimmune diseases such as MG, where autoantibodies to AChR are present in the circulation. (4) The anergy induced is antigen-specific, and long-lasting or permanent. (5) The B cells used in this strategy are unlikely to be immunogenic because they are autologous. (6) Although we are developing this strategy for the treatment of MG, it should be applicable to the treatment of virtually any immune disorder for which the antigen is known.

REFERENCES

1. WALKER, M. B.Treatment of myasthenia gravis with physostigmine. 1934. Lancet **1:** 1200–1201.
2. GROB, D., N. G. BRUNNER & T. NAMBA. 1981. The natural course of myasthenia gravis and effect of therapeutic measures. Ann. N.Y. Acad. Sci. **377:** 652–669.
3. DRACHMAN, D. B. 1992. Treatment of myasthenia gravis. *In* Current Therapy in Rheumatology and Immunology. 4th Edition. A Fauci, Ed. 347–352. B. C. Decker. Philadelphia, PA.
4. MICHELS, M., R. HOHLFELD, H-P HARTUNG, K. HEININGER, U. A. BESINGER & K. V. TOYKA. 1988. Myasthenia gravis: Discontinuation of long-term azathioprine. Ann. Neurol. **24:** 798.
5. DRACHMAN, D. B., S. DE SILVA, D. RAMSAY & A. PESTRONK. 1987. Humoral pathogenesis of myasthenia gravis. Ann. N.Y. Acad. Sci. **505:** 90–105.
6. VINCENT, A. 1980. Immunology of acetylcholine receptors in relation to myasthenia gravis. Physiol. Rev. **60:** 756–824.
7. ENGEL, A. G. 1984. Myasthenia gravis and myasthenic syndromes. Ann. Neurol. **16:** 519–534.

8. LINDSTROM, J. 1985. Immunobiology of myasthenia gravis, experimental autoimmune myasthenia gravis, and Lambert-Eaton syndrome. Annu. Rev. Immunol. **3:** 109–131.
9. KAO, I. & D. B. DRACHMAN. 1977. Myasthenic immunoglobulin accelerates acetylcholine receptor degradation. Science **196:** 527–529.
10. STANLEY, E. F. & D. B. DRACHMAN. 1978. Effect of myasthenic immunoglobulin on acetylcholine receptor of intact mammalian neuromuscular junctions. Science **200:** 1285–1287.
11. APPEL, S. H., R. ANWYL, M. W. MCADAMS & S. ELIAS. 1977. Accelerated degradation of acetylcholine receptor from cultured rat myotubes with myasthenia gravis sera and globulins. Proc. Natl. Acad. Sci. USA **74:** 2130–2134.
12. ALMON, R. R., C. G. ANDREW & S. H. APPEL. 1974. Serum globulin in myasthenia gravis: inhibition of α-bungarotoxin binding to acetylcholine receptors. Science **186:** 55–57.
13. DRACHMAN, D. B., R. N. ADAMS, J. F. JOSIFEK & S. G. SELF. 1982. Functional activities of anti-AChR autoantibodies and clinical severity of myasthenia gravis. N. Engl. J. Med. **307:** 769–775.
14. TOYKA, K. V., D. B. DRACHMAN, D. E. GRIFFIN, A. PESTRONK, J. A. WINKELSTEIN, K. H. FISCHBECK & I. KAO. 1977. Myasthenia gravis: study of humoral immune mechanisms by passive transfer to mice. N. Engl. J. Med. **296:** 125–131.
15. ENGEL, A. G. & K. ARAHATA. 1987. The membrane attack complex of complement at the endplate in myasthenia gravis. Ann. N.Y. Acad. Sci. **505:** 326–332.
16. MARON, R., R. ZERUBAVEL, A. FRIEDMAN & I. R. COHEN. 1983. T lymphocyte line specific for thyroglobulin produces or vaccinates against autoimmune thyroiditis in mice. J. Immunol. **131:** 2316–1322.
17. MOKHTARIAN, F., D. E. MCFARLIN & C. S. RAINE. 1984. Adoptive transfer of myelin basic protein-sensitized T cells produces chronic relapsing demyelinating disease in mice. Nature **309:** 356–358.
18. HOLOSHITZ, J., A. MATITIAU & I. R. COHEN. 1984. Arthritis induced in rats by cloned T lymphocytes responsive to myocobacteria, but not to collagen type II. J. Clin. Invest. **73:** 211–215.
19. LENNON, V. A., J. M. LINDSTROM & M. E. SEYBOLD. 1976. Experimental autoimmune myasthenia gravis: cellular and humoral immune responses. Ann. N.Y. Acad. Sci. **274:** 283–299.
20. ABRAMSKY, O., A. AHARONOV, C. WEBB & S. FUCHS. 1975. Cellular immune response to acetylcholine receptor rich fraction in patients with myasthenia gravis. Clin. Exp. Immunol. **19:** 11–16.
21. RICHMAN, D. P., J. P. ANTEL, J. W. PATRICK & B. G. W. ARNASON. 1979. Cellular immunity to acetylcholine receptor in myasthenia gravis: Relationship to histocompatibility type and antigenic site. Neurology **29:** 291–296.
22. NOGUCHI, S. A., D. B. DRACHMAN, R. N. ADAMS & R. L. HIRSCH. 1980. Lymphocyte responsiveness to acetylcholine receptor in rats with experimental autoimmune myasthenia gravis. Ann. Neurol. **8:** 32–37.
23. HOHLFELD, R., K. V. TOYKA, L. L. MINER, S. L. WALGRAVE & B. M. CONTI-TRONCONI. 1988. Amphipathic segment of the nicotinic receptor alpha subunit contains epitopes recognized by T lymphocytes in myasthenia gravis. J. Clin. Invest. **81:** 657–660.
24. FUJII, Y. & J. LINDSTROM. 1988. Specificity of the T cell immune response to acetylcholine receptor in experimental autoimmune myasthenia gravis. J. Immunol. **140:** 1830–1837.
25. BROCKE, S., C. BRAUTBAR, L. STEINMAN, O. ABRAMSKY, J. ROTHBARD, D. NEUMANN, S. FUCHS & E. MOZES. 1988. *In vitro* proliferative responses and antibody titers specific to human acetylcholine receptor synthetic peptides in patients with myasthenia gravis and relation to HLA class II genes. J. Clin. Invest. **82:** 1894–1900.
26. MELMS, A., S. CHRESTEL, B. C. SCHALKE, H. WEKERLE, A. MAURON, M. BALLIVET & T. BARKAS. 1989. Autoimmune T lymphocytes in myasthenia gravis. Determination of target epitopes using T lines and recombinant products of the mouse nicotinic acetylcholine receptor gene. J. Clin. Invest. **83:** 785–790.

27. PROTTI, M. P., A. A. MANFREDI, C. STRAUB, X. D. WU, J. F. HOWARD & B. M. CONTI-TRONCONI. 1990. Use of synthetic peptides to establish anti-human acetylcholine receptor $CD4^+$ cell lines from myasthenia gravis patients. J. Immunol. **144:** 1711–1720.
28. MELMS, A., G., MALCHEREK, U. GERN, H. WIETHOLTER, C. A. MUELLER, R. SCHOEPFER & J. LINDSTROM. 1992. T cells from normal and myasthenic individuals recognize the human acethlcholine receptor: Heterogeneity of antigenic sites on the α-subunit. Ann. Neurol. **31:** 311–318.
29. NEWSOM-DAVIS, J., G. HARCOURT, N. SOMMER, D. BEESON, N. WILLCOX & J.-B. ROTHBARD. 1989. T cell reactivity in myasthenia gravis. J. Autoimmunity **2** (Suppl. 2): 101–108.
30. PALACE, J., C. M. WILES & J. NEWSOM-DAVIS. 1991. 3,4-diaminopyridine in the treatment of congenital (hereditary) myasthenia. J. Neurol. Neurosurg. Psychiatry **54:** 1069–1072.
31. KUNCL, R. W., R. N. ADAMS & D. B. DRACHMAN. 1988. Inhibition of methyltransferase reduces the turnover of acetylcholine receptors. Proc. Natl. Acad. Sci. USA **85:** 4032–4036.
32. KUNCL, R. W., D. B. DRACHMAN, I. WITTSTEIN, R. ADAMS, W. WIGGINS, O. AVILA, A. PESTRONK, K. MCINTOSH, D. MAURATH & S. DESILVA. 1991. Novel therapy of experimental myasthenia gravis by reducing endocytosis of acetylcholine receptors. Neurology **41:** 154.
33. HEININGER, K., M. HENDRICKS & K. V. TOYKA. 1985. Myasthenia gravis: a new semi-selective procedure to remove acetylcholine receptor autoantibodies from plasma. Plasma Ther. Transfus. Technol. **6:** 771–775.
34. DRACHMAN, D. B., R. N. ADAMS, K. MCINTOSH & A. PESTRONK. 1985. Treatment of experimental myasthenia gravis with cyclosporin A. Clin. Immunol. Immunopathol. **34:** 174–188.
35. TINDALL, R. W. A., J. A. ROLLINS, J. T. PHILLIPS, R. G. GREENLEE, L. WELLS & G. BELENDIUK. 1987. Preliminary results of a double-blind, randomized-placebo-controlled trial of cyclosporine in myasthenia gravis. N. Engl. J. Med. **316:** 719–724.
36. DESILVA, S., J. E. BLUM, K. R. MCINTOSH, S. ORDER & D. B. DRACHMAN. 1988. Treatment of experimental myasthenia gravis with total lymphoid irradiation. Clin. Immunol. Immunopath. **48:** 31–41.
37. PESTRONK, A. & D. B. DRACHMAN. 1980. Dimethyl sulfoxide reduces titers of anti-receptor antibody in experimental myasthenia gravis. Nature **288:** 733–734.
38. CRAIN, E., K. R. MCINTOSH, G. GORDON, A. PESTRONK & D. B. DRACHMAN. 1989. The effect of thalidomide on experimental autoimmune myasthenia gravis. J. Autoimmunity **2:** 197–202.
39. PESTRONK, A., D. B. DRACHMAN, R. TEOH & R. N. ADAMS. 1983. Combined short-term immunotherapy cures experimental autoimmune myasthenia gravis. Ann. Neurol. **14:** 235–241.
40. KILLEN, J. A. & J. M. LINDSTROM. 1984. Specific killing of lymphocytes that cause experimental autoimmune myasthenia gravis by ricin toxin-acetylcholine receptor conjugates. J. Immunol. **133:** 2549–2553.
41. STERZ, R. K., G. BIRO, K. RAJKI, G. FILIPP & K. PEPER. 1985. Experimental autoimmune myasthenia gravis: can pretreatment with ^{125}I-labeled receptor prevent functional damage at the neuromuscular junction? J. Neuroimmunol. **7:** 97–105.
42. SHIZURU, J. A., C. TAYLOR-EDWARDS, B. BANKS, A. R. GREGORY & C. G. FATHMAN. 1988. Immunotherapy of the nonobese diabetic mouse: treatment with an antibody to T-helper lymphocytes. Science **240:** 659–662.
43. BOGEN, S., E. MOZES & S. FUCHS. 1984. Induction of AChR-specific suppression: an *in vitro* model of antigen-specific immunosuppression in myasthenia gravis. J. Exp. Med. **159:** 292–304.
44. PACHNER, A. R. & F. S. KANTOR. 1987. Suppressor T-cell lines and hybridomas in murine myasthenia. Ann. N.Y. Acad. Sci. **505:** 619–627.
45. MCINTOSH, K. R. & D. B. DRACHMAN. 1986. Induction of suppressor cells specific for acetylcholine receptor in experimental autoimmune myasthenia gravis. Science **232:** 401–403.

46. McIntosh, K. R., D. B. Drachman & R. W. Kuncl. 1989. Antigen-specific suppressor macrophages induced by culture with cyclosporin A plus acetylcholine receptor. J. Neuroimmunol. **25:** 75–89.
47. Whitham, R. H., A. A. Vandenbark, D. N. Bourdette, Y. K. Chou & H. Offner. 1990. Suppressor cell regulation of encephalitogenic T cell lines: Generation of suppressor macrophages with cyclosporin A and myelin basic protein. Cell. Immunology **126:** 290–303.
48. Waldor, M. S., Sriram, H. O. McDevitt & L. Steinman. 1983. *In vivo* therapy with monoclonal anti-IA antibody suppresses immune responses to AChR. Proc. Natl. acad. Sci. USA **80:** 2713–2717.
49. Aharoni, R., D. Teitelbaum, R. Arnon & J. Puri. 1991. Immunomodulation of experimental allergic encephalomyelitis by antibodies to the antigen-Ia complex. Nature **351:** 147–150.
50. Adorini, L. 1992. Inhibition of T cell activation by MHC blockade: A possible strategy for immunointervention in autoimmune diseases. J. Autoimmunity **5** (Suppl. A): 73–81.
51. Schonbeck, S. & D. S. Dwyer. 1984. Regulation of the immune response to acetylcholine-receptor after injection of anti-idiotypic antibodies. Immunobiology **168:** 184.
52. De Baets, M. H. & P. J. C. Van Breda Vriesman. 1985. Immunomodulatory effects of anti-idiotypic antibodies in experimental autoimmune myasthenia gravis. Fed. Proc. Fed. Am. Soc. Exp. Biol. **44:** 607.
53. Agius, M. A. & D. P. Richman. 1986. Suppression of development of experimental autoimmune myasthenia gravis with isogeneic monoclonal antiidiotypic antibody. J. Immunol. **137:** 2195–2198.
54. Souroujon, M. C. & S. Fuchs. 1987. Antiidiotypic antibodies in the regulation of experimental autoimmune myasthenia gravis. Ann. N.Y. Acad. Sci. **505:** 676–682.
55. Chiller, J. M., G. S. Habicht & W. O. Weigle. 1970. Cellular sites of immunologic unresponsiveness. Proc. Natl. Acad. Sci. USA **65:** 551–556.
56. Weiner, H. L., Z. J. Zhang, S. J. Khoury, A. Miller, A. Al-Sabbagh, S. A. Brod, O. Lider, P. Higgins, R. Sobel, R. B. Nussenblatt & D. A. Hafler. 1991. Antigen-driven peripheral immune tolerance: Suppression of organ-specific autoimmune diseases by oral administration of autoantigens. Ann. N.Y. Acad. Sci. **636:** 227–232.
57. DeSilva, S., K. R. McIntosh, J. E. Blum, S. Order, D. Mellits & D. B. Drachman. 1990. Total lymphoid irradiation and antigen-specific tolerance: Treatment for experimental myasthenia gravis? J. Neuroimmunol. **29:** 93–103.
58. Waldmann, H., S. Qin & S. Cobbold. 1992. Monoclonal antibodies as agents to reinduce tolerance in autoimmunity. J. Autoimmunity **5** (Suppl. A): 93–102.
59. McIntosh, K. R. & D. B. Drachman. 1992. Tolerance to acetylcholine receptor induced by AChR-coupled syngeneic cells. J. Neuroimmunol. **38:** 75–84.
60. Oki, A. & E. Sercarz. 1985. T cell tolerance studied at the level of antigenic determinants. J. Exp. Med. **161:** 897–911.
61. Das, M. K. & J. Lindstrom. 1991. Epitope mapping of antibodies to acetylcholine receptor α subunits using peptides synthesized on polypropylene pegs. Biochemistry **30:** 2470–2477.
62. Cantrell, D. A. & K. A. Smith. 1984. The interleukin 2 T cell system: A new cell growth model. Science **224:** 1312–1316.
63. Waldmann, T. A. 1986. The structure, function, and expression of interleukin-2 receptors on normal and malignant lymphocytes. Science **232:** 727–732.
64. Williams, D. B., K. Parker, P. Bacha, W. Bishai, M. Borowski, F. Genbauffe, T. Strom & J. R. Murphy. 1987. Diphtheria toxin receptor binding domain substitution with interleukin 2: genetic construction and properties of a diphtheria toxin-related interleukin-2 fusion protein. Protein Eng. **1:** 493–498.
65. Bacha, P., D. B. Williams, C. Waters, J. M. Williams, J. R. Murphy & T. B. Strom. 1988. Interleukin-2 receptor targeted cytotoxicity: interleukin-2 receptor-mediated action of a diphtheria toxin-related interleukin-2 fusion protein. J. Exp. Med. **167:** 612–622.

66. YAMAIZUMI, M., E. MEKADA, T. UCHIDA & Y. OKADA. 1978. One molecule of diphtheria toxin fragment A introduced into a cell can kill the cell. Cell **15:** 245–250.
67. BALCER, L. J., K. R. MCINTOSH, J. C. NICHOLS & D. B. DRACHMAN. 1991. Suppression of immune responses to acetylcholine receptor by interleukin 2-fusion toxin: *in vivo* and *in vitro* studies. J. Neuroimmunol. **31:** 115–122.
68. BACHA, P., S. FORTE, N. KASSAM, J. THOMAS, D. AKIYOSHI, C. WATERS, J. NICHOLS & M. ROSENBLUM. 1990. Pharmacokinetics of the recombinant fusion protein DAB 486-IL2 in animal models. Cancer Chemother. Pharmacol. **26:** 409–414.
69. KIRKMAN, R. L., P. BACHA, L. V. BARRETT, S. FORTE, J. R. MURPHY & T. B. STROM. 1989. Prolongation of cardiac allograft survival in murine recipients treated with a diphtheria toxin-related interleukin-2 fusion protein. Transplantation **47:** 327–330.
70. KELLEY, V. E., P. BACHA, O. PANKEWYCA, J. C. NICHOLS, J. R. MURPHY, & T. B. STROM. 1988. Interleukin-2 diphtheria toxin fusion protein can abolish cell-mediated immunity *in vivo*. Proc. Natl. acad. Sci. USA **85:** 3980–3984.
71. COHEN, I. R., A. BEN-NUN, J. HOLOSHITZ, R. MARON & R. ZERUBAVEL. 1983. Vaccination against autoimmune disease with lines of autoimmune T lymphocytes. Immunol. Today **4:** 227–230.
72. ANDERSSON, L. C., H. BINZ & H. WIGZELL. 1976. Specific unresponsiveness to transplantation antigens induced by autoimmunization with syngeneic, antigen-specific T lymphoblasts. Nature **264:** 778–780.
73. BEN-NUN, A., M. WEKERLE & I. R. COHEN. 1981. Vaccination against autoimmune encephalomyelitis with T-lymphocyte line cells reactive against myelin basic protein. Nature **292:** 60–61.
74. NEILSON, E. G. & S. M. PHILLIPS. 1982. Suppression of interstitial nephritis by auto-anti-idiotypic immunity. J. Exp. Med. **155:** 179–189.
75. MARON, R., R. ZERUBAVEL, A. FRIEDMAN & I. R. COHEN. 1983. T lymphocyte line specific for thyroglobulin produces or vaccinates against autoimmune thyroiditis in mice. J. Immunol. **131:** 2316–2322.
76. HOLOSHITZ, J., Y. NAPARSTEK, A. BEN-NUN & I. R. COHEN. 1983. Lines of T lymphocytes induce or vaccinate against autoimmune arthritis. Science **219:** 56–58.
77. KAHN, C. R., K. R. MCINTOSH & D. B. DRACHMAN. 1990. T-cell vaccination in experimental myasthenia gravis: a double-edged sword. J. Autoimmunity **3:** 659–669.
78. LIDER, O., N. KARIN, M. SHINITZKY & I. R. COHEN. 1987. Therapeutic vaccination against adjuvant arthritis using autoimmune T cells treated with hydrostatic pressure. Proc. Natl. Acad. Sci. USA **84:** 4577–4580.
79. LIDER, O., T. RESHEF, E. BERAUD, A BEN-NUN & I. R. COHEN. 1988. Antiidiotypic network induced by T cell vaccination against experimental autoimmune encephalomyelitis. Science **239:** 181–183.
80. ACHA-ORBEA, H., D. J. MITCHELL, L. TIMMERMANN, D. C. WRAITH, G. W. TAUSCH, M. K. WALDOR, S. S. ZAMVIL, H. O. MCDEVITT & L. STEINMAN. 1988. Limited heterogeneity of T cell receptors from lymphocytes mediating autoimmune encephalomyelitis allows specific immunointervention. Cell **54:** 263–273.
81. VANDENBARK, A. A., Y. K. CHOU, D. N. BOURDETTE, R. WHITHAM, G. A. HASHIM & H. OFFNER. 1992. T cell receptor peptide therapy for autoimmune disease. J. Autoimmunity **5** (Suppl A): 83–92.
82. HOWELL, M. D., S. T. WINTERS, T. OLEE, H. C. POWELL, D. J. CARLO & S. W. BROSTOFF. 1989. Vaccination against experimental allergic encephalomyelitis with T cell receptor peptides. Science **246:** 668–670.
83. BERRIH-AKNIN, S., S. COHEN-KAMINSKY, D. NEUMANN, D. SAFAR, B. EYMARD, C. GAND, P. LEVASSEUR, S. FUCHS & J.-F. BACH. 1988. Cellular aspects of myasthenia gravis. Immunol. Res. **7:** 189–199.
84. PROTTI, M. P., A. A. MANFREDI, X. D. WU, L. MOIOLA, J. F. HOWARD & B. M. CONTI-TRONCONI. 1991. Myasthenia gravis. T epitopes on the subunit of human muscle acetylcholine receptor. J. Immunol. **146:** 2253–2261.
85. SCHWARTZ, R. H. 1985. T-lymphocyte recognition of antigen in association with gene products of the major histocompatibility complex. Annu. Rev. Immunol. **3:** 237–261.

86. ROITT, I., J. BROSTOFF & D. MALE. 1986. Immunology. 8.5–8.7. C. V. Mosby Co. St. Louis.
87. JANEWAY, C. A., J. RON & M. E. KATZ. 1987. The B cell is the initiating antigen-presenting cell in peripheral lymph nodes. J. Immunol. **138:** 1051.
88. RON, J. & J. SPRENT. 1987. T cell priming *in vivo:* a major role for B cells in presenting antigen to T cells in lymph nodes. J. Immunol. **138:** 2848.
89. ASHWELL, J. D. 1988. Are B lymphocytes the principal antigen-presenting cells *in vivo*? J. Immunol. **140:** 3697.
90. ROCK, L. L., B. BENACERRAF & A. K. ABBAS. 1984. Antigen presentation by hapten-specific B lymphocytes. 1. Role of surface immunoglobulin receptors. J. Exp. Med. **160:** 1102.
91. LANZAVECCHIA, A. 1985. Antigen specific interactions between T and B cells. Nature **314:** 537.
92. CHESNUT, R. W. & H. M. GREY. 1986. Antigen presentation by B cells and its significance in T-B interactions. Adv. Immunol. **39:** 51.
93. CASTEN, L. A. & S. K. PIERCE. 1988. Receptor-mediated B cell antigen processing. J. Immunol. **140:** 404–410.
94. SNIDER, D. P. & D. M. SEGAL. 1987. Targeted antigen presentation using crosslinked antibody heteroaggregates. J. Immunol. **139:** 1609–1616.
95. SNIDER, D. P. & D. M. SEGAL. 1989. Efficiency of antigen presentation after antigen targeting to surface IgD, IgM, FcRII, and B220 molecules on murine splenic B cells. J. Immunol. **143:** 59–65.
96. LINDSTROM, J. 1984. Use of monoclonal antibodies in the study of myasthenia gravis. *In* Monoclonal Antibodies: Probes for the study of autoimmunity and immunodeficiency. G. Eisenbarth & B. Haynes, Eds.: 91–109. Academic Press. New York.
97. SPDP. Heterobifunctional reagent. Pharmacia Fine Chemicals. Uppsala, Sweden.
98. REIM, J., K. R. MCINTOSH, S. MARTIN & D. B. DRACHMAN. 1992. Specific immunotherapeutic strategy for myasthenia gravis: targeted antigen-presenting cells. J. Neuroimmunol. **41:** 61–70.
99. JENKINS, M. K. & R. H. SCHWARTZ. 1987. Antigen presentation by chemically modified splenocytes induces antigen-specific T cell unresponsiveness *in vitro* and *in vivo*. J. Exp. Med. **165:** 302–319.
100. SCHWARTZ, R. H. 1990. A cell culture model for T lymphocyte clonal anergy. Science **248:** 1349–1356.
101. GILBERT, K. M., K. D. HOANG & W. O. WEIGLE. 1990. Th1 and Th2 clones differ in their response to a tolerogenic signal. J. Immunol. **144:** 2063–2071.
102. LEA, T., F. VARTDAL, C. DAVIES & J. UGELSTAD. 1985. Magnetic monosized polymer particles for fast and specific fractionation of human mononuclear cells. Scand. J. Immunol. **22:** 207–216.
103. GILLIS, S. & D. A. SMITH. 1977. Long term culture of tumor-specific cytotoxic cells. Nature **268:** 154–156.

Seronegative Myasthenia Gravis

Evidence for Plasma Factor(s) Interfering with Acetylcholine Receptor Function[a]

A. VINCENT,[e] Z. LI, A. HART, R. BARRETT-JOLLEY, T. YAMAMOTO, J. BURGES,[b] D. WRAY,[b,c] N. BYRNE, P. MOLENAAR,[d] AND J. NEWSOM-DAVIS[b]

Department of Clinical Neurology
University of Oxford
Oxford, United Kingdom

[b]*Department of Pharmacology*
Royal Free Hospital School of Medicine
London, United Kingdom

[c]*Department of Pharmacology*
University of Leeds
Leeds, United Kingdom

[d]*Department of Pharmacology*
University of Leiden
Leiden, the Netherlands

INTRODUCTION

It has long been recognized that about 10–15% of patients with typical features of generalized myasthenia gravis (MG) do not have detectable serum anti-AChR antibody, as measured by immunoprecipitation of [^{125}I]α-bungarotoxin (α-BuTx)-labeled human acetylcholine receptor (AChR). A few of these patients may become anti-AChR positive (seropositive, SPMG) over a period of time,[1] but there remains a substantial number of persistently seronegative (SNMG) patients.

Although anti-AChR is undetectable, other autoantibodies and autoimmune disorders may be present,[2] and SNMG patients frequently respond to plasma exchange and immunosuppressive therapy such as prednisolone and azathioprine,[2,3] clearly indicating that their disorder of neuromuscular transmission is mediated by humoral factors. Here we review the various approaches that we have used to investigate humoral immune mechanisms in generalized SNMG.

CLINICAL OBSERVATIONS

SNMG must first be distinguished from congenital myasthenic syndromes. The latter usually present at birth or within the first two years, do not respond to

[a] J. Burges and R. Barrett-Jolley were supported by Action Research for the Crippled Child.

[e] Address for correspondence: Neurosciences Group, Institute of Molecular Medicine, John Radcliffe Hospital, Oxford OX3 9DU, UK.

TABLE 1. Comparison of Seropositive and Seronegative Myasthenia Gravis; Clinical Features in 480 Patients with Generalized Disease

	Seropositive	Seronegative
Total number	436	44
M : F	1 : 3	1 : 1
Age of onset		
<10	4%	19%***
<20	28%	38%
20–40	58%	52%
40–60	11%	5%
>60	14%	10%
IIa[a]	28%	29%
IIb	50%	52%
III	17%	17%
IV	5%	2%

*** $p < 0.001$.

[a] Roman numerals indicate severity of symptoms: IIa, mild; IIb, moderate; III, severe; IV, chronic severe.

plasma exchange or to immunosuppression, and are due to congenital or inherited abnormalities of neuromuscular transmission (see Vincent *et al.* and Engel *et al.*, this volume).

Out of 480 consecutive patients with typical symptoms and clinical evidence (positive tensilon test or >10% decrement) of generalized MG, 44 were seronegative in our hands (TABLE 1), and of these, 19% presented before the age of 10 compared to only 4% of SPMG patients ($p < 0.001$). The sex distribution was approximately equal, contrasting with a predominance of females in seropositive cases. In other respects the clinical features did not generally differ from those in SPMG, and in particular the proportion of patients presenting within each clinical (Osserman) classification was very similar.

There were, however, highly significant differences in thymic pathology (TABLE 2). Whereas the majority (58%) of SPMG patients showed hyperplasia of the thymus, only 17% of SNMG patients had a hyperplastic thymus, and in the remaining 83% the thymus was described as atrophied, involuted, or normal. Moreover, there were no cases of thymoma among SNMG patients compared to 28% of SPMG cases.

TABLE 2. Comparison of Seropositive and Seronegative Myasthenia Gravis; Thymic Pathology

	Seropositive	Seronegative
Number available	286	18
Hyperplasia	59%	17%***
Thymoma	28%	0%**
Atrophy/involuted/normal	13%	83%***

** $p < 0.01$.
*** $p < 0.001$.

PATHOLOGY OF NEUROMUSCULAR TRANSMISSION IN SNMG

We had the opportunity to investigate intercostal muscle biopsies from several patients with SNMG (for more detail of the methods, see Vincent *et al.*, this volume). In four out of six biopsies, miniature end-plate potential (mepp) amplitudes were significantly reduced (TABLE 3), whereas end-plate [^{125}I]α-BuTx binding was reduced in five of seven. Mepp amplitudes, however, did not correlate well with α-BuTx binding; for instance in one case α-BuTx binding was normal whereas mepp amplitudes were reduced, and in another case the reverse was found. Quantal content was reduced, substantially in one case, in each of the three biopsies in which it was measured (TABLE 3).

TABLE 3. Results from Intercostal Muscle Biopsies

Patient	End-plate AChR, α-BuTx sites $\times$ 10^{-7}	Mepp amplitude, mV: Normal Solution	Mepp amplitude, mV: Eserine Present	Quantal Content
1	0.4	0.43 ± 0.01 (2)	0.68 ± 0.10 (4)	—
2	1.4 ± 0.4 (4)[a]	—	0.47 ± 0.13 (4)	—
3	0.59 ± 0.08 (3)	0.33 ± 0.03 (7)	—	—
4	0.54 ± 0.08 (6)	0.86 ± 0.16 (5)	—	45 ± 17 (4)
5	0.86 ± 0.1 (9)	0.69 ± 0.15 (3)	—	56 (5)
6	undetectable	0.41 ± 0.02 (3)	—	–
7	0.23 ± 0.03 (5)	—	—	27 ± 6 (5)
control	1.2 ± 0.1 (7)	0.84 ± 0.07 (4)	1.05 ± 0.08 (12)	79 ± 9 (15)

[a] Results are mean ± SD; (no. of observations).

PATHOGENIC FACTORS IN SNMG PLASMA

We showed previously that injection of plasma Ig from SNMG patients into mice, at 60 mg/day for three days, resulted in a defect in neuromuscular transmission: a significantly increased decrement in the compound muscle action potential of the diaphragm, during phrenic nerve stimulation at 3 Hz in the presence of *d*-tubocurarine (dTC; 0.8 μg/mL), compared to the decrement in diaphragms from control Ig-treated animals.[3]

When SPMG Ig was passively transferred to mice,[4] a reduction in end-plate AChRs was found, mepp amplitudes were reduced, and Ig could be detected bound to AChR extracted from the mouse leg muscle.[5,6] When SNMG Ig was injected, however, there was only a marginal reduction in the number of end-plate AChRs in the mice and no evidence of Ig bound to the extracted AChRs.[3]

Further investigations were made in mice injected with SNMG Ig or plasma followed by microelectrode recordings from the diaphragm muscle. After 15 days injection of plasma, there were significant reductions in mepp amplitude (FIG. 1). There were also reductions in quantal content in mice treated with three out of six of the plasma preparations. Furthermore the amount of dTC required to just block neuromuscular transmission (in order to measure quantal content) was significantly less in mice injected with each SNMG plasma compared to control plasma, indicating impairment of neuromuscular transmission (Burges *et al.*, in preparation) and confirming the increased sensitivity to dTC shown by Mossman *et al.*[3] Also in agreement with previous findings,[3] despite the reduction in mepp amplitude, there was no detectable reduction in leg muscle [^{125}I]α-BuTx binding, nor was there any detectable Ig bound to extracted AChRs. In further experiments, the crude Ig fraction from three SNMG plasmas was injected for three days, and reductions in mepp amplitude, quantal content, and dTC concentration required for neuromuscular block were seen, though to a lesser extent (FIG. 2).

EFFECT OF SNMG PLASMA ON ACETYLCHOLINE METABOLISM

In mice treated with two SNMG plasma preparations, there were significant reductions in mepp amplitudes without any reduction in quantal content or leg AChR. One possible explanation would be a reduction in the amount of ACh

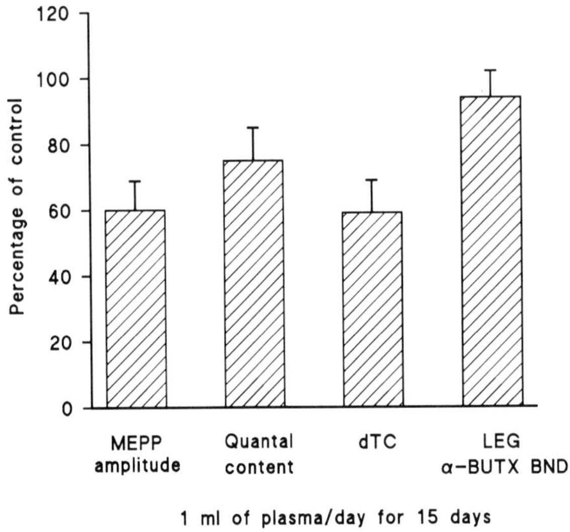

FIGURE 1. Effect of 15 days of SNMG plasma injections. The FIGURE shows mepp amplitudes, quantal contents, dTC concentrations that were required to block neuromuscular transmission, and leg muscle AChR numbers ([^{125}I]α-BuTx binding) in mice injected for 15 days with 1 mL of plasma/day. Mean results (+SEM) from six SNMG preparations are expressed as a percentage of results in control-injected muscles. Mepp amplitudes, quantal content, and dTC concentrations were significantly reduced overall; individually 5/6 SNMG plasmas reduced mepp amplitudes, and 3/6 reduced quantal content.

in each synaptic vesicle. We therefore measured in detail ACh metabolism in diaphragms of mice treated with one SNMG Ig preparation and compared the results to control Ig-treated muscles (TABLE 4).

There were no significant differences in ACh content, resting or KCl-evoked ACh release, or the activity of cholinesterase or choline acetyltransferase. A moderate (27%) reduction in ACh release induced by stimulation of the phrenic nerve at 3 Hz was observed, but this was not statistically significant (TABLE 4; P. Molenaar unpublished observations, 1986; Burges *et al.*, in preparation).

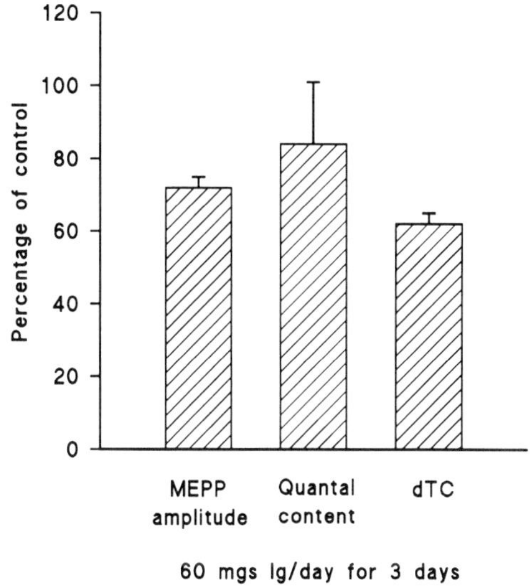

FIGURE 2. Effect of three days of SNMG Ig injections. The FIGURE shows mepp amplitudes, quantal contents, and concentrations of d-tubocurarine (dTC) that were required in mice injected with approximately 60 mg/day of SNMG Ig. Mean results (+SEM) from animals treated with three different SNMG preparations are expressed as a percentage of result in control-injected muscles.

EFFECT OF PLASMA AND PLASMA FACTORS ON AChR FUNCTION IN CULTURED MUSCLE CELLS

In order to investigate further the nature of the defect in neuromuscular transmission in SNMG we have established an assay for AChR function, using the TE671 rhabdomyosarcoma cell line, which expresses human AChR.[8,9] Incubation of the cells in carbachol (0.5–1.0 mM) in the presence of $^{22}Na^+$ results in uptake of $^{22}Na^+$, which can be measured by gamma or scintillation counting.[10] This uptake was re-

TABLE 4. Acetylcholine Metabolism in Mouse Muscle after 15 Days Injection of SNMG Plasma

	Mepp Amplitude mV	Fresh ACh[a] Content in EDL (pmol)	Evoked ACh Release (pmol/min)[c] 3 Hz Stimulation	50 mM KCl	CHAT[a,b] pmol/h	AChE[a] nmol/h
Control Ig[a]	0.83	17.8	0.89	1.66	1250	960
SNMG Ig	0.42	17.3	0.65[d]	1.68	1400	1020

[a] Measured in extensor digitorum longus (EDL).
[b] CHAT = choline acetyltransferase.
[c] Measured from phrenic nerve diaphragm preparation. For methods see ref. 7.
[d] Not significantly reduced.

TABLE 5. Effects of SNMG Plasma on AChR Function in TE671 Cells[a]

		ACh-Induced Currents Percentage of Initial Current	Carbachol-Induced ^{22}Na Flux Percentage Control
Seronegative plasmas	1	122 (1)	81.5 (2)
	2	141 (1)	ND[b]
	3	56 ± 56 (8)	52 ± 17 (4)
	4	38 ± 20 (11)	43 ± 5 (4)
	5	107 ± 1 (2)	96 ± 6 (3)
	6	59 ± 33 (4)	31 ± 34 (6)
Healthy control plasmas		111 ± 16 (6)	99 ± 5 (7)

[a] All results are mean ± SD; (no. of observations).
[b] Not determined.

duced by up to 70% when cells were preincubated in the presence of about 50% of the SNMG plasmas at 1 : 10 dilution,[11] and the same SNMG plasmas also inhibited the ACh-induced current in whole-cell voltage clamped TE671 cells[12] (TABLE 5). As in the passive transfer experiments reviewed above, however, there was insignificant loss of AChR, as measured by [^{125}I]α-BuTx binding, and no evidence of Ig bound to the extracted AChRs.[11] Moreover, this inhibition of AChR function was found in the IgG-depleted fraction which contains other immunoglobulins (TABLE 6) and copurified with IgM.[11] Recently we have attempted, with little success, to demonstrate binding of IgM to AChR on the surface of the TE671 cells by using the bifunctional reagent BS_3 to cross-link it to its target before extraction, followed by immunoprecipitation with anti-Ig (FIG. 3).

EFFECT OF OTHER SUBSTANCES ON CARBACHOL-INDUCED $^{22}Na^{\dagger}$ FLUX

Alternative possibilities are that a non-Ig factor binds to the AChR, or that the target for the plasma factor, whether Ig or non-Ig, is another component of the muscle membrane. We have found that a number of substances can inhibit carbachol-induced $^{22}Na^+$ flux into the TE671 cells; the lectins concanavalin A and wheat germ agglutinin, which are known to bind to *Torpedo* AChR, inhibit flux, the latter doing so without affecting the binding of [^{125}I]α-BuTx to the AChR (FIG.

TABLE 6. Effects of SNMG Plasma and Ig Fractions on AChR Function in TE671 Cells[a]

	$^{22}Na^+$ Flux as Percentage of Results in PBS or Control Plasma (n = 12)
Plasma	68 ± 24%
IgG	101 ± 12%
IgG-depleted	52 ± 25%

[a] TE671 cells were preincubated in SNMG plasma or plasma fractions for two hours before measurement of carbachol-induced $^{22}Na^+$ flux. IgG was purified by affinity chromatography on Protein G Sepharose (Pharmacia Ltd).

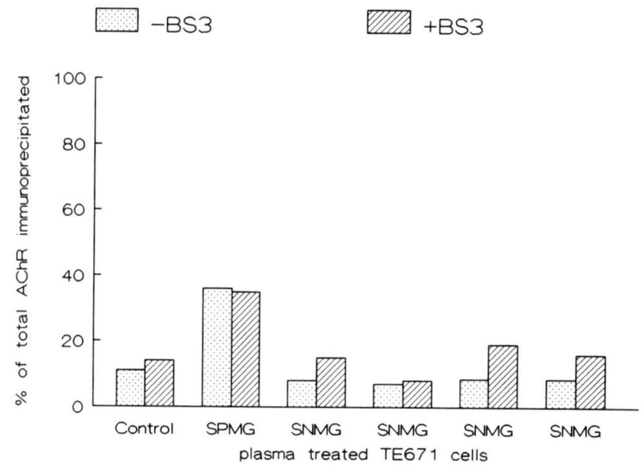

FIGURE 3. Effect of the cross-linking agent BS_3. TE671 cells treated with SNMG, SPMG, or control plasma for 2 hours at room temperature were exposed to 5 mg/mL BS_3 for 30 min, quenched in 5 mM glycine, and incubated in [^{125}I]α-BuTx to label surface AChR. The cells were extracted, and [^{125}I]α-BuTx-AChRs with bound or cross-linked Ig attached were precipitated with anti-Ig and counted on a gamma counter. Results are expressed as a percentage of total [^{125}I]α-BuTx-AChR extracted. Four different SNMG plasmas were tried.

4; Li *et al.*, in preparation) and thus mimicking the effect of SNMG plasma. Although plasma from healthy controls does not affect the flux (*e.g.* TABLE 5), we previously found that plasma and IgG-depleted plasma from some patients with demyelinating neuropathies produced a substantial inhibition,[11] suggesting the possibility that antiganglioside antibodies might be involved. Indeed, peanut

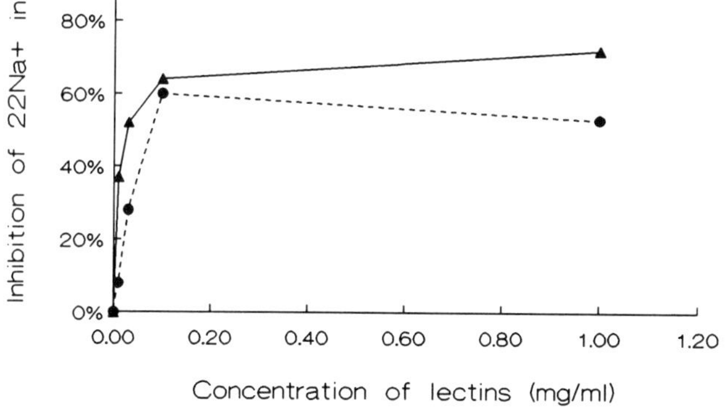

FIGURE 4. Effect of lectins on carbachol-induced $^{22}Na^+$ flux. TE671 cells were exposed to concanavalin A (Con A) and wheat germ agglutinin (WGA) at the relevant concentration in PBS for 1 h before performing the flux assay. Results are expressed as a percentage of flux in PBS-treated cells. —▲—, Con A; ---●--- WGA.

agglutinin and cholera toxin, both of which are thought to bind to GM1 gangliosides, inhibited the flux (data not shown).

DISCUSSION

Now that the immunopathological mechanisms in typical seropositive MG are relatively well understood, attention is being directed towards those in patients who are seronegative. Soliven *et al.*[2] have recently reviewed the serological and clinical data in 221 MG patients. Of the 41 who were negative for anti-AChR antibodies, 8 had congenital myasthenic syndromes and another 8 had purely ocular symptoms. No significant differences from SPMG patients were found in the remaining 25 SNMG patients with generalized weakness, although there was a tendency towards a younger age of onset, particularly among females; and no thymoma cases were present. In our study of 480 patients with acquired generalized MG, there were 44 cases (8%) designated as seronegative (anti-AChR <0.2 nM), and a relatively high proportion of these (8/44 compared with 17/436, $p < 0.001$) presented before the age of 10 (TABLE 1).

Among the 18 thymectomized patients, there were no cases of thymoma ($p < 0.01$), very few with thymic hyperplasia, and a greatly increased incidence of atrophied, involuted, or normal thymus ($p < 0.001$; TABLE 2). These differences are supported by an investigation of cellular function in cultured thymic tissue.[13] These authors found significantly fewer germinal centers and less IgG produced by thymic lymphocytes in culture. There were T-cell areas, however, similar to those found in SPMG thymus, that were not found in controls. None of the SNMG thymuses produced anti-AChR antibody.

In spite of these differences in age of onset and thymic pathology between SNMG and SPMG, however, the distribution of each group of patients within the Osserman classification was almost identical, indicating that the absence of detectable anti-AChR antibody does not associate with less severe disease (TABLE 1).

There have been few detailed studies of biopsied muscle from SNMG patients. A loss of end-plate AChRs, however, and reduced mepp amplitudes have been reported in a few cases.[14] These changes are similar to those found in SPMG muscle (for example[15]). In the seven SNMG intercostal muscle biopsies we have examined, there has been a poor correlation between the reductions in mepp amplitude and end-plate AChR numbers (TABLE 3) among these patients, in contrast to that found in typical SPMG. Furthermore, there is evidence for a decreased quantal content in some patients.

There seems little doubt that seronegative myasthenia gravis is due to the presence of a circulating plasma immune factor; the response to plasma exchange and immunosuppression is similar to that found in seropositive MG patients, and a defect in neuromuscular transmission can be passively transferred to mice by injection of SNMG immunoglobulins. In contrast to SPMG, however, the injected mice showed very little reduction in end-plate AChR and no evidence of Ig bound to the extracted AChRs.[3]

These results have now been confirmed and extended by microelectrode studies of injected mice. After 15 days of SNMG plasma injections, there were electrophysiological abnormalities, either reduced mepp amplitudes, reduced quantal content, or both (FIG. 1; Burges *et al.*, in preparation). Similar effects, though to a lesser extent, were seen after three days injection of immunoglobulin. Thus these experiments suggest both pre- and postsynaptic effects of a plasma Ig factor.

We have previously shown that the TE671 cell line, which expresses human AChR, provides a convenient system for looking at AChR function[10] and a source of human AChR that is antigenically similar to that found in denervated muscle.[9] SPMG plasma and IgG preparations inhibited the carbachol-induced $^{22}Na^{+}$ flux into these cells;[10,11] this results in loss of cell-surface AChR and detectable Ig bound to extracted AChR. Plasma and IgG-depleted plasma fractions from a proportion (about 50%) of SNMG patients also inhibit the flux,[11] and the ACh-induced currents, into voltage-clamped TE671 cells[12] (TABLE 4), but in these cases there is minimal loss of AChR and no detectable Ig bound.[11] One possible explanation for the lack of detectable Ig bound to the AChR after incubation of TE671 cells in the presence of SNMG plasma would be the existence of a low-affinity, IgM anti-AChR antibody that dissociated during or after extraction. For this reason we attempted to attach the putative IgM to its target (presumably the AChR) by using a bifunctional cross-linking agent, BS_3. Even under these conditions, however, the extracted AChRs had very little Ig attached (FIG. 3).

Another possibility is that the factor in SNMG plasma, although immune-related, is not an immunoglobulin. We have found that several non-Ig substances can affect AChR function in the TE671 cells, including concanavalin A and wheat-germ agglutinin, two lectins that are known to bind to AChR (FIG. 4). It is conceivable that high molecular weight, lectin-like molecules that copurify with IgM could be responsible for the inhibition of AChR function that we observe in SNMG plasma. On the other hand there is increasing evidence that a proportion of SNMG plasma does contain immunoglobulins that bind to a cultured mouse cell line[16] or factors that cause modulation or reduction in $[^{125}I]\alpha$-BuTx binding sites on human muscle cells in about 30% of SNMG sera (V. A. Lennon, personal communication). Thus an alternative explanation is that they bind to some other component of the muscle surface. We investigated the possibility that binding to gangliosides might indirectly affect AChR function, because we had found that plasma from patients with demyelinating peripheral neuropathies also reduced $^{22}Na^{+}$ flux.[11] Both cholera toxin and peanut agglutin, which are known to bind to GM1 gangliosides, reduced flux (Vincent, in preparation) indicating the potential of other Ig or non-Ig factors to inhibit indirectly AChR function. Cholera toxin is thought to exert its effects on cellular function by raising intracellular cyclic AMP; interestingly, the inhibition by one SNMG plasma of ACh-induced currents in voltage-clamped whole TE671 cells (patient 4, TABLE 5) was prevented by buffering internal Ca^{2+} with 10 mM EGTA, indicating that it involves a Ca^{2+}-dependent process.[12]

In conclusion, we have confirmed the importance of plasma factor(s) in the pathogenesis of SNMG, but their exact identity and target(s) are still unclear. Our results suggest that non-IgG factors, possibly IgM, in a proportion of SNMG patients inhibit AChR function, but these factors may act indirectly, rather than by binding to the AChR itself. The factors involved in the remaining patients, whose plasma does not inhibit AChR function in TE671 cells, are unknown. Plasmas from several of these patients were effective in passive transfer.[3] Some plasmas had both pre- and postsynaptic actions, and one only affected quantal content (FIG. 1 and Burges *et al.*, in preparation), altogether implying heterogeneity in the mechanisms underlying SNMG. It is clearly important that assays should be established that can demonstrate the existence of Ig or non-Ig factors interacting with muscle or neuronal surface antigens, thus enabling more extensive studies on their incidence and disease specificity to be performed.

ACKNOWLEDGMENTS

We are grateful to the Medical Research Council of Great Britain for financial support.

REFERENCES

1. VINCENT, A. & J. NEWSOM-DAVIS. 1985. Acetylcholine receptor antibody as a diagnostic test for myasthenia gravis: results in 153 validated cases and 2967 diagnostic assays. J. Neurol. Neurosurg. Psychiatry **48:** 1246–1252.
2. SOLIVEN, B. C., D. J. LANGE, A. S. PENN *et al.* 1988. Seronegative myasthenia gravis. Neurology **38:** 514–517.
3. MOSSMAN, S., A. VINCENT & J. NEWSOM-DAVIS. 1986. Myasthenia gravis without acetylcholine receptor antibody: a distinct disease entity. Lancet **i:** 116–119.
4. TOYKA, K. V., D. B. DRACHMAN, D. E. GRIFFIN, A. PESTRONK, J. A. WINKELSTEIN, K. H. FISCHBECK & I. KAO. 1977. Myasthenia gravis: study of humoral immune mechanisms by passive transfer to mice. N. Engl. J. Med. **296:** 125–131.
5. WILSON, S., A. VINCENT & J. NEWSOM-DAVIS. 1983. Acetylcholine receptor turnover in mice with passively transferred myasthenia gravis. I. Receptor degradation. J. Neurol. Neurosurg. Psychiatry **46:** 377–382.
6. MOSSMAN, S., A. VINCENT, & J. NEWSOM-DAVIS. 1988. Passive transfer of myasthenia gravis by immunoglobulins: lack of correlation between AChR with antibody bound, acetylcholine receptor loss and transmission defect. J. Neurol. Sci. **84:** 15–28.
7. MOLENAAR, P. C. 1990. *In* Neuromuscular transmission: basic and applied aspects. A. Vincent and D. Wray, Eds. Manchester University Press. Manchester, UK.
8. LUTHER, M. A., R. SCHOEPFER, P. WHITING *et al.* 1989. A muscle acetylcholine receptor is expressed in the human cerebellar medulloblastoma cell line TE671. J. Neurosci **9:** 1082–1096.
9. WALKER, R., A. VINCENT & J. NEWSOM-DAVIS. 1988. Immunological and pharmacological heterogeneity of alpha-bungarotoxin binding sites extracted from TE671 cells. J. Neuroimmunol. **19:** 149–157.
10. LANG, B., G. RICHARDSON, J. REES, A. VINCENT & J. NEWSOM-DAVIS. 1988. Plasma from myasthenia gravis patients reduces acetylcholine receptor agonist-induced Na^+ flux into TE671 cell line. J. Neuroimmunol. **19:** 141–148.
11. YAMAMOTO, T., A. VINCENT, T. A. CIULLA, B. LANG, I. JOHNSTON & J. NEWSOM-DAVIS. 1991. Seronegative myasthenia gravis: A plasma factor inhibiting agonist-induced acetylcholine receptor function copurifies with IgM. Ann. Neurol. **30:** 550–557.
12. BARRETT-JOLLEY, R., N. G. BYRNE, J. NEWSOM-DAVIS & A. VINCENT. 1992. Inhibition of ACh-currents by human plasma in a whole-cell voltage-clamped human muscle cell line. J. Physiol. Lond. **446:** 267P.
13. WILLCOX, N., M. SCHLUEP, M. A. RITTER & J. NEWSOM-DAVIS 1991. The thymus in seronegative myasthenia gravis patients. J. Neurol. **238:** 256–261.
14. PESTRONK, A., D. B. DRACHMAN & S. B. SELF. 1985. Measurement of junctional acetylcholine receptors in myasthenia gravis: clinical correlates. Muscle & Nerve **8:** 245–251.
15. ITO, Y., R. MILEDI, A. VINCENT & J. NEWSOM-DAVIS. 1978. Acetylcholine receptors and end-plate electrophysiology in myasthenia gravis. Brain **101:** 345–368.
16. BROOKS, E. B., A. R. PACHNER, D. B. DRACHMAN & F. S. KANTOR. 1990. A sensitive rosetting assay for detection of acetylcholine receptor antibodies using BC_3H-1 cells: Positive results in "antibody-negative" myasthenia gravis. J. Neuroimmunol. **28:** 83–93.

A Clinical Therapeutic Trial of Cyclosporine in Myasthenia Gravis[a]

RICHARD S. A. TINDALL,[b] J. THEODORE PHILLIPS,
JULIA A. ROLLINS, LIESELOTTE WELLS,
AND KAREN HALL

Department of Neurology
University of Texas Southwestern Medical Center
Dallas, Texas

INTRODUCTION

Cyclosporine A is a cyclic undecapeptide and fungal metabolite that has been shown effective in preventing and suppressing the expression of autoimmune disorders induced in experimental animals, and in inducing and maintaining tolerance in cardiac, renal, and hepatic transplantation.[1-3] In a noncytotoxic fashion, it inhibits predominantly T lymphocyte–dependent functions with differential effects on T-cell subpopulations. It has been shown to reversibly inhibit clonal expansion of activated T-helper cells while allowing the activation and expression of T-suppressor lymphocytes. T cell–dependent antibody responses are so inhibited.[4] Following internalization, cyclosporine binds to a distinct cytosolic protein, cyclophilin, which is an immunophilin with *cis-trans* prolyl isomerase activity, resulting in the formation of an inhibitory complex that blocks transcription directed by the factor NF-AT.[5] This factor is essential for early T-cell gene activation. FK506 and rapamycin appear to act in an analogous fashion.[6] By inhibiting cytokine gene expression at a pretranscription level, the production and secretion of IL-2 and other lymphokines and of IL-2 receptor is blocked, resulting in the well-defined inhibitory effects on $CD4^+$ T cells.[4,5]

Myasthenia gravis (MG) is a T-cell dependent chronic autoimmune disorder due to a persistent but limited break in self-tolerance.[7] Both the induction and expression of experimental autoimmune MG (EAMG) can be blocked or suppressed by cyclosporine.[8] An earlier controlled therapeutic trial for patients with recent onset progressive generalized MG demonstrated that cyclosporine was effective at suppressing the manifestations of the disease, although nephrotoxicity was an important limitation.[9,10]

The conventional therapy for generalized MG has been thymectomy and/or corticosteroid therapy.[11] Many patients, however, will subsequently require continuing corticosteroid therapy for years, leading to a substantial degree of cumulative toxicity. In an effort to reduce or eliminate corticosteroid dosage in a population of corticosteroid-dependent MG patients, a therapeutic clinical trial of cyclosporine was undertaken.

[a] This work was supported by Grants from the Muscular Dystrophy Association, Sandoz Research Institute, and the General Clinical Research Center of the National Institutes of Health (Grant M01-RR-00633).

[b] Address correspondence to Richard S. A. Tindall, M.D., 930 Afton Road, San Marino, CA 91108.

Based on preliminary data demonstrating that a split dosage format (b.i.d.) resulted in reduced peak serum drug concentrations with increased trough levels, and reduced nephrotoxicity, the dosage in this trial was reduced and divided (5 mg/kg/day; split dosage) in comparison to the earlier study (6 mg/kg/day; once daily).

METHODS

The design of this therapeutic trial was randomized, placebo-controlled, and double-blinded. Men and women over the age of 18 years, with or without prior thymectomy, who had generalized MG with prominent clinical symptoms, despite continuing moderate or high-dose corticosteroid therapy, and who were seropositive for antiacetylcholine receptor antibody titer, were eligible for the study. The diagnosis of MG was based on accepted clinical findings, seropositivity, and, where indicated, repetitive nerve stimulation, demonstrating a decrement in conduction. Patients were considered steroid-dependent if improvement followed initiation of steroid therapy, and if, despite taking a dosage of ≥30 mg of prednisone (or equivalent) every other day, weakness was present on exam and the patient experienced unacceptable (greater) symptoms on a lower dosage. Patients with uncontrolled or poorly controlled hypertension and severe hepatic or renal disease were excluded.

The duration of the study was 6 months based upon a time-to-maximal benefit of 3.6 months found in an earlier study (at 6 mg/kg/day). Patients who qualified for this study were randomly assigned (random number table) to treatment with either placebo or cyclosporine. Cyclosporine was initiated at a dosage of 5 mg/kg per day in a divided dosage format; one half after breakfast and one half after dinner, approximately 10 hours apart. The dosage was adjusted by the unblinded investigator (J. T. Phillips) on the basis of whole blood levels of the drug to maintain the morning trough levels between 300 and 500 ng/mL (whole blood RIA), and to maintain serum creatinine values at <2.0 mg/deciliter (<177 μmol/liter). To maintain blinding, the unblinded investigator changed dosages on 2 or more patients simultaneously. The dosage was adjusted monthly by the blinded investigator (R. S. A. Tindall) on the basis of any adverse reactions. In the event of a conflict in dosage adjustments between the blinded and unblinded investigator, the unblinded investigator's adjustments were followed or the patient would be removed from the study.

Patients were evaluated monthly with a measurement of disease severity (Quantified Myasthenia Gravis Scale score or MG score),[9] antireceptor antibody titer,[9] steroid dosage, and Mestinon dosage. At each evaluation, the patients were tested more than 12 hours after their last dose of anticholinesterase therapy to assess the actual degree of weakness. Steroid withdrawal was initiated beginning with visit 4 (2 months), with a reduction of 10 mg if the dosage was 60 mg every other day or less, and 20 mg if the dose was 80 to 100 milligrams. This schedule of very rapid tapering was based upon the results of an earlier trial indicating the beneficial effects of cyclosporine to be evident within 2 months. On visits 5 through 8 (3–6 months), the dosage was reduced by 10 mg per visit (per month). If weakness increased following reduction in steroid dosage, the dosage was increased. Patients were withdrawn from the study if treatment failure occurred. Treatment failure was defined as progression to insufficiency of respiration (grade 3) or swallowing (grade 3) that was not responsive to an increase in steroid dosage to the patient's entry level dosage. If the patient was declared a treatment failure, the patient's steroid dosage could be subsequently increased to a sufficient level to control the symptoms or be started on additional therapy. A doubling of the creatinine value from entry or the occurrence of persistent side effects unresponsive to dosage reduction was defined as drug failure. Protocol violation occurred if any patient stopped the study medicine for more than 5 days.

The primary measures of efficacy were entry to exit change in the MG score, antiacetylcholine receptor antibody titer, and dosage of steroid medication. Secondary measures of efficacy were dosage reduction of anticholinesterase medication, and number of treatment or drug failures by group. Safety was assessed by physical and laboratory examination, evaluation of vital signs and adverse reactions, and direct measures of hepatic and renal function. Both parametric and nonparametric methods were used to analyze the measures of efficacy and safety. Change in MG scores was analyzed by the Wilcoxon Rank Sum method. Change in antireceptor antibody titers was expressed as the percent change in titer and analyzed by the analysis of variance (ANOVA) method. Change in steroid dosage was analyzed as a percent change from entry using ANOVA. To determine the extent of correlation between measures of efficacy, correlation coefficients were calculated by the Pearson Product-Moment method. For all methods of analysis, the p values were calculated as 2-tailed; values of ≤ 0.05 were considered significant.

RESULTS

Thirty-nine patients were entered in the study and randomly assigned to receive either cyclosporine (20 patients) or placebo (19 patients) (TABLE 1). At entry, the two groups did not differ significantly in age, sex, disease severity (MG score), antireceptor antibody titer, steroid dosage, or prior therapy (thymectomy). The mean duration of prior steroid therapy was 36.9 months for placebo and 39.8 months for cyclosporine. All patients had undergone at least two attempts (unsuccessful) to reduce steroid dosage below 30 mg per 48 hours, one of which occurred within 6 months of entry. Steroid dosage had been tapered to the presumed lowest tolerated dose, and that was maintained for at least 30 days prior to admission. Ten patients in the placebo arm and 11 patients in the cyclosporine arm had previously taken azathioprine in addition to steroid medication; duration of prior azathioprine therapy was 29.1 months in the placebo arm and 26.2 months in the cyclosporine arm. Azathioprine had been discontinued at least 60 days prior to entry because of failure to improve and/or inability to reduce steroid dosage. The placebo group was taking significantly more Mestinon at entry: mean total daily Mestinon (anticholinesterase; 60 mg tablet) dosage of 344 mg (range 0–1170) compared to 161 mg (range 0–750) for cyclosporine ($p = 0.03$). The dosage of Mestinon (180 mg time span) tablets, however, did not differ significantly between the groups at entry. The cyclosporine group had a significantly lower forced vital capacity at entry, with a mean of 2.36 liters compared to 3.09 liters for placebo ($p = 0.01$). The mean disease severity score (entry MG score) did not differ significantly: 12.37 for the placebo group and 11.50 for the cyclosporine group ($p = 0.669$). Because the entry disease severity score (assessed from anticholinesterase therapy) and entry antibody titers of the two groups did not differ significantly, the groups were considered to have clinically equivalent disease severity, despite differences in 60 mg (but not 180 mg) Mestinon dosage and forced vital capacity.

TABLE 1. Cyclosporine and Steroid-Dependent MG[a]

Data by Treatment Group

Patient No.	Age	Sex	Entry MG[b]	Exit MG[b]	Entry Steroid[c]	Exit Steroid[c]	Outcome
Group 1: Placebo							
01	37	F	18	16	80	65	completed
03	39	M	10	8	60	10	completed
06	70	M	15	18	60	30	completed
07	70	M	3	3	40	30	completed
09	37	F	16	20	40	40	completed
12	44	F	13	13	32	32	completed
14	27	M	7	17	100	100	treatment failure
15	25	M	5	11	40	40	completed
17	49	F	10	7	80	80	completed
19	31	F	8	6	40	10	completed
21	34	F	4	3	80	0	completed
24	29	F	12	8	80	60	completed
25	47	F	22	25	50	50	completed
26	39	F	19	17	100	60	completed
29	71	F	19	16	100	80	completed
31	25	F	19	23	80	80	treatment failure
33	69	M	3	3	40	10	completed
34	35	M	21	26	80	80	treatment failure
38	53	M	11	10	40	20	completed
Group 2: Cyclosporine							
02	23	F	2	0	80	80	protocol violation
04	46	F	4	5	50	20	completed
05	34	F	7	1	40	20	completed
08	75	F	13	11	40	40	completed
10	51	F	15	10	30	2.5	completed
11	41	F	6	9	50	50	completed
13	40	F	10	1	52	30	completed
16	34	F	31	14	105	60	completed
18	61	F	9	6	35	20	completed
20	35	F	10	7	40	24	completed
22	67	F	12	11	40	10	completed
23	75	F	10	6	40	30	completed
27	87	F	7	9	40	30	completed
28	34	M	18	19	100	60	completed
30	78	M	13	8	40	20	completed
32	70	F	12	9	160	130	drug failure
35	64	M	8	2	40	20	completed
36	81	F	17	18	40	20	completed
37	68	F	9	6	50	20	completed
39	57	M	17	7	100	0	completed

[a] There were three treatment failures in the placebo group and no treatment failures in the cyclosporine group. There were no drug failures in the placebo group, and there was one drug failure in the cyclosporine group. This drug failure occurred in a patient receiving high-dose steroid therapy who experienced a recurrence of her cutaneous *Herpes simplex* infection. Pain and infection were not controlled with acylovir. Therefore, the study medication (cyclosporine) was discontinued. Her symptoms subsequently resolved. One patient left the study due to protocol violation; this patient became pregnant despite an approved form of birth control. She subsequently delivered a normal, healthy infant.

[b] Quantified MG score; higher score = greater weakness (see ref. 9).

[c] Steroid dosage in mg per 48 hours.

TABLE 2. Cyclosporine and Steroid-Dependent MG

Analysis of Measures of Efficacy

Measure	Placebo Group			Cyclosporine Group			P value[e]
	Min.	Max.	Mean ± SD	Min.	Max.	Mean ± SD	
Change QMG[a,b] score	−10.0	4.0	−0.79 ± 3.74	−3.0	17.0	3.55 ± 4.68	0.004
Percent change AB titer[c]	−540	63.6	−28.1 ± 130.8	−77.8	100	34.8 ± 42.4	0.01
Percent change steroid dosage[d]	0.0	100	29.6 ± 33.5	0.0	100	43.2 ± 27.3	0.12

[a] QMG: Quantified Myasthenia Gravis score (MG score).
[b] Change MG score: entry minus exit score; positive = reduced score = improvement.
[c] Percent change AB titer: entry minus exit antireceptor antibody titer divided by entry titer, times 100; positive = reduced titer.
[d] Percent change steroid dosage: total steroid dosage per 48 hours at entry minus exit divided by entry dosage, times 100; positive = reduction in dosage.
[e] ANOVA or Wilcoxon Rank Sum; 2-tailed.

Change in Mestinon dosage, however, and forced vital capacity were not used as an outcome variable.

When all patients had reached six months of treatment or an earlier end point, an analysis of the primary outcome measures was performed (TABLE 2). For the first two measures, the cyclosporine group demonstrated significantly greater increase in strength and reduction of symptoms ($p = 0.004$) and greater reduction in antireceptor antibody titer ($p = 0.01$) than did the placebo group. There were three treatment failures in the placebo group and no treatment failures in the cyclosporine group; there were no drug failures in the placebo group, and there was one drug failure in the cyclosporine group (TABLE 3). For the third outcome measure, percentage reduction in steroid dosage at 6 months, the group difference was not significant.

A variation in the MG score of 1 or 2 points between consecutive visits was

TABLE 3. Cyclosporine and Steroid-dependent MG, at Six Months or End of Study

Analysis of Measures of Efficacy

	Placebo Group	Cyclosporine Group
QMG:[a]		
Better	9	15
Same	3	0
Worse	7	5
Improving ≥ 4 points QMG	2/19	8/20
Completing 6 months	16	18
Treatment failure	3	0
Drug failure	0	1
Protocol violation	0	1

[a] Quantified MG score.

noted to occur without a trend being established. An improvement of 4 points or more, however, was associated with a sustained change and was seen in only 2 of 19 placebo patients compared to 8 of 20 cyclosporine patients. A comparison of mean change in MG score from baseline (percentage) at monthly intervals indicated an initial improvement in both groups during the first two months (FIG. 1). Thereafter, the placebo group gradually returned to baseline, whereas continued improvement was noted in the cyclosporine group. An analysis of the components of the MG score indicated a differential sensitivity to clinical changes. As shown in FIGURE 2, for the vision composite (ptosis and diplopia) of the MG score, the magnitude of improvement was greater in the cyclosporine group, and the difference in the two arms of the study was significant by 8 weeks ($p < 0.05$). A comparison of other composites of the total MG score indicated that the facial composite (ptosis, diplopia, facial weakness, chewing, and swallowing) gave re-

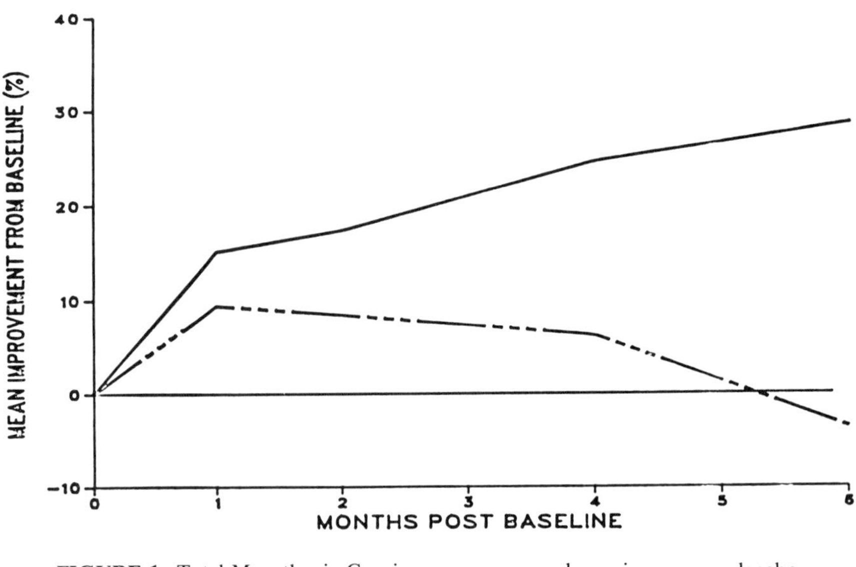

FIGURE 1. Total Myasthenia Gravis score. ——, cyclosporine; -------, placebo.

sults similar to the visual composite. Improvement in the nonfacial composite (arm and leg outstretch times, grip, forced vital capacity, and head lift) was significantly greater for the cyclosporine group, although the percentage improvement was less than for vision or facial (data not shown).

Steroid dosage expressed as the median percent decrease from entry (median value used because of the wide range of doses) is shown in FIGURE 3 as a comparison of the two arms of the study. Once reduction in steroid dosage was initiated (2 months), the cyclosporine group demonstrated steady dose reduction, whereas the placebo group did not. As noted, at 6 months this difference was not significant, although it increased during the next 6 and 18 months. The mean reduction (improvement) in antireceptor antibody titer, expressed as the percent change in log titer, is shown in FIGURE 4 for the two groups in the study. At 6 months, the

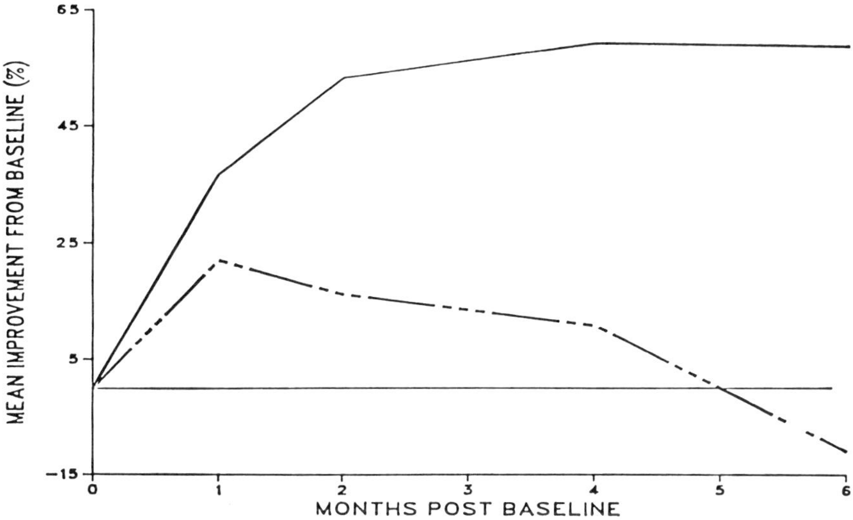

FIGURE 2. Vision composite. ——, cyclosporine; -·—·-, placebo.

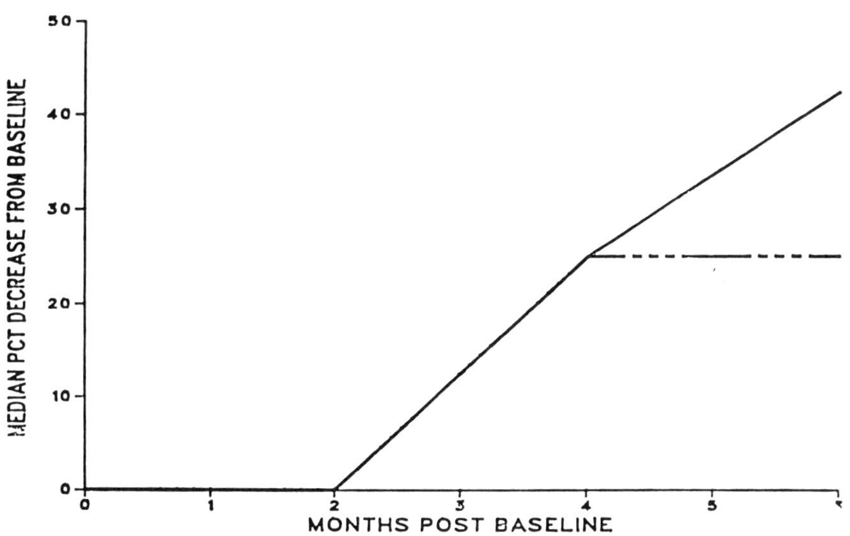

FIGURE 3. Daily prednisone dose. ——, cyclosporine; -·—·-, placebo.

cyclosporine group experienced a statistically significant mean difference in reduction of titer compared to placebo.

The correlation between changes in MG score and antireceptor antibody titer was statistically significant. Using Pearson product moment coefficient of correlation, the change (entry-exit) in total MG score and in antibody titer had a positive correlation ($p = 0.01$). The vision composite ($p = 0.03$) and facial composite ($p = 0.03$) correlated best, whereas the nonfacial composite (all measures except diplopia, ptosis, face, chewing, and swallowing) only approached significance ($p = 0.067$). These correlations were found in the cyclosporine arm, whereas in the placebo arm no significant correlations were seen, as no consistent trend was observed.

The measures of safety consisted of the exit BUN, creatinine, and creatinine clearance, as well as a change from entry to exit in these three parameters (TABLE

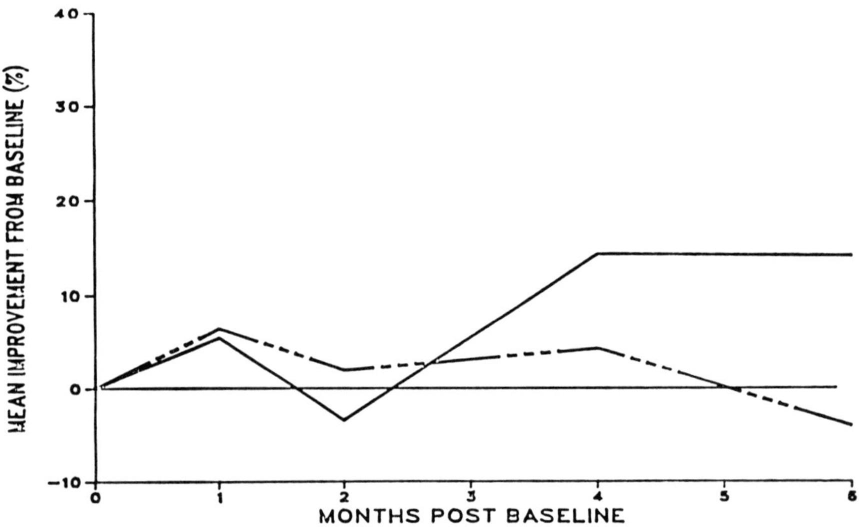

FIGURE 4. Log AAR antibody titer. ——, cyclosporine; ------, placebo.

4). There was a small but significant increase in BUN in the cyclosporine group. There was no significant difference, however, in exit creatinine and creatinine clearance or in the entry to exit change in these two measures when the cyclosporine group was compared to the placebo group. Thus, there was no clinically significant renal toxicity evident in this study using 5 mg/kg/day in a divided dosage with maintenance of 12-hour trough levels at 300–500 ng/mL. Although side effects were reported in both groups, there was no significant difference in the incidence of nausea, hypertension, headache, episodes of infection, paresthesias, gingival hyperplasia, alteration in taste, increased hair growth, or palpitations of sufficient magnitude to require dosage reduction during the first 6 months. Many of the patients in this study had preexisting hypertension primarily due to concurrent steroid therapy and were receiving antihypertensive medications

TABLE 4. Cyclosporine and Steroid-dependent MG

	Measures of Safety						
	Placebo Group			Cyclosporine Group			
Measure	Min.	Max.	Mean ± SD	Min.	Max.	Mean ± SD	P value[a]
Exit BUN	6	55	17 ± 10	10	40	21.3 ± 8	0.003
Exit creatinine	0.9	2.7	1.2 ± 0.4	0.8	1.8	1.2 ± 0.3	NS[b]
Exit creatinine clearance	26	181	85 ± 42	37	153	79.5 ± 36	NS
Change in BUN	−5.0	+17.0	+1.3 ± 4.8	−4.0	+20.0	+6.5 ± 5.6	0.003
Change in creatinine	−0.9	+0.7	+0.06 ± 0.22	−0.1	+0.6	+0.15 ± 0.22	NS
Change in creatinine clearance	−36.0	+34.0	+9.2 ± 31.8	−57.0	+29.0	+17.3 ± 32.2	NS

[a] ANOVA, 2-tailed.
[b] Not significant.

at entry. Increased systolic or diastolic pressures occurred in both groups and were easily managed by the addition of a calcium channel blocking antihypertensive medication and/or dosage reduction of the study medication.

A comparison of cyclosporine dosages and whole blood concentrations observed during the current study and an earlier higher dosage trial is shown in TABLE 5. A larger initial dosage was associated with a shortened time to maximal improvement. Lower initial dosage was associated with less nephrotoxicity, longer time to maximal improvement, but a similar maintenance dosage at 6 months.

For the 18 patients in the cyclosporine arm, all were permitted to continue medication for up to 3 years. TABLE 6 summarizes the outcome. By 36 months, 55% of patients had stopped cyclosporine. Nephrotoxicity accounted for only 10 percent of patients prematurely leaving the study. Protocol violation (15%),

TABLE 5. Cyclosporine and Steroid-dependent MG

	Present Study		Initial Study[9]	
Cyclosporine Group	Range	Mean	Range	Mean
Minimum dosage (mg/kg/d)	2–5	3.9	2–6	3.8
Maximum dosage (mg/kg/d)	5–8	5.7	6–8	6.5
Dosage at maximal improvement (mg/kg/d)	2–7	4.4	ND[a]	
Dosage at 6 months (mg/kg/d)	3–8	4.8	2–6	4.4
Time to maximal improvement (months)	2–6	4.6	1–6	3.6
Time to maximal reduction in AChR Ab titer (months)	2–6	4.8	3–8	4.8

[a] Not determined.

headache and psychiatric symptoms (10%), infection (5%), and gastrointestinal symptoms (10%) were also responsible. Forty-five percent of patients tolerated cyclosporine well and were willing to continue therapy beyond 3 years. Patients randomized to placebo were switched to cyclosporine after the blinded phase (6 months) if they chose to continue in the study; their outcomes were similar to the cyclosporine arm.

DISCUSSION

The results of this study suggest that cyclosporine was effective in the treatment of myasthenia gravis. Our initial study population was one of recent onset, largely elderly patients, who had not previously been treated with thymectomy or immunosuppressive therapy.[9,10] The present study population involved a more severely ill group of patients, both younger and older patients, most with prior thymectomy, and all of whom were dependent upon long-term (mean > 3 years) steroid therapy for symptomatic control. Many (54%) had previously received combined steroid and azathioprine therapy without improvement. In this patient population, cyclosporine resulted in a significant improvement in two measures of efficacy: increased strength and decreased symptoms, and a reduction in antireceptor antibody titer. Unfortunately, group mean differences for the third measure, reduction in steroid dosage, did not reach significance by 6 months. Data presented in TABLE 6 indicate that reduction in steroid dosage was not maximal at 6 months, but between 12 and 24 months. The study protocol may have been too short (6 months) to detect a statistically and clinically meaningful reduction in steroid dosage. The effect of cyclosporine was reversible with a return of weakness upon cessation of cyclosporine. The dosage of steroid medication could usually be tapered to 5 or 10 mg/48 hours with sustained control of symptoms. Those patients who were able to discontinue their steroid medication ultimately (for 1 to 6 months) restarted it when symptoms again increased.

TABLE 6. Cyclosporine and Steroid-dependent MG

Cyclosporine Group: Status at 6, 12, and 24 Months
Number at Entry = 20

	6 Months	12 Months	24 Months
Number completing	18	14	9
Steroid: Percent Reduction	43	60	69
Discontinued[a] due to:			
Protocol violation (15%)	1		2
Failure to improve (5%)		1	
Renal toxicity (10%)		1	1
Infection (5%)	1		
GI symptoms (10%)		2	
Psychiatric symptoms (5%)			1
Headaches (5%)			1

[a] Percentage of patients entered and randomized to cyclosporine who discontinued the medication for the indicated problem.

Although long-term (12–24 months) accumulating nephrotoxicity did occur in 10% of the patients, the results of the present study are more acceptable in this regard than the initial study. Cyclosporine nephrotoxicity appears to be due to both endothelial-vascular injury and to interstitial fibrosis.[12] Even at the low dosage utilized in this protocol (3 to 5 mg/kg/day, split), some degree of cumulative interstitial fibrosis did occur. With careful (monthly) monitoring and prompt reduction or cessation of cyclosporine, this adverse reaction was infrequent and not severe. The frequency of other side effects over a 24-month period was higher than anticipated; 20% of patients discontinued therapy due to psychiatric symptoms, headache, or gastrointestinal complaints.

After the conclusion of this study, a European study comparing low dosage cyclosporine to azathioprine, rather than to placebo, was reported.[13] Both drugs were equally efficacious in producing clinical improvement after a 12-month trial. Although side effects of the medications differed, the extent of nephrotoxicity in that study was minimal and considered clinically acceptable. Despite five decades of therapeutic advances in the treatment of myasthenia, a substantial number of patients remain dependent for control of their symptoms upon long-term immunosuppressive medications.[11] Recent studies have confirmed the following in myasthenia gravis: chronic T-cell activation with elevation of serum IL-2 and soluble IL-2 receptor proportional to disease severity; extensive heterogeneity of V beta T cell–receptor gene usage between and within individual patients; and B-cell somatic hypermutation with resultant marked diversity in the fine specificity of antireceptor antibodies between and within individual patients.[14–18] Thus future therapies directed at blockade or deletion of specific T- or B-cell clones by synthetic peptides or toxin-coupled anticlonotypic antibodies do not appear therapeutically promising. Three therapeutic avenues remain available—two pragmatic and one investigational. First, the use of better, less toxic, and more effective combination therapies with existing immunosuppressive medications (*e.g.*, prednisone, azathioprine, cyclosporine, and cyclophosphamide) or derivative/analogous medications (*e.g.*, FK506, rapamycin, chlorambucil) must be perfected. Second, the role of thymectomy must be clarified by a controlled and blinded (blinded rater) clinical trial with a sufficient number of patients to allow subsequent multivariate analysis to identify prognostic/predictive factors regarding probable need for subsequent immunosuppressive medication(s) and disease duration (decades). Such factors are not now known, and the decision regarding thymectomy is frequently based more on opinion than fact. Third, recognition that T-cell tolerance to self-antigens is maintained throughout life by the continual interaction of processed self-antigen in a major histocompatibility complex (MHC) (HLA,B,D) groove with specific T-cell receptors, in the absence of a costimulatory signal (*e.g.*, IL-2), provides a potential mechanism to re-establish tolerance to the acetylcholine receptor. Induction of suppression or anergy by autologous B cells that have been induced to process *Torpedo* receptor (by cross-linking a monoclonal anti-*Torpedo* acetylcholine receptor to an anti-mouse IgM) and that are then used to "stimulate" naive autologous T cells in the absence of IL-2 has been demonstrated in a murine EAMG model.[19] A human $DR4^+$ T-lymphocyte clone known to react with a portion of the alpha-subunit showed reduced antigen-specific proliferation following preincubation with a complex of soluble DR4w4-alpha–peptide (class II MHC solubilized by detergent complexed with alpha-peptide 138–167) for 18 hours.[20] The therapeutic potential of anergy induction using soluble MHC-receptor peptides or autologous B cells linked to receptor remains to be demonstrated, though such antigen-specific noncytoxic approaches are a natural extension of current laboratory models.

The proper place of cyclosporine in the treatment of chronic human autoimmune disorders such as myasthenia is still not firmly established. Based on the present study and previous single center studies, one may conclude that cyclosporine is an effective immunosuppressant agent for the treatment of myasthenia (as initial therapy or as a second agent), that improvement is reversible following cessation of the drug, and that clinical benefit (if it occurs) is evident by 6 months and maximal at 12 months. Although a meaningful reduction in steroid dosage may be achieved by 12 months, a small dosage of steroids will probably need to be continued, for example, 5 mg daily. Cumulative nephrotoxicity does occur despite reduced dosage, and with long-term therapy other side effects are as limiting as nephrotoxicity. When tolerated, long-term (>4 years) therapy is feasible. Confirmation of these findings must await the analysis and report of two large multicenter clinical trials already completed.

SUMMARY

We randomly assigned 39 patients with steroid-dependent generalized myasthenia gravis to treatment with cyclosporine (5 mg/kg per body weight in divided doses) or placebo. Duration of treatment was 6 months. Patients were evaluated monthly. Primary measures of efficacy were quantified strength testing, anti-human acetylcholine receptor antibody titer, and dosage of corticosteroid medication. At the end of the study, patients in the cyclosporine group had significantly greater improvement in strength ($p = 0.004$) and a reduction in antireceptor antibody titer ($p = 0.01$). Percentage reduction of steroid medication was greater in the cyclosporine group, although the difference was not statistically significant ($p = 0.12$). There were no treatment failures, and there was one drug failure in the cyclosporine group. In the placebo group, there were three treatment failures. No significant nephrotoxicity was noted at this dosage during the first 6 months. During the subsequent 18 months of open-label therapy, continued reduction in steroid dosage occurred. Cumulative side effects, however, caused 35% of patients to discontinue the medication; 10% did so secondary to slowly progressive nephrotoxicity.

REFERENCES

1. Wartburg, A. von & R. Traber. 1986. Chemistry of the natural cyclosporin metabolites. Prog. Allergy **38:** 28–45.
2. Nussenblatt, R. B., H. C. Gunn, B. Ryffel & J. F. Borel. 1986. Experimental autoimmunity. Prog. Allergy **38:** 159–180.
3. Kahan, B. D. 1985. Cyclosporine: the agent and its actions. Transplant. Proc. **17**(4): Suppl. **1:** 5–18.
4. Ryffel, B., K. Tammi, A. Grieder & A. D. Hess. 1985. Effects of cyclosporine on human T cell activation. Transplant. Proc. **17:** 1268–1270.
5. Flanagan, W. M., F. Corthesy, R. J. Bram & G. R. Crabtree. 1991. Nuclear association of a T-cell transcription factor blocked by FK-506 and cyclosporine A. Nature **352:** 803–807.
6. Thomson, A. W. 1991. The immunosuppressive macrolides FK-506 and rapamycin. Immunol. Lett. **29:** 105–111.
7. Vincent, A. 1980. Immunology of acetylcholine receptors in relation to myasthenia gravis. Physiol. Rev. **60:** 756–824.
8. Drachman, D. B., R. N. Adams, K. McIntosh & A. Pestronk. 1985. Treatment

of experimental myasthenia gravis with cyclosporine A. Clin. Immunol. Immunopathol. **34:** 174–188.
9. TINDALL, R. S. A., J. R. ROLLINS, J. T. PHILLIPS *et al.* 1987. Preliminary results of a double-blind, randomized placebo-controlled trial of cyclosporine in myasthenia gravis. N. Engl. J. Med. **316:** 719–724.
10. TINDALL, R. S. A., J. T. PHILLIPS, J. A. ROLLINS *et al.* 1988. Cyclosporine in the treatment of myasthenia gravis. Monogr. Allergy **25:** 135–147.
11. ENGEL, A. G. 1984. Myasthenia gravis and myasthenic syndromes. Ann. Neurol. **16:** 519–534.
12. KOPP, J. B. & P. E. KLOTMAN. 1990. Cellular and molecular mechanisms of cyclosporine nephrotoxicity. J. Am. Soc. Nephrology **1**(2): 162–179.
13. SCHALKE, B. C. G., L. KAPPOS, E. ROHRBACH *et al.* 1988. Cyclosporine A vs. azathioprine in the treatment of myasthenia gravis: final results of a randomized, controlled double-blind clinical trial. Neurology **38**(Suppl I): 135.
14. VOSKUHL, R. 1992. The mechanism of T-cell activation in myasthenia gravis. Neurology **40**(Suppl 1): 184.
15. MICHELS, M., K-W. PFLUGHAUPT, B. SCHALKE, K. TOYKA & H-P. HARTUNG. 1992. Increased serum levels of soluble IL-2 receptors in myasthenia gravis. From conference sponsored by the New York Academy of Sciences entitled Myasthenia Gravis and Related Disorders: Experimental and Clinical Aspects. Abstract.
16. CONFALONIERI, P., C. ANTOZZI, O. SIMONCINI, F. CROSTI, F. CORNELIO & R. MANTEGAZZA. 1992. Soluble interleukin-2 receptor levels are increased in myasthenia gravis. From conference sponsored by the New York Academy of Sciences entitled Myasthenia Gravis and Related Disorders: Experimental and Clinical Aspects. Abstract.
17. THEOFILOPOULOS, A. N., R. S. BALDERAS, R. BACCALA & D. H. KONO. 1993. T-cell receptor genes in autoimmunity. Ann. N.Y. Acad. Sci. This volume.
18. HEIDENREICH, F., A. VINCENT & J. NEWSOM-DAVIS. 1988. Differences in fine specificity of anti-acetylcholine receptor antibodies between subgroups of spontaneous myasthenia gravis of recent onset, and of penicillamine induced myasthenia. Autoimmunity **2:** 31–37.
19. REIM, J., K. R. MCINTOSH & D. DRACHMAN. 1993. Acetylcholine receptor presentation by B cells using heterobifunctional antibody conjugates. Ann. N.Y. Acad. Sci. This volume.
20. NICOLLE, M. W., A. VINCENT, S. SHARMA, B. NAG, N. WILLCOX & J. NEWSOM-DAVIS. 1993. An *in vitro* model for disease-specific immunotherapy in myasthenia gravis using soluble MHC class II bound to AChR-derived peptide. Ann. N.Y. Acad. Sci. This volume.

Clinical Improvement of Myasthenia Gravis by Treatment with a Chimeric Anti-CD4 Monoclonal Antibody

R. ÅHLBERG,[a] Q. YI,[a] R. PIRSKANEN,[b] G. MATELL,[b]
C. SWERUP,[c] P. RIEBER,[d] G. RIETHMÜLLER,[d] G. HOLM,[a]
AND A. K. LEFVERT[a]

[a]*Department of Medicine*
Karolinska Hospital
Stockholm, Sweden S-104 01

[b]*Department of Neurology*
[c]*Department of Clinical Neurophysiology*
South Hospital
Stockholm, Sweden

[d]*Institute of Immunology*
University of Munich
Munich, Germany

INTRODUCTION

Monoclonal antibodies (MAb) against the CD4 antigen on T cells have been effective in the treatment of both experimental autoimmune diseases, including experimental autoimmune myasthenia gravis,[1] and human autoimmune disorders, such as rheumatoid arthritis.[2–4] We here report the clinical and immunological effects of a chimeric anti-CD4 MAb in a patient with myasthenia gravis (MG).

PATIENT AND METHODS

A 57-year-old woman with severe myasthenic symptoms was treated for seven days with 25 mg/day of a chimeric (human constant regions and murine variable regions) anti-CD4 MAb (M-T412 Centocor). The therapeutic effect was assessed by muscle function tests and by single-fiber EMG. T-cell activation was studied in peripheral blood mononuclear cells (PBMC) by enzyme-linked immunospot assay for enumeration of IL-2–producing cells and by thymidine incorporation assay. PBMC were cultured in the presence or absence of a human antireceptor antibody (2-115-19), an antiidiotypic antibody (2-115-58),[5] or rIL-2. Peripheral blood immunophenotyping was performed by double-staining immunofluorescence and analyzed on a FACScan flow cytometer.

RESULTS

Clinical Evaluation

A clinical and electrophysiological improvement was evident on the fourth day of treatment, was most pronounced during the second to eighth weeks, and was still present at the five-month follow-up. A progressive deterioration started three months after initiation of the treatment (FIG. 1). She had a moderate fever for two days during the treatment and a common cold 10 weeks later. No other side effects or infections were observed.

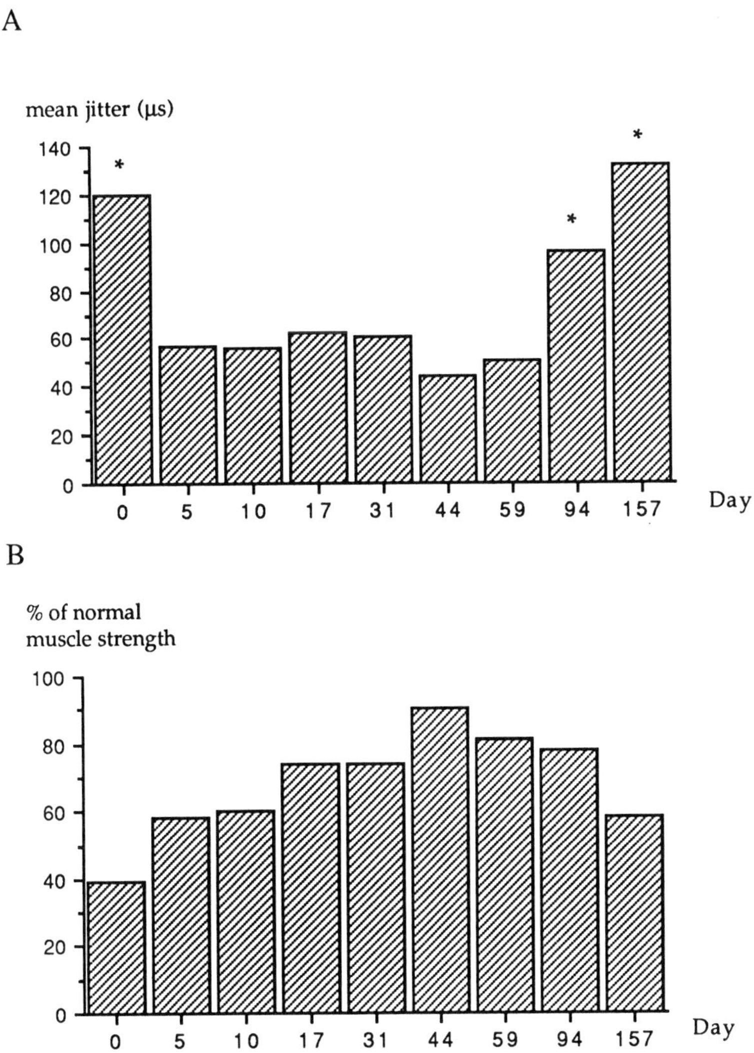

FIGURE 1. A. Changes in the mean jitter of single-fiber EMG (SF-EMG) examined on the right extensor digitorum muscle. The recordings with mean jitter above 90 μs showed significant blocking (*). ▨, SF-EMG. **B.** Changes in the arm muscle strength in percent of normal muscle strength in a patient with severe myasthenia gravis treated with a seven-day course (starting day one) of a chimeric anti-CD4 MAb. ▨, muscle strength.

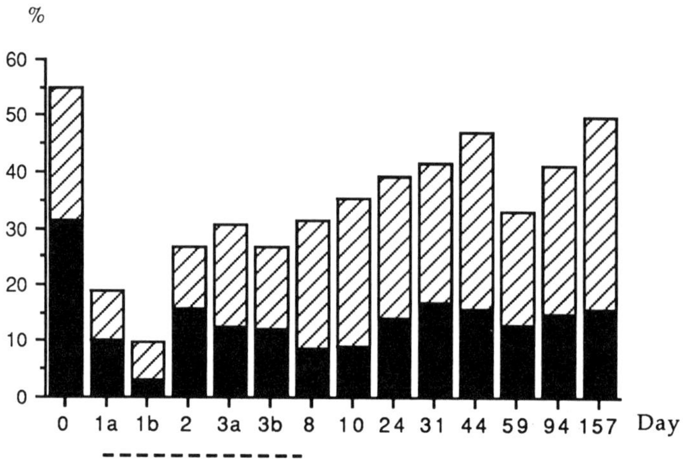

FIGURE 2. Changes in the lymphocyte subsets in a patient with severe myasthenia gravis who was treated (- - -) with a seven-day course of a chimeric anti-CD4 MAb. The percentage of CD4 and CD8 immunofluorescence-positive cells was determined before, during (a = two hours and b = five hours after injections), and after the treatment. ▨, CD8; ■, CD4.

Lymphocyte Subsets

Treatment with the anti-CD4 MAb resulted in rapid decrease in the percentage of $CD4^+$ and $CD8^+$ cells (FIG. 2). The $CD4^+$ cells recovered to 50% of pretreatment value at the five-month follow-up. The CD4/CD8 ratio became inverse five hours after the first antibody injection and was persistently decreased during five months of follow-up. There were no major alterations in the subpopulations of $CD4^+$ cells as evidenced by double staining with anti-CD29 or anti-CD45RA MAb.

T-cell Function

IL-2 production and T-cell proliferation was abolished during the anti-CD4 treatment, both spontaneously and in response to antireceptor antibody, antiidiotypic antibody, and rIL-2, all which induced T-cell stimulation prior to treatment. During several months after completion of therapy, T cells were spontaneously more activated than before therapy. The idiotypic and antiidiotypic antibody induced further T-cell activation during the five-month follow-up, whereas rIL-2 induced an increase in the number of IL-2–producing cells, but no further increase in thymidine incorporation.

Acetylcholine Receptor Antibodies (AChRec-Ab) in Serum

The anti-CD4 treatment did not induce any major changes in the level of serum AChRec-Ab.

DISCUSSION

The intravenous administration of chimeric anti-CD4 MAb induced a clinical and electrophysiological improvement of the disease and a striking immunomodulation, including a rapid substantial depletion of $CD4^+$ cells and T-cell activation after termination of the treatment. A similar rapid reduction of $CD4^+$ cells induced by anti-CD4 MAb treatment has also been reported in rheumatoid arthritis with a duration varying from 24 hours[3] to more than two months.[4] Our findings support the hypothesis that $CD4^+$ lymphocytes play an important role in the pathogenesis of MG and suggest that chimeric anti-CD4 MAb could be helpful in the treatment of severe MG resistant to conventional therapy.

REFERENCES

1. Christadoss, P. & M. J. Dauphinee. 1986. Immunotherapy for myasthenia gravis: a murine model. J. Immunol. **136:** 2437–2440.
2. Herzog, C., C. Walker, W. Müller, P. Rieber, C. Reiter, G. Rietmüller, P. Wassmer, H. Stockinger, O. Madic & W. J. Pichler. 1989. Anti-CD4 antibody treatment of patients with rheumatoid arthritis. J. Autoimmunity **2:** 627–642.
3. Reiter, C., B. Kakavand, E. P. Rieber, M. Schattenkirchner, G. Rietmüller & K. Krüger. 1991. Treatment of rheumatoid arthritis with monoclonal CD4 antibody M-T151. Arthritis Rheum. **34:** 525–536.
4. Horneff, G., G. R. Burmester, F. Emmrich & J. R. Kalden. 1991. Treatment of rheumatoid arthritis with an anti-CD4 monoclonal antibody. Arthritis Rheum. **34:** 129–140.
5. Yi, Q., R. Åhlberg & A. K. Lefvert. 1992. T cells with specificity for idiotypic determinants on human monoclonal autoantibodies in myasthenia gravis. Res. Immunol. **143:** 149–156.

Searching for Prognostic Factors at the Time of Thymectomy for Myasthenia Gravis

Correlation between Outcome after Long-term Follow-up and Immunophenotype of Thymic and Peripheral Blood Lymphocytes

LUCA DURELLI,[c] GIULIANO MAGGI,[a]
CATERINA CASADIO,[a] ROBERTO FERRI,
BRUNO FERRERO, GUGLIELMO POCCARDI,[b]
ANNA DUTTO, AND LODOVICO BERGAMINI

Clinica Neurologica and
[a]Chirurgia Toraco-polmonare dell'Università di Torino
[b]Divisione di Ematologia Ospedale Mauriziano
Torino, Italy

A role for the thymus in the pathogenesis of myasthenia gravis (MG) is strongly supported both by the excellent outcome reported by many surgical series after thymectomy[1] as well as by structural and phenotypic immunologic alterations suggesting that the gland, enriched by mature B and T cells capable of responding to antigens, is the site of an active immune reaction.[2,3] The study of thymic and peripheral blood (PB) lymphocyte phenotypes and of their changes after thymectomy and the analysis of the possible correlations between lymphocyte phenotypes and clinical outcome may further clarify the pathogenetic role of thymus in MG. We therefore started a prospective study of the correlations between clinical outcome after thymectomy, lymphocyte phenotypes (studied with dual-color immunofluorescence (IF) and with an extensive panel of monoclonal antibodies) both in thymus and PB, and antiacetylcholine receptor (AChR) antibody (Ab) titers.

Twenty MG patients were followed for 3–4 years after thymectomy. The mean age was 40.7 ± 15.0 years (range, 20–70); 14 patients had hyperplastic thymuses and 6 had thymomas. Normal thymus was obtained from 10 age-matched controls undergoing cardiac surgery. Lymphocyte phenotypes were analyzed on thymic cell suspensions. Patients were regularly seen every 2–6 months for 3–4 years after thymectomy. The clinical status was evaluated and graded according to Osserman's classification; blood was drawn for phenotypic analysis and anti-AChR Ab titration at each visit. Anti-AChR Ab titer was determined using a preparation of AChR from fetal calf muscle, labeled with [^{125}I]alpha-bungarotoxin,

[c] Send all correspondence to Luca Durelli, M.D., Clinica Neurologica, Via Cherasco, 15, I-10126 Torino, Italy.

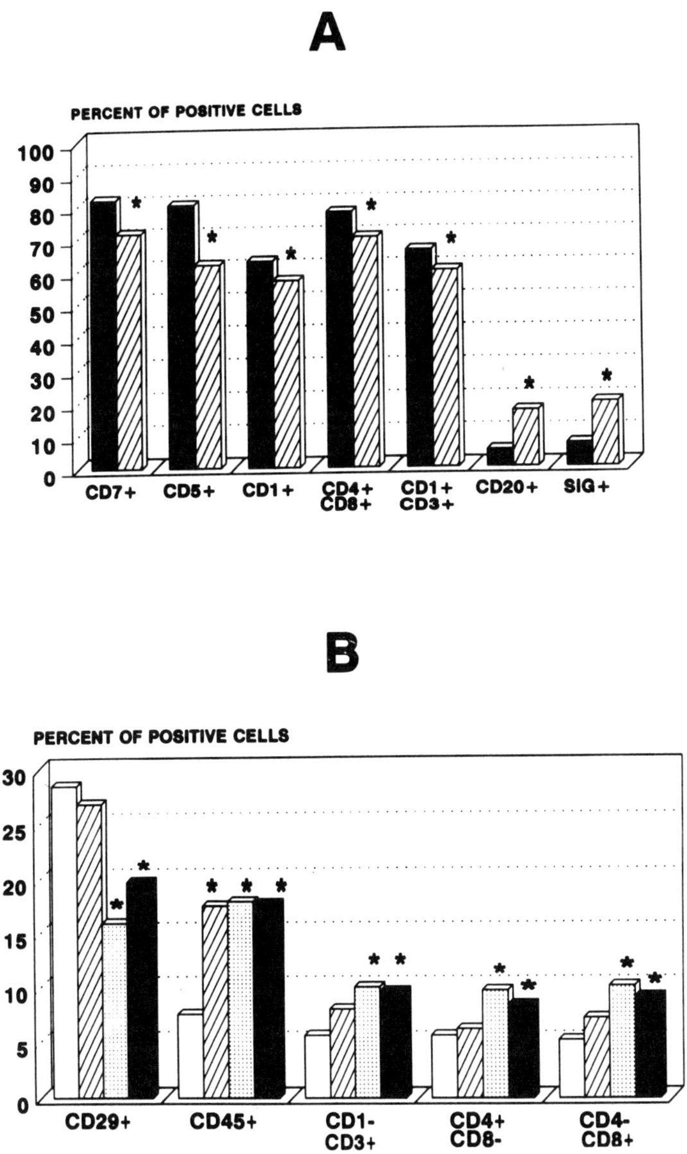

FIGURE 1. Subsets of lymphocytes in thymic cell suspensions of cardiopathic control subjects and MG patients. Bars represent the mean percent expression of lymphocyte differentiation antigens identified with single- or dual-color direct immunofluorescence. **A:** The phenotypes corresponding to the first steps of the intrathymic T-lymphocyte differentiation process ($CD7^+$, $CD5^+$, $CD1^+$, $CD4^+CD8^+$, and $CD1^+CD3^+$) were decreased in MG thymus, whereas those corresponding to B cells ($CD20^+$, SIG^+) were increased in MG thymus. ■, controls (10); ▨, MG (20); * significantly different from controls. **B:** The phenotypes of mature lymphocytes ($CD45^+$, $CD1^-CD3^+$, $CD4^+CD8^-$, and $CD4^-CD8^+$) were in-

and using rabbit anti-human IgG as an immunoprecipitating agent. The cardiopathic controls were seen 6–12 months after surgery, and the PB lymphocyte phenotypes of 40 healthy age-matched volunteers were prospectively studied every 6–12 months for 3 years.

The percent expression of T-cell lineage phenotypes that are expressed during the first steps of intrathymic T-cell differentiation ($CD1^+$, $CD5^+$, $CD7^+$, $CD4^+CD8^+$, $CD1^+CD3^+$) (FIG. 1A) were significantly decreased, whereas those corresponding to the most mature T lymphocytes ($CD1^-CD3^+$, $CD4^+CD8^-$, $CD4^-CD8^+$, and $CD45RA^+$), and those identifying mature B cells ($CD20^+$, SIG^+), were significantly increased in MG thymus (FIG. 1A and B). When data from patients who improved after thymectomy were separated from those of patients who did not, it appeared clear that, within thymuses not associated with thymoma, the significant increase of mature T phenotypes could be ascribed almost totally to the thymuses of improved patients (FIG. 1B). In addition, in MG thymuses not associated with thymoma, statistically significant correlations could be calculated between phenotypes corresponding to mature T lymphocytes, suggesting, in those thymuses, a parallel increase of many different antigens identifying mature T cells.

In the PB, $CD4^+CD45RA^+$, $CD8^+CD57^+$, and $CD4^+CD8^+$ T cells were significantly increased, whereas $CD16^+$, $CD4^+CD29^+$, and $CD4^+LEU8^-$ were decreased before thymectomy (TABLE 1). Most PB lymphocyte phenotypes rapidly normalized 4–6 months after thymectomy and remained normal without any correlation to the clinical status after surgery (TABLE 1). This was not the case for percent T $CD4^+CD8^+$ that, after a decrease to normal values in the months after thymectomy, began increasing again 3–4 years after thymectomy only in unimproved patients. The anti-AChR Ab titer, that was high in most patients before thymectomy, slowly decreased within 1–2 years after surgery, more remarkably in improved patients.

The thymus is open to two-way traffic with the PB, and lymphocytic traffic is increased in the MG thymus, where immune-activated mature T and B cells, specifically and actively involved in the production of anti-AChR Ab, have been demonstrated.[2,3] The mature lymphocytes that we showed increased in MG thymus by IF analysis, and by the study of the reciprocal correlation of many lymphocyte differentiation antigens, these lymphocytes may well be PB lymphocytes involved in an active immune response. The increase of mature lymphocytes was particularly pronounced in the thymuses of patients who improved after thymectomy: the more the thymus is involved in an active immune response, the greater the probability that thymectomy will be clinically useful.

In the PB, most immunologic alterations normalized after surgery without any correlation to clinical status, except for the percentage of T $CD4^+CD8^+$ that increased in unimproved patients 3–4 years after thymectomy. This small subset

---

creased in MG thymuses not associated with thymoma, and this increase could be ascribed to the thymuses of patients who improved (IMP) after thymectomy. These mature lymphocytes may partly be peripheral blood (PB) lymphocytes migrated into an immunologically activated thymus. This hypothesis is in keeping with the decrease of $CD29^+$ T lymphocytes in MG thymus. CD29 was studied with fluorescein-conjugated monoclonal antibodies and therefore seen only on those T lymphocytes expressing the antigens at high density ($CD29^+$ HIGH) (immature thymocytes and PB memory T lymphocytes). The decrease of $CD29^+$ HIGH suggests an increase of those T lymphocytes expressing CD29 at low density (mature thymocytes or virgin PB lymphocytes). □, controls (10); ▨, not IMP (5); □, IMP (9); ■, total (14); * significantly different from controls.

TABLE 1. Peripheral Blood Lymphocyte Differentiation Antigens before and after Thymectomy

		MG Patients				
			After Thymectomy			
					More than 30 Mos.	
Antigens	Normals (n = 40)	Before Thymectomy (n = 20)	4–6 Mos. (n = 20)	12–24 Mos. (n = 18)	Improved (n = 12)	Not Improved (n = 6)
Lymphocyte families						
$CD3^+$	1769 ± 448	1867 ± 355	1842 ± 442	1727 ± 357	1767 ± 344	1614 ± 325
$CD4^+$	1138 ± 315	1337 ± 287	1217 ± 294	1105 ± 328	1277 ± 222	1192 ± 309
$CD8^+$	701 ± 208	638 ± 162	679 ± 179	649 ± 84	574 ± 122	603 ± 109
$CD16^+$	284 ± 143	141 ± 83[a]	189 ± 79	278 ± 93	292 ± 84[b]	226 ± 75[b]
$CD20^+$	224 ± 66	294 ± 171	274 ± 96	232 ± 57	196 ± 61	207 ± 49
$CD4^+$ family						
% $CD45RA^+$	45.2 ± 7.3	65 ± 10[a]	47 ± 14[b]	49 ± 8	49 ± 7[b]	45 ± 15[b]
% $CD29^+$	39.8 ± 7.3	28 ± 7[a]	38 ± 8[b]	36 ± 7	40 ± 7[b]	46 ± 12[b]
% $LEU8^-$	26.6 ± 7.5	22 ± 7[a]	25 ± 4	26 ± 4	24 ± 5	24 ± 8
$CD8^+$ family						
% $CD11b^-$	73.5 ± 6.9	73 ± 10	70 ± 10	74 ± 6	73 ± 4	72 ± 4
% $CD57^+$	15.3 ± 12.8	28 ± 13[a]	22 ± 12	28 ± 21	27 ± 18	24 ± 12
% $CD4^+CD8^+$	0.83 ± 0.45	3.68 ± 2.2[a]	1.28 ± 1.04[b]	0.69 ± 0.3	0.46 ± 0.4[b]	1.21 ± 0.6[b]

[a] Significantly different from normal values.
[b] Significantly different from baseline values.

of PB lymphocytes is composed of mature T cells ($CD1^-$, $WT31^+$) that can generate *in vitro* either $CD4^+CD8^-$ or $CD4^-CD8^+$ T cells and possess suppressor and cytotoxic functions. These double-positive PB lymphocytes have been recovered *in vivo* from sites undergoing immune reactions and have been, already, demonstrated increased in the PB of MG patients.[4] It may be a T subset involved in immunosurveillance mechanisms.

REFERENCES

1. Durelli, L., G. Maggi, C. Casadio, R. Ferri, S. Rendine & L. Bergamini. 1991. J. Neurol. Neurosurg. Psychiatry **54:** 406–411.
2. Leprince, C., S. Cohen-Kaminsky, S. Berrih-Aknin, B. Vernet-Der Garabedian, D. Treton, P. Galanaud & Y. Richard. 1990. J. Immunol. **145:** 2115–2122.
3. Sommer, N., N. Willcox, G. C. Harcourt & J. Newsom-Davis. 1990. Ann. Neurol. **28:** 312–319.
4. Matsui, M., H. Fukuyama, I. Akiguchi & M. Kameyama. 1989. J. Neurol. **236:** 329–335.

Infections Occurring in Hospitalized Myasthenia Gravis Patients

C. A. ELECHI, A. SHAH, AND R. P. LISAK

Department of Neurology
Wayne State University School of Medicine
4201 St. Antoine 6E UHC
Detroit, Michigan 48201

INTRODUCTION

Patients with myasthenia gravis (MG) are often hospitalized and may be concurrently receiving treatment with corticosteroids and immunosuppressive agents, such as azathioprine and plasmapheresis. They are therefore at risk of acquiring nosocomial infections.

Previous studies have described a possible association between the occurrence of infections and the course and outcome of MG.[1,2] We could find no studies that examined the relationship between the reason for hospitalization and the occurrence of infections.

METHODS

We retrospectively examined the hospital records of patients with generalized MG who were admitted to the Detroit Receiving and Harper Hospitals between 1985 and 1991. The charts were scrutinized for the following: (1) reason for admission; (2) infections existing at the time of admission; (3) infections occurring during the hospital stay; (contamination or colonization was carefully distinguished from actual infection); (4) steroid use as treatment of MG; (5) azathioprine or similar agents for treating MG; (6) length of hospital stay; (7) use of ventilatory assistance; and (8) the need for plasmapheresis.

RESULTS

TABLE 1 summarizes the salient clinical features. There were 171 total admissions among 65 patients with generalized MG. Sixty-seven admissions were due primarily to worsening of MG symptoms, whereas 104 were for reasons other than worsening of MG symptoms. There were 23 documented serious infections (pneumonia, septicemia, or brain abscess) in both groups.

Eleven of 67 (16%) patients admitted due to MG worsening had gram-negative pneumonias, all acquired during their hospital stay. None of these admissions was precipitated by an infection. Nine of these eleven patients required ventilatory assistance during their hospitalization. Their average length of stay (LOS) in the hospital was 38.8 days.

Twelve serious infections occurred among patients admitted for other reasons (12%), of which only 4 were acquired in the hospital. There were 3 septicemias,

TABLE 1. Summary of Clinical Features

	MG Exacerbations	Other Reasons
Total admissions	67	104
Serious infections	11 (16%)	12 (12%)
Acquired in hospial	11	4
Mean hospital stay	38.8	11.1
Ventilatory assistance	9	4

one brain abscess, and 8 pneumonias (2 pneumococcal, one *Hemophilus influenzae,* one atypical, and 4 with unknown organisms). The infection itself was the primary reason for admission in 8 of these 12 patients. The mean length of stay was 11.1 days.

Three of 8 patients (37%) who had had an infection prior to admission were on corticosteroid therapy compared to 8 of 15 patients (53%) acquiring an infection in the hospital.

Sixteen patients were receiving treatment with azathioprine. Seven episodes of serious infections occurred in 6 patients receiving azathioprine. Two episodes (pneumococcal pneumonia and septicemia) occurred prior to hospitalization.

CONCLUSION

Patients admitted because of MG exacerbations have a higher risk of acquiring gram-negative pneumonia than patients admitted for other reasons. Corticosteroid therapy may further increase this risk. Azathioprine does not appear to bestow any excessive risk.

On the whole patients admitted because of exacerbations of MG were more ill than patients admitted for other reasons. This was reflected in the average length of hospital stay and also in the proportions of patients requiring ventilatory support. Both of these factors may have contributed to the relatively high frequency of gram-negative pneumonia acquired in the hospital among the patients admitted because of exacerbation of myasthenia gravis.

REFERENCES

1. Rowland, L. P., P. F. A. Hoeffer, H. Aranow & H. H. Merritt. 1956. Fatalities in myasthenia gravis. Neurology **6:** 307–326.
2. Oosterhuis, H. J. G. H. 1964. Studies in myasthenia gravis. Part 1. A clinical study of 180 patients. J. Neurol. Sci. **1:** 512–516.

Therapies for Exacerbation of Myasthenia Gravis

The Mechanism of Action of Intravenous High-Dose Immunoglobulin G

BRUNO FERRERO,[b] LUCA DURELLI,
ROBERTO CAVALLO, ANNA DUTTO, GIUSEPPE AIMO,[a]
FERNANDO PECCHIO,[a] AND BRUNO BERGAMASCO

Clinica Neurologica dell'Università di Torino
[a]Laboratori Baldi e Riberi
Ospedale Molinette
Torino, Italy

Lymphocytes and their immunoglobulin (Ig) product are in a dynamic equilibrium, up- and down-regulated by a network of idiotypes (Id) and anti-Id.[1] Anti-Id antibodies (Ab) may specifically affect: (a) Ab activity through a direct F(ab)-mediated interaction with Id of the F(ab) portion of circulating Ig, probably disturbing Ag-Ab binding with a rapid reduction of the affinity of the Ab; and (b) Ab synthesis through a F(ab)-mediated interaction with B and T cell surface Id with a slowly occurring inhibition of lymphocyte activation and Ig production. Anti-Id Abs are natural components of human serum, and they have been demonstrated in commercial IgG preparations. Intravenous (iv) high-dose IgG (HIgG) has been used in the treatment of MG since 1982 with encouraging results.[2] MG patients with rapidly worsening bulbar symptoms or respiratory insufficiency have been treated with 7S IgG at 0.2–0.4 g/kg/d for 5 consecutive days. The rapid (beginning 4–5 days after starting iv HIgG) improvement of severe MG symptoms, observed in about 85% of cases, was usually not correlated with the decreased anti-AChR Ab titers observed in less than 50% of cases. In order to elucidate the possible mechanisms of the immunosuppressive action of iv HIgG in MG, we studied, in 15 MG patients, the effects of such treatment both on anti-AChR Ab titer and on AChR-Ab binding avidity. All patients presented with acute severe worsening of MG, with bulbar symptoms, or respiratory insufficiency, shortly after disease onset and before thymectomy. All patients were treated only with iv HIgG (intact 7S polyvalent IgG, 0.4 g/kg/day for five consecutive days) and anticholinesterase drugs. Anti-AChR Ab titer was measured using a preparation of AChR from fetal calf muscle,[3] labeled with [^{125}I] alpha-bungarotoxin, and using rabbit anti-human IgG or protein A–bearing *Staphylococcus aureus* (Staph-A) as immunoprecipitating agents. Muscle strength was assessed using a quantified MG score calculated from standardized endurance tests and vital capacity. A rapidly occurring (3–4 days after starting iv HIgG) objective clinical improvement was observed in 13/15 patients. During treatment, the concentration of serum IgG increased to 200% of baseline value. Anti-AChR Ab titer, determined using anti-human IgG as immuno-

[b] Send all correspondence to Bruno Ferrero, M.D., Clinica Neurologica, Via Cherasco, 15, I-10126 Torino, Italy.

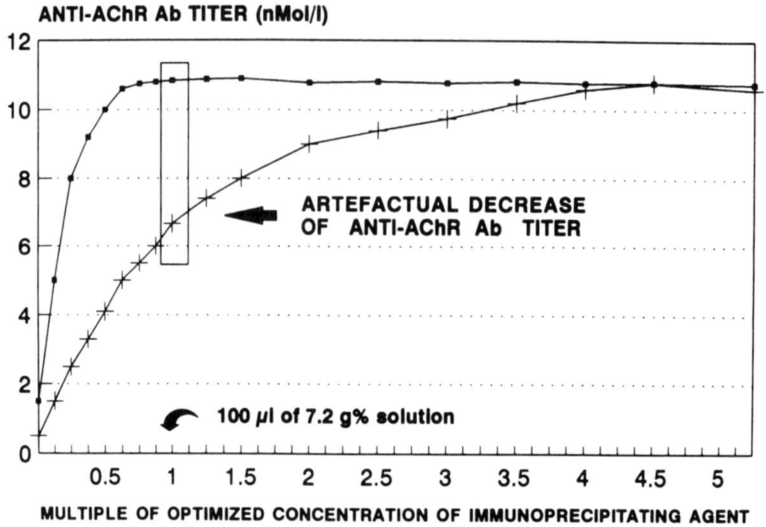

FIGURE 1. Titration curve of anti-AChR Abs using different concentrations of rabbit anti-human IgG in one MG patient before and after iv HIgG treatment. Before treatment, the amount of 100 μL of a 7.2 g % solution of anti-human IgG (concentration optimized to precipitate all IgG present in 5 μL of normal serum) precipitated all the anti-AChR Abs, and the subsequent part of the curve showed a plateau. After treatment, the same amount of anti-human IgG precipitated only a part of anti-AChR Abs present in the serum, and the increase of the amount of the immunoprecipitating agent resulted in a proportional increase of anti-AChR Ab titer, reaching a plateau at a concentration of the immunoprecipitating agent 5-fold higher than that used before treatment. The abnormally high serum IgG concentration after iv HIgG probably aspecifically consumed the immunoprecipitating agent, resulting in an artefactual decrease of anti-AChR Ab titer if a too-low concentration of immunoprecipitating agent was used. —■—, before iv HIgG; +, after iv HIgG.

precipitating agent, appeared to be reduced at the same time. In this first set of experiments, the amount of anti-human IgG was optimized to the lowest concentration precipitating all Ab present in 5 μL of normal serum (usually 100 μL of a 7.2% solution). Before iv HIgG, increasing the concentration of anti-human IgG above the optimized concentration did not result in a further increase of anti-AChR Ab titer (FIG. 1). On the contrary, after iv HIgG, the increase of the immunoprecipitating agent above the optimized concentration resulted in a proportional increase of anti-AChR Ab titer, reaching a plateau at an anti-human IgG concentration 5–6-fold higher than the optimized concentration (FIG. 1). Serum IgG, abnormally increased after iv HIgG, probably nonspecifically consumed the immunoprecipitating agents, yielding an artefactual decrease of anti-AChR Ab titer if too-low an amount of rabbit anti-human IgG was used. Anti-AChR Ab titer (determined with an excess of immunoprecipitating agent) began decreasing 10 days after starting iv HIgG and remained low for 30 days, returning to baseline 45–60 days after starting iv HIgG (FIG. 2). The rapidity of the occurrence of clinical improvement (3–4 days after starting iv HIgG) does not correlate with the delayed inhibition of anti-AChR Ab synthesis and, therefore, points to another possible mechanism of action, that is, the direct idiotype-antidiotype interaction between autoantibody and iv HIgG. We therefore studied the kinetics of binding anti-AChR Ab to the receptor. The use of Staph-A rather than anti-human IgG as the immunoprecipitating agent allows a detailed analysis of the kinetics of binding of anti-AChR Ab to the receptor, which can be quantified as the time to 50% of maximal binding.[4] The interaction of Staph-A and the AChR-Ab complex is essentially instantaneous: this immunoadsorbent was therefore added after different intervals (15, 30, 45, and 60 min, and 2, 4, 6, and 12 hr) following receptor and anti-AChR Ab incubation, and the number of bungarotoxin binding sites precipitated by the sera and the immunoadsorbent at each different interval was calculated. Binding avidity of serum anti-AChR Ab was significantly reduced a few days after starting iv HIgG treatment (FIG. 2) in good chronological association with clinical improvement. No significant side effects occurred in our patients.

Intravenous HIgG seems to be a rapidly effective and safe therapy to be used in rapidly deteriorating MG patients with bulbar symptoms. The therapy of choice of myasthenic crises is plasma exchange that rapidly lowers anti-AChR Ab titer, yielding a rapid clinical improvement probably associated with the regeneration of AChR. Plasma exchange, however, is not technically feasible in all emergency care units and in all patients, and it has been associated with rare, but at times fatal, side effects. Intravenous HIgG may be a safe alternative in treatment of myasthenic crises. The immunosuppressive action of iv HIgG seems to be associated with specific idiotype-antiidiotype interactions: both a direct binding between circulating autoantibodies and iv HIgG (disturbing the binding to AChR) as well

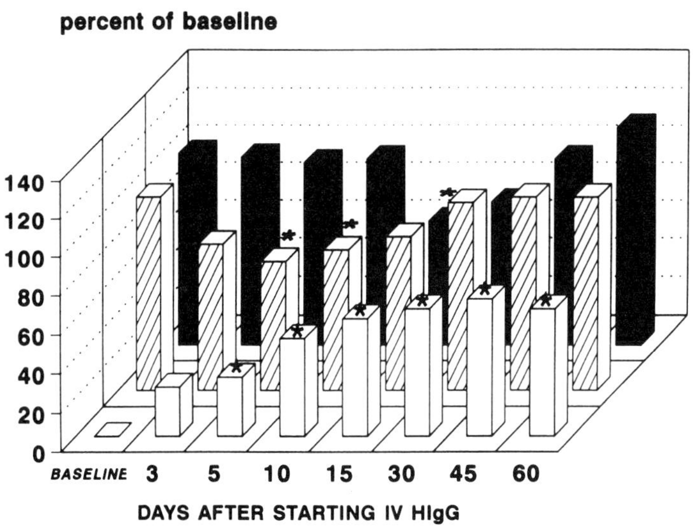

FIGURE 2. Clinical status, anti-AChR Ab binding avidity, and titer in 15 MG patients before and after iv HIgG treatment. Each bar represents the mean value. Clinical improvement is expressed as 100-percent decrement of the baseline MG standardized score. Anti-AChR Ab binding avidity and titer are expressed as a percent of baseline value. Clinical improvement began a few days after starting iv HIgG in good chronological association with the decrease of anti-AChR Ab binding avidity. Anti-AChR Ab titer decreased later. □, clinical improvement; ▨, binding avidity; ■, antibody titer; ★, significantly different from baseline.

as between iv HIgG and lymphocytes (inhibiting autoantibody production) can be hypothesized.

REFERENCES

1. JERNE, N. K. 1974. Ann. Immunol. Inst. Pasteur (Paris) **125C:** 373–389.
2. ARSURA, E. 1991. Immunol. Updates **3:** 1–5.
3. GOTTI, C., R. MANTEGAZZA & F. CLEMENTI. 1984. Neurology **34:** 374–377.
4. TINDALL, R. S. A., M. KENT & L. WELLS. 1981. J. Immunol. Methods **45:** 1–14.

Single-Fiber EMG and Repetitive Stimulation of the Same Muscle in Myasthenia Gravis

JAMES M. GILCHRIST,[a] JANICE M. MASSEY,[b] AND DONALD B. SANDERS[b]

[a]*Department of Neurology*
Rhode Island Hospital
Brown University
593 Eddy Street
APC 689
Providence, Rhode Island 02903

[b]*Division of Neurology*
Duke University Medical Center
Durham, North Carolina

INTRODUCTION

Presumably, the same pathophysiologic abnormality (impulse blocking) is tested by repetitive nerve stimulation (RNS) and single-fiber EMG (SFEMG),[1] though SFEMG has been shown to be more sensitive in diagnosing myasthenia gravis[2] (MG). The relationship, however, between SFEMG and RNS, and in particular, jitter, impulse blocking, and decrement, has not been examined in the same muscle in a large group of patients with MG. Decrement to RNS should be coexistent with impulse blocking seen during SFEMG. If so, blocking should also be seen in some myasthenic muscles without significant decrement, if the percentage of blocking fibers is less than the 10% decrement that must be seen to assure that technical factors have not artifactually reduced the size of the response. To test this hypothesis we identified a group of patients with MG who had RNS and SFEMG studies of the same muscle.

METHODS

Data were collected from patients referred to the EMG laboratories at Duke University Medical Center and Rhode Island Hospital. The diagnosis of MG was based on historical, clinical, pharmacologic, and immunologic findings. Only patients with RNS and SFEMG of the same muscle performed during a single examination were included. RNS with decrement of at least 10% of M-wave area or amplitude between the first and fourth response, and SFEMG with increased jitter or blocking,[3] were abnormal.

RESULTS

RNS and SFEMG were performed on the same muscle in 50 muscles of 46 patients with MG (TABLE 1). The maximum decrement to 3–5 Hz stimulation pre-

TABLE 1. RNS and SFEMG of the Same Muscle

		SFEMG +	SFEMG −	Blocking +	Blocking −	Mean Percent Blocking	Range
RNS	+	18	0	18	0	53.3	12–100%
	−	25	7	19	13	27.8	6–67%
Total		50		50		41.2	6–100%

and postexercise tetanization of the tested muscle, the percentage of action potential pairs with increased jitter and blocking, and the mean consecutive difference (MCD) were compared. We found significant decrement was found only in muscles with increased jitter and impulse blocking on SFEMG. Sixty percent of patients with less than 10% decrement had blocking seen on SFEMG. Increased jitter, blocking, and mean MCD were each correlated with maximum decrement ($R > 0.60$, $p < 0.0001$) (FIG. 1).

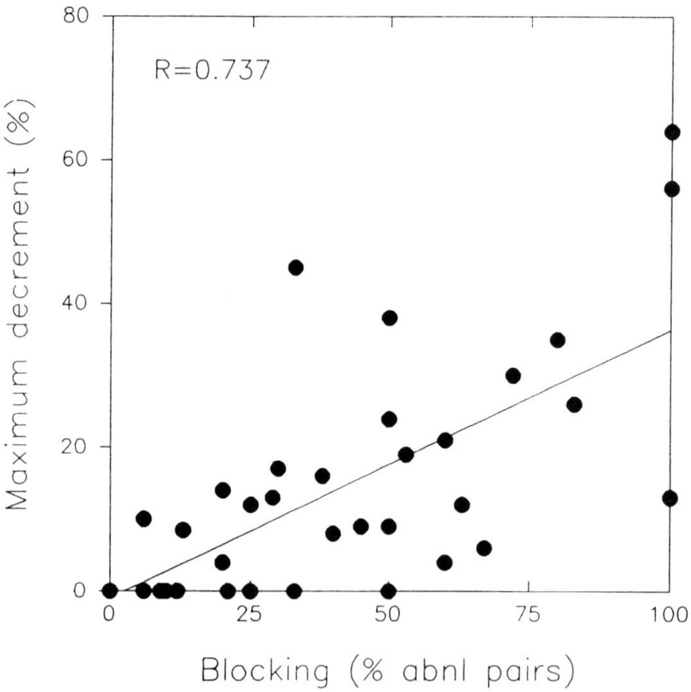

FIGURE 1. Maximum decrement before or after maximum voluntary contraction versus the percentage of blocking pairs seen with SFEMG in 50 muscles in which RNS and SFEMG were performed in 46 patients with myasthenia gravis. R is the correlation coefficient, ($p < 0.0001$).

CONCLUSION

The decrement to RNS results from the same physiologic phenomenon as impulse blocking on SFEMG, and SFEMG is more sensitive at detecting disordered neuromuscular transmission given its ability to detect blocking at levels below the resolution of RNS and increased neuromuscular jitter when there is not blocking.

REFERENCES

1. Schwartz, M. S. & E. Stalberg. 1975. Single fibre electromyographic studies in myasthenia gravis with repetitive nerve stimulation. J. Neurol. Neurosurg. Psychiatry **38:** 678–682.
2. Sanders, D. B. & J. F. Howard. 1986. AAEE minimonograph #25: single-fiber electromyography in myasthenia gravis. Muscle & Nerve **9:** 809–819.
3. Ad Hoc Committee of the AAEM Special Interest Group on Single Fiber EMG. 1992. Single fiber EMG reference values: a collaborative effort. Muscle & Nerve **15:** 151–161.

Analysis of HLA DPB1 Allele in Patients with Myasthenia Gravis in Japan

TERUMI HORIKI, JUNKO MORIUCHI,
HIDETOSHI INOKO, KAZUYUKI MORITA,
YUKINOBU ICHIKAWA, YUKITO SHINOHARA,
KIMIYOSHI TSUJI, AND SHIGERU ARIMORI

Departments of Internal Medicine IV, V,
and Transplantation II
Tokai University School of Medicine
Kanagawa, Japan 259-11

INTRODUCTION

The association of HLA-B8, DR3 with myasthenia gravis (MG) in Caucasians is now well established.[1] A study of Japanese MG patients, however, demonstrated a high frequency of HLA-DR9 and DR53 in females with early onset.[2] Moreover, analysis of the restriction fragment length polymorphism (RFLP) of the DR53-positive patients revealed significant differences between the patients and controls when a DQB cDNA probe was used, indicating the important role of the DQB gene on the disease process of MG.[2]

Recently, a new technique for HLA class II genotyping by digestion of polymerase chain reaction (PCR)-amplified DNA with allele-specific restriction endonucleases (PCR-RFLP method) has been developed.[3] In this study, we used the PCR-RFLP method for HLA-DPB1 typing to analyze the distribution of DPB1 alleles in MG patients. Serological typing of HLA-A, B, C, and DR antigens was also performed.

TABLE 1. Determination of DPB1 Allele by PCR-RFLP Method

genomic DNA + primers
(5′ DPB101 ; GTGAAGCTTTCCCCGCAGAGAATTAC)
(3′ DPB201 ; CACCTGCAGTCACTCACCTCGGCGCTG)
↓
96°C, 1 min (denaturation)
↓
62°C, 1 min (primer annealing)
↓
72°C, 2 min (extent the annealed primers by Taq DNA polymerase)

} 30 cycles

↓
digestion with restriction endonuclease
(Bsp1286I, BssH II, Sau96I, DdeI, EcoNI, Fok I, RsaI)
↓
electrophoresis

TABLE 2. Phenotype Frequencies (pf) and Relative Risk (RR) of HLA Antigens and DPB1 Alleles in Myasthenia Gravis

Antigen/ Allele	MG Total			MG Female Total			MG Female ≦30			MG Female 30<			MG Male Total			Control	
	n	pf (%)	RR	n	pf (%)	RR	n	pf (%)	RR	n	pf (%)	RR	n	pf (%)	RR	n	pf (%)
DR9	62	28.1	1.5	45	31.1	1.7	22	50.0[a]	3.8	23	13.0	0.6	17	41.1	1.2	114	21.1
DR53	62	74.2	1.6	45	80.0	2.2	22	95.5[b]	11.8	23	65.2	1.1	17	59.3	0.8	114	64.0
DPB1*0201	43	58.1	2.5	35	60.0[c]	2.7	18	83.3[d]	9.1	17	35.3	1.0	8	50.0	1.8	45	35.6

[a] p = 0.00749; corrected p value (Pc) not significant (NS).
[b] p = 0.00435; Pc < 0.05.
[c] p = 0.0236; Pc NS.
[d] p = 0.000752; Pc < 0.02.

PATIENTS AND METHODS

The HLA-A, B, C, and DR phenotypes of 62 patients and 114 controls were determined by the microcytotoxicity method. Determination of DPB1 alleles of 43 patients and 45 controls was as shown in TABLE 1.

RESULTS AND DISCUSSION

The distribution of HLA-A, B, and C antigens in MG patients did not differ significantly from those of the controls. The strong association of DR9 and DR53 with early-onset female patients has been confirmed (TABLE 2). The PCR-RFLP analysis of the DPB1 allele demonstrated a high frequency of DPB1*0201 in early-onset female patients compared to the controls (83.3% vs. 35.6%, $p < 0.005$, relative risk = 9.1). Because no linkage disequilibrium has been shown between genes coding for DPB1*0201 and DR9 or DR53, these results indicate an additive effect of certain allelic combinations of HLA class II genes in the disease process of myasthenia gravis.

REFERENCES

1. NAEIM, F., J. C. KEESEY, C. HERRMANN, C. HERRMANN, JR., J. LINDSTORM, E. ZELLER & R. L. WALFORD. 1978. Association of HLA B8, DRW3, and anti-acetylcholine receptor antibodies in myasthenia gravis. Tissue Antigens **12**(5): 381–386.
2. MORITA, K., J. MORIUCHI, H. INOKO, K. TSUJI & S. ARIMORI. 1991. HLA class II antigens and DNA restriction fragment length polymorphism in myasthenia gravis in Japan. Ann. Neurol. **29**(2): 168–174.
3. MAEDA, M., N. URYU, N. MURAYAMA, H. ISHII, M. OTA, K. TSUJI & H. INOKO. 1990. A simple and rapid method for HLA-DP genotyping by digestion of PCR-amplified DNA with allele-specific restriction endonucleases. Hum. Immunol. **27**(1): 111–121.

Myasthenia Gravis Is Associated with Two DQα/β Heterodimers

I. KHALIL,[a] S. BERRIH-AKNIN,[c] V. LEPAGE,[b]
M. N. LOST,[b] P. GAJDOS,[d] D. CHARRON,[b] J. HORS,[a]
AND L. DEGOS[a]

[a]*INSERM U 93*

[b]*Laboratoire d'Immunologie et d'Histocompatibilité*
Hôpital Saint-Louis
Paris, France

[c]*CNRS UA1159*
Hôpital Marie Lannelongue
Le Plessis-Robinson, France

[d]*Hôpital Raymond Poincarré*
Garches, France

Myasthenia gravis (MG) is an autoimmune disease in which antibodies to the acetylcholine receptors play a crucial role in the pathogenesis.[1] The thymus gland is pathologically changed. Most HLA associations in Caucasian MG patients have been previously observed after subgrouping patients according to thymic morphology. The DR3-DQw2 haplotype has been found positively associated with MG with thymic hyperplasia, whereas DR2(Dw2)-DQw1 was found among female MG patients with thymoma.[2] Recently it has been reported that the DQB1*0201 allele was associated with MG with thymic hyperplasia.[3] To investigate further the correlations between HLA class II alleles and MG disease and to examine whether different subgroups of myasthenia patients display different HLA associations, sequence-specific oligonucleotide (SSO) analysis was used to identify the frequencies of HLA-DRB1, B3, B4, DQA1, DQB1, and DPB1 alleles in 47 Caucasian MG patients and in 105 healthy controls. Altogether, 69 alleles were defined, and none of them was found significantly associated with the disease. Subgrouping MG patients according to thymic morphology, an increased frequency of DQA1*0501 allele was found in MG with thymic hyperplasia (90% vs. 46% in controls, relative risk (RR) = 11.7, Pc < 0.001). To investigate the possibility that MG could be associated with a common amino acid sequence shared by DQA1 or DQB1 alleles, DQA1 alleles were divided into DQA1*0100 and DQA1*0X. DQA1*0100 corresponds to DQA1*0101, 0102, and 0103 alleles, sharing a common amino acid sequence at the 42-85 position. The remaining DQA1 alleles are considered as 0X. Similarly, the DQB1 alleles were divided into DQB1*0201, DQB1*0301 alleles, and DQB1*0Y. The latter corresponds to DQB1 alleles other than DQB1*0201 and 0301. The frequencies of the DQA1 and DQB1 genotypes in patients and controls showed that (TABLE 1) the highest risk for developing the disease was found in individuals carrying the heterozygote DQA1*0100/DQA1*0X together with either DQB1*0301/DQB1*0Y or DQB1*0201/DQB1*0Y genotype (72% vs. 29% of controls, RR = 6.2, PC < .001). Individuals carrying DQA*0100/0100-DQB*0Y/0Y or DQA1*0X/0X-DQB1*0201 and/or 0301 genotype seem to be protected. These results suggest the requirement of both DQA*0100 and DQB1*0201 or DQB1*0301 alleles in conferring susceptibility to MG disease.

TABLE 1. Frequencies of the Different DQA1 and DQB1 Genotype Combinations among MG Patients and Controls

Genotype		MG Patients n = 47	Controls n = 105		
DQB1*/DQB1*	DQA1*/DQA1*	%	%	RR	Pc[d]
0201/0Y[b]	0100[a]/0X[c]	34	12	3.7	0.01
0301/0Y	0100/0X	36	17	3	0.04
0201/0301	0100/0X	02	04		—
0Y/0Y	0100/0X	04	13		—
0Y/0Y	0100/0100	06	25		—
0201/0Y	0X/0X	0	06		—
0301/0Y	0X/0X	06	06		—
0201/0201	0X/0X	0	05		—
0301/0301	0X/0X	06	07		—
0201/0301	0X/0X	04	10		—

[a] DQA1*0100 : DQA1*0101, 0102, and 0103 alleles sharing a common aa squence (42-85).
[b] Y : DQB1 alleles other than DQB1*0201 and DQB1*0301.
[c] X : remaining DQA1 alleles (DQA1*0201, 0301, 0401, 0501, and 0601).
[d] Pc = P corrected.

Based on the known linkage disequilibrium between DQA1 and DQB1 alleles, DQA1*0100 is usually located in the *trans* position with regard to DQB1*0201 and DQB1*0301 alleles. The expression, at the cell surface, of such *trans* DQ combinations has been previously demonstrated in DR1-DQw1/DR7-DQw2 cell lines.[4] DQA1*0100-DQB1*0201 and DQA1*0100-DQB1*0301 *trans* combinations seem to be positively associated with MG independent of thymic morphology. Whether these *trans* DQ combinations represent the same or different peptides is not established.

REFERENCES

1. LINDSTROM, J. M., M. E. SEYBOLD, V. A. LENNON, S. WHITTINGHAM & D. D. DUANE. 1976. Antibody to acetylcholine receptor in myasthenia gravis: prevalence, clinical correlates, and diagnostic value. Neurology (Minneap.) **26:** 1054–1059.
2. CARLSSON, B., J. WALLIN, R. PIRSKANEN, G. MATELL & E. I. E. SMITH. 1990. Different HLA-DR-DQ associations in subgroups of idiopathic myasthenia gravis. Immunogenetics **31:** 285–290.
3. SPURKLAND, A., N. E. GILHUS, K. S. RONNINGEN, J. A. AARLI & F. VARTDAL. 1991. Myasthenia gravis patients with thymus hyperplasia and myasthenia gravis patients with thymoma display different HLA associations. Tissue Antigens **37:** 90–93.
4. CHARRON, D. J., V. LOTTEAU & P. TURMEL. 1984. Hybrid HLA-DC antigens provide molecular evidence for gene trans complementation. Nature **312:** 157–159.

How Reliable Are Imaging Procedures in Detecting Residual Thymus after Previous Thymectomy?

PETER KORNFELD,[a,b] AVRAHAM MERAV,[a]
SONDRA FOX, AND KAREN MAIER

[a]*Mount Sinai School of Medicine*
New York, New York

Departments of Medicine and Surgery
Englewood Hospital
Englewood, New Jersey

Five cases of recalcitrant generalized myasthenia gravis (MG) of long duration are presented, who underwent transsternal reexploration of chest after having

TABLE 1. Summaries of Clinical Histories

Patient	Classification (Osserman)	P.A. and Lateral Chest X Rays	CAT Scan of Chest GE CT9800, Quick Unit	MRI GE Sigma 1.5 Tesla Unit	Clinical Response to Reexploration	Thymic Pathology
D.S. F Age 35 MG 13 yr	II B	0[a]	0	+[b]	+	55 grams resected GC[c] 0
G.W. F Age 61 MG 12 yr	II B	0	0	0	+	44 grams resected GC +
F.G. F Age 42 MG 12 yr	IV	0	0	0	0	49 grams resected GC +
W.S. F Age 35 MG 13 yr	II B	0	0	0	0	37.9 grams resected GC +
R.E. F Age 32 MG 5 yr	II B	0	0	0	0	12.1 grams resected GC 0

[a] Negative.
[b] Positive.
[c] GC = germinal centers.

[b] Current address: Department of Medicine, Stanford University School of Medicine, Stanford, CA 94305.

previously undergone transcervical thymectomy. Preoperative chest X rays, CAT scans, and MRI examinations were performed in all five patients. Except for one MRI in a single patient, all failed to demonstrate imaging evidence of residual thymic tissue. Yet, significant amounts of thymic tissue (12–55 g) were found in all five patients. Three out of five resected thymuses contained many germinal centers. Only two out of five patients demonstrated significant clinical improvement within two years of reexploration (see TABLE 1). Apparently duration of MG and presence of germinal centers bore no relation to clinical results in this small series. We, therefore, recommend that the decision to operate or reoperate be based on clinical findings rather than on imaging results alone.[1–15]

REFERENCES

1. OSSERMAN, K. E., P. KORNFELD, E. COHEN, G. GENKINS, H. MENDELOW, H. GOLDBERG, H. GOLDBERG, H. WINDSLEY & C. I. KAPLAN. 1958. Studies in myasthenia gravis. A review of 282 cases. Arch. Intern. Med. **2:** 72–81.
2. OSSERMAN, K. E. & G. GENKINS. 1971. Studies in myasthenia gravis. Review of a twenty year experience in over 1,200 patients. Mt. Sinai J. Med. **38:** 497–537.
3. PAPATESTAS, A. E., G. GENKINS, S. H. HOROWITZ & P. KORNFELD. 1976. Thymectomy in myasthenia gravis: pathologic, clinical, and electrophysiologic correlations. Ann. N.Y. Acad. Sci. **274:** 555–573.
4. JARETZKI, A., III, M. BETHEA, M. WOLFF, R. E. LOVELACE, A. S. PENN, M. R. OLARTE & L. ROWLAND. 1978. Total thymectomy in myasthenia gravis. Anatomical and surgical considerations N.Y. Sate J. Med. **78:** 1094–1097.
5. JARETZKI, A., III, A. S. PENN, D. S. YOUNGER *et al.* 1988. Maximal thymectomy for myasthenia gravis. J. Thorac. Cardiovasc. Surg. **95:** 747–757.
6. COOPER, F. O., M. B. AL-JILAIHAWA, F. G. PEARSON & J. G. HUMPHREY. 1988. An improved technique to facilitate transcervical thymectomy for myasthenia gravis. Am. Thorac. Surg. **45:** 242–247.
7. PAPATESTAS, A. E., G. GENKINS & P. KORNFELD. 1981. Comparison of the results of the transcervical and transsternal thymectomy in myasthenia gravis. Ann. N.Y. Acad. Sci. **377:** 766–778.
8. GENKINS, G., A. E. PAPATESTAS, S. H. HOROWITZ & P. KORNFELD. 1975. Early thymectomy in myasthenia gravis. Am. J. Med. **58:** 517–524.
9. MOLNAR, J. & A. SZOBOR. 1990. Myasthenia gravis: Effect of thymectomy in 425 patients. Eur. J. Cardiothorac. Surg., **4:** 8–14.
10. KAYE, A. D., R. JANNSEN, P. H. ARGER *et al.* 1983. Mediastinal computed tomography in myasthenia gravis. J. Comput. Tomogr. **7:** 273–279.
11. GOLDBERG, R. E., J. R. HAAGA & B. S. YULISH. 1987. Serial CAT scans in thymic hyperplasia. J. Comput. Assisted Tomogr. **11**(3): 539–540.
12. MOORE, N. R. 1989. Imaging in myasthenia gravis. Clin. Radiol. **40:** 115–116.
13. SLOTHES, S., K. F. R. NEUFANG & F. W. HAUPST. 1989. Value of radiologic imaging of the thymus gland in myasthenia gravis. Röntgen Bl. **42:** 455–461.
14. OLANOW, C. W., A. S. WECHSLER, M. SIROTKIN-ROSES, J. STAJICK & A. D. ROSES. 1987. Thymectomy as primary therapy in myasthenia gravis. Ann. N.Y. Acad. Sci. **505:** 595–606.
15. GENKINS, G., P. KORNFELD, A. E. PAPATESTAS, A. N. BENDER & R. J. MATTA. 1987. Clinical experience in more than 2,000 patients with myasthenia gravis. Ann. N.Y. Acad. Sci. **505:** 500–514.

An *in Vitro* Model for Disease-specific Immunotherapy in Myasthenia Gravis Using Soluble MHC Class II Bound to AChR-derived Peptide[a]

M. W. NICOLLE, A. VINCENT, S. SHARMA,[b] B. NAG,[b]
N. WILLCOX, AND J. NEWSOM-DAVIS

Neurosciences Group
Institute of Molecular Medicine
John Radcliffe Hospital
University of Oxford
Oxford OX3 9DU, United Kingdom

[b]*Anergen Incorporated*
301 Penobscot Drive
Redwood City, California 94063

INTRODUCTION

Although MG is known to be mediated by antibodies against the acetylcholine receptor (AChR), propagation of MG is thought to depend on B-cell help provided by disease-specific T-helper cells (T_H). Previous attempts to anergize peripheral disease-specific T_H cells include models in which the relevant peptide is presented by MHC molecules either on the surface of fixed antigen-presenting cells (APC) or on planar membranes.[1] In these models other surface accessory molecules are modified or lacking, or the necessary cytokines are not produced. This results in provision of the first signal for T-cell activation: TCR engagement by MHC-bound peptide. The second signal, necessary for complete T-cell activation, is not adequately provided, and functional inactivation (anergy) results. The aim of our *in vitro* model is to develop and define the mechanisms for a disease-specific immunotherapy in MG in which we attempt to inactivate antigen-specific T cells. An animal model of experimental allergic encephalomyelitis (EAE) produced *in vivo* and *in vitro* evidence that this approach might work in MG.[2] Inactivating T_H cells in MG should remove the help necessary for anti-AChR antibody production.

MATERIAL AND METHODS

We preincubated clonal CD4$^+$ T cells from a well-characterized clone[3] with a soluble complex of MHC II (DR4Dw4) bound to the relevant peptide epitope (p138-167) from the AChR α-subunit. After preincubation with complex, or media as a control, for varying periods of time, the cells were stimulated with APCs and

[a] M. Nicolle is supported by a Canadian MRC Fellowship.

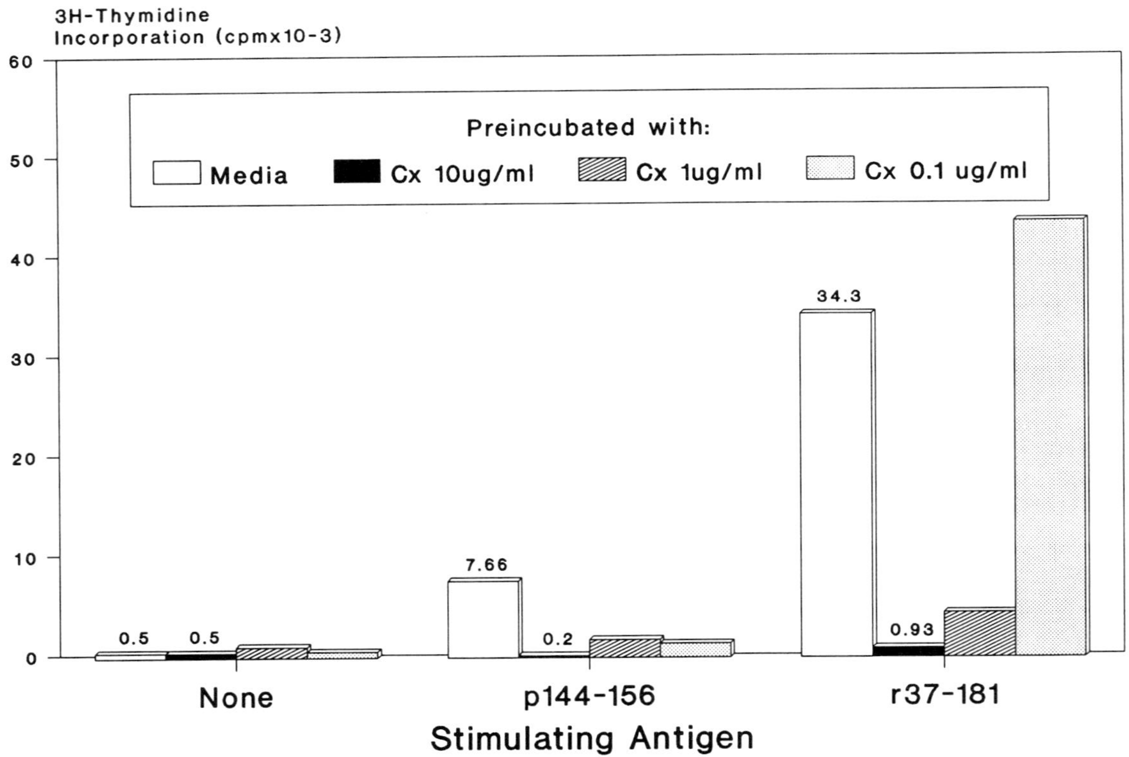

FIGURE 1. T cells were incubated with medium or complex, at either 10, 1, or 0.1 μg/mL, for 5 days before washing and counting. Three days after the addition of APCs and stimulating antigen to 2×10^4 T cells, [^{3}H]thymidine was added, and the cells were harvested 18–24 hours later. Proliferation, as measured by incorporated radioactivity (cpm $\times 10^{-3}$), is shown on the Y axis. Results are shown for stimulation with media (None), peptide 144-156, or recombinant 37-181 (both of the AChR α-subunit). SDs were all less than 15 percent.

antigens. Proliferation was assessed using a standard assay measuring [^{3}H]thymidine incorporation.

RESULTS

Incubation of T cells with complex overnight, or for several days (FIG. 1), substantially reduced the subsequent proliferative response to antigen. Shorter periods of incubation were ineffective. Incubation with irrelevant complexes was ineffective (FIG. 2). Incubation of non-AChR, non-DR4–restricted clones with complex had no effect, and incubation with free peptide alone inhibited proliferation but required concentrations that were 10^3 moles higher (data not shown). Cell counts using Trypan blue and chromium release assays show that cell death occurs in the presence of the complex (data not shown).

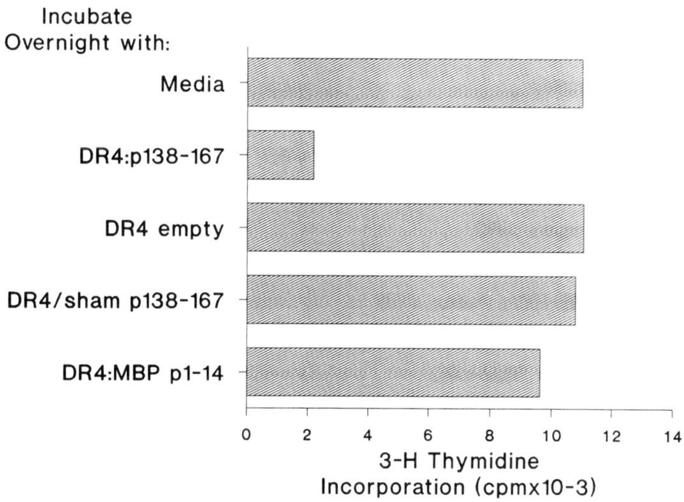

FIGURE 2. T cells were incubated overnight with one of the following: DR4 : p138-167 complex; DR4 empty; DR4 and sham p138-167 (unbound); or DR4 : MBP 1-14 complex. APCs and stimulating antigen (recombinant 37-181) were added, and proliferation was assessed as described for FIG. 1. SDs were all less than 15 percent. MBP p1-14 = p1-14 of myelin basic protein.

CONCLUSIONS

We have shown that preincubation of CD4$^+$ T cells, raised against recombinant peptides of the AChR α-subunit, with a soluble complex of the restricting MHC II haplotype and peptide epitope substantially reduces subsequent antigen-specific proliferation. We postulate that the complex provides the first signal for partial or transient T-cell activation. If a second signal is not provided within a finite time, the T cell becomes refractory to further stimulation and, in this model, may actually be killed. The nature of the second signal that can "rescue" the T cells from the inactivating effects of the complex is unknown. It is provided by APCs (irradiated PBLs) but not by IL2 (data not shown). This *in vitro* model may provide a basis for rational formulation of a specific immunotherapy in MG, similar to that shown for EAE in an animal model. This therapy would have the advantage that it appears to be selective for disease-specific T cells and is nontoxic to other T cells.

REFERENCES

1. Quill, H. &. R. H. Schwartz. 1987. J. Immunol. **138**(11): 3704–3712.
2. Sharma, S., B. Nag, X. M. Su, D. Green, E. Spack, B. R. Clark & S. Sriram. 1991. Proc. Natl. Acad. Sci. USA **88:** 11465–11469.
3. Ong, B., N. Willcox, P. Wordsworth, D. Beeson, A. Vincent, D. Altmann, J. S. S. Lanchbury, G. C. Harcourt, J. I. Bell & J. Newsom-Davis. 1991. Proc. Natl. Acad. Sci. USA **88:** 7343–7347.

HLA Haplotype in Myasthenia Gravis (MG) of Japanese Children

Correlative Studies between MG Patients and Their Normal Siblings and Parents

Y. NOMURA, M. EHARA, K. MATSUKI,[a] T. JUJI,[a] AND M. SEGAWA

Segawa Neurological Clinic for Children
2-8 Surugadai, Kanda
Chiyoda-ku, Tokyo 101, Japan

[a]Blood Transfusion Service
Tokyo University Hospital
Tokyo, Japan

Childhood MG in Japan has a correlation to specific HLA class II antigens: DR9, DRw13, DQw1, and DQw3 (see ref. 1 for previous nomenclature). The purpose of this report is to examine the genetic influences with regard to the onset of MG in children by comparing those frequently occurring class II antigens, genotypes, and haplotypes in children with MG, their non-MG siblings, and their parents.

Forty-one MG patients with onset in childhood, 22 non-MG siblings, and 49 parents (41 mothers and 8 fathers) were subjected to this study. These are the same subjects and data source reported previously,[1] though DQ antigens are expressed as a new denomination. In this study, incidences of HLA class II antigens and genotypes observed frequently in childhood-onset MG in our previous report were compared with these three groups. Furthermore, incidences of DR-DQ haplotypes that increased in childhood MG were compared in individuals whose haplotypes were all made clear and informative: 32 childhood-onset MG patients, 18 non-MG siblings, and 25 parents (17 mothers and 8 fathers). Incidences of these haplotypes were also assessed in accordance with the ages of onset of MG and compared with MG patients and non-MG siblings.

HLA class II antigens observed frequently in childhood MG were also detected in higher incidences in their non-MG siblings and parents (TABLE 1a). These incidences, however, were less than those in MG patients. The frequency of the DR9 antigen of non-MG siblings and parents was also significantly higher compared to normal controls (<0.005). By contrast, the DRw13 frequency of parents was slightly higher compared to normal controls (<0.05), but it showed no significant differences with non-MG siblings. Statistically significant differences were observed with DR9, DRw13, and DQw6 between MG patients and non-MG siblings, and with DR9 and DQw9 between MG patients and parents. The differences were more remarkable when the incidences of genotypes of these antigens in MG patients and non-MG siblings, and in MG patients and parents, were compared (TABLE 1b).

DR-DQ haplotypes that were frequently observed in MG were significantly higher in MG patients than in non-MG siblings (TABLE 2a), whereas a combination of these haplotypes showed significant differences between MG patients and parents (TABLE 2b). The differences of incidences of particular haplotypes or of their combinations between MG patients and non-MG siblings showed specific age dependency, for example, DR9-DQw9/DRw13-DQw6 was significantly higher in patients with MG onset before 3 years of age, as well as MG onset between ages 6 and 12 years.

There were some MG patients who shared an identical haplotype with their non-MG siblings; these patients included two sets of monozygotic twins and one set of dizygotic twins. One family had normal parents who had identical MG-specific haplotypes to their MG son. On the contrary, in one case of familial MG with involvement of two siblings, each showed different HLA haplotypes.

Thus the parents and non-MG siblings of childhood-onset MG patients also showed the significantly increased childhood MG-related class II antigens, but levels were not as high as the patients'. It was also shown that the specific DR and DQ genotypes were significantly increased in MG patients compared with their parents or non-MG siblings. These results suggest that childhood-onset MG consists of a genetically specific group that is under both parental influences. However, the presence of the non-MG siblings, including two monozygotic twins with the same haplotypes and normal parents with "MG haplotype," suggests the roles for environmental or other exogenous factors and also the age-dependent influences of these factors for development of MG during childhood.

TABLE 1.

a. Incidences of DR and DQ Antigens in Children with MG, Their Non-MG Siblings, and Their Parents[a]

	MG Patients N = 41	Non-MG Siblings N = 22	Parents N = 49
DR9	34 (82.9)	13 (59.1)*	27 (55.1)**
DRw13	17 (41.5)	4 (18.2)*	14 (28.6)
DQw6	27 (65.9)	8 (36.4)*	26 (53.1)
DQw9	33 (80.5)	13 (59.1)	26 (53.1)**

( ): %

Japanese control: DR9 60 (27%)
N = 220 DRw13 36 (16%)

b. Incidences of Frequently Observed DR and DQ Genotypes in Children with MG, Their Non-MG Siblings, and Their Parents[a]

	MG Patients N = 41	Non-MG Siblings N = 22	Parents N = 49
DR9-w13/DQw6-w9	14 (34.1)	2 (9.1)**	4 (8.2)**
DR9,- /DQw6-w9	20 (48.8)	2 (9.1)**	10 (20.4)**
DR9,- /DQw9,-	34 (82.9)	11 (50.0)**	26 (53.1)**

( ): %

[a] Statistical differences were compared between MG patients and non-MG siblings, and also between MG patients and parents.

* $p < 0.05$

** $p < 0.01$

TABLE 2.

a. Frequencies of DR-DQ Haplotypes in Childhood MG, Their Non-MG Siblings, and Their Parents[a]

	MG Patients N = 32	Non-MG Siblings N = 18	Parents N = 25
Bw61-DR9	12 (37.5)	2 (11.1)*	5 (20.0)
DR9-DQw9	26 (81.3)	9 (50.0)*	12 (48.0)**
DRw13-DQw6	12 (37.5)	2 (11.1)*	6 (24.0)

( ): %

b. Frequencies of DR-DQ Haplotype Combination in Childhood MG, Their Non-MG Siblings, and Their Parents[a]

	MG Patients N = 32	Non-MG Siblings N = 18	Parents N = 25
DR9-DQw9/ DRw13-DQw6	10 (31.3)	1 (5.6)*	2 (8.0)*
DR9-DQw9/ —, DQw6	5 (15.6)	2 (11.1)	0 (0.0)*

( ): %

[a] Statistical differences were compared between MG patients and non-MG siblings, and also between MG patients and parents.

* $p < 0.05$

** $p < 0.01$

REFERENCE

1. Matsuki, K., T. Juji, K. Tokunaga, M. Takamizawa, H. Maeda, M. Soda, Y. Nomura & M. Segawa. 1990. HLA antigens in Japanese patients with myasthenia gravis. J. Clin. Invest. **86:** 392–399.

Electrophysiological Characteristics in Seronegative Myasthenia Gravis[a]

SHIN J. OH

Department of Neurology
The University of Alabama at Birmingham
The Veterans Administration Medical Center
UAB Station
Birmingham, Alabama 35294

There have been several papers addressing the differences in clinical features between AChR-Ab positive and AChR-Ab negative myasthenia gravis (MG) patients.[1-3] The electrophysiological differences, however, between these two groups have not been studied. We are reporting here the electrophysiological characteristics of seronegative and seropositive myasthenic patients.

MATERIAL AND METHODS

This study is based on 140 MG patients (61 males and 79 females ranging in age from 19 to 79 years) seen in the UAB MG clinic during the past 11 years. All these patients were symptomatic and had objective evidence of weakness at the time of evaluation. The severity of MG was graded according to a modified Osserman's classification at the time of evaluation: I, ocular; IIA, mild generalized; IIB, moderate generalized; and IIC, severe generalized.

The AChR Ab was assayed following a standard immunoprecipitation method using human AChR. When the titer was above 0.05 nmol [^{125}I]alpha-bungarotoxin AChR bound per liter of serum, the test was considered positive. When the AChR-Ab assay was negative at least twice over the follow-up period, patients were classified seronegative.

The repetitive nerve stimulation (RNS) test on the abductor digiti quinti (ADQ), flexor carpi ulnaris (FCU), orbicularis oculi, trapezius, and deltoid muscles and the single fiber EMG (SFEMG) on the extensor digitorum communis muscle were performed, and an interpretation was made following the standard methods.[4]

RESULTS

Among 140 MG patients, the AChR-Ab assay was positive in 85 patients (73%), whereas the RNS test and the SFEMG were positive in 110 (79%) and 126 (91%), respectively. Between the seronegative and seropositive groups, there was no significant difference in sex, age, distribution of weakness, or severity of disease

[a] This study was supported by a Grant from the Alabama Chapter of the National MG Foundation.

(TABLE 1). Mild MG (I and IIA) was found in 31 (82%) of seronegative patients and in 65 (64%) of seropositive patients, suggesting that seropositive patients have a tendency to have more severe MG. This difference, however, was not statistically significant ($p = 0.094$).

There was a statistically significant difference between seronegative and seropositive groups in the overall abnormality rate of the RNS test ($p = 0.025$) (TABLE 1). No significant difference was found, however, in the abnormality rate on the

TABLE 1. Clinical and Electrophysiological Differences in Seropositive and Seronegative MG

Parameters	Seronegative (N = 38[a])	Seropositive (N = 102)	Significance Level
Clinical Features:			
Sex: Male	19[a]	42	p = 0.349
Female	19	60	
Age (yr):	51.4	46.6	
Symptoms:			
Oculobulbar	30	92	p = 0.367
Limb girdle	30	86	p = 0.454
MG type:			p = 0.093
I	8	16	
IIA	23	49	
IIB	4	32	
IIC	3	5	
Electrophysiological findings:			
Repetitive Nerve Stimulation:			
Abnormal	25 (65.8%)	85 (83.3%)	p = 0.025
Distal muscle	22	66	p = 0.458
Proximal muscle	3	19	
Normal	18	16	
ADQ: Abnormal	19	57	p = 0.605
Type 1[b]	9	30	p = 0.371
2	4	20	
3	5	6	
4	1	1	
Normal	19	45	
FCU: Abnormal	16	53	p = 0.379
Type 1	6	27	p = 0.502
2	6	21	
3	3	4	
4	1	1	
Normal	20	47	
Single-fiber EMG:			
Abnormal	35 (92.1%)	93 (91.1%)	p = 0.861
Normal	3	9	
Mean MCD (μs)	71.34 ± 53.66	91.53 ± 71.11	p = 0.119
PP[c] > 53 μs (%)	38.11 ± 30.86	51.05 + 39.67	p = 0.070
PP with blocking (%)	15.45 ± 22.58	20.00 ± 27.35	p = 0.361

[a] Number represents number of patients unless specified otherwise.

[b] Type 1, decrement at the low rate of stimulation (LRS) alone. Type 2, decrement at LRS and high rate of stimulation (HRS). Type 3, decrement at HRS alone. Type 4, posttetanic exhaustion alone.

[c] PP, potential pairs.

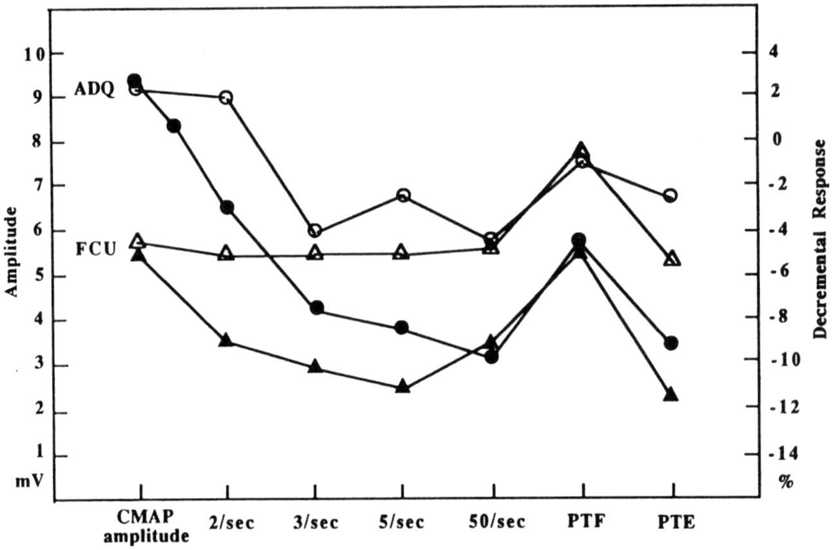

FIGURE 1. Mean responses on the RNS test on the ADQ (● ○) muscle and FCU muscle (▲ △) in the seronegative (○ △) and seropositive (▲ ●) MG patient. PTF (posttetanic facilitation); 5/sec stimulation immediately after 50/sec stimulation. PTE (posttetanic exhaustion); 5/sec stimulation 4 minutes after the 50/sec stimulation. CMAP = compound muscle action potential.

distal or proximal RNS tests in the type of RNS abnormalities on the ADQ and FCU muscles, or in the frequency of SFEMG abnormalities, mean MCD (mean consecutive difference) values, percentage of SF potential pairs with MCD > 53 μs, or with blocking. The decremental responses at the low rates of stimulation on the RNS tests in the ADQ and FCU muscles were worse in the seropositive group (FIG. 1).

DISCUSSION

Seven to 33% of MG patients were persistently AChR-Ab negative.[1] One study suggested that the pathogenesis of seronegative MG is immune-mediated, but that the pathogenic autoantibody interferes with neuromuscular transmission by binding to non-AChR determinants.[5] One study showed no significant difference in the clinical features between these two groups,[1] but another report showed that "milder MG restricted to the ocular muscle" is more likely in seronegative MG.[2] Our study showed that RNS abnormalities are significantly more common in seropositive MG. This finding is most likely due to the nonsignificant trend toward more severe MG in the seropositive group as shown by our study. One previous study showed a significant difference in the abnormality rate of the RNS test.[1]

REFERENCES

1. SOLIVEN, B. C., D. J. LANGE, A. S. PENN, D. YOUNGER, A. JARETZKI, R. E. LOVELACE & L. P. ROWLAND. 1988. Seronegative myasthenia gravis. Neurology **38:** 514–517.

2. SANDERS, D. B., J. F. HOWARD, J. M. MASSEY, M. MIHOVILOVIC, C. W. OLANOW, A. D. ROSES & C. R. STEWART. 1987. Seronegative myasthenia gravis. Ann. Neurol. **22:** 126.
3. BIRMANNS, B., T. BRENNER, O. ABRAMSKY & I. STERINER. Seronegative myasthenia gravis: clinical features, response to therapy and synthesis of acetylcholine receptor antibodies *in vitro*. 1991. J. Neurol. Sci. **102:** 184–189.
4. OH, S. J. 1988. Electromyography: Neuromuscular Transmission Studies. Williams & Wilkins. Baltimore, MD.
5. MOSSMAN, S., A. VINCENT & J. NEWSOM-DAVIS. 1986. Myasthenia gravis without acetylcholine-receptor antibody: a distinct disease entity. Lancet **1:** 116–118.

3,4-Diaminopyridine in Lambert-Eaton Myasthenic Syndrome and Myasthenia Gravis[a]

DONALD B. SANDERS, JAMES F. HOWARD JR.,[b] AND JANICE M. MASSEY

Duke University Medical Center
Box 3403
Durham, North Carolina 27710

[b]*University of North Carolina Medical School at Chapel Hill*
Chapel Hill, North Carolina

The aminopyridines increase chemical transmission at central and peripheral synapses by increasing transmitter release. 3,4-diaminopyridine (DAP) has been used for several years in Europe to treat diseases that affect neuromuscular transmission[1] but has only recently been investigated for this purpose in this country.[2] We report here the current results of an ongoing study of the efficacy and safety of DAP in Lambert-Eaton myasthenic syndrome (LEMS) and myasthenia gravis (MG).

METHODS

In patients with LEMS, malignancy was sought and, if found, treated before entry into this study. The diagnosis of MG was based on clinical criteria, including a history of fluctuating weakness that improves after administration of cholinesterase inhibitors and the demonstration of abnormal neuromuscular transmission by EMG tests. Pregnant women and patients with seizures or cardiac, renal, or hepatic disease were excluded. DAP was obtained as a powder from Regis Chemical Company and was packaged in capsules with lactose by the Duke University Pharmacy. Patients received increasing doses of DAP in hospital to detect immediate side effects and to establish an effective dose for the subsequent double-blinded phase. Patients then received blinded placebo or DAP for one week, then crossed over to the alternate. Some patients received nicotinamide during both arms of the blinded phase to mimic and mask the side effects of DAP. The severity of weakness was measured by a quantitative MG (QMG) score, a standardized assessment of function of muscle groups typically involved in MG and LEMS.[3] Treatment efficacy was assessed by comparing QMG scores on the placebo and DAP arms of the blinded, crossover trial. Patients could continue taking DAP chronically if they improved more than three QMG points on DAP, or exceptionally, if they had only symptomatic improvement on DAP. Pyridostigmine was

[a] This work was supported by the GCRC Program Grant, NCRR, NIH, and USPHS Grant FD-U-000460.

added during this phase if not previously being taken, and the DAP dose was adjusted to produce maximum clinical benefit.

RESULTS

Seven of ten patients with LEMS improved more than 3 QMG points on DAP (FIG. 1). QMG scores after DAP were significantly less than after placebo in this group ($p < .01$). Two of eight patients with acquired MG improved at least 3 QMG points on DAP ($p = .07$ for this group). Therapeutic responses were seen with doses of 5 to 15 mg DAP (tid). The addition of pyridostigmine produced greater

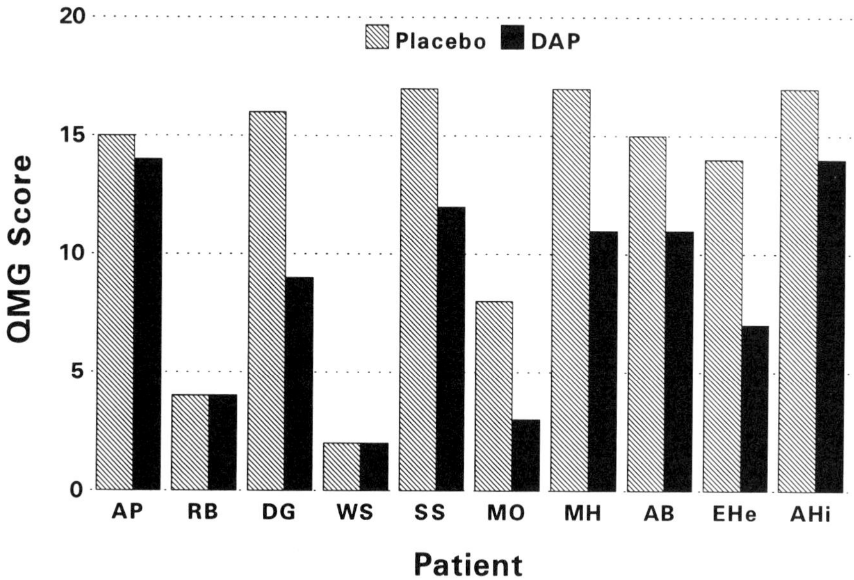

FIGURE 1. Quantitative MG (QMG) scores on placebo and 3,4-diaminopyridine (DAP) in 10 patients with Lambert-Eaton myasthenic syndrome.

symptomatic improvement than did DAP alone in most patients and permitted reduction of the DAP dose.

SIDE EFFECTS

Sixteen patients have taken DAP chronically (15 to 50 mg/day) for 4 to 45 months (average 22 months). Side effects were identified by symptom questionnaire, physical examination, blood tests of renal, hepatic, and hematopoietic function, EKGs, EEGs, and endocrine function tests performed serially before and during administration of DAP. Seizures occurred in two patients taking

100 mg DAP/day (one had toxic blood levels of theophylline). Other side effects were transient digital and perioral paresthesias after doses of 10 mg or more of DAP.

CONCLUSIONS

In ten patients with LEMS, DAP produced a significant improvement in strength as assessed by a quantitative scoring system. In patients with acquired MG there was a variable response, some patients having dramatic, functionally significant improvement, but others having none. The only significant side effects of DAP were seizures that occurred in two patients taking 100 mg/day. Optimum clinical response was usually obtained with DAP doses of 30 to 50 mg/day, especially if pyridostigmine was given concomitantly.

DAP appears to be a safe, effective treatment in patients with LEMS and in some patients with acquired myasthenia gravis.

REFERENCES

1. LUNDH, H., O. NILSSON & I. ROSEN. 1983. Novel drug of choice in Eaton-Lambert syndrome. J. Neurol. Neurosurg. Psychiatry **46:** 684–685.
2. MCEVOY, K. M., A. J. WINDEBANK, J. R. DAUBE & P. A. LOW. 1989. 3,4-Diaminopyridine in the treatment of Lambert-Eaton myasthenic syndrome. N. Engl. J. Med. **321:** 1567–1571.
3. TINDALL, R. S. A., J. A. ROLLINS, J. T. PHILLIPS, R. G. GREENLEE, I. WELLS & G. BELENDIUK. 1987. Preliminary results of a double-blind, randomized, placebo-controlled trial of cyclosporine in myasthenia gravis. N. Engl. J. Med. **316:** 719–724.

Relapses in Patients with Myasthenia Gravis Treated with Azathioprine

R. WILENSKY, B. DWYER, AND R. F. MAYER[a]

Department of Neurology
University of Maryland School of Medicine
Baltimore, Maryland 21201

Generalized myasthenia gravis (MG) is a chronic relapsing autoimmune disease characterized by weakness and fatigability of skeletal muscles secondary to a defect in neuromuscular transmission (NMT). The defect is NMT results from decreased sensitivity of the postsynaptic membrane[1] secondary to the effects of acetylcholine receptor antibodies.[2] Because patients with MG have thymic abnormalities and impaired immunoregulation, therapy has been directed at altering these dysfunctions such as total thymectomy or corticosteroids (CST). But in some patients, recurrent MG signs occur or undesirable side effects result from long-term CST therapy. Azathioprine (AZA) has been reported to be effective in MG,[3] and this study was undertaken to assess its long-term effects in patients with chronic MG.

CLINICAL DATA

Thirty-one patients (median age 59 years) with mild (48%), moderate (45%), or fulminating (7%) MG were treated with daily AZA for 1.5 to 8 years. Median daily dose of AZA was 150 mg given in 3 equal doses with meals. Clinical status was graded 0 to 4 (asymptomatic to severe generalized weakness with respiratory failure). Of these patients, 58% had a thymectomy, 58% were on CTS, and 71% were on cholinesterase inhibitors (AChE-I).

RESULTS

At one year the mean clinical grade of the patients had significantly improved (TABLE 1). Seventy-six percent of the patients improved, and of the 7 not improved, 6 were females under the age of fifty. Eighty-nine percent of the patients were able to reduce the dose of CST. Although no serious complications occurred, multiple side effects were observed, especially hematologic changes (33%), gastrointestinal symptoms (26%), and infections (21%). Six patients withdrew due to rash, anemia, nausea, or cost.

[a] Address correspondence to Dr. Richard Mayer, Department of Neurology, Room N4W46, University of Maryland Hospital, 22 S. Greene Street, Baltimore, MD 21201.

TABLE 1. Clinical Grade of Patients Treated with Azathioprine

Number of Patients	Treatment Years	Grade[a]
26	Onset	2.7 ± 0.6
26	1	1.2 ± 0.7[b]
23	2	1.2 ± 0.9[b]
19	4	1.2 ± 0.8[b]
11	6	1.0 ± 0.6[b]

[a] Mean values ± 1 SD.
[b] All values significant, $p < 0.01$, compared with onset.

Reviewing the clinical course of 23 of these patients (TABLE 2) revealed that 70% had had at least one relapse, defined as a change of one clinical grade. However, 96% of these patients had had at least one relapse prior to AZA therapy. During the first year of AZA therapy, 23% of the relapses occurred; 48% occurred during the second through fourth years, and 29% occurred after 4–6 years. The relapses were precipitated by medication reduction (35%), by infection (26%), following surgery (16%), or by personal stress (13%). Patients were able to discontinue AChE-I drugs (55%) and discontinue (37%) or reduce (63%) the dose of CTS. Thirty-nine percent of the patients have been able to reduce the dose of AZA, and two patients have stopped AZA after 6 years and remained in remission off medication.

DISCUSSION

Long-term AZA therapy, 150 mg daily, is effective and well-tolerated by most MG patients. AZA may be more effective in patients over the age of 50 since 6 out of 7 of our nonresponders were young females. The presence of macrocytosis, anemia, or leukopenia was not a reliable indicator of adequate AZA dosage or a predictor of a positive response to therapy. A moderate number of side effects did occur on AZA therapy, and it was necessary to follow blood counts and liver enzymes, and to monitor for infections.

Relapses do occur during AZA therapy and may follow reduction of the medication.[3] Therefore the dosage should be reduced slowly over months to years when the patient is in a stable remission after 2 years of therapy. Recurrent MG signs, if severe, usually respond to plasma exchange, but pyridostigmine or restoring

TABLE 2. Clinical Data of Myasthenia Gravis (MG) Patients with Relapses on Azathioprine (AZA) Therapy

Number and sex: 23; 10 males, 13 females
Ages: 34–88 years, 62.3 ± 14.5 years[a]
Thymectomy 61%; prednisone 83%
Duration MG: 3.5–25 years, 11.9 ± 6.4 years[a]
Duration MG when AZA started: 6.9 ± 6.4 years[a]
Duration AZA therapy: 1.5–8 years, 4.7 ± 2 years[a]
Number of relapses before AZA therapy: 0–19, 3.5 ± 4.5[a]
Number of relapses on AZA therapy: 0–7, 1.4 ± 1.6[a]

[a] Mean values ± 1 SD.

the dose of AZA is often sufficient in reducing mild signs. Relapses of MG symptoms have been reported many months after discontinuation of AZA,[4] and patients need follow-up for years.

REFERENCES

1. Albuquerque, E. X., J. E. Rash, R. F. Mayer & J. R. Satterfield. 1976. An electrophysiological and morphological study of the neuromuscular junction in patients with myasthenia gravis. Exp. Neurol. **51:** 536–563.
2. Engel, A. G. 1984. Myasthenia gravis and myasthenic syndromes. Ann. Neurol. **16:** 519–534.
3. Mertens, H. G., G. Hertel, P. Reuther & K. Ricker. 1981. Effect of immunosuppressive drugs (azathioprine). Ann. N. Y. Acad. Sci. **377:** 691–699.
4. Hohlfeld, R., K. V. Toyka, U. A. Besinger, B. Gerhold & K. Heinenger. 1985. Myasthenia gravis: reactivation of clinical disease and of autoimmune factors after discontinuation of long-term azathioprine. Ann. Neurol. **17:** 238–242.

Immunosuppressive Treatments

Their Efficacy on Myasthenia Gravis Patients' Outcome and on the Natural Course of the Disease[a]

FERDINANDO CORNELIO,[b] CARLO ANTOZZI,[b] RENATO MANTEGAZZA,[b] PAOLO CONFALONIERI,[b] EMILIA BERTA,[c] DIONISIO PELUCHETTI,[c] ANGELO SGHIRLANZONI,[c] AND FOLCO FIACCHINO[d]

[b]Neuromuscular Research Center
[c]Neurological Department
[d]Intensive Care Unit
Istituto Nazionale Neurologico "C.Besta"
Milan, Italy

The natural course of myasthenia gravis (MG) has been extensively investigated, with particular attention to the years 1940 to 1960 and 1960 to 1980.[1] The former period preceded the introduction of steroid treatment and positive pressure ventilation. Morbidity and mortality of MG were significantly different in the 1940–1960 group versus the 1960–1980 group. Grob *et al.* stated, "The improvement was attributed mainly to better respiratory care, and to a lesser extent to corticosteroid, both of which increased survival during the early period of maximal severity allowing patients to enter the subsequent years of spontaneous improvement or stability."[1]

The main goal of any therapeutic approach, however, a stable and complete remission, is still far from being achieved, as demonstrated by similar percentages of true remission in different epidemiological studies on large series of MG patients reported during the last two decades.[1–5] The evaluation of the efficacy of therapeutic measures adopted in MG has often relied upon retrospective or uncontrolled studies, due to obvious ethical reasons. The lack of standardization in the assessment of the clinical outcome and in definition of what is meant by "stable remission" makes the comparison between different experiences particularly difficult. This lack of standardization might explain the conflicting results on the efficacy of thymectomy and of the various immunosuppressive therapies. The increased use of corticosteroids, other immunosuppressive drugs, and plasma exchange, however, seem to be responsible for higher percentages of patients achieving pharmacological remission or a significant clinical improvement.[6–8]

Unfortunately, the etiological agent of MG is unknown, but a considerable amount of data on the immunopathogenesis of the disease is now available.[9] These data represent the basis for the investigation of new and more specific

[a]The support of IMG (International Myasthenia Gravis Association), AIM (Associazione Italiana per la Lotta contro la Miastenia Grave), and Telethon Italia (to R. Mantegazza) is gratefully acknowledged.

immunosuppressive treatments. In this regard, a critical reassessment of the conventional immunosuppressive therapies is still necessary. We report our experience on the use of corticosteroids, azathioprine, and plasma exchange in a large sample of MG patients.

PATIENTS AND METHODS

We selected from our series a group of 297 MG patients: only patients with at least three years of clinical follow-up were chosen (the mean follow-up was 6.9 years). The diagnosis of MG was performed according to clinical, pharmacological, and neurophysiological criteria.

All patients were diagnosed and treated by the same group of neurologists according to the same therapeutic decision flow chart shown in FIGURE 1. As far as the pharmacological treatment was concerned, patients affected by ocular MG were treated with anticholinesterase inhibitors (anti-AChE) and, in case of failure to control ocular symptoms, with prednisone (0.5 mg/kg body weight/day), gradually shifted to an alternate-day regimen and then slowly tapered to the lowest maintenance dose. Generalized bulbar and respiratory MG patients were submitted

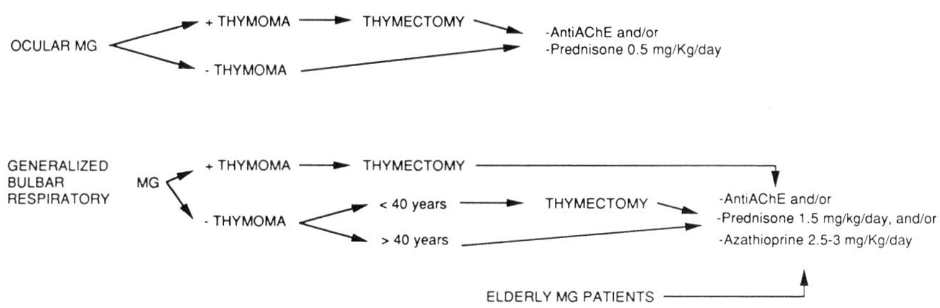

FIGURE 1. Therapeutic decision flowchart.

to anti-AChE and immunosuppressive treatment with prednisone (1.5 mg/kg body weight/day) until a significant and stable clinical improvement was obtained. Medication was then shifted to every other day and was gradually tapered (about 10% every 5–6 weeks) to the lowest effective dose. Azathioprine was added to the above schedule in case of poor responsiveness to corticosteroids, severe clinical relapse during steroid tapering, or intolerance to prednisone due to side effects.

Thymectomy was performed in ocular MG patients only in case of thymoma; generalized MG patients were operated on if age at onset was less than 40 years, or at any age in case of thymoma.

Therapeutic plasma exchange (TPE) was always used as an emergency procedure in patients affected by the most severe forms of the disease, that is, bulbar and respiratory MG. The apheretic schedule consisted of two alternate-day exchanges with removal of 80–100% of the estimated plasma volume calculated on the basis of the patients' body weight and venous hematocrit. Five percent albumin supplemented with single-donor plasma (data until 1990) and, more recently, 5%

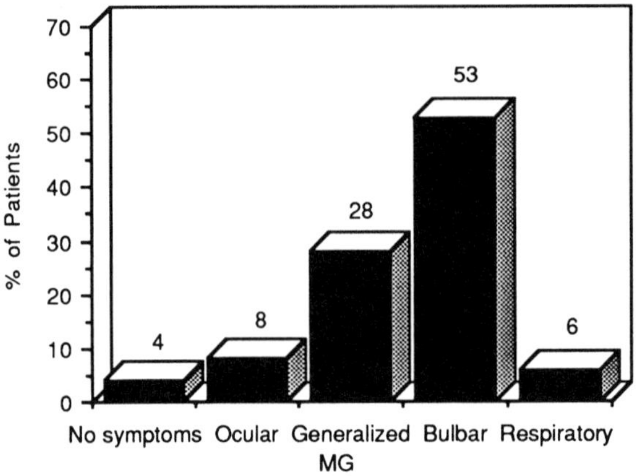

FIGURE 2. Clinical compromise at first observation.

albumin alone were used as replacement fluids. TPE was always performed in association with immunosuppressive drugs. The clinical assessment of the efficacy of TPE was performed seven days after the first exchange in order to minimize the effect of the ongoing immunosuppressive treatment, particularly corticosteroids, on the outcome evaluation. The clinical conditions of the patients before plasmapheresis were as follows: bulbar MG, 69%; bulbar and respiratory MG, 20%; respiratory compromise needing assisted ventilation, 11 percent. The clinical assessment of the patients was always performed according to our MG score, as reported elsewhere.[10]

The following outcomes were identified on the basis of the clinical score: asymptomatic, improved, and unchanged. Patients were considered in remission when the complete absence of symptoms and signs of MG was observed for at least one year without any ongoing treatment. Pharmacological remission was defined as remission with patients still taking anti-AChE and/or immunosuppressive drugs. We evaluated the clinical compromise, outcome, and ongoing treatments after 3, 6, and 9 years of follow-up.

RESULTS

The clinical compromise recorded at first observation is shown in FIGURE 2, and its distribution was asymptomatic MG, 4%; ocular MG, 8%; generalized MG, 28%; bulbar MG, 53%; and respiratory MG, 6 percent. The distribution of patients receiving anti-AChE alone or with immunosuppressive drugs (either prednisone or its association with azathioprine) is indicated in FIGURE 3: at first observation, 50% of patients were taking anti-AChE only, whereas 38% were already immunosuppressed; the percentage of immunosuppressed patients increased up to 58% after 3 years and remained stable during the follow-up (58% at 3 years, 65% at 6 years, and 66% at 9 years). The mean duration of immunosuppression in the whole sample of MG patients was 5.2 years.

The clinical status of the patients (expressed according to a modified Osserman classification) at different times of the follow-up is shown in FIGURE 4. The percentage of asymptomatic patients increased from 4% at first observation to 27%, 25%, and 23% at 3, 6, and 9 years of clinical follow-up, respectively. The percentage of patients showing ocular signs remained stable during the follow-up (8% at first observation, 11% at 3, 9% at 6, and 7% at 9 years). An increase in the percentage of patients affected by generalized MG was noted from 28% at first observation up to 42% at 3 years; the percentage observed remained stable at 3 and 6 years (44% and 48%, respectively). A parallel reduction of patients showing bulbar MG was noted from 53% at first observation to 20% at 3 years, 19% at 6, and 22% at 9 years. Respiratory MG patients (*i.e.*, those requiring assisted ventilation) were 6% at first observation and 0.3% at 3 years.

Whenever the clinical outcome was considered, the percentage of asymptomatic, improved, and unchanged patients remained stable at each step of the clinical follow-up, as shown in FIGURE 5 (asymptomatic patients: 27%, 35%, and 23%; improved 45%, 45%, and 53%; unchanged 29%, 30%, and 24%, at 3, 6 and 9 years, respectively). The percentage of complete remission was 7% at 3 years and remained stable during the follow-up. One hundred twenty patients were treated with prednisone alone (FIG. 6). The percentage of patients achieving pharmacological remission was 15% at 3 years, 18% at 6 years, and 32% at 9 years. Much improved and improved patients were 63%, 54%, and 54%, respectively. The mean duration of corticosteroid treatment was 5.23 years (range 1–14 years).

One hundred two patients were treated with the association of corticosteroids and azathioprine (FIG. 6). The percentage of patients in pharmacological remission was 13% at 1 year, 26% at 3 years, and 22% afterwards. Considering the positive outcomes (pharmacological remission and improved together), the percentages of improved patients increased from 41% to 62% and 71%, respectively. The mean duration of treatment with the association of the two drugs was 3.3 years.

TPE was always performed in patients showing severe bulbar impairment. One hundred six patients were submitted to a total of 194 exchange cycles consisting of two exchanges every other day each (FIG. 7). A significant improvement was noted in 67% of the cycles performed; in 29% the outcome was unchanged, and in 4% the outcome worsened. The mean duration of assisted ventilation was 9 days, significantly reduced when compared to nonexchanged patients in which

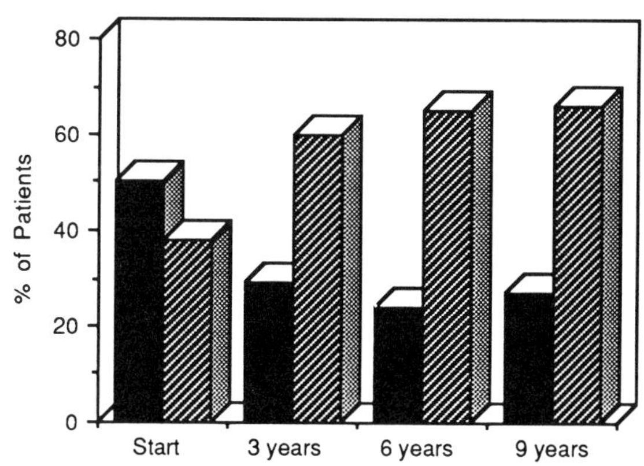

FIGURE 3. Distribution of patients treated with AChE-inhibitors or with immunosuppressive drugs (prednisone alone or prednisone in association with azathioprine). ■, anti-AChE; ▨, immunosuppression.

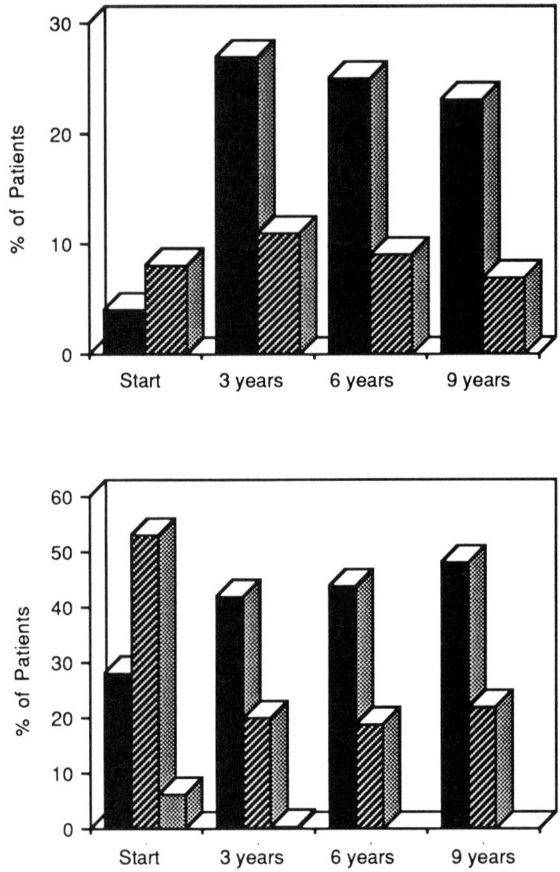

FIGURE 4. Clinical status of the patients at different time points of the clinical follow-up (modified Osserman classification). Top: ■, no symptoms; ▨, ocular MG. Bottom: ■, generalized MG; ▨, bulbar MG; ▩, respiratory MG.

the mean duration was 19.3 days. TPE was always performed in patients treated with immunosuppressive drugs or in whom immunosuppression was started together with plasmapheresis. The incidence of a clinical relapse within one month after TPE was negligible inasmuch as a significant deterioration was noted in only 4% of the cycles, whereas the incidence of relapses was equally distributed along the clinical follow-up (FIG. 7).

Thymectomy was performed in 152 out of 297 patients (51%): at the end of follow-up, the percentage of patients achieving complete remission was 12% in the thymectomized group and 13% in the nonthymectomized patients.

DISCUSSION

The analysis of our experience regarding a large group of MG patients confirms that conventional immunosuppressive treatment is effective in a large proportion of MG patients. As shown in FIGURE 4, we observed a gradual shift towards the less severe forms of the disease, with a parallel reduction of patients showing severe bulbar impairment. The use of a standardized immunosuppressive regimen significantly improved the patients' clinical conditions inasmuch as the majority of cases were asymptomatic or improved. When improvement was expressed in terms of clinical outcome (*i.e.*, asymptomatic, improved, and unchanged), however, the percentage of asymptomatic patients and that of improved and unchanged cases remained stable during the follow-up period up to 9 years. This is noteworthy because immunosuppressive drugs, even though effective, were not able to influence the clinical course in a significant proportion of cases (about 25%). Moreover, in our patients the clinical improvement over a 9-year follow-up was not accompanied by a substantial increase in the percentage of patients free of immunosuppressive treatment. This finding indicates that a stable control of the disease can be obtained, but long-term (several years) immunosuppression is necessary.

No definite conclusions can be drawn on whether prednisone alone is better than with its association with azathioprine. Nevertheless, the association was effective in patients showing a poor response to corticosteroids or in those cases requiring a more rapid tapering of prednisone due to intolerance or side effects, as we already reported elsewhere.[10]

Plasma exchange has definitely acquired a specific role in the treatment of severe MG patients.[11] Since no clear-cut indications were available with regard to the optimal apheretic schedule, considering blood volume processed and number of exchanges to be performed in each patient, we treated our patients according to a short standardized protocol. As already reported in a smaller

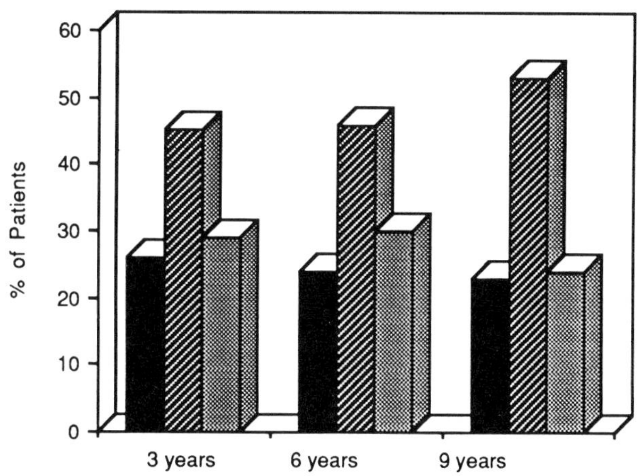

FIGURE 5. Clinical outcome at 3, 6, and 9 years of follow-up. ■, asymptomatic; ▨, improved; ▩, unchanged.

group of MG patients,[12] our approach was effective in the majority of cases, suggesting that the rapid removal of plasma in a short period of time induces a significant clinical improvement. Indeed, our apheretic approach was effective also in patients with respiratory insufficiency: in 84% of the cycles, performed in these cases, spontaneous breathing was resumed within seven days after TPE, and the mean duration of mechanical ventilation was significantly shortened when compared to the mean of a pre-TPE period. Longer apheretic schedules are not necessary to improve the clinical conditions, even in the most severe forms of the disease. Plasma exchange was always associated with immunosuppressive drugs and was not followed by an early clinical deterioration, as demonstrated by the low incidence of worsening within one month after the exchange treatment.

We cannot make any definite conclusion about the influence of thymectomy. In a recent multicenter study on the prognosis of MG, we found that the only

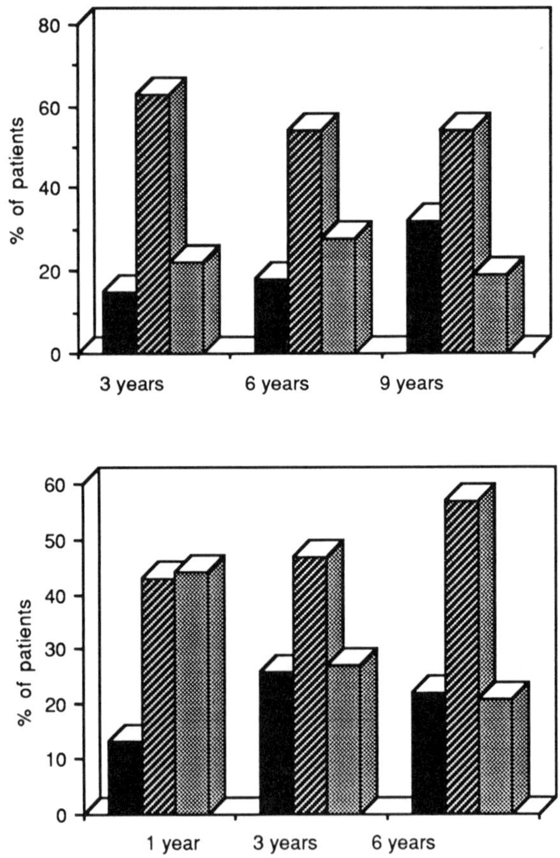

FIGURE 6. Upper panel: clinical outcomes in patients treated with prednisone alone. Lower panel: clinical outcomes in patients treated with prednisone and azathioprine. ■, pharmacological remission; ▨, improved; ▩, unchanged, worsened.

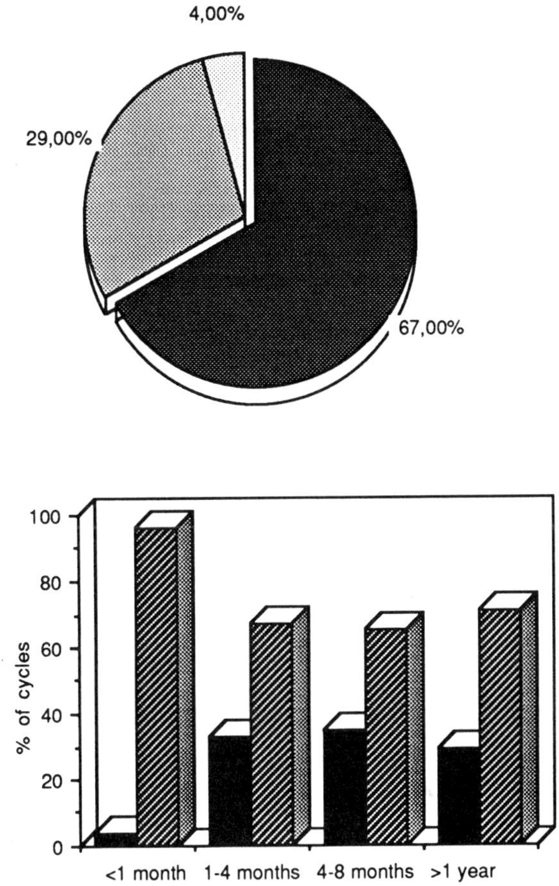

FIGURE 7. Upper panel: clinical outcomes after TPE. ▩, much improved, improved; ▨, unchanged; □, worsened. Lower panel: incidence of clinical relapses after TPE. ■, relapses; ▨, stabilizations.

variables that correlated to the chance of complete remission were younger age at onset, lower severity of symptoms at onset, and shorter disease duration at diagnosis, whereas the chance of complete remission in thymectomized versus nonsurgical patients was of borderline significance after multivariate analysis.[9] These factors, as well as the occurrence of spontaneous remission, must be considered in assessing the real efficacy of thymectomy, because the suggested delayed effect of surgery may not be related to thymectomy itself. Moreover, short disease duration may be a cue for prompt therapeutic intervention, resulting in a better prognosis.

Altogether, the conventional therapeutic strategies we used were able to improve significantly the clinical course of the majority of our patients. Nevertheless, about 25% of our patients showed little or no improvement using any of the immunosuppressive treatment on a long clinical follow-up. This group of patients represents a major demand for a better and more specific immunosuppressive treatment. The immunobiological characteristics associated with the disease will help us to understand the mechanisms by which the final effectors, that is, the anti-AchR antibodies, are produced and are of pathogenic relevance. In turn, nonresponder patients' characteristics might be of critical importance for explaining poorly characterized immunological events.

REFERENCES

1. GROB, D., N. G. BRUNNER & T. NAMBA. 1981. The natural course of myasthenia gravis and effect of therapeutic measures. Ann. N.Y. Acad. Sci. **377:** 652–669.
2. GROB, D., E. L. ARSURA, N. G. BRUNNER & T. NAMBA. 1987. The course of myasthenia gravis and therapies affecting outcome. Ann. N.Y. Acad. Sci. **505:** 472–499.
3. OSSERMAN, K. E. & G. GENKINS. 1971. Studies in myasthenia gravis: review of a twenty years experience in over 1200 patients. Mt. Sinai J. Med. **38:** 497–537.
4. MANTEGAZZA, R., E. BEGHI, D. PAREYSON, C. ANTOZZI, D. PELUCHETTI, P. TONALI, A. EVOLI, E. RICCI, A. P. BATOCCHI, C. ANGELINI, G. F. MICAGLIO, G. MARCONI, R. TAIUTI, L. BERGAMINI, L. DURELLI, & F. CORNELIO. 1990. A multicenter follow-up study of 1152 patients with myasthenia gravis in Italy. J. Neurol. **237:** 339–344.
5. SGHIRLANZONI, A., D. PELUCHETTI, R. MANTEGAZZA, F. FIACCHINO & F. CORNELIO. 1984. Myasthenia gravis: prolonged treatment with steroids. Neurology **34:** 170–174.
6. DAU, P. C. 1981. Response to plasmapheresis and immunosuppressive drug therapy in sixty myasthenia gravis patients. Ann. N.Y. Acad. Sci. **377:** 700–708.
7. MATELL, G. 1987. Immunosuppressive drugs: Azathioprine in the treatment of myasthenia gravis. Ann. N.Y. Acad. Sci. **505:** 588–594.
8. BEGHI, E., C. ANTOZZI, A. P. BATOCCHI, F. CORNELIO, V. COSI, A. EVOLI, M. LOMBARDI, R. MANTEGAZZA, M. L. MONTICELLI, G. PICCOLO, P. TONALI, D. TREVISAN & M. ZARRELLI. 1991. Prognosis of myasthenia gravis: a multicenter follow-up study of 844 patients. J. Neurol. Sci. **106:** 213–220.
9. LINDSTROM, J., D. SHELTON & Y. FUJII. 1988. Myasthenia gravis. Adv. Immunol. **42:** 233–284.
10. MANTEGAZZA, R., C. ANTOZZI, D. PELUCHETTI, A. SGHIRLANZONI & F. CORNELIO. 1988. Azathioprine as a single drug or in combination with steroids in the treatment of myasthenia gravis. J. Neurol. **235:** 315–453.
11. Consensus Conference. 1986. The utility of therapeutic plasmapheresis for neurological disorders. J. Am. Med. Assoc. **256:** 1333–1337.
12. ANTOZZI, C., M. GEMMA, B. REGI, E. BERTA, P. CONFALONIERI, D. PELUCHETTI, R. MANTEGAZZA, F. BAGGI, M. MARCONI, F. FIACCHINO & F. CORNELIO. 1991. A short plasma exchange protocol is effective in severe myasthenia gravis. J. Neurol. **238:** 103–107.

Treatment Strategies in Myasthenia Gravis

GABRIEL GENKINS, MARK SIVAK,
AND PAUL I. TARTTER

Myasthenia Gravis Division
Mount Sinai Medical Center
One Gustave L. Levy Place
New York, New York 10029-6575

This year marks the fortieth anniversary of the founding of the Myasthenia Gravis Clinic at the Mount Sinai Hospital. This term is used to denote a patient cohort, largely of private patients, which have been followed since 1952. Additionally, through the efforts of the late Dr. A. E. Papatestas, we periodically reviewed data of patients treated by Dr. Geoffrey Keynes in Great Britain (318 patients). This type of longitudinal experience in one medical institution has resulted in the formulation of strategies in therapy of myasthenia gravis. The total number of patients followed is now 2,308 and include the performance of 1,108 thymectomies (TABLE 1).

DIAGNOSIS

Clinical history and physical examination form the major initial critical building blocks. In the milder cases (groups I, IA, and IIA) of the Osserman classification, determination of the acetylcholine receptor antibody (AChR) has not been particularly helpful in that it is often found to be within normal limits at presentation. General laboratory data is always obtained, including thyroid function tests and determination of antinuclear antibodies (TABLE 2).

Chest X-rays in multiple views and CT scans with contrast are routinely performed in an effort to detect thymic abnormalities. It should be noted that the latter procedure is helpful when positive but may be negative even in the face of small thymomas.

The intravenous Tensilon® test (3–4 mg of edrophonium chloride in bolus injection) can be most helpful. It is, however, in the realm of electrophysiological testing that the most specific data is obtained by use of a three-stage screening.

First, after a minimum of 12 hours without anticholinesterase medication, supramaximal percutaneous stimulation at the three/second is employed (Jolly test). If negative, single-fiber testing is performed looking for jitter and blocking phenomena. The third phase consists of administration of a provocative test using "regional" curare administration. If all these determinations are negative, the patient may still demonstrate pure ocular myasthenic symptoms, to be followed on anticholinesterase medication alone (primarily pyridostigmine, Mestinon®). Repeat electrophysiological testing is performed on a six-month basis or more often if generalized symptoms develop earlier.

Specific education in the use of these agents is given to all patients and, in general, the use of time span medication is avoided, reserved primarily for night-

TABLE 1. Experience to 1992

	All Patients	Thymectomized Patients
Mount Sinai Medical Center	2308	972
Patients followed at Mount Sinai Hospital but operated on elsewhere		136
Total		1108
New End Hospital	319	266

time administration in cases with nocturnal respiratory distress or inability to swallow the first morning dose. All patients found to have evidence of generalized myasthenia, even if by electrophysiological testing alone, receive recommendations to undergo thymectomy if no other medical contraindications exist.

THERAPY

Anticholinesterases

Mestinon® remains the premier drug of this class of compounds in terms of effectiveness, smoothness of action, and predictability. Patient involvement in the pharmacodynamics of the drug is essential in that dosage is variable and usually requires both time and dose adjustment over a twenty-four hour period, with variables such as menstrual cycle, infections, and emotional trauma considered. With greater awareness, cholinergic crisis has become rarer, yet occurrence of overdose with muscarinic side effects is common.

Edrophonium as both a diagnostic agent and monitor of adequacy and anticholinesterase levels is extremely useful and most effective in proper doses. A few patients cannot tolerate Mestinon, but rather require Mytelase® (ambenonium) as a substitute.

Steroids

In general, we reserve the use of steroids for patients with severe symptoms interfering with the activities of living or those with debilitation bulbar manifesta-

TABLE 2. Clinical Classifications in the Myasthenia Gravis Registry

	Patient Diagnosed Before 1972				Patients Diagnosed 1972 and After			
Class	No Thymectomy		Thymectomy		No Thymectomy		Thymectomy	
		Thymoma		Thymoma		Thymoma		Thymoma
IIB	228	(14)	147	(43)	12	—	125	(29)
III	102	(17)	94	(26)	4	—	45	(10)
IV	39	(5)	38	(10)	1	—	2	(2)

TABLE 3. Thymectomy and Patient Age

Age	No Thymoma	Thymoma	Total
Thymectomy			
50 years and over	251	100	351
Less than 50 years	652	105	757
Total	903	205	1108
No thymectomy (age at diagnosis)			
50 years and over	417	17	434
Less than 50 years	731	35	766
Total	1148	52	1200

tions. All patients on steroids are covered by H2 antagonists and antituberculous therapy, where indicated. If clinical condition forbids thymectomy, a course of ACTH (10 days ACTH gel) is given to improve clinical status. This is preferred to prednisone in that it is of great effectiveness and does not commit the patient to long-term steroid therapy.

Additionally, the conjoint intramuscular administration of gamma globulin (10 cc) every three weeks seems to potentiate prednisone, allowing for lower doses (20–30 mg) of the latter on an alternate day dosage with good results.

Plasmapheresis

This procedure represents an alternative or adjuvant form of therapy in the clinically severe myasthenic and is attended by a high level of benefit. We generally perform pheresis by an initial "loading" of 12–15 liters of exchange (6–8 treatments) on an alternate-day basis, followed by serial once-per-week treatments. Although many such severely ill patients are being treated by prednisone and/or specific immunotherapy, we have not been able to demonstrate additive benefits through the use of Imuran® (azathioprine) or steroids. Early experience with different absorptive columns, such as Staph A (Prosorba©), are yet inconclusive. Seronegative patients may benefit from pheresis in line with Dr. Angela Vincent's observations.

TABLE 4. Thymectomy Technique

	Patients Diagnosed	
	Before 1972	1972 and After
Transcervical	236	544
Transsternal	196	55
Chamberlain	4	20
Thoracotomy	5	14
Biopsy	5	3
Transthoracic	3	1
Partial split	1	0

Specific Chemotherapy

Azathioprine in doses of 2.5–3.5 mg/kg has been the primary agent used with stringent blood sampling monitoring. Effects were not seen for at least eight weeks and are usually not seen for three to six months. Qualitatively, the older myasthenic (above age 45) seems to respond best. Serious hepatic, pancreatic, and bone marrow complications have been observed. Cytoxan and methotrexate were employed with a few early patients with a higher incidence of problems. Experience with cyclosporin has been too limited to comment upon.

CRISIS

The myasthenic crisis, that is, respiratory and swallowing difficulties have been attacked in many ways. In general, we find treatment by intubation, nasogastric feeding, and sedation for 72–96 hours to be extremely helpful while discontinuing all anticholinesterase treatment.

Prednisone is continued with H2 antagonist coverage, and clinical search for inciting factors (*i.e.*, infection, thyroid disease) is undertaken. Most patients respond well to such a "hibernation" program without need for additional specific treatment. When the latter is necessary, administration of ACTH courses has been highly effective. A few patients have been treated by the use of intravenous immunoglobulins, in massive amounts, with good results. Applying the same methodology to patients with severe myasthenia gravis undergoing major surgical manipulations, either emergent or elective, such as abdominal surgery, heart valve replacement, or CNS surgery has been very fruitful.

THYMECTOMY

Although never subjected to the true scientific type of prospective studies recommended in the past, thymectomy has empirically been adopted by most treatment centers as the single most effective treatment strategy in this disease. Indeed, there seems to be general agreement that all patients with generalized myasthenia gravis should undergo this procedure as early as possible in their course. We have become convinced over time that (1) progression to severe disease is halted; (2) the incidence of thymomas decreases; (3) a high proportion of total disease remissions occurs; (4) the older patient (over 45) with no thymic tumors has the best statistical results; (5) prepubertal patients should not be operated upon because of oncogenic risk, unless life-threatening symptoms exist; (6) all thymomas should be removed on discovery, together with the thymus, even when there is no evidence of myasthenia clinically, although thorough electrophysiological testing will usually reveal a myasthenic deficit (TABLE 3).

After an initial experience of 196 cases of transsternal thymectomies, we have changed to the transcervical approach exclusively, except in some cases of thymoma or reexploration. The reduced morbidity and mortality in these procedures, coupled with more retractors (Cooper), seem appropriate justification for this view. Reoperation has been performed, as illustrated, with generally good results on those patients in the early part of the series whose clinical results seemed unsatisfactory. Most patients are discharged within 36 hours of admission.

TABLE 5. Rethymectomies in Mg Registry

Type	Number
Transsternal—Transcervical	11
Transcervical—Transcervical	19
Thoracotomy (no thymectomy)—Transsternal	1[a]
Thoracotomy (no thymectomy)—Transcervical	7[a]
Recurrent thymoma	3[a]
Transcervical—Transsternal	5
Other	1
Total	47
Pathology at Rethymectomy	**Number**
Residual thymus	12
No thymic tissue	15
Germinal centers[b]	5
No germinal centers	2
Thymoma	9
Recurrent thymoma	3
Unknown	1

[a] Thymus not removed at first procedure (thymomectomy).
[b] Germinal centers found at thymectomy.

THYMOMA

Because no effective chemotherapy currently exists for this tumor, benign in microscopic appearance, but malignant in behavior, we believe in the earliest complete extirpation possible. When the tumor is recurrent, low dose (2500 R) radiotherapy is highly effective for the thymocyte component alone. The epithelial component seems totally radiotherapy-resistant. The current remaining controversy regarding the surgical technique for thymectomy will, we believe, also be settled in time. The work of Dr. G. Maggi in Italy and centers in St. Louis, California, and Canada seem to point the way (TABLE 4).

In our hands, early thymectomy has proved to be an incalculable boon for this group of patients. No other form of treatment, in spite of our far greater understanding of pathophysiologic mechanisms, has matched its success. Performance of surgery early in the course of the disease, prior to the development of multiple germinal centers, seems most quickly to result in remission. Indeed, the older patients (> age 50), with no ability to form such structures, have the best clinical result (TABLE 5). Of continuing interest is the association of myasthenia

TABLE 6. Mount Sinai Medical Center Series: Extrathymic Neoplasia

	Number of Neoplasms	Number of Patients
Mount Sinai Series	244	2308
New End Hospital Series	200	319
Total	264	2627

gravis and extrathymic neoplasia. Prior to thymectomy the patients so afflicted represent 10% of the Myasthenic cohort. There is disproportionate frequency of lymphoma, leukemia, and breast carcinoma. Following thymectomy, particularly when disease remission occurs, the incidence of such tumors drops far below normal levels (TABLE 6).

Conclusion

ROBERT L. RUFF

Neurology Service
Department of Veterans Affairs Medical Center
10701 East Boulevard
Cleveland, Ohio 44106

MYASTHENIA GRAVIS

Myasthenia gravis is an autoimmune disease in which the body begins to attack the acetylcholine receptors at the end-plate of the neuromuscular junction. The acetylcholine receptors are responsible for chemical communication between nerves and muscle fibers that permits muscle contraction to occur when the nerve signals the muscle. When this communication is compromised, as occurs in myasthenia gravis, fatigue and weakness occur. There are two elements to the attack: a humoral component, produced by antibodies directed against the acetylcholine receptors, and a cellular component, in which inflammatory cells attack and destroy the membrane containing the acetylcholine receptors.

Several presentations in this volume focus on aspects of cellular and humoral immunology that directly relate to myasthenia gravis and other related diseases. The presenters discuss how thymus-derived immune cells, called T cells, are produced in the thymus, and how those cells can recognize specific regions of proteins, called epitopes, from foreign tissue and, in some cases, native epitopes, such as the acetylcholine receptors, which are normal parts of our bodies. These discussions suggest several new ways of abating the cell-mediated immune response in myasthenia gravis by specifically targeting activated T cells or, more specifically, the group of T cells that are responsible for the attack on the acetylcholine receptors. In addition, several presentations suggest different ways for inducing anergy or tolerance and T cells to acetylcholine receptors.

Two types of antibodies are produced by a class of lymphocytes called B cells: IgG and IgM. A subset of B cells that has a specific CD5 surface marker, called $CD5^+$ B cells, is associated with several autoimmune diseases. Discussions of the production of immunoglobulins by B cells raises the possibility that the B cells responsible for producing antibodies against acetylcholine receptors might be identified and either destroyed or suppressed.

The thymus gland, located in the chest in front of the heart, plays a major role in the genesis of myasthenia gravis, and several speakers focus on the changes that occur in the thymus gland of some patients with myasthenia gravis. In addition, several speakers and posters indicate that components of the acetylcholine receptors or acetylcholine receptor–like proteins are present in the thymus and may serve to initiate or perpetuate autoimmune myasthenia gravis. The discussion of the potential roles of the thymus in myasthenia gravis provides a basis for understanding why thymectomy can benefit some patients.

The acetylcholine receptor in skeletal muscle is a large protein molecule with five subunits. The alpha-subunit binds the neurotransmitter acetylcholine. Each receptor contains two alpha-subunits along with a beta-, delta-, and either an epsilon- or a gamma-subunit. Acetylcholine receptors from mature innervated muscle fibers contain the epsilon-subunit, whereas receptors with the gamma-subunit are found on immature and denervated muscle fibers. Several presentations describe recent findings about the amino acid composition of the subunits and the three-dimensional structure of the acetylcholine receptor. Most of the antibodies are directed against portions of the alpha-subunit. Several presentations describe the regions of the alpha-subunit that are targets for antibodies. Humans have two variants of the alpha-subunit, and it appears that one of the variant alpha-subunits is disproportionately expressed in patients with myasthenia gravis. It will be interesting to find out if the overexpression of one type of alpha-subunit is important in the genesis of myasthenia gravis or if this overexpression represents an adaptive response of muscle to the autoimmune attack. The reduction in the number of acetylcholine receptors at the end-plate is due, in part, to receptors being pulled back (internalized) into the muscle fiber when antibodies bind to the receptor. One presentation suggests a possible strategy to reduce internalization of receptors, thereby reducing the degree to which neuromuscular transmission is compromised in myasthenia gravis.

Double vision and droopy eyelids, ptosis, are frequent symptoms and, in some patients, the only symptoms of myasthenia gravis. These symptoms suggest that the extraocular muscles, which control eye and eyelid movements, are targeted in myasthenia gravis. One presentation focuses on the relationship between the unique structure of extraocular muscles and their involvement in myasthenia gravis. Some patients with ocular myasthenia gravis have antibodies that selectively target extraocular muscle fibers and do not react with other muscle fibers. Further characterization of the eye muscle–specific antibodies and the epitopes that these antibodies recognize will enable us to develop specific strategies for treating patients with ocular myasthenia gravis.

Members of local chapters of the Myasthenia Gravis Foundation often note that there may be subsets of patients who have unusual manifestations of myasthenia gravis or who respond to treatment differently. Several patient care groups in the United States and other countries have begun to address the variability of presentation and response by establishing large-scale patient registries. As these registries expand they will enable patients and practitioners to get a better idea of the nationwide and regional incidence and prevalence of myasthenia gravis as well as to compare the efficacy of different treatment modalities among different groups of patients in different regions.

The final section of this *Annal* is devoted to a discussion of the different therapies available to treat myasthenia gravis. The therapies currently used include drugs, such as pyridostigmine to potentiate neuromuscular transmission; thymectomy; plasmapheresis; immunoglobulin administration; and immunosuppressive chemotherapy.

LAMBERT-EATON MYASTHENIC SYNDROME AND OTHER DISORDERS OF NEUROMUSCULAR TRANSMISSION

The Lambert-Eaton myasthenic syndrome is an autoimmune disorder of neuromuscular transmission in which antibodies are produced that react with the nerve terminal and interfere with the release of the neural transmitter acetylcholine. About half of the patients with myasthenic syndrome have a small-cell lung cancer. Patients with myasthenic syndrome have generalized weakness, due to compromise of neuromuscular transmission, and other problems, such as difficulty urinating, impaired production of saliva, and constipation due to compromise of neural transmission to the bladder, various glands, and the gastrointestinal tract.

This section focuses on the role of calcium channels in releasing acetylcholine from nerve terminals, the targets of the antibodies directed toward the nerve terminals, and strategies for treating Lambert-Eaton myasthenic syndrome. The concentration of calcium in cells is extremely tightly regulated. The free intracellular concentration of calcium is less than one ten thousandth of the concentration found in the extracellular fluid that surrounds cells. Almost all cells use changes in intracellular calcium to trigger enzymatic processes or other cellular functions, such as release of neural or hormonal transmitters. Due to the different needs of cells, there are many types of calcium channels that are found in different cell types. There appear to be two specific types of calcium channels, N-type and P-type, that are associated with nerve terminals that connect to muscle fibers. Data are presented indicating that the antibodies in Lambert-Eaton myasthenic syndrome may be directed towards N-type and P-type channels. One group found that some of the antibodies from patients with Lambert-Eaton myasthenic syndrome bind to a specific calcium-channel subunit.

The novel findings about the types of calcium channels that are responsible for acetylcholine release and specificity of the antibodies in Lambert-Eaton myasthenic syndrome pave the way for developing more effective treatments for patients with Lambert-Eaton myasthenic syndrome. One group presents very encouraging results in treating Lambert-Eaton myasthenic syndrome by using a drug that blocks potassium channel activity in nerve terminals. Blocking potassium channels allows the nerve terminal calcium channels to remain open longer, thereby enabling the remaining calcium channels to conduct more calcium current into the nerve terminal to stimulate transmitter release. In addition, high-dose immunoglobulin therapy may also benefit patients with Lambert-Eaton myasthenic syndrome as it did in some patients with myasthenia gravis.

Some patients with weakness that resembles that seen in myasthenia gravis have a congenital disorder of neuromuscular transmission, also called congenital myasthenia or congenital myasthenic syndrome. Congenital disorders of neuromuscular transmission differ from myasthenia gravis and Lambert-Eaton myasthenic syndrome in that congenital disorders of neuromuscular transmission are not caused by antibodies directed against the neuromuscular junction. Congenital disorders of neuromuscular transmission represent a collection of congenital disorders that effect the structure or function of the nerve terminal, the synaptic cleft, or the end-plate. This *Annal* contains a symposium on congenital myasthenic syndromes that characterizes several different congenital myasthenic syndromes. These disorders result from genetic mutations that affect the structure and function of different parts of the neuromuscular junction. In this symposium specific mutations in the acetylcholine subunits and in the acetylcholinesterase molecule are defined that are responsible for specific congenital myasthenic syndromes.

In summary, this volume defines new aspects of the immune disorders in myasthenia gravis and Lambert-Eaton myasthenic syndrome, presents new strategies for treating myasthenia gravis and Lambert-Eaton myasthenic syndrome, clarifies the role of calcium channels in the release of acetylcholine from nerve terminals and the calcium channel targets of the antibodies in Lambert-Eaton myasthenic syndrome, provides the first major symposium on congenital myasthenic syndromes that defines the genetic defects in some of the congenital myasthenic syndromes, and serves as a forum for exchange of information among researchers worldwide.

Subject Index

Index of Contributors